Edited by Carlos Bo

Air Pollution Modeling and Its Application XVII

Carlos Borrego
Department of Environment and Planning
University of Aveiro
Aveiro, Portugal
PORTUGAL
borrego@ua.pt

Ann-Lise Norman
Department of Physics and Astronomy
University of Calgary
Calgary, Alberta T2N 1N4
CANADA
annlisen@phas.ucalgary.ca

Cover illustration: Photo Courtesy of Adolf Ebel

Library of Congress Control Number: 2005939180

ISBN-10: 0-387-28255-6 eISBN-10: 0-387-68854-4
ISBN-13: 978-0-387-28255-8 eISBN-13: 978-0-387-68854-1

Printed on acid-free paper.

9 8 7 6 5 4 3 2 1

springer.com

Preface

In 1969 the North Atlantic Treaty Organisation (NATO) established the Committee on Challenges of Modern Society (CCMS). The subject of air pollution was from the start, one of the priority problems under study within the framework of various pilot studies undertaken by this committee. The organization of a periodic conference dealing with air pollution modeling and its application has become one of the main activities within the pilot study relating to air pollution. The first five international conferences were organized by the United States as the pilot country; the second five by the Federal Republic of Germany; the third five by Belgium; the next four by The Netherlands; and the next five by Denmark; and with this one, the last three by Portugal.

This volume contains the papers and posters presented at the 27th NATO/CCMS International Technical Meeting on Air Pollution Modeling and Its Application held in Banff, Canada, 24-29 October 2004. The key topics at this ITM included: Role of Atmospheric Models in Air Pollution Policy and Abatement Strategies; Integrated Regional Modeling; Effects of Climate Change on Air Quality; Aerosols as Atmospheric Contaminants; New Developments; and Model Assessment and Verification.

104 participants from North and South America, Europe, Africa and Asia attended the 27th ITM. The conference was jointly organized by the University of Aveiro, Portugal (Pilot Country) and by The University of Calgary, Canada (Host Country). A total of 74 oral and 22 poster papers were presented during the conference.

Invited papers by Stephen Dorling of the United Kingdom (Climate Change and Air Quality), Tim Oke of Canada (Siting and Exposure of Meteorological Instruments at Urban Sites), and Greg Yarwood of the United States (Particulate Matter Source Apportionment Technology (PSAT) in the CAMx Photochemical Grid Model) were presented.

We wish to thank the volunteers who assisted with the presentations and registration including Shannon Fargey, Vivian Wasiuta, Astrid Sanusi, Tara Allan, and Naisha Lin. We also wish to thank the Scientific Committee members and the Banff Centre and University of Aveiro staff involved in coordinating the meeting and preprint volume.

On behalf of the Scientific Committee and as Organizers and Editors, we would like to express our gratitude to all participants who made the meeting successful. The efforts of the chairpersons and rapporteurs were appreciated.

Special thanks are due to the sponsoring institutions:

Department of Physics and Astronomy and Environmental Science Program and The Faculty of Science at The University of Calgary
Meteorological Service of Canada
University of Aveiro, Portugal
EURASAP: European Association for the Sciences of Air Pollution
GRICES: Office for International Relations in Science and Higher Education, Portugal
NATO/CCMS: Committee on the Challenges of Modern Society

The next conference in this series will be held in 2006 in Leipzig, Germany

Ann-Lise Norman
Local Conference Organizer
Canada

Carlos Borrego
Scientific Committee Chairperson
Portugal

The Members of the Scientific Committee for the 27th NATO/CCMS International Technical Meetings on Air Pollution Modeling and Its Application

G. Shayes, Belgium
D. Syrakov, Bulgaria
D. Steyn, Canada
S.-E. Gryning, Denmark
N. Chaumerliac, France
E. Renner, Germany
G. Kallos, Greece
D. Anfossi, Italy
A. L. Norman, Canada
P. Builtjes, The Netherlands
T. Iversen, Norway
C. Borrego, Portugal (Chairman)
J. M. Baldasano, Spain
S. Incecik, Turkey
B. Fisher, United Kingdom
F. Schiermeier, United States
S. Trivikrama Rao, United States

History of NATO/CCMS Air Pollution Pilot Studies

Pilot Study on Air Pollution: International Technical Meetings (ITM) on Air Pollution and Its Application

Dates of Completed Pilot Studies:

1969-1974 Air Pollution Pilot Study (Pilot Country United States)
1975-1979 Air Pollution Assessment Methodology and Modelling (Pilot Country Germany)
1980-1984 Air Pollution Control Strategies and Impact Modelling (Pilot Country Germany)

Dates and Locations of Pilot Study Follow-Up Meetings:

Pilot Country – United States (R. A. McCormick, L. E. Niemeyer)

Feb 1971 – Eindhoven, The Netherlands
First Conference on Low Pollution Power Systems Development

Jul 1971 – Paris, France
Second Meeting of the Expert Panel on Air Pollution Modeling

All of the following meetings were entitled NATO/CCMS International Technical Meetings (ITM) on Air Pollution Modelling and Its Application

Oct 1972 – Paris, France – Third ITM
May 1973 – Oberursel, Federal Republic of Germany – Fourth ITM
Jun 1974 – Roskilde, Denmark – Fifth ITM

Pilot Country – Germany (Erich Weber)

Sep 1975 – Frankfurt, Federal Republic of Germany – Sixth ITM
Sep 1976 – Airlie House, Virginia, USA – Seventh ITM
Sep 1977 – Louvain-La-Neuve, Belgium – Eighth ITM

Aug 1978 – Toronto, Ontario, Canada – Ninth ITM
Oct 1979 – Rome, Italy – Tenth ITM

Pilot Country – Belgium (C. De Wispelaere)

Nov 1980 – Amsterdam, The Netherlands – Eleventh ITM
Aug 1981 – Menlo Park, California, USA – Twelfth ITM
Sep 1982 – Ile des Embiez, France – Thirteenth ITM
Sep 1983 – Copenhagen, Denmark – Fourteenth ITM
Apr 1985 – St. Louis, Missouri, USA – Fifteenth ITM
Apr 1987 – Lindau, Federal Republic of Germany – Sixteenth ITM
Sep 1988 – Cambridge, United Kingdom – Seventeenth ITM
May 1990 – Vancouver, British Columbia, Canada – Eighteenth ITM
Sep 1991 – Ierapetra, Crete, Greece – Nineteenth ITM

Pilot Country – Denmark (Sven-Erik Gryning)

Nov 1993 – Valencia, Spain – Twentieth ITM
Nov 1995 – Baltimore, Maryland, USA – Twenty-First ITM
Jun 1997 – Clermont-Ferrand, France – Twenty-Second ITM
Sep 1998 – Varna, Bulgaria – Twenty-Third ITM
May 2000 – Boulder, Colorado, USA – Twenty-Fourth (Millenium) ITM

Pilot Country – Portugal (Carlos Borrego)

Oct 2001 – Louvain-la-Neuve, Belgium – Twenty-Fifth ITM
May 2003 – Istanbul, Turkey – Twenty-Sixth ITM
Oct 2004 – Banff, Canada – Twenty-Seventh ITM

Contents

ROLE OF ATMOSPHERIC MODELS IN AIR POLLUTION POLICY AND ABATEMENT STRATEGIES

EFFECTS OF CLIMATE CHANGE ON AIR QUALITY

AEROSOLS AS ATMOSPHERIC CONTAMINANTS

NEW DEVELOPMENTS

MODEL ASSESSMENT AND VERIFICATION

POSTERS
ROLE OF ATMOSPHERIC MODELS IN AIR POLLUTION POLICY AND ABATEMENT STRATEGIES

INTEGRATED REGIONAL MODELING

EFFECTS OF CLIMATE CHANGE ON AIR QUALITY

NEW DEVELOPMENTS

MODEL ASSESSMENT AND VERIFICATION

Role of Atmospheric Models in Air Pollution Policy and Abatement Strategies

1
A Photochemical Screening Tool Based on a Scale Analysis of Ozone Photochemistry

Bruce Ainslie and D. G. Steyn*

1. Introduction

Controlling ground level ozone, a relatively recent problem but now pervasive in many urban centers, requires linking knowledge about its formation (science) to choices society makes about present and future economic development and social behaviour (policy). But the future is unknown and can follow many courses, and so future air quality is inherently unpredictable and modeling it an inexact exercise.

Uncertainties about the future and the complexity of ozone formation also present a tension for modelers: on the one hand, a model is needed that is sensitive to the intricacies of meteorology, photochemistry and emissions while on the other hand a model is needed that can be used to investigate many possible futures. This tension leads to a spectrum of modeling techniques: at one end are comprehensive Eulerian grid models, such as Models-3/CMAQ (EPA, 2004), which represent the most complete way of describing ozone formation, but due to their complexity and costs, preclude examining large numbers of possible futures; and at the other end, less exact models having quick execution times, like the SOMS model (Venkatram, 1994), containing many approximations but nonetheless include the main processes in ozone formation.

To this end, this paper presents a simple model for ozone formation, capable of characterizing a region's sensitivity to VOC and NOx emissions, which serves as a screening tool for comprehensive Eulerian modeling.

2. Model Development

The model arises from research addressing two related issues:

- How much detail is necessary to capture ozone sensitivity to VOC and NOx emissions in a particular region?

* Atmospheric Science Programme, The University of British Columbia, Vancouver, B.C., Canada

- Given that a model is only as strong as its weakest link, what level of detail is justified in each of the various modeling components?

Through an exploration of these ideas, an air quality model has been developed that incorporates the most important meteorological and photochemical processes with an emissions inventory based on a limited number of sources and whose purpose is to characterize ozone sensitivity to emission. While the aim of the research is to formulate such a model under general conditions, a specific location, the lower Fraser Valley B.C., during a specific ozone episode, has been used to develop and demonstrate the model.

2.1. Modeling Domain and Design Day

The Lower Fraser Valley (LFV), a roughly triangular valley that spans the Canada/U.S.A. border along the 49th parallel, is flanked by the Coast Mountain Ranges to the North, the Cascades Ranges to the South and the Strait of Georgia to the West (Steyn et al., 1997) (see Figure 1). It narrows from a width of 100 km at its western edge to a few kilometers at is eastern boundary some 90 km inland. In the north-south direction, it extends from the major metropolitan area of Vancouver to the town of Bellingham in Northern Washington State. While the

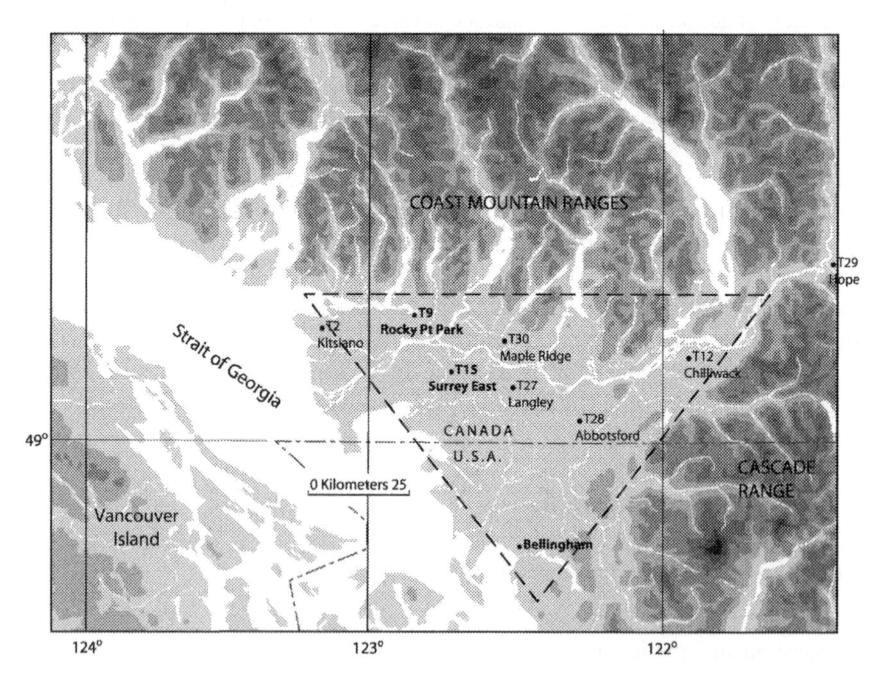

FIGURE 1. The Lower Fraser Valley, B.C. with topographic relief. The roughly triangular shape of the valley is outlined by the heavy dashed line. The valley, much narrower through its eastern reaches, has its head at the town of Hope. Also identified are a number of air quality stations in the region.

valley floor is flat, valley walls rise steeply to over 2000 m in the North and 1200 m in the South. The valley has a population of 2.4 million, the majority living in Vancouver and its surrounding communities.

Although the LFV has generally good air quality, it occasionally experiences high ozone concentrations during summertime anticyclonic weather. Elevated ozone levels arise from local production, and with the presence of sea-breeze circulations during summertime fair-weather conditions, highest concentrations generally occur up-valley.

While the aim of this project is to study ozone formation under generic conditions, for the present analysis, observations from a 'design day' were used to characterize the meteorology. The selected day was part of a minor ozone episode during August 2001 that exhibited many features typical of episodes in this region: it began with the appearance of a 500-hPa ridge off the coast and a surface thermal trough extending up from Southwestern B.C on August 8th and by the 10th, stagnant conditions prevailed-winds at Vancouver International airport reached only 3.6 ms^{-1} and temperatures in the valley exceeded 30°C – with maximum hourly ozone concentrations reaching the National Ambient Air Quality Objective (NAAQO) acceptable limit of 82 ppb.

2.2. *Emissions*

One of the first modeling tasks was to develop an emissions inventory of sufficient detail to characterize the spatial and temporal variability of the emissions sources and sufficient flexibility to quickly allow the calculation of new emissions fields for different future scenarios. The inventory considered only VOC and NOx emissions and all sources were classified using three main categories (mobile, area and point sources) and a limited number of sub-categories. The inventory was based on published regional yearly emission totals, which were used to produce hourly emissions using general monthly, daily and hourly temporal emission profiles. Spatial allocation was handled using six general land use categories (urban, transportation, airport, agricultural, water and forest) which formed a non-overlapping covering of the entire domain and four additional hybrid land use categories (shipping, commercial, developed and green space) formed by the union of two or more of the main categories. Table 1 lists emission sources and typical domain wide emission totals for a typical summer day.

Comparisons of hourly emissions totals with a detailed, episode specific inventory, shows the simple model captures temporal and spatial variability of the emissions fields very well (not shown). A high level of detail was not needed to characterize emissions because, in many cases, this detail has been subsumed in the calculation of yearly totals. For instance, with mobile sources (the region's dominant emissions source), yearly totals are based on detailed emission modeling programs making use of the region's fleet characteristics, driving patterns and road network use. The contributions of these factors are directly included in the yearly totals and indirectly in the hourly values through the downscaling of the yearly totals. In essence, a model of the complex emission models is developed using a simple set of scalings.

TABLE 1. Emissions categories and domain wide fluxes of VOC and NOx for a typical summer day based on the simple emissions inventory.

General Category	Sub-category	VOC(tonnes/day)	NOx (tonnes/day)
Mobile	LDV	79	70
	HDV	3	46
	Marine	32	35
	Rail	0	9
	Aircraft	2	3
	Construction	21	35
	Agriculture	2	12
Area	Space Heating	0	0
	Evaporative	52	–
	Forests	235	44
	Livestock	6	–
	Crop	32	–
	Miscellaneous	5	–
Point		13	17
Total		484	270

2.3. *Meteorology*

The influence of meteorology on ozone formation was captured using a series of box models, each advected for a single day, by the mean winds, over the emission fields. Model trajectories were determined from hourly wind fields produced by interpolating wind observations from 53 measuring stations within the model domain on August 10th. Mixing depths (h) for each box were determined using a simple slab model (Gryning and Batchvarova, 1990) based on a TKE budget within the mixing height. This model incorporates the influence of horizontal advection by the mean winds, subsidence and surface heat flux fluxes on mixing heights. The resulting equation for mixing height was solved over the entire domain using a Bott advection scheme with 60-second time step and 500 m spatial resolution.

2.4. *Photochemistry*

Ozone photochemistry was incorporated into the model by means of a scaling analysis of a chemical mechanism. The OZIPR trajectory model was used to simulate a smog chamber using a modified version of the SAPRC90 chemical mechanism (Jiang et al., 1997). The Buckingham Pi method of dimensional analysis was used to express the relevant variables in terms of dimensionless groups. The analysis followed the same methods given by Ainslie and Steyn (2003) and showed dimensionless maximum ozone concentration to be described by a product of powers of dimensionless NOx concentration, dimensionless temperature ($\theta(T)$) and a similarity relationship (f) dependent on the ratio of VOC to NOx (R) only:

$$\frac{[O_3]_{\max}}{j_{av}/k_{NO}} = \left(\frac{[NOx]_{eff}}{j_{av}/k_{NO}}\right)^a \theta(T)^b f(R) \tag{1}$$

where j_{av} is the average rate of NO_2 photolysis and k_{NO} is the NO-O_3 titration rate constant. The similarity relationship was parameterized using two Weibull distributions requiring five parameters: one to normalize the model output, two to describe the shape of the Weibull distributions and two for the location their intersection. Parameters were found using a curve fitting scheme after first Weibull transforming the model output. NOx and VOC concentrations were calculated at each hour based on a budget of emissions rates and deposition loses as well as precursor entrainment/detrainment as column heights varied. Effective initial concentrations, required by the similarity relationship, were calculated by weighting the hourly concentrations by the cumulative NO_2-photolysis rate to which they were exposed.

2.5. Initial Boundary Layer Concentrations

To account for pollutant build up in the boundary layer before the simulation day, a set of pre-conditioning simulations was performed. In these, precursors were emitted into the boundary layer and advected by the mean winds without chemical reactions until steady state concentrations were reached. Final steady state concentrations were found to depend on the choice of background concentration used to initialize the pre-conditioning simulations. These values were chosen so that resulting boundary layer concentrations were consistent with available observations and resulting **peak** ozone concentrations were consistent with recent observed episodic values.

2.6. Baseline Ozone Plume

Figure 2 shows the ozone plume based on the simple model (A) along with the observed plume (B) found by interpolating measured concentrations from 20

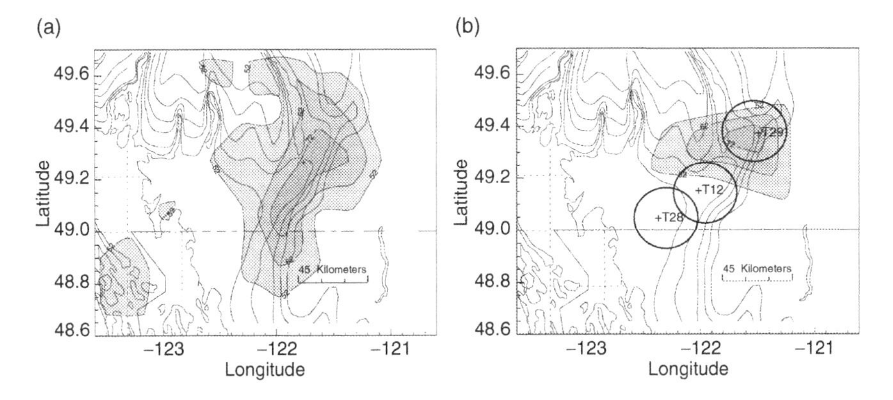

FIGURE 2. Modeled (a) ozone concentrations and observed (b) ozone concentrations (in ppb) at 1800 PST on August 10th. Concentrations over the mountainous regions are uncertain and may differ from the interpolated values given by isopleths. To highlight plume uncertainty away from the valley, the three most easterly measurement stations (T29, T12 and T28) are marked with 20 km radii of influence around each.

monitoring stations. The modeled plume, centered between monitoring stations at Chilliwack (T12) and Hope (T29), has a maximum of 81 ppb. This can be compared with the observed plume, centered 12 km northeast, with maximum of 82 ppb. While the modeled plume covers a much greater area, it is difficult to characterize the spatial extent of the observed plume due to a lack of measuring stations away from the valley axis. To mark this uncertainty, a conservative 20-km radius of influence is indicated around the three most easterly stations.

3. Model Validation

In a departure from previous modeling studies, model validation was not through point by point analysis of model output versus observations but through a high level comparison of model sensitivity to meteorological conditions and emission levels using a range of different modeling techniques and observations. In order to judge the general spatial structure of the ozone plume, comparisons were based on the shape and location of the 52-ppb isopleth. This isopleth was chosen because it represents the NAAQO maximum 1-hour desirable limit for ozone concentration. Comparisons were also based on ability of model to correctly predict maximum concentration to fall into one of three broad categories: below 52 ppb, between 52 and 82 ppb and above 82 ppb, again based on NAAQO guidelines.

3.1. Sensitivity to Meteorological Conditions

A previous classification and regression tree statistical analysis (CART) by Burrows et al. (1995) was used to identify the key meteorological variables influencing ozone formation in the LFV. In a series of sensitivity studies, these variables were individually perturbed about their baseline state (using ranges determined by the CART analysis) with the resulting ozone plume compared to the original (baseline) plume. Response of the modeled ozone plume to changes in temperature, actinic flux, wind speed and mixing heights showed: decreasing temperatures by 5°C reduces peak ozone concentration by about 15% with little change in plume location; decreasing the actinic flux (by using fluxes appropriate to September 10th) shifts the 52-ppb isopleth 10 km eastward and lowers concentrations by about 15%; increasing wind speeds by a factor of 1.5 decreases concentrations by about 10% and shifts plume 20 km eastward and using mixing heights calculated by an Eulerian grid model, which produced a generally deeper mixed layer, reduces maximum concentrations by 5% while leaving the plume in the virtually the same location. These results are consistent with observations and previous studies.

3.2. Sensitivity to Emissions

Next, the sensitivity of the modeled plume to emissions was tested using backcasted emissions levels from 1985 (GVRD, 2003); a period when episodic ozone concentrations were higher (Vingarzan and Taylor, 2003) and plume location

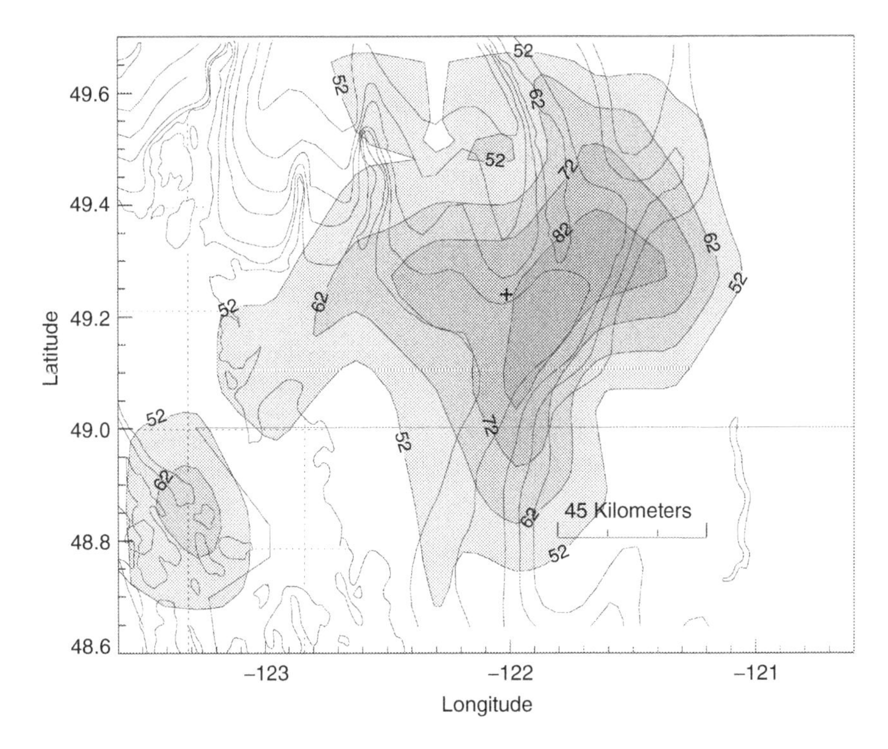

FIGURE 3. Ozone concentration (in ppb) using 1985 emissions levels. Location of 52-ppb centroid given by '+'.

generally closer to the coast (Joe et al., 1996). The resulting plume, shown in Figure 3, has concentrations exceeding 82 ppb, 52-ppb isopleth covering a larger area and plume centroid 20 km further west. These results are consistent with past observations but are likely lower estimates of past concentrations under the modeled conditions.

Nevertheless, the sensitivity studies suggest the main processes controlling ozone formation have been captured to first order and that the model can be used as a screening tool to explore the region's sensitivity to precursor emissions.

4. Domain-Wide Ozone Isopleths

To explore the region's sensitivity to anthropogenic emissions, a series of 225 simulations was performed for a matrix of VOC and NOx emissions levels ranging from 1.7 to 0.2 times year 2000 anthropogenic levels. To include the dependence of initial boundary layer concentrations on emissions, a library of concentration fields was calculated in advance of the simulations. Boundary layer pre-conditioning simulations were performed for emissions rates of 0.2, 0.5, 1.0 and 1.5 times year 2000 levels. For each of the subsequent 225 simulations, that

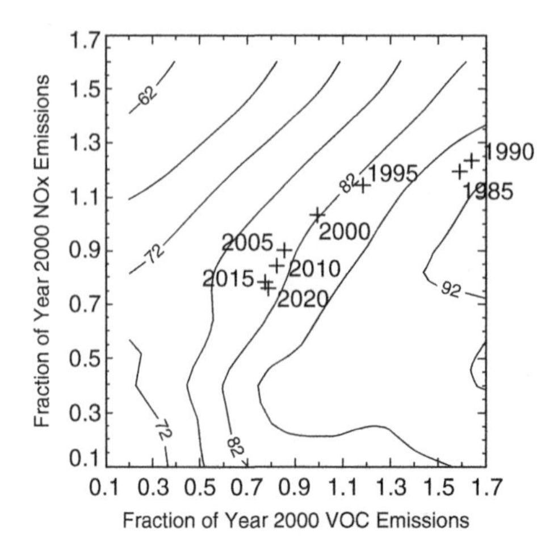

FIGURE 4. Domain-wide ozone isopleths (in ppb) as a function of anthropogenic VOC and NOx emissions. Also included are maximum ozone concentrations for the years 1985 through 2020 based on forecasted and backcasted emissions inventories (GVRD,2003).

simulation's initial boundary layer concentrations were interpolated from the library values based on the it's emission levels.

In Figure 4 the resulting isopleths of maximum ozone concentration as a function of year 2000 anthropogenic VOC and NOx emissions is given. Also located on the figure are inventory estimates for the years 1985 to 2020, in 5-year increments, based on published forecasts and backcasts emission inventories (GVRD, 2003). The figure suggests: episodic ozone concentrations were higher in the past; the region is VOC-limited; greater reductions in VOC than NOx emissions since 1985 have made NOx emission controls a less effective strategy and expected future emissions reductions may not significantly improve episodic ozone concentrations.

5. Discussion and Conclusions

This paper outlines the development of a simple air quality model through the integration of a simple meteorological model and simple emissions inventory and a scaling analysis of ozone photochemistry. The model captures the spatial structure of the ozone plume under typical episode conditions. Model response to wind speed, temperature, mixing heights, actinic fluxes and historic emissions levels is consistent with previous observations and analyses.

Despite realistic appearance of ozone plume, many qualifications must be stressed. First, while the intent was to examine ozone formation under generic conditions, all simulations were based on the meteorology taken from a single day. Also, the general plume response appears conservative and in many cases shows little spatial variability. The pre-conditioning experiments do not consider the effects of nighttime chemistry and decoupling of the stable boundary layer from the residual layer. While these effects likely influence magnitude of the plume response, it appears they don't alter the direction of response – justifying the model's use as a screening tool.

References

Ainslie, B and Steyn, D.G., 2003. Scaling analysis of ozone precursor relationships. *In Air Pollution Modeling and Its Application XVI*, Borrego and Incecik (eds.), Kluwer Academic Press, New York, pp.321-328.

Burrows, W. R., Benjamin, M., Beauchamp, S., Lord, E. L., McCollor, D. and Thomson B., 1995. CART decision-tree statistical analysis and prediction of summer season maximum surface ozone for the Vancouver, Montreal and Atlantic regions of Canada. *Journal of Applied Meteorology* 34, 1848-1862.

Environmental Protection Agency (EPA), 2004. Introduction to the CMAQ model can be found at: http://www.epa.gov/asmdnerl/models3/cmaq.html

Greater Vancouver Regional District (GVRD), 2003. Forecast and backcast of the 2000 emissions inventory for the Lower Fraser Valley airshed 1985-2020. Technical Report, GVRD Policy and Planning Department, Burnaby B.C.

Jiang, W., Singelton, D.L., Hedley, D.L., McLaren, M.,1997. Sensitivity of ozone concentrations to rate constants in a modified SAPRC90 chemical mechanism used for the Canadian Lower Fraser Valley ozone studies. *Atmospheric Environment* 31, 1195-1208.

Joe, H., Steyn, D. G. and Susko, E., 1996. Analysis of trends in tropospheric ozone in the Lower Fraser Valley, British Columbia. *Atmospheric Environment* 30, 3413-3421.

Steyn, D. G., Bottenheim, J. W. and Thomson, R. B., 1997. Overview of tropospheric ozone in the Lower Fraser Valley, and the Pacific 93 field study. Atmospheric Environment 31, 2025-2035.

Venkatram, A., Karamchandani, P., Pai, P. and Goldstein, R., 1994. The development and application of a simplified ozone modeling system (SOMS). Atmospheric Environment 28, 3665-3678.

Vingarzan, R. and Taylor, B., 2003. Trend analysis of ground level ozone in the greater Vancouver/Fraser Valley area of British Columbia. *Atmospheric Environment* 37, 2159-2171.

A Photochemical Screening Tool Based on a Scale Analysis of Ozone Photochemistry

Speaker: B. Ainslie

Questioner: R. Bornstein

Question: What has caused the inland movement of the peak ozone as precursor emissions have been reduced?

Answer: As the region has lowered its VOC emissions faster than NOx emissions, we find the reactivity of the airmass has decreased which translates into an ozone peak later in the day and further inland.

Questioner: P. Builtjes

Question: Can you indicate the limitations of the screening model, would it be valid for other areas, or for larger domains like the whole of Canada?

Answer: Originally, we hoped to build a generic model that would be applicable anywhere, but as we proceeded, we found the model we built made use of many properties specific to the LFV. For example, "a model for an urban area in Eastern North America would have to contend with higher background concentrations and greater advection of precursors from outside the modeling domain – two features which are not significant on the West Coast of Canada."

Questioner: D. Cohan
Question: How robust is your modeled finding of VOC limitation of ozone formation in the Lower Fraser Valley, and is it consistent with other studies of the region?
Answer: There have been a few studies which suggest the region is VOC limited. In the last 2 decades two refineries have been closed which were major VOC sources. Our ambient air is often from the Pacific Ocean which is quite clean and there are few deciduous forests in the region (they are mainly coniferous). However, these results are based on a very simplified model, so I hope to test them against model output from comprehensive grid modeling studies this winter.

2
Modeling and Analysis of Ozone and Nitrogen Oxides in the Southeast United States National Parks

Viney P. Aneja*, Quansong Tong, Daiwen Kang, and John D. Ray

1. Introduction

High O_3 episodes are observed in several eastern US national parks, among which the Great Smoky Mountains National Park by far of the fastest increase in frequency of exceedance days (days when any 8-hour average O_3 concentration exceeding 85 ppbv). It has been well established that the southeast US rural areas are characterized with strong biogenic VOCs emissions and that O_3 production in this region is mostly NOx-limited during summer time. Thus, understanding the contribution of nitrogen oxides to O_3 formation during transport and for local photochemistry is essential to predict what effects the planned reductions in NO_x emissions from large point sources might have on observed O_3 concentrations at these southeast national parks.

Our specific interests in this study are: 1) to quantify the relative importance of point sources and mobile sources to total nitrogen oxides emissions; 2) to identify origins of air masses associated with high levels of nitrogen oxides and O_3; 3) to quantify contributions of individual chemical and physical processes, i.e., chemistry, transport, emission, and deposition, to the budget of production and removal of nitrogen oxides and O_3 in the southeast national parks.

2. Methodology

2.1. Measurements

O_3, NO_y, NO, SO_2, and CO, and meteorological parameters were measured at two enhanced monitoring sites in the southeast United States (TVA, 1995; Olszyna et al. 1998). The Great Smoky Mountains (GRSM) site (35°41′48″N, 83°36′35″W, 1243 m ASL) is located on a ridge in the Great Smoky Mountains

* Viney P. Aneja, Quansong Tong, and Daiwen Kang, North Carolina State University, Raleigh, NC 27695; John D. Ray, Air Resource Division, National Park Science, Denver, CO 80225

National Park. The Mammoth Cave (MACA) site (37°13′04″N, 86°04′25″W, 230 m ASL) is located in Mammoth Cave National Park. Samples of O_3, SO_2, and CO were analyzed using O_3 Model 49, SO_2 Model 43S, and CO Model 48S monitors from Thermo Environmental Instruments, Incorporated (TEII). More detailed information of instruments, experimental techniques, and data QA/QC procedures can be found from Olszyna et al. (1998) and TVA AQ/QC manual.

2.2. *Emission Source Strength: Point vs. Mobile Sources*

According to US EPA, more than 90% of the anthropogenic NO_x ($NO_x = NO + NO_2$) emissions in the United States are from either mobile or point sources (EPA, 1997). Mobile sources emit high levels of CO, but relatively low levels of SO_2, while the reverse is true for point sources. Therefore, it is possible to valuate the relative emission strength using observed pollutant data and the techniques of regression analysis and of emission inventory analysis. The method is described in detail as follows,

Regression Analysis

NO_y is taken as the response variable in a multiple linear regression analysis and the combination of CO and SO_2 as factors. The mathematical expression of the fitted model is:

$$[NO_y] = \alpha[SO_2] + \beta[CO] + \delta \tag{1}$$

where α and β represent the linear coefficients between $[NO_y]$ and $[SO_2]$ and [CO]; δ represents the intercept. After the coefficients, α and β are parameterized, the model can be used to quantify relative contributions from mobile and point sources by plugging in measured CO and SO_2 concentrations. The terms, α, β, and δ in Eq. (1) are estimated using validated measurement data from MACA and GRSM.

Emission Inventory Analysis

Emission inventory analysis obtains the ratio (indicated by x) of NO_y from point sources ($NO_y|_p$) to mobile sources ($NO_y|_m$) from the division of two factors:

$$x = \frac{NO_{y|p}}{NO_{y|m}} = \left[\frac{NSR_p}{NCR_m}\right]\left(\frac{\mu^*[SO_2]}{[CO]-[CO]_{bg}}\right) \tag{2}$$

where NSR_p and NCR_m represent the molar ratios of NO_y to SO_2 from point sources and NO_y to CO from mobile sources, respectively. $[CO]_{bg}$ is background CO concentration independent of local processes. An adjusting parameter, μ, is introduced to account for the fraction of SO_2 oxidized into sulfate before arriving at the receptor site. Thus,

Fraction of NO_y attributed to mobile sources = 1/(1+x) (3-a)
Fraction of NO_y attributed to print sources = x/(1+x) (3-b)

2.3. *Trajectory-Cluster Analysis*

Origin of air masses approaching a receptor site is investigated by a combination of cluster analysis (Dorling, et al., 1992) of Hybrid Single-Particle Lagrangian Integrated Trajectories (HY-SPLIT) model output (Draxler, 1997) and emission source categorization based on EPA emission inventory (EPA, 2001). This method aims to maximize inter-group variance and to minimize within-group variance. The algorithm chosen in this study is the one proposed by Dorling (1992).

2.4. *MAQSIP Model*

A comprehensive 3-dimentional Eulerian grid model called Multiscale Air Quality SImulation Platform (MAQSIP) (Odman and Ingram, 1996) is employed to study the production and removal processes of nitrogen oxides and O_3 in the southeast US national parks. One of MAQSIP's attributes is a truly modular platform where physical/chemical processes are cast into modules following the time-splitting approach. Each process module operates on a common concentration field, making it possible to conduct process budget analysis for each modeled species (Kang et al., 2003). In this study, we analyze process budgets for modeled nitrogen species and O_3.

3. Results and Discussion

3.1. *Source Apportionment of Nitrogen Oxides*

Both methods discussed in section 2.2 have been used to quantify the relative contribution of point sources to total nitrogen oxides emissions. Relative contribution of point source to NO_y are given in Table 1 for each season at GRSM and MACA as well as results from previous studies for other locations in Eastern US sites (Stehr et al., 2000).

TABLE 1. Estimation of contribution of point sources to total NOy emissionat Eastern US sites.

Sites	Periods	Regression Analysis	Emission Inventory Analysis
GRSM, TN[a]	Spring (MAM)	23%	34%
	Summer (JJA)	16%	22%
	Fall (SON)	20%	21%
	Winter (DSF)	30%	38%
	All Data	23%	26%
SHEN, VA[b]	September-Dec.	29±5%	30±8%
MACA, KY[a]	Spring (MAM)	25%	53%
	Summer (JJA)	21%	40%
	Fall (SON)	29%	42%
	Winter (DSF)	33%	47%
	All Data	27%	45%
Wye, MD[b]	Sept.-Dec.	11±5%	16±4%
Arendiville, PE[b]	June-September	21±3%	26±6%

[a] 95% confidence interval on the mean; [b] Stehr. et al., 2000.

Both analyses show a lower fraction of NO_y from point sources in summer and fall, and higher in winter and spring. Point sources contribute 16–22% and 30–38% to total reactive nitrogen oxides in summer and winter at GRSM, and 21–40% and 33–47% at MACA.

3.2. Influence Areas Based on Trajectory-Clustering Analysis

Individual trajectories in 1996 were calculated using the HSPLIT-4 model for those days when simultaneous measurements of O_3 and NO_y were available at GRSM. The whole trajectory set was then clustered into seven groups using the clustering algorithm discussed by Dorling (1992). A seed trajectory, i.e. a trajectory that can minimize the average Root Mean Square Deviation (RMSD) for all other trajectories within a cluster, was selected out of each cluster. Figure 1 illustrates the resulting seed trajectories from each cluster. Each seed trajectory is labeled with the direction best describing the relative position of its origin and general path to the receptor site. The number in parenthesis shows the percentage of trajectories assigned to a particular cluster. Emission density of NO_y, taken from the EPA county-based emission inventory (EPA, 2001), is displayed as the background in Figure 1 to provide a visual indicator of regional emission levels that

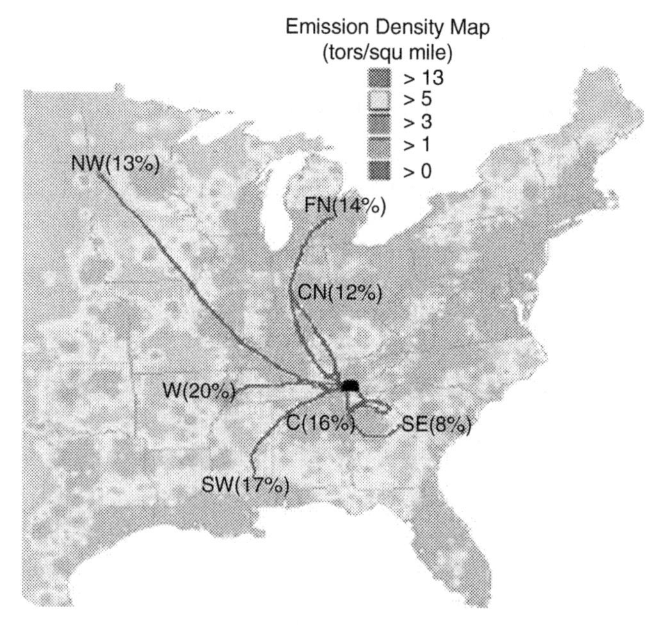

FIGURE 1. Average trajectory of air masses approaching the Great Smoky Mountain (GRSM) National Park, TN. Each average trajectory is labeled with the relative position of its origin to the receptor site. The number in the parenthesis shows the percent of the trajectories assigned to a particular cluster. The background is emission density of NO_y taken from EPA (1999).

might be loaded into those air parcels following a cluster trajectory. Figure 1 shows that air masses from west (W, 20%) and southwest (SW, 17%) sweep GRSM most frequently, while pollutants transported from the eastern half (i.e., Eastern, Northeast, or Southeast) have limited influence (< 10%) on air quality in the Great Smoky Mountain National Park.

3.3. Process Budget Analysis of Nitrogen Oxides by MAQSIP

Process budget analysis using MAQSIP shows that although the major contributions to NO_x at the two locations come from horizontal advection (58–62%) and local emissions (38–42%), and the primary removal process for NO_x is local chemistry (92–95%), the magnitude of each process at MACA is more than 3 times as large as that at GRSM, which matches the observational results (1.2-4.8 times, Tong et al., submitted manuscript, 2004). However, 84% of NO_z, the oxidized products of NO_x, is the result of local chemistry at MACA, which only accounts for 32% at GRSM; the rest comes from transport. If we compare the process budgets of NO_z with that of O_3 (Figure 2, recreated from Kang et al., 2003), the similarity among the locations as well as the contributions of each individual process at each location is apparent, especially at the positive side, i.e. the production or accumulation side. For instance, chemistry contributions of 32% and 84% to NO_z correspond to 26% and 80% to O_3 at GRSM and MACA, respectively. Similarity between NO_z and O_3 process budgets further demonstrates that NO_z can be used as evidence of close association between nitrogen oxides and effective O_3 production at these rural locations.

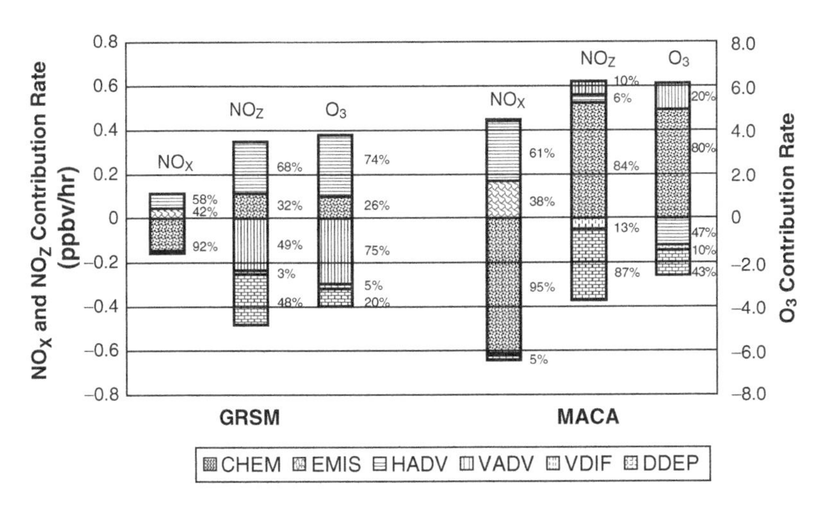

FIGURE 2. Process budget analysis of NO_x, NO_z, and O_3 using MAQSIP in the chosen domain during the 1995 summer time.

4. Conclusions

This study focuses on both observation-based and modeling analyses in elucidating source attribution, influence area, and process budget of reactive nitrogen oxides at two rural southeast national parks. Using two independent observation-based techniques, we demonstrate that point sources contribute a minimum of 23%~26% and 27%~45% of total NO_y at GRSM and MACA respectively. The influencing area, or origin of nitrogen oxides, is investigated using trajectory-cluster analysis. The result shows that air masses from western and southwest regions sweep over GRSM most frequently, while pollutants transported from the eastern half has limited influence (< 10%) on the air quality in the Great Smoky Mountain National Park. Processes budget analysis using MAQSIP reveals similarities between NO_z and O_3 process budgets, which serve as further evidences of close association between nitrogen oxides and effective O_3 production at these rural locations.

Acknowledgements This research was funded by the National Park Service, Air Resources Division, cooperative agreement #4000-7-9003. The authors thank Dr. Rohit Mathur in MCNC for his help in the modeling and data processing, and Ms. Binyu Wang at North Carolina State University for her assistance in statistical analysis. Thanks to Bob Carson, Scott Berenyi, and Jim Renfro (NPS), and to Ken Olszyna (TVA) for their efforts in collecting the measurement data.

References

Dorling, S.T., T.D. David, and C.E. Pierce, 1992. Cluster analysis: a technique for estimating the synoptic meteorological controls on air and precipitation chemistry – method and applications. Atmos. Env., 26A, 2575-2581.

Draxler, R.R., 1997. Description of the HYSPLIT_4 modeling system. National Oceanic and Atmospheric Administration technical Memorandum ERL ARL-224.

Environmental Protection Agency, 1997. National air pollutant emission trends, 1900-1996, EPA-454/R-97-011, Washington, D.C., Dec.

Kang, D., Aneja, V.P., Mathur, R., and Ray, J., 2003. Nonmethane hydrocarbons and ozone in the rural southeast United States national parks: A model sensitivity analysis and its comparison with measurement, J. of Geophys. Res., 108(D10), 4604, doi: 10.1029/2002JD003054.

Olszyna, Kenneth J.; William, J. Parkhurst; and James F. Meagher. 1998. Air chemistry during the 1995 SOS/Nashville intensive determined from level 2 network. J. Geophys. Res. 103, 31, 143-31, 153, December 20.

Tennessee Valley Authority (TVA), SOS Nashville / Middle Tennessee Ozone Study, vol. 2, 1995. Level 2 operations manual, Environ. Res. Cent., Muscle Shoals, Ala.

Modeling and Analysis of Ozone and Nitrogen Oxides in the Southeast United States National Parks

Speaker: Tong

Questioner: A. Ebel
Question: Budget calculations have been carried out for different processes, among them horizontal and vertical advection. What is the net effect of both in the examples presented in the paper?
Answer: The net effects of horizontal and vertical advections can be calculated from figure 2 in our full manuscript, as well in the corresponding figures in my presentation.

Questioner: S. T. Rao
Question: Have you looked at ozone data in the post-1998 period to assess whether ozone has been improving in the southeast? As you know, there are substantial decreases in the NOx emissions in the post 2001 period. It would be good to examine the linkages between emission reductions and ozone concentrations.
Answer: This has been covered in another one of our papers, which talks about ozone observations after 1998 from the same sites. It has been reported that NOx cut so far has not contributed too much to O_3 reductions, since the benefits of decreasing emissions of NOx are partially offset by an increase in the O_3 production efficiency of remaining NOx.

Questioner: D. Cohan
Question: In regressing NO_y against SO_2 and CO, how do you account for the fact that the ratios of these pollutants change during transport from an emissions source to a receptor?
Answer: We have applied an oxidizing parameter to account for the pollutants change during transport in the emission inventory analysis. For regression analysis, we assume the coefficients obtained by regression analysis of real-world measurements have taken into account the changes due to transformations during transport.

Questioner: R. Bornstein
Question: Could the reduction in reactive organics be postponing the location of ozone max to the Parks, and thus have caused the increase in "bad" days from the '80's to the '90's?
Answer: It depends. Our budget analysis shows that while local photochemistry is important at Mammoth Cave national park, O_3 in the Great Smoky Mountains national park is dominated by transport and regional O_3 background. Apparently both sites are subject to upwind emissions, particularly that of NOx.

3 An Investigation of Local Anthropogenic Effects on Photochemical Air Pollution in Istanbul with Model Study

Umit Antepliogl u[*], Selahattin Incecik[**], and Sema Topcu[**]

1. Introduction

Urban ozone is a major pollutant produced by various sources as well as urban traffic through photochemical transformation of nitrogen oxides, carbon monoxide, and volatile organic compounds. Ozone pollution in urban areas is a complex problem involving both atmospheric diffusion processes, chemical reactions and transport. The superimposition of chemical production and physical processes leads to episodic level of photochemical air pollution under favorable meteorological conditions and abundance of precursors.

The air quality of Istanbul has been a major concern since the early 1980s. The city has experienced several air pollution problems in 1980s. Usage of poor quality lignite was banned in late 1993. The fuel switching from coal to natural gas has gradually improved the air quality (Topcu et al., 2003). Today SO_2 and TSP levels are below the national air quality standards. However, a new air pollution type has appeared in Istanbul that is the photochemical pollution. Surface ozone concentrations are increasing in the city depending on increasing numbers of cars that use mostly gasoline and poor dispersion conditions. Ozone episodes are frequently observed when anticyclonic pressure systems are in the vicinity of Istanbul.

A series of ozone studies have been conducted in Istanbul in the last half decade. Previous works are mainly focused on analyses of the ozone and its precursor monitored data (Topcu and Incecik, 1999; Topcu and Incecik, 2002; Topcu and Incecik. 2003). It is not easy to obtain the inner relations among different processes and to identify which process is dominant solely by observation studies due to the limited number of measurements available. In order to understand photochemical smog formation and its nonlinear characteristics and then to apply emission reduction strategies, special measurements and models are needed.

[*] Umit Antepliogl u, Division of Meteorology, Kandilli Observatory, Bosphorus University, Istanbul, Turkey.

[**] Selahattin Incecik and Sema Topcu, Department of Meteorology, Istanbul Technical University, Maslak, Istanbul, Turkey.

There are several studies in understanding these process in Europe and Asian cities Tirabasi et al., (2001); Blond et al., (2003), Andreani-Aksoyoglu, (2004) and Huang et al., (2004).

The first attempt in understanding of spatial and temporal distributions of ozone and its simulations was made by Anteplioglu (2000) using SAIMM hydrostatic meteorological model and an urban airshed model (UAM-V) in Istanbul. Furthermore, a new attempt was made using non-hydrostatic PSU/NCAR Mesoscale Meteorological Model (MM5) (Anteplioglu et al., 2002). Besides, CAMx as an Eulerian Photochemical model which is one of the best models in this field, was used in model studies.

In this study, the levels of NOx emissions, which is the principal ozone precursor and ozone concentrations covering the urban areas of the city, were investigated based on measurements and both wind and temperature fields in the area for a worst photochemical period (22-26 June) in 2003 ozone season.

2. Region

Istanbul is one of the world's largest metropolitan areas, containing nearly 12 million inhabitants. The city (41°N,29°E) is located on both continents, Asia and Europe. About 40 percent of the city lives in the Asian part of the city and the most part of the business centers are in the European side. The region and air quality stations are given in Fig.1. Anthropogenic emissions based on traffic have increased significantly in the last decade. The dominant pollution sources are domestic heating, traffic and industry in the city. The city is connected by two bridges.

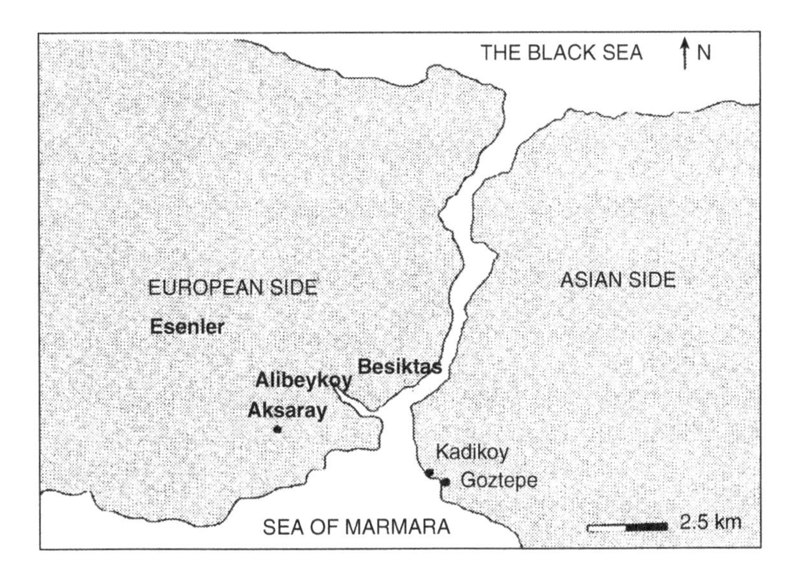

FIGURE 1. The region and air quality stations.

These are Bosphorus and Fatih Sultan Mehmet Bridges. The Bosphorus Bridge carries the heaviest vehicular volumes of the bridges due to its location in the city center. According to Highway report (2003) the number of cars passing over the Bosphorus Bridge is 1,945,650 in July 2002 and 1,766,995 in January 2003. About 80% of the vehicles passing over the Bosphorus Bridge are cars. Approximately 70,000 cars use Bosphorus Bridge every day. 98% of the total cars use gasoline and the other is diesel. The climate is basically influenced by The Black Sea and Sea of Marmara and The Mediterranean Sea. The complex topography of the region is characterized by small valleys, hills and the Bosphorus wide channel. During summer anti-cyclonic conditions, local circulation is completely driven by thermal contrasts between land and Sea of Marmara and The Black Sea. During the sunny summer days when the chemical air mass becomes stagnant under low wind, ozone pollution episodes are observed over a wide area.

Surface hourly ozone concentrations are provided by the IBB Environmental Protection Office. Unfortunately, only two urban monitoring sites data are available. Additionally, there are also 5 monitoring sites including NOx and CO measurements are used.

3. Episode Analysis

Generally mesoscale and regional factors dominate the episode characteristics in the region. Weak morning surface winds, early morning stable conditions and higher precursor concentrations cause higher ozone concentrations. These conditions are most favorable for ozone formation. On the other hand, intense solar radiation causes convective conditions and increases instability. Intense solar radiation and other favorable ozone formation conditions might be the reason for the June episodes. Time series of solar radiation in Istanbul indicate that maximum levels of solar radiation have been occurred in June 2003. The impact of the meteorological conditions as well as solar radiation, wind and cloud cover on high ozone in the region are given in Table 1. The high ozone days are assumed as concentrations of exceeding 100 $\mu g/m^3$. Table 1 indicates the number of days of ozone concentrations exceeding 100 $\mu g/m^3$ and related meteorological conditions. Maximum levels of ozone have been in June 2003 for both Aksaray and Kadiköy sites. Recently, motor vehicles are the major source of a number of air pollutants like CO, NOx, HCs, lead and VOCs depending on the number of cars.

TABLE 1. The number of days of ozone concentrations exceeding 100 $\mu g/m^3$ in the summer months of 2003.

	Aksaray		Kadikoy	
Month/ Station	number of days/data	number of days of O_3> 100 $\mu g/m^3$	number of days/data	number of days of O_3>100 $\mu g/m^3$
June	29	11	30	5
July	28	3	14	0
August	29	3	8	2

The number of cars exceeded 1.7 million in the city. NOx (NO + NO_2) concentrations are one of the major ozone precursors based on traffic emissions. Monitoring data of NOx are provided from five locations in the city. Four of them belong to the European side of the city. Time series of the NOx data for June, July and August are shown in Fig.2. Aksaray monitoring station seems to be maximum NOx potential location in the city. The secondary maximum levels of NOx have occurred in Besiktas. These figures may reflect the traffic volume in the city. In this study, 22-26 June 2003 was selected as an episode depending on the high ozone concentrations which are observed. Daily maximum ozone concentrations in June, 2003 are given in Fig.3. Maximum ozone concentrations in both monitoring locations of the city are higher than the other days. In order to understand

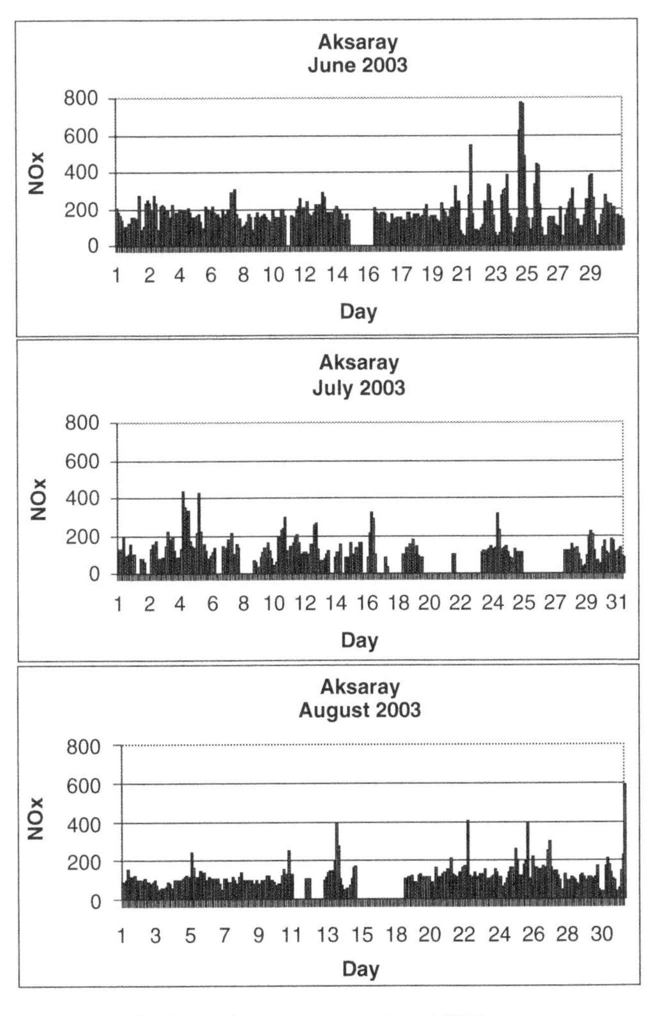

FIGURE 2. Time series of NOx in the summer months of 2003.

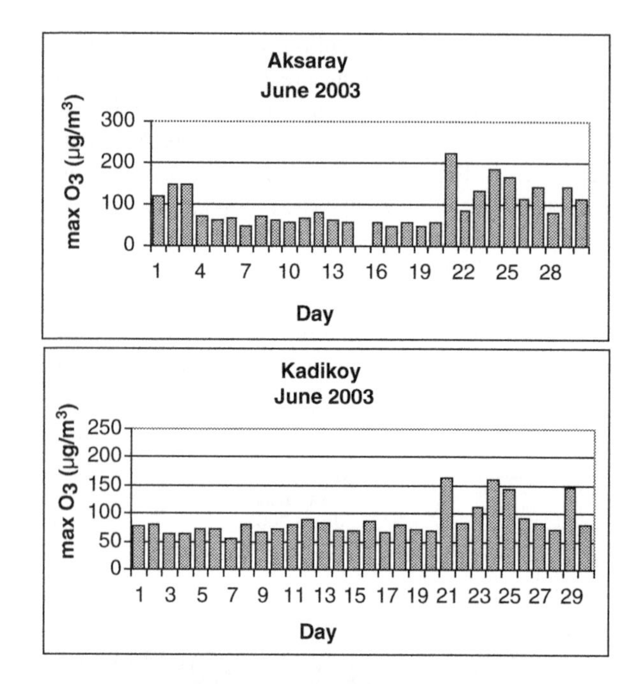

FIGURE 3. Daily ozone maximum concentrations in both sites of the city in June 2003.

the impact of ozone precursors on the ozone levels depending on the traffic hours, we have used the episode period days starting two days before and two days after the episode.

4. Model Simulations

4.1. MM5 Simulation

MM5 is a mesoscale meteorological model developed by PSU/NCAR. The MM5 meteorological pre-processor was run to obtain three dimensional meteorological fields. The area where MM5 was used was larger than the area where concentrations were evaluated. The model rebuilds three-dimensional wind and potential temperature fields. In this model, we have used MRF for boundary layer, Schultz microphysics for moisture scheme, KF2 for cloud scheme and NOAH land-surface scheme for the soil respectively. MM5 is configured with four two way interactive nested grids. Three domains are used. The outmost domain covers the most of Europe to account for the pressure gradients and frontal passages. The first domain has 148×170 grid points with a horizontal grid distance of 27km. The second domain has 88×94 grid points with a horizontal resolution of 9km. The innermost domain encompasses the whole Marmara region and has a 91×139 grid points with a horizontal resolution of 3.0km. In the vertical direction 30 sigma levels were defined unequally from ground to top of the model 100 mb. The first ten layers are included in the atmospheric boundary layer. The episode simulated by the photochemical model occurred in June 19-21, 2003, during a period characterized by clear sky condition and light wind speed in the morning hours. The modeling system is run in a box on the Istanbul urban area. Meteorological input is obtained by MM5 meteorological pre-processor. The initial and boundary concentrations of species are obtained from the default values.

Land–Sea breezes influence ozone concentrations at coastal areas. In European cities, the highest ozone concentrations take place in summer under stable high-pressure systems with clear skies (Andreani-Aksoyoglu et al., 2004). During these episodes, ozone levels are well above the international guidelines. Model results for potential temperature fields and wind fields for 19-21 of June are shown in Figs. 4-6. These figures represent the results of first level of the model in midday

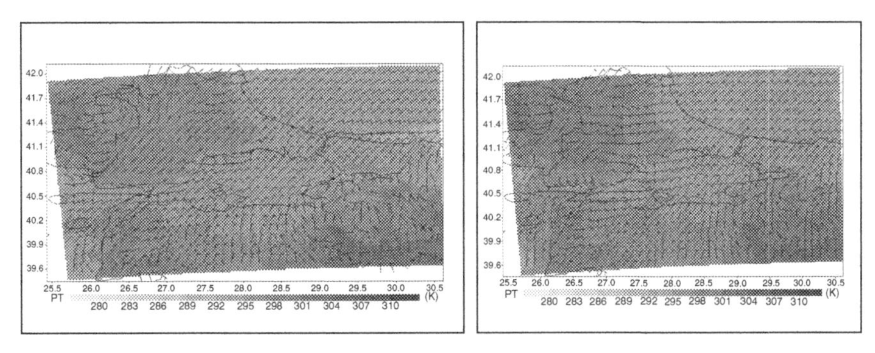

FIGURE 4. MM5 Potential temperatures and surface wind fields for June 19th at 10 UTC (left) and 14 UTC (right).

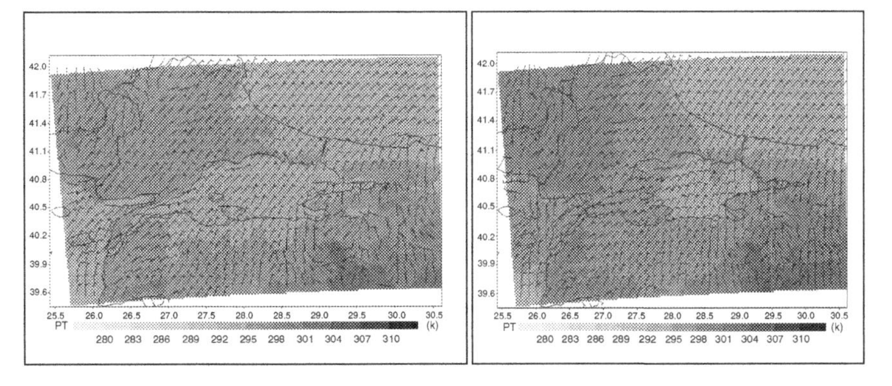

FIGURE 5. MM5 Potential temperatures and surface wind fields for June 20th at 10 UTC (left) and 14 UTC (right).

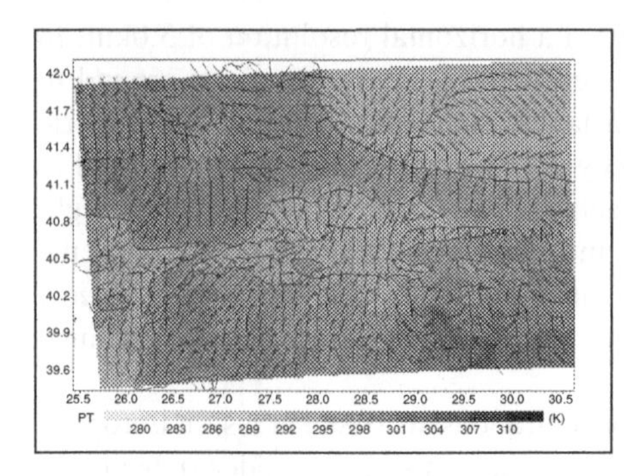

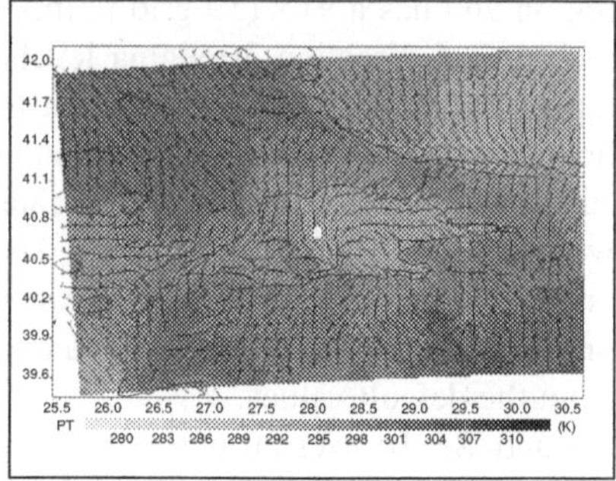

FIGURE 6. MM5 Potential temperatures and surface wind fields for June 21st at 10 UTC (left) and 14 UTC (right).

hours for 10 and 14 UTC. The effects of surface heating which is an important driving force in inducing the sea/land breeze circulation can be seen in the model results. The circulation is clearly appearing during intense solar heating hours and then disappearing at the late afternoon hours. Synoptic pattern was characterized by an anticyclonic system during this period. The surface winds are dominated by weak WSW winds.

The overall picture from 19-21 of June 2003 which is the beginning of episode period, simulations show a clear signal with sea-breeze circulation during the episode dominated by mesoscale activities in the region. We hoped that our limited understanding of urban ozone formation will be improved by using non-hydrostatic MM5 meso-scale meteorological model and numerical simulations.

4.2. CAMx Simulations

The available hourly ozone and its precursors data, the potential of the highest photochemical activity, a case has been selected for simulations by CAMx (Environ, 2002). CAMx is an Eulerian photochemical model for integrated assessment of air pollution (gaseous and particulates). The model requires inputs to describe photochemical conditions, surface characteristics, initial and boundary conditions, emission rates and meteorological fields over the modeling system. Model domain consists of 22 vertical layers up to 6000 m and 2 km horizontal resolution, with 60^*60 grids. Chemical mechanisms based on Carbon bond mechanism. June 2003 has been used for initialization. There is no emission inventory yet at national and city level for a range of pollutants, including NOx and VOCs. Therefore, the only traffic emissions except local roads in the city are used. Model simulations of NO and NO_2 for June 21 at 10 and 14UTC are illustrated in Figs 7 and 8.

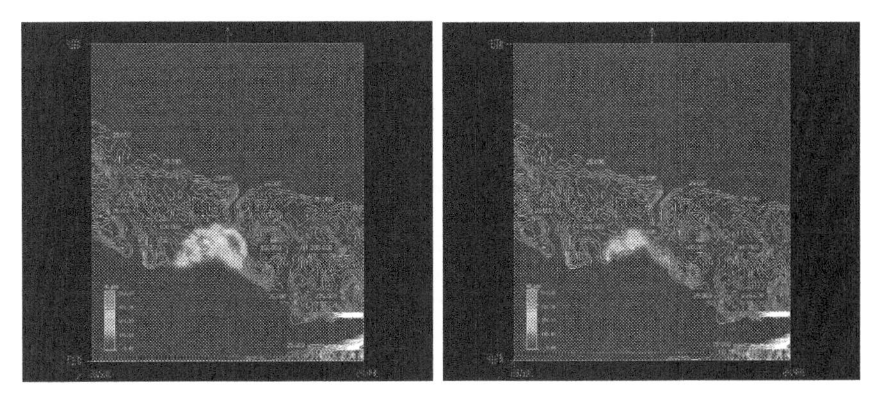

FIGURE 7. NO for June 21st 10 UTC (left) and 14 UTC (right).

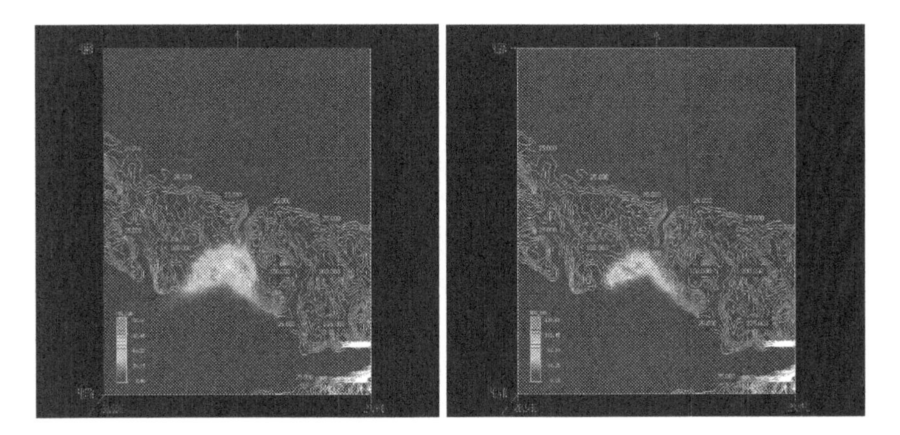

FIGURE 8. NO_2 for June 21st 10 UTC (left) and 14 UTC (right).

5. Conclusions

In this study we reported on the ozone conducive days for a selected period. The realistic evaluations of flow fields and potential temperature by MM5 simulations have been shown over Istanbul and surrounding areas. The findings of the study are in agreement with what is generally known about the environmental conditions for ozone pollution, namely the importance of sunlight and motor vehicle traffic. Results strengthen the expected flow characteristics over complex terrain and ozone concentrations. Its realistic representations by the model will influence the success of the wind field simulations. Realistic emission inventory is very important for proper modeling results. Emissions for the restricted area are much lower than the whole area. Therefore updated emission inventories are needed for future studies. Emissions of chemical complex in Izmit bay should be added to the improved emission data.

References

Andreani-Aksoyoglu S., A.S.H.Prevot, U.Baltensperger, J.Keller, and J.Dorman, 2004, Modeling of formation and distribution of secondary aerosols in the Milan area, *J.Geophysical Research*, 109, D05306,1-12.

Antepliogl U., 2000, Investigation of surface ozone by photochemical – dynamic model in Istanbul (in Turkish), PhD Thesis, Istanbul Technical University

Antepliogl, U, Topcu, S., and Incecik,S., 2002, An application of a photochemical model for urban airshed in Istanbul, Air Pollution Modeling and Its Application, C.Borrego and G.Schayes, Ed., Kluwer Academic / Plenium Publishers, New York, pp 167-175.

Blond,N., L.Bel and R.Vautard, 2003, Three-dimensional ozone data analysis with an air quality model over the Paris area, *J. of Geophysical Research*, 108 D.23,4744.

Environ, 2002, User's Guide, Comprehensive Air Quality Model with Extensions (CAMx), Environ International Corporation, California.

Highways General Directorate, 2003, Bosphorus Bridge, Highway Traffic Report (in Turkish)

Huang J.P., J.C.H.Fung, A.K.H.Lau and Y.Qin, 2004, Numerical study of ozone episode formation in Hong Kong, 13th World Clean Air and Environmental Protection Congress, 22-27 August, 2004.

Tirabasi, T., M.Deserti, L.Passni and A.Kerschbuma, 2001, Photochemical smog evaluation in an urban area for environmental management, ISEE/RC'2001, Moscow, Russia, September,26-29, 2001.

Topcu S and S.Incecik, 1999, An evaluation of the preliminary measurements of surface ozone levels (in Turkish), National Conference on Air Pollution and Control, September 27-29, 1999, Izmir (281-285).

Topcu, S, and Incecik, S., 2002, Surface ozone measurements and meteorological influences in the urban atmosphere of Istanbul, *Int. J Environment and Pollution*,17, 390-404.

Topcu, S., Incecik, S. and Unal, Y.S., 2003, The influence of meteorological conditions and stringent emission control on high TSP episodes in Istanbul, *Environmental Sciences and Pollution Research*, 8, 24-32.

Topcu, S., and Incecik, S., 2003, Characteristics of surface ozone concentrations in urban atmosphere of Istanbul: A case study, *Fresenius Environmental Bulletin*. 12.

An Investigation of Local Anthropogenic Effects in Istanbul with Model Study

Speaker: S. Incecik

Questioner: P. Builtjes

Question: Are your calculated NOx concentrations similar to the NOx observations you showed?

Answer: Yes, calculated NOx concentrations are similar to the NOx measurements taken at the both side of the city.

4
Forecasting Urban Meteorology, Air Pollution and Population Exposure (European FUMAPEX Project)

A. Baklanov, N. Bjergene, B. Fay, S. Finardi, A. Gross, M. Jantunen, J. Kukkonen, A. Rasmussen, A. Skouloudis, L. H. Slørdal, and R. S. Sokhi

1. Introduction

The main problem in forecasting urban air pollution is the prediction of episodes with high pollutant concentrations in urban areas, where most of the well-known methods and models based on in situ meteorological measurements, fail to realistically produce the meteorological input fields for the urban air pollution (UAP) models.

UAP models in operational urban air quality information and forecasting systems, as a rule, use simple in-situ meteorological measurements which are fed into meteorological pre-processors (Fig. 1, dash line). Lacking an adequate description of physical phenomena and the complex data assimilation and parameterisations of numerical weather prediction (NWP) models, these pre-processors do not achieve the potential of NWP models in providing all the meteorological fields needed by modern UAP models to improve urban air quality forecasts.

During the last decade substantial progress in NWP modelling and in the description of urban atmospheric processes has been achieved. Modern nested NWP models are utilising land-use databases down to hundred meters resolution or finer, and are approaching the necessary horizontal and vertical resolution to provide weather forecasts for the urban scale. In combination with the recent scientific developments in the field of urban sublayer atmospheric physics and the enhanced availability of high-resolution urban surface characteristics, the capability of the NWP models to provide high quality urban meteorological data will therefore increase.

Despite the increased resolution of existing operational numerical weather prediction models, urban and non-urban areas mostly contain similar sub-surface, surface, and boundary layer formulations. These do not account for specifically urban dynamics and energy exchange and their impact on the numerical simulation of the atmospheric boundary layer and its various characteristics (e.g. internal boundary layers, urban heat island, precipitation patterns). Additionally NWP models are not primarily developed for air pollution modelling and their results need to be

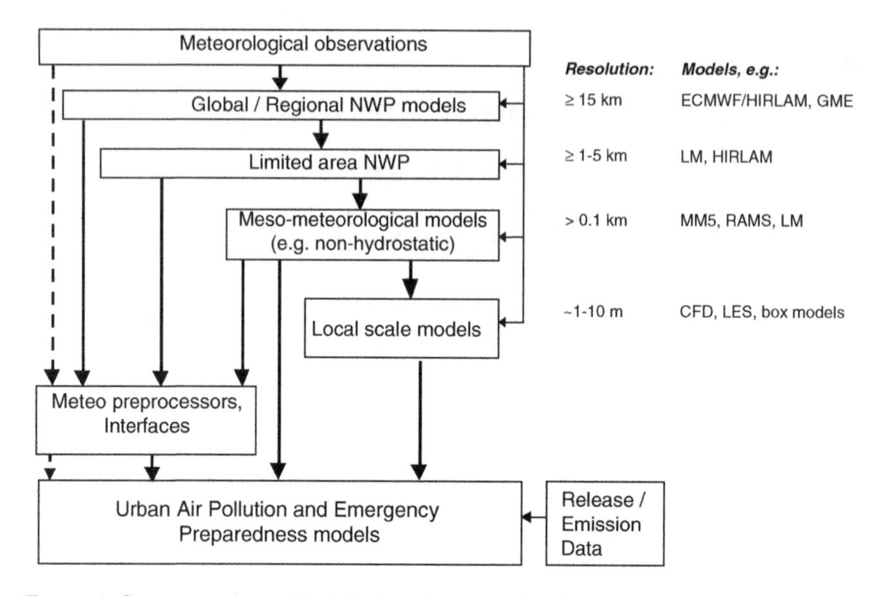

FIGURE 1. Current regulatory (dash line) and suggested (solid line) ways for systems of forecasting of urban meteorology for urban air quality information and forecasting systems.

designed as input to urban and meso-scale air quality models. Therefore the situation in urban air quality information and forecasting systems is changing and requires a revision of the conventional conception of UAP forecasting.

2. Project Objectives and Implementation

In response to the above research needs, a research project on Integrated Systems for Forecasting Urban Meteorology, Air Pollution and Population Exposure (FUMAPEX, project web-site: http://fumapex.dmi.dk) was initiated by the COST 715 community, and funded under the Fifth Framework Programme, Subprogram: Environment and Sustainable Development, Key Action 4: City of Tomorrow and Cultural Heritage. FUMAPEX started in November 2002, will continue for a period of 3 years and involves 16 contractors and 6 subcontractors from 10 European countries. It is also a member of the EC CLEAR cluster.

The main objectives of FUMAPEX are to improve meteorological forecasts for urban areas, to connect NWP models to UAP and population exposure models, to build improved Urban Air Quality Information and Forecasting Systems (UAQIFS), and to demonstrate their application in cities subject to various European climates. The FUMAPEX scheme for the improvement of meteorological forecasts in urban areas, interfaces and integration with UAP and population exposure (PE) models for UAQIFS is presented in Figures 1 and 2.

The improvement of urban meteorological forecasts will also provide information to city managers regarding other hazardous or harmful urban climates

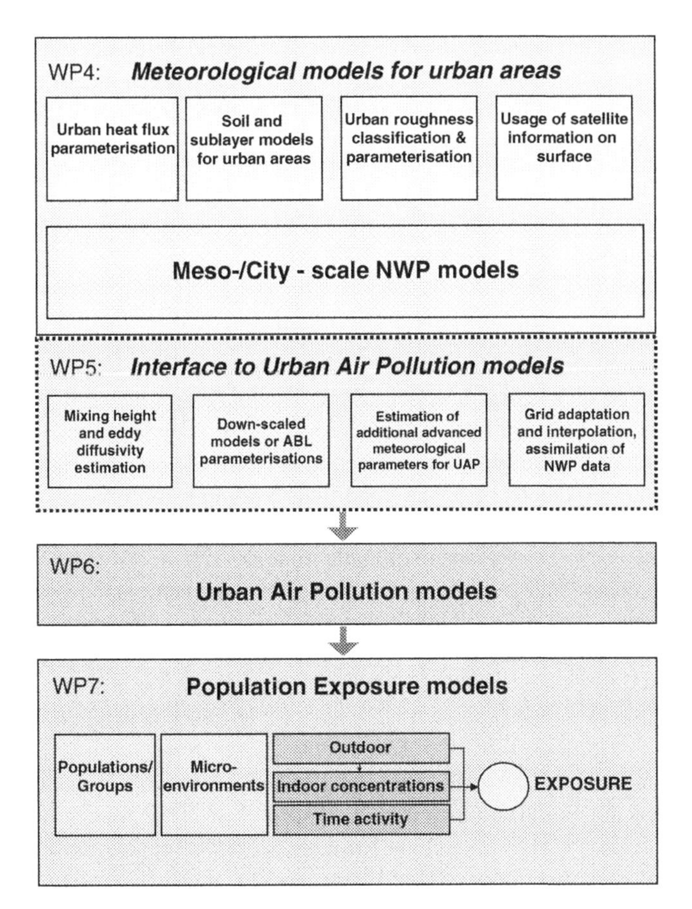

FIGURE 2. FUMAPEX scheme of the improvements of meteorological forecasts (NWP) in urban areas, interfaces and integration with UAP and population exposure models for urban air quality information forecasting and information systems (UAQIFS).

(e.g. urban runoff and flooding, icing and snow accumulation, high urban winds or gusts, heat or cold stress in growing cities and/or a warming climate). Moreover the availability of reliable urban scale weather forecasts could be of relevant support for the emergency management of fires, accidental toxic emissions, potential terrorist actions etc..

In order to achieve the goal of establishing and implementing an improved new urban air quality information forecasting and information system to assist sustainable urban development, the following steps are being undertaken:

1. improve predictions of the meteorological fields needed by UAP models by refining resolution and developing specific parameterisations of urban effects in NWP models,
2. develop suitable interface from NWP to UAP models,

3. validate the improvements in NWP models and meteorological pre-processors by evaluating their effects on the UAP models against urban measurement data,
4. apply the improved meteorological data to UAQIFS, emergency preparedness and population exposure models and compare and analyse the results, and
5. link meteorologists / NWP modellers with UAP scientists and UAQIFS end-users.

The project involves the following main stages.

Classification of air pollution episodes focusing on relevant meteorological variables

1. Identification and classification of various types of air pollution episodes in cities located in different European climatic and geographic regions.
2. Key pollutants relevant to EU Air Quality Directives and Daughter Directives (EC/96/62; EC/99/30) will be selected for different regions/city characteristics.
3. Classification of meteorological conditions leading to pollution episodes and identification of the more relevant meteorological parameters to define these conditions in various European climatic regions.
4. Compilation and analysis of existing datasets of concentration and meteorological data measured during UAP episodes in different European climatic and geographic regions.

Improvement of the quality of urban meteorological forecasting for UAP and PE models

1. Improvement of urban weather forecasts and calculation of key meteorological parameters for pollution episodes. A hierarchy of NWP models from large-scale global circulation models to local-scale obstacle-resolving meteorological models will be employed (see Figure 1).
2. Improvement of boundary layer formulations and parameterisations, and the physiographic data description of urban areas.
3. Development of assimilation techniques with satellite remote sensing data in NWP.
4. Development of interfaces to connect NWP to UAP models.

Verification of the improved NWP, UAP and population exposure models

1. Evaluation of improved urban meteorological forecast models based on UAP episodes.
2. Estimation of sensitivity of UAP models to uncertainties in meteorological input data.
3. Evaluation of the impact of the improved output of the urban air quality models on simulations of an urban population exposure model.

Application of Urban Air Quality Information Forecasting and Information Systems (UAQIFS) and emergency systems

1. Integration of the improved NWP, UAP and PE models into UAQIFSs.
2. Implementation of the new improved UAQIFS in air quality forecasting mode to be applied in four target cities, in urban management or public health and

planning mode in one target city, and of the emergency preparedness system in one target city.
3. The six target cities for testing the improved systems implementations in association with end users are: Oslo (Norway), Turin (Italy), Helsinki (Finland), Castellon/ Valencia (Spain), Bologna (Italy), and Copenhagen (Denmark).

3. Current Urban Meteorology Achievements

3.1. Testing the Quality of Different Operational NWP Systems for Urban Areas

The focus of this aspect of the project is on the description of existing forecasting systems and the evaluation of their capability to forecast key meteorological parameters in urban areas. Partners in FUMAPEX use different operational NWP, or research mesoscale models, for providing the meteorological input data for the UAP models in the UAQIFSs (Fay, 2003; Fay *et al*, 2004). Therefore the tasks comprise:

1. the description and comparison of the selected operational NWP models
2. the harmonised analysis and evaluation of the model simulations with original, and increased model resolution, but unaltered physics for agreed cities and episodes, and
3. the controlled inter-comparison of the simulations of the various models.

The NWP and meso-meteorological models considered for urbanisation include: 1. DMI-HIRLAM model (DMI); 2. Lokalmodell (LM) (DWD); 3. MM5 model (CORIA, DNMI, UH); 4. RAMS model (CEAM, Arianet); 5. Topographic Vorticity-Mode Mesoscale (TVM) Model (UCL); 6. Finite Volume Model (FVM) (EPFL); 7. SUBMESO model (ECN); 8. aLMo model (MeteoSwiss).

A model overview (Fay, 2003) was completed for operational mesoscale NWP models, plus established research mesoscale models. It contains detailed information on all model aspects including information on the different interfaces, or pre-processors, converting the NWP output data into input data for UAP models.

A model comparison design study on model comparison and evaluation (Fay, 2003a) deals in detail with the choice of cities and episodes, and the various theoretical and applied aspects of evaluation methodology for episode, as well as for longer-term evaluation. The choice of cities, the proposed evaluation strategy, and the harmonised use of the GRADS visualisation software and of the MMAS evaluation tool (FMI) had been discussed and agreed upon.

With focus on winter and spring episodes in Helsinki, simulations were performed using the operational NWP/mesoscale models HIRLAM, LM, MM5 and RAMS (Fay *et al.*, 2004). Results show improvements with increasing model resolution (down to 1.1 km) but the need for adapted external parameters and urbanised parameterisations was apparent. Harmonised model evaluation and comparison was discussed in detail (Neunhäuserer *et al.*, 2004) and will be performed for all target cities.

3.2. Improvement of Parameterisation of Urban Atmospheric Processes and Urban Physiographic Data Classification

The following urban features can influence the atmospheric flow, microclimate, turbulence regime and, consequently, the transport, dispersion, and deposition of atmospheric pollutants within urban areas: (1) local-scale non-homogeneities, sharp changes of roughness and heat fluxes, (2) the building effect in reducing wind velocity, (3) redistribution of eddies, from large to small, due to buildings, (4) trapping of radiation in street canyons, (5) effect of urban soil structure on diffusivities of heat and water vapour, (6) anthropogenic heat fluxes, including the urban heat island effect, (7) urban internal boundary layers and the urban mixing height (MH), (8) effects of pollutants (including aerosols) on urban meteorology and climate, (9) urban effects on clouds and precipitation.

Accordingly the following aspects of urban effects was considered in improved urban-scale NWP models: higher spatial grid resolution and model downscaling, improved physiographic data and land-use classification, calculation of effective urban roughness and urban heat fluxes, urban canopy and soil sub-models, MH in urban areas.

Since these involve many complexities the FUMAPEX project has decided to concentrate on three main steps, or levels of complexity in NWP urbanisation (Baklanov, 2003):

1. corrections to the surface roughness for urban areas (Baklanov and Joffre, 2003) and to the heat fluxes (by adding an extra urban heat flux, e.g., via heat/energy production/use in the city, heat storage capacity and albedo change) in the existing non-urban physical parameterisations of the surface layer in higher resolution NWP models with improved land-use classification. Furthermore, an analytical model for wind velocity and diffusivity profiles inside the urban canopy is suggested (Zilitinkevich and Baklanov, 2004).
2. improvement and testing of a new flux aggregation technique, suggested by the Risø National Laboratory in co-operation with DMI (Hasager *et al.*, 2003) for urban areas. Recently this module was coupled to the DMI-HIRLAM model for non-urban areas. The approach can be extended for urban canopies as well. However, experimental data are needed to verify parameterisations for urban areas.
3. implementation of special physical parameterisations for the urban sublayer into the NWP models. It is considered for incorporation into both the HIRLAM and LM models a new urban module, developed within FUMAPEX (Baklanov and Mestayer, 2004), and based on two urban submodels: (i) the BEP urban surface exchange parameterisation, developed by EPFL team (Martilli *et al.*, 2002); (ii) the SM2-U urban area soil submodel, developed by ECN team (Dupont *et al.*, 2003a,b).

At the current stage of progress the following examples of 'urbanisation' within models have been completed (Baklanov and Joffre, 2003; Baklanov and Mestayer, 2004).

1. Increased resolution and surface data bases for the roughness length calculation in NWP models was tested in DMI-HIRLAM, LM, MM5 and RAMS models. For instance, DMI-HIRLAM was run with a horizontal grid resolution of 1.4 km and 1 km land-use classification using 21 land classes and urban subclasses. DWD tested the LM2LM that provides initial and boundary values of the local model with coarser resolution needed to calculate the LM with higher resolution. Using the LM2LM, 3 level nested LM forecasts (7, 2.8, 1.1km) were performed.

2. Modified parameterizations and algorithms for roughness parameters in urban areas based on the morphometric method were developed. Urban database analysis for mapping morphometric and aerodynamic parameters is tested using the ESCOMPTE data (Mestayer *et al.*, 2004) and the Copenhagen study.

3. Improved models for urban roughness sublayer simulation, including: (i) effective roughness and flux aggregation techniques, (ii) effect of stratification on the surface resistance over very rough surfaces, (iii) roughness lengths for momentum, heat, and moisture, (iv) wind and diffusivity profiles inside the urban canopy, were suggested.

4. It was shown that the roughness length depends on the atmospheric temperature stratification, new parameterisations for the effect of stratification on the surface resistance over very rough surfaces were suggested. The roughness lengths for momentum, temperature and moisture are different for urban areas. Possible parameterisations for the scalar roughness length for urban areas were recommended for urban-scale NWP models, but they need to be verified and improved.

5. Experimental studies of urban roughness inhomogeneity effects on the urban boundary layer development has been undertaken by the University of Hamburg and the results have been made available for model verification.

6. EPFL improved the BEP (Martilli) module to simplify its introduction in NWP models and tested it using the BUBBLE measurements. The computer code as a separate module, suitable for implementation into NWP models, was prepared. The improved BEP module, which combines the thermal and dynamical effect of the urban canopy, was introduced in the TVM and HIRLAM models.

7. ECN further developed a 1-dimensional, 2-dimensional and box version of the SM2-U model and implemented SM2-U and a simplified Martilli parameterisation in the MM5 model. SM2-U in SUBMESO was tested vs. the CLU-ESCOMPTE Marseilles data.

8. The sea surface temperature data, obtained from NOAA satellite high-resolution images, were incorporated into the RAMS model by CEAM. The land categories (from CORINE and PELCOM datasets) were reclassified following the USGS categories.

3.3. Development of Interface between Urban Scale NWP and UAP Models

The possibility of obtaining reliable air quality forecasts in urban areas depends on the appropriate exploitation of technical features of meteorological and air quality models. For reaching this objective, the communication between models has to be physically consistent and finalised to suit practical applications. This specifically includes the need to incorporate the improved description of the urban boundary layer introduced in NWP models.

The tasks that have to be covered by interface modules have been analysed to identify: which physical variables have to be processed or estimated, which computational methods are normally used and which kinds of improvements are desired to better exploit the new features of parameterisations and urbanised meteorological models that are under development in the project (Finardi, 2003). The air quality model types that have been individuated and analysed cover almost all the models that can be used to build UAQIFS depending on the different European climatic and pollutant emission conditions (statistical, steady state, Lagrangian and Eulerian models). The possible and desirable improvements for each model interface class have been identified in a preliminary way.

Guidelines have been elaborated to help the construction of improved interface modules from operational NWP models to UAP models (Finardi, 2004). General items like meteorological fields interpolation and downscaling, evaluation of boundary layer scaling parameters, turbulence modelling, evaluation of dispersion parameters and use of meteorological data for emissions related computations have been described and discussed in their general aspects and with examples taken from the different partners experience and activities. The document contains a critical review of the limits of parameterisations currently employed for urban areas and suggestions for possible improvements. Special attention has been devoted to boundary layer scaling parameters and MH evaluation.

Relevant FUMAPEX References

Baklanov, A., A. Rasmussen, B. Fay, E. Berge and S. Finardi 2002. Potential and Shortcomings of Numerical Weather Prediction Models in Providing Meteorological Data for Urban Air Pollution Forecasting. *Water, Air and Soil Poll., Focus*, 2(5-6), 43-60.

Baklanov, A., A. Gross, J. H. Sørensen 2004. Modelling and forecasting of regional and urban air quality and microclimate. *J. Computational Technologies*, 9: 82-97.

Baklanov, A., J. H. Sørensen, S. C. Hoe, B. Amstrup, 2003. Urban meteorological modelling for nuclear emergency preparedness. Submitted to *Journ. Envir. Radioactivity.*

Bruinen de Bruin, Y., Hänninen, O., Carrer, P., Maroni, M., Kephalopoulos, S., Scotto di Marco, G., Jantunen, M. 2004. Simulation of working population exposures to carbon monoxide using EXPOLIS-Milan microenvironment concentration and time activity data. *Journal of Exposure Analysis and Environmental Epidemiology* 14(2), 154-163.

Chenevez, J., A. Baklanov, J. H. Sørensen 2004. Pollutant transport schemes integrated in a numerical weather prediction model: Model description and verification results. *Meteorological Applications,* 11(3), 265-275.

Dupont, S., I. Calmet, P. G. Mestayer and S. Leroyer 2003a. Parameterisation of the Urban Energy Budget with the SM2-U Model for the Urban Boundary-Layer Simulation. Submitted to *Boundary Layer Meteorology*, September 2003.

Dupont, S. E. Guilloteau, P. G. Mestayer, E. Berthier and H. Andrieu 2003b. Parameterisation of the Urban Water Budget by Using SM2-U model. Submitted to *Applied Meteorology*, August 2003.

Elperin, T., Kleeorin, N., Rogachevskii, I. and Zilitinkevich, S., 2003: Turbulence and coherent structures in geophysical convection. *Quart, J. Roy. Met. Soc.* In press.

Hasager, C. B. Nielsen, N. W., Boegh, E., Jensen, N. O., Christensen, J. H, Dellwik, E. and Soegaard, H. 2003. Effective roughnesses calculated from satellite-derived land cover maps and hedge information and used in a weather forecasting model. *Boundary Layer Meteorol.*, 109: 227-254.

Hänninen, O., Kruize, H., Lebret, E., Jantunen, M. 2003. EXPOLIS simulation model: PM2.5 application and comparison with measurements in Helsinki. *Journal of Exposure Analysis and Environmental Epidemiology* 13(1): 74-85.

Hänninen, O. O., Lebret, E., Tuomisto, J. T., and Jantunen, M. 2004. Characterization of Model Error in the Simulation of PM2.5 Exposure Distributions of the Working Age Population in Helsinki, Finland. *JAWMA*. In press.

Hänninen, O. O., Lebret, E., Ilacqua, V., Katsouyanni, K., Künzli, N., Srám, Radim J., and Jantunen, M. J. 2004. Infiltration of ambient PM2.5 and levels of indoor generated non-ETS PM2.5 in residences of four European cities. *Atmospheric Environment* 38(37): 6411-6423.

Kitwiroon, N., R. S. Sokhi, L. Luhana and R. M. Teeuw 2002. Improvements in air quality modeling by using surface boundary layer parameters derived from satellite land cover data. *Water, Air and Soil Pollution, Focus*, 2, 29-41.

Kukkonen, J., Partanen, L., Karppinen, A., Ruuskanen, J., Junninen, H., Kolehmainen, M., Niska, H., Dorling, S., Chatterton, T., Foxall, R. and Cawley, G. 2003. Extensive evaluation of neural network models for the prediction of NO2 and PM10 concentrations, compared with a deterministic modelling system and measurements in central Helsinki. *Atmospheric Environment* 37(32), pp. 4539-4550.

Kukkonen, J., Pohjola, M. A., Sokhi, R. S., Luhana, L., Kitwiroon, N., Rantamäki, M., Berge, E., Odegaard, V., Slørdal, L.H., Denby, B. and Finardi, S. 2004. Analysis and evaluation of local-scale PM10 air pollution episodes in four European cities: Oslo, Helsinki, London and Milan. *Atmos. Environ.* (accepted)

Martilli, A., A. Clappier, and, M. W. Rotach 2002. An urban surfaces exchange parameterisation for mesoscale models. *Boundary Layer Meteorol.*, 104, 261-304.

Mestayer, P., R. Almbauer, O. Tchepel 2003. Urban Field Campaigns, Air quality in cities, N. Moussiopoulos, editor. Springer Verlag Berlin Heidelberg (ISBN3-540-00842-x), 2003, pp.51-89

Mestayer, P. G., P. Durand, P. Augustin, et al. 2004. The Urban Boundary Layer Field Experiment over Marseille UBL/CLU-Escompte, Experimental Set-up and First Results. *Boundary-Layer Meteorology*, in press.

Neunhäuserer, L, Fay B., Baklanov A., Bjergene N., Kukkonen J., Palau J. L., Perez-Landa G., Rantamaki M., Rasmussen A., Valkama I. 2004. Evaluation and comparison of operational NWP and mesoscale meteorological models for forecasting urban air pollution episodes – Helsinki case study. 9th Intern. Conference on Harmonisation within Atmospheric Dispersion Modelling for Regulatory Purposes, 1-4 June 2004, Garmisch-Partenkirchen, Germany.

Pohjola, M. A., J. Kukkonen, M. Rantamäki, A. Karppinen, E. Berg 2004. Meteorological evaluation of a severe air pollution episode in Helsinki on 27–29.12.1995. *Boreal Envi.t Research*, 9: 75-87.

Zilitinkevich, S. and A. Baklanov, 2002. Calculation of the height of stable boundary layers in practical applications. *Boundary Layer Meteorology*, 105(3): 389-409.

Zilitinkevich, S. and A. Baklanov 2004. A simple model of turbulent mixing and wind profile within urban canopy. Manuscript to be submitted to *Boundary Layer Meteorology*.

FUMAPEX Reports

Baklanov, A. (ed.) 2003. FUMAPEX Integrated Systems for Forecasting Urban Meteorology, Air Pollution and Population Exposure – Project Kick-off Meeting and First Progress Report. Copenhagen, DMI, Denmark. DMI Sci. Report 03-12, ISSN 0905-3263. April 2003, 140p.

Baklanov, A. and S. Joffre (eds.) 2003. Improved Models for Computing the Roughness Parameters of Urban Areas. / Baklanov, A., P. Mestayer, M. Schatzmann, S. Zilitinkevich, A. Clappier, etc. D4.4 FUMAPEX Report, DMI Sci. Report 03-19, ISBNnr.: 87-7478-495-1, 51 p.

Baklanov, A. and P. Mestayer (eds.) 2004. Improved parameterisations of urban atmospheric sublayer and urban physiographic data classification. D4.1, 4.2 and 4.5 FUMAPEX Report, April 2004, DMI Sci. Report: #04-05, ISBN nr. 87-7478-506-0.

Baklanov, A. and S. Zilitinkevich (eds.) 2004. Parameterisation of nocturnal UBL for NWP and UAQ models. D4.6 FUMAPEX Report. DMI Sci. Report 04-08, ISBN: 87-7478-510-9, 71p.

Eastwood, S., V. Ødegaard and K. H. Midtbø 2004. Algorithms for assimilation of snow cover. D4.3 FUMAPEX Report, September 2004, Norwegian Meteorological Institute, Oslo, Norway. 21 p.

Batchvarova, E., R. Sokhi, J. L. Palau, et al. 2003. Comparison and analysis of currently employed meteo-rological approaches for modelling urban air pollutants & Identification of gaps in met data required by UAP models for characterising urban BL. FUMAPEX D2.3-2.4 Report.

Fay, B. 2003. Overview of NPW/wind field models in FUMAPEX. D3.1 FUMAPEX report, 29 p.

Fay, B. 2003a. Design of model comparison study. D3.2 report for FUMAPEX, 26 p.

Fay, B., Neunhäuserer, L., Pérez-Landa, et al. 2004. Model simulations and preliminary analysis for three air pollution episodes in Helsinki. D3.3 report for FUMAPEX, 60 p.

Finardi, S. (ed.) 2003. Definition of NWP models output products, UAP models input needs and gaps to be filled. FUMAPEX report for M5.1, Arianet, Italy.

Finardi, S. (ed.) 2004. Guidelines for the construction of the interfaces from operational NWP models to UAP models. FUMAPEX report for D5.1, Arianet, Italy, January 2004, 47 p.

Hänninen, O. O., Karppinen, A., Valkama, I., Kousa, A., Kukkonen, J., and Jantunen, M. 2003. Refined and validated population exposure models. FUMAPEX project report deliverable D7.1.

Hänninen, O. O. 2004. Probabilistic and deterministic time-location-activity models for urban populations. D7.2 FUMAPEX Report, KTL, Finland.

Hänninen, O., Karppinen, A., Valkama, I., Kousa, A., Koskentalo, T., and Jantunen, M. 2004. Integration of population exposure models to the urban air quality information and forecasting system in Helsinki. FUMAPEX project report deliverable D7.3.

Ødegaard, V. 2004. Guidelines to the study of sensitivity of UAP forecasts to meteorological input. FUMAPEX Deliverable D6.1 Report, Met.no, May 2004.

Slördal, L. H. (ed.) 2003. FUMAPEX: Guidelines of output from UAQIFSs as specified by end-users. FUMAPEX D8.1 report. NILU Report OR 2/2004, Ref: U-102144.

Sokhi, R. S., N. Kitwiroon and L. Luhana 2003. FUMAPEX Datasets of Urban Air Pollution Models and Meteorological Pre-processors. D2.1-2.2 report for FUMAPEX, 41 p.

Valkama, I., J. Kukkonen 2003. Identification and classification of air pollution episodes in terms of pollutants, concentration levels and meteorological conditions. FUMAPEX D1.2 report, FMI.

Valkama, I. (ed.) 2004. Identification and evaluation of the practical measures taken during episodes, in the selected European cities. FUMAPEX Deliverable D1.3 Report, 26 April 2004.

Forecasting Urban Meteorology, Air Pollution and Population Exposure: European FUMAPEX Project.

Speaker: A. Gross (Baklanov)

Questioner: G. Kallos

Question: You presented here emission inventories at street resolution and you mentioned that you calculated emissions according to the population movement. To me, such a resolution is not supported by the key that the uncertainties due to the change in population or due to "per capita" emissions are higher than the horizontal differentiation. Therefore, the information you claim to gain through this approach is missing within the "noise". Simple statements i.e. a better agreement with observations are not adequate

Answer: The emission inventory includes vehicular emissions from approximately 5000 line sources within the area. We have computed the concentrations on a receptor grid, which contains approximately 10000 receptor points. The receptor point network covers the whole area, and the largest grid intervals are 200 m. A more densely-spaced grid was applied in the Helsinki downtown area and in the vicinity of the main roads and streets, the smallest grid interval was 20 m. The variable receptor grid is required in order to evaluate isoconcentration curves with adequate accuracy from the computed data.

The amount of receptor points is limited by the computational resources. Computations of the present study contain an hourly time series of one year (8760 meteorological and emission situations). The dispersion equations therefore had to be solved separately of the order of $5\times10^3\times10^4\times10^4=5\times10^{11}$ times. The chemical transformation computations increase the computational time even further. The computations regarding PM2.5 therefore required 24 h CPU time on a supercomputer.

Questioner: H. Schmidt

Question 1: I did not understand the slide showing daytime activity in Helsinki and what the numbers on the slide represent. Is this something more than daytime population? How was this population determined?

Answer: We combined the home coordinates together with the information on the number of inhabitants at each home location and the time spent at home during the day. Correspondingly, we combined the workplace coordinates, the number of personnel and the time spent at the workplace. We have utilised the time-microenvironment activities produced in the EXPOLIS study, which is applicable for the adult population from 25 to 55 years of age.

The population activities at other locations (other than for home and workplaces) were also evaluated. The number of persons carried by vehicular traffic was evaluated based on the predicted traffic flow information. E.g., in the case of buses, the number of persons and the time they spent in each street section is predicted directly by the EMME/2 model.

Question 2: How detailed is the building dimension information in the model?
Answer: In general for the urban module we use high-resolution urban databases, resolving individual building structure. Several European countries (e.g., Denmark, France, Switzerland) have such databases for big cities with a horizontal resolution up to 5-25 meters. However, the use of any real 'obstacle resolving' NWP and dispersion models for the city-scale is not possible in the forecasting mode due to high computational time demands. Therefore, urban database analysis for mapping morphometric and aerodynamic parameters is realized for the resolution of the city-scale NWP/meteorological models in use (e.g., 1.4 km for Copenhagen), so the building structure is represented for each grid-cell by different parameters, like the average building surfaces, heights of buildings (average and % of different heights), building height vs. width ratios, main street directions, etc. This approach is tested by FUMAPEX partners for Marseille using the ESCOMPTE data, for Basel using the BUBBLE data and for Copenhagen.

For the specific study of Helsinki the NWP model in use is not urbanized, the dispersion models applied are Gaussian and these do not generally take into account the influence of individual buildings on the atmospheric dispersion (although the UDM-FMI model allows for the influence of the source building itself on the dispersion). However, the terrain in the area is relatively flat and the average height of the buildings is fairly low (most buildings are lower than 15–20 m). There is only a moderate number of street canyons in the area. We have used the roughness length of 0.7 m in the numerical computations. The computations presented can be extended using nested street canyon dispersion model (OSPM) computations. Such computations have been performed for some specific locations. However, there is no database for Helsinki that would contain detailed building information within the whole of the metropolitan area.

5
Models-3/CMAQ Simulations to Estimate Transboundary Influences on Ozone and Particulate Matter Concentrations Over Ontario in Spring – Summer 1998

An. Chtcherbakov[1], R. Bloxam, D. Yap, D. Fraser, N. Reid, and S. Wong

1. Introduction

This study describes the application of atmospheric modelling to assess transboundary transport from the USA and the influence of natural emissions sources for the 12 most populated sub regions across Ontario, Canada. The impact of Ontario emissions was determined by the difference between model runs including all emissions and model runs with Ontario's anthropogenic NO_x, SO_2, VOC and primary particulate matter emissions shut off.

The modelling has been done for the entire May through September 1998 period. Since there were many days with high ozone and/or high $PM_{2.5}$ that summer, the model runs provide estimates of the variations in Ontario's impact from episode to episode and from the spring to the summer.

2. Methodology

The MODELS-3 modelling system MM5/SMOKE/CMAQ was run for 153 days of continuous simulations from May through September 1998.

The MODELS-3 system was developed by the US EPA and its main components are:

- The Fifth-Generation NCAR / Penn State Mesoscale Model, version 3.6 (MM5) – a limited-area, nonhydrostatic, terrain-following sigma-coordinate

[1] Ontario Ministry of the Environment, 125 Resources Road, Etobicoke, ON, M9P 3V6, Canada, : Andrei.Chtcherbakov@ene.gov.on.ca

model designed to simulate or predict mesoscale meteorological conditions (PSU/NCAR, 2002);

- US EPA Community Multiscale Air Quality modelling system, version 4.3 (CMAQ) – a modelling system for urban to regional scale air quality simulation of tropospheric ozone, acid deposition, visibility and fine particulate matter (Byun and Ching, 1999);
- Sparse Matrix Operator Kernel Emissions Modelling System (SMOKE) (Houyoux *et al.*, 2003) – an emission processor designed to convert annual emission inventory data to the temporal and spatial resolution needed by an air quality model – CMAQ.

Simulations were made for the domain shown in Figure 1, which covers northeastern U.S.A. and southern Ontario and Quebec. This domain includes all the major emission sources that could have influenced ozone, fine particulate matter and their precursors' concentrations in Ontario to any significant extent. The model resolution was 36 km with 50 grid cells in the east-west direction, 42 in the south-north direction and with 15 vertical levels that go up to 100 kPa. This grid size is sufficient to spatially allocate emissions to resolve transboundary influences.

The simulation procedure was as follows:

The MM5 model was run to produce meteorological fields from May through September of 1998. This time period included many high ozone and $PM_{2.5}$ episodes, some of which were stagnation events while others involved more rapidly

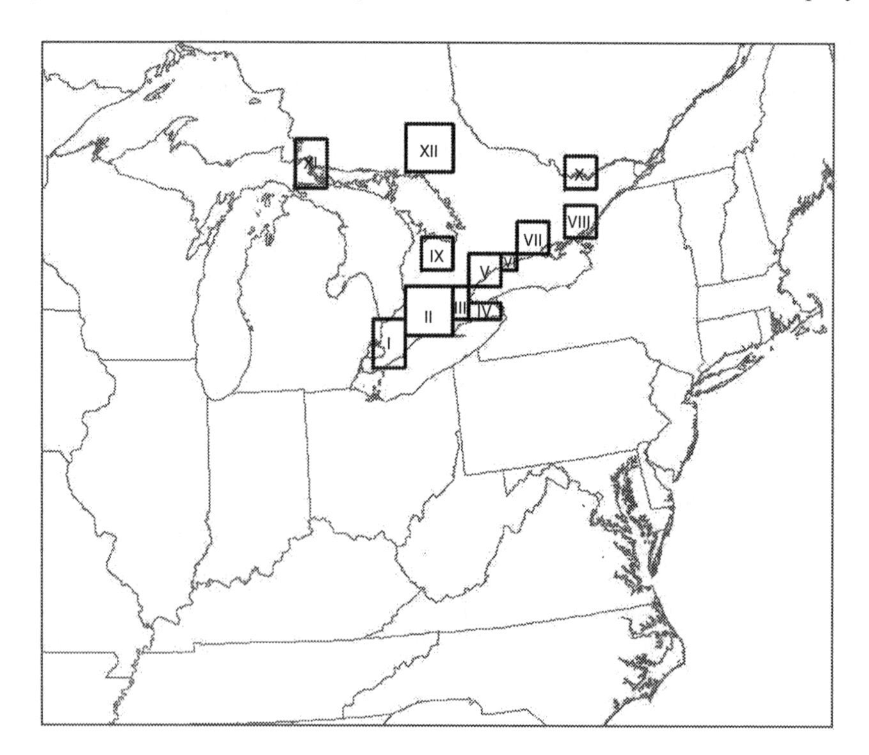

FIGURE 1. Modelling domain and Ontario sub-domains.

transported air masses. The relative importance of Ontario emissions versus transboundary influences could be expected to change for different meteorological conditions. Modelling for the 5-month period could show the variability in the contributions of Ontario's emissions to episode concentrations. Both spring and summer months were selected to model the impact of seasonal variations in emissions and chemistry on transboundary transport.

SMOKE was used for emission processing. Emission inventories from 1995 were used for Canada and from 1996 for the US. The original area source emission inventories were modified to reflect a more realistic seasonal variation of NH_3 emissions, especially for late spring and early summer when they are significantly larger than annual average values.

The emission rates and their spatial and temporal distributions agreed well with reported emission data (OME, 2003) and with previous emission estimates (Chtcherbakov *et al.*, 2003).

Figure 1 shows the 12 sub-domains, representing major Ontario communities, used in assessing transboundary influences. Modelled ground-level concentrations were extracted each day for all 12 sub-domains and 24-hour averages were calculated for aerosol species and 8-hour averaged maximum concentrations for ozone in each grid cell. All concentrations were then spatially averaged over each sub-domain's grid cells.

3. Model Evaluation

Comparisons between modelled and measured concentrations have been made on a month-by-month basis for time-series and frequency distributions of ozone and $PM_{2.5}$. Overall, the modelled time-series and frequency distributions compared well with observed results (Figure 2). The frequency distributions for ozone systematically showed more days with observed 8-hour maxima below 40 ppb for most sub-domains. This systematic over prediction of modelled ozone on cleaner days is likely due to local urban nitrogen oxide emissions reacting with ozone to reduce ozone concentrations. When modelled ozone concentrations were above 60 ppb the frequency distributions usually agreed well with observations.

Comparisons of the observed and modelled $PM_{2.5}$ frequency distributions showed very similar patterns with modelled peak concentrations generally lower than measured values.

Discrepancies in modelled/observed concentrations were caused by:

- Differences between modelled and actual meteorological fields; the most important parameters are wind speed and direction, cloud cover, position and evolution of the low- and high-pressure systems and atmospheric fronts;
- Natural events, which cannot be captured by model e.g. particulate matter transported from wild forest fires;
- Physical and chemical limitations of the model itself, the model's spatial resolution, and quality of input emissions and boundary conditions; and
- Problems with the monitoring data.

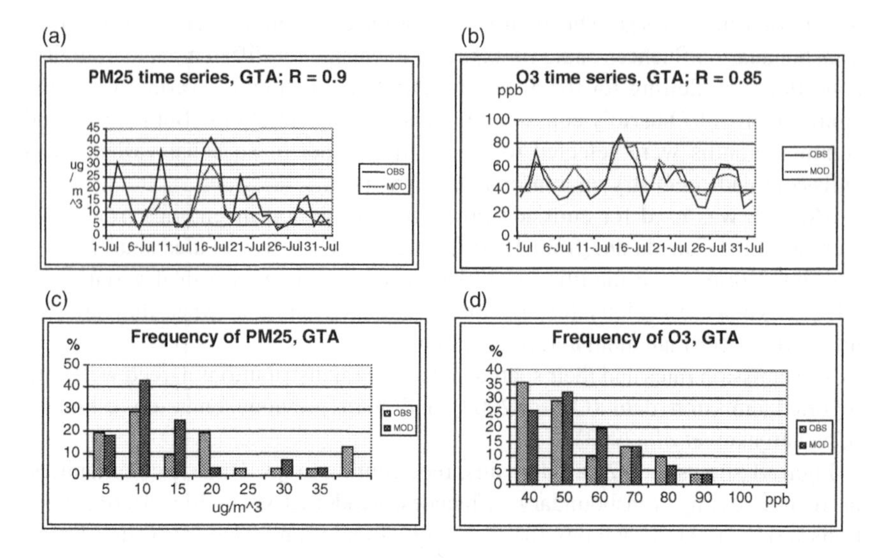

FIGURE 2. Time-series (a) and (b), and frequency distribution (c) and (d) of ozone and PM25 modelled and observed concentrations in the Greater Toronto Area (GTA) in July 1998.

Correlation coefficients were calculated for each month and each sub-domain for which observed data was available. The typical range of the coefficients was 0.7~0.8, with maximum values of 0.9 and minimum of 0.5. Good correlations were found for both ozone and $PM_{2.5}$ and for each of the sub-domains. There were no significant seasonal or spatial patterns in the coefficients' values. The average values ranged from 0.8 in southwestern Ontario to about 0.7 in southeastern and central Ontario.

Although there were some discrepancies between modelled and observed data for individual days and episodes, the differences appear to be random due to small shifts in the meteorological fields. The model/monitoring comparisons indicate that the model performed well overall.

4. Modelling Results

The number of "dirty" days when concentrations of ozone and/or $PM_{2.5}$ exceed the Canada Wide Standards (CWS): 65 ppb for ozone and 30 $\mu g/m^3$ for $PM_{2.5}$ were calculated for each month using monitored data. At the same time, the number of modelled "dirty" days was evaluated using a lower criterion for $PM_{2.5}$ of 20 $\mu g/m^3$. The lower criterion for $PM_{2.5}$ was chosen to select a large enough number of higher concentration days to see the variations in the contribution of Ontario's emissions on higher $PM_{2.5}$ concentrations. Also, the modelled concentrations analyzed were the averages over each sub-domain. There can be higher concentrations at sites within the area.

The number of days with observed high concentration of ozone (> 65 ppb) varied across Ontario decreasing from 35~40 at southwestern Ontario to 12 in the eastern area. Modelled data show the same pattern: about 30 in southwestern Ontario dropping to 10 in southeastern part of the province.

The number of days with observations above the CWS criteria was usually much higher for ozone than for $PM_{2.5}$. Selecting concentrations greater than 20 $\mu g/m^3$ as high modelled $PM_{2.5}$ values, the number of days with high PM2.5 concentrations for southwestern Ontario and the Greater Toronto Area (GTA) ranged from about 10 to 20; this is about the same number of days with observed values over the CWS criterion (30 $\mu g/m^3$). There were no days with modelled $PM_{2.5}$ concentrations above 20 $\mu g/m^3$ in northern regions of Ontario.

Ozone concentrations on "dirty" days were usually well above the criterion with averages in the 72~77 ppb range. For the cleaner days, the ozone concentrations were generally between 45 and 55 ppb with larger values occurring in June-August. Ozone concentrations in northern communities on these cleaner days were about 5 to 8 ppb lower than found for southwestern Ontario.

Typical modelled $PM_{2.5}$ concentrations on "dirty" days were in the range of about 23~30 $\mu g/m^3$ across Ontario. Mean $PM_{2.5}$ concentrations on cleaner days were about 8~9 $\mu g/m^3$ for southwest Ontario through to the GTA with lower concentrations of 6 $\mu g/m^3$ in eastern Ontario and the lowest concentrations at about 3~4 $\mu g/m^3$ in the northern communities.

Table 1 presents the spatial and temporal variations of sulphates and nitrates, which are very important components of fine particles. For higher concentration days, the modelled chemical composition of fine $PM_{2.5}$ had sulphate at usually 40 to 60% of the total mass. This is what would be expected for the warm season (PCAFM, 2001). The predicted concentrations of sulphate on days with $PM_{2.5}$ above 20 $\mu g/m^3$ were similar in magnitude across the southern part of Ontario; however there were fewer high concentration days at sub-domains east of the GTA. On cleaner days sulphate concentrations were generally predicted to decrease from southwestern to eastern Ontario.

TABLE 1. Average sulphate and nitrate concentrations ($\mu g/m^3$) on days with $PM_{2.5}$ > 20 $\mu g/m^3$.

		Domain			
Month	Species	I	V	VII	X
May	SO_4	12.3	10.7	N/A	N/A
	NO_3	1.3	6.6	N/A	N/A
June	SO_4	14.3	7.9	13.0	N/A
	NO_3	0.4	5.2	0.3	N/A
July	SO_4	13.7	12.5	14.4	N/A
	NO_3	0.6	0.7	0.3	N/A
Aug	SO_4	15.0	8.0	15.5	14.9
	NO_3	0.3	1.6	0.2	0.2
Sept	SO_4	12.9	10.1	15.9	12.1
	NO_3	0.7	2.4	0.6	0.3

The percentage of nitrate was higher in May than July or August in part because of larger ammonia emissions. Formation of ammonium nitrate is more likely at lower temperatures, which occur more often in May. The location relative to agricultural regions and animal feed lot locations affects the ammonia emissions and thus the formation of nitrate. The highest nitrate concentrations were registered in the Niagara peninsula, which is a farming area, and the GTA located downwind from agricultural regions of southwestern Ontario. In the GTA nitrate concentrations on "dirty" days dropped from 6 in May to 1 $\mu g/m^3$ in July and August. Further to the east nitrate concentrations sharply decreased to about 0.5 $\mu g/m^3$ further to the east in May and from to 0.2 $\mu g/m^3$ in August.

The contribution of Ontario's emissions to $PM_{2.5}$ concentrations depends on which constituents are significantly affecting the total mass. When nitrate is a major contributor to the total mass the percentage impact of Ontario's emissions would usually be larger.

Shutting down Ontario emissions had a different degree of impact on ozone and $PM_{2.5}$ concentrations in the sub-domains depending on their distances from the border, position relative to prevailing airflows and Ontario's emissions and the extent of urbanization. Communities near the border showed little change in the ozone concentrations with 0 to 2% reductions on days, which exceed the CWS. Reductions of $PM_{2.5}$ were larger but still only reached about 5% on high concentration days. This response reflects a dominant transboundary influence from U.S. emissions in this region (Table 2). On cleaner days, the impact of Ontario's emissions on $PM_{2.5}$ was larger with reductions of 15 to 20% when these emissions were shut off.

In the GTA, the impact of zeroing out Ontario's emissions was quite different. There was often a disbenefit for ozone concentrations on "cleaner" days (increases of up to 5~7%) that can be explained by less scavenging of ozone by NO emissions in the urban Greater Toronto Area. On days with higher ozone concentrations, eliminating Ontario's emissions resulted in reductions of ozone concentrations in most cases. Sub regions downwind of the Toronto area showed the largest reductions in ozone at about 10~15% averaged over all dirty days.

TABLE 2. Percent impact of shutting off Ontario emissions on ozone and $PM_{2.5}$ concentrations.

		Species					
		Ozone			$PM_{2.5}$		
Month	Domain	Average	Dirty	Clean	Average	Dirty	Clean
May	I	−2.7	−1.9	−2.8	−19.6	− 5.9	−22.3
	V	3.1	−17.1	4.5	−53.7	− 46.8	−56.2
June	I	−0.8	−0.7	−0.8	−15.2	− 4.9	−15.5
	V	0.1	−14.9	1.7	−60.8	− 57.6	−62.0
July	I	−2.5	−1.1	−3.0	−18.4	− 6.8	−19.6
	V	−0.7	−6.4	0.3	−57.3	− 35.8	−60.5
August	I	−4.8	−1.4	−6.0	−23.7	− 2.3	−24.4
	V	−2.4	−4.6	−2.2	−58.5	− 54.9	−59.9
September	I	−1.1	−1.2	−1.1	− 18.9	−5.7	−20.4
	V	7.2	−2.5	7.9	−54.7	−38.7	−58.7

When modelled $PM_{2.5}$ concentration were high, zeroing out Ontario emissions resulted in 47–57% reductions in the GTA for May and June with reduction of 35–55% in other months. The larger ammonium nitrate concentrations in May and June were greatly reduced when Ontario's emission were shut off. Reductions of about 55–60% in $PM_{2.5}$ were found for cleaner days in the GTA for all months.

On the higher concentration days there was large day by day variability in the modelled reduction for the sub domains. The average reduction in $PM_{2.5}$ for the GTA was about 37% but the range was from 20% to 67% reduction on these days. The range of response for ozone in the GTA on dirty days was from 5.1% to a benefit of 27.7%.

5. Conclusions

The contribution of Ontario's anthropogenic emissions to ozone and $PM_{2.5}$ concentrations varies with location across the province, season of the year and with meteorology.

Modelled results showed good overall agreement with observations with some cases where the timing and/or spatial extent of an episode were offset from what actually occurred.

Zeroing out Ontario's emissions showed small impacts in southeastern Ontario with larger influences on $PM_{2.5}$ in the GTA area and on ozone east of the GTA.

References

Byun D.W. and J.K.S. Ching, 1999: Science Algorithms of the EPA Models-3 Community Multiscale Air Quality (CMAQ) Modeling System. EPA/600/R-99/030. http://www.epa.gov/asmdnerl/models3/doc/science/science.html

Chtcherbakov A., R. Bloxam, S. Wong, P.K. Misra, M. Pagowski, J. Sloan, S. Soldatenko, and X.. Lin, 2003, 'MODELS-3/CMAQ evaluation during high particulate episodes over Eastern North America in summer 1995 and winter 1998', Int. J. Environment Pollution, Vol. 20, Nos. 1–6, pp.242–255.

Houyoux M., J. Vukovich, J. E. Brandmeyer, C. Seppanen, A. Holland, 2003: SMOKE USER MANUAL, Version 2.0; http://www.cep.unc. edu/empd/products/smoke/SMOKEDOCS.shtml#manual

OME, 2003, Ontario Ministry of the Environment. Fast Reference Emission Document: Version5, (2003) P1BS1531e03

PCAFM, 2001, Precursor Contributions To Ambient Fine Particulate Matter in Canada. 2001. A report by the Meteorological Service of Canada.

PSU/NCAR, 2002, PSU/NCAR Mesoscale Modeling System Tutorial Class Notes and Users' Guide: MM5 Modeling System Version 3, NCAR, 2002 http://www.mmm.ucar.edu/mm5/

6
Cost-Optimized Air Pollution Control Using High-Order Sensitivity Analysis

Daniel S. Cohan and Armistead G. Russell*

1. Introduction

Beyond the associated harm to human health,[1] non-attainment of air quality standards can substantially hamper the economy of a region[2] and its access to federal funds. The costs of emissions control are substantial as well and vary greatly among various options. Thus, much is at stake in reducing air pollution in a cost-efficient manner.

Alternative goals can be considered in the optimization of air pollution control strategies. The development of regulatory attainment plans can be abstracted as a constrained optimization problem of attaining air quality standards at minimal cost. One may also consider how to minimize regional pollutant concentrations or potential population exposure subject to a budget constraint. Because ozone forms from complex nonlinear interactions involving nitrogen oxides (NO_x) and volatile organic compounds (VOC),[3] the impact of a control measure will depend on which pollutant is reduced, the location of reduction, and variable factors such as meteorology. Thus, ozone sensitivity must be considered along with cost in evaluating cost-effectiveness.

With the onset of more stringent U.S. National Ambient Air Quality Standards (NAAQS) for ozone, many mid-sized metropolitan regions have been designated non-attainment and for the first time must develop control plans.[4] Here we consider Macon, Georgia, as a case study of the options facing one such area. The U.S. Environmental Protection Agency (EPA) in 2004 designated Bibb County and portions of Monroe County as non-attainment[4] based on ozone observations in Bibb County, Macon's only regulatory monitor. We develop a comprehensive menu of potential control measures, and link this menu with concentration-emission sensitivities computed from a photochemical model to determine the least-cost approach to ozone attainment in Macon. We compare the least-cost

* Daniel S. Cohan, School of Earth & Atmospheric Sciences, Georgia Institute of Technology, Atlanta, GA 30332 USA, dcohan@eas.gatech.edu, Phone: 1-404-385-4565. Fax: 1-404-894-8266. Armistead G. Russell, School of Civil & Environmental Engineering, Georgia Institute of Technology, Atlanta, GA 30332.

attainment strategies with those optimized for other goals, such as reducing regional concentrations or potential population exposure to elevated ozone.

2. Methods

2.1. Emissions Abatement Options

Where possible, estimates of the costs and emissions reductions associated with each NO_x and VOC control measure were obtained from AirControlNET v. 3.2.[5] Annual costs represent the sum of operational costs plus the amortized value of capital costs expressed in Year 2000 U.S. dollars. Here we apply AirControlNET to a projected 2010 National Emissions Inventory, the closest in the software to our 2007 target year.

Because AirControlNET focuses primarily on emissions from area and point sources, separate analysis was conducted for potential controls of mobile sources, both on-road and non-road. Estimates of costs and effectiveness were taken from available literature[6-11] and applied to the Fall-Line Air Quality Study (FAQS) 2007 inventory.[12-13] Taken together, the menu of measures represents the potential to control about 20-35% of NO_x and VOC, but with marginal costs increasing rapidly beyond 15-20% reductions (Figure 1). VOC tends to be slightly cheaper to control than NO_x, but marginal costs depend much more on the fraction reduced than on the precursor category.

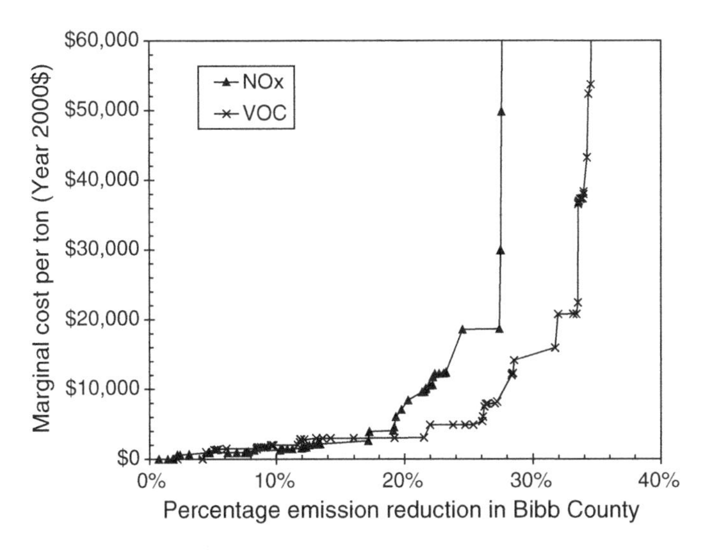

FIGURE 1. Marginal cost of emission reduction in Bibb County, indicative of cost curves for other regions.

2.2. *Photochemical Model Simulations*

Two air pollution episodes—1-19 August 1999 and 11-19 August 2000—were simulated with the Community Multiscale Air Quality (CMAQ) model[14] v. 4.3. Both episodes contain several days representative of ozone exceedances in Macon.[15] The first two days from each episode were discarded as initialization periods. The nested domain has 13 vertical layers and covers the southeastern U.S. with 12-km resolution and northern Georgia with 4-km resolution (Figure 2). Initial and boundary conditions were supplied by simulations on a 36-km grid covering the eastern U.S. Meteorology simulations are described in detail elsewhere.[16-17] Separate simulations were conducted using base year and Year 2007 emissions inventories developed for FAQS.[12-13] Base year simulations have been evaluated extensively relative to observations.[13]

To evaluate the impact of emissions controls on ozone, we used a high-order sensitivity analysis feature, the Decoupled Direct Method in Three Dimensions (DDM-3D).[18-19] DDM-3D computes the sensitivity of concentrations to perturbations in model parameters and inputs, utilizing the equations that compute concentrations in the underlying model. Taylor expansions of first- and second-order DDM-3D sensitivity coefficients[19] accurately capture the response of the underlying model even for large perturbations.[13]

CMAQ-DDM-3D was applied with 4-km resolution to compute the sensitivity of ozone to Year 2007 projected emissions of NO_x and VOC from each of 9 Georgia regions (Table 1): the 7 counties (each considered separately) which comprise the Macon-Warner Robins Combined Statistical Area; the "Macon Buffer", the 12 mostly rural counties bordering the Macon region; and the Atlanta region, defined here as the 20 counties designated in 2004 as non-attainment for 8-hour ozone.[4] Sensitivities to NO_x from two coal-fired power plants that are the largest point sources near Macon, Plant Scherer in Monroe County (Macon

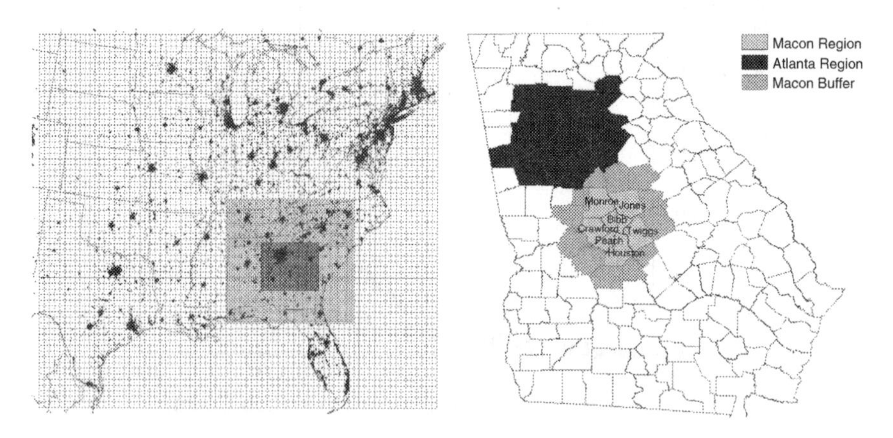

FIGURE 2. Modeling domain with 36, 12, and 4-km resolution and the Georgia regions for sensitivity analysis.

TABLE 1. Various metrics of 8-hour ozone sensitivity to Year 2007 emissions.

	Daily emissions during episodes (tpd)		Sensitivity near monitor[a] (10^{-3}ppb/tpd)		Spatial sensitivity[b] (10^{-3}ppb/tpd)	Population-weighted sensitivity[b] (10^3ppb-persons/tpd)
	NO_x	VOC	$s^1_{O3,NOx}$	$s^1_{O3,VOC}$	$s^1_{O3,NOx}$	$\Psi_{O3\geq 85ppb,NOx}$
Macon Region						
Bibb County	25	27	295.3	17.3	8.3	25.0
Crawford County	2	2	36.7	5.5[c]	10.0	10.8
Houston County	21	15	60.4	1.8	10.0	14.0
Jones County	5	5	80.7	5.5[c]	8.3	16.3
Monroe County	8	5	84.3	5.5[c]	7.0	12.8
Peach County	5	5	115.2	4.0	11.2	20.2
Twiggs County	4	3	44.5	5.5[c]	10.0	10.0
Plant Scherer	99	NC	49.4	NC	3.9	7.1
Macon Buffer	48	41	16.8	0.7	9.6	7.4
Plant Branch	60	NC	9.4	NC	3.8	3.8
Atlanta Region	537	401	8.8	0.7	7.2	46.1
Rest of Georgia	706	NC	3.5	NC	5.9	5.1
Alabama	949	NC	2.3	NC	2.8	1.9
North Carolina	944	NC	0.7	NC	0.9	0.5
South Carolina	702	NC	1.1	NC	2.0	0.8
Tennessee	850	NC	2.1	NC	2.1	6.3

[a] Sensitivity of cell in 7 × 7 box around Macon monitor with daily maximal O_3, averaged over 2 episodes (days with $O_3 < 70$ppb excluded).[20]
[b] Evaluated on 4-km resolution domain, averaged over all days; 85-ppb threshold is applied only to Ψ.
[c] 4 counties grouped together for VOC sensitivity calculations.

region) and Plant Branch in Putnam County (Macon Buffer), were computed separately. Sensitivities were computed on the 12-km grid with respect to NO_x from "Rest of Georgia", consisting of all other counties in the state, and from Alabama, South Carolina, and the within-domain portions of Tennessee and North Carolina. Preliminary modeling showed VOC from these regions negligibly impacts Macon ozone.

3. Results and Discussion

3.1. Ozone Attainment in Macon

The new NAAQS mandate that yearly 4th highest 8-hour ozone concentrations, averaged over 3 years, must remain below 85 ppb. Under the EPA attainment demonstration method,[20] concentrations are simulated near non-attaining monitors for historical episodes under base year and projected emission rates. The ratio of future to base modeled ozone is known as the "relative reduction factor (RRF)." Sufficient emission reductions must be identified such that the product of RRF (evaluated in 4-km modeling based on the daily maximal 8-hour ozone within a 7×7 cell box around the monitor) and the design value (computed from ozone observations in the three years straddling the base year) is below 85 ppb. We deviate slightly from the EPA method[20] by not rounding or truncating model results because the sensitivities considered in subsequent analysis are continuous.

Because control measures may reduce NO_x, VOC, or both, and because impact depends on emission location, a metric was developed to facilitate comparison. Measures were ranked based upon their annual cost per change in ozone as approximated by first-order sensitivities:

$$cost - effectiveness\,(\$/ppbO_3) = \frac{Annual\ Cost}{s^1_{O_3,NO_x} \cdot R(NO_x) + s^1_{O_3,VOC} \cdot R(VOC)} \quad (1)$$

In Eq. 1, s^1 is the first-order sensitivity of ozone to an incremental short ton per day (tpd) of emissions, and R is the reduction in emissions associated with a measure. This accounts for the impact of reductions in both VOC and NO_x without the use of an arbitrary ratio or scaling. Second-order terms for NO_x slightly enhance ozone impact in the final analysis,[13] but cannot be considered in the initial ranking because their importance in Taylor expansions depends on the cumulative reduction.[19] Total impact on ozone is computed as the sum of impacts from controls in each region, ignoring cross-sensitivities.

The cost-effectiveness metric is driven both by the net cost of a measure, which varies from zero to tens of thousands of dollars per ton (Figure 1), and by the sensitivity to emissions from that region (Table 1). Sensitivities depend strongly on the species emitted and on the proximity of the source to the monitor. Macon ozone is much more sensitive to NO_x than to VOC as is typical in the biogenic VOC-rich southeastern U.S.[3] Ozone near the Bibb monitor is several times more sensitive to emissions from within the county than to neighboring counties, and far less sensitive to emissions from elsewhere in the state. It is also less than 60% as sensitive to each ton of NO_x from the two elevated point sources, Scherer and Branch, than to other emissions from the corresponding regions, reflecting lower ozone production efficiencies in concentrated NO_x plumes.[21] Despite its relatively small per-ton impacts on Macon, Atlanta's large emission rates and its own non-attainment status lead to high interest in control options there. Macon ozone is relatively insensitive to emissions from neighboring states (Table 1). All sensitivities exhibit large day-to-day variability driven by meteorological conditions.

We apply the EPA attainment demonstration method[20] to the two episodes, with 2007 as the target year. NO_x and VOC emissions are projected to decline about 20-30% between the base years and 2007 as stationary source controls associated with an earlier Atlanta State Implementation Plan[22] (SIP) and the NO_x SIP Call[23] along with cleaner on-road and non-road vehicles more than offset growth. Thus, ozone is modeled to be 12-15% lower in 2007. Although the two episodes are only one year apart and have similar emission rates, the base year design value for the 1999 episode (based on 1998-2000 observations) is 7 ppb higher than the 2000 episode (based on 1999-2001). Thus, a hypothetical Year 2007 demonstration for the 2000 episode would indicate attainment, whereas the 1999 episode would indicate the need for 6.4 ppb of additional controls. Joint consideration with equal weight for each episode indicates that 2.7 ppb of ozone reduction is needed.

Control measures were selected in order of cost-effectiveness until cumulative ozone reduction near the Macon monitor achieved the threshold. For the combined episodes, the necessary 2.7 ppb reduction would require annual expenditures of only $750,000, mostly for low-cost NO_x controls for industrial sources and local locomotives in the Macon region. Measures assumed to have zero net cost—continuation of lower-emitting Powder River Basin (PRB) coal at Scherer, a seasonal burning ban, parking pricing,[11] replacement of water heaters, and the planned closure of Brown & Williamson Tobacco—achieve 2.1 ppb of the reduction, mostly due to the PRB coal. Because almost all ozone reduction results from controls within the Macon region, excluding extra-regional controls only marginally raises the price tag. Modest reductions beyond the 2.7 ppb threshold could be achieved relatively affordably, primarily through additional controls on local industry.

If the 1999 episode is considered separately, minimal attainment costs would soar to $72.6 million annually to meet the higher 6.4 ppb threshold. Participation from neighboring Georgia regions would be essential, as full implementation of available control measures within the Macon region would not quite achieve the necessary reduction. The least-cost approach would include SCRs at all four Scherer units and one Branch unit and a wide array of stationary and mobile source controls throughout the Macon counties and to a lesser extent elsewhere in Georgia (Figure 3). As may be expected for an attainment-optimized strategy, largest ozone reductions would occur within the Macon region.

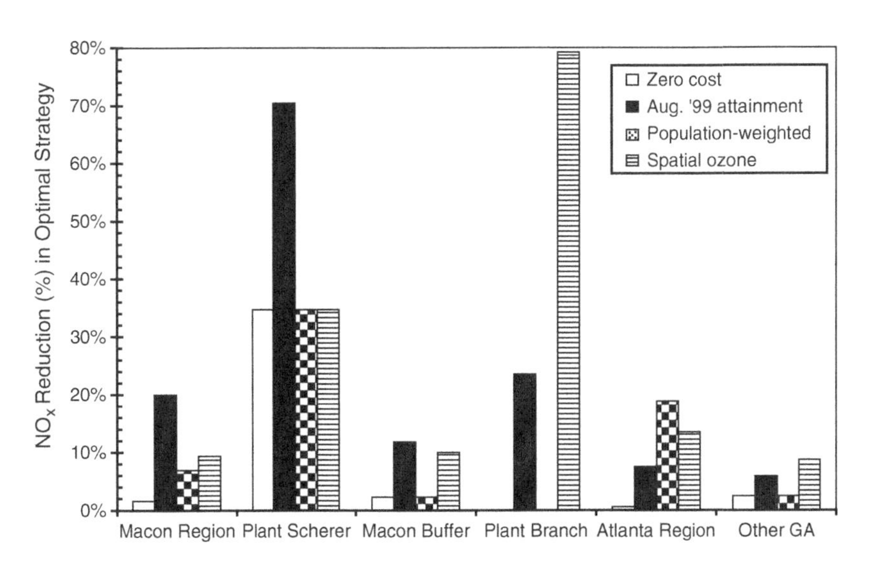

FIGURE 3. NO_x reductions for zero-cost measures, least-cost Macon attainment (1999 episode), and the strategies that would minimize population-weighted and spatial ozone at the same cost as the attainment strategy.

3.2. Alternative Metrics

Is the least-cost attainment strategy well suited to achieving other goals? For simplicity we consider only two metrics averaged over all modeled days: (1) the domain-wide average sensitivity of 8-hour ozone and (2) a potential exposure metric, Ψ, which weights s by population where 8-hour ozone is at least 85 ppb under default 2007 emissions:

$$\Psi\,(ppb - persons O_3/tpd) = \sum_{O_3 \geq 85ppb} s_{O_3,E}(i,j) \cdot Population\,(i,j) \qquad (2)$$

Here, $s_{O3,\,E}$ is the per-ton sensitivity of 8-hour ozone in cell (i,j) to source E. Population is from the 2000 U.S. Census. The spatial metric is a proxy for ozone accumulation efficiency, and $\Psi_{O3 \geq 85ppb,E}$ quantifies contribution to high ozone in populated areas.

Spatial sensitivity declines with NO_x emissions intensity,[3,21] but the range among fine domain regions is narrow relative to the variation in emissions density (Table 1). Sensitivities to "Rest of Georgia" and other states are lower because their impact occurs mainly outside the fine domain. On a population-weighted basis, Atlanta emissions are by far the most important per-ton, reflecting Atlanta's denser population and higher ozone.

We re-rank emission control measures by applying Ψ in place of s^1 in Eq. 1. As before, total impact for each level of cumulative emission reduction is assessed by incorporating 2nd-order NO_x sensitivities in Taylor expansions.[19] Given an annual budget constraint of $72.6 million, the minimal Macon attainment cost for the 1999 episode, the maximal impact on Ψ would be 6.32 million ppb-persons (Figure 4). This Ψ-optimized strategy would devote 98% of control expenditures to Atlanta NO_x (Figure 3), largely for control technologies at coal-fired power plants. By contrast, the optimal Macon attainment strategy achieves 3.23 million ppb-persons impact. The gap would be more pronounced except that both strategies include all zero-cost options, which account for 0.60 million ppb-persons of total impact. However, the gap is accentuated by the case-specific situation of mid-sized city attainment near a much larger city. Similar consideration of the spatial-average metric shows that the $72.6 million attainment strategy for the 1999 episode would reduce domain-wide ozone by 1.04 ppb, compared with 1.38 ppb at equal cost by optimizing for this metric (zero-cost measures yield 0.30 ppb). Because per-ton spatial impacts have a narrower range than the other metrics (Table 1), rankings under this metric are driven more by per-ton NO_x control costs than by source region.

3.3. Other Considerations

Rankings by cost-effectiveness should not be the only criterion for selection of attainment strategies. Political will and regulatory structure play major roles in the ability to implement cost-optimized strategies.[24] Cost estimates are inherently subject to alternative assumptions and some measures yield ancillary impacts. A crucial unknown in attainment-oriented control strategy formulation is future

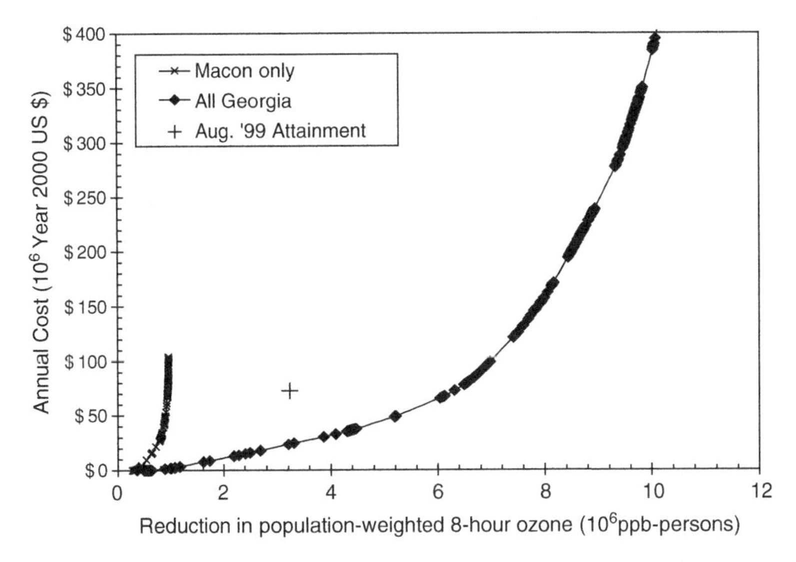

FIGURE 4. Least-cost reduction in population-weighted 8-hour ozone above 85 ppb threshold, averaged over all modeled days, by controls in Macon or anywhere in Georgia. The statewide menu totals 11.1×10^6 ppb-persons for \$1.14 billion. "+" denotes the least-cost attainment strategy for Macon monitor (1999 episode).

meteorological conditions. Observed ozone design values have fluctuated about 10 ppb over years with similar emissions rates and despite 3-year smoothing. Such fluctuations are comparable to the impact of the entire Georgia control menu, complicating efforts to assure future attainment.[25]

All of these nuanced considerations require the oversight of informed policymakers and cannot be decided based on cost-effectiveness alone. However, the linkage of costs with sensitivities has been shown to be a valuable approach for informing policy decisions. Since sensitivities can differ widely even among nearby source regions, their incorporation along with costs facilitates the selection of efficient strategies. Cost-optimized strategies such as those developed here provide a starting point to discuss appropriate actions, conduct additional assessments, and identify viable strategies.

Acknowledgements The Fall-line Air Quality Study was funded by Georgia Department of Natural Resources (DNR). However, views expressed herein are those of the authors alone. Yongtao Hu and Di Tian (Georgia Tech (GIT)) modeled meteorology and emissions. Alper Unal (GIT) created the Georgia graphic. We thank Larry Sorrels (U.S. EPA) for guidance to AirControlNET; Allen Greenberg (Federal Highway Administration) and Randall Guensler (GIT) for conversations about mobile source controls; Steve Ewald (Georgia Power) for information about Scherer; and Jim Boylan (DNR) for helpful discussions.

References

1. B. Brunekreef and S. T. Holgate, 2002. Air pollution and health, *Lancet* **360**, 1233-1242.
2. J. V. Henderson, 1996. Effects of air quality regulation, *American Economic Review* **86**, 789-813.
3. S. Sillman, 1999. The relation between ozone, NO_x and hydrocarbons in urban and polluted rural environments, *Atmos. Environ.* **33**, 1821-1845.
4. US EPA, 2004. 8-hour ozone National Ambient Air Quality Standards, *Federal Register* **69**, 23,858-23,951.
5. E.H. Pechan & Associates, 2003. AirControlNET Version 3.2 Development Report, prepared for U.S. EPA.
6. C. S. Weaver and D. B. McGregor, 1995. Controlling locomotive emissions in California: Final Report, prepared for California Air Resources Board.
7. W. T. Davis and T. L. Miller, 2004. Estimates of potential emissions reductions for the Nashville ozone early action compact area, prepared for Nashville Dept. of Environment and Conservation.
8. J. R. Kuzmyak, 2002. Cost-effectiveness of congestion mitigation and air quality strategies, in: *The Congestion Mitigation and Air Quality Improvement Program: Assessing 10 Years of Experience*, Transportation Research Board Special Report 264, Washington: National Academy Press.
9. T. Litman, 1997. Distance-based insurance as a TDM strategy, *Transportation Quarterly* **51**(3), 119-137.
10. E.H. Pechan & Associates, 2002. Control strategy cost analyses for the Southern Appalachian Mountain Initiative, prepared for Southern Appalachian Mountain Initiative.
11. D. C. Shoup, 1997. Evaluating the effects of cashing out employer-paid parking: Eight case studies, *Transport Policy* **4**, 201-216.
12. A. Unal, D. Tian, Y. Hu, A. G. Russell, 2000. August Emissions Inventory for Fall Line Air Quality Study (FAQS), prepared for Georgia Dept. of Natural Resources (2003).
13. Y. Hu, D. S. Cohan, M. T. Odman, and A. G. Russell, 2004. Air quality modeling of the August 11-20, 2000 episode for the Fall Line Air Quality Study, prepared for Georgia Dept. of Natural Resources.
14. D. W. Byun and J. K. S. Ching, 1999. Science algorithms of the EPA Models-3 Community Multiscale Air Quality (CMAQ) modeling system, U.S. EPA/600/R-99/030.
15. ICF Consulting/Systems Applications International, 2002. Ozone episode selection analysis for urban areas in northern Georgia and northern Alabama (1995-2001), prepared for Southern Company.
16. Y. Hu, M. T. Odman, and A. G. Russell, 2003. Meteorological modeling of the first base case episode for the Fall Line Air Quality Study, prepared for Georgia Dept. of Natural Resources.
17. D. Tian, Y. Hu, and A. G. Russell, 2004. Meteorological modeling of the 1999 August episode for the Fall-Line Air Quality Study, prepared for Georgia Dept. of Natural Resources.
18. Y.-J. Yang, J. Wilkinson, and A. G. Russell, 1997. Fast, direct sensitivity analysis of multidimensional photochemical models, *Environ. Sci. Technol.* **31**, 2859-2868.
19. A. Hakami, M. T. Odman, and A. G. Russell, 2003. High-order, direct sensitivity analysis of multidimensional air quality models, *Environ. Sci. Technol.* **37**, 2442-2452.
20. U.S. EPA, 1999. Draft guidance on the use of models and other analyses in attainment demonstrations for the 8-hour ozone NAAQS, EPA-454-R-99-004.

21. T. B. Ryerson et al., 2001. Observations of ozone formation in power plant plumes and implications for ozone control strategies, *Science* **292**, 719-723.
22. GA Dept. of Natural Resources, 2001. State Implementation Plan for Atlanta ozone non-attainment area.
23. U.S. EPA, 2004. Interstate Ozone Transport: Response to Court Decisions on the NO_x SIP Call, NO_X SIP Call Technical Amendments, and Section 126 Rules; Final Rule. Federal Register **69**, 21603-21648.
24. NRC, 2004. Air Quality Management in the United States, Washington: National Academies Press.
25. D. P. Chock, S. L. Winkler, and S. Pezda, 1997. Attainment flip-flops and the achievability of the ozone air quality standard, J. Air Waste Manage. Assoc. **47**, 620-622.

Cost-Optimized Air pollution Control Using High-Order Sensitivity Analysis

Speaker: D. Cohan

Questioner: G. Yarwood

Question: Ozone sensitivities to NOx depend upon the magnitude of emissions, as you have demonstrated. Are you concerned that a large source could have zero, or even negative, sensitivity and thus not be targeted for control by the method you have described?

Answer: My concern is that sensitivities are computed accurately and that policy judgments are founded upon sound science and a thorough consideration of relevant implications. Inclusion of second-order sensitivities obviates some of the problem that you raise because, given the concavity of ozone-NOx response, a second-order Taylor expansion will for some large sources show that major emission reductions would have a beneficial impact even if the incremental sensitivity is zero or negative. If a NOx control would indeed have negligible impact on ozone, then that fact should be considered in policy decisions.

Questioner: C. Hogrefe

Question: You showed that the results of your sensitivity towards different source categories depended on the episode you chose to model. Do you have a feeling for how long you would have to model to reduce the sensitivity of your results to episode selection?

Answer: For sensitivities of concentrations in one region to emissions from the same region, modeling a couple of weeks appears sufficient for average sensitivities to converge to a fairly representative value. To capture representative sensitivities of concentrations in one region to emissions from another region or point source would require significantly longer and climatologically representative periods because these are much more driven by highly variable transport patterns.

We have looked at different years, and the results are consistent, though not identical. These episodes were chosen as being representative of the types of meteorologies that lead to elevated ozone, but not necessarily the highest levels.

Thus, an additional question (which also should be explored) is if you focus on the very highest ozone levels, alone, how does that change the results.

Questioner: R. Yamartino

Questions:

1. How big an effect did inclusion of the second derivatives have on your results (i.e. versus just the 1st order approach)?

Answer: For individual power plants, where SCR technology makes it feasible to reduce NOx emissions by 80% or more, inclusion of second derivatives raises the modeled impact of control by about one-fourth and thus improves the relative cost-effectiveness of these options. For the broader emissions regions, much smaller percentage reductions are possible from the identified control menu and thus nonlinearity of response is less significant.

2. Did you include cross terms in with your second derivatives (eg. $\delta^2 C / \delta X \delta Y$ or just diagonal $\delta^2 C / \delta X^2$)?

Answer: Cross-term second derivatives were not considered in this study, but we have examined them elsewhere and found them to be smaller than diagonal terms. Second-order terms were included here only for NOx, as second-order VOC sensitivities were found to be small.

Questioner: D. Anfossi

Question: Which is the order of magnitude of the uncertainties (or the error bar) in your calculated results?

Answer: Uncertainties were not rigorously quantified in this study. The greatest uncertainties are in the costs and impacts of individual measures, and thus the specific measures comprising a "least-cost" strategy are subject to debate. However, because cost-per-ton is likely overestimated for some sources and underestimated for others, the overall curves of ozone impact v. cost are less uncertain than the rankings of individual measures.

Questioner: S. T. Rao

Question: Have you started looking at how these optimization techniques can be used to address multi pollutant strategies? In this regard it would be useful to look as the least cost options to control both ozone and PM2.5 not ozone concentrations first and PM2.5 later. As you know NOx and VOC affect both ozone and PM2.5 concentrations.

Answer: We have considered this as an excellent avenue for further research, and it could be facilitated by the recent incorporation of aerosol processes into CMAQ-DDM by S. Napelenok (Georgia Tech). Multi-pollutant optimization could be conducted as a straightforward extension of the current work by developing a common metric (e.g., monetized health impacts) by which to evaluate perturbations in multiple pollutants. A more complex alternative would be to conduct a multi-objective optimization problem in which separate goals are set for each pollutant.

7

Seasonal Evaluation of EU Road Traffic Emission Abatement Strategies on Photochemical Pollution in Northern Italy

Giovanna Finzi, Veronica Gabusi, and Marialuisa Volta*

1. Introduction

Road traffic is widely recognised to be a significant and increasing source of photochemical pollution precursors. EU Directives in force up to 2010 require substantial NO_x and VOC emission decreases. Such reductions can improve or further deteriorate air quality because of the emission mix and the photochemical regime characterizing the area under study.

The comprehensive modelling system GAMES (Gas Aerosol Modelling Evaluation System) (Volta and Finzi, 2004) has been used to evaluate the seasonal impact of the EU Directive abatement strategies on road traffic emissions in Northern Italy. In this area frequent stagnating meteorological conditions and elevated Mediterranean solar radiation, associated with critical anthropogenic emissions, regularly cause high ozone level episodes, especially during summer months. Several experimental campaigns (Vecchi and Valli, 1999), (Neftel et al., 2002) as well as modelling studies (Silibello et al., 1998), (Baertsch-Ritter et al., 2003), (Gabusi and Volta, 2004) have been carried out to investigate photochemical pollution in such complex domain.

The seasonal impact of the EU Directive on road traffic emissions scheduled for 2010 has been investigated as follows:

- the future emission scenario has been assumed to be equal to the base-case one (i.e. the 1996 summer season) except for road traffic sources; it has been evaluated under the hypothesis of a fleet emission trend in accordance to the EU Directives implementation;
- the impact of the 2010 emission scenario has been assessed performing simulations fed both by the 1996 and 1999 summer season meteorological fields.

* Department of Electronics for Automation, University of Brescia, Via Branze, 38, 25123 Brescia, Italy.

2. The Modelling System

Modelling simulations have been carried out by means of GAMES system, allowing to perform long-term homogenous gas-phase and multi-phase simulations.

The gas-phase configuration includes the 3D meteorological processor CALMET (Scire et al., 1990), the emission model POEM-PM, (Carnevale et al., 2004) and the photochemical transport model CALGRID (Yamartino et al., 1992).

- The diagnostic meteorological pre-processor CALMET provides (1) 3D wind and temperature fields, (2) turbulence parameters, merging background fields provided by regional model with measurements.
- The emission processor POEM-PM has been designed to provide present and alternative emission fields by means of an integrated top-down and bottom-up approach.
- The photochemical model CALGRID is an Eulerian three-dimensional model. It implements an accurate advection-diffusion scheme in terrain following coordinates with vertical variable spacing; a resistance-based dry deposition algorithm takes into account pollutant properties, local meteorology and terrain features. The chemical module implements SAPRC-90 mechanism (Carter, 1990), including 54 chemical species with 129 reactions and the QSSA (Quasy Steady State Approximations) solver (Hessvedt et al., 1978) for the integration of kinetic equations.

The performance assessment of the integrated modelling system as for the base case simulation over the Northern Italy domain (Figure 1) is described in detail in Gabusi et al. (2003), Pirovano et al. (2003) and Gabusi et al. (2004), including the estimate of the indicators required in the EC Directive 2002/3 as well as the recommended US EPA indexes.

3. The 1996 and 2010 Emission Field Estimation

Emission fields used in these modelling applications have been provided by POEM-PM model, recommended by the Italian National Environmental Agency (APAT) (Deserti et al., 2001). POEM-PM implements an integrated top-down and bottom-up approach allowing the assessment of both real and alternative emission scenarios, the harmonization of non-homogenous emission data. It implements a NMVOC lumping algorithm for both structure and molecule approaches and provides emission fields suitable for mesoscale and local scale multiphase modelling application. The road transport, agriculture and biogenic emissions have been estimated by means of a bottom-up approach while emissions of the other sectors have been disaggregated, modulated and split on the basis of last available Italian CORINAIR database (1994).

In particular, the road transport emissions are obtained by the following procedure:

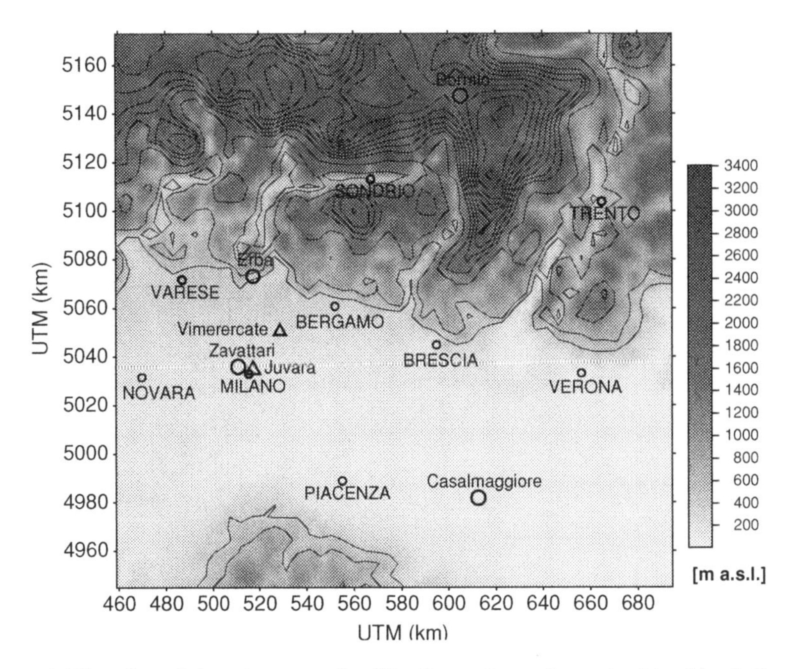

FIGURE 1. The selected domain orography. The figure shows the main cities (dots), the air quality (triangles) and the meteorological (circles) measurement stations referred in this work.

- The national vehicle fleet (number of vehicles and age distribution) is split in the CORINAIR road traffic categories. The driving condition parameters (annual mileage per vehicle class, annual mileage per road class, average vehicle speed) are estimated on the basis of the recorded data.
- The yearly national emissions have been computed for the estimated vehicle fleet by means of the software COPERT III (Kouridis et al., 2000); this model approach is based on four main information classes: the fuel consumption (per fuel type and per vehicle category), the vehicle park (number of vehicles per vehicle category, age distribution of the vehicle park per vehicle category), the driving conditions (annual mileage per vehicle class, annual mileage per road class, average speed of vehicles) and the emission factors (per vehicle class, per production year, per road class). The model includes the emission factors for future EU regulations scheduled up to 2010.
- The emission scenarios are spatially and temporally resolved by the POEM-PM. The spatial distribution is supported by the output of a traffic model, starting from the regional road transport network and the traffic load data.

Table 1 shows the diffuse annual emission in the simulation domain. It can be noticed that transport is the main source of NO_x emissions, representing the 75% of total diffuse emissions, as well as for CO (63%) and NMVOC (41%).

TABLE 1. Total diffuse emissions over the domain [ton/year].

Sector	NO_x	CO	SO_2	NMVOC
Industry	48,076	185,824	55,920	31,459
Solvent use	–	–	–	165,005
Road transport	273,850	1,176,955	38,544	171,825
Waste	10,469	495,712	5,551	25,120
Agricultural	85	4,540	–	209
Biogenic	36,370	–	–	28,844
Total	**368,850**	**1,863,031**	**100,015**	**422,462**

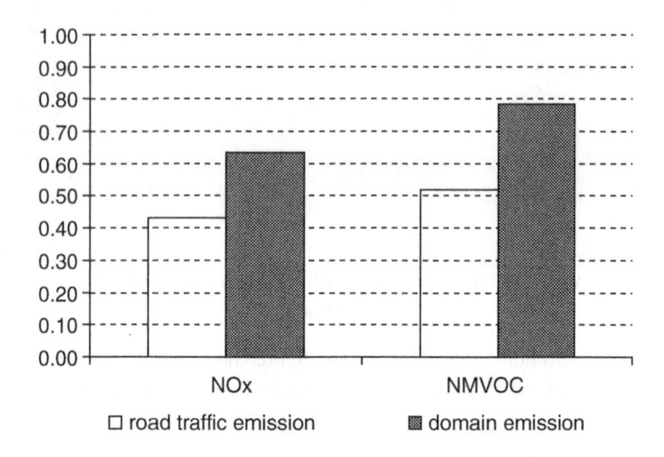

FIGURE 2. The emission ratio between the 2010 and the base case scenario.

The 2010 emission scenario has been assumed to be the base-case except for road traffic sources, assigned in accordance to the EU Directives implementation. Following the above-mentioned procedure, the first step implies the forecasting of the vehicle fleet up to 2010. The procedure involves the evaluation of the national fleet growth and the estimation of the vehicle age trends, as described in Volta and Finzi (1999).

Such estimates indicate that present EU Directives on the road traffic emission will yield reductions of about 60% on NO_x and 50% on NMVOC (Figure 2).

As a consequence of the emission sectors distribution overall the domain, it comes out an estimated total reduction amount of about 40% for NO_x and 20% for NMVOC.

4. Meteorological and Chemical Characterization of the 1996 and 1999 Summer Seasons

To provide a comprehensive evaluation of the two selected summer seasons, the meteorological and chemical data recorded by local network have been analysed and processed.

Temperature statistics and wind roses have been computed to characterize the meteorological conditions occurred during 1996 and 1999 summer seasons. Table 2 reports temperature mean values and standard deviations for three stations representative of Alpine, urban and plan profiles respectively. Lower mean values together with higher standard deviations characterise 1996 with respect to warmer 1999 summer, so evidencing a higher temperature variability in the former season.

As for the wind features, wind roses for mountain, foothill and plan monitoring stations have also been processed. Figure 3 shows the 1996 and 1999 summer wind roses computed for Erba station, well representing seasonal piedmont wind circulation, characterized by generally low wind speeds and a significant amount of calms (54.03% in 1996 and 59.54% in 1999 summer months). The wind roses comparison evidences that 1996 summer period is characterized by local breezes regimes induced by complex orography, while during 1999 wind prevailing directions are set along the Po Valley axis†.

Ozone critical episodes distribution in 1996 and 1999 summer seasons is summarized in Tables 3 and 4. Such information, obtained processing measured

TABLE 2. Temperature statistics.

	Mean		Standard Deviation	
Station	1996	1999	1996	1999
Bormio (Alpine region)	13.6	14.2	5.9	5.7
Zavattari (Milan)	19.0	20.3	5.2	4.9
Casalmaggiore (Po Valley)	21.2	21.3	5.7	5.3

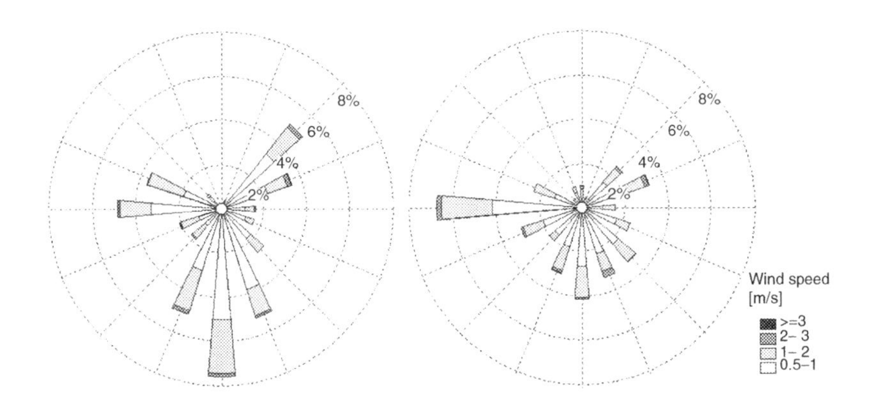

FIGURE 3. Wind roses at Erba station: 1996 (left) and 1999 (right) summer season.

† Figures 2 and 3 (page 124 in Preprints) have been replaced with current Table 2 and Figure 3 in order to meet the objections raised in the discussion following the oral presentation. Concerning comments have been consequently modified in the text.

TABLE 3. Exceedances of the threshold for information of the public (1h concentration > 90 ppb) during summer season on a year basis.

	No. of stations	Stations with exceedance	Days with exceedance	Average exceedance per station [hour]	Max. obs. conc. [ppb]
1996	33	32	74 (40%)	90.8	219
1999	48	47	71 (39%)	67.0	173

TABLE 4. Exceedances of the threshold for human health (55 ppb average between 12 and 20) during summer season on a year basis.

	Stations with exceedance	Days with exceedance	Average exceedance per station [days]
1996	33	138 (75%)	58.1
1999	48	140 (77%)	69.7

ozone concentrations, allow to point out the intensity and the extension of critical ozone episodes in the domain. In 1996 a higher number of days with exceedances of 90 ppb and the highest maximum ozone value have been recorded, while the human health threshold has been more frequently exceeded in 1999.

5. Emission Scenarios Impact Assessment

GAMES system has been performed for the 1996 base case and 2010 scenarios, using two different meteorological dataset: the 1996 and 1999 summer season meteorological fields. The modelling simulations results over long-term periods are summarized by means of proper seasonal indicators (e.g. Jonson et al., 2001), as mean ozone and AOT40 patterns.

In Figure 4 ozone daytime mean concentrations are presented, in term of differences between the 2010 and the base case scenario, feeding the model with 1996 and 1999 meteorological fields respectively. The percentage differences range from −16% to +28%. It should be observed that the two maps present a similar distribution pattern, mainly in the central part of the domain where big urban areas are prevailing.

Such analysis highlights that a large part of the Po Valley, where major urban areas are located, is VOC sensitive as already pointed out by Gabusi and Volta (2004). Actually the simulated ozone fields exhibit increasing concentrations compared to the base case ones in urban areas, where high ozone levels are related to anthropogenic emissions, mainly from road transport. As for the Alpine region, it may be observed that, accounting for 1996 meteorological data, the model simulations estimate a decrease in O3 mean value, while accounting for 1999 meteorological fields, this area seems mainly influenced by transboundary pollution phenomena.

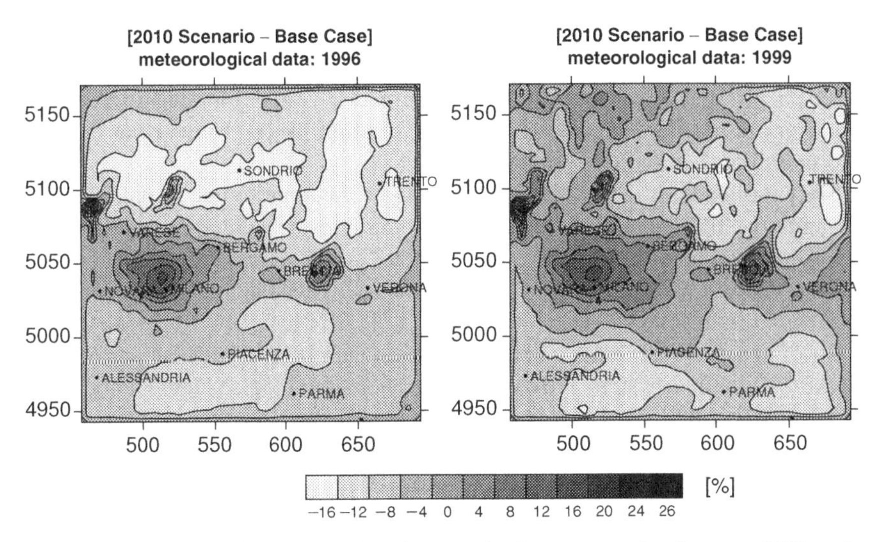

FIGURE 4. Calculated differences [%] of ozone daytime mean value between 2010 emission scenario and the reference case.

The simulated mean daily pattern, computed averaging hourly concentrations over the 6-months period, in correspondence of two monitoring stations is presented in Figure 5. Both the stations are located in the Milan metropolitan area, characterized by high traffic emissions. As for the 2010 scenario, the estimated ozone behaviours in term of trends are similar for both stations, despite the different meteorological fields used in the modelling simulations. The EU traffic emission strategies, mainly focused on measures leading to a prevailing reduction of NO_x emissions, give rise to an ozone concentrations increase in the main metropolitan area.

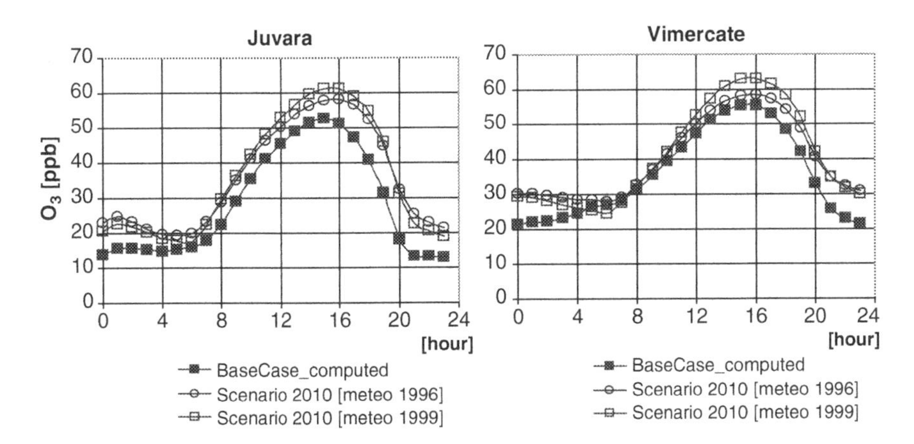

FIGURE 5. Ozone mean day pattern for two urban stations located in the Milan metropolitan area.

6. Conclusions

GAMES modelling system has been performed to assess long-term impact of emission scenarios on photochemical pollution over Lombardia Region (Northern Italy).

Since the road traffic has been estimated to be the main source of pollutants in the area, the EU Directive on road traffic emissions scheduled up to 2010 has been evaluated on the area. The estimated total emission reduction comes out to be nearly 40% for NO_x and 20% for NMVOC, with even higher abatement rates located in urban areas.

The impact of the 2010 emission scenario has been evaluated performing GAMES modelling system, fed by two different meteorological data set: the 1996 and the 1999 summer seasons (April-September).

The modelling evaluation of the EU traffic emission abatement strategies has pointed out the following evidence. (1) Accounting different meteorological data set over a long-term period does not produce significant differences in simulated pollution patterns. (2) As a consequence of the VOC-limited photochemical behaviour over a large part of the domain, the EU pollution abatement measures for road traffic emissions seem to be inadequate or at least not sufficient for the complexity of the considered area.

Acknowledgements This research was mainly supported by MIUR and University of Brescia. The authors are grateful to Guido Pirovano and Cesare Pertot (CESI) for their valuable cooperation in the frame of EUROTRAC2-SATURN and CityDelta Project.

References

Baertsch-Ritter, N., Prévôt, A., Dommen, J., Andreani-Aksoyoglu, S., Keller, J., 2003, Model Study with UAM-V in the Milan area (I) during PIPAPO: simulations with changed emissions compared to ground and airborne measurements. *Atmospheric Environment,* 37, pp. 4133–4147.

Carnevale, C., Gabusi, V., Volta, M., 2004, POEM-PM: an emission model for secondary pollution control scenarios, *Environmental Modelling and Software*, in press.

Carter, W., 1990, A detailed mechanism for the gas-phase atmospheric reactions of organic compounds. *Atmospheric Environment,* 24A, 481–518.

Deserti, M., Angelino, E., Finzi, G., Pasini, A., Zanini, G., Gabusi, V., Volta, M., Minguzzi, E., 2001. Metodi per la pre-elaborazione di dati di ingresso ai modelli tridimensionali di dispersione atmosferica. *Tech. Rep. CTN ACE 5/2001*, Agenzia Nazionale per la Protezione dell'Ambiente (APAT).

Gabusi, V., Finzi, G., Pertot, C., 2003, Performance assessment of long-term photochemical modelling system. *International Journal of Environment and Pollution,* 20, pp. 64–74.

Gabusi, V., Volta, M., 2004, Seasonal modelling assessment of ozone sensitivity to precursors in Northern Italy. *Atmospheric Environment*, in press.

Hessvedt, E., Hov, O., Isakesen, I., 1978, Quasy Steady State Approximation in Air Pollution Modeling. *Internation Journal of Chemical kinetics*, 10, 971– 994.

Jonson, J., Sundet, J., Tarrason, L., 2001, Model calculation of present and future levels of ozone and ozone precursors with a global and a regional model. *Atmospheric Environment,* 35, pp. 525–537.

Kouridis, C., Ntziachristos, L., Samaras, Z., 2000, COPERT III Computer programme to calculate emissions from road transport – User manual, *European Environment Agency Technical report No 50.*

Neftel, A., Spirig, C., Prévôt, A., Furger, M., Stuz, J., Vogel, B., Hjorth, J., 2002, Sensitivity of photooxidant production in the Milan basin: an overview of results from a EUROTRAC-2 Limitation of Oxidant Production field experiment. *Journal of Geophysical Research,* 107 (D22).

Pirovano, G., Pertot, C., Gabusi, V., Volta, M., 2004, Evaluating seasonal model simulations of ozone in Northern Italy, in: *Air Pollution Modeling and Its Application XVI,* Kluwer Academic / Plenum Publishers, New York, pp. 171–178.

Scire, J., Insley, E., Yamartino, R., 1990, Model formulation and User's Guide for the CALMET meteorological model. *Tech. Rep. A025-1,* California Air Resources Board, Sacramento, CA.

Silibello, C., Calori, G., Brusasca, G., Catenacci, G., Finzi, G., 1998. Application of a photochemical grid model to Milan metropolitan area. *Atmospheric Environment,* 32, pp. 2025–2038.

Vecchi, R., Valli, G., 1999. Ozone assessment in the southern part of the Alps. *Atmospheric Environment*, 33, 97–109.

Volta M., G. Finzi, 1999, Evaluation of EU Road Traffic Emission Abatement Strategies in Northern Italy by a Photochemical Modelling System, *Eurotrac Newsletter*, 21, pp. 29–34.

Volta, M. and G. Finzi, 2004: GAMES, a new comprehensive gas aerosol modeling system, *Environmental Modelling and Software*, in press.

Yamartino, R., Scire, J., Carmichael, G., Chang, Y., 1992, The CALGRID Mesoscale Photochemical Grid Model – I. Model Formulation. *Atmospheric Environment,* 26A, pp. 1493–1512.

8
Risk Based Approaches to Assessing the Environmental Burden of Acid Gas Emissions

Bernard Fisher*

1. The Problem

This paper discusses simple risk-based ways of choosing locations where reductions in airborne emissions should be made, based on exceedences of environmental objectives. A major regulatory issue is how emissions should be managed within national emission limits, at the same time minimising exceedences of environmental objectives. As air quality models become more detailed, complex environmental optimisation becomes harder to manage. At the same time decisions involve other qualitative factors besides meeting air quality standards alone. In this paper a practical method is introduced for dealing with the risk posed by acid and nutrient deposition and exposure to particles based on describing the source-receptor relationship in terms of a Green's function. The aim of the research is to produce a map (or maps) showing the relative importance of source locations for emissions from a specific type of source. The method allows one to decide where would be the best place to reduce current emissions. The examples are chosen to illustrate the approach to environmental decision making and in later examples show how social/demographic factors may be taken into account.

2. Risk Based Approach to Decision Making

Although national air quality objectives, in terms of concentrations and deposition, have been laid down, decisions regarding the management of emissions in the event of exceedences are not straightforward. Decisions have to be made about the most cost-effective and most equitable ways of ensuring improvement, taking account of the uncertainty in current methods. Decisions focusing on exceedences alone can be very sensitive to underlying assumptions, whereas methods accepting that the transition from acceptable to unacceptable air quality

* Risk and Forecasting, Environment Agency, Kings Meadow House, Kings Meadow Road, READING RG1 8DG UK Bernard.Fisher@environment-agency.gov.uk

is gradual, incorporate a more realistic representation of knowledge (Fisher, 2003). This is important in situations when the assessment is subject to considerable uncertainty. Recent studies have shown the uncertainty in model predictions for acid deposition (Abbott, Hayman, Vincent, Metcalfe, Dore, Skeffington, Whitehead, Whyatt, Passant and Woodfield, 2003) and dispersion calculations (Hall, Spanton, Dunkerley, Bennett and Griffiths, 2000a, Hall, Spanton, Dunkerley, Bennett and Griffiths, 2000b) based on sensitivity studies, Monte-Carlo simulation and scenario analysis. The degree of uncertainty would be regarded as quite large, broadly a factor of two i.e. changing input values within their bounds of uncertainty generally leads to a range of predictions within a range of two of a central value.

For acid deposition the environmental objective is given by the critical load usually described at a 1km × 1km spatial scale in the UK. The acid deposition to a grid square should not exceed the critical load of that grid square. For air quality the environmental objectives are set in terms of the frequency of occurrence of concentrations which should not be exceeded in specified years. This leads to the setting of objectives for the total national emissions of the main acidifying species, sulphur oxides, nitrogen oxides and reduced nitrogen compounds for each country in Europe.

3. Choice of Penalty Function

Accepting that the deposition and critical load have some uncertainty associated with them, one can define a membership function for the set of unacceptable conditions where the critical load is exceeded based on the ratio of deposition to critical load which is fuzzy, and similar criteria could be applied to air quality. A strict criterion would be a step function with a vertical line where the deposition/critical load equals one.

Figure 1 shows pictorially an example of the membership function of a fuzzy set describing low acid deposition or low concentrations, namely good environmental quality. The critical loads approach is equivalent to maximising the membership function over all grid squares. Similarly achieving air quality objectives is equivalent to maximising a function describing good air quality over all grid

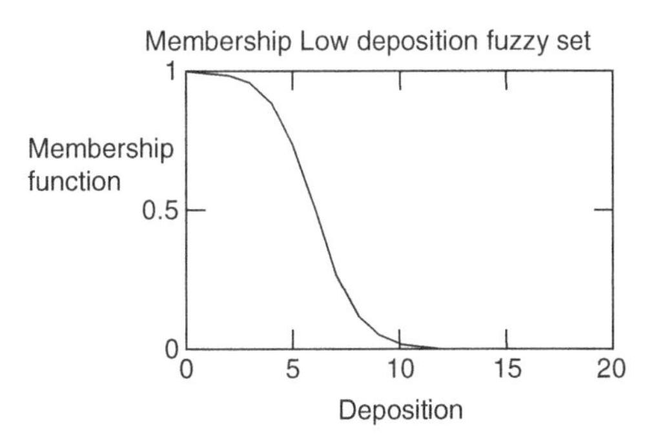

FIGURE 1. Example of fuzzy membership functions $\mu(x)$.

squares. For a strict membership function representing a sharp cutoff, improvement would arise from reducing deposition in areas near to where the deposition approximately equals the critical load. There would no advantage in reducing exceedences in areas where the deposition is very much greater than the critical load, since the membership function is still equal to 1 in these regions.

If the intention is to reduce the membership of the fuzzy sets where gross exceedences occur, one should aim to minimise the function e^x, where x is the ratio of deposition to critical load. Reducing this membership function emphasises the regions of greatest exceedence. This is the kind of penalty function P(x) used in this paper.

For acid deposition the optimum choice of location of a point source depends on the deposition of sulphur oxides, nitrogen oxides and reduced nitrogen from the source over the UK. A new source should be placed at a location where the exponential penalty function is a minimum. Similarly the greatest effect of a reduction in source strength occurs when this is a maximum, so one should reduce sources where the penalty function is a maximum. The penalty function can be written explicitly as:

$$P(x) = \exp\left(\iint_{UK} D(y,x)/CL(y)\, d^2 y \right) \tag{1}$$

where x is the location of the source, whose total deposition at point y somewhere in Britain is D(y,x), and CL(y) is the critical load at position y. The choice of location x depends on evaluating the integral $\iint_{UK} D(y, x)/CL(y) d^2y$.

This integral depends on a source-receptor matrix, or a Green's function, through D(y,x). For most semi-empirical regional transport models one can write the deposition at point y subject to a source distribution Q(x) in a region A as:

$$D(y, x) = G(y, x)\, Q(x) + D_B(y) \tag{2}$$

where G(y,x) is a Green's function, multiplicative factors involving dry and deposition factors have been neglected and $D_B(y)$ is the background deposition from all other sources. The ratio of the deposition from the point source at x to the critical load integrated over the region is

$$\begin{aligned} \iint_A D(y,x)/CL(y)\, d^2 y &= \iint_A \iint_A G(y,z)\, Q(x)\, \delta(x-z)/CL(y)\, d^2 z d^2 y \\ &= \iint_A \iint_A G'(z,y)\, Q(x)\, \delta(x-z)/CL(y)\, d^2 y d^2 z \qquad (3) \\ &= \iint_A G'(x,y)\, Q(x)/CL(y)\, d^2 y \end{aligned}$$

where G′ is the adjoint of G. The optimum location x for a new point source is when the change in penalty function P(x) because of the extra source Q(x) is a minimum. Assuming that the extra source makes a relatively small contribution to P, a first order expansion is a good approximation. Since

$$\frac{\delta P}{\delta Q(x)_{Q(x)=0}} = \iint_A G'(x,y)/CL(y)\,d^2y P(Q(x)=0) \tag{4}$$

the best location to locate the source, so as to minimise the increase in possible critical load exceedence, occurs when

$$\iint_A G'(x,y)/CL(y)\,d^2y \tag{5}$$

is a minimum. This is equivalent to the minimum of the adjoint of the deposition from a distribution of sources of strength 1/CL(y). A single calculation, assuming a distribution of source strengths equal to the inverse of the critical load, gives the deposition. The minima in the deposition field are the optimum source locations. Similar arguments could be applied to sources of particles. However in this case the criterion would be total exposure defined below, where n(y) is the population distribution

$$\iint_A G'(x,y)\,n(y)\,d^2y \tag{6}$$

The treatment used here does not fully take account of the sub-grid scale process nor the spatial variation in some factors, such as deposition velocity, although it can take into account of the dependence of wet deposition on the orographic enhancement of rain, and the enhanced deposition of particulate in hill cloud. The Green's function G and its adjoint have been determined from a simple semi-empirical transport model. It is a linear solution consisting of a series of exponentially decaying functions over travel distance of the primary and combined secondary forms of sulphur oxides, nitrogen oxides and reduced nitrogen species. The solution is derived from the diffusion equation to describe vertical mixing with a stochastic treatment of precipitation scavenging (Fisher, 1987). The adjoint is equal to the Green's function apart from a reflected wind direction weighting (north-west winds have a weighting equal to south-east winds etc). Analytical quasi-linear solutions can be justified in the current application given that one has to define the source-receptor relationship between every 1km × 1km source square to every other 1kmx × 1km receptor square over an area of 700km by 1000km.

4. Application to Multiple Criteria

One of the main advantages of fuzzy sets is that can be used to aggregate sets describing different criteria. One can generalise the single criterion approach and consider the intersection of individual membership functions $\mu_i(x)$, defined in various ways, to express multi-criteria conditions. For acid deposition this requires S and N deposition (N deposition is taken to be the sum of the deposition of oxidised and reduced nitrogen compounds) to fall within the boundaries of a region of acceptability defined in terms of S and N deposition. The boundaries are defined by the maximum sulphur acidity $CL_{max}S$, the maximum nitrogen acidity $CL_{max}N$ and the maximum nutrient nitrogen $CL_{nut}N$. There are three conditions on S and N, any one of which may lead to a critical load exceedence: $S_{dep}>CL_{max}S$, $S_{dep}+N_{dep}>CL_{max}N$ and $N_{dep}>CL_{nut}N$. In reality it cannot be possible to assign precise numbers to the three limiting values: $CL_{max}S$, $CL_{max}N$ and $CL_{nut}N$ and one may consider a simple generalisation of the penalty function above:

$$P(x) = \exp\left(\iint_{UK} \frac{S_{dep}(y,x)}{CL_{max}S(y)} + \frac{S_{dep}(y,x) + N_{dep}(y,x)}{CL_{max}N(y)} + \frac{N_{dep}(y,x)}{CL_{nut}N(y)} d^2y \right) \quad (7)$$

If the sensitivity of the dominant soil type in each 1km×1km square is modified by one of five vegetation types: acid grassland, calcareous grassland, heathland, deciduous woodland/forest, coniferous woodland/forest, there are five critical load functions per 1km × 1km grid square. If the penalty function above is adopted, then by taking the harmonic mean of the five ecosystem specific critical load conditions (so that high critical load values are not weighted highly) one can reduce the number of individual optimisations.

Simplified approaches have been used to define criteria for the acceptability of VOCs in terms of their ozone production (Derwent and Nelson, 2003). For particles we adopt a more speculative aggregation than usually applied in air quality modelling. Although it is desirable to reduce particle concentrations as much as possible, it is not possible to eliminate them, as there is no alternative to road transport in many regions of the country. One may consider an environmental criterion based on a fuzzy relation between primary particle concentrations from area sources, population density and average distance travelled person by road transport on minor roads. One may consider using PM_{10} concentration, population density n and average travel distance T to define good environmental quality with different criteria in urban areas compared with rural areas. As an example, the following criteria were adopted.

If population is rural then very low pollution and low travel ⇒ good quality.
If population is urban then low pollution and very low travel ⇒ good quality.

The combined criterion using the usual fuzzy representations of the logical AND and OR functions is the maximum of:

$$\min(\min(\mu_{very\ low}(PM), \mu_{low}(T), \mu_{rural}(n)) \text{ and } \min(\min(\mu_{low}(PM), \mu_{very\ low}(T), \mu_{urban}(n))$$

where the fuzzy membership functions, $\mu_{very\ low}$, μ_{low} and μ_{rural} etc, are somewhat arbitrary but can be chosen plausibly in terms of the average PM_{10} concentrations, annual average journey length per person and average population density.

5. Examples of Results

In this paper some preliminary results are given. These are all expressed in a similar form showing the weighting of a certain kind of source relative to a base case. The base cases are as follows:

- a poultry farm emitting ammonia, assuming a critical load for nitrogen $CL_{nut}N$ based on heathland scaled relative to a farm in the East Anglia. For a low–level source, local effects (Fig 2) are dominant as seen by the local variability.

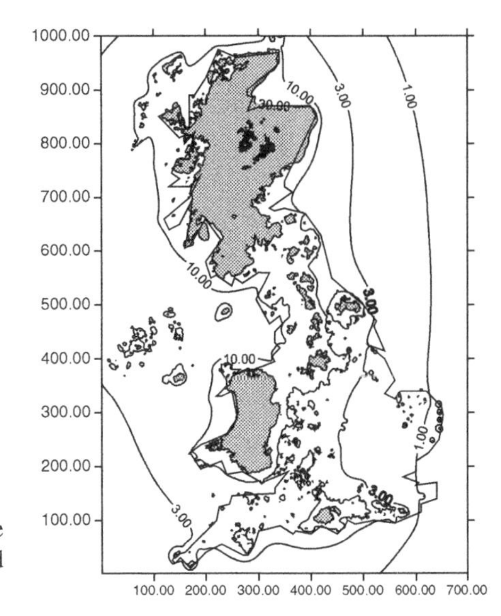

FIGURE 2. Relative weighting of source locations for a poultry farm, critical load map moorland.

- a urban development near London, emitting extra NOx, assuming a critical load for nitrogen $CL_{max}N$ based on aggregated critical loads. The area where development should be avoided is central Scotland (Fig 3).

Figure 2 contains much detailed structure, as the source is near ground-level, producing high concentrations which deposit readily in the near-field. In figure 3 the local structure is not so pronounced because neither the dry or wet deposition

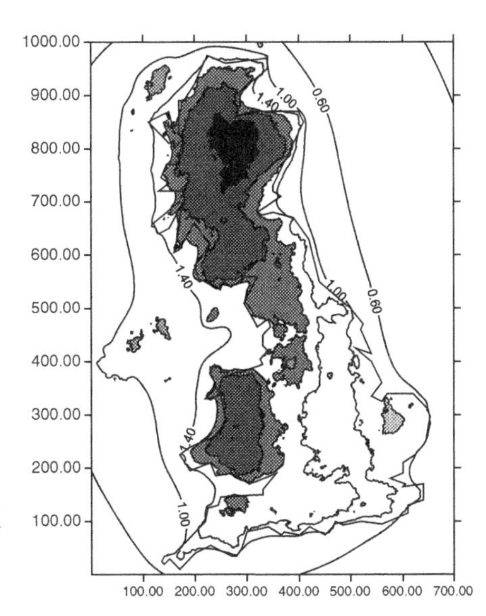

FIGURE 3. Relative weighting of location of urban NOx emissions harmonised over critical loads for acidic and nutrient nitrogen oxides, no orographic enhancement.

of nitrogen oxides is rapid in the near-field. The outcome of the calculation is that the optimum location is somewhere in the south-east of the country. The calculation is based on a number of assumptions, such as that the preferred location of the source is not dependent on the background deposition from other sources. Another assumption is that the extra source makes only a small extra contribution, so that the penalty function can be approximated by a first order expansion.

A full optimisation is possible using Markov Chain Monte Carlo methods in which the emissions are changed at random and the space of possible states is investigated, but not described here. However it would only be practical using a simple Green's function to describe the source-receptor relationship. Comprehensive models which are often used to investigate particle concentrations would impose too heavy a computational burden. Of course comprehensive models have other uses such as to parameterise complex processes. Both kinds of models have problems looking at relationships on a very local scale where nesting of models and emissions inventories is required.

In Figure 4 the method is used to determine the optimum position of a low-level PM_{10} emission, so as to minimise the exposure of population. Not surprisingly locations around major population centres, particularly London in the south-east of the country, should be avoided. In another example (Fig 5), fuzzy aggregation, as briefly outlined above, is used to find the preferred location of a new area source emitting PM_{10}, or the preferred location where emission reductions would have greatest benefit, based on aggregated fuzzy criteria dependent on (1) different standards in rural and urban areas, (2) population exposure to area sources, (3) average travel distance and (4) population density. It illustrates how concentrations derived from an air quality model may be combined with social considerations to obtain a weighting. The weighting is expressed in relative terms and does not equate to a concentration, or density. Higher values denote less

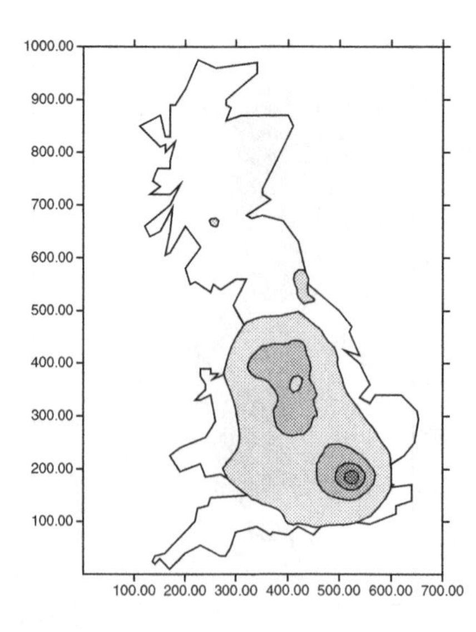

FIGURE 4. Relative weighting of source positions according to exposure of population to primary PM_{10} emissions.

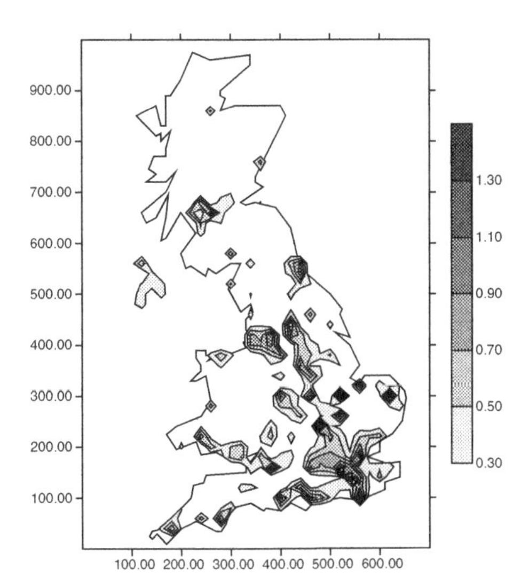

FIGURE 5. Environmental quality expressed as a weighting based on PM_{10} concentration, average travel distance, and population density.

favourable areas in terms of the chosen environmental quality i.e. poorer environmental quality. The map still emphasises urban areas, but could be used to identify rural areas, or urban areas, whose environmental quality was not proportionate to similar demographic areas.

6. Conclusions

The direct approach to critical loads is to sum all sources and then compare with the critical load. This paper demonstrates an alternative approach. Sources have a footprint of deposition. Their impact depends on the source location relative to the critical load. Assuming a simple source-receptor relation one can estimate for a typical sources, on a scale of 1km×1km, the optimum source location for the UK. The application to a single source shows that the effect of a unit of emission is dependent on location.

The paper also illustrates how results from air quality models and social considerations may be combined in the decision making process to minimise exposure of the population to PM_{10}, and also to take into consideration other factors in a planning decision such as population density and travel by car. It illustrates how risk based approaches can be used in combination with an air quality model to represent regional vulnerability to environmental influences. The method could be extended to incorporate other factors in environmental decision making.

Acknowledgements The views expressed in this paper are those of the author and are not necessarily those of the Environment Agency.

References

Abbott J, Hayman G, Vincent K, Metcalfe S, Dore T, Skeffington R, Whitehead P, Whyatt D, Passant N, and Woodfield M, 2003, Uncertainty in acid deposition modelling and critical load assessments, *Environment Agency R&D Technical Report P4-083(5)/1*

Derwent R G and Nelson N, 2003, Development of a reactivity index for the control of the emissions of organic compounds, *Environment Agency R&D Technical Report P4-105 RC8309*

Fisher B E A, 2003, Fuzzy environmental decision making: Applications to Air Pollution, *Atmospheric Environment*, 37, 1865-1877.

Fisher B E A, 1987, case study of sulphur transport modelling over Great Britain, 16th Int Technical Meeting on Air Pollution Modelling and its Application, Lindau, Germany April 1987.

Hall D J, Spanton A M, Dunkerley F, Bennett M, and Griffiths R F, 2000a, A review of dispersion model inter-comparisons studies using ISC, R91, AERMOD and ADMS, *Environment Agency Technical Report P353*

Hall D J, Spanton A M, Dunkerley F, Bennett M, and Griffiths R F, 2000b, An inter-comparison of the AERMOD, ADMS and ISC dispersion models for regulatory applications, *Environment Agency Technical Report P362*

Risk Based Approaches to Assessing the Environmental Burden of Acid Gas Emissions

Speaker: B. Fisher

Questioner: R. Dennis

Question: It appears that critical load cost functions will place emissions preferably in the SE. But this is where there is a high population. Thus health effects stemming from putting now emissions (e.g. ammonia) in the SE (to mitigate critical loads) will be exacerbated. Has this conflict been seen? Is it being addressed?

Answer: The issue raised is an example of the potential conflicts that may arise when addressing simultaneously two issues, namely acid deposition and the population exposure to particles with different effects footprints. Though not addressed directly in the paper, the conflict has been considered and the author thinks it could be tackled using a fuzzy aggregation technique. This could be used to define a joint penalty function which combines weightings for both acid deposition and particle exposure, with weighting factors chosen to reflect the degree of harm. This is necessarily somewhat subjective, but could be used to judge the preferred location of a source which has two possible harmful effects in differing geographical areas.

9
Assessment of Different Land use Development Scenarios in Terms of Traffic Flows and Associated Air Quality

Filip Lefebre[1], Koen De Ridder[1], Stefan Adriaensen[1], Liliane Janssen[1], Luc Int Panis[1], Stijn Vermoote[1], Jiri Dufek[2], Annett Wania[3], Jacky Hirsch[3], Christiane Weber[3], and Annette Thierry[4]

1. Introduction

Compact and polycentric city forms are associated with minimal consumption of land and energy, and are often promoted as the more sustainable and hence preferred mode of urban development. In this context, a series of numerical simulations was performed to evaluate the impact of two urban development scenarios on air quality and related human exposure.

The area that was selected consists of a highly urbanised region in the Ruhr area, located in the north-western part of Germany in central North Rhine-Westphalia with a total population in excess of 5.5 million. The choice for this particular area was mainly motivated by its size and importance, as well as its conversion potential. Two distinct scenarios were selected. The first is referred to as 'urban sprawl' and is characterized by a significant increase in built-up surface. This scenario supposes a continuation of the current process of people leaving the highly occupied central part of the study area to settle in the greener surroundings. In the second scenario, referred to as 'satellite cities', persons and jobs were displaced to five existing towns located near the core of the urban area. Models dealing with land use, traffic flows, and atmospheric dispersion were applied, first under conditions representative of the urbanised area as it is today. Subsequently, the urban development scenarios were implemented using spatial modelling techniques, and the impact of the scenarios with respect to air quality was evaluated, including an estimate of human exposure to air pollution and the associated external costs.

[1] Flemish Institute for Technological Research (Vito), Mol, Belgium;
[2] Transport Research Institute, Brno, Czech Republic;
[3] Université Louis Pasteur, Strasbourg, France;
[4] Danish Town Planning Institute, Copenhagen, Denmark

2. Land Use and Traffic Flows

Land use maps for the current situation were derived from satellite imagery of the Landsat Thematic Mapper (TM) instrument, using the 'stepwise discriminant analysis' image classification technique (e.g., Sabins, 1997) together with ground truth data. From Figure 1 it can be seen that built-up land use types cover a significant proportion of the study domain.

Apart from the land use characteristics, two other types of geographical information were required, that is, the spatial distribution of people and jobs (both quantities are important drivers of traffic flows, hence are required for the traffic model, see below). The spatial distribution of population was obtained from census information, and this information was aggregated to a limited number of zones to suit the requirements of the traffic model. The spatial distribution of jobs was obtained using data provided for a certain number of administrative zones of the area. These data were also processed towards the traffic zones.

After the establishment of the maps for the reference case, spatial modelling techniques were applied to simulate changes in land use, population, and job density, according to the urban-sprawl and the satellite-city scenarios outlined above. As far as the re-distribution of land use is concerned, spatial simulations were done using the so-called 'potential model', which models the potential for transition to a given land use type of all the grid cells in the domain (see, e.g., Weber and Puissant, 2003). In our study, a natural area has a high probability of being converted into a built-up area if it is mainly surrounded by already existing built-up areas, and otherwise it does not change its state. Already built-up categories remain, and

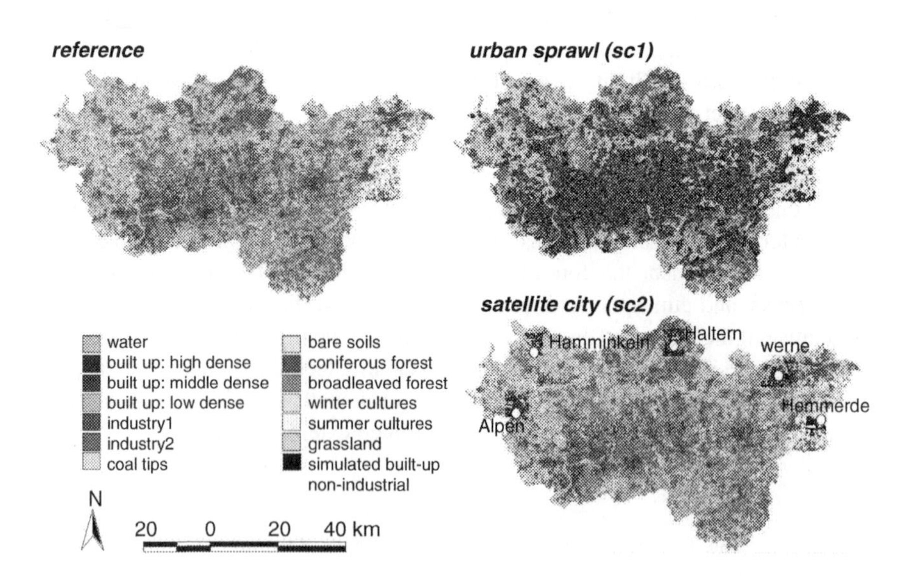

FIGURE 1. Land use categories of the reference state and the two scenarios. The dark band extending in the east-west direction in the central portion of the domain corresponds with the urbanised areas.

newly created built-up areas are always of the non-industrial type, while existing industrial areas do not change. The results of this procedure are contained in Figure 1. The main result here is that, for the urban-sprawl situation, the urbanised area in the study domain increases by almost 75%, hence land consumption is rather drastic. For the satellite-city scenario, urban land use changes are much lower, around 9%. The resulting land use maps for each scenario were then used to model the corresponding spatial distribution of people and jobs.

Traffic flow modelling was done with the AUTO model (De Ridder et al., 2004), which employs the traditional four-step methodology, and which contains sub-models that deal with trip generation, trip distribution, mode choice and traffic assignment. Trip generation or travel demand models calculate, based on the spatial distribution of population, jobs, and activities, the amount of travel that a population will undertake, distinguishing between different zones and using information regarding their characteristics to determine the number of trips that will originate or end in each zone. The first step in the traffic modelling activities was the creation of a simplified network for the study domain, consisting of highways and other major roads. Apart from using this network, the traffic model also subdivides the study area into different zones (typically a few hundred).

The number of trips to and from each zone (i.e., the traffic volume) was calculated for a 24-hour period. For the reference situation, the number of car trips was calculated from the number of inhabitants and jobs as well as the average number of daily car trips per inhabitant or per job. As a next step, matrices were created containing the traffic volumes between zones in the area. This was done using a so-called gravity model, which calculates the number of trips between any two zones following the number of trips produced in each zone as well as the number of trips attracted to each zone, the probability of travel between two zones decreasing with their interdistance. Information regarding the traffic relations between individual zones was then used to calculate the traffic relations on individual road segments. To obtain realistic spatial distribution of traffic loads for the reference case, a calibration procedure was carried out using data from the most recent traffic census.

In order to simulate the effect of the scenarios on traffic volumes and their spatial distribution, the traffic origins and destinations for each zone were recalculated based on the changes in land use, number of inhabitants and number of jobs. The road network and the traffic zones were taken identical as for the reference state. The main result of the traffic simulations is that, owing mainly to the increased average travel distance to get from home to work, passenger car traffic increased by 17% for the urban-sprawl scenario, and by 15% for the satellite-city scenario.

3. Air Quality and Impact Assessment

The methodology to compute regional-scale air quality is based on computer simulations with the atmospheric dispersion model AURORA (Mensink *et al.*, 2001), which receives relevant meteorological data from the Advanced Regional Prediction System (ARPS), a non-hydrostatic mesoscale atmospheric model developed at the University of Oklahoma (Xue *et al*., 2000, 2001). AURORA contains modules representing transport, and (photo-)chemistry, and has nesting capabilities. The chemistry module is the Carbon-Bond IV-model, which lumps together different chemical species into single components in order to reduce the computing-time.

An advanced land surface scheme (De Ridder and Schayes, 1997) was incorporated in AURORA to account for the impact of land use changes on atmospheric circulations and pollutant dispersion. The surface scheme calculates the interactions between the land surface and the atmosphere, including the effects of vegetation and soils, on the surface energy balance, and was specifically adapted to better represent urban surfaces.

A three-week period, 1-20 May 2000, was selected to perform the simulations on. This period was characterized by the presence of a blocking anticyclone over southern Scandinavia, producing weak south-easterly winds, clear skies, and moderately high temperatures over the Ruhr area. The nice weather ended abruptly on the 17th, when a cold front swept over the area. Emissions from industry, shipping and building heating were obtained from the 'Landesumweltamt Nordrhein-Westfalen', the local environmental administration. Traffic-related emissions were calculated using the MIMOSA model (Mensink et al., 2000; Lewyckyj et al., 2004), which uses the COPPERT III methodology to calculate geographically and temporally distributed traffic emissions using traffic information (including fluxes of vehicles and their speeds) from the AUTO traffic flow model (see above). Apart from anthropogenic emissions, biogenic emissions from forests (isoprene) were also calculated. The simulations carried out here focused on ground-level ozone and fine particulate matter, both pollutants being recognised as having major effects on human health.

Air quality simulations performed with the AURORA model were validated by comparing model results with available observations from two stations in the study domain (Figure 2). Even though the simulations overestimate the ozone peak concentrations rather systematically by up to a few ten percent, the diurnal cycle as well as the behaviour of the model over the entire three-week period is rather satisfactory, and well in line with results produced by models of this type. In particular the difference of nighttime concentrations between the two locations, which is due to the titration effect (reduction of ozone by traffic-related NO emissions) caused by the intenser traffic at Bottrop, is well captured by the model, meaning that the spatial distribution of traffic emissions as well as the chemical processes accounted for in the model perform correctly.

After the successful completion of the simulations for the reference case, the AURORA model was run on the urban-sprawl and satellite-city scenarios established previously, using the modified land use characteristics as well as the correspondingly modified traffic flows as inputs. The calculated pollutant emissions, largely traffic-related in this area, underwent increases of the same order as the increases of the traffic flows themselves.

The simulated change of ground-level ozone and of primary particulate matter is shown in Figure 3. With respect to ozone, the largest changes are seen to occur

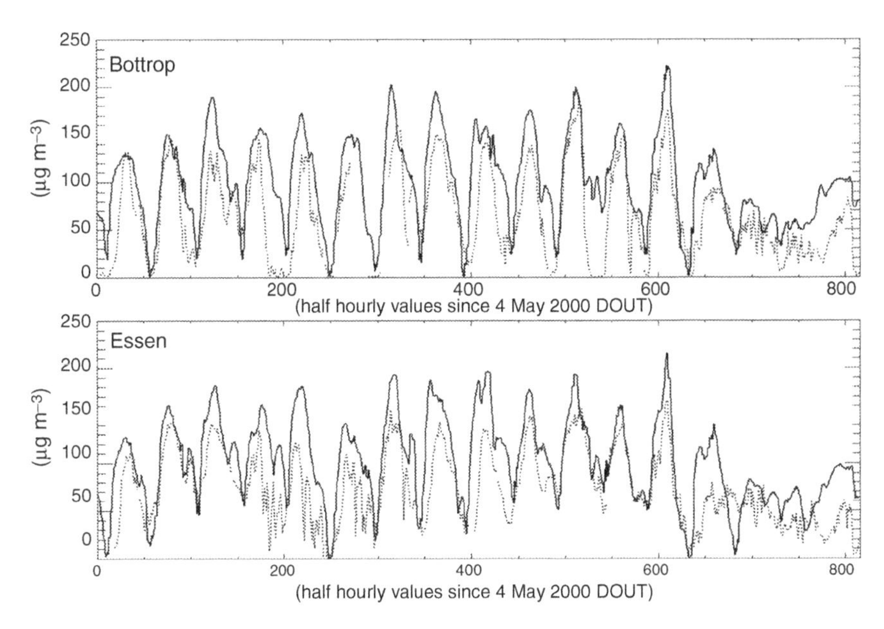

FIGURE 2. Simulated (solid line) as compared to observed (dotted line) ground-level ozone concentrations for the stations Bottrop (upper panel) and Essen (lower panel).

for the urban-sprawl scenario. Owing to the dominating south-easterly wind direction during this episode, an increased ozone plume is simulated north-west of the agglomeration. The titration effect, on the other hand, slightly depresses ozone concentrations in the central portion of the domain, i.e., where the highest population densities occur. As a result, the average exposure to ozone pollutants (calculated as the average of the concentrations, spatially weighted with population density) remained almost unchanged – they increased by 0.3% – between the reference case and the urban-sprawl scenario (Figure 4). Also in the satellite-city scenario the changes are minimal (decrease by 0.45%), despite the increased domain-average emissions.

With respect to fine particulate matter, the effect of the scenarios is perhaps not so clear (Figure 3). Whereas the satellite-city scenario clearly exhibits local spots of (a very modest) increase of this pollutant, the concentration patterns in the urban-sprawl case appear almost unaltered. A detailed analysis shows that there is a slight overall increase of domain-average concentration. However, the effects on human exposure to this pollutant are not so straightforward: whereas one would intuitively associate increased emissions and the ensuing increased domain-average concentrations with increased human exposure values, the contrary is seen to occur. Indeed, a detailed analysis shows that the urban-sprawl scenario results in an exposure *reduction* of 5.7%, and a reduction of 1.4% for the satellite-city case (Figure 4). The dominant driver of these exposure changes appears to be the movement of people from locations with high to locations with

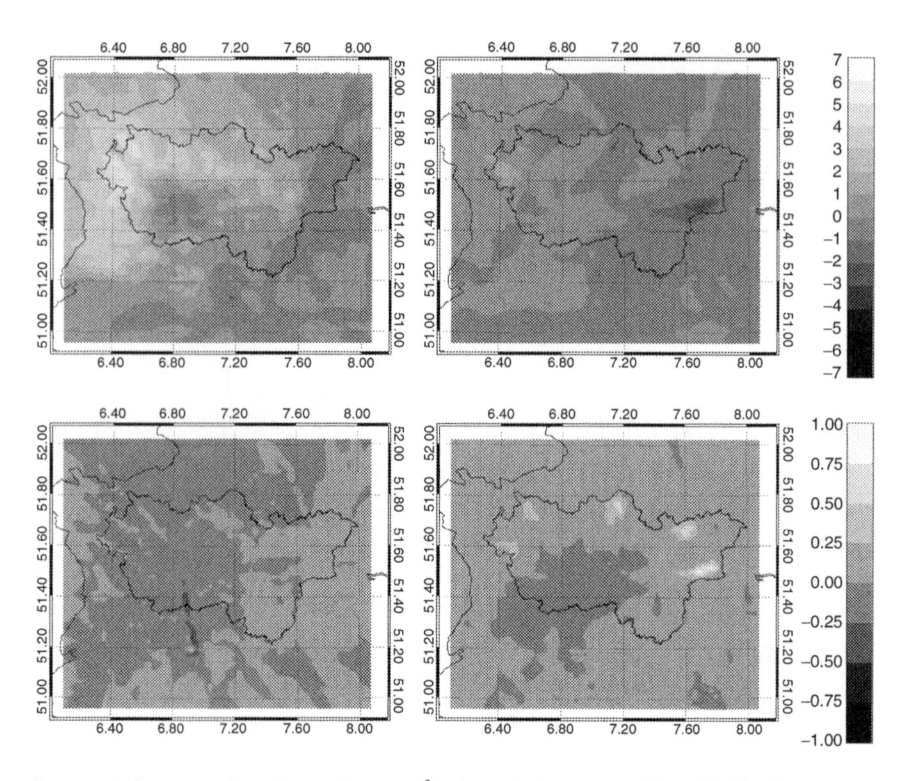

FIGURE 3. Concentration change (in μg m^{-3}) of ozone (upper panels) and PM_{10} (lower panels) for scenario 1 (left panels) and scenario 2 (right panels). Positive values indicate an increase of the considered scenario compared to the reference situation.

lower particulate matter concentrations. Stated otherwise, the global exposure decreases when a portion of the population moves from the relatively polluted conurbation to less-polluted areas.

The air pollution-related public health damage, together with changes in CO_2 emissions, were used for the calculation of the damage costs using the ExternE methodology (see, e.g., Friedrich and Bickel, 2001). The emissions in CO_2 increase in both scenarios, though slightly more so for the urban sprawl scenario, when compared to the reference state. This results in higher damage costs related to global change. The changes in damage costs of public health related to primary particulate matter and ground-level ozone are determined by the changes in exposure which reflect the combined effect of changes in concentrations and changes in population at a specific location in the study area. Because of the dominant effect of relocating people to less polluted areas, the urban development patterns presented in both scenarios result in a positive effect in the public health damage costs, especially for the urban sprawl scenario. In the calculation of the total damage costs, the effects of exposure changes to particulate matter were dominant, owing to the severe health impacts attributed to this pollutant. As a result, the

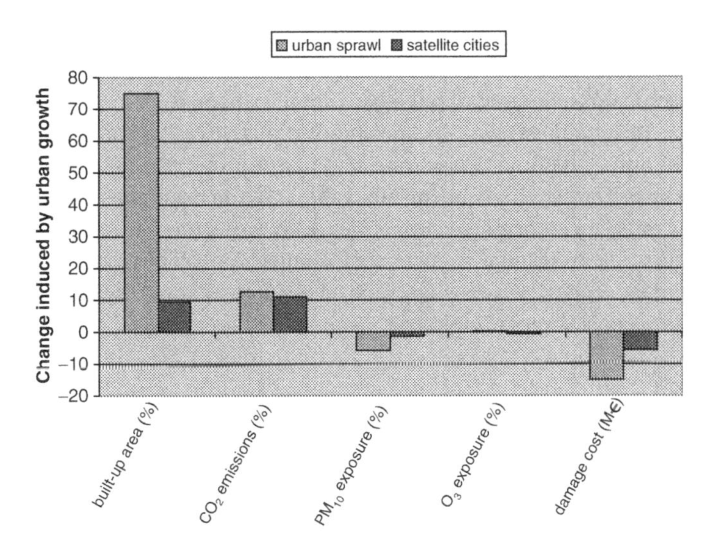

FIGURE 4. Overview of the changes induced by the urban-sprawl and satellite-city scenarios, for the indicators listed on the horizontal axis. All changes are expressed as a percentage increase compared to the reference case, except the damage cost, which is expressed in millions of Euros.

total avoided damage costs range between 5.6 M€; and 15.0 M€; for respectively, the satellite city scenario and the urban sprawl scenario when compared to the reference state (Figure 4).

4. Conclusions

Compact and polycentric city forms are associated with minimal consumption of land and energy, hence they are often promoted as being the more sustainable and hence preferred mode of urban development. In this context, a series of numerical simulations was performed to evaluate the impact of two urban development scenarios – urban sprawl and the creation of satellite cities – on air quality and related human exposure.

Working on a highly urbanised area in the German Ruhrgebiet, models dealing with satellite data processing (to map land use and relevant socio-economic indicators), traffic flows, pollutant emission and their dispersion in the atmosphere were applied, under conditions representative of the urbanised area as it is today. A fair agreement was obtained between simulated and observed pollutant concentrations.

Subsequently, urban sprawl and satellite-city scenarios were defined and implemented using spatial modelling techniques. The resulting updated maps of the area were then used as input for the traffic and atmospheric dispersion simulations, which showed that total traffic kilometres and associated emissions increased by up to almost 17%. As a consequence, the domain-average pollutant concentrations also increased, though by a smaller amount.

However, despite these global concentration increases, an analysis of human exposure to atmospheric pollution revealed that both scenarios considered here lead to *lower* rather than higher exposure values. While not contesting the evident advantages of compact or polycentric cities with respect to a host of sustainability indicators, these results indicate that compact/polycentric cities may also induce adverse effects, which should be taken into account by policy makers when making choices regarding urban development scenarios.

Acknowledgements This work was carried out as a part of the project *Benefits of Urban Green Space* (BUGS, more information available from *http://www.vito.be/bugs*) and supported by the EU under the Fifth Framework Programme. Meteorological simulations were performed using the Advanced Regional Prediction System (ARPS), a non-hydrostatic mesoscale meteorological model developed at the Centre for Analysis and Prediction of Storms, University of Oklahoma (*http://www.caps.ou.edu/ARPS*). Finally, we greatefully acknowledge the department Landesumweltamt NRW (Germany) for providing emissions data.

References

De Ridder, K., and G. Schayes, 1997. The IAGL land surface model. *Journal of Applied Meteorology*, **36**, 167-182.

De Ridder et al., 2004. Benefits of Urban Green Space. Case study in the German Ruhra area and dezcription of the methodology. Part II: Urban/regional scale. Vito report 2004/IMS/R/163, 57 pp., available on request from koen.deridder@vito.be.

Friedrich R., and P. Bickel, 2001. Environmental external costs of transport. Springer-Verlag, Berlin, Heidelberg, New York, 326 pp.

Lewyckyj, N., Colles, A., Janssen, L. and Mensink, C., 2004. MIMOSA: a road emission model using average speeds from a multi-modal traffic flow model, **in:** R. Friedrich and S. Reis (Eds.): Emissions of air pollutants, Measurements, Calculations and Uncertainties, pp. 299-304, Springer Verlag, Berlin, Heidelberg, New York.

Mensink, C., De Vlieger, I. and Nys, J., 2000. An urban transport emission model for the Antwerp area, *Atmospheric Environment*, **34**, 4595-4602.

Mensink C, K. De Ridder, N. Lewyckyj, L. Delobbe, L. Janssen, P. Van Haver, 2001. Computational aspects of air quality modelling in urban regions using an optimal resolution approach (AURORA). *Large-scale scientific computing – lecture notes in computer science*, **2179**, 299-308.

Sabins, F.F., 1997. Remote sensing, principles and interpretations, Freemann, 432 p.

Weber, C., and A. Puissant, 2003. Urbanization pressure and modeling of urban growth: example of the Tunis Metropolitan Area, *Remote Sensing of Environment*, **86**, 341-352.

Xue, M., K. K. Droegemeier, and V. Wong, 2000. The Advanced Regional Prediction System (ARPS) – A multiscale nonhydrostatic atmospheric simulation and prediction tool. Part I: Model dynamics and verification. *Meteorology and Atmopheric Physics*, **75**, 161-193.

Xue, M., K. K. Droegemeier, V. Wong, A. Shapiro, K. Brewster, F. Carr, D. Weber, Y. Liu, and D.-H. Wang, 2001. The Advanced Regional Prediction System (ARPS) – A multiscale nonhydrostatic atmospheric simulation and prediction tool. Part II: Model physics and applications. *Meteorology and Atmopheric Physics*, **76**, 134-165.

Assessment of Different Land Use Development Scenarios in Terms of Traffic Flows and Associated Air Quality

Speaker: C. Mensink (Lefebre)

Questioner: S. T. Rao

Question: Have you run the meteorological model to reflect the changing land-use options on the meteorological fields and then used these modified fields in simulating air quality?

Answer: This was indeed the strategy followed. However, we found that in the end the main effect of land use changes on atmospheric pollutant concentrations was caused by the altered emission patterns rather than by the altered atmospheric flow.

Questioner: D. Cohan

Question: I would expect that traffic emissions in your "satellite cities" scenario would be highly dependent on the proportion of residents who find employment within their city rather than commuting to other satellite locations. What assumptions were made regarding how both traffic demand and road infra-structure might respond to the land use scenarios? Do you assume that road capacity expands to accommodate traffic demand, or that congested conditions develop on corridors where traffic demand increases?

Answer: Jobs were distributed uniformly over industrial and urban land use, the latter to represent jobs in the service and commercial sectors. The traffic induced by the spatial arrangement of persons and jobs was calculated using a gravity model, where the 'attractiveness' of each zone increases with the number of jobs. No changes to road infrastructure were made. Even though relatively simplistic, this method does account for the increased average travel distance for persons living in the satellite cities but still having their jobs in the core city. However, future research should include more realistic spatial modelling of land use, and its connections with population density and jobs.

Questioner: P. Builtjes

Question: What would your final advice to the authorities in the Ruhr area be?

Answer: One the one hand our results indicate that urban sprawl induces lower personal exposure to poor air quality. But on the other hand, urban sprawl leads to (irreversible) consumption of open land. Therefore, our advice is to avoid urban sprawl, and at the same time reduce high personal exposure by decreased car use/a better public transportation system, combined with changes to the small-scale structure of the urban fabric, e.g., reducing densely habitated city quarters near busy roads.

Questioner: R. Bornstein

Question: One factor to consider in the compact vs. diffuse cities discussion is the availability of open spaces e.g. the western U.S. vs. Israel or Belgium?

Answer: This is correct. Urban sprawl leads to lower exposure to PM10 than a compact city and to lower overall damage costs, but enhances the (irreversible) consumption of open land. The latter is not expressed in terms of damage costs. Therefore other considerations should be taken into account: what is the value of availability of open spaces ? How is this evaluated ? This of course can be different in countries where open space is limited like in Belgium or where it is more available like in the western U.S.

Questioner: W. Jiang

Comments: To make a decision on city planning issue. Other social and economical aspects need to be considered in additional to the air quality studies such as this.

Answer: I agree with this. However, expressing the environmental damage in terms of external costs is already an attempt to consider environmental damage economically. In this way urban planning issues like e.g. traffic measures can then be evaluated economically through a cost-benefit analysis. Social aspects should be taken into account as well. A multi-criteria analysis could be a way to bring all these aspects together into the decision making process.

10 Concentrations of Toxic Air Pollutants in the U.S. Simulated by an Air Quality Model

Deborah J. Luecken and William T. Hutzell*

1. Introduction

The U.S. Environmental Protection Agency is examining the concentrations and deposition of air pollutants that are known or suspected to cause cancer or other serious health effects in humans. These "air toxics" or "hazardous air pollutants" (HAPs) include a large number of chemicals, ranging from non reactive (i.e. carbon tetrachloride) to reactive (i.e. formaldehyde), exist in gas, aqueous, and/or particle phases and are emitted from a variety of sources. Some HAPs, such as formaldehyde and xylene, also play an important role in the production of ozone and particulate matter. In addition, concentrations of air toxics are required over both shorter (days) as well as longer (a year) time scales in order to analyze health risks resulting from exposure to these compounds. These requirements challenge the current capabilities of numerical air quality models beyond their needs for other pollutants, such as ozone.

Most previous assessments of risks from HAPs have used Gaussian plume dispersion models to predict concentrations, while ignoring or simplifying the atmospheric chemistry that affects the concentrations of these pollutants (i.e. Rosenbaum et al., 1999). Several HAPs, such as formaldehyde and acetaldehyde, can be produced in the atmosphere in greater quantities than they are directly emitted, so it is critical to adequately characterize this complex chemistry. A 3-D photochemical grid model is better suited to account for atmospheric chemistry, including the time-varying changes in radical concentrations that affect the ambient concentrations of HAPs.

We have modified a numerical air quality model to simulate the concentration of toxic air pollutants over large spatial and temporal scales. The application described here focuses on a subset of HAPSs that exist in the gas phase. We describe the development and testing of a chemical mechanism appropriate for

* U.S. Environmental Protection Agency, MD E243-03, Research Triangle Park, NC. Luecken.deborah@epa.gov

HAPs; the incorporation of this chemistry and physics into a chemical transport model; and analysis of the model results.

2. Model Description

2.1. Model Platform, Domain and Meteorology

The Community Multi-Scale Air Quality Model (CMAQ) version 4.3 (Byun and Ching, 1999, Byun and Schere, 2004) was the base air quality model used for this application. In order to provide predictions for a domain that sufficiently covers the continental U.S., the domain extends at least 450 km beyond the US borders in all directions. This domain includes 153 horizontal east-west and 117 north-south grid cells, and 15 vertical layers from the surface to $1.0E^4$ Pa (~12 km.) Simulations were performed on an IBM SP2 for the full year of 2001 with a 10-day spinup period.

Meteorology for the simulation was calculated with the Penn State/NCAR Mesoscale Model (MM5) (*http://box.mmm.ucar.edu/mm5/*), version 3.6.1. The simulation for 2001 meteorology consisted of 34 vertical layers, using ACM parameterization for the PBL, Kain-Fritsch cumulus parameterization and 4DDA nudging (Alpine Geophysis, 2003).

2.2. Modeled HAPs and Chemical Mechanism

The toxic pollutants simulated represent the gas-phase HAPs that EPA has identified, under the Urban Air Toxics Program, to pose the highest risk to the U.S. population. To calculate concentrations of HAPs, we started with a Carbon Bond 4 (CB4) mechanism (Gery et. al, 1989) with cloud chemistry and minor modifications (Gipson and Young, 1999). The new mechanism, CB4_TX1P, accounts for the additional production and decay of air toxics, while retaining the full chemistry and radical cycling of the mechanism. Toxic species were added to the mechanism either by 1) integrating species production and decay into the full mechanism, or 2) calculating chemical decay at each time step based on the current model conditions, but with no feedback to the mechanism.

In the first instance, the full chemical mechanism was modified by changing two existing CB4 model species (FORM and ALD2) so that they simulate only formaldehyde and acetaldehyde, and adding 7 model species, listed as having feedback in Table 1. Including primary-only species quantifies the role of atmospheric chemical production on the total concentrations, as opposed to atmospheric transport of direct emissions of these species. Reactions which originally produced ALD2 were modified to produce either acetaldehyde or higher aldehydes, depending on the reactants. Acrolein and 1,3-butadiene were added to CB4_TX1P using reaction rates from Carter (2000) and product distributions corresponding to those from mapping the species to CB4 model species ([2.0]OLE for 1,3-butadiene, [0.5]OLE+[1.0]ALD2 for acrolein), with product coefficients scaled to the reaction rates. Production of acrolein from 1,3-butadiene reactions were added based on the product yields identified in SAPRC99 (Carter, 2000).

Under the second criteria, sixteen species were added to the chemical mechanism with no feedback to the chemical mechanism (Table 1). Because these species are present in small quantities or are relatively non-reactive, they do not affect the overall radical balance and chemistry, therefore their effect on the chemistry was not included. Their concentrations were updated at each chemical time step based on the current radical and environmental conditions. By including them as decay-only, however, the computational requirements of the model were significantly reduced. These species were included in all subsequent transport, advection, and deposition calculations of CMAQ.

TABLE 1. Species added to CB4, with and without feedback to chemistry calculations.

Species	Feedback	Reactions	Definition
Formaldehyde	Yes	Photolysis, OH, NO_3, O	Formaldehyde
Form-surrogate	Yes	Photolysis, OH, NO_3, O	Species which are not formal de-hyde but are mapped in CB4 as formaldehyde
Prim. formaldehyde	Yes	Photolysis, OH, NO_3, O	Formaldehyde from direct emissions only
Acetaldehyde	Yes	Photolysis, OH, NO_3, O	Acetaldehyde plus internal olefins that react immediately to form acetaldehyde
Higher aldehydes	Yes	Photolysis, OH, NO_3, O	Aldehydes with more than 2 carbons
Prim. acetaldehyde	Yes	Photolysis, OH, NO_3, O	Acetaldehyde from direct emissions only
Acrolein	Yes	Photolysis, OH, O_3, NO_3, O	Acrolein
Prim. Acrolein	Yes	Photolysis, OH, O_3, NO_3, O	Acrolein from direct emissions only
1,3-butadiene	Yes	OH, NO_3, O_3, O	1,3-butadiene from direct emissions only
Naphthalene	No	OH, O_3, NO_2, NO_3	
1,3-dichloropropene	No	OH, O_3	
Quinoline	No	OH, O_3, NO_2	
Vinyl chloride	No	OH, NO_3	
Acrylonitrile	No	OH, O_3, NO_3,	
Trichloroethylene	No	OH	
Benzene	No	OH	
1,2-dichloropropane	No	OH	
Ethylene oxide	No	OH	
1,2-dibromoethane	No	OH	
1,2-dichloroethane	No	OH	
Tetrachloroethylene	No	OH	
Carbon tetrachloride	No	OH	
Dichloromethane	No	OH	
1,1,2,2-tetrachloroethane	No	OH	
Chloroform	No	OH	

2.4. *Emissions*

The simulation included hourly emissions of all relevant organic and inorganic species. Emissions of HAPs in Table 1 were from the 1999 National Emission Inventory (NEI) v3 (www.epa.gov/ttn/chief/net/index.html). Other emissions were from the 1999 NEI v2. The 1999 inventory was chosen because it was the best inventory available at the time of simulation, and the differences between 1999 and 2001 were not large. Biogenic VOC emissions were from BEIS v3.11 (www.epa.gov/asmdnerl/biogen.html). Emissions were integrated and processed using the SMOKE processing software.

3. Results and Discussion

3.1. *Concentration Distributions*

The resulting annual concentrations of formaldehyde at the surface are displayed in Figure 1. One noticeable characteristic of this figure is the area of higher concentrations in the Southeast. Formaldehyde is emitted directly as well as produced in the atmosphere via chemical reaction with almost every other VOC in the atmosphere. Formaldehyde is higher in summer than in winter, but the summer to winter ratio varies across the U.S., with highest values in the Midwest (7-10), slightly lower values in the Northeast (5-7) and lowest values in the West (< 4).

Figures 2 and 3 display the fraction of total formaldehyde resulting from atmospheric production during winter (Dec., Jan., Feb.) and summer (June, July, Aug). In both seasons, over half the total formaldehyde is due to production in the atmosphere, but the fraction is higher in summer than in winter. The importance of atmospheric production varies across the domain and by season. In summer, higher photolysis rates, temperatures and biogenic emissions

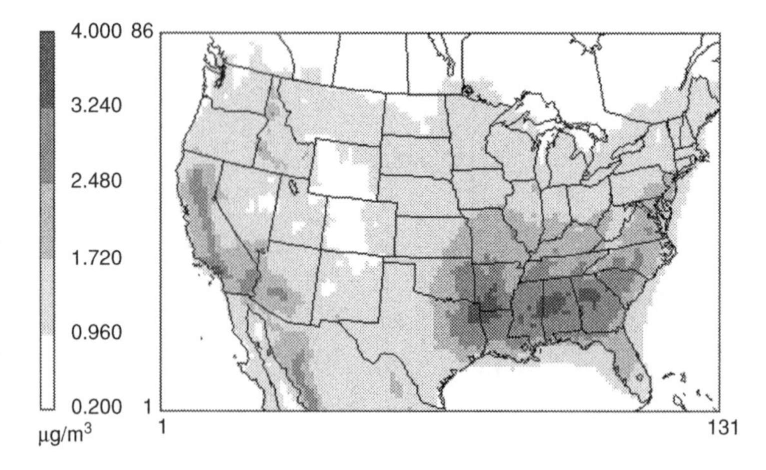

FIGURE 1. Concentration of total formaldehyde. Annual average, μg/m³.

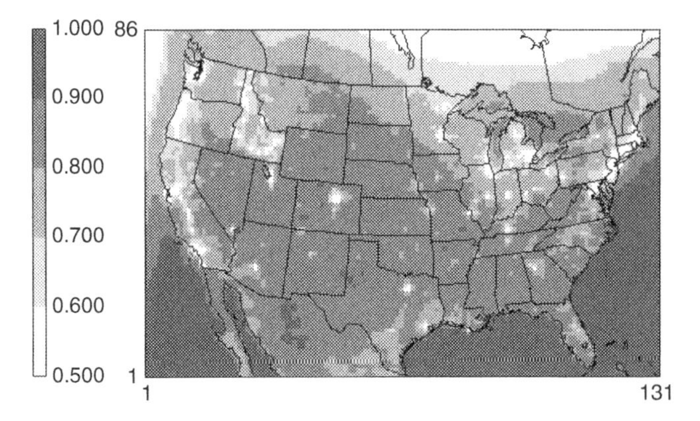

FIGURE 2. Fraction of total formaldehyde due to atmospheric formation in winter. Three-month averages.

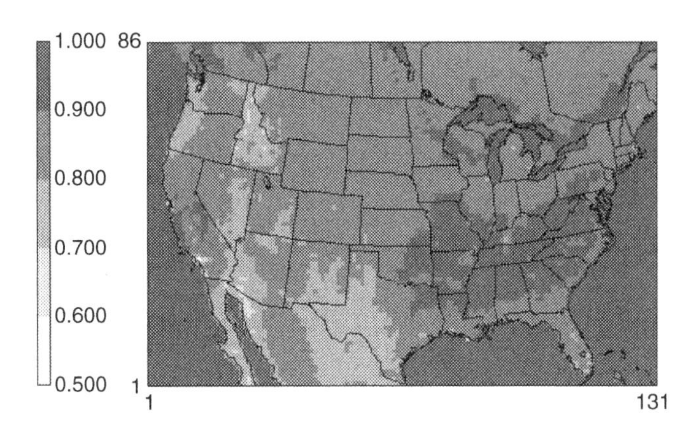

FIGURE 3. Fraction of total formaldehyde due to atmospheric formation in summer. Three-month averages.

contribute to the observed high formaldehyde concentrations. Acetaldehyde concentrations show similar patterns and behavior to formaldehyde.

Annual benzene concentrations are presented in Figure 4. Benzene behaves differently from formaldehyde because it is less reactive (half life of about 6 days in summer, vs. 2 hours for formaldehyde) and is not produced in the atmosphere. The concentration patterns reflect the distribution of emissions. Benzene in summer is less than half of its winter concentrations over most of the U.S.. The major loss process for atmospheric benzene is reaction with the OH radical, which is lower in winter. Other factors, such as increased emissions, may also increase winter benzene concentrations.

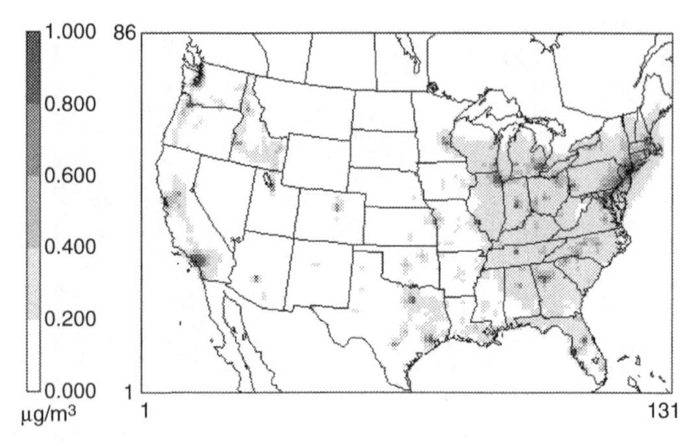

FIGURE 4. Concentration of total benzene. Annual average, µg/m^3.

Acrolein distributions are similar to benzene, but acrolein has slightly different sources as well as competition between atmospheric production (from 1,3-butadiene) and decay. Figure 5 shows that atmospheric production accounts for about 30-40% of the total acrolein concentrations, although this varies spatially and temporally.

3.2. Comparison of Modeled to Observed Concentrations

We compared model results with HAP concentrations measured in the U.S. in 2001. Our primary source of observational data was 35 monitors at 8 cities from the Air Toxics Pilot Study (Battelle, 2003), supplemented with 11 monitors from the Urban Air Toxics Monitoring Program (Eastern Research Group, 2002). Comparisons between point measurements made by monitors and volume-average concentrations from grid models such as CMAQ are difficult to

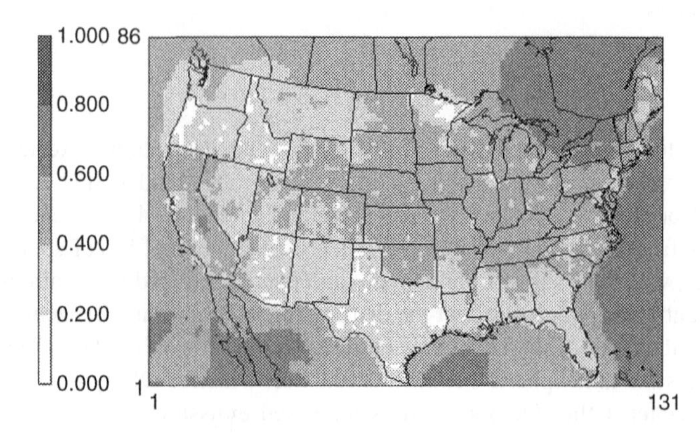

FIGURE 5. Fraction of total acrolein due to atmospheric formation. Annual averages.

interpret because there is a high degree of environmental variability within the 1296 km^2 area represented by one model cell. To compensate for short-term variability, we focus on monthly-averaged values. Figure 6 displays scatter plots of observed vs. modeled concentrations for formaldehyde and benzene.

The model tends to slightly underpredict the formaldehyde measurements, especially at the highest measured values. Overall, about 56% of the modeled values are within a factor of two of the observations. However, the model does a much better job in the spring and summer, with 72% and 62% of the predictions matching within a factor of 2, versus 43% and 47% in the fall and winter. Greater dependence of formaldehyde concentrations on atmospheric production in the spring and summer may be a factor for the better model performance in the warmer months. Differences in meteorological model performance in the warmer months may also play a role. The majority of monitors that are severely underpredicted by the model are located at the St. Louis, MO, Salt Lake City, UT, and Grand Junction, CO sites. There can be large differences in measurements between multiple monitors sited in the same city, and the model predicts some of these monitors well and others poorly (such as the River Rouge monitor in Detroit). Overall, the relative bias is –0.47, although this varies among the states, from +0.48 to –0.82, with the larger biases at the UT and CO sites.

CMAQ also tends to underpredict benzene concentrations, with the greatest differences occurring at the monitors in Salt Lake City, UT and Grand Junction, CO, at all four Mississippi sites and at the Yellow Freight monitor in Detroit. Overall, 60% of the model predictions fall within a factor of 2, with slightly better prediction in spring and summer (67% and 62%) versus fall and winter (54% and 48%). The overall relative bias is –0.54, with individual state bias ranging from +0.48 to –0.95.

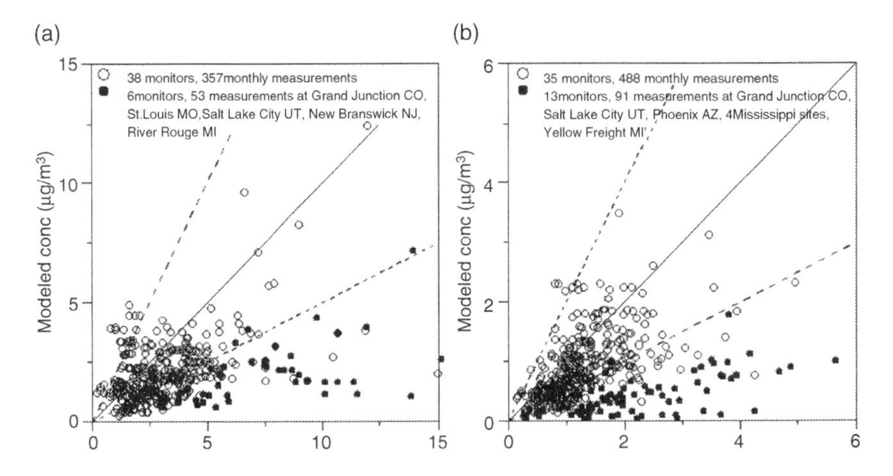

FIGURE 6. Comparison of observations with model predictions of (a) formaldehyde and (b) benzene. Monthly averaged concentrations (µg/m^3) for all months reporting data. The 1:1, 1:2 and 2:1 lines are also shown.

4. Summary

The CMAQ model has been adapted to model concentrations of air toxics across the continental U.S. for the year of 2001. A large portion of the modeled values are within a factor of 2 of the observations.

Formaldehyde concentrations across the continental U.S. are largely due to production in the atmosphere from other VOCs. While direct emissions of formaldehyde play a role, especially in urban areas in winter, their influence is generally smaller than atmospheric production. This has implications for the development of strategies to control toxic concentrations of formaldehyde – control efforts must identify the contributing VOCs, whether toxic or not. Isoprene emitted from biogenic sources can be a major source of formaldehyde, which complicates control efforts. Formaldehyde is approximately 5 times larger in summer than winter months, due to enhanced formation rates, increased emissions of biogenic VOCs and increased volatilization of organics.

Benzene concentration distributions are influenced primarily by direct emissions of benzene, because there is no gas phase production. It is critical to obtain accurate and complete emission inventories in order to predict benzene concentrations and test the results of control strategies. High density source areas of benzene, which result in "hot spots" of concentration, are not predicted well by the model, which distributes emissions uniformly within a grid cell. The role of OH in benzene decay, and the substantial diurnal and seasonal variation in OH concentrations indicate that accurately accounting for atmospheric OH is essential for benzene predictions. Concentrations are larger in winter than summer largely due to increased OH in summer.

Acrolein concentrations have a significant secondary contribution, but the majority of acrolein is from direct emissions. Acrolein is modulated by OH concentrations in two ways: it is lost through chemical reaction, but it is produced through reaction of OH with 1,3-butadiene. A complete description of OH radical concentrations is also necessary in order to accurately predict atmospheric acrolein concentrations.

Disclaimer The research presented here was performed under the Memorandum of Understanding between the U.S. Environmental Protection Agency (EPA) and U.S. Department of Commerce's National Oceanic and Atmospheric Administration (NOAA) and under agreement number DW13921548. Although it has been reviewed by EPA and NOAA and approved for publication, it does not necessarily reflect their policies or views.

References

Byun, D.W. and Ching, J.K.S. (eds), 1999, *Science Algorithms of the EPA Models-3 Community Multiscale Air Quality (CMAQ) Modeling System*. EPA-600/R-99/030, U.S. Environmental Protection Agency, Research Triangle Park, NC.

Byun, D. and Schere, K.L., 2004, Review of the governing equations, computational algorithms, and other components of the Models-3 Community Multiscale Air Quality (CMAQ) Model. Submitted to Applied Mechanics Reviews.

Battelle Memorial Institute and Sonoma Technology, Inc., 2003. Phase II: Air Toxics Monitoring Data: Analyses and Network Design Recommendations. Final report prepared for Lake Michigan Air Directors Consortium, December 19, 2003. Available from http://www.ladco.org/toxics.html

Carter, W.P.L., 2000, *Documentation of the SAPRC-99 Chemical Mechanism for VOC Reactivity Assessment.* Final Report to California Air Resources Board Contract No. 92-329, and 95-308. May, 2000. Available at http://pah.cert.ucr.edu/~carter/absts.htm#saprc99

Eastern Research Group, 2002. 2001 Urban Air Toxics Monitoring Project (UATMP). EPA 454/R-02-010. Final report prepared for EPA (OAQPS) under contract number 68-D99007, October, 2002.

Gery, M.W., Whitten, G.Z., Killus, J.P., and Dodge, M.C., 1989, A photochemical kinetics mechanism for urban and regional scale computer modeling, *J. Geophys. Res.* **94**:12925-12956.

Gipson, G.L. and Young, J.O., 1999, Gas-phase chemistry, in: *Science Algorithms of the EPA Models-3 Community Multiscale Air Quality (CMAQ) Modeling System*, Byun, D.W. and J.K.S. Ching (eds), EPA-600/R-99/030, U.S. Environmental Protection Agency, Research Triangle Park, NC, pp.8-1 to 8-99.

McNally, D., 2003, Annual *Application of MM5 for Calendar Year 2001.* Report to U.S. EPA by Alpine Geophysis, Arvada, CO, March 31, 2003.

Rosenbaum, A.S., Axelrad, D.A., Woodruff, T.J., Wei, Y.H, Ligocki, M.P. and Cohen, J.P., 1999, National estimates of outdoor air toxics concentrations, *J. Air Waste Manage. Assoc.* **49**: 1138-1152.

Concentrations of Toxic Air Pollutants in the U.S. Simulated by and Air Quality Model

Speaker: D.J. Luecken

Questioner: M. Sofiev

Question: A serious difficulty in simulating the many toxic pollutants originates from their stability in various environmental media: soil, water, canopies, etc.. A secondary emission flux from these media can be very important for e.g. polychlorinated biphenyls and, in particular, for mercury. How are you going to handle these processes?

Answer: It is becoming apparent that accounting for the re-emission of mercury is essential for accurately predicting the concentrations. While the current version of CMAQ can only account for atmospheric processes and deposition of mercury, our goal in the near future is to be able to describe mercury emissions as a bi-directional process. A parallel project at EPA is investigating ways to describe the interactions among the atmosphere and multiple other environmental media, using mercury as a first test case.

Questioner: R. Bornstein

Question: Have you tried to use some technique that interpolates and extrapolates (both the model results and observational) concentration gradients to obtain "point" values for use in your model validation efforts?

Answer: We have not done any interpolation or extrapolation with the observed data or the modeled values. While we have multiple monitors in some cities, most of them fall within one model grid but are not representative of the grid. In the case of the Rhode Island monitors, for example, five monitors fall within one corner of a CMAQ grid. So I agree that it would be useful to examine some alternative methods other than just straight grid to monitor comparisons.

Questioner: R. San Jose

Question: Have you run CMAQ with SASPR99 for benzene simulation instead of adapting the BCIV for obtaining benzene?

Answer: We have current simulations using SAPRC99, and the results for benzene are very similar. There are, however, some differences for the aldehydes, where differences between the two mechanisms are larger.

11
A Numerical Study of Recirculation Processes in the Lower Fraser Valley (British Columbia, Canada)

Alberto Martilli[1] and Douw G. Steyn

1. Introduction

The Lower Fraser Valley (LFV, Figure 1) spanning the Canada/USA border at 49° N is a roughly triangular valley with its westward end being the shoreline of the Strait of Georgia. The LFV contains the city of Vancouver, and its satellite communities with a total population of two million persons, mostly in the Canadian part of the valley. In the last two decades, the LFV has experienced several air pollution episodes. Primary as well as secondary pollutant concentrations are unusually high for such relatively small population. Several experimental campaigns (Pacific93, Pacific2001, Steyn et al. 1997), have been organized to investigate the physical mechanisms responsible for such elevated air pollutant levels. It has been hypothesized that the particularly low boundary layer height in the valley and the recirculation processes, induced by mesoscale circulations, play an important role.

2. Meteorology

The atmospheric motions in the region are the result (in cases of low synoptic forcing), of the combination of mountain/valley winds (in the main valley as well as in the tributaries), slope flows induced by mountain ridges up to 2000 m height a. s. l., land/sea breezes, and channeled flows in the Georgia Straight between Vancouver Island and the Mainland. This complex situation has been studied with the mesoscale atmospheric model MC2 (version 4.9.1, Laprise et al., 1997) for four days (10-13 of August 2001) of the Pacific2001 field experiment. In those days the highest ozone concentrations of the whole campaign were recorded.

---

[1] Alberto Martilli, and Douw G. Steyn, Atmospheric Programme, Department of Earth and Ocean Sciences, The University of British Columbia, 6339 Stores Road, Vancouver, BC V6T1Z4, Canada.

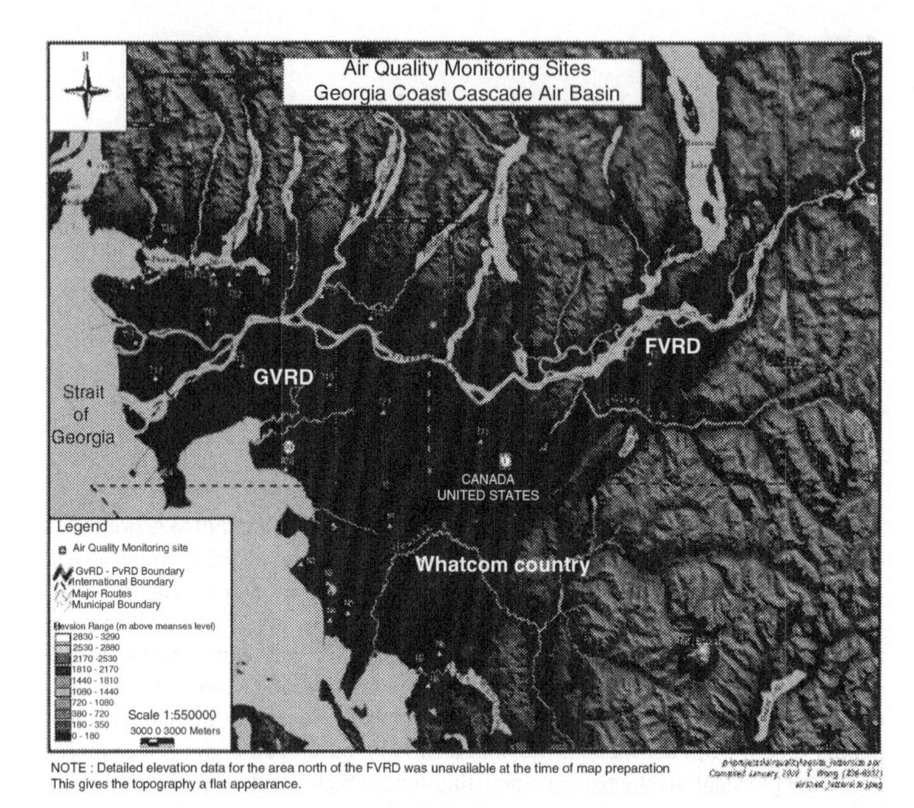

FIGURE 1. Map of the Lower Fraser Valley.

3. Modelling Technique

The technique used to investigate the recirculation consists of modeling the dispersion of several passive tracers emitted from the city of Vancouver, close to the coast line (where the strongest emissions are located), for different periods of the day (e.g. one tracer from 600 to 1800 of the 10th of August, a second tracer from 1800 of the same day to 600 of the 11th, etc.). The tracer is transported and diffused on the same eurlerian grid of the meteorological model, and with the same time-step. Evidence of recirculation of pollutants is that the modeled emissions return over the city, after emissions have stopped. Besides the inherent scientific interest in the phenomena of mesoscale atmospheric recirculation, the technique can be used to calculate air mass age and emission times. This information can be useful to define abatement strategies, in order to reduce (or prevent the increase) of future pollutant levels (population in the region is increasing at the rate of forty thousand new inhabitants per year).

4. Numerical Results

An analysis of the results shows the existence of three main recirculation processes: 1) a day-to-night recirculation, where tracers emitted during daytime and pushed inland by sea-breezes, valley winds and upslope flow, are transported back towards the coast by down-slope flows and mountain winds at night (Fig. 2); 2) a night-to-day recirculation, where tracers emitted during night and pushed over the sea by land breezes, are brought back over land during daytime by sea breezes (Fig. 3); 3) a day-to-day recirculation where tracers emitted during daytime are transported vertically by up-slope flows, and stored in a reservoir layer. Those tracers are then fumigated back to the ground the following day (Fig. 4).

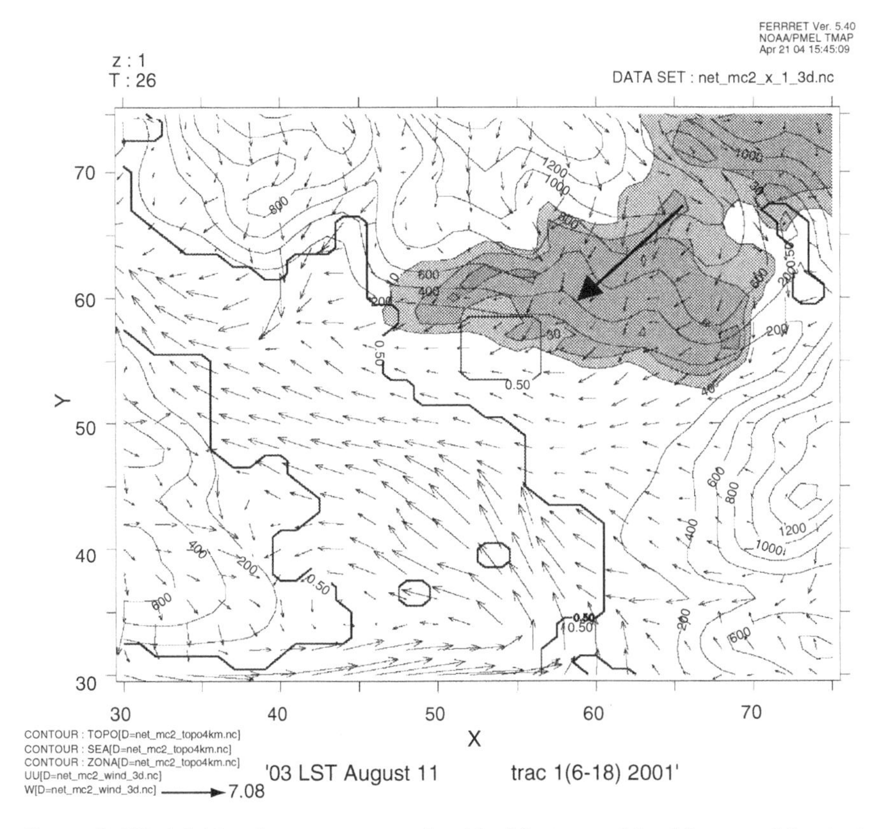

FIGURE 2. Wind field and tracer concentration (shade) at ground level from model output at 0300 LST. Downslope flows push the tracer emitted during daytime back toward the city.

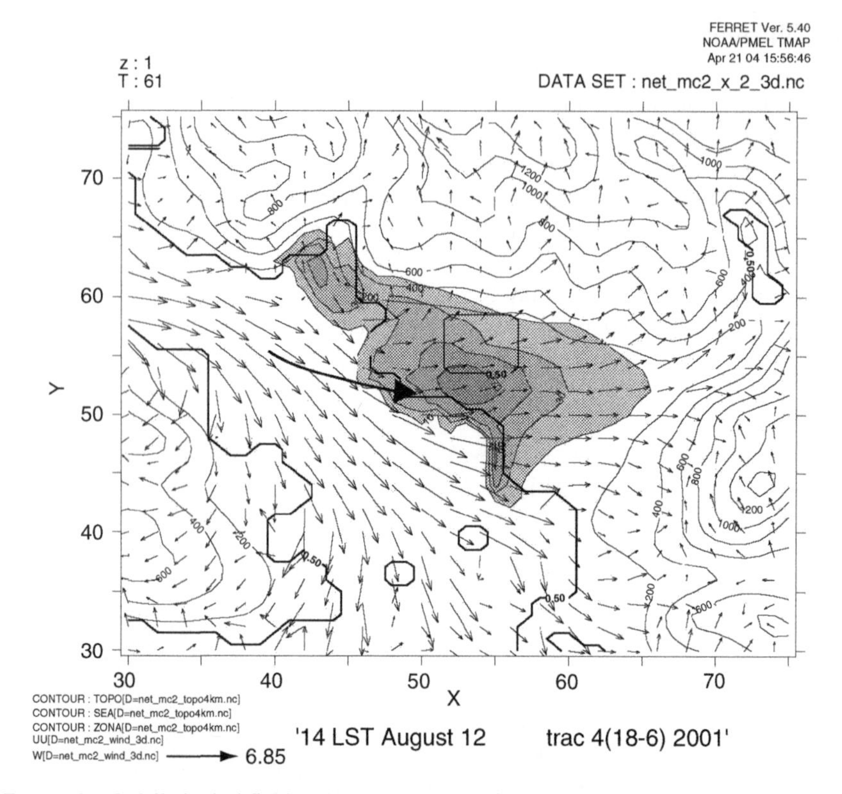

FIGURE 3. Modelled wind field and tracer concentration at ground level at 1400 LST. Sea breeze pushes the tracer emitted during night back to the land.

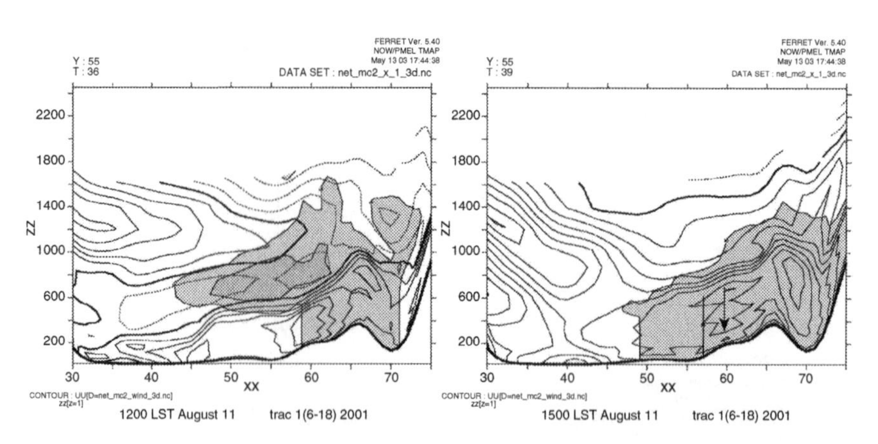

FIGURE 4. East-West vertical cross-section through the Georgia Basin of tracer concentration (shaded) at 1200 LST and 1500 LST. The thick solid line at the bottom is the topography. The tracer has been emitted between 600 and 1800 of the day before, and it has been transported in an elevated reservoir layer by up-slope flows and sea-breeze return current (1200 PST, left panel). Later, when the convective boundary layer increases, it is fumigated down to the ground (right panel, 1500 PST).

References

Laprise, R., C. Caya, G. Bergeron and M. Giguere, 1997: The formulation of the Andre Robert MC2 (Mesoscale Compressible Community) Model. *Atmosphere Ocean*, **35**, 195-220.

Steyn, D. G., J. W. Bottenheim, and R. B. Thompson, 1997: Overview of Tropospheric Ozone in the Lower Fraser Valley, and the Pacific '93 Field Study, *Atmospheric Environment*, **31**, 2025-2035.

A Numerical Study of Recirculation Processes in the Lower Fraser Valley

Speaker: A. Martilli

Questioner: B. Terliuc
Question: Does the model account for the vertical transport of the tracer?
Answer: Yes, it is a full 3-D motion.

12
A Preliminary Estimate of the Total Impact of Ozone and PM2.5 Air Pollution on Premature Mortalities in the United States

Denise L. Mauzerall and Quansong Tong

Abstract

Our objective is to estimate the excess mortalities resulting from ambient present-day concentrations of O_3 and PM2.5 in the United States. We use the U.S. Environmental Protection Agency's Community Multi-scale Air Quality (CMAQ) version 4.3 model to simulate present levels of air pollution. We then remove anthropogenic emissions and repeat the simulations. Using epidemiological dose-response functions we use the increase in pollution levels between the two simulations to estimate total mortalities resulting from exposure to O_3 and PM2.5. We estimate that in 1996 107,000 and 162,000 additional mortalities occurred in the United States due to increased exposure to ambient O_3 and PM2.5 concentrations respectively. The total mortalities incurred, 269,000 deaths, is more than ten times larger than the 23,000 mortalities that are predicted to be averted in 2010 from implementation of the 1990 CAAA. Our analysis indicates that tens of thousands of lives could be saved by substantial further improvements in U.S. air quality.

1. Introduction

Epidemiological studies of the health impacts of air pollution have quantified mortality and morbidity dose-response relationships resulting from exposure to O_3, PM2.5 (particulate matter less than 2.5μm in diameter) and other pollutants. A comprehensive summary of the dose-response functions from these studies is available in [*EPA*, 2003]. In addition, integrative analyses have examined the benefits to human health achieved by the implementation of the Clean Air Act (CAA) and its amendments (CAAA). The health benefits from the CAAA are substantial.

Denise Mauzerall and Quansong Tong, Science, Technology and Environmental Policy program, Woodrow Wilson School of Public and International Affairs, Princeton University, Princeton, NJ 08544.

23,000 premature deaths due to exposure to PM2.5 are estimated to be avoided in 2010 due to implementation of CAAA regulatory measures [*EPA*, 1999]. In light of the large benefits obtained from regulatory control measures, we wish to estimate the remaining benefits that could be attained with further controls on emissions of air pollutants.

To estimate the maximum possible benefits achievable with further controls on emissions, we estimate the total effects on mortalities of current air pollution levels. To make this estimate, we conduct two continental-scale air pollution computer model simulations. The first simulation includes all present-day anthropogenic emissions; the second simulation removes all anthropogenic emissions and provides a baseline of air unaffected by anthropogenic pollution. For simplicity, we call the simulation without anthropogenic emissions "pre-industrial". We use the concentration difference between the present and "pre-industrial" simulations to estimate the anthropogenic contribution to present levels of O_3 and PM2.5. In addition, we use the concentration difference in dose-response functions obtained from epidemiological studies to calculate the premature mortalities that result from exposure to present levels of O_3 and PM2.5.

In section 2 of this paper we describe the model frame-work we use for our analysis including a description of the atmospheric chemistry model and emissions used for the present-day and "pre-industrial" simulations. In section 3 we show O_3 and PM2.5 concentrations in January and July of the present simulation and the difference between the present and "pre-industrial" simulations. In section 4 we summarize the dose-response functions we use to evaluate the impact of present-day levels of O_3 and PM2.5 on mortality and morbidity. Section 5 quantifies mortalities and morbidities resulting from exposure to O_3 and PM2.5. Section 6 concludes with the benefits possible from additional controls on the emission of air pollutants and improvements in air quality.

2. Atmospheric Chemistry Model Configuration

2.1. CMAQ/MM5

We use the U.S. Environmental Protection Agency's Community Multi-scale Air Quality (CMAQ) version 4.3 model with a domain extending over the continental United States [*Byun and Ching*, 1999]. The model uses a Lambert Conformal projection with grid boxes of $(36km)^2$ and 12 vertical layers extending on a sigma co-ordinant system from the surface to 100mb. CMAQ is driven by meteorology from the Pennsylvania State University/National Center for Atmospheric Research (NCAR) Mesoscale Model version 5 (MM5). The Carbon Bond IV chemical mechanism with aqueous chemistry and aerosols (CB4-aq-ae3 module) is used with a 5-minute time-step.

Two sets of CMAQ simulations were conducted. MM5 simulations for January and July 1996 are used to drive CMAQ simulations from 1-31 January and 1-31 July for both sets of simulations. The first set used emissions for January and July 1996 and represents present-day conditions. Anthropogenic and biogenic emissions were obtained from the 1995 U.S. EPA national emission inventory (NEI) and from the Biogenic Emissions Inventory System version 3 (BEIS3) model, respectively. The second set removed anthropogenic emissions and repeated the simulations for January and July; this set is intended to represent "pre-industrial" conditions or maximum improvement in air quality possible with pollution controls. The first 5-days of each simulation are used for model spin-up. Boundary conditions for the January and July 1996 CMAQ simulations are obtained from monthly mean results of the global chemical transport model, Model of Ozone and Related Tracers, version 2 (Mozart-2) using emissions designed to represent the 1990s. MOZART-2 has been described and evaluated in [*Horowitz et al.*, 2003]. The version we use has been updated to include aerosols. The "pre-industrial" CMAQ boundary conditions were obtained from a MOZART-2 simulation in which anthropogenic emissions were also removed. Initial conditions for both the present and "pre-industrial" simulations use uniform low concentrations.

2.2. Emissions

The difference between present and "pre-industrial" July emissions of O_3 precursors, NO_x ($NO_x = NO + NO_2$) and CO include only anthropogenic emissions and are shown in Figure 1a and 1b. Biogenic emissions (isoprene plus terpenes) are identical in both the present and "pre-industrial" simulations. July mean emissions are shown in Figure 1c.

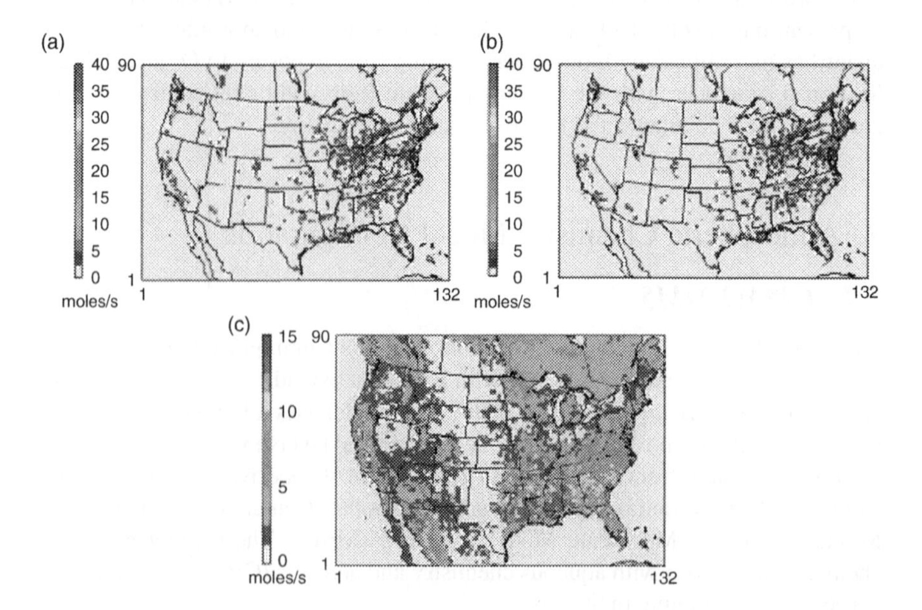

FIGURE 1. (a) July Mean NO_x Emissions: Present – Preindustrial. (b) July Mean CO Emissions: Present – Preindustrial. (c) July mean biogenic emissions of isoprene plus terpenes. Identical for both present and "pre-industrial" simulations.

2.3. *Model Evaluation*

The July 1996 simulated O_3 concentrations are evaluated in [Tong et al., this publication]. Evaluation of the simulated PM concentrations is on-going.

3. Results – Increase in O_3 and PM2.5 Concentrations due to Present levels of Anthropogenic Emissions

Figure 2a shows the maximum mean daily O_3 concentration in July 1996 across the continental United States and Figure 2b shows the increase in maximum mean daily O_3 concentrations due to anthropogenic emissions. O_3 concentrations in the "pre-industrial" simulation (not shown) are about 25 ppbv across the United States.

Figure 3a (3c) shows the monthly mean PM2.5 concentration in July (January) 1996 across the United States and Figure 3b (3d) shows the increase in monthly mean concentrations due to anthropogenic emissions. PM2.5 concentrations in the "pre-industrial" simulation (not shown) are less than 3 ug/m^3 across the United States.

4. Evaluation of Health Impacts

Many epidemiological studies have indicated that both short- and long-term exposure to air pollutants have an adverse impact on human health and mortality rates [*EPA*, 2003]. These studies fall into two categories – those which examine acute short term effects (time-series studies) and those which examine chronic long-term effects (cohort studies). Time-series studies quantify the association between short-term time-varying exposures and population-averaged health effects. They are easier to execute and are hence more numerous. However, because they are only able to account for short-term exposure, they have the disadvantage of likely underestimating the total effect of exposure to air pollution. Cohort studies follow

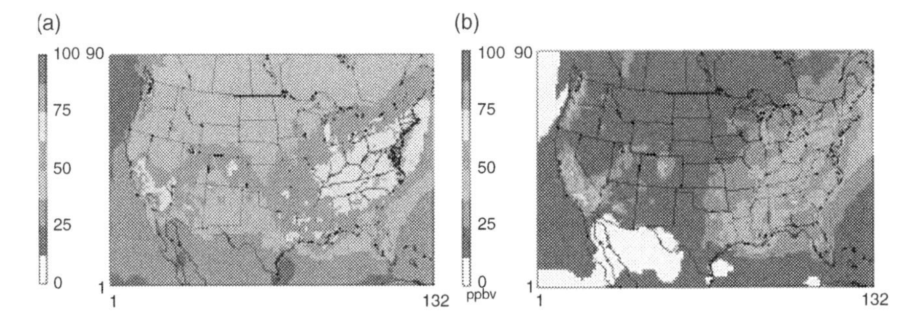

FIGURE 2. (a) Maximum mean daily O_3 concentration in July 1996. (b) Increase in maximum mean daily O_3 concentrations from "pre-industrial" to present time due to anthropogenic emissions.

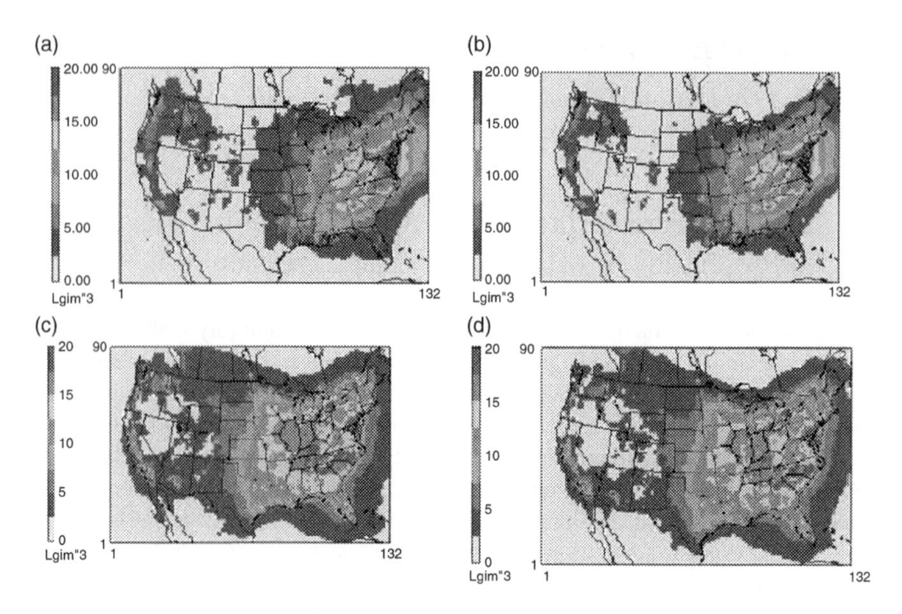

FIGURE 3. (a) Present (**July** 1996) daily mean PM2.5. (b) Increase in **July** daily mean PM2.5 from "pre-industrial" to present due to increased anthropogenic emissions. (C) Present (**January** 1996) daily mean PM2.5. (d) Increase in **January** daily mean PM2.5 from "pre-industrial" to present due to increased anthropogenic emissions.

groups over long periods of time and estimate the health effects of chronic long-term exposures. They are more expensive and difficult to execute; as a result few studies of this type have been conducted. Cohort studies include the effects of the pollution to which the cohorts were exposed over the entire period and are believed to be more indicative of the full effect of exposure to the pollution. For both types of study the change in the incidence of mortality (or respiratory disease) is estimated using the following function:

$$\Delta M = Y_o\left(1 - e^{\beta \Delta O_3}\right) * population \tag{1}$$

Here ΔM is the increase in mortalities resulting from exposure to the difference between current and "pre-industrial" levels of air pollution (indicated by ΔC). Y_o is the baseline incidence of daily non-accidental deaths per person of any age. The concentration response coefficient, β, is derived from the relative risk of exposure which is commonly reported in epidemiological studies.

It is difficult to separate the health effects of one pollutant from another. In order to avoid double-counting, we estimate mortalities resulting from exposure to present levels of O_3 and PM2.5 concentrations (two pollutants whose concentrations are not well correlated) as indicative of the total effects of present air pollution levels. Table 1 includes a description of the parameters from the two epidemiological studies we use in this analysis.

Total additional mortalities in the United States due to exposure to surface O_3 produced from anthropogenic emissions in January (not shown) and July 1996 relative to mortalities that would occur due to O_3 concentrations in an environment

TABLE 1. Epidemiological studies used in this analysis.

Pollutant - averaging interval (unit)	β (95% CI)	Age Group	Health Endpoint	Study Type	Reference
PM2.5 – annual average (ug/m3)	0.58 (0.2-1.04)	Adults, 30+	Mortality – chronic	Cohort	[*Pope et al.*, 2002]
O3 -1-hr. daily max (ppb)	0.051 (0.04-0.06)	All ages	Mortality – acute	Meta-analysis of time-series studies	[*Steib et al.*, 2002]

without anthropogenic pollution are estimated to be 3,200 and 14,600 persons, respectively (Figure 4). To approximate an annual effect of O_3 on mortalities in the United States, we average the January and June mortality rates and multiply by 12 obtaining 106,800 additional mortalities (Figure 5).

Total additional mortalities in the United States due to exposure to PM2.5 produced from anthropogenic emissions among those over the age of 30 in January and July 1996 relative to mortalities that would occur due to PM2.5 pollution in an environment without anthropogenic pollution are estimated to be 15,300 and 11,700 persons, respectively. To approximate an annual effect of PM2.5 on mortalities in the United States, we average the January and June mortality rates and multiply by 12 obtaining 162,000 additional mortalities.

5. Conclusions

Our simulations indicate that exposure of the U.S. population to ambient O_3 and PM2.5 in 1996 resulted in approximately 269,000 additional mortalities. This value is more than ten times larger than the mortalities that are predicted to be

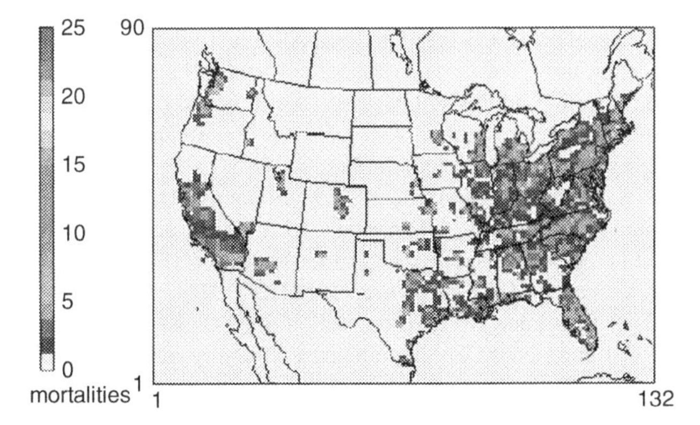

FIGURE 4. Additional mortalities resulting from enhancements in daily maximum **July** O_3 concentrations due to total anthropogenic emissions. Units: Mortalities / (36 km)^2 gridbox. Maximum value in red square = 312.

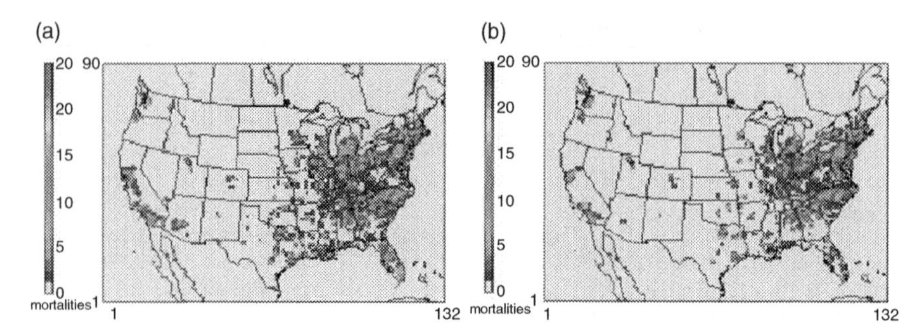

FIGURE 5. (a) Additional mortalities resulting from enhancements in **January 1996** monthly mean PM2.5 concentration due to total anthropogenic emissions. Units: Mortalities/ $(36\ km)^2$ gridbox. Maximum value in red square = 615. (b) Additional mortalities resulting from enhancements in **July 1996** monthly mean PM2.5 concentration due to total anthropogenic emissions. Units: Mortalities/$(36\ km)^2$ gridbox. Maximum value in red square = 662.

averted in 2010 from implementation of the 1990 CAAA. Our analysis indicates that tens of thousands of lives could be saved by substantial further improvements in U.S. air quality.

Acknowledgements We thank Curt Hilligas of the Office of Information Technology at Princeton University for computer system support. We are pleased to acknowledge funding from the Grace Foundation.

References

Byun, D.W., and J.K.S. Ching, 1999. Science Algorithms of the EPA Models-3 Community Multiscale Air Quality (CMAQ) Modeling System, U.S. Environmental Protection Agency.

EPA, 1999. U.S., Benefits and Costs of the Clean Air Act: 1990-2010, U.S. Environmental Protection Agency, http://www.epa.gov/air/sect812/copy99.html

EPA, 2003. U.S., Benefits and Costs of the Clean Air Act 1990 – 2020: Revised Analytical Plan For EPA's Second Prospective Analysis, U.S. Environmental Protection Agency, http://www.epa.gov/air/sect812/blueprint.html

Horowitz, L.W., S. Walters, D.L. Mauzerall, L.K. Emmons, P.J. Rasch, C. Granier, X.X. Tie, J.F. Lamarque, M. Schultz, G.S. Tyndall, J.J. Orlando, and G.P. Brasseur, 2003. A global simulation of tropospheric ozone and related tracers: Description and evaluation of MOZART, version 2., *J. Geophys. Res.*, *108* (D24), 4784.

Pope, C.A., III, R.T. Burnett, M.J. Thun, E.E. Calle, D. Krewski, K. Ito, and G.D. Thruston, 2002. Lung Cancer, Cardiopulmonary Mortality and Long-Term Exposure to Fine Particulate Air Pollution, *J. Amer. Med. Assoc.*, *287* (9), 1132-1141.

Steib, D.M., S. Judek, and R.T. Burnett, 2002. Meta-Analysis of Time-Series Studies of Air Pollution and Mortality: Effects of Gases and Particles and the Influence of Cause of Death, Age, and Season, *J. Air & Waste Manage. Assoc.*, *52*, 470-484.

13 Application of a Comprehensive Acid Deposition Model in Support of Acid Rain Abatement in Canada

Michael D. Moran[1]

1. Introduction

Acid deposition has been an environmental and policy concern in North America since the 1970s. To date there have been two national initiatives in Canada to address the acid deposition problem through SO_2 emission reductions: (a) the 1985 *Eastern Canada Acid Rain Program* between the Canadian federal government and the seven eastern provinces (e.g., Environment Canada, 1994); and (b) the *Canada-Wide Acid Rain Strategy for Post-2000*, signed in 1998 between the federal government and all 12 Canadian provinces and territories (e.g., Federal/Provincial/Territorial Ministers of Environment and Energy, 1998; Environment Canada, 2002). In addition, in 1991 the Canadian and U.S. federal governments signed the *Canada-United States Air Quality Agreement*, Annex 1 of which committed each country to specific SO_2 and NO_x emission reductions (Government of Canada and Government of the United States of America, 1991, 2002).

The development and assessment of Canadian strategies to reduce acid deposition has been supported by acid deposition modelling. Most recently, the Acid Deposition and Oxidant Model (ADOM), a comprehensive three-dimensional, regional-scale, episodic Eulerian transport/chemistry model, has been used by Environment Canada to predict future-year annual wet deposition and ambient air concentration fields in eastern North America for 24 SO_2 and NO_x emission control scenarios for both Canada and the U.S., including various emission control strategies for the *Canada-Wide Acid Rain Strategy for Post-2000* and proposed Canadian and U.S. acid-deposition legislation (Moran, 2004).

This paper illustrates the role played by ADOM in support of acid rain abatement in Canada by describing the application of ADOM for two realistic SO_2 emission reduction scenarios. The first scenario represents the impact after 2010 of the "first generation" of Canadian and U.S. acid deposition legislation from the

[1] Michael D. Moran, Air Quality Research Branch, Meteorological Service of Canada, 4905 Dufferin Street, Downsview, Ontario, M3H 5T4, Canada; Mike.Moran@ec.gc.ca (e-mail); 1-416-739-5708 (fax).

late 1980s; the second scenario represents the impact of additional "second-generation" emission reductions in both countries. Key metrics used for assessing and comparing the impact of different control strategies are the magnitude, extent, and locations of critical-load exceedances or gaps.

2. Model and Methodology Description

2.1. *Model Description*

ADOM is a comprehensive, three-dimensional, regional-scale, episodic Eulerian transport-chemistry model. It is comprehensive in the sense that it has been designed to consider all of the important processes that govern the fate of acidifying pollutants and oxidants in the atmosphere. These processes include pollutant emission, atmospheric transport and diffusion, gas-phase chemistry, aqueous-phase chemistry, cloud mixing and scavenging, and dry and wet deposition.

The acid-deposition version of ADOM has usually been run for a 33 by 33 uniform horizontal grid with 127-km grid spacing and 12 logarithmically-spaced vertical layers up to 10 km. The model domain covers eastern and central North America (e.g., Figure 2). Hourly emissions of 18 chemical species are considered both from area sources and from roughly 3,000 individual point sources. Hourly gridded meteorological fields, including wind, temperature, vertical diffusivity, mixed-layer height, cloud cover, cloud type, and stratiform and cumuliform precipitation that have been obtained from a combination of observations and numerical weather prediction model forecasts are input for each hour of ADOM simulation (e.g., Venkatram et al., 1988).

The gas-phase chemistry mechanism consists of 114 reactions amongst 47 species, including isoprene. The aqueous-phase chemistry mechanism consists of 25 reactions amongst 13 species. Two cloud modules are used, a cumuliform module to represent small-scale (i.e., subgrid-scale) convective clouds and precipitation, and a stratiform module to represent large-scale clouds and precipitation. Both types of cloud serve as aqueous-phase chemical reactors, scavenge pollutants within cloud, and mix pollutants vertically. Below-cloud scavenging of particles and four soluble gases by precipitation is also represented. Dry deposition of 12 gaseous species and particulate sulphate, nitrate, ammonium, and crustal material is parameterized using species-specific resistance formulas that account for variations in surface properties associated with changing land-use category and for diurnal and seasonal changes in these surface properties (e.g., Stockwell and Lurmann, 1989; Fung et al., 1991; Karamchandani and Venkatram, 1992; Padro et al., 1993).

2.2. *Episode Aggregation*

To estimate annual concentration and deposition fields with ADOM, which is a detailed and computationally demanding short-term model, an episode-aggregation technique is used. Specifically, ADOM is run for 33 three-day events or "episodes" drawn from all seasons from the years 1988 and 1990. Aggregated annual estimates are then calculated for a given quantity by combining the 33 mean or accumulated three-day fields predicted by ADOM for that quantity using pre-calculated, episode-specific, gridded weighting-factor fields.

Episode aggregation is a semi-empirical technique (i.e., partly dependent on data) because the gridded weighting-factor fields needed to combine ADOM episode predictions are pre-determined from historical gridded meteorological fields and from historical air-chemistry and precipitation-chemistry measurements at stations (Brook et al., 1995a,b; Environment Canada, 1997). In order to create the gridded weighting-factor fields from scattered station measurements, a horizontal interpolation technique called kriging was used (e.g., Finkelstein, 1984; Schaug et al., 1993). One limitation of this approach, however, is that kriging can only estimate values within the confines of the convex hull (the area that is defined by drawing connecting lines between every pair of stations and then deleting all interior lines) determined by the geographic distribution of measurement stations on the periphery of the measurement network. Hence, the spatial coverage of the aggregated ADOM annual fields is constrained by the spatial coverage of the historical weighting-factor fields, even though the "raw" ADOM-predicted fields cover the entire ADOM domain (e.g., Figure 2).

2.3. *Performance Evaluation*

ADOM performance has been evaluated against both precipitation-chemistry measurements and surface and upper-air air-chemistry concentration measurements for a number of species and time and space scales (e.g., Macdonald et al., 1993; Li et al., 1994; Moran, 1998). The overall model performance of ADOM, along with that of RADM, a comprehensive acid-deposition model used by the U.S. EPA, was also assessed by an international, external peer-review panel. The panel concluded that they would have confidence in the ability of ADOM and RADM to represent total sulphur and nitrogen loading of the atmosphere and deposition at seasonal and annual time scales and regional space scales (e.g., U.S. EPA, 1995).

2.4. *Wet Deposition Calibration*

A final adjustment step is performed for two ADOM predicted annual fields to minimize the impact of statistical fluctuations resulting from the relatively small number of episodes used in the episode aggregation (33) to represent the full range of meteorological conditions contributing to the long-term transport and chemical climatology of eastern North America. The ADOM annual SO_4 wet deposition field estimated by episode aggregation for each future-year scenario is multiplied by the ratio of the observed 1986-1990 mean annual SO_4 wet deposition field to the predicted annual SO_4 wet deposition field for the ADOM 1989 base-case simulation (scenario "BASE89"). The same adjustment step is performed for the annual NO_3 wet deposition field using the corresponding observed 1986-1990 mean annual NO_3 wet deposition field. No calibration was performed for the predicted future-year annual ambient concentration fields (SO_2, SO_4, NO_x, and t-NO_3), on the other hand, because of a lack of sufficient measurements for these species for the 1986-1990 period.

This adjustment procedure is equivalent to multiplying the observed base-case field by the ratio of the future-year to base-case predicted fields, and it has the advantage of making use of available observations to augment model predictions. It also emphasizes ADOM's prediction of relative changes as opposed to absolute changes for different scenarios. Other agencies have used this calibration procedure as well; for example, some U.S. EPA documents recommend this same approach and refer to the ratios of model-predicted scenario to model-predicted base-case fields as relative-reduction-factor fields (e.g., U.S. EPA, 1999).

2.5. Critical-Load Exceedance Calculation

The primary environmental measure that has been used to evaluate future-year deposition fields predicted by ADOM for ecosystem sustainability is the sulphate-wet-deposition critical-load field for eastern Canadian aquatic ecosystems. This field is shown in Figure 1 plotted on a grid with 42.3-km-by-42.3-km grid spacing

To identify areas with SO_4 wet deposition critical-load exceedances, that is, with wet deposition values greater than critical load, the predicted scenario annual SO_4 wet deposition field on the ADOM 127-km-by-127-km grid is first horizontally interpolated to the higher-resolution, 42.3-km-by-42.3-km critical-load grid by bilinear interpolation. The critical-load field is then subtracted from the scenario SO_4 wet deposition field.

Any positive values correspond to areas where the SO_4 wet deposition is predicted to remain higher than the critical load value for the same area, that

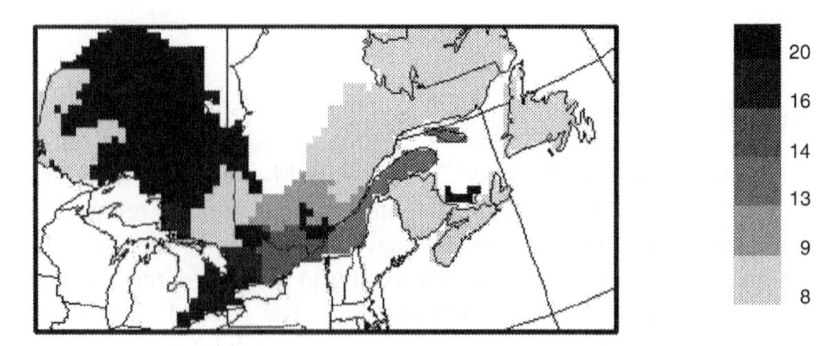

FIGURE 1. Plot of wet-sulphate-deposition critical-load field for eastern Canada at 5% lake-sacrifice level in units of kg SO_4 ha^{-1} yr^{-1} (from Environment Canada, 1997).

is, higher than can be neutralized by natural weathering processes. Negative critical-load-exceedance values, on the other hand, indicate sustainable acid-deposition levels.

3. Emission Scenario Descriptions

Emission scenarios can generally be divided into (a) simpler "rollback" scenarios, in which emissions from all sources in a particular geographic region or even the entire modelling domain are reduced by the same relative amount and (b) more complicated "realistic" scenarios, in which emission reductions are targeted by source type (e.g., coal-fired power plants, heavy-duty diesel trucks) and by jurisdiction. The two ADOM SO_2 emission reduction scenarios considered here belong to the second category.

3.1. Scenario "CCUSA2"

Scenario "CCUSA2" (abbreviation of "Canadian Controls and U.S.A. 1990 CAAA Title IV Phase 2 controls") is a future-year SO_2 emission control scenario that was designed to quantify the impact of (a) the 1985 *Eastern Canada Acid Rain Program* and (b) the acid deposition provisions (i.e., Title IV, Phases 1 and 2) of the 1990 U.S. *Clean Air Act Amendments* (CAAA) (e.g., U.S. EPA, 1990). The nominal projection year for this scenario is 2010, since that is the year in which the U.S. CAAA SO_2 control program will be fully implemented (the 1985 ECARP SO_2 reductions having been fully implemented by 1994: e.g., Environment Canada, 1994).

Scenario "CCUSA2" is a realistic scenario in that SO_2 emission reductions are modelled at the smokestack level for seven smelters and 11 power plants in Canada and for 110 power plants in the U.S. The U.S. power plant emissions are based on SO_2 emission allowances for 2010 from version 3.11 of the U.S. EPA's National Allowance Data Base, and no SO_2 emissions trades are considered. Total Canadian SO_2 emissions in the ADOM domain are reduced by 28% total U.S. SO_2 emissions in the ADOM domain are reduced by 29% from 1989 levels for this scenario (Table 1). See Environment Canada (1997) and Moran (2004) for more details about this scenario.

TABLE 1. Total SO_2 emissions within ADOM domain for the 1989 base case and two future-year emission control scenarios.

		SO_2 Emissions (Ktonnes yr^{-1})			
Scenario Name	Nominal Year	Canada	U.S.	Total	Fraction of "BASE89"
BASE89	1989	2,688	17,511	20,199	1.00
CCUSA2	2010	1,939	12,446	14,385	0.71
PST2010F	2030	1,520	6,204	7,724	0.38

3.2. *Scenario "PST2010F"*

The name "PST2010F" is an abbreviation of "Post-2010 Scenario F". This semi-realistic scenario considers further SO_2 emission reductions beyond the "CCUSA2" scenario. In Canada, the additional SO_2 reductions are due to implementation of the *Canada-Wide Acid Rain Strategy for Post-2000*, which requires further reductions in SO_2 emissions from four eastern Canadian provinces: Ontario; Quebec; New Brunswick; and Nova Scotia. All four provinces have committed to further 50% reductions in SO_2 emissions relative to their 1985 ECARP caps of 885, 500, 175, and 189 Ktonnes/yr, respectively (Environment Canada, 2002).

Major SO_2 point sources are targetted for much of these additional reductions in this scenario. In Ontario SO_2 emissions from two smelters in Sudbury are reduced by 50%, SO_2 emissions from a foundry in Wawa are reduced by 100%, and SO_2 emissions from the three largest coal-fired power plants in Ontario (Nanticoke Generating Station, Lambton GS, Lakeview GS) are reduced by 35%, 69%, and 100%, respectively. SO_2 emissions from all other Ontario sources are reduced by 36%. In Quebec 92% of the SO_2 emission reductions are imposed on a single smelter at Rouyn. In both New Brunswick and Nova Scotia, the largest SO_2 emissions reductions are made at a handful of power plants. The other two Canadian provinces that were party to the 1985 ECARP, Manitoba and Newfoundland, are assumed to remain at their 1994 caps (550 and 45 Ktonnes/yr, respectively), and Saskatchewan is assumed to remain at 1985 levels. Finally, emissions from all U.S. SO_2 sources are assumed to be reduced across the board by 50% from the estimated 2010 levels considered in scenario "CCUSA2".

The net result of these various SO_2 control measures is a reduction in SO_2 emissions on the ADOM domain of 46% relative to the "CCUSA2" scenario and 62% relative to the 1989 base scenario (cf. Table 1). See Moran (2004) for more details about this scenario.

4. Results

Figure 2 shows plots of calibrated annual SO_4 wet deposition patterns for the two ADOM scenarios. The greater impact of the larger reduction in SO_2 emissions for the "PST2010F" scenario is evident. For example, the 20 kg/ha/yr isopleth extends into southwestern Ontario for the "CCUSA2" scenario (Figure 2a) but has vanished completely for the "PST2010F" scenario (Figure 2b). The 10 kg/ha/yr isopleth can also be seen to retreat westward across the Maritimes. Maximum annual SO_4 wet deposition is reduced from 25.9 kg SO_4/ha/yr in the "CCUSA2" scenario to 16.6 kg/ha/yr in the "PST2010F" scenario. The corresponding annual SO_2 and SO_4 ambient air concentration fields for these two scenarios are presented in Moran (2004).

Figure 3 is based upon Figures 1 and 2. It shows the SO_4 wet deposition critical-load exceedance fields for eastern Canada for the same two scenarios. An

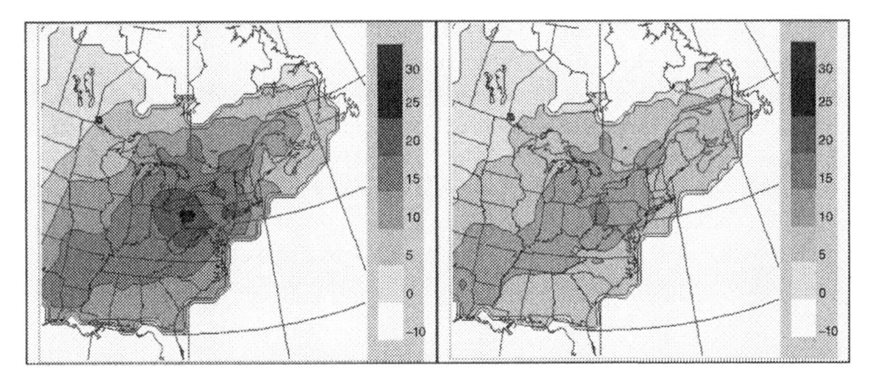

FIGURE 2. Plots of calibrated annual SO_4 wet deposition patterns in units of kg SO_4/ha/yr for ADOM scenarios (a) CCUSA2 and (b) PST2010F.

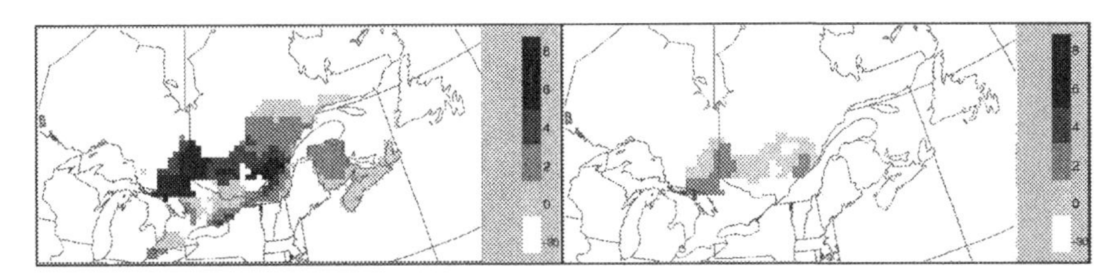

FIGURE 3. Plots of annual SO_4 deposition critical-load exceedance fields (units of kg SO_4/ha/yr) for ADOM scenarios (a) CCUSA2 and (b) PST2010F.

important finding of Environment Canada (1997) and Moran (1997) was that much of eastern Canada would remain in exceedance of critical loads even after the significant reductions in both Canadian and U.S. SO_2 emissions considered in the "CCUSA2" scenario (Figure 3a). ADOM predicts that the additional SO_2 emission reductions modelled in the "PST2010F" scenario will reduce both the extent of the area in exceedance of critical load and the magnitude of the exceedances themselves (Figure 3b). However, a band of critical-load exceedances remains across central Ontario and central Quebec. Only one ADOM scenario has been considered to date in which the critical-load exceedance field virtually vanishes, and that scenario required a 75% reduction in both Canadian and U.S. SO_2 emissions relative to the "CCUSA2" scenario (Moran, 2004).

5. Conclusions

ADOM, a comprehensive, regional-scale, Eulerian transport/chemistry model, has been used to predict future-year annual wet deposition and ambient air concentration fields in eastern North America for a number of SO_2 and NO_x emission control scenarios. The application of ADOM for two realistic SO_2 emission control scenarios, the first representing the effect of the "first generation" of Canadian and U.S. acid deposition legislation from the late 1980s and the second the effect of further "second-generation" SO_2 emission reductions in both countries, was described. The model predicts that sulphate critical loads for aquatic ecosystems will still be exceeded in central Ontario and Quebec even for a scenario with reductions in SO_2 emissions considerably larger than those already required by legislation. Other ADOM scenarios presented elsewhere support this finding, suggesting that even larger SO_2 emission reductions will be required to achieve critical load in eastern Canada.

Acknowledgements The contributions of Dr. J. Kaminski, formerly with ARM Consultants and now with WxPrime Corporation, in running the ADOM model for numerous emission scenarios over the past ten years is greatly appreciated.

References

Brook, J.R., Samson, P.J., and Sillman, S., 1995a, Aggregation of selected three-day periods to estimate annual and seasonal wet deposition totals for sulfate, nitrate, and acidity. Part I. A synoptic and chemical climatology for eastern North America, *J. Appl. Meteor.*, **34**, 297-325.

Brook, J.R., Samson, P.J., and Sillman, S., 1995b, Aggregation of selected three-day periods to estimate annual and seasonal wet deposition totals for sulfate, nitrate, and acidity. Part II. Selection of events, deposition totals, and source-receptor relationships, *J. Appl. Meteor.*, **34**, 326-339.

Environment Canada, 1994, *1994 Annual Report on the Federal-Provincial Agreements for the Eastern Canada Acid Rain Program*, Cat. EN40-11/29-1994E, ISBN 0-662-23665-3, Gatineau, Quebec, 16 pp. [Available from http://www.ns.ec.gc.ca/reports/pdf/acid_rain_94_e.pdf].

Environment Canada, 1997, *The 1997 Canadian Acid Rain Assessment: Volume 2. Atmospheric Science Assessment Report*, February, Meteorological Service of Canada, Environment Canada, Downsview, Ontario, 296 pp.

Environment Canada, 2002, *2001 Annual Progress Report on The Canada Wide Acid Rain Strategy for Post-2000*, Cat. En40-11/39-2001, ISSN 1488-948X, December, Gatineau, Quebec, 16 pp. [Available from http://www.ccme.ca/assets/pdf/acid_rain_e.pdf].

Federal/Provincial/Territorial Ministers of Environment and Energy, 1998, *The Canada-Wide Acid Rain Strategy for Post-2000: Strategy and Supporting Document*, October 19, 1998, Halifax, Nova Scotia, 20 pp. [Available at http://www.ccme.ca/initiatives/climate.html?category_id=31].

Finkelstein, P.L., 1984, The spatial analysis of acid precipitation data. *J. Clim. Appl. Meteor.*, **23**, 52-62.

Fung, C.S., Misra, P.K., Bloxam, R., and Wong, S., 1991, A numerical experiment on the relative importance of H_2O_2 and O_3 in aqueous conversion of SO_2 to SO_4^{2-}, *Atmos. Environ.*, **25A**, 411-423.

Government of Canada and Government of the United States of America, 1991, *Agreement between the Government of Canada and Government of the United States of America on*

Air Quality, March 13, 1991, Ottawa, Canada. [Text available at http://www.ijc.org/rel/agree/air.html].

Government of Canada and Government of the United States of America, 2002, *Canada–United States Air Quality Agreement 2002 Progress Report*, 64 pp. [See http://www.ec.gc.ca/air/qual/2002/index_e.html or http://www.epa.gov/airmarkets/usca/2002report.html].

Karamchandani, P.K. and Venkatram, A., 1992, The role of non-precipitating clouds in producing ambient sulfate during summer: results from simulations with the Acid Deposition and Oxidant Model (ADOM), *Atmos. Environ.*, **26A**, 1041-1052.

Li, S.-M., Anlauf, K.G., Wiebe, H.A., Bottenheim, J.W., and Puckett, K.J., 1994, Evaluation of a comprehensive Eulerian air quality model with multiple chemical species measurements using principal component analysis, *Atmos. Environ.*, **28**, 3449-3461.

Macdonald, A.M., Banic, C.M., Leaitch, W.R., and Puckett, K.J., 1993, Evaluation of the Eulerian Acid Deposition and Oxidant Model (ADOM) with summer 1988 aircraft data, *Atmos. Environ.*, **27A**, 1019-1034.

Moran, M.D., 1997, Evaluation of the impact of North American SO_2 emission control legislation on the attainment of SO_4 critical loads in eastern Canada, Paper 97-TA28.01, *Proc. 90th AWMA Annual Meeting*, June 8-13, Toronto, Canada, Air & Waste Management Association, Pittsburgh.

Moran, M.D., 1998, Operational evaluation of ADOM seasonal performance with surface data from the Eulerian Model Evaluation Field Study, *Proc. 10th AMS/AWMA Joint Conf. on Applications of Air Pollution Meteorology*, Jan. 11-16, Phoenix, American Meteorological Society, Boston, pp. 404-408.

Moran, M.D., 2004, Current and proposed emission controls: how will acid deposition be affected?, in: *2004 Canadian Acid Rain Science Assessment*, Environment Canada, Downsview, Ontario [In review].

Padro, J., Puckett, K.J., and Woolridge, D.N., 1993. The sensitivity of regionally averaged O_3 and SO_2 concentrations to ADOM dry deposition velocity parameterizations, *Atmos. Environ.*, **27A**, 2239-2242.

Schaug, J., Iversen, T., and Pedersen, U., 1993, Comparison of measurements and model results for airborne sulphur and nitrogen compounds with kriging, *Atmos. Environ.*, **27A**, 831-844.

Stockwell, W.R. and Lurmann, F.W., 1989, *Intercomparison of the ADOM and RADM gas-phase chemistry mechanisms*. Report prepared for the Electric Power Research Institute, June, Electric Power Research Institute, Palo Alto, California, 260 pp.

U.S. EPA, 1990, Full text of 1990 Clean Air Act Amendments, U.S. Environmental Protection Agency [Available at *http://www.epa.gov/oar/caa/caaa.txt].*

U. S. EPA, 1995, *Acid Deposition Standard Feasibility Study: Report to Congress*, Report No. EPA-430/R-95-001a, U. S. Environmental Protection Agency, Research Triangle Park, North Carolina. October, 242 pp.

U.S. EPA., 1999, *Draft Guidance On The Use Of Models And Other Analyses In Attainment Demonstrations For The 8-Hour Ozone NAAQS*. U.S. Environmental Protection Agency Report No. EPA-454/R-99-004, May, 168 pp. [See *http://www.epa.gov/ttn/naaqs/ozone/* eac/gd19990531_epa-454_r-99-004.pdf].

Venkatram, A., Karamchandani, P.K., and Misra, P.K., 1988, Testing a comprehensive acid deposition model, *Atmos. Environ.*, **22**, 737-747.

Application of a Comprehensive Acid Deposition Model in Support of Acid Rain Abatement in Canada

Speaker: M. Moran

Questioner: C. Mensink

Question: In your presentation you focus on the contribution of sulphates to acid rain deposition in Canada. Does that mean that nitrates and ammonium do not pose any problems with respect to acid deposition and critical loads?

Answer: No, it does not. Initial efforts in eastern North America to reduce acid deposition focussed on SO_2 *emission reductions due to the dominant contribution of sulphur to acid deposition in the 1970s and 1980s. The success of controls on* SO_2 *emissions coupled with little change in* NO_x *and* NH_3 *emissions in eastern North America and a greater understanding of the contribution of nitrogen to forest critical loads means that emissions of nitrogen species must also be considered in acid-deposition abatement plans.*

Questioner: B. Fisher

Question: Given the relatively small changes in total emissions of SO_2 ***in eastern North America (at least relative to Europe) over the past 20 years, is there any evidence of trends in measurements which are consistent with emission scenarios?***

Answer: Total emissions of SO_2 *in North America decreased by nearly 40% between 1980 and 2000, a significant decline though less than the 60% decrease in* SO_2 *emissions reported for the EMEP area over the same 20-year period. There is considerable evidence from both* $SO_4^=$ *wet concentration and deposition measurements and* SO_2 *and* SO_4 *air concentration measurements in eastern North America of concurrent decreases in atmospheric sulphur levels over this period, especially in the main source regions.*

14
Modeling Source-Receptor Relationships and Health Impacts of Air Pollution in the United States

Quansong Tong*, Denise Mauzerall, and Robert Mendelsohn

1. Introduction

The objective of this study is to examine the source-receptor relationships and health impacts of air pollution resulting from emissions from individual states in the continental United States. Air pollution is a widespread problem of spatially and temporally varying magnitude. Over 115 million U.S. individuals are exposed to air pollution levels in excess of one or more health-based ambient standards in 1996 (EPA, 1996). Policy decisions can benefit from a quantification of source-receptor relationships and resulting health and environmental damages. The damage or environmental assessment processes, however, often suffer from a lack of scientific knowledge and tools.

To help address this problem, we have assembled an integrated assessment model that links emissions of major air pollutants to their chemical transformation and transport, to human exposure and finally to resulting health impacts. The architecture of the assembled integrated assessment model is shown in Figure 1.

Section 2 of this paper describes our integrated assessment modeling system. Section 3 describes the model simulations we conduct in our source-receptor and health analysis. Section 4 presents an evaluation of the model performance against available measurement data. Section 5 describes our preliminary results on the source-receptor relationship between NO_x emissions and O_3 concentration distribution. Section 6 summarizes the resulting excess premature mortalities due to O_3 exposure in the continental United States. Conclusions are presented in Section 7.

* Quansong Tong, Denise Mauzerall, Woodrow Wilson School, Princeton University, Princeton, NJ 08544.
Robert Mendelsohn, School of Forest & Environmental Studies, Yale University, New Haven, CT.

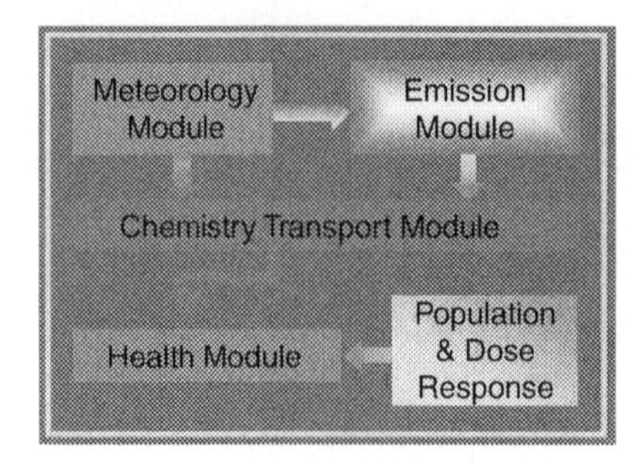

FIGURE 1. Architecture of the integrated assessment model.

2. Description of the Integrated Assessment Modeling System

We assemble a variety of models from different scientific disciplines to perform an integrated assessment study. Anthropogenic and natural emissions are processed using the MCNC SMOKE model (http://www.cep.unc.edu/empd/products/smoke/index.shtml). Meteorology is provided by NCAR/PSU MM5 (Grell et al., 1993). US EPA Models-3/Community Multiscale Air Quality (CMAQ) model (Byun and Ching, 1999) is used to calculate ambient concentrations of pollutants using emissions from SMOKE and meteorology from MM5 as input. Human exposure is calculated from concentration distributions of pollutants obtained from CMAQ, and population census data. Finally, the effects of air pollution on human health are estimated using epidemiological concentration response functions on the calculated exposure.

The simulation domain covers the 48 contiguous US states and parts of Southern Canada and Northern Mexico, with western and eastern borders in the Pacific and Atlantic Oceans, respectively. The domain is divided into 132 columns by 90 rows with a horizontal grid resolution of 36 km. There are 12 vertical layers extending from surface to approximately 200 mb. The surface layer is about 38 m. Lambert map projection is applied for horizontal domain, which has a center at 100W and 40N, a lower left corner at (122.69W, 22.20N), and aupper right at (66.42W, 51.39N). A vertically varying lateral boundary condition is used for both baseline and perturbation simulations.

3. Simulation Design

To quantify the source-receptor relationship between NO_x emissions and resulting O_3 production over the continental United State, we conduct emissions-perturbation simulations for each state. These simulations include (1) a standard simulation (baseline emissions), (2) perturbation simulations in which NO_x emissions are increased from each of the 48 states, one state per scenario. The difference between a perturbation run and the baseline run represents the marginal contributions of NO_x

emissions from a single state to total ozone production over the modeling domain. To conserve the original emission patterns inside a state, we implement the emission perturbation as follows,

$$NO_{ij} = \frac{E_T^* NO_{ij}^* f_i}{\sum_{i=1}^{M} \sum_{j=1}^{N} [(NO_{ij} + NO2_{ij})^* f_i]} \tag{1-a}$$

$$NO2_{ij} = \frac{E_T^* NO2_{ij}^* f_i}{\sum_{i=1}^{M} \sum_{j=1}^{N} [(NO_{ij} + NO2_{ij})^* f_i]} \tag{1-b}$$

where E_T is the total increase of emissions from that state, M is number of grid cells inside the state, f is the fraction of grid cell i located inside the state (f ranges from 0 to 1), and N is the number of layers in which NO_x is emitted. The total increase in NO_x emissions (E_T) for each perturbation is 10 moles/sec (equivalent to 27.3 tons per day assuming 90% NO_x emitted in the form of NO), distributed across each state in proportion to emissions in the baseline simulation.

4. Model Evaluation

The ability of CMAQ to capture the main features of surface O_3 and its variations over the continental United States is assessed by comparing model output with hourly O_3 measurements obtained from the Aerometric Information Retrieval System (AIRS). We use the statistical measures recommended by US EPA (1991) which are shown in Table 1.

Prediction-observation pairs are often filtered based on a cut-off value for observed data, as suggested by the evaluation guidance developed by US EPA (1991). Those pairs below the cut-off values are excluded from the analysis since most air quality models are designed to simulate high O_3 episodes that are linked to adverse impacts on human health. We have adopted a cut-off value of 40ppbv

TABLE 1. Definitions of the US EPA recommended statistical measures (EPA, 1991).

Measures	Definition[1]
Mean Normalized Bias (MNB)	$\frac{1}{N}\sum_{i=1}^{N}\frac{C_{mod}(i) - C_{obs}(i)}{C_{obs}(i)}$
Mean Normalized Error (MNE)	$\frac{1}{N}\sum_{i=1}^{N}\frac{\lvert C_{mod}(i) - C_{obs}(i)\rvert}{C_{obs}(i)}$
Unpaired Peak Prediction Accuracy (UPPA)	$\frac{C_{mod}(i)_{max} - C_{obs}(i)_{max}}{C_{obs}(i)_{max}}$

[1] N is the number of hourly data pairs. C_{mod} and C_{obs} are the predicted and measured data pairs, and C_{max} is the maximum hourly value of measurement or modeling data.

based on EPA recommendations (40-60ppbv) and previous work (e.g., 40ppbv, Kang et al., 2003).

Although there is no objective criterion set for performance standards, US EPA has suggested informal criteria for regulatory modeling practices: ±5 to ±15% for MNB, 30 to 35% for MNE, and ±15 to ±20 for UPPA (Kang et al., 2003). With the cut-off value of 40 ppbv, we obtain an overall MNB value of –6%, MNE of 25%, and UPPA of –12%, all of which lie within the US EPA suggested ranges.

5. Source-Receptor Relationship between O_3 and NO_x Emissions

The 49 simulations are designed to quantify how much and where O_3 perturbations will be for identical perturbations of NO_x emissions from each state. Figure 2 illustrates monthly averaged daily maximum differences between perturbation and baseline runs for four states. The difference of daily maxima is required for the concentration-response functions used to determine the health impacts of exposure to O_3. The selected states in Figure 2 represent the West, Mid-West, Southeast and Northeast regions, respectively.

Increased emissions in all four states lead to O_3 increases in downwind states. The effect of emissions from one state on downwind states is largest in the north-east and

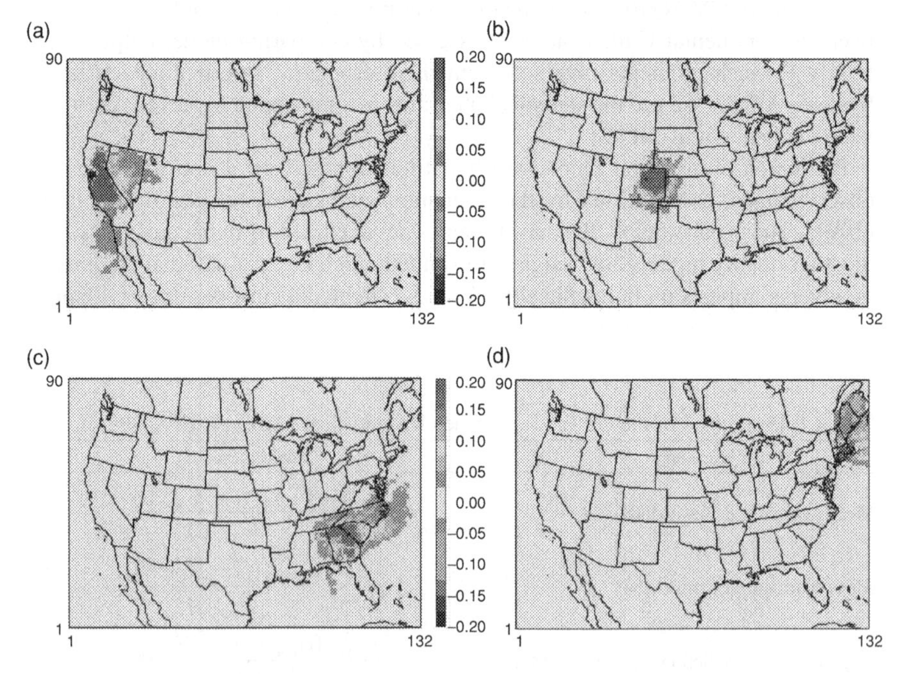

FIGURE 2. Monthly means of the difference in daily maximum O_3 between baseline and perturbation simulations with increased emissions for (a) California; (b) Colorado; (c) Georgia; (d) Connecticut in July 1996.

south-east. Except in Georgia where biogenic hydrocarbon emissions are high, O_3 destruction occurs in locations with large emissions.

6. Resulting Exposure and Health Impacts

Estimating the health impacts of air pollution exposure requires three inputs: 1) air pollution concentrations; 2) population distributions; 3) concentration-response (C-R) relationships. Concentration distributions are obtained from CMAQ model results as described in previous sections. Population data are compiled at the census block group level, with about 2.9 million block groups for the United States. A mapping tool has been developed to project block data onto grid cells of different resolution and coordinate projections (Figure 3).

The concentration-response functions are taken from published epidemiological literature (e.g., Stieb et al., 2002). We use a Poisson regression approach to link daily deaths with O_3 concentrations,

$$M_i = Y_{0_i} * \left(1 - e^{\Delta C_i * \ln RR}\right) * P_i \qquad (2)$$

Here i represents the ith grid cell in the domain; M is mortality, Y_0 is non-accidental mortality in a corresponding grid cell. ΔC is the dimensionless concentration change of air pollutants of concern (here O_3). RR is relative risk for a unit change of concentration reported by epidemiological studies. P is population in a grid cell.

Total mortality change resulting from O_3 exposure is shown in Figure 4 for the same states as in Figure 2.

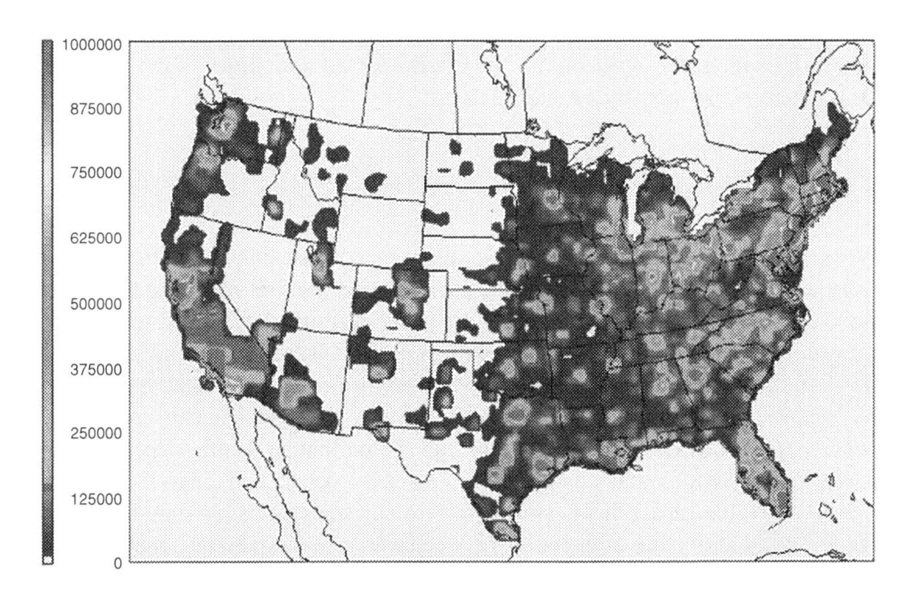

FIGURE 3. Distribution of US Population in 1999.

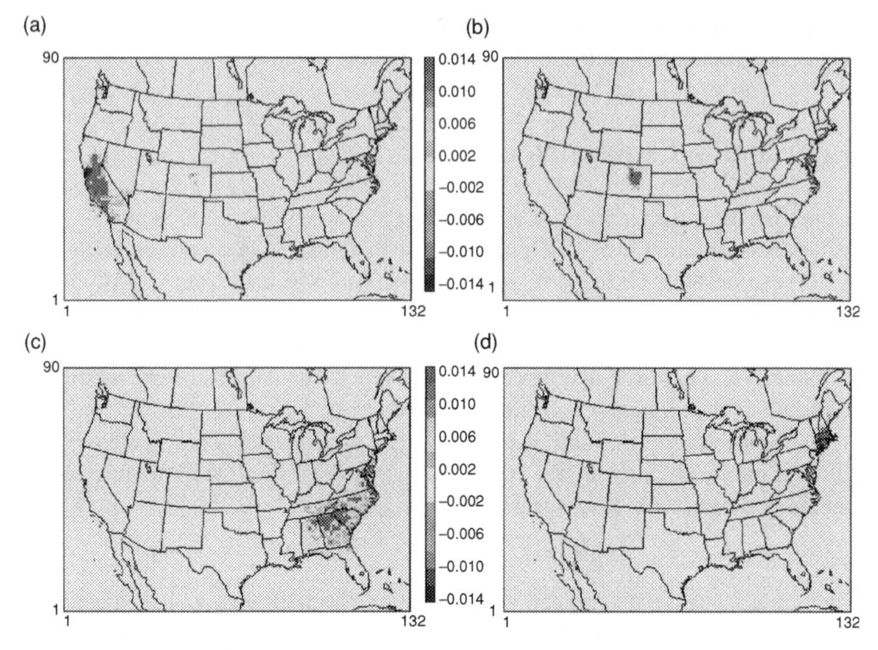

FIGURE 4. Mortality change resulting from O_3 exposure in July 1996.

The sum of resulting mortalities across the continental United States for July 1996 for NO_x emission increases in California, Colorado, Connecticut, and Georgia are –0.8, 1.7, 0.4, and 4.3, respectively. Given identical emission increase (846.3 tons NO_x in July 1996), our results show that, depending on the state from which NO_x was emitted, the effect on total mortalities of the resulting O_3 production can vary by a factor of 10.

7. Conclusions

We have assembled an integrated assessment model that links emissions of primary air pollutants to their chemical transformation and transport and to human exposure and resulted health impacts. Here we examined the health impacts of exposure to O_3 resulting from identical quantities of NO_x emitted from individual states in the continental United States. We demonstrate that the same quantity of NO_x emissions can produce different quantities of O_3 which can cross state boundaries and can have very different impacts on human health, depending on where the emissions and concentration increases occur. Our analysis demonstrates the value of a full integrated assessment analysis using state-of-the-art models to quantify the connection between emissions, ambient concentration distributions, exposure, and health impacts.

Acknowledgements This research was funded by the Grace Foundation. The authors thank Dr. Carey Jang at US EPA for his help in the modeling and data processing, and Dr. Curtis W. Hillegas at Princeton Office of Information Technology for his assistance with computational resources.

References

Byun, D.W., Ching, J.K.S. (Eds.), 1999. Science algorithms of the EPA Models-3 Community Multiscale Air Quality (CMAQ) Modeling System. EPA Report No. EPA-600/R-99/030.

Grell, G.A., J. Dudhia, and D.R. Stauffer, 1993: A description of the Fifth-Generation Penn State/NCAR Mesoscale Model (MM5). NCAR Technical Note, NCAR/TN-398+IA.

Kang, D., Aneja, V.P., Mathur, R., and Ray, J., Nonmethane hydrocarbons and ozone in the rural southeast United States national parks: A model sensitivity analysis and its comparison with measurement, J. of Geophys. Res., 108(D10), 4604, doi: 10.1029/2002 JD003054, 2003.

Stieb, D.M, S. Judek, R.T. Burnett, 2002. Meta-analysis of time series studies of air pollution and mortality: effects of gases and particles and the influence of cause of death, age, and season. Journal of the Air & Waste management Association, 52:470-484.

US Environmental Protection Agency (EPA), 1991: Guidance for regulatory application of the Urban Airshed Model, USEPA Report No. EPA-450/4-91-013.

US Environmental Protection Agency (1996) *EPA Green Book: Nonattainment Areas for Criteria Pollutants*. http://www.epa.gov/oar/oaqps/greenbk

Modeling Source-Receptor Relationships and Health Impacts of Air Pollution in the United States

Speaker: Q. Tong

Questioner: M. Sofiev

Question: Source receptor relationships should be also a subject for changes due to meteorological variability. Probably they are somewhat more stable than the concentrations themselves, but to what extent? Do you have any estimates of variability of these relationships?

Answer: Variability of meteorological variables plays and important role in shaping the source-receptor relationships. While our focus is on the relationships between O_3 concentrations and NOx emissions, our simulations are run for one month to allow time for meteorological conditions to change and to include several different meteorological episodes that are representative of the summer season.

Questioner: R. Bornstein

Question: Do you account for population movements, such as urban inflow during daytime work periods?

Answer: These movements are not accounted for in this study. The resolution of our modeling domain in 36 km by 36 km. We assume the grid cell is large enough that the majority of the population assigned to each box is not traveling into

another box, although this assumption may not be entirely defensible for certain locations.

Questioner: S. Daggupaty

Question: Models predict concentrations of individual chemicals. The module that you are presenting is concentration-response module to infer health impacts. Generally toxic limits of individual chemicals are available for humans. My question is: How do you evaluate health impacts due to polluted air that consists of several chemicals instead of one?

Answer: This paper only addresses O_3 production and the resulting human exposure to O_3. Our next step is to simulate PM production and population exposure. Evaluating health impacts due to several chemicals (such as O_3 and PM) will produce different magnitudes of impacts, although this treatment has been a subject of considerable debate. Some argue that including several chemicals leads to double-counting, unless the original epidemiological study has been designed to evaluate multiple-pollutant relationships.

Questioner: A. Chtcherbakov

Question: How did you assess the migration of the population in concentration-health impact relationships?

Answer: See above. This problem will be of greater importance at higher model resolutions.

Integrated Regional Modeling

15 Evaluation of Local Ozone Production of Chamonix Valley (France) During a Regional Smog Episode

Eric Chaxel, Guillaume Brulfert, Charles Chemel, and Jean-Pierre Chollet*

1. Introduction

During the summer 2003 a POVA Intensive Observation Period (IOP) aimed at determining the sources of airborne pollutants and monitoring their concentrations in two French Alpine valleys: the Chamonix and the Maurienne valleys (see figure 1 for geographic location). The Pollution of Alpine Valleys (POVA) program was launched in 2000 after the traffic interruption under the Mont-Blanc that followed the tragic accident in the tunnel. The Mont-Blanc tunnel was reopened at the end of 2002 and caused the high duty vehicle traffic (about 1100 trucks per day) to be back in the Chamonix valley. The summer 2003 IOP took place from 5 to 12 July in the Chamonix valley. A high ozone event occurred from 5 to 14 July at regional scales and was well characterised by measurements at rural monitoring sites. To better understand the particular atmospheric circulation, and to study the chemical reactions of airborne pollutants within the valleys, mesoscale modelling is applied. For meteorological calculation, the fifth generation PSU/NCAR Mesoscale Model (MM5) was used at scales ranging from 27 to 1 km. MM5 was coupled with the Chemistry Transport Model (CTM) CHIMERE at regional scales and with the CTM TAPOM at a one-kilometre resolution. Simulations were performed for the period 5-12 July 2003 with different emission sets aiming at studying the impact of the international road traffic in the valley on airborne pollutant concentrations.

2. Methodology

Before describing atmospheric processes within the valley such as slope winds, thermal inversions and ozone production, a good description of the meteorological synoptic situation must be performed with regional models. MM5 was chosen

* E. Chaxel, G. Brulfert, C. Chemel, and J.P. Chollet, Laboratoire des Ecoulements Géophysiques et Industriels, Université Joseph Fourrier, Institut National Polytechnique de Grenoble et Centre National de la Recherche Scientifique, BP 53, 38041 Grenoble Cedex 9, France. eric.chaxel@hmg.inpg.fr

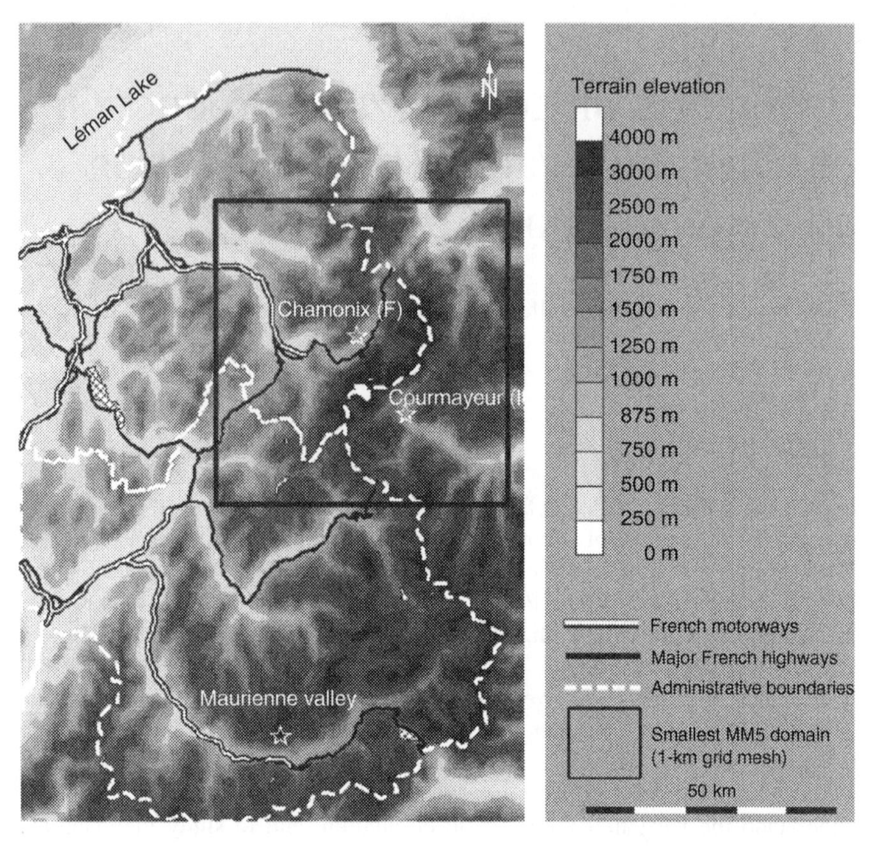

FIGURE 1. Geographical locations of major roads in the study area and of the small simulation domain.

to be used both at the regional and local scales. The two CTMs CHIMERE (Vautard et al., 2003) and TAPOM (Clappier, 1998; Gong et Cho, 1993) are then powered with MM5 meteorological fields. The interaction between the different models and domains is presented on figure 2.

2.1. *Meteorological Calculation*

The fifth-generation PSU/NCAR mesoscale model (MM5) is a nonhydrostatic code which allows meteorological calculations at various scales with a two-way nesting technique described by Grell et al. (1994). For our simulations four different domains are used as shown in figure 2. The smallest domain has a 1-km grid mesh and is geographically located on figure 1. The coarsest domain is powered with the ECMWF gridded analysis and first guess with a 0.5° resolution. Vertically MM5 uses 27 sigma-pressure levels with thickness ranging from 35 m at the ground to 2000 m at 15000 m. The top of the model is at the pressure 100 hPa.

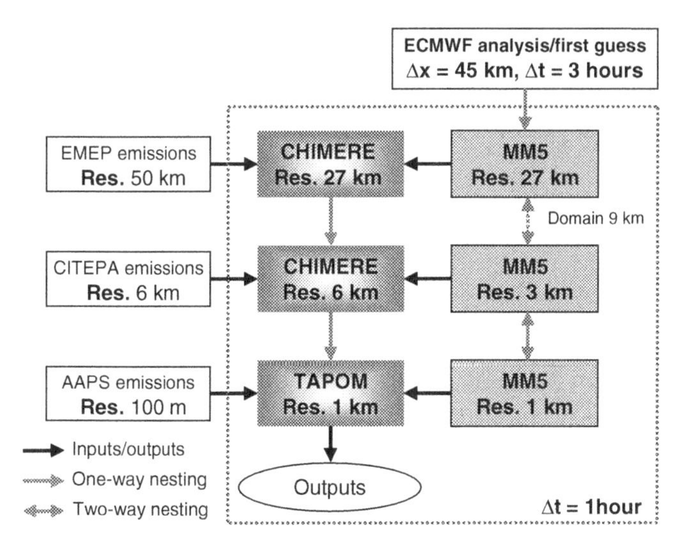

FIGURE 2. Schematic of the simulation system and its main inputs.

In this configuration the planetary boundary layer (PBL) is described with about 15 layers from 0 to 2000 m agl. PBL height calculation is performed by the MRF scheme described by Hong and Pan (1996).

2.2. *Chemistry Transport Calculation*

As shown on figure 2, the eulerian code CHIMERE is used in two configurations: on a domain with a 27-km grid mesh covering Central Europe (called continental) and on a domain with a 6-km grid mesh covering Rhône-Alpes region (called regional). CHIMERE has 8 hybrid sigma-p levels with a model top set at 500 hPa (equivalent to 5600 m amsl). The chemical mechanism MELCHIOR implemented in the CHIMERE code calculates concentrations of 44 species following a set of 116 reactions. MELCHIOR was developed by Derognat (1998) based on EMEP mechanism. CHIMERE is powered off-line with MM5 meteorological fields (see the model technical documentation for details). The cloud effects and dry deposition are taken into account in ozone production and destruction. Both domains run from 2 to 14 July 2004. Results for 20 MELCHIOR species matching RACM species from the regional domain are then used at the boundaries and on top of the TAPOM model. TAPOM runs from 5 to 12 July 2004.

TAPOM is an eulerian model that deals with transport, diffusion, chemistry and dry deposition in a 1×1 km grid mesh with 12 terrain-following vertical levels. The Regional Atmospheric Chemistry Mechanism (RACM) (Stockwell et al., 1997) implemented in TAPOM resolves chemistry with 77 species and a set of 237 reactions. Chemical concentrations calculated on the regional domain of CHIMERE are linearly interpolated on the TAPOM levels.

Nested domains at continental and regional scales provide a good description of the temporal variation of background concentrations of ozone and of other secondary species. Over this one-week simulation CHIMERE correctly deals with the important smog episode that occurred at continental scale during the period 5-12 July 2003.

2.3. Emission Inventories

As shown in figure 2, the EMEP emission inventory of 2001 for NOx, SOx, CO, NMVOC and NH_3 on a 50-km grid mesh is used in the continental domain (at a 27-km resolution) of CHIMERE model by recalculating emissions following the land use. The EMEP emission inventory is described by EEA (2003). To perform chemistry-transport calculations at 6-km resolution in the regional domain of CHIMERE for the Rhône-Alpes region a 6-km resolution emission inventory from CITEPA (Bouscaren et al., 1999) is used for NOx, CO, NMVOC and SOx. At the smallest scale an accurate emission inventory based on the CORINAIR methodology and developed in the framework of the POVA program by Air de l'Ain et des Pays de Savoie with a 100-m resolution is distributed on the 1-km grid mesh used for the TAPOM calculation. Biogenic emissions are recalculated by CHIMERE following land use and temperature and for the small domain are included in the 100 m × 100 m inventory provided for the POVA program.

2.4. Definition of the Emission Scenarios

Beside the base case two scenarios for emissions are considered. The base case (BC) takes into account all the emissions for 2003. The first scenario (S1) aims at determining which part of ozone is produced by the vehicles using the Mont-Blanc tunnel between France and Italy. This international traffic is mainly located along the RN 205 road. Emissions are computed based on 1-km emissions by assigning multiplicative factors to each type of sources. The second scenario (S2) aims at determining which part of ozone is produced by all the sources in the valley, biogenic and anthropogenic and then all emissions are set to zero. For the first scenario road counting realized by local agencies were used to accurately determine the part of the international traffic in the total traffic.

3. Results

3.1. Description of the Base Case (BC)

The validation of the meteorological fields from MM5 was realised using ground station data, a tethered balloon, an instrumented cable car and an UHF windprofiler radar located in Chamonix. MM5 correctly describes the two valley wind systems and the wind switch between the systems that occurs at 0600 GMT and 2000 GMT as shown on figure 3.

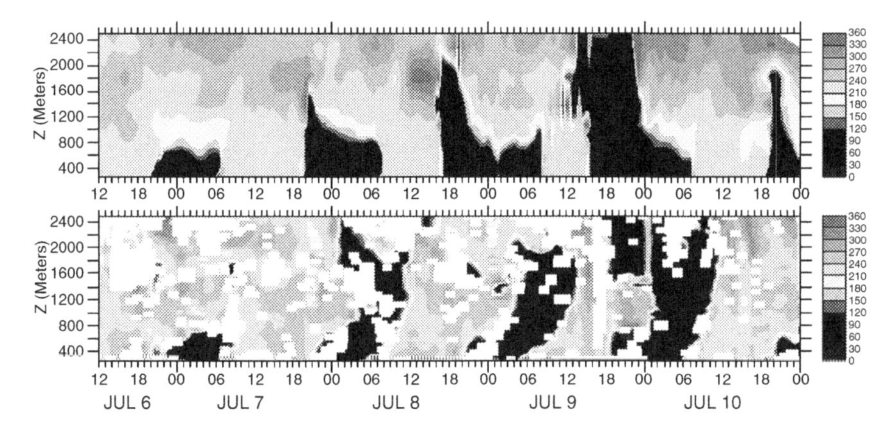

FIGURE 3. Wind direction (in degrees clockwise from north) modelled by MM5 (up) and measured with UHF radar (down) from 250 to 2500 m above the ground at the vertical of Chamonix from 6 to 10 July. The two wind switches occur every day at 0600 UTC and 2000 UTC.

As shown on the figure 3 the southwesterly wind resulting from slope winds and blowing in the valley axe from 0600 UTC to 2000 UTC has a vertical extent of 1300 m agl. During the night the northeasterly valley wind blows from 2000 UTC to 0600 UTC and has a vertical mean extend of 500 m agl at the vertical of Chamonix. Figure 4 shows that on 8 July morning MM5 calculates PBL heights that are twice greater than heights determined from the available measurements. However the maximum PBL height of 1300 m agl that is reached at 1400 UTC is well reproduced.

The validation of chemical–transport calculation for the base case on figure 5 shows very good agreement with hourly measurements of ozone at rural and ranged sites. Two comparisons are shown on figure 6 and figure 7 respectively for

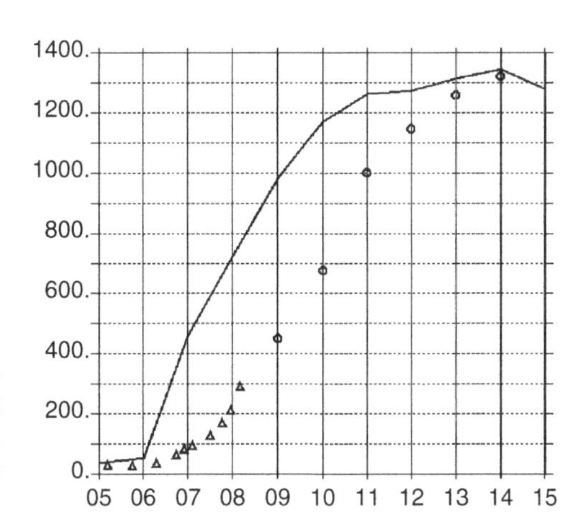

FIGURE 4. Determination of PBL height growth during the morning with MM5 (line), tethered balloon (triangles) and windprofiler (dots) on 8 July. Timescale is UTC.

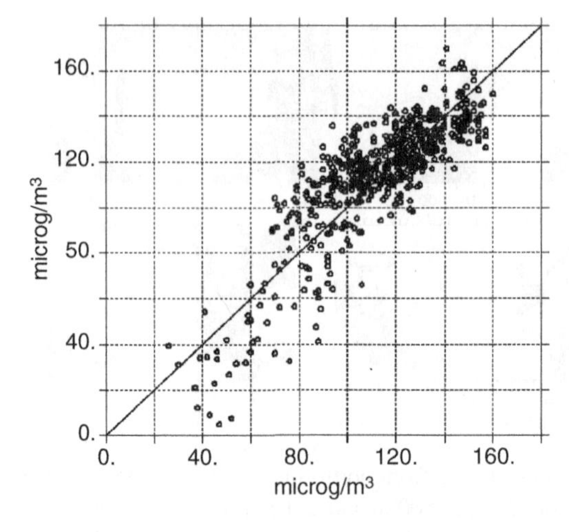

FIGURE 5. Modelled (y axis) versus observed (x axis) hourly concentrations of ozone (in μg/m^3) at two rural sites (les Houches, Argentière) and two high-altitude sites (plan de l'Aiguille, col des Montets).

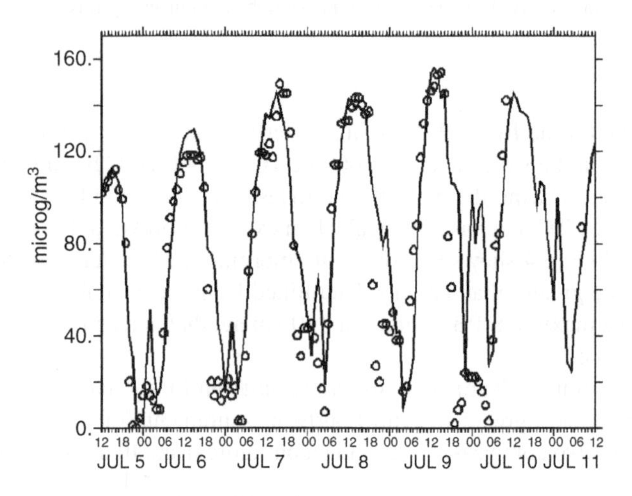

FIGURE 6. Measured (dots) and modelled (line) hourly concentrations of ozone at the suburban station Clos de l'Ours from 8 to 12 July 2003.

the suburban site Clos de l'Ours (1100 m amsl) and for the high-altitude site col des Montets (1465 m amsl). For other pollutants such as nitrogen dioxide, temporal variations are slightly less satisfactory described at source-influenced sites but still in agreement with the measurements at background sites located in the Chamonix urban area and at ranged sites.

TAPOM results and measurements both show different temporal evolutions of ozone concentrations following the part of the domain and the typology of the monitoring site. The ranged site col des Montets (on figure 7) catches very well the increase of ozone concentrations in the free troposphere from 5 July to 9 July. After 9 July the ozone concentration does not exceed 140 μg/m^3. At sites located

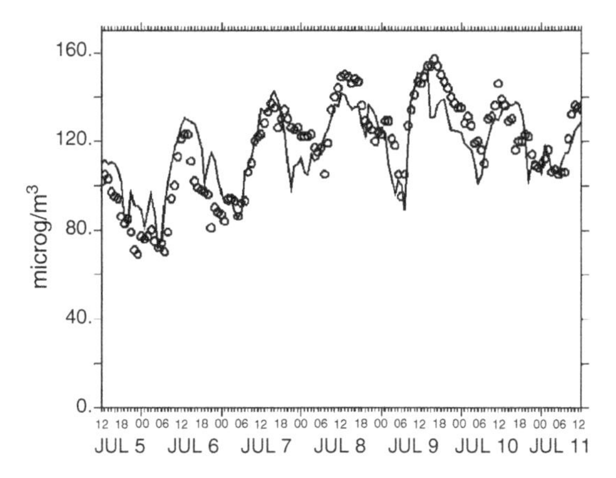

FIGURE 7. Measured (dots) and modelled (line) hourly concentrations of ozone at the ranged station *col des Montets* from 8 to 12 July 2003.

at the bottom of the valley, as Clos de l'Ours on figure 6, diurnal variations of ozone concentrations are well modelled with low values at night characterising source-influenced sites.

During the day from 0600 UTC to 2000 UTC (as stated by figure 3) the valley wind system is composed of up-slope winds generated by convection and of the resulting main valley wind blowing in the central part of the valley with a mean speed of 5 m/s. This system is performing a very effective dispersion and a fast transport of ozone precursors from the sources to higher altitude areas. The combination of the main valley wind and of slope winds causes the formation of two ozone plumes downwind of the Chamonix urban area. The production of ozone in the plumes is enhanced by the availability on the slopes of the valley of large amounts of biogenic VOC such as terpenes. The simulated production rate of ozone in the central part of the valley is 2 $\mu g/m^3/h$ whereas it reaches 5 $\mu g/m^3/h$ on the south face of Aiguilles Rouges mountains. This transport starts when the valley wind sets up at 0600 UTC and is really effective from 0800 UTC to noon. The calculated ozone peak in the plumes reaches 170 $\mu g/m^3$ on 9 July while the background is 150 $\mu g/m^3$ what represents a net production in the valley of 20 $\mu g/m^3$.

The BC simulation highlights that the ozone maximums are linked with the ozone background level because of intensive mixing that brings air from free troposphere. With sunset the convection stops and the valley wind weakens progressively to become null at 1800 UTC. The atmosphere becomes stable and primary pollutants such as NOx are accumulated in the surface layer. Titration of ozone by NO and dry deposition cause the ozone concentrations to rapidly decrease at urban and suburban sites. On figure 6 the concentration reaches 0 $\mu g/m^3$ at 1900 UTC on 9 July at the site Clos de l'Ours. At ranged sites such as col des Montets ozone concentration are under the control of long distance transport and dry deposition and so shows low-amplitude variations.

3.2. Impact of International Traffic on Ozone Production (scenario S1)

The simulation performed without the emissions of the tunnel gives very similar results as the base case during the day because of the strong dependency of ozone concentrations in the valley with the regional background. However the simulation shows that the two ozone plumes downwind of Chamonix have lower concentrations. Depending on the day the difference between the base case and the case without tunnel emissions is in the range 2-7 μg/m^3 as shown on figure 8.

3.3. Impact of all Local Sources on Ozone Production (scenario S2)

In order to determine the total production of ozone in the valley a simple way to proceed is to set all the emissions of the valley to zero. In this case the processes that efficiently control the variation of ozone concentration are: long distance transport by advection, mixing, UV radiation and dry deposition. Depending on the day the difference between ozone maximums in the base case and in the scenario S2 is in the range 8-27 μg/m^3 as shown on figure 8 what is roughly 5 to 15% of ozone peak. The figure also shows that the local production in the domain was the most effective on 9 July. This day was characterised by very weak winds that enhanced ozone production. Although ozone production in the model domain is clearly detected by the simulation (figure 8) some monitoring sites do not experience a reduction in ozone with scenario S2. Table 1 shows that ozone concentration is even higher in scenario S2 at the 4 sites located upwind of Chamonix and in the Chamonix urban area. This result is explained by the NOx-saturated (VOC-limited) regime of ozone production in this part of the valley as shown by Brulfert et al. (2003). Table 1 shows also that only sites located downwind of Chamonix experience ozone reduction except on 8 July when all sites experience reduction.

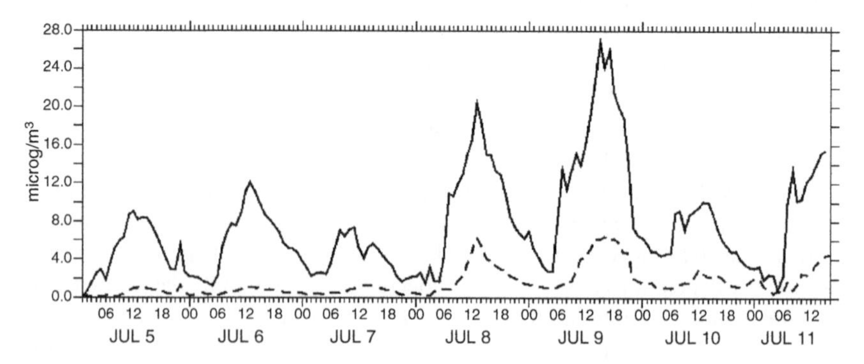

FIGURE 8. Difference between the ozone peaks in BC and in S1 (dashed line) and in BC and in S2 (solid line). Positive values indicate that the peak values of BC are always greater than values calculated for S1 and S2.

TABLE 1. Maximum reduction of hourly ozone concentrations (in μg/m^3) at monitoring sites by applying S2. (+) indicates augmentation, (–) indicates diminution.

Site	Typology	Downwind of Chamonix	Sat. 05/07	Sun. 06/07	Mon. 07/07	Tues. 08/07	Wed. 09/07	Thu. 10/07	Fri. 11/07
Houches	Rural	no	–5 (+)	–5 (+)	–3 (+)	1 (–)	–6 (+)	–4 (+)	–2 (+)
Bossons	Traffic	no	–2 (+)	–3 (+)	–1 (+)	4 (–)	–8 (+)	–3 (+)	–2 (+)
Clos de l'O.	Suburban	no	–2 (+)	–3 (+)	–1 (+)	4 (–)	–8 (+)	–2 (+)	–1 (+)
Chamonix	Centre	no	–2 (+)	–3 (+)	–1 (+)	6 (–)	–3 (+)	–2 (+)	2 (–)
Plan de l'A	Ranged	yes	4 (–)	3 (–)	2 (–)	10 (–)	10 (–)	3 (–)	8 (–)
Bouchet	Suburban	yes	0 (–)	–2 (+)	0	7 (–)	5 (–)	1 (–)	3 (–)
Argentière	Rural	yes	4 (–)	3 (–)	2 (–)	12 (–)	18 (–)	6 (–)	12 (–)
Montets	Ranged	yes	6 (–)	5 (–)	3 (–)	14 (–)	14 (–)	9 (–)	10 (–)

The ozone production in the valley only occurs in the region downwind of Chamonix except days when large amounts of photochemical secondary species regionally transported from outside to the valley modify the ozone production regime to NOx-limited (VOC-saturated). Anywhere else at the valley bottom destruction occurs even during the day.

4. Conclusion

This numerical study based on a chain of nested models highlighted the processes of production and destruction of tropospheric ozone in the very narrow valley of Chamonix (France) for the period 5-12 July 2003. Two pure academic cases of emission reduction in the valley showed that ozone concentrations were regionally controlled during the day whereas during the night under stable conditions the valley was totally decoupled from synoptic conditions. The precursors emitted by the Chamonix urban area and by the traffic are rapidly dispersed during the day and two plumes were modelled downwind of Chamonix. The net ozone production in the valley during the day accounts for 5 to 15% of the ozone peaks which are in the range 110-170 μg/m^3 depending on the day. The impact of the tunnel accounts for about 5% of these peaks. A dependency of ozone production regime with regional background of photochemical secondary species has been observed with simulations on 8 July but must be confirmed by a three-dimension field investigation. The aim of this study was also to evaluate the ability of MM5, which is a regional model dedicated to prediction, to provide meteorological fields for a CTM with a 1-km grid mesh. Even if MM5 fields gives good results for ozone at this scale it remains too coarse to resolve fine scale processes that control concentrations of primary pollutants such as nitrogen oxides. Additional simulations with MM5 at a 500-m scale showed that the model was unable to work at such scales in a so complex terrain because processes such as slope winds were resolved twice, explicitly and by the PBL scheme, and causes unrealistic instabilities to occur near the ground. The use of the MM5 model in our modelling applications will be limited at a 1-km scale until further improvements.

Acknowledgements The POVA program is supported by Région Rhône Alpes, ADEME, METL, MEDD. Meteorological data are provided by ECMWF, emission inventories by EMEP, CITEPA, GIERSA and Air de l'Ain et des Pays de Savoie.

References

Bouscaren, R., Riviere, E. and Heymann, Y., 1999, *Réalisation d'un Inventaire d'Emissions Simplifiées pour la Ville de Grenoble*, in Convention n°98338001, CITEPA 392.

Brulfert, G., Chaxel, E., Chemel, C. and Chollet, J.P., 2003, Numerical simulation of air quality in Chamonix valley, use of different chemistry indicators, *14th IUAPPA International Conference Air quality*, Dubrovnik, Croatia, pp 661-667.

Clappier, A., 1998, A correction method for use multidimensional time splitting advection algorithms: application to two and three dimensional transport, *Mon. Wea. Rev.* **126**: 232-242.

Derognat, C., 1998, *Elaboration d'un Code Chimique Simplifié Applicable à l'Etude de la Pollution Photooxydante en Milieu Urbain et Rural*, Diploma thesis report, Université Pierre et Marie Curie, France.

European Environmental Agency (EEA), 2003, *EMEP/CORINAIR Emission Inventory Guidebook – 3rd edition*, European Environmental Agency.

Gong, W. and Cho, H.-R., 1993, A numerical scheme for the integration of the gas phase chemical rate equations in a three-dimensional atmospheric model, *Atmos. Env.* **27A**: 2147-2160.

Grell, G.A, Dudhia, J. and Stauffer, J., 1994, *A Description of the Fifth-Generation Penn State/NCAR Mesoscale Model (MM5).* NCAR Tech Note TN-398, 122 pp.

Hong, S.-Y. and Pan, H.-L., 1996, Nonlocal boundary layer vertical diffusion in a medium-range forecast model, *Mon. Wea. Rev.* **124**: 2322-2339.

Stockwell, R., Kirchner, F., Kuhn, M. and Seefeld, S., 1997, A new mechanism for atmospheric chemistry modelling, *J. of Geophys. Res.* **102**(D22): 25847-25879.

Vautard, R., Martin, D., Beekmann, M., Drobinski, P., Friedrich, R., Jaubertie, A., Kley, D., Lattuati, M., Moral, P., Neininger, B. and Theloke, J., 2003, Paris emission inventory diagnostics from ESQUIF airborne measurements and chemistry transport model, *J. Geophys. Res.* **108**(D17): 8564.

Evaluation of Local Ozone Production of Chanonix Valley (France) During a Regional Smog Episode

Speaker: E. Chaxel

Questioner: D. Steyn

Question: In the map Riviera modeling study, we had substantial difficulty capturing topographic shading (or shadowing) effects. Given the step topography in your domain, how did you deal with this problem?

Answer: No sensitivity tests have been made by now on the impact of the shadowing effect on the dynamical calculation since MM5 does not include a radiation module taking into account the topographic shading. About these effects they are thought to cause the MM5 model to overestimate the PBL height during the morning and the evening transitions. For our air quality applications this problem has to be tackled especially in winter. The impact of shading on photochemistry has been studied and has very little effect on ozone concentration.

Questioner: P. Builtjes

Question: You are using Chimere down to 6×6 km ? And then use TAPON down to 1×1 km? Why do you not use Chimere down to 1×1 km? Going from one model to another model often creates problems.

Answer: We did not use CHIMERE down to 1×1 km because the CHIMERE model has never been used at such a fine resolution and on the other hand the actual version of the CHIMERE model can not have a model top higher than 500 mb and the highest point in our 1×1 km gridcell domain was around the pressure 500 mb (Mont-Blanc summit) what makes the model in its original version unusable. Future versions of CHIMERE may use a higher top level. The main problem from going to CHIMERE to TAPOM is the use of different chemical mechanisms. TAPOM uses RACM mechanism and CHIMERE uses MELCHIOR mechanism. These two mechanisms are different in a such extent that RACM is a bit more detailed than MELCHIOR considering the NMVOC splitting but the main photochemical species such as ozone, PAN, formaldehyde or nitric acid are individual species and are present in the two mechanisms.

16 Alternative Approaches to Diagnosing Ozone Production Regime

Daniel S. Cohan, Yongtao Hu, and Armistead G. Russell*

1. Introduction

Effective formulation of control strategies requires knowledge of the responsiveness of ozone to emissions of its two main precursors, nitrogen oxides (NO_x) and volatile organic compounds (VOC). While responsiveness depends nonlinearly on an array of spatially and temporally variable factors, a large body of research has sought to classify ozone formation into categories of chemical regime (Sillman, 1999). In NO_x-limited regimes, ozone increases with increasing NO_x and exhibits only slight sensitivity to VOC; in VOC-limited (or NO_x-saturated) regimes, ozone increases with VOC and exhibits slight or even negative sensitivity to NO_x. Transitional conditions of dual sensitivity also occur. Classification of ozone production regime helps determine whether NO_x or VOC emissions should be targeted more aggressively in strategies to reduce ozone.

In addition to photochemical modeling of ozone responsiveness, "indicator ratios" have been sought to diagnose ozone production regime based on observable concentrations and to corroborate atmospheric models (e.g., Sillman, 1995). Regardless of diagnosis method, two factors hinder the usefulness of any single classification. First, because ozone production is nonlinear, response to large changes in emissions may not scale linearly from incremental sensitivity. Second, because ozone forms downwind of emission sources, response to domain-wide emissions may not reflect response to local emissions.

Here, we compare modeled ozone sensitivities to NO_x and VOC during a summertime air pollution episode in the southeastern U.S. with modeled values of three widely used indicator ratios. By modeling the response of ozone to both large and infinitesimal perturbations of both local and region-wide emissions, we examine how ozone sensitivity depends on the size and scope of the emission perturbation. Lessons are drawn regarding the usefulness of ozone production regime diagnosis.

* Daniel S. Cohan, School of Earth & Atmospheric Sciences, Georgia Institute of Technology, Atlanta, GA 30332, dcohan@eas.gatech.edu, Phone: 404-385-4565. Fax: 404-894-8266. Yongtao Hu and Armistead G. Russell, School of Civil & Environmental Engineering, Georgia Institute of Technology, Atlanta, GA 30332.

2. Methods

2.1. Direct, High-Order Sensitivity Analysis

The Decoupled Direct Method in Three Dimensions (DDM-3D) provides an efficient method for computing the sensitivity of modeled concentrations to perturbations in model parameters and inputs (Yang et al., 1997). It computes sensitivities simultaneously with concentrations, utilizing the transport and chemistry mechanisms of the underlying model. Hakami et al. (2003) extended DDM-3D to compute higher-order sensitivities.

We have implemented 2nd-order DDM-3D into the Community Multiscale Air Quality (CMAQ) model v. 4.3 (Byun and Ching, 1999) with the SAPRC-99 chemical mechanism (Carter, 2000). First-order sensitivity coefficients, $s_{ij}^{(1)} = \partial C_i/\partial p_j$, represent the local sensitivity or "slope" of species i with respect to input parameter p_j whose unperturbed value is P_j. Perturbations in p_j are considered by defining a scaling variable, ε_j, with a nominal value of 1 such that $p_j = \varepsilon_j P_j = (1+ \Delta\varepsilon_j)P_j$. Second-order sensitivities, $s_{i,j1,j2}^{(2)} = \partial^2 C_i/(\partial p_{j1}\partial p_{j2})$, represent the 2nd derivative of the species-parameter relationship. In this paper we present sensitivity coefficients S_{ij} semi-normalized to the size of the unperturbed input field:

$$S_{ij}^{(1)} = P_j \frac{\partial C_i}{\partial p_j} = P_j \frac{\partial C_i}{\partial(\varepsilon_j P_j)} = \frac{\partial C_i}{\partial \varepsilon_j} \quad (1)$$

For certain comparisons, we divide S by the coincident ozone concentrations; thus, for example, a concentration-normalized ozone-to-NO_x sensitivity of 0.15 (unitless) means that a 1% reduction in NO_x emissions would reduce ozone by 0.15%.

Due to nonlinearity, accurate approximation of response to large perturbations requires consideration of 2nd-order sensitivity via Taylor expansion (Hakami et al., 2003):

$$(\mathbf{C}|_{p_j = (1+\Delta\varepsilon_j)P_j} - \mathbf{C}|_{p_j = P_j}) \approx \Delta\varepsilon_j \mathbf{S}^{(1)} + \frac{1}{2}\Delta\varepsilon_j^2 \mathbf{S}^{(2)} + \textit{higher order terms} \quad (2)$$

Note the 2nd-order term scales with $\Delta\varepsilon^2$, and thus its relative importance increases with the size of the perturbation. The accuracy of CMAQ-DDM has been rigorously demonstrated by comparison to finite difference calculations for a variety of brute force perturbations (Hu et al., 2004). Thus we have high confidence in the ability of DDM-3D to capture response of the underlying model.

2.2. Species Indicator Ratios

Numerous studies have suggested metrics by which the relative concentrations of species could indicate whether ozone formed under primarily NO_x– or VOC-limited conditions. Here we focus on three indicator ratios that were introduced by Sillman (1995): (1) H_2O_2/HNO_3, (2) $HCHO/NO_y$, and (3) O_3/NO_z. NO_y is total reactive nitrogen and NO_z is the sum of NO_x reaction products, or NO_y-NO_x. All three ratios are expected to be higher in NO_x-limited regimes and lower in

VOC-limited regimes. The first ratio compares the concentration of H_2O_2, the major sink for odd hydrogen radicals (OH and HO_2) in NO_x-limited regimes, with the concentration of HNO_3, the major sink for odd hydrogen in VOC-limited regimes (Sillman, 1995). The ratio $HCHO/NO_y$ serves as a reactivity-weighted proxy for the VOC/NO_x ratio, because HCHO is a product of reactions of VOC with OH (Sillman, 1995). The rationale for O_3/NO_z is more complex, and supposes that the quantity is approximately proportional to photochemical production of odd hydrogen divided by loss of odd nitrogen (Sillman, 1995; Kleinman et al., 1997). The ability of O_3/NO_z to diagnose regime weakens when reactions other than ozone photolysis provide significant sources of odd hydrogen (Sillman and He, 2002).

2.3. Model Episode

CMAQ-DDM is applied to model ozone and its sensitivity to precursor emissions in the southeastern U.S. during the August 11-20, 2000 air pollution episode. The first two days are discarded as model initialization. Emissions, meteorology, and photochemical modeling methodology for the episode are presented extensively elsewhere (Unal *et al.*, 2003; *Hu et al.*, 2003 and 2004). Agreement of modeled and observed concentrations is well within U.S. EPA benchmarks. We focus on results in the 12-km resolution Fall-Line Air Quality Study (FAQS) nest, which covers Georgia and neighboring states. We define "domain-wide emissions" as anthropogenic emissions within the outer 36-km domain that spans from Texas to Maine. Sensitivities to domain-wide emissions are modeled by using 36-km resolution results as boundary concentrations and sensitivities for the 12-km nest; sensitivities to local emissions are modeled on the 12-km domain only, with boundary concentrations from the 36-km domain but boundary sensitivities set to zero.

3. Results and Discussion

3.1. Sensitivity – Indicator Ratio Correlations

For each day and ground-level grid cell, indicator ratios and DDM-3D first-order sensitivities are evaluated at the time of peak hourly ozone. Most of the domain exhibits significantly positive ozone sensitivity ($S^{(1)} \geq 5$ ppb) only to NO_x, but some urban centers show significant sensitivity to VOC and negative sensitivity to NO_x.

As expected, sensitivity to NO_x tends to increase and sensitivity to VOC tends to decrease with H_2O_2/HNO_3 (Figure 1). Similar patterns are observed for $HCHO/NO_y$ (not shown). The response of NO_x sensitivity can be described as asymptotic in each case, as NO_x-inhibition (i.e., negative sensitivity) occurs only at very low values of the ratios but sensitivities plateau as the ratios increase. Normalized sensitivity to NO_x peaks at about 0.5, whereas VOC sensitivity peaks at less than 0.15, indicative of predominantly NO_x-limited ozone conditions.

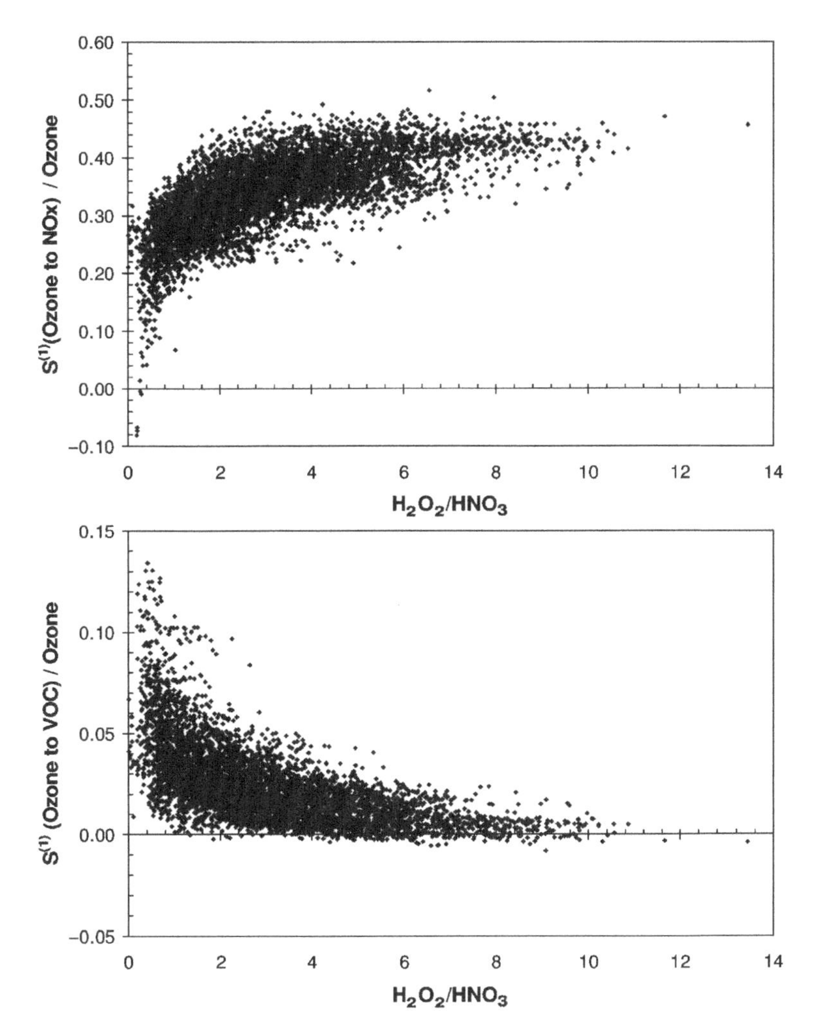

FIGURE 1. First-order sensitivity coefficients, normalized by ozone concentrations, of ozone response to anthropogenic NO_x (top) and VOC (bottom), plotted against concurrent concentration ratio H_2O_2/HNO_3 in CMAQ-DDM simulations. Each data point corresponds to a grid-cell-day at the hour of its daily peak ozone. For plotting purposes, only every fourth point is shown.

NO_x-inhibition is confined to locations with H_2O_2/HNO_3 less than 0.3 and $HCHO/NO_y$ less than 0.5 (Table 1). NO_x sensitivity consistently exceeds VOC sensitivity whenever H_2O_2/HNO_3 is greater than 0.7 or $HCHO/NO_y$ is greater than 0.8. The paucity of VOC-limited locations in the FAQS domain precludes clear definition of regime thresholds. However, the approximate transition zones indicated by the bounds on NO_x inhibition and greater VOC sensitivity are consistent with the thresholds suggested by earlier studies (Table 1). We note that

TABLE 1. Maximal indicator ratios in our simulations for (1) NO_x inhibition and (2) VOC-sensitivity exceeding NO_x-sensitivity, and ozone regime thresholds suggested elsewhere. Ozone tends to be NO_x-limited above thresholds, and can be VOC-limited below.

	Max. ratio for NO_x-inhibition	Max. ratio for $S_{O3,VOC}^{(1)} > S_{O3,NOx}^{(1)}$	Sillman (1995)	Lu and Chang (1998)	Sillman and He (2002)
H_2O_2/HNO_3	0.3	0.7	0.25–0.67	0.8–1.0	0.23–0.54
$HCHO/NO_y$	0.5	0.8	0.2–0.39	0.5–0.7	NA
O_3/NO_z	NA	NA	6–14	25–30	8–20

even when indicator ratios are below the threshold defined by NO_x-inhibition, a majority of grid cells have greater sensitivity to NO_x than to VOC.

The ratio O_3/NO_z fails to delineate ozone production regime in our simulations. This is consistent with studies that have found O_3/NO_z to be a weaker indicator of sensitivity than other ratios (Sillman and He, 2002). Two factors hinder the usefulness of the O_3/NO_z ratio here. First, the ratio almost always exceeds 10 at the time of peak-hour ozone in our simulations, within the ranges of thresholds reported elsewhere (Table 1). Second, abundant biogenic VOC provides a source of odd hydrogen other than ozone photolysis, undermining the premise of the O_3/NO_z ratio.

Unfortunately, few observations of the ratio species are available for comparison with model results on the 12-km domain. Measurements of H_2O_2 and HNO_3 are unavailable, and HCHO was measured only sporadically at two stations. NO_x, NO_y, and O_3 were concurrently measured at only 2 stations, both in Atlanta suburbs. At the time of daily peak ozone, observed O_3/NO_z at these stations ranged from 11 to 35 during the episode, compared to the modeled range of 11-19 at corresponding grid cells.

3.2. *Size of Perturbation*

Though we have so far considered sensitivity to incremental perturbations, regimes have traditionally been characterized by the response of ozone to arbitrary percentage reductions in emissions of each precursor. Given the concave-down response of daytime ozone to NO_x, greater nonlinearity increases the extent to which ozone reductions from large NO_x reductions exceed those indicated by scaling from incremental sensitivity.

Our 2nd-order results show ozone-NO_x response to be most nonlinear where HNO_3 is highest. As a result, the range of response to NO_x indicated by Figure 1 would narrow if large-scale response rather than incremental sensitivity was considered. The relationship between NO_x sensitivity and H_2O_2/HNO_3, though consistently positive, becomes more muddled as larger perturbations are considered. In fact, 2nd-order Taylor approximations (Eq. 2) indicate that in all grid cells with negative incremental sensitivity to NO_x, ozone would actually decline if domain-wide NO_x emissions were reduced by 50% or more.

3.3. *Location of Perturbation*

For urban and regional pollution control strategy formulation, ozone sensitivity to local emissions can be more pertinent than its response to domain-wide emissions. If a receptor is diagnosed as NO_x– (VOC–) limited with respect to domain-wide emissions, will NO_x (VOC) controls necessarily be the most effective local response? We investigate this question by applying DDM-3D to compute the sensitivity of ozone to emissions from two regions: the Atlanta region, defined here as the 13 non-attainment counties at the time of the episode, and the 7-county Macon-Warner Robbins Combined Statistical Area.

VOC emissions enhance ozone production primarily within 100 km of the source, so sensitivity to local VOC closely tracks sensitivity to domain-wide emissions. Intense NO_x emissions, however, often generate an initial decline in concentrations before forming ozone downwind (Ryerson et al., 2001). Figure 2 compares the first-order sensitivities of Macon ozone to domain-wide NO_x emissions and to emissions from within the region. Sensitivity to domain-wide emissions is consistently higher than sensitivity to regional emissions, and the gap between the lines indicates the positive sensitivity to emissions from outside the region. In Macon, sensitivities to local emissions remain positive during the afternoon, the time of highest ozone concentrations that determine regulatory attainment. On a 24-hour average basis, however, the sensitivity to local emissions is virtually zero in Macon despite strongly positive sensitivity to domain-wide emissions.

Within the Atlanta region, sensitivity of peak-hour ozone at a particular location and day to domain-wide emissions is a poor predictor of that location's sensitivity to Atlanta emissions (Figure 3). While sensitivity to domain-wide

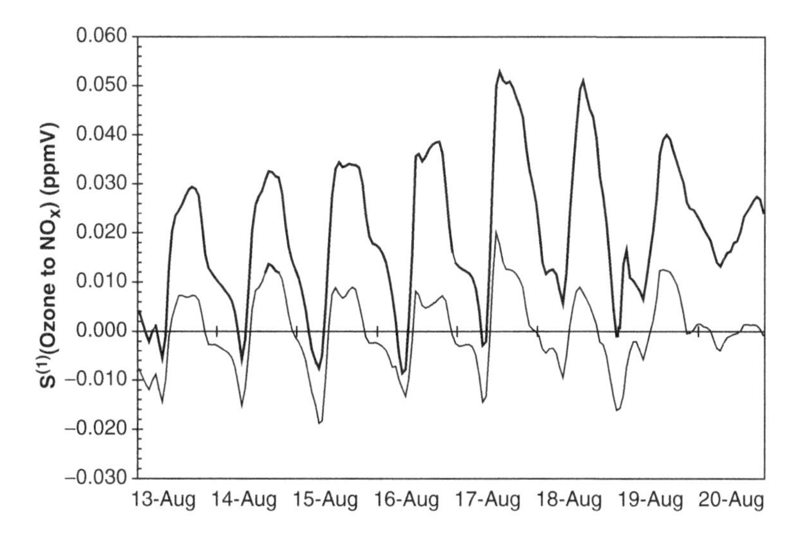

FIGURE 2. First-order sensitivity of Macon ozone to domain-wide (bold line) and local (thin line) NO_x.

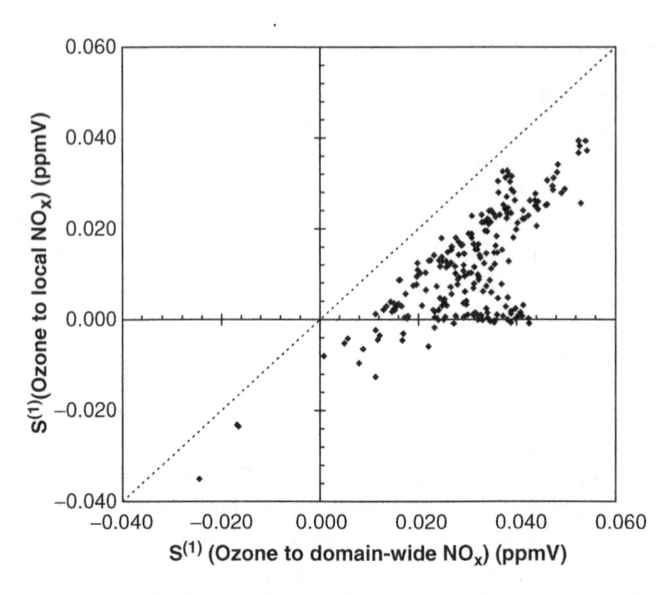

FIGURE 3. First-order sensitivity of daily peak-hour ozone to Atlanta region NO_x emissions plotted against concurrent sensitivity to domain-wide emissions. Each point represents a grid-cell-day within the Atlanta region.

emissions provides an upper-bound for sensitivity to local emissions, the local emissions impact is often negative or negligible at locations with positive sensitivity to domain-wide emissions. As shown in Figure 4, sensitivity to non-Atlanta NO_x emissions exhibits an asymptotic response to the H_2O_2/HNO_3 ratio

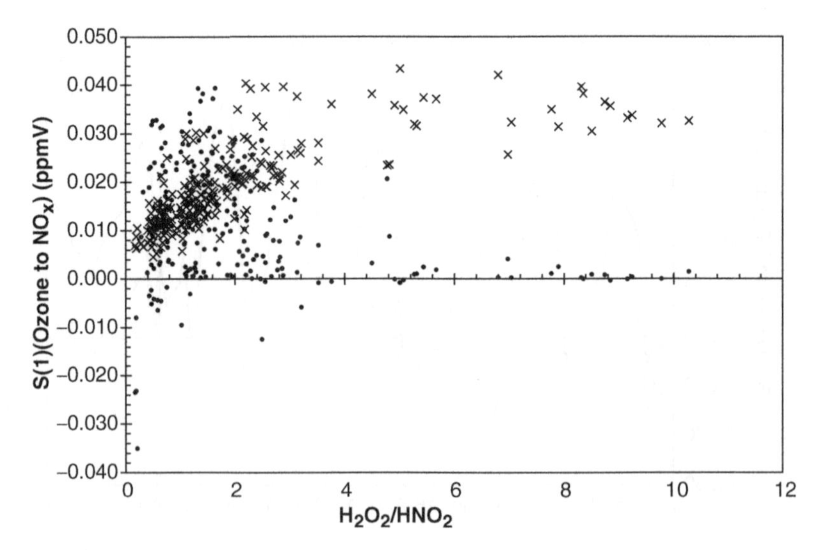

FIGURE 4. First-order sensitivity of daily peak-hour ozone at Atlanta region grid cells to local NO_x emissions (dots) and to other-than-local NO_x emissions (x's), plotted against concurrent H_2O_2/HNO_3 ratios.

analogous to Figure 1, but sensitivity to local emissions declines to near zero as the ratio increases. These patterns suggest an additional meaning of indicator ratios—a signal of the origin of ozone in urban areas. As before, high H_2O_2/HNO_3 ratios signify NO_x-limited ozone production and low ratios signify NO_x-saturation. Because the Atlanta plume has a much higher NO_x/VOC ratio than the biogenic VOC-laden domain, ozone would more likely have formed under NO_x-limited conditions if it originated upwind of the region rather than from Atlanta emissions, and high H_2O_2/HNO_3 serves as a proxy for such conditions.

4. Conclusions

CMAQ-DDM results have shown that ozone in the southeastern U.S. is predominately governed by NO_x-limited conditions during the episode, consistent with earlier studies of the VOC-rich region (e.g., Chameides et al., 1992). Significant sensitivity to VOC occurs only in urban centers, but the dense populations and high ozone concentrations of these areas make them important from a policy perspective. The concentration ratios H_2O_2/HNO_3 and $HCHO/NO_y$ have been shown to be useful but not definitive indicators of ozone sensitivity to domain-wide emissions. Ozone sensitivity to NO_x tends to increase asymptotically with each ratio, and significant sensitivity to VOC is confined to low values of the ratios. The ratio O_3/NO_z is not a meaningful predictor of ozone sensitivity during this episode. The associations between ozone sensitivities and species indicator ratios during the episode are consistent with those reported in earlier studies, but the paucity of VOC-limited conditions hinders definitive quantification of threshold values here.

We have identified and explored two factors which complicate the usefulness of ozone production regime categorization. First, ozone production is nonlinear and the degree of nonlinearity varies spatially and temporally. Ozone is found to be most nonlinear when H_2O_2/HNO_3 and $HCHO/NO_y$ are low, so trends of sensitivity with these ratios will diminish as the size of perturbation increases. Second, and more significant for air pollution policy, are the distinctions between response to domain-wide and local emissions. For VOC, domain-wide sensitivity is strongly indicative of the impact of local controls. For NO_x, however, local control may elicit a positive, negative, or neutral impact even when ozone exhibits strong positive sensitivity to domain-wide NO_x. Diagnosing the ozone production regime at a given receptor, whether by indicator ratio or by modeled sensitivity to domain-wide emissions, is therefore insufficient to determine whether local NO_x control would be beneficial. In localities whose NO_x/VOC emissions ratios differ markedly from surrounding areas, species ratios may be recast as indicators of the upwind versus local character of ozone formation.

Acknowledgements This work was conducted as part of the Fall-Line Air Quality Study with funding from the Georgia Department of Natural Resources.

References

Carter, W.P.L., 2000, Documentation of the SAPRC-99 chemical mechanism for VOC reactivity assessment. Final Report to California Air Resources Board, Contract No. 92-329 and 95-308.

Chameides, W.L., et al., 1992, Ozone precursor relationships in the ambient atmosphere, *J. Geophys. Res.* **97**, 6037-6055.

Hu, Y., Odman, M. T., and Russell, A. G., 2003, Meteorological modeling of the first base case episode for the Fall Line Air Quality Study. Prepared for Georgia Department of Natural Resources.

Hu, Y., Cohan, D. S., Odman, M. T., and Russell, A. G., 2004, Air quality modeling of the August 11-20, 2000 episode for the Fall Line Air Quality Study. Prepared for Georgia Department of Natural Resources.

Kleinman, L. I., Daum, P. H., Lee, J. H., Lee, Y.-N., Nunnermacker, L. J., Springston, S. R., Newman, L., Weinstein-Lloyd, J., and Sillman, S., 1997, Dependence of ozone production on NO and hydrocarbons in the troposphere, *Geophys. Res. Lett.* **24**, 2299-2302.

Lu, C.-H., and Chang, J. S., 1998, On the indicator-based approach to assess ozone sensitivities and emission features, *J. Geophys. Res.* **103**, 3453-3462.

Ryerson, T. B, et al., 2001, Observations of ozone formation in power plant plumes and implications for ozone control strategies, *Science* **292**, 719-723.

Sillman, S., 1995, The use of NO_y, H_2O_2 and HNO_3 as indicators for O_3-NO_x-VOC sensitivity in urban locations. *J. Geophys. Res.* **100**, 14,175-14,188.

Sillman, S., 1999, The relation between ozone, NO_x and hydrocarbons in urban and polluted rural environments. *Atmos. Environ.* **33**, 1821-1845.

Sillman, S., and D. He, 2002, Some theoretical results concerning O_3-NO_x-VOC chemistry and NO_x-VOC indicators, *J. Geophys. Res.* **107**, doi: 10.1029/2001JD001123.

Unal, A., Tian, D., Hu, Y., and Russell, A. G., 2003. 2000 August Emissions Inventory for Fall Line Air Quality Study (FAQS). Prepared for Georgia Department of Natural Resources.

Yang, Y.-J., Wilkinson, J., and Russell, A. G., 1997, Fast, direct sensitivity analysis of multidimensional photochemical models, *Environ. Sci. Technol.* **31**, 2859-2868.

Alternative Approaches to Diagnosing Ozone Production Regime

Speaker: D. Cohan

Questioner: S. Andreani Aksoyoglu

Question: Did you include particulate nitrate in the H_2O_2/HNO_3 ratio?

Answer: No, we considered only gas phase concentrations in the ratio.

Questioner: R. Yamartino

Question: How much effort is involved in putting the DDM into an existing photochemical AQM?

Answer: Incorporation of DDM requires considerable effort because it must account for all major processes simulated by the AQM. It took me about $1\frac{1}{2}$ years to incorporate DDM into CMAQ and assure its accuracy.

17 Analysis of Seasonal Changes of Atmospheric Aerosols on Different Scales in Europe Using Sequentially Nested Simulations

Adolf Ebel, Michael Memmesheimer, Elmar Friese, Hermann J. Jakobs, Hendrik Feldmann, Christoph Kessler, and Georg Piekorz*

1. Introduction

Long-term simulations of atmospheric pollutants containing aerosols as an essential component may conveniently be used to support the assessment of the role of suspended particulate matter for the atmospheric environment on various temporal and spatial scales. They can help to bridge gaps of our knowledge and thus to broaden the range of applications of observations in various respects. This holds, among others, for the distribution of aerosols and its variability in space and time, their generation, emission and resulting source-receptor relationships, their spectral distribution and possible composition. For instance, the finest particle fraction cannot yet efficiently be measured by existing observational networks. Regarding the possibility that this component is mainly responsible for adverse impacts of aerosols on human health, simulations of monitored components – usually total suspended particulates (TSP), PM_{10} or $PM_{2.5}$ – may be employed to establish useful quantitative relationships between these parameters and the finer particle mode of the aerosol spectrum and apply it to health impact studies.

Though impressive advances have been achieved in the field of aerosol dynamics, chemistry and transport modeling which enable studies and applications as mentioned above, it is obvious that considerable uncertainties still exist for the simulation of particulate matter in the atmosphere. Our knowledge of aerosol sources and certain aspects of the formulation of particle formation, transformation and aerosol-gas phase interaction may serve as examples. A major problem of aerosol model application and development is the fact that

* Adolf Ebel, Michael Memmesheimer, Elmar Friese, Hermann J. Jakobs, Hendrik Feldmann, Christoph Kessler, and Georg Piekorz, Rhenish Insitute for Environmental Research at the University of Cologne, 50931 Cologne, Germany. Christoph Kessler also at Ford Research Center, 52072 Aachen, Germany.

comprehensive evaluation of simulated particle distributions in space and time can only be carried out with rather unspecific measurements of PM_{10} (or TSP, occasionally used in this study when not enough PM_{10} data is available) and $PM_{2.5}$. Obviously, considerably more research is needed for more reliable assessment of the specific role of atmospheric aerosols for air quality and their long- and short-term on the environment in general and human health in particular. Model results as presented in this paper should be regarded as a contribution to this task.

In the following sections we begin with a brief characterization of the EURAD model system which has been employed for long-term simulations introduced in Section 3. Evaluation studies and selected results focusing on seasonal effects are discussed in Section 4 and 5, respectively.

2. The Model System

The EURAD system consists of a chemical transport model (EURAD CTM2; Hass, 1991), the Meso-scale Meteorological Model, Version 5 (MM5; Grell et al. 1993) and the EURAD Emission Model (EEM, Memmesheimer et al., 1991) providing gridded hourly emission data as input to the CTM. The basic model domain covers the major part of Europe with coarse resolution. For application to smaller domains the method of sequential nesting is used (Jakobs et al., 1995). Such downscaling helps to improve the treatment of initial and lateral boundary conditions on smaller scales for less well observed atmospheric species and, of course, allows the simulation of fine structures of tracer distributions as well as meteorological fields. Meteorological initial and boundary conditions are obtained from ECMWF or NMC analyses and from weather forecasts in the case of air quality predictions. Emission input data have been derived from inventories provided by EMEP, GENEMIS, TNO or regional sources, depending on the details of air quality simulations.

Nested calculations as discussed in this study have been carried out with resolution of 125 km (so-called coarse grid, covering Europe), 25 km (nest 1, Central Europe) and 5 km (nest 2, states in the FRG, e.g. North Rhine-Westphalia (NRW)). For special applications to local problems a resolution of 1 km may be employed. The standard model version has 23 layers between the surface and 100 hPa, where 15 are found below an altitude of 3000 m. The height of the lowest layer is about 40 m.

The chemical mechanism applied in CTM2 is either RADM2 (Stockwell et al., 1990) or RACM (Stockwell et al., 1997). The latter one is always used when secondary organic aerosols are included in the simulations. 63 and 77 chemical species are treated in the gas phase calculations of RADM2 and RACM, respectively. The aerosol module attached to the CTM is MADE2 (Modal Aerosol Dynamic Model for Europe in EURAD, Version 2). It combines a modal aerosol dynamic mechanism with inorganic aerosol equilibrium thermodynamics in the system NH_4^+–NO_3^-–$SO_4^{2-}$$H_2O$ (Ackermann et al., 1998;

Binkowski and Roselle, 2003), a cloud-aerosol interaction module (Friese et al., 2000) and a Secondary Organic Aerosol Module (SORGAM; Schell et al., 2001). The output of the CTM which is mainly used for air quality studies consists of hourly values for ozone, NO_x, NO_2, NO, SO_2, CO, NH_3, TSP, PM_{10}, $PM_{2.5}$ and the secondary inorganic aerosol components NH_4^+, NO_3^- and SO_4^{2-}. Furthermore, information about anthropogenic particles (ANTH), elemental carbon (EC), organic carbon (OC), biogenic secondary organic aerosols (BSOA) and anthropogenic secondary organic aerosols (ASOA) may be obtained from MADE2 (see Fig. 1).

3. Long-Term Simulations

Long-term simulations with the EURAD system which can be used for the analysis of seasonal effects on the behavior of air pollutants are available for the years 1997 and 2002. Furthermore, operational air quality forecasts have been carried out which also give insight into seasonal changes. As a special application studies of the sensitivity of annual variations to emission reductions may be mentioned. An example of long-term calculations of TSP at a rural North German station with corresponding observations and estimated composition is shown in Fig. 1. Long-term numerical aerosol simulations with EURAD are still too sparse for establishing a general climatology of the model domain. Yet they can well be used to study typical seasonal effects in different regions, e.g. the relative importance of BSOA and ASOA in summer in Scandinavia and Central Europe. In this study the main emphasis is put on the analysis of the reliability of aerosol simulations and their dependence on season. The aerosol emission data used for 1997 and 2002 coarse grid ands nest 1 simulations is based on the TNO inventory for 1995 (TNO, 1997). A specific emission inventory has been employed for 1997 and 2002 nest 2 calculations covering NRW.

4. Evaluation Studies

Evaluation of the long-term simulations is mainly carried out with available observations from monitoring networks which provide TSP, PM_{10} and/or $PM_{2.5}$ data. Though of limited information content regarding composition and vertical distribution, they may conveniently be used to assess the reliability of horizontal and temporal changes at a large range of scales. Furthermore, these integral parameters play a key role for assessing air quality in the framework of EU directives and therefore deserve special attention in models employed for this purpose. Various methods have been applied. Only some examples of evaluation studies can briefly be addressed in this paper due to lack of space.

Comparison of calculated and observed time series is a convenient way to check the performance of a model with regard to long-term simulations. Fig. 1 gives the impression that the EURAD model frequently underestimates aerosol mass during

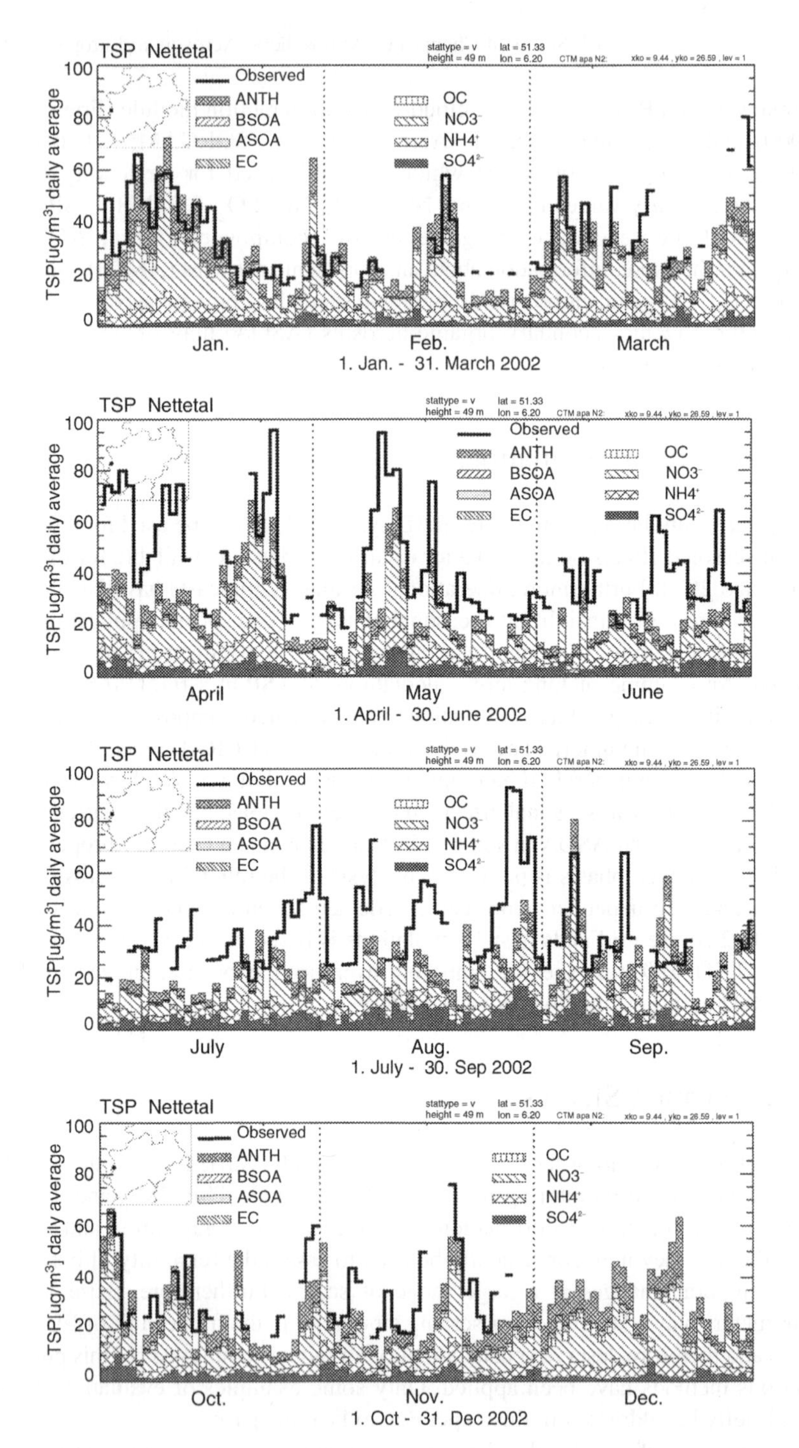

FIGURE 1. Simulated (columns) and observed (thick line) time series of daily TSP averages for the year 2002. The modeled composition is also shown (ANTH anthropogenic fraction of primary aerosols; BSOA, ASOA biogenic and anthropogenic secondary organic aerosols; EC elemental carbon; OC organic carbon; $NH4^+$, NO_3^- and SO_4^{2-} inorganic secondary components).

episodes with high measured values and that this more frequently occurs during summer than during the other seasons. Due to the lack of observations it is not possible to validate the calculated composition also shown for curiosity in the figure. The seasonal dependence of aerosol prediction quality is clearly evident from the scatter plots exhibited in Fig. 2. Measurements from NRW and nest 1 calculations with 25 km resolution have been used. The upper part shows results for PM_{10} from January till June 1997 and the lower one for TSP from July till December 2002 (with a smaller number of presently available observations, but the same distribution characteristics as PM_{10} from autumn to winter in 1997). Comparing the results for nest 1 with those for nest 2 (not shown) it is found that

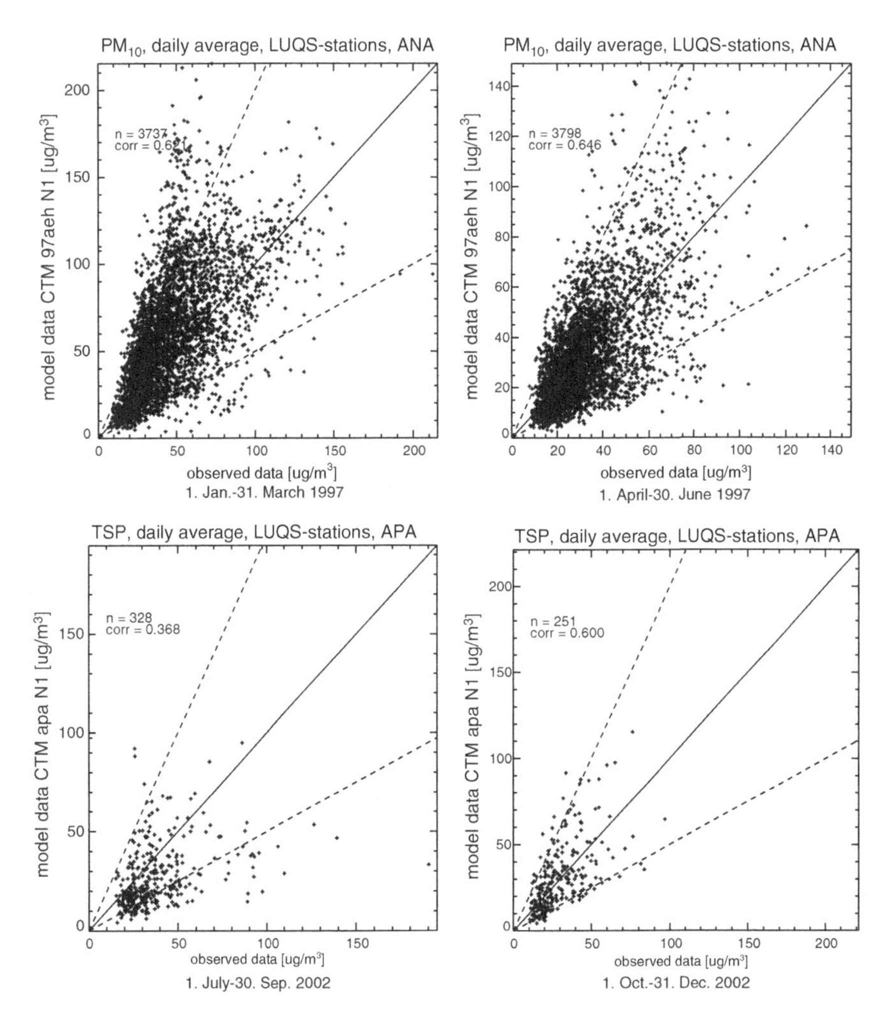

FIGURE 2. Comparison of observed and simulated atmospheric aerosol for NRW, nest 1 calculations. Upper panels show PM10 in 1997, January–March (left) and April– June (right). Lower panels show TSP in 2002, July– Sept. (left) and Oct.– Dec. (right).

a tendency exists to underestimate stronger daily maxima increases with increasing resolution. It seems to be a persistent seasonal feature that the modeled summer values exhibit a clear negative bias. The monthly hit rates for 50% accuracy found for NRW in 1997 and 2002 for PM_{10} and TSP, respectively, are shown in Fig. 3. They range between 50 and 89% for the coarse grid, 49 and 92% for nest 1 with 25 km horizontal resolution and 40 and 85% for nest 2 with 5 km resolution. Better correspondence of PM10 estimates than simulated TSP with observations as seen in Fig. 3 is expected since the model is underestimating the large particle fraction. It should be noted that a special emission scenario for NRW has been employed in the latter case leading to an improvement of the simulations in the autumn and winter months of 1997 and to a notable decrease of accuracy during the summer months in correspondence with the strong summer bias seen in the scatter plots (Fig. 2). The question why this improvement does not appear in 2002 is still under investigation.

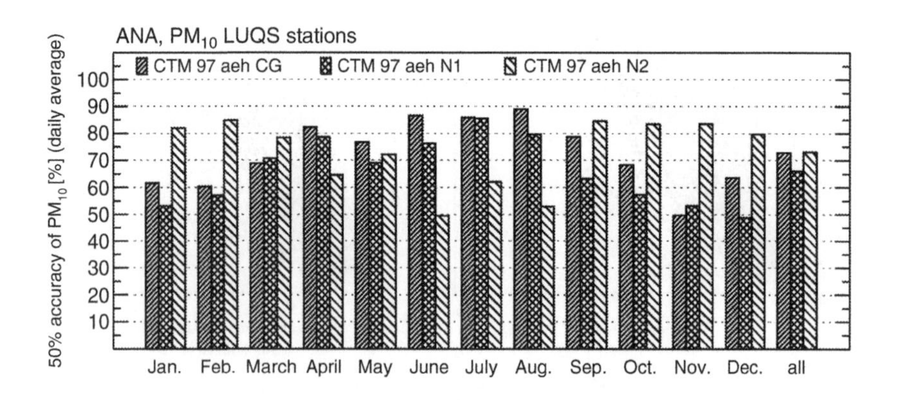

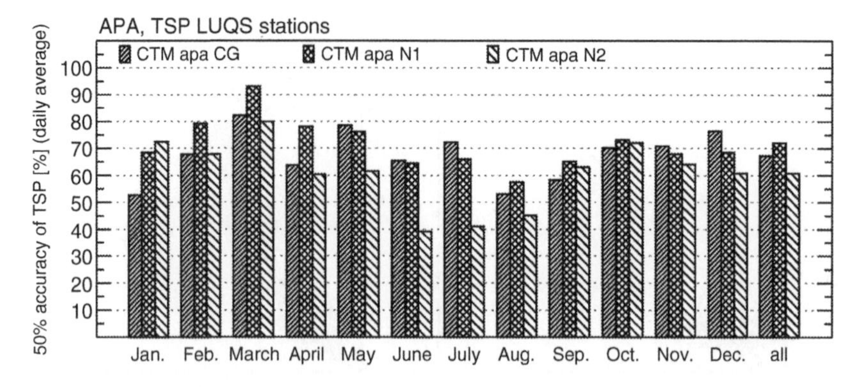

FIGURE 3. Hit rates for 50% accuracy of PM_{10} (1997, upper panel) and TSP (2002, lower panel), North Rhine-Westphalia. Results for coarse grid (CG), nest 1 (N1) and nest 2 (N2) calculations.

5. Discussion and Conclusions

CTMs like EURAD including advanced aerosol modules have proven to be a valuable and necessary tool for air quality assessment and control, though comparisons of observations and calculations as presented in Section 4 show that improvements of numerical simulations are still possible and necessary. The apparent summer deficit of modeled aerosol mass may partly be attributed to a shortcoming of the aerosol emission inventory not taking into account the generation of specific emissions like mineral dust during dry and hot episodes. The tendency of the EURAD system to underestimate aerosol maxima quite often in summer still has to be investigated. It could also partly originate from the applied emission inventory.

Composition of aerosols could become one of the most valuable and relevant model outputs. Better knowledge of composition is particularly needed for detailed impact studies regarding health and ecological effects. Yet a comprehensive

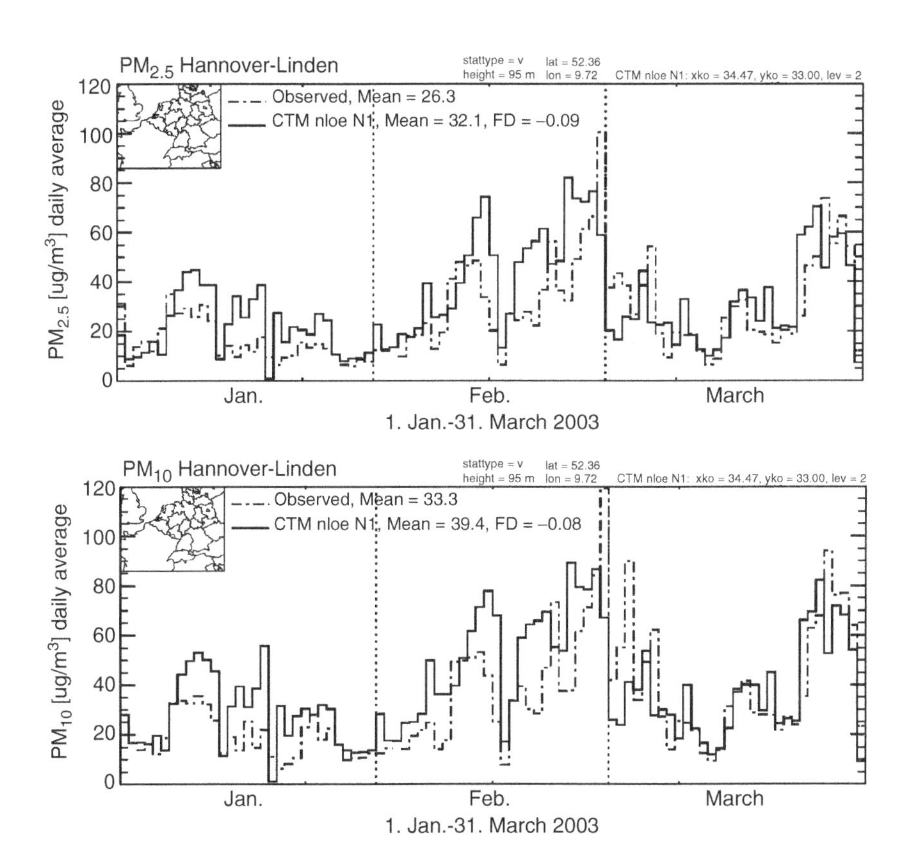

FIGURE 4. $PM_{2.5}$ and PM_{10} forecast (continuous line) with the EURAD system compared to observations (broken line) at the North German station Hannover-Linden. January–March 2003.

evaluation of this property is still difficult due to the lack of sufficient and suitable observations. Nevertheless, first principal studies are possible and have been carried out. For instance, the differences between the behavior of anthropogenic and biogenic secondary organic aerosols with respect to origin, chemistry and transport can well be analyzed. Furthermore, specific features of space and time dependence of aerosol modes (nucleation, Aitken) and their dependence on meteorological conditions including the presence of clouds have been investigated. Confining the simulations to primary and secondary inorganic aerosols the computation time needed by the EURAD model is short enough that PM_{10} and/or $PM_{2.5}$ can be included in air quality predictions. An example is shown in Fig. 4. Neglecting secondary organic aerosols appears to be acceptable during seasons and episodes when the SOA formation potential is low.

Acknowledgement The development of the aerosol module MADE2 of the EURAD system and its application has been funded in the framework of various projects by the federal (BMBF) and state government (MWF/NRW) and by the Landesumweltamt (LUA, State Environmental Agency) of NRW. The authors also gratefully acknowledge that meteorological data has kindly be provided by ECMWF via DWD and by NCEP, emission data by EMEP, LUA and TNO and of TSP and PM data by the state environmental agencies of North Rhine-Westphalia and Lower Saxony.

References

Ackermann, I. J., Hass, H., Memmesheimer, M., Ebel, A., Binkowski, F. B., and Shankar, U., 1998, Modal Aerosol dynamics model for Europe: Development and first applications. *Atmos. Environm.*, **32**, 2891–2999, 1998.

Binkowski, F.S., and Roselle, S.J., 2003, Models-3 Community Multiscale Air Quality (CMAQ) model aerosol component, 1. Model description, *J. Geophys. Res.*, **108**, NO. D6, 4183, doi:10.129/2001JD001409,2003.

Friese, E., Memmesheimer, M., Ackermann, I. J., Hass, H., Ebel, A., and Kerschgens, M. J., 2000, A study of aerosol/cloud interactions with a comprehensive air quality model. *J. Aerosol Sci.*, **31**, Suppl 1, 54–55.

Grell, A. G., Dudhia, J., and Stauffer, D. R., 1993, A description of the fifth-generation PennState/NCAR Mesoscale Model (MM5), *NCAR Techn. Note, NCAR/TN-398+1A*.

Hass, H., 1991, Description of the EURAD Chemistry Transport Model, version 2 (CTM2), *Mitteil. Inst. Geophys. Meteor., Universitaet zu Koeln*, no. 83.

Jakobs, H.J., Feldmann, H., Hass, H., and Memmesheimer, M., 1995, The use of nested models for air pollution studies: an application of the EURAD model to a SANA episode. *J. Appl. Met.*, **34**, 1301–1319.

Memmesheimer, M., Tippke, J., Ebel, A., Hass, H., Jakobs, H. J., and Laube, M., 1991, On the use of EMEP emission inventories for European scale air pollution modelling with the EURAD model. In: *EMEP Workshop on Photo-oxidant Modelling for Long-Range Transport in Relation to Abatement Strategies*, Berlin, April 1991, pp. 307–324.

Schell, B., Ackermann, I. J., Hass, H., Binkowski, F. S., Ebel, A., 2001, Modeling the formation of secondary organic aerosol within a comprehensive air quality modeling system. *J. Geophys. Res.*, **106**, 28275–28293.

Stockwell, W. R., Middleton, P., and Chang, J. S., 1990, The second generation regional acid deposition model chemical mechanism for regional air quality modelling, *J. Geoph. Res.*, **95**, 16 343–16367.
Stockwell, W. R., Kirchner, F., and Kuhn, M., 1997, A new mechanism for regional atmospheric chemistry modeling. *J. Geophys. Res.*, **102**, D22, 25847–25879.
TNO, 1997, Particulate matter emissions (PM10, PM2.5, PM < 0.1) in Europe in 1990 and 1993, *TNO Report TNO-MEP* R96/472 (see also http://www.air.sk/tno/cepmeip/)

Analysis of Seasonal Changes of Atmospheric Aerosols on Different Scales in Europe

Speaker: A. Ebel

Questioner: W. Jiang
Question: Could you comment a little more on your first conclusion item on reliability of modelled results vs data?
Answer: The comparison of simulated and measured aerosol mass and composition has been carried out assuming that the observations are correct. This is, of course, not true. Measurements may exhibit various degrees of uncertainty, depending on availability and quality of instruments, environmental conditions etc. Comprehensive model evaluation and reliability estimates should take into account the quality of data.

Questioner: P. Bhave
Question: I noticed that the maximum PM_{10} concentrations are observed in the fall. What is the cause of this feature and is it representative of Westphalia only or the larger region.
Answer: Anticyclonic weather conditions frequently occur in early fall in Central Europe and neighbouring regions. I suggest that they have caused high PM_{10} concentrations in the years when the measurements which were shown during the presentation have been made. Since high pressure systems usually cover large areas it may be concluded that such temporal maxima occur in larger regions. Yet it should be noted that a long-term analysis of TSP observations in North Rhine-Westphalia did not reveal significant regular (climatologic) maxima of aerosol concentrations in any season (Luftqualität in Nordrhein-Westfalen, LUQS-Jahresbericht 1999, pp. 101–131, Landesumweltamt Nordrhein-Westfalen, Essen, 2001).

18 Interaction Between Meteorological and Dispersion Models at Different Scales

Eugene Genikhovich[*], Mikhail Sofiev[†], and Irina Gracheva

1. Introduction

A general understanding of the interface between meteorological and dispersion models is that the latter ones should assimilate the information coming from meteorological model with minor internal processing, mainly oriented at computation of some surrogate indices like stability parameters or mixing layer height. However, the situation is rarely that simple. The dispersion model (DM) as a client of the numerical weather prediction (NWP) meteorological model (MM) requires significantly higher spatial resolution that that of NWP and, possibly, it should also vary in space and time because concentrations of atmospheric pollutants are usually much more irregular than meteorological variables. This brings about the problems of multi-scale nested modelling, boundary conditions, agreement between the scales and several other issues sometimes referred as a downscaling problem and its influence on accuracy of the DM at various scales.

The second set of issues touched in this paper is related to the problem of self-consistency of the meteorological fields provided to dispersion models and their compatibility with the physical parameterizations of the DMs themselves.

There are several key areas for harmonization of meteorological and dispersion models: (i) effects of natural variability and scales of processes should be adequately represented, (ii) extra variables required by dispersion models should be derived from the input, (iii) representations of various processes in NWP and dispersion models have to be made coherent, (iv) technical problems and associated losses in accuracy during transfer of large amounts of data should be taken into account, (v) when a feedback from the dispersion model is important for the NWP model, an appropriate two-way interface should be set.

[*] E.Genikhovich, I.Gracheva Main Geophysical Observatory, Karbysheva str. 7, 194021, St. Petersburg, Russia; *ego@main.mgo.rssi.ru*

[†] M.Sofiev Finnish Meteorological Institute, Sahaajankatu 20 E, 00880 Helsinki Finland; *mikhail.sofiev@fmi.fi*

2. Downscaling Using a Perturbation Approach: An Example of the Blasius Problem

There are several ways to perform downscaling of meteorological fields. The most common one comprises a set of sequential runs of one (or several) meteorological models with an increasing resolution and reducing sizes of the nested domains. The key limitation of the method originates from a problem of synchronization of the inner and outer fields at the boundaries of the smaller domain – also known as a problem of transparent boundary conditions. Indeed, if some discrete field satisfies a set of equations (including the continuity equation that is an absolute imperative for the dispersion models), its interpolation to any other grid – whether denser or scarcer – will not satisfy these equations any more. Various attempts to cope the issue, including all sorts of Deivis-relaxation constructions usually introduce their own problems barely reducing the influence of the main one.

Below, we demonstrate another approach that, being also not ideal, shows good performance in comparison with "standard" interpolation-based methodologies. The approach is based on a consideration of the small-scale problem as a high-frequency perturbation of the fields obtained at the coarser scale.

Let's consider a classical Blasius problem, which is a two-dimensional idealized case of formation of a boundary layer (BL) over a flat terrain ($z=0,\ x>0$) from the originally homogenous wind field $u = u_\infty$:

$$\begin{cases} u\dfrac{\partial u}{\partial x} + w\dfrac{\partial u}{\partial z} = \nu\dfrac{\partial^2 u}{\partial z^2} \\ \dfrac{\partial w}{\partial x} + \dfrac{\partial w}{\partial z} = 0 \end{cases}, \quad \begin{array}{l} u\big|_{x=0} = u_\infty,\ u\big|_{z=\infty} \to u_\infty \\ u\big|_{z=0} = 0,\ w_{z=0} \to 0 \end{array}$$

The solution of this problem can be found using the stream-function approach, when the stream function, ψ, is introduced to satisfy the following requirements:

$$u = \frac{\partial \psi}{\partial z};\ w = -\frac{\partial \psi}{\partial x}$$

Such an approach is preferable for the purposes of the dispersion modeling because it ensures the mass consistency (see next section); its analogue in 3D case is the vorticity description (Schayes et al., 1996).

Let's also modify the problem by making a "hole" in the plane, which creates a high-frequency perturbation in the originally smooth fields. Behavior of the BL around this hole will be the subject for the downscaling analysis (Figure 1).

The problem is then solved two times – with high-resolution grid (an "exact" field) and with the perturbation downscaling approach (Figure 2). All fields are made non-dimensional here using u_∞ as the wind-speed scale and the height of the computational domain as the length scale. A comparison of the fields is shown in Figure 3, which also presents comparison of the interpolated coarse-grid

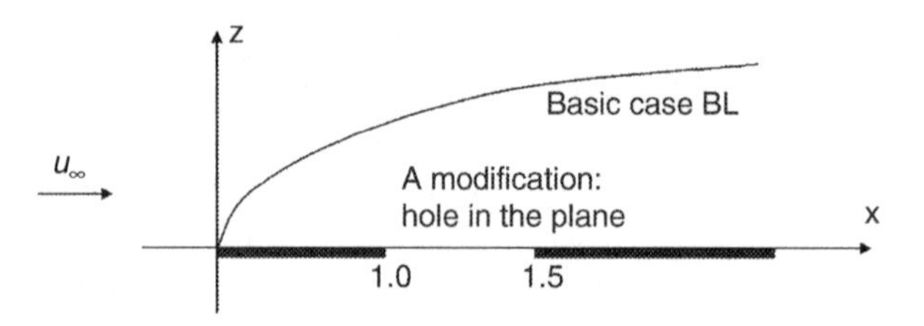

FIGURE 1. A Scheme of Blasius problem and its modification for the current study.

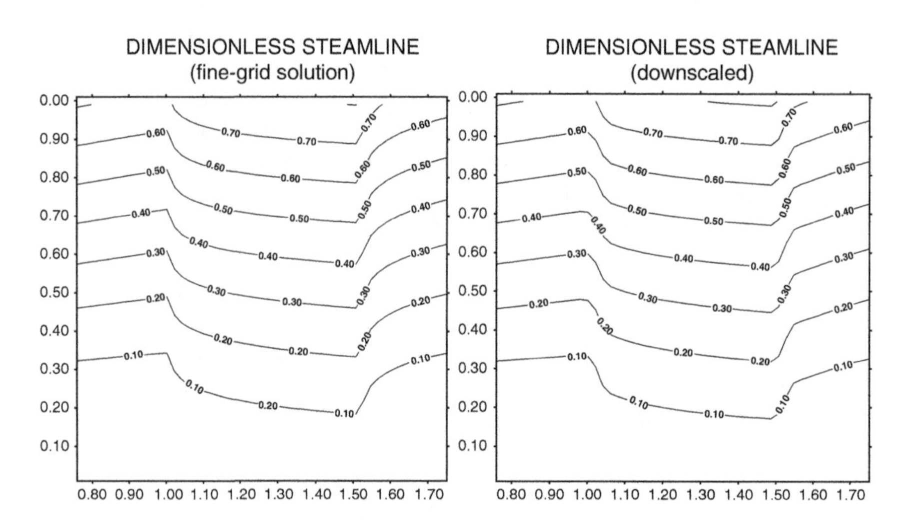

FIGURE 2. Wind fields for the problem solutions: "exact" high-resolution field (left panel), downscaled field (right panel).

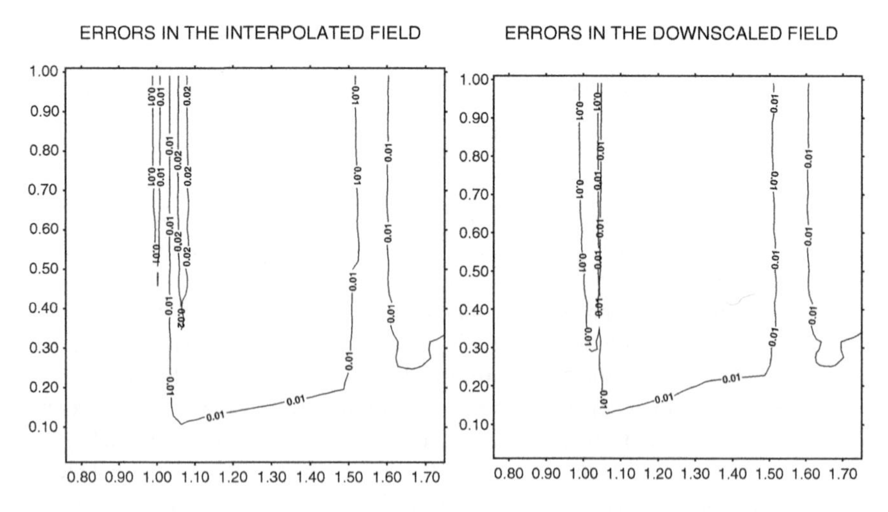

FIGURE 3. Errors of interpolation (left panel) and downscaling via perturbation computations (right panel).

field and the exact solution. It is well seen that the perturbation approach shows much better performance, especially in the areas with high gradients.

3. Self-Consistency of Meteorological Fields and Dynamical and Physical Parameterizations of Dispersion Models

As mentioned above, significant errors in dispersion calculations could appear as a result of downscaling, if the self-consistency and mutual coherence (transparent boundary conditions) of the both coarse and fine grids are not ensured. It is easy to show that interpolation of the fields near the boundary between the outer coarse and inner fine grids do not provide the self-consistency to either of them. In application to the transport equation, it results in violation of the mass consistency because the interpolated fields do not satisfy to the continuity equation. Since the natural variability of the meteorological and pollution fields depends on spatial resolution, the downscaled parameters should, in principle, resolve the corresponding variability, which is inside the internal uncertainty of the coarse fields. Taken together with probable non-linearity of the involved processes, this again poses the problem of consistency and mutual agreement of the coarse and fine grids. A potential cost of disagreements at this stage is illustrated by Cameron-Smith and Connell (2003). The problem of coherent representation of natural variability and scales of the processes leads to significant challenges. However, there is one more dimension for harmonization: chemical and physical processes, including advection and diffusion. As shown by Gong (2003) and Cameron-Smith and Connell (2003), violation of continuity equation can result in dramatic consequences, such as 100% error in total mass in air of some chemical or two orders of magnitude local peaks of concentrations, appearing and disappearing in a chaotic manner. They can show up if the downscaled wind fields are not solenoidal, *i.e.*, do not satisfy the continuity equation to ensure the local mass conservation. In such a case, fictitious sources and/or sinks of pollutants distributed over the computational domain are introduced in the model.

The aforementioned effects can originate from non-conservative advection (as often happens in meteorological models), from an inconsistency between the schemes in meteorological and dispersion models, from limited accuracy of the data transfer and from grid interpolation and transformation. One of the efficient, though expensive, recipes here is to explicitly restate the continuity equation in the meteorological fields inside the dispersion model right before using the fields. Residuals should be distributed in the horizontal wind components as they are less sensitive to such a disturbance. Problems of harmonization of other physical and chemical processes can be illustrated by the following example. Let at some time an air volume limited by the grid cell borders contain a certain amount of liquid cloud water, with some species dissolved in it. Let at the next time step the liquid water content becomes zero due to cloud microphysical processes, which are computed in meteorological model. What should then be done with the dissolved species? It is easy to show that virtually all ways of handling such situations lead to errors, which significantly depend on the mutual relation of timescales – chemical and microphysical.

4. Re-Stating the Boundary Layer Integrated Parameters from the Meteorological Fields

One of the most important extra variables used by the dispersion models are the scaling parameters for the similarity theory applications to the atmospheric boundary layer. These variables, such as Monin-Obukhov length L, friction velocity u^*, temperature scale T^* and convection velocity scale w^*, can be obtained from vertical profiles of wind and temperature, or from the sensible heat flux H, which, however, is not always in a standard output of the NWP models. A well-known iteration approach for computing L, u^*, T^* does not always provide good estimates due to problems with convergence (especially in stable conditions) and the necessity to perform numerical differentiations of slowly varying parameters. The scope of corresponding problems is illustrated on Figure 4 that reproduced the fields of friction velocity computed with HIRLAM (right-hand panel) and estimated with a "standard" M-O similarity approach. A one-step approach introduced by Genikhovich and Osipova (1984) and Groisman and

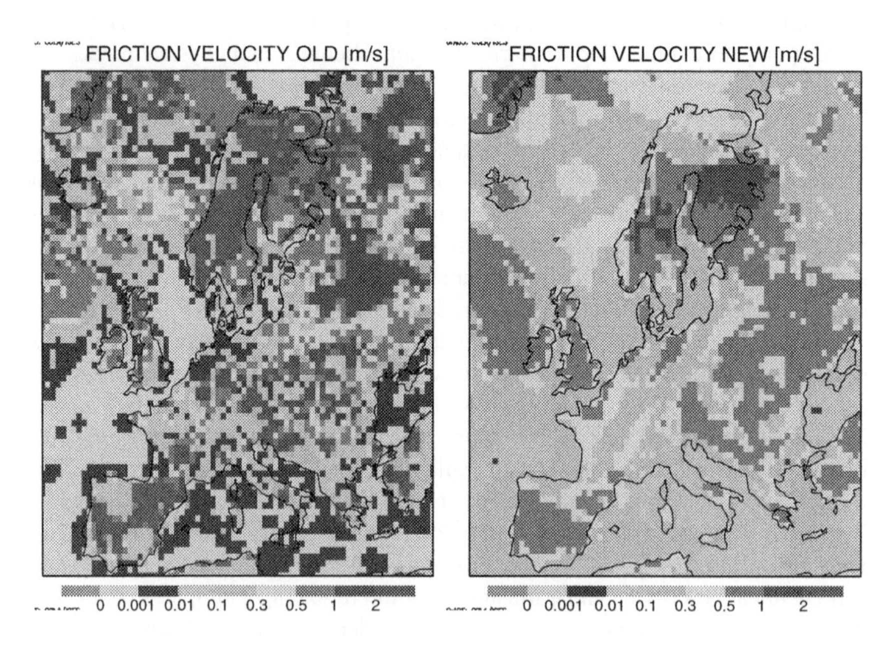

FIGURE 4. Friction velocity obtained from the iterative (left panel) and non-iterative (right panel) methods re-stating the similarity parameters.

Genikhovich (1997) can be used to provide robust estimates of parameters listed above. The method is based on the following formulation for eddy diffusivity:

$$K_z = \left\{ 0.5\kappa \int_0^2 \frac{[(dU/dz)^2 - \sigma\beta d\theta/dz]^{5/4}}{(dU/dz)^2 - 0.5\sigma\beta d\theta/dz} dz \right\}^2 .$$

Here z is the characteristic height inside the surface layer (z ~ 1 m), $\sigma = 10$ and $\kappa = 0.4$ are dimensionless constants, $\beta = g/T_0$ is the buoyancy parameter and g = 9.81 m s^{-2}. From this equation, it is straightforward to get the required scaling parameters:

$$u^* = \sqrt{K_z(z_c)\partial U(z_c)/\partial z};\ H = -\frac{c_p\rho}{\Pr} K_z(z_c)\partial T(z_c)/\partial z;$$

$$T^* = \frac{H}{c_p\rho u^*};\ L = -\frac{c_p\rho(u^*)^3}{\kappa\beta H};\ w^* = u\left(\frac{h_{ABL}}{-\kappa L}\right)^{1/3}.$$

Its performance is illustrated in Figure 5 by comparison of the sensible surface heat fluxes computed with HIRLAM (right-hand panel) and estimated with the aforementioned robust parameterization scheme (left-hand panel).

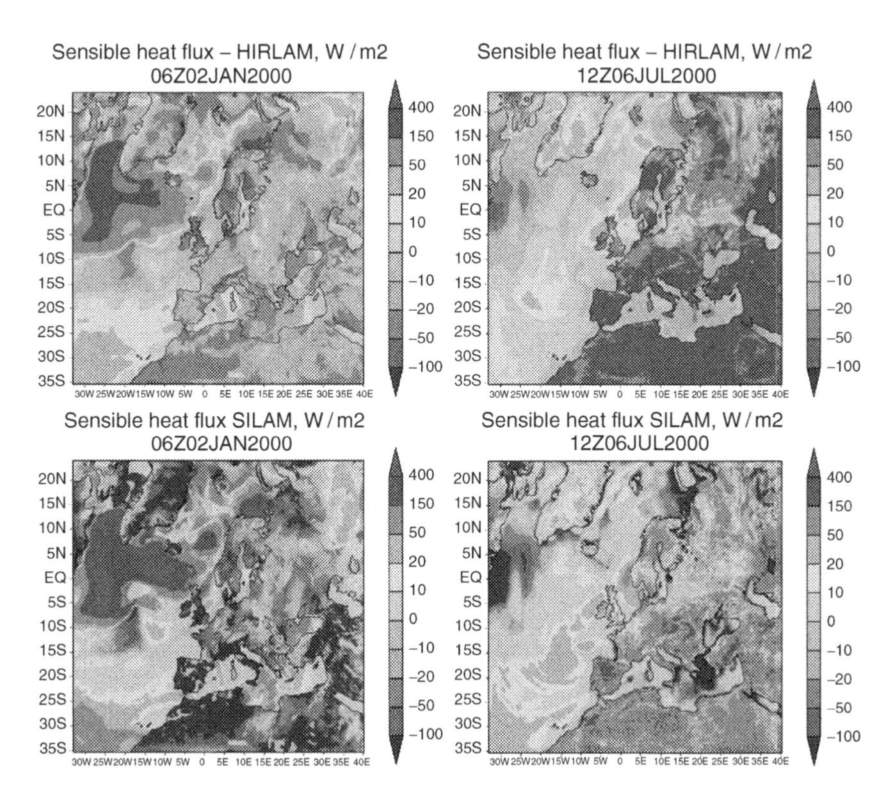

FIGURE 5. Re-stated and HIRLAM-internal sensible heat fluxes.

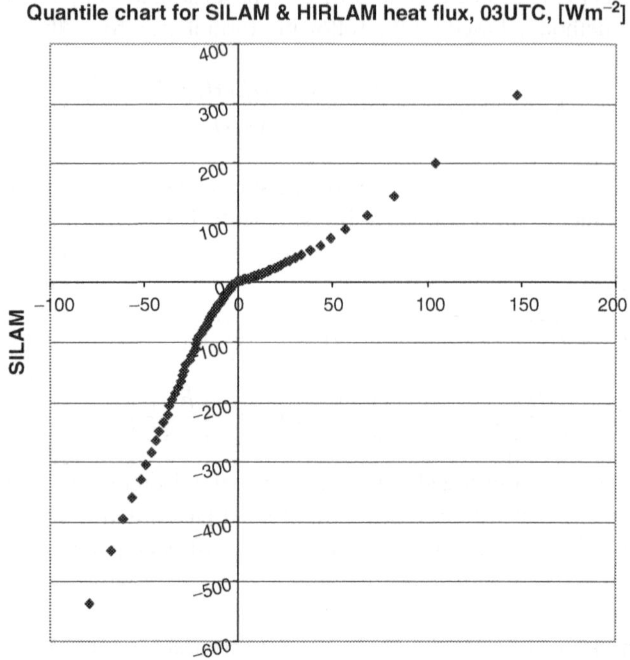

FIGURE 6. Quantile chart of HIRLAM sensible heat flux (x-axis) and the re-stated one with the above methodology (y-axis).

An integrated assessment of the methodology is on-going and is not yet completed. A specific problem on this way is that the sensible heat flux is not an explicitly verifiable variable of HIRLAM, which makes its accuracy unknown. There can be also some systematic differences between the treatment of this variable in NWP models and by the above methodology.

A possibility of such disagreement is hinted by the quantile-quantile chart in Figure 6, where the different treatment of stable and unstable conditions is quite evident. However, the stability type is determined exactly in the same manner (the chart passes through the (0,0) point), and the overall disagreement is within a factor of a few times, which is not too bad keeping the above-mentioned uncertainties in mind.

5. Conclusions

The most straightforward approach to handle at least some of the above difficulties is an "integrated" one when transport and transformation equations are integrated simultaneously with the NWP model. It requires, however, significant computational resources and a higher resolution that it not always possible to achieve. In addition, this method does not solve problems of methodological compatibility of the models. Therefore, in many cases the "interfaced" approach is used when results of the NWP modelling are transferred to dispersion model with downscaling, if necessary. In this case, however, the downscaling procedures could introduce additional errors in computed concentration fields. If one write the dynamic system of equation in the vorticity form (Shcayes et al., 1996), it becomes obvious that these procedures should properly account for the effects of baroclinicity in the outer flow as well as for correct description of distribution of the second derivatives of the wind velocity field. That is why the Hermite polynomials seem to be more appropriate for interpolation of the outer fields in the computational domain than the Lagrange ones.

References

Cameron-Smith, P., Connell, P. 2003. Pressure fixers: conserving mass in chemical transport models by adjusting the winds. Internet publication, *http://asd.llnl.gov/pfix/*

Genikhovich E.L. 2003. Indicators of performance of dispersion models and their reference values. *Int. J. Environment and Pollution,* Vol. 20, Nos. 1–6, pp. 321–329.

Genikhovich, E.L., Osipova, G.I. 1984. On determination of the turbulent eddy diffusivity from standard meteorological observation data. *Trudy GGO (Proc. Main Geoph. Observatory*), iss. 479, pp. 62–69 (in Russian).

Genikhovich E., Ziv, A., Iakovleva, E., Palmgren, F., Berkowicz, R. 2004. Joint analysis of air pollution in street canyons in St. Petersburg and Copenhagen (accepted for publication in *Atmospheric Environment).*

Gifford 1958. Statistical plume model, in: Frenkiel, F.N., and Sheppard, P.A. (Eds.): Atmospheric Diffusion and Air Pollution, Acad. Press, NY, pp. 143–164.

Gong, W., Makar, P.A., Moran, M.D. 2004. Mass-conservation issues in modelling regional aerosols. In: Air Pollution Modelling and its Application XVI (Eds. C. Borrego, S. Incecik), Kluwer, Acad./Plenum Press, pp. 317–325.

Groisman, P., Genikhovich, E. 1997. Assessing surface-atmosphere interaction using former Soviet Union standard meteorological network data. Part 1: Method, *Journal of Climate*, v. 10, No 9, pp. 2154–2183.

Kiselev, V.B., Gorelova, V.V. 1979. The study of the atmospheric diffusion equation by means of perturbation theory for functionals. *Trudy GGO (Proc. Main Geoph. Observatory*), iss. 436, pp. 37–42 (in Russian).

Schayes, G., Thunis, P., Bornstein, R. 1996. Topographic vorticity-mode mesoscale-β model. Part I: Formulation. *Journal of Applied Meteorology*, v. 35, pp. 1815–1823.

Interaction Between Meteorological and Dispersion Models at Different Scales

Speaker: M. Sofiev

Questioner: T. Castelli Silvia

Question: As far as regards the surface layer parameterization problem, have you performed an intercomparison with the well known Louis(1970)

parameterization, which also, as your proposed method, allows avoiding iterative procedures and it is based on surface layer similarity theory?
Answer: No, we did not apply that very method. Maybe, in the future we will compare the performance of different approaches. Our first idea was just to speed-up the evaluation of the similarity parameters on the basis of the explicitly verified meteorological variables – wind and temperature profiles (contrary to non-verified heat fluxes). The next problem, which came a bit later and which is getting more and more stress is to check the compatibility of the meteorological models output with the "standard" assumptions used in the dispersion modelling – in particular, the Monin-Obukhov theory. This problem appeared non-trivial because dispersion models do not have full information about the internal meteorological fields. As we just saw, the resulting differences between "meteorological" and "dispersion" interpretations of the same profiles can be quite different – and we do not think that this is only due to simplifications used in our method. There are more fundamental problems of interfacing the meteorological data to dispersion models that may be responsible for large part of the deviations.

19 Modeling Photochemical Pollution in the Northeastern Iberian Peninsula

P. Jiménez, O. Jorba, R. Parra, C. Pérez, and J. M. Baldasano*

1. Introduction

The Western Mediterranean Basin (WMB) is surrounded by high coastal mountains and is influenced by the Mediterranean sea. In summer it becomes isolated from the traveling lows and their frontal systems. The meteorology and the origin of the air masses arriving at the Iberian Peninsula are highly influenced by the Azores high pressure system which is located over the Atlantic Ocean and that intensifies during the warm season inducing very weak pressure gradient conditions all over the region. A number of studies have shown that during this period, layering and accumulation of pollutants such as ozone and aerosols were taking place along the northeastern Iberian Peninsula (NEIP) (Millán *et al.*, 1992, 1997). Baldasano *et al.* (1994) and Soriano *et al.* (2001) combined a numerical approach with an elastic-lidar sounding campaign to study the circulatory patterns of air pollutants over Barcelona (NEIP) in a typical summertime situation, where pollutant layers were, formed by the return flow of the breeze and forcings caused by the complex orography combined with the compensatory subsidence. This was confirmed and detailed in a regular basis by Pérez *et al.* (2004). The complex topography of the NEIP induces a extremely complicated structure of the flow because of the development of α- and β-mesoscale phenomena that interact with synoptic flows. The characteristics of the breezes have important effects in the dispersion of pollutants emitted. In addition, the flow can be even more complex because of the non-homogeneity of the terrain, the land-use and the types of vegetation. In these situations, the structure of the flow is extremely complicated because of the superposition of circulations of different scale. In this study, a numerical approach was adopted in order to study the formation and dynamics of photochemical pollutants over the NEIP with a third-generation air

* Pedro Jiménez, Oriol Jorba, René Parra, Carlos Pérez and José M. Baldasano, Laboratory of Environmental Modeling, Department of Engineering Projects, Universitat Politècnica de Catalunya (UPC), SPAIN.

quality model (MM5-EMICAT2000-CMAQ) applying high resolution, and an evaluation of the results for O_3 was performed during the photochemical episode of August 13-16, 2000.

2. Models

In order to have a detailed picture of the local mesoscale processes that contribute to the episodes of photochemical pollution in the NEIP, a numerical simulation of meteorological fields was run with the PSU/NCAR Mesoscale Model 5 (MM5), version 3, release 4 modeling system (MMD/NCAR, 2001), for the period 13-16 August 2000. Four nested domains were selected (Figure 1), which essentially covered Europe (Domain 1, D1), the IP (Domain 2, D2), the NEIP (Domain 3, D3) and the Catalonia area (Domain 4, D4).

A one-way nesting approach was used. The vertical resolution was of 29 σ-layers for all domains, the lowest one situated approximately at 10 m AGL and 19 of them below 1 km AGL. The upper boundary was fixed at 100 hPa. Initialization and boundary meteorological conditions were introduced with analysis data of the ECMWF. Data at 1-degree resolution were available (100-km approx. at the working latitude) at the standard pressure levels every 6 hours. The physics options used for the simulations were: the Mellor-Yamada scheme as used in the Eta model for the boundary layer parameterization, the Anthes-Kuo and Kain-Fritsch cumulus scheme, the Dudhia simple ice moisture scheme, the cloud-radiation scheme, and the five-layer soil model. Comparisons with measurements were done in order to evaluate the reliability of the model results, and to validate the model under weak synoptic conditions over the Iberian Peninsula. Reasonable agreement was produced between model results and observations (Jorba *et al.*, 2003).

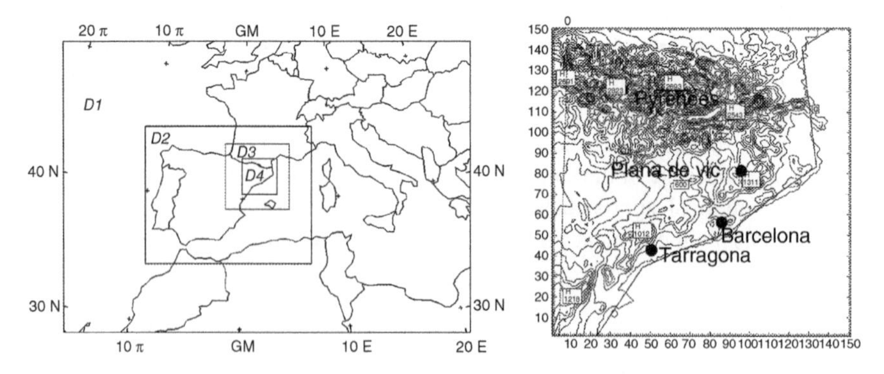

FIGURE 1. Four-nested domain configuration. D1: 35×50 72-km cells. D2: 61×49 24-km cells. D3: 93×93 6-km cells. D4: 136×136 2-km cells.

The high resolution (1-h and 1 km^2) EMICAT2000 emission model (Parra, 2004) has been applied in the NEIP. This emission model includes the emissions from vegetation, on-road traffic, industries, energy facilities and domestic-commercial solvent use. Biogenic emissions were estimated using a model that takes into account local vegetation data (land-use distribution and biomass factors) and meteorological conditions (surface air temperature and solar radiation) together with emission factors for native Mediterranean species and cultures (Parra *et al.*, 2004). On-road traffic emission includes hot exhaust, cold exhaust and evaporative emissions (Parra and Baldasano, 2004). Industrial emissions include real records of some chimneys connected to the emission control network of the Environmental Department of the Catalonia Government (Spain), and the estimated emissions from other industrial installations.

The chemical transport model used to compute the concentrations of photochemical pollutants was Models-3/CMAQ (Byun and Ching, 1999). The initial and boundary conditions were derived from a one-way nested simulation covering a domain of 1392 × 1104 km^2 (Domain 2) centered in the Iberian Peninsula, that used EMEP emissions corresponding to year 2000. A 48-hour spin-up was performed in order to minimize the effects of initial conditions. The chemical mechanism selected for simulations, after a comparison of the performance of diverse photochemical mechanisms (Jiménez et al., 2003) was CBM-IV, including aerosols and heterogeneous chemistry. Horizontal resolution considered was 2 km, and 16-sigma vertical layers cover the troposphere.

3. Modeling Case

Modeling was conducted for the photochemical pollution event in the NEIP that took place on August 13-16, 2000. Values over the European threshold of 180 $\mu g\ m^{-3}$ for ground-level O_3 are attained (189 $\mu g\ m^{-3}$ in the area of Vic). The domain of study covers a squared area of 272 × 272 km^2 centered in the NEIP. The complex configuration of the zone comes conditioned by the presence of the Pyrenees mountain range (altitudes over 3000m), the influence of the Mediterranean Sea and the large valley canalization of Ebro river. That produces a sharp gradient in the characteristics of the area.

This episode corresponds to a typical summertime low pressure gradient with high levels of photochemical pollutants over the Iberian Peninsula. The day was characterized by a weak synoptic forcing, so that mesoscale phenomena, induced by the particular geography of the region, would be dominant. This situation is associated with weak winds in the lower troposphere and high maximum temperatures. Under this weak synoptic forcing, strong insolation promotes the development of prevailing mesoscale flows associated with the local orography, while the difference of temperature between the sea and the land enhances the development of sea-land breezes (Barros *et al.*, 2003). Since similar behavior of pollutants were observed each day of the episode, representative results for August 14, 2000 are depicted.

4. Results

4.1. Transport of Pollutants

The Iberian Peninsula was dominated by the Azores anticyclone during the episode of August 13-16, 2000, with very low pressure gradients. The sea-breeze regime, developed within all the western Mediterranean coast, induced an anticyclonic circulation over the NEIP with compensatory subsidence over the region (Millán *et al.*, 1992).

Figure 2a shows the ozone patterns on August 14, 2000 at 0600 UTC. The maintenance of the anticyclonic circulation over the NEIP at this time from the previous day is remarkable. At night, the whole eastern Iberian coast presented down-slope winds over the mountains and general offshore breeze flows. The canalization between the Pyrenees and the Central Massif introduced northwestern flows into the Mediterranean. At low levels, this canalization plays an important role, because it is the only pass bringing fresh air into the WMB (Gangoiti *et al.*, 2001). The offshore flows produce a drainage of pollutants towards the coast through the river valleys. As the day advanced, a well developed sea-breeze regime established along all the domain with breeze circulation cells up to 2000 m height,

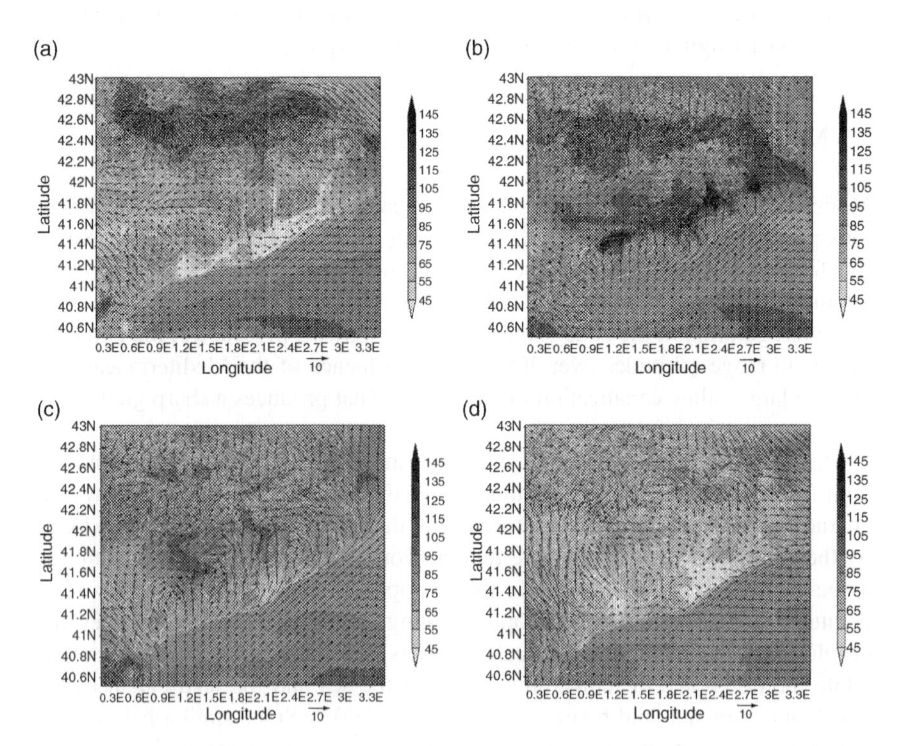

FIGURE 2. Wind field vectors and ozone concentrations (μg m^{-3}) over the northeastern Iberian Peninsula on August 14, 2000 at 0600 UTC (a), 1200 UTC (b), 1600 UTC (c) and 2000 UTC (d).

over the mixing height (800 m) (Sicard *et al.*, 2003). From 0800 UTC, the onshore winds are well developed along all the eastern coast, intensifying the anticyclonic circulation and deflecting to the east the flow between the Pyrenees and the Central Massif. Katabatic winds weaken and a clear breeze regime is observed, with the development of anabatic and valley winds. In the southern part of the domain, the intensity of the breeze is not capable to overcome the littoral mountain range, and therefore a local maximum downwind the industrial area of Tarragona appeared (Figure 3) before noon, with ozone levels above 160 $\mu g\ m^{-3}$.

At noon, pollutants departing from the Barcelona area, as the main emitter zone, and the line coast, are transported inland following the breeze front (Figure 2b), since they arrive at Plana de Vic (70-km downwind Barcelona), where calm winds prevail and allow ozone and its precursors to accumulate, meeting values over the European Information Threshold of 180 $\mu g\ m^{-3}$.

In this area, the flow is conditioned by the presence of the pre-Pyrenees and the northwesterly wind aloft, allowing the creation of re-circulation patterns that allow the accumulation of ozone. This pattern keeps constant during the afternoon (Figure 2c), but the breeze gains intensity and adds to the upslope winds transporting pollutants over the pre-littoral mountain ranges. Around 1900-2000 UTC (Figure 2d), the photochemical activity ceases, the sea-breeze regime loses intensity and winds in the coast weaken. Inland, in the east part of the domain, strong south winds are observed, which translates into a transport of ozone that adds to the quenching procedure of photochemical ozone by its reaction with nitrogen monoxide. At 2300 UTC, inland winds calm with the development of a weak land-breeze, with drainages in the valleys and katabatic winds.

4.2. *Regional Re-Circulation*

The strength of the sea breeze and the very complex orography of the eastern Iberian coast produce several injections due to orographic forcing. As the sea breeze front advances inland reaching different mountain ranges, orographic

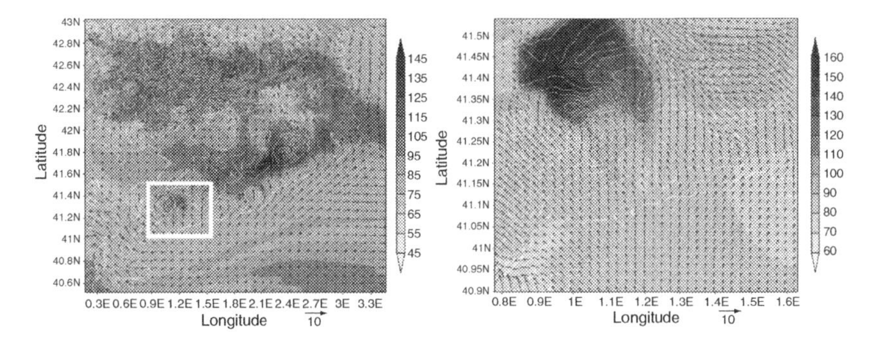

FIGURE 3. Wind field vectors and ozone concentrations ($\mu g\ m^{-3}$) over the northeastern Iberian Peninsula on August 14, 2000 at 1100 UTC, for the whole domain (left) and the industrial area of Tarragona (right).

injections occur at different altitudes. By early afternoon, the sea-breeze front has passed over the coastal mountain range and reached the central plain (~100km from the coast), overwhelming the upslope winds in the opposite direction (Jorba *et al.*, 2003). Strong thermally driven convections appear at the central plain injecting air masses up to 3000-4000 m (Figure 4). Once the air masses are injected at these altitudes, they are incorporated into the dominant synoptic flux and are transported towards the coast. Here, the compensatory subsidence of the Iberian thermal low incorporates the polluted air masses into the breeze cell, provoking a positive feedback. Once at low levels over the Mediterranean, some air masses may re-circulate over the sea with a possible return to the coast on the following days (Millán *et al.*, 1997). If the orographic injections are strong enough, part of these pollutants are injected over the mixing layer and extracted out from the domain because of the northwesterly winds aloft. Backtrajectory analysis by means of HYSPLIT trajectory model (Draxler *et al.*, 1998) from the MM5 high resolution domain has been applied in the study area

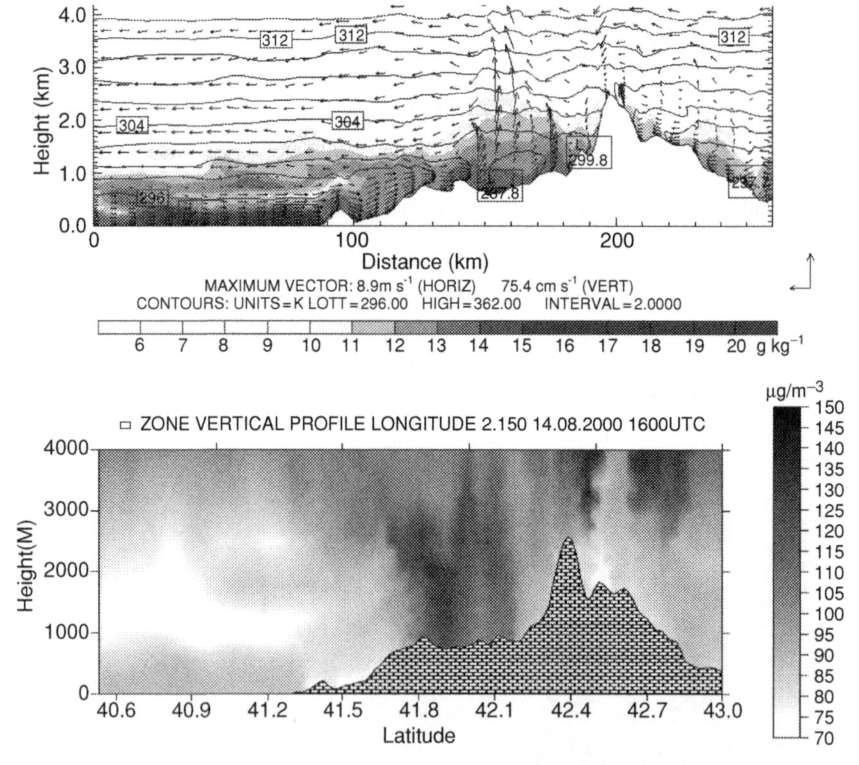

FIGURE 4. Vertical profile of the wind fields, potential temperature and mixing ratio for August 14, 2000 (up) at 1600 UTC; and thermal and mechanic orographic injections of ozone (µg m^{-3}) at 1600 UTC for a constant longitude of 2.150 E (profile Barcelona-Pyrenees).

in order to explain the main pathways of the polluted air masses over the city of Barcelona for August 14, 2000.

Figure 5 plots four representative backtrajectories arriving over the city of Barcelona at 1500 and 1900 UTC between 500 and 1500 m asl. Air injections are produced at the Garraf (littoral mountain) and the Montserrat (pre-littoral) mountain up to at least 1300 and 2100 m respectively with later return over Barcelona sinking by compensatory subsidence, re-circulating on the following days.

4.3. Model Evaluation

Ground-level O_3 simulations were compared to measurements from 48 surface stations in Catalonia, located in both urban and rural areas. The O_3 bias is negative on each day, ranging from −2.1% on the first day of simulation until −14.3% on August 15. That suggests a slight tendency towards underprediction; however, EPA goals of ± 15% are met (US EPA, 1991). This negative bias may suggest that the O_3-production chemistry may not be sufficiently reactive. The modeled episode peak (189 μg m^{-3}) is well-captured by the model. Peak accuracy is overestimated on the first and last day of simulations (14.4% and 5.2%, respectively) and underestimated on the central days of the episode (−3.8% and −11.7%). The mean normalized error increases from August 13 until August 16 (16.8% to 26.7%), mainly to deviations in meteorological predictions that enlarge with the time of simulation.

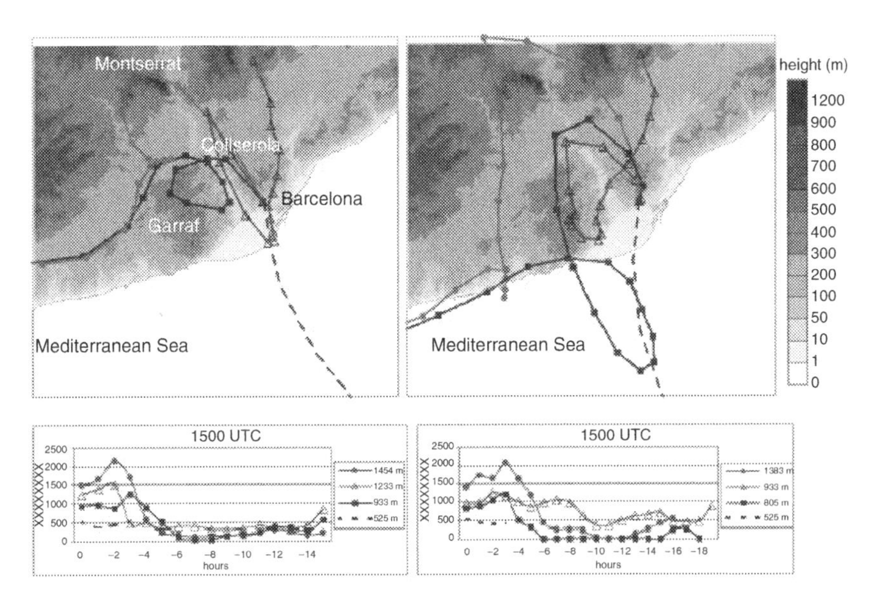

FIGURE 5. HYSPLIT kinematic back trajectories arriving over the city of Barcelona on August 14, 2000; (left) at 15 UTC and (right) at 19 UTC.

5. Conclusions

A high-resolution simulation with MM5-EMICAT2000-CMAQ air quality model was carried out in order to assess the dynamics of air pollutants over the NEIP for a representative summertime episode of photochemical pollution with stagnant synoptical conditions. Results revealed that the combination of mesoscale wind flows, emissions and very complex local orography is crucial in the production and transport of photochemical pollutants within the domain. Maximum ozone levels in the NEIP are located downwind from Barcelona (Vic area) and the industrial zone of Tarragona (1-hr peaks above 160 $\mu g\ m^{-3}$). The combination of sea breezes, upslope winds and important valley canalizations transports pollutants inland from coastal areas, following the advance of the sea-breeze front and causing high levels of photochemical pollutions.

Air masses departing mainly from the coast act as photochemical reactors as they move northeast with the sea breeze, whose direction is conditioned by the presence of an anticyclone over the Mediterranean. Simultaneously, due to orographic forcings, the sea breeze flow, reinforced by upslope winds, injects part of these pollutants in aloft returns flows that move back towards the sea. A fraction of the air masses is transported over the mixing layer out from the domain, extracting pollutants injected in a medium-long range transport depending on the upper flow over the region. The other fraction is forced to descend by subsidence over the east Iberian coast, where it is re-circulated and plays a fundamental role in the increment of levels of photochemical pollutants. Finally, this work depicts the circulation and formation pattern of ground-level ozone and its three-dimensional complexity in the area of study.

Acknowledgements This work was developed under the research contract REN2003-09753-C02 of the Spanish Ministry of Science and Technology. The Spanish Ministry of Education and Science is also thanked for the FPU doctoral fellowship hold by P. Jiménez. The authors gratefully acknowledge E. López for the implementation of EMICAT2000 into GIS.

References

Baldasano, J.M., Cremades, L., Soriano, C., 1994. Circulation of Air Pollutants over the Barcelona Geographical Area in Summer. Proceedings of Sixth European Symposium Physico-Chemical Behaviour of Atmospheric Pollutants. Varese (Italy), 18-22 October, 1993. Report EUR 15609/1: 474-479.

Barros, N., Toll, I., Soriano, C., Jiménez, P., Borrego, C., Baldasano, J.M., 2003. Urban Photochemical Pollution in the Iberian Peninsula: the Lisbon and Barcelona Airsheds. Journal of the Air & Waste Management Association, **53**, 347-359.

Byun, D.W., Ching, J.K.S. (Eds.), 1999. Science algorithms of the EPA Models-3 Community Multiscale Air Quality (CMAQ) Modeling System. EPA Report N.

EPA-600/R-99/030, Office of Research and Development. U.S. Environmental Protection Agency, Washington, DC.

Draxler, R.R., Hess, G.D., 1998. An overview of the Hysplit_4 modelling system for trajectories, dispersion, and deposition. Australian Meteorological Magazine 47, 295-308.

Gangoiti, G., Millán, M.M., Salvador, R., Mantilla, E., 2001. Long-range transport and re-circulation of pollutants in the western Mediterranean during the project regional cycles of air pollution in the west–central Mediterranean area. Atmospheric Environment 35, 6267-6276.

Jiménez, P., Dabdub, D., Baldasano, J.M., 2003. Comparison of photochemical mechanisms for air quality modeling. Atmospheric Environment, 37, 4179-4194.

Jorba, O., Gassó, S., Baldasano, J.M., 2003. Regional circulations within the Iberian Peninsula east coast. 26th Int. Tech. Meeting of NATO-CCMS on Air Pollution Modelling and its application, Istanbul, Turkey.

Millán, M.M., Artiñano, B., Alonso, L., Castro, M., Fernandez-Patier, R., Goberna, J., 1992. Mesometeorological cycles of air pollution in the Iberian Peninsula. Air Pollution Research Report 44. Commission of the European Communities. Brussels, Belgium. 219 pp.

Millán, M.M., Salvador, R., Mantilla, E., 1997. Photooxidant Dynamics in the Mediterranean Basin in Summer: Results from European Research Projects. Journal of Geophysical Research, 102 (D7), 8811-8823.

MMMD/NCAR, 2001. PSU/NCAR Mesoscale Modeling System Tutorial Class Notes and User's Guide: MM5 Modeling System Version 3.

Parra, R., 2004. Development of the EMICAT2000 model for the estimation of air pollutants emissions in Catalonia and its use in photochemical dispersion models. Ph.D. Dissertation, Polytechnic University of Catalonia (Spain).

Parra R., Baldasano, J.M., 2004. Modeling the on-road traffic emissions from Catalonia (Spain) for photochemical air pollution research. Weekday – weekend differences. In: 12th International Conference on Air Pollution (AP'2004), Rhodes (Greece).

Parra, R., Gassó, S., Baldasano, J.M., 2004. Estimating the biogenic emissions of non-methane volatile organic compounds from the North western Mediterranean vegetation of Catalonia, Spain. The Science of the Total Environment (In Press).

Pérez, C., Sicard, M., Jorba, O., Comerón, A., Baldasano, J.M., 2004. Summertime re-circulations of air pollutants over the north-eastern Iberian coast observed from systematic EARLINET lidar measurements in Barcelona. Atmospheric Environment (In Press).

Sicard M., Pérez C., Comerón A., Baldasano J.M, Rocadenbosch F., 2003. Determination of the Mixing Layer Height from Regular Lidar Measurements in the Barcelona Area. Remote Sensing of Clouds and the Atmosphere VIII (Editors: K.P. Schäfer *et al.*) Proc. Of SPIE, Vol. 5235: 505-516

Soriano, C., Baldasano, J.M., Buttler, W.T., Moore, K., 2001. Circulatory Patterns of Air Pollutants within the Barcelona Air Basin in a Summertime situation: Lidar and Numerical Approaches. Boundary-Layer Meteorology, **98** (1), 33-55.

US EPA, 1991. Guideline for Regulatory Application of the Urban Airshed Model. US EPA Report No. EPA-450/4-91-013. Office of Air and Radiation, Office of Air Quality Planning and Standards, Technical Support Division. Research Triangle Park, North Carolina, US.

Modeling Photochemical Pollution in the Northeastern Iberian Peninsula

Speaker: J. Baldasano

Questioner: S. Andreani Aksoyoglu
Question: Why is the predicted peak O_3 at 12:00? Can you explain why it is early?
Answer: It is not early, the units of time are indicated in UTC, two hours before in summer of the official time.

20 Modeling the Weekend Effect in the Northeastern Iberian Peninsula

Pedro Jiménez, René Parra, Santiago Gassó, and José M. Baldasano*

1. Introduction

The chemistry of ozone (O_3) and its two main precursors, nitrogen oxides (NO_x) and volatile organic compounds (VOCs) represents one of the major fields of uncertainty in atmospheric chemistry. The ozone weekend effect refers to a tendency in some areas for ozone concentrations to be higher on weekends compared to weekdays, despite emissions of VOCs and NO_x that are typically lower on weekends due to different anthropogenic activity. This phenomenon was first reported in the United States in the 1970s (Cleveland *et al.*, 1974; Lebron, 1975) and has been since reported mainly in the U.S. and Europe. Higher weekend ozone tends to be found in urban centers, while lower weekend ozone is found in downwind areas. Altshuler *et al.* (1995) have suggested that the weekend effect is related to whether ozone formation is VOC- or NO_x-sensitive, with higher weekend ozone occurring is VOC-sensitive areas. Despite this, there is a high uncertainty in the causes of the weekend effect, and six hypothesis have been set (CARB, 2003): (1) NO_x reduction; (2) NO_x timing; (3) carryover near the ground; (4) carryover aloft; (5) increased weekend emissions; (6) increased sunlight caused by decreased soot emissions.

Emission inventories for each day of the week are needed to help determine the causes of the ozone weekend effect. In this work, a day-specific hourly emissions inventory is used for stationary, area and on-road sources in order to help assess the ozone weekend effect observed in as complex terrain as the northeastern Peninsula. The hypothesis of changing mass and timing of emissions, ozone quenching and carryover are analyzed, discarding the hypothesis that have been proved not to have importance on the weekend effect, such as increased emissions or increased sunlight on weekends. An evaluation of the performance of the

* Pedro Jiménez, René Parra, Santiago Gassó and José M. Baldasano, Laboratory of Environmental Modeling, Department of Engineering Projects, Universitat Politècnica de Catalunya (UPC), SPAIN.

model is also considered, comparing results with data from air quality stations and analyzing the up-to-date hypothesis about the ozone weekend effect.

2. Modeling Case

Modeling was conducted for a photochemical pollution event in the Western Mediterranean Basin that took place on August 13-16, 2000. Two non-labor days (August 13 and 15) and two working days (August 14 and 16) are considered in order to evaluate the O_3 weekend effect. Values over the European threshold of 180 μg m^{-3} for ground-level O_3 are attained. The domain of study (Figure 1) covers a square area of 272 × 272 km^2 centered on the northeastern Iberian Peninsula. The very complex configuration of the region is conditioned by the presence of the Pyrenees mountain range (altitudes over 3000m), the influence of the Mediterranean Sea and the large valley canalization of Ebro river. That produces a sharp gradient in the characteristics of the area.

The weekend effect results reported here are for to the episode of August 13-16, 2000, which corresponds to a typical summertime low pressure gradient with high levels of photochemical pollutants over the Iberian Peninsula. This situation is related to a decrease in air quality. The day was characterized by a weak synoptic forcing, so that mesoscale phenomena, induced by the particular geography of the region, would be dominant. This situation is associated with weak winds in the lower troposphere, cloudless skies and high maximum temperatures. Under this weak synoptic forcing, strong insolation promotes the development of prevailing mesoscale flows associated with the local orography, while the difference of temperature between the sea and the land enhances the development of sea-land breezes (Barros *et al.*, 2003).

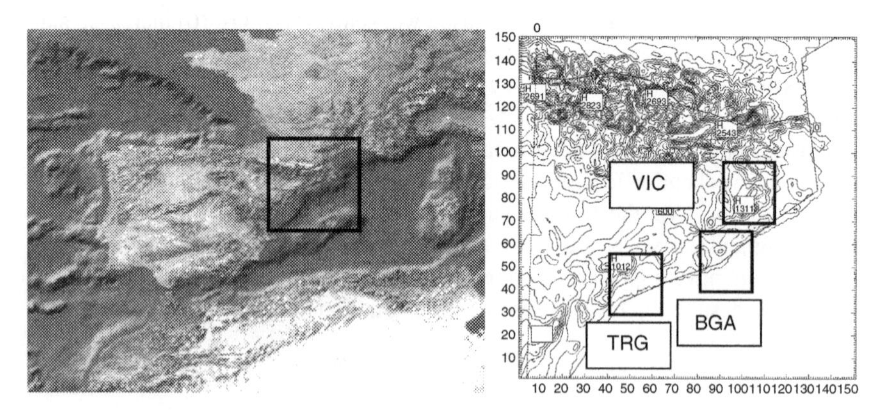

FIGURE 1. Geography of the areas where main O_3 problems in the northeastern Iberian Peninsula are located: Barcelona Geographical Area (BGA, urban), Plana de Vic (VIC, downwind) and Tarragona (TRG, industrial).

3. Models

MM5 numerical weather prediction model (MMMD/NCAR, 2001) provided the dynamic meteorological parameters. MM5 physical options used for the simulations were: Mellor-Yamada scheme as used in the Eta model for the PBL parameterization; Anthes-Kuo and Kain-Fritsch cumulus scheme; Dudhia simple ice moisture scheme, the cloud-radiation scheme, and the five-layer soil model. Initialization and boundary conditions for the mesoscale model were introduced with analysis data from the European Center of Medium-range Weather Forecasts global model (ECMWF).

The high resolution (1-h and 1 km^2) EMICAT2000 emission model has been applied in the northeastern Iberian Peninsula. This emission model includes the emissions from vegetation, on-road traffic, industries and emissions by fossil fuel consumption and domestic-commercial solvent use. Biogenic emissions were estimated using a model that takes into account local vegetation data (land-use distribution and biomass factors) and meteorological conditions (surface air temperature and solar radiation) together with emission factors for native Mediterranean species and cultures (Parra *et al.*, 2004). On-road traffic emission includes hot exhaust, cold exhaust and evaporative emissions (Parra and Baldasano, 2004). Industrial emissions include real records of some chimneys connected to the emission control web of the Environmental Department of the Catalonia Government (Spain), and the estimated emissions from other industrial installations.

The chemical transport model used to compute the concentrations of photochemical pollutants was Models-3/CMAQ (Byun and Ching, 1999). The initial and boundary conditions were derived from a one-way nested simulation covering a domain of 1392×1104 km^2 centered in the Iberian Peninsula, that used EMEP emissions corresponding to year 2000. A 48-hour spin-up was performed in order to minimize the effects of initial conditions. The chemical mechanism selected for simulations was CBM-IV (Gery *et al.*, 1989), including aerosols and heterogeneous chemistry. Horizontal resolution considered was 2 km, and 16-sigma vertical layers cover the troposphere.

4. Results

4.1. Weekday/Weekend Emissions within EMICAT2000

Existing gridded inventories used as input to air quality models typically assume weekday patterns and lack of accurate estimates of emissions on weekends (Marr and Harley, 2002). We used EMICAT2000 inventory emission model (Parra, 2004) that takes into account main weekday/weekend differences on ozone precursors emissions profiles due mainly to variations in on-road traffic emissions. Figure 2 shows some samples of hourly traffic profiles both for weekdays and weekends for highways and road stretches in the northeastern Iberian Peninsula.

Weekday profiles have higher percent usage about 0700 UTC and 1700 UTC. There are drops at midday and the lower percentages are present during night time and first hours of the early morning. Weekend profiles have similar shapes, but the higher values during the morning are displaced to 0900 UTC and the maximum values in afternoon are higher in relation to weekday values. Figure 2 also indicates weekday/weekend urban traffic profiles for Barcelona city. Weekend profiles have important traffic percentages during night time and the first hours of early morning. On weekends, there is an average 60% reduction of heavy-duty traffic in the northeastern Iberian Peninsula. Moreover, heavy-duty vehicles average 22% and 14% of traffic fleet on highways and roads respectively.

4.2. Evidence of the Ozone Weekend Effect

An analysis of ozone weekday/weekend differences was carried out by averaging concentrations in the whole domain of the northeastern Iberian Peninsula for August 13-15 (non-labor days) and 14-16 (weekdays). Ambient O_3 data indicates concentrations up to 189 μg m^{-3} on weekends and 177 μg m^{-3} during weekdays, both measured in the industrial zone of Alcover, downwind of the city of Tarragona. A significant weekend increase in ozone weekend concentrations is observed in urban areas of the domain (Barcelona and Tarragona, mainly), where peaks increase by over 30 μg m^{-3} (Figure 3). In the case of Barcelona city, simulations yield differences over 50 μg m^{-3} (increment of +66% on weekends). This value is supported by observations, where increments from 81 μg m^{-3} (weekdays) to 125 μg m^{-3} (weekends) are measured at air quality stations in the area (increment of +54% on weekends). This behavior is also stated for average daily values (Figure 3), where both simulations and observations provide growths of around 14 μg m^{-3} (+21%) for the ozone weekend effect.

On the other hand, areas downwind the city of Barcelona exhibit the inverse trend of the weekend effect. O_3 reductions of about 20 μg m^{-3} (−10%) on weekends are detected in peak-O_3 values in Vic for both measurement and simulations;

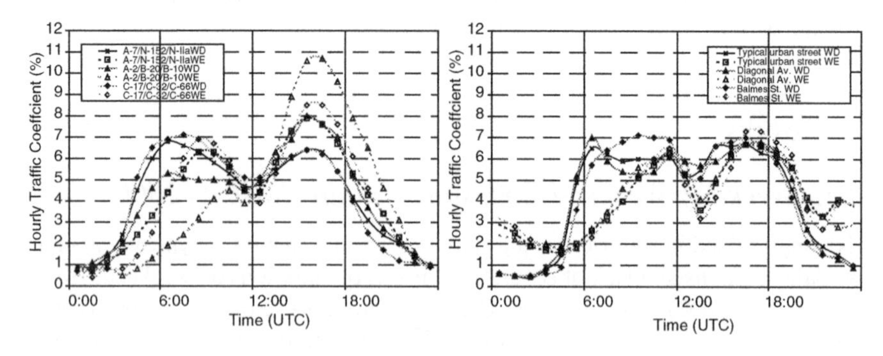

FIGURE 2. Weekday/weekend traffic profiles for (left) main roadways in the northeastern Iberian Peninsula; and (right) streets in the city of Barcelona (WD, weekdays, solid, black pointers; WE, weekends, dashed, white pointers).

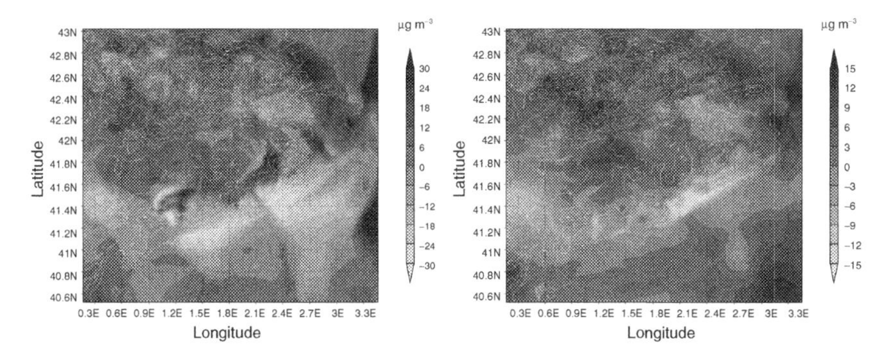

FIGURE 3. Differences in weekday/weekday ozone (µg m^{-3}) for (left) 1-hr daily maximum concentration; and (right) average daily values. Higher weekend values are observed in upwind areas (white) whereas a reverse weekend effect is observed downwind (black).

and this reduction is also observed for daily-average levels. Background air quality stations don't display a significant weekend effect since pollutants in these areas are consequence of medium-range transport. Several factors likely contribute to the lower weekend O_3 in downwind areas, including the upwind shift in O_3 peaks caused by reduced NO_x inhibition, and reduced O_3 production in the downwind areas in response to lower anthropogenic emissions. These effects can also be described in terms of the upwind areas being VOC-sensitive and the downwind areas being NO_x-sensitive.

4.3. *Mass and Timing of Emissions*

Reduction of heavy-duty traffic and hourly variations imply different profiles for O_3 precursor emissions on weekends. On weekdays, NO_x and VOCs emissions were 236 t d^{-1} and 172 t d^{-1}, respectively. On weekends, NO_x and VOCs emissions were 184 t d^{-1} and 179 t d^{-1}. On weekends, traffic from heavy-duty vehicles during all hours undergoes a substantial reduction (60% in average), and also variations of on-road traffic. They both imply differences in emission profiles. Total NO_x emission on weekends are 22% lower than weekdays. The evolution of ozone precursor emissions during the weekdays and weekends of August shows a bimodal profile for both NO_x and VOC (Figure 4).

On weekdays, NO_x peak occurs about 0700UTC and 1700 UTC, but on weekends the first peak is lower and displaced to the right (around 0900-1000UTC). Therefore, the timing of NO_x emitted on weekends causes the midday emissions to produce O_3 more efficiently compared with the NO_x emitted on weekdays. NO_x-reduction, in combination with the NO_x-timing, contribute to the ozone weekend effect. For VOCs, similar timing as in the case of NO_x is observed, but first peaks have equivalent magnitudes and the second peak on weekends is higher. Total VOC emissions on weekends are slightly higher (4%) than weekdays. This highly

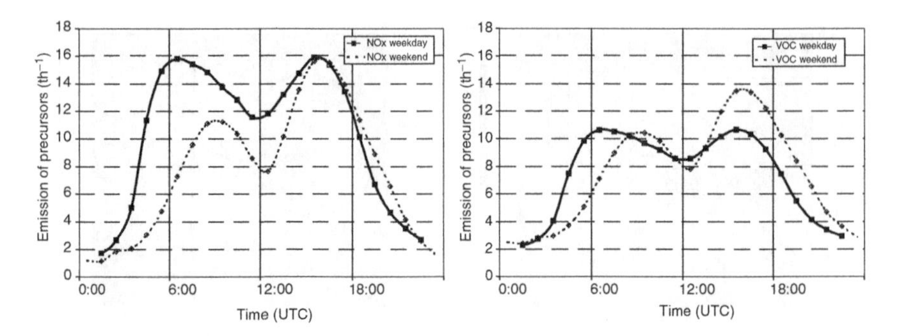

FIGURE 4. Hourly emission of (left) NO_x and (right) VOCs (both in t h^{-1}) on weekday (solid, squares) and weekend (dashed, diamonds) of August in the northeastern Iberian Peninsula.

influences the NO_x/VOCs ratio. On weekdays, this ratio reaches 2.0-2.7, meanwhile on weekends, lower NO_x emissions provide values for the considered ratio under 1.0. Values of NO_x/VOCs are 40% less on weekends.

The model simulations show that CO and NO_x ground-level concentrations on weekends are lower than on weekdays, as shown in (Figure 5) (–19% and –41%, respectively, for average values and –39% and –61% for 1-hr peaks) but the higher proportional reduction of NO_x makes ozone-forming photochemistry more active on weekends compared to weekdays (lower NO_x/VOC ratios). This pattern is also present in measured ambient concentrations of precursors. CO and NO_x are also significantly averagely reduced by –37% and –23%, respectively.

The cause of this phenomenon is that reductions in traffic on weekends, according to the emission model, imply higher NO_x-than VOC-reductions. This reduction of precursors is also observed downwind of the emitting sources. If we

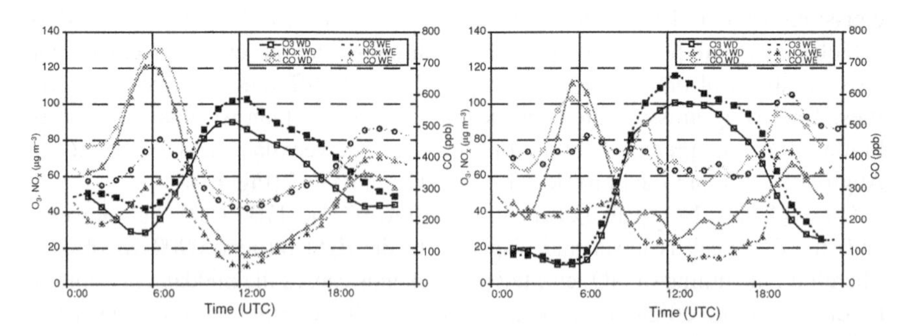

FIGURE 5. Averaged hourly weekday (solid) and weekend (dashed) ozone (black, squares) and precursors (NO_x, triangles, dark grey; CO, circles, light grey) for the northeastern Iberian Peninsula during the episode of August 13-16, 2000; (left) simulations; and (right) measurements.

take into account modeling results, ambient NO_x decreases on weekends are important in downwind areas (over –30%), but the influence of VOCs precursors (represented by CO) are not as important (reductions of –4% for simulated maximum and average levels).

4.4. *Ozone Quenching*

Emissions of O_3 precursors are greater during the morning on weekdays than on weekends, as stated before. The higher values of NO in NO_x emissions destroys (quenches) more of the available ground-level ozone in pervasive emission areas such as cities, according to the titration reaction in which NO and O_3 combines to produce NO_2 and oxygen ($NO + O_3 \rightarrow NO_2 + O_2$). When VOCs are present, they participate in chain reactions that convert NO to NO_2 without using up an O_3 molecule (Atkinson, 2000). Thus, the amount and kind of hydrocarbons determine the ratio of NO_2 to NO. Measured and simulated ground-level precursors indicate that ambient levels of NO_x are substantially lower on weekend mornings than on weekdays (around –40% both in upwind and downwind areas for average NO_x levels) and smaller reductions of VOCs levels on weekends (reductions of only –4% for average simulated CO levels and –2% for measured CO), and therefore the potential for ozone quenching considerably decreases since NO_x/VOC ratios are lower on weekends. The higher proportional presence of VOCs oxidizes NO to NO_2 and thus NO does not contribute to the ozone titration reaction. Furthermore, in the study case, both ambient data and simulations depict that the NO_2/NO relationship is higher for almost all daylight hours on weekends compared to weekdays at all upwind locations, so the amount of NO available for quenching ozone near the surface is smaller on weekends than on weekdays.

4.5. *Carryover Contribution to the Weekend Effect*

Increased VOC and NO_x emissions from traffic on weekend nights may carry over and lead to greater O_3 formation after sunrise on the following day (Yarwood *et al.*, 2003). The importance of carryover because spatial/temporal source-receptor relationships was investigated by changing weekend emissions into weekday emissions for the whole period of study. As derived from simulations, pollutant carryover is a negligible factor because differences in precursor concentrations during the carryover period have only a small effect on precursor concentrations and ratios during the O_3 accumulation period. Higher O_3 concentrations are depicted and noon in the case of considering weekend-weekday profiles instead of not considering weekend-weekday differences, since mass and timing of emissions have been changed. However, no differences are observed on labor-days, and that implies that increases in the O_3 effect are dominated by same-day emission changes. The additional nighttime emissions appear to be much less that the additional fresh emissions from traffic than occurs in the morning. Therefore, ozone precursors that carryover do not appear to be a significant cause in the study case for the northeastern Iberian Peninsula ozone weekend effect.

5. Conclusions

Day-of-week emission inventories are needed to support air quality models that simulate the ozone weekend effect. A day-specific hourly emissions inventory considering day-to-week variations in emissions that is used for stationary, area and on-road sources has been developed in the framework of the EMICAT2000 emission model. This emission model has been coupled with MM5-CMAQ to conduct a study of the weekend effect of ozone and its precursors with very high spatial resolution in as complex terrain as the northeastern Peninsula.

A significant weekend increase in ozone weekend concentrations is simulated in urban areas (as Barcelona city). On the other hand, areas downwind the city of Barcelona reduce or even reverse the weekend effect. Background air quality stations don't reflect a significant weekend effect since pollutants in these areas are the consequence of short to medium range transport. Several factors likely contribute to the lower weekend O_3 in downwind areas, including the upwind shift in O_3 peaks caused by reduced NO_x inhibition, and reduced O_3 production in the downwind areas in response to lower anthropogenic emissions. These effects can also be described in terms of the upwind areas being VOC-sensitive and the downwind areas being NO_x-sensitive. Reduction of heavy-duty traffic and hourly variations imply different profiles for O_3 precursor emissions on weekends. On weekends, traffic from heavy-duty vehicles undergoes a substantial reduction. The shift of 1-2 hours in the peaks of precursor emissions at weekends causes the midday emissions to produce O_3 more efficiently compared with the NO_x emitted on weekdays. Model simulations and air quality stations measurements for precursors depict that CO and NO_x ground-level concentrations on weekends are lower than those corresponding to weekdays. The higher proportional reduction of NO_x makes ozone-forming photochemistry more active on weekends compared to weekdays (lower NO_x/VOC ratios). Finally, pollutant carryover is a negligible factor because differences in precursor concentrations during the carryover period would have only a small effect on precursor concentrations and ratios during the O_3 accumulation period.

The fact that this modeling system performed well in describing the weekday/weekend differences in ozone levels helps supporting the use of this air quality model for future scientific and air quality planning applications in very complex terrains, since this approach provides a novel contribution to the analysis of the weekend effect in the Western Mediterranean Basin.

Acknowledgements This work was developed under the research contract REN2003-09753-C02 of the Spanish Ministry of Science and Technology. The Spanish Ministry of Education and Science is also thanked for the FPU doctoral fellowship hold by P. Jiménez. The authors gratefully acknowledge E. López for the implementation of EMICAT2000 into a GIS system. Air quality stations data and information for implementing industrial emissions were provided by the Environmental Department of the Catalonia Government (Spain).

References

Altshuler, S.L., Arcado, T.D., Lawson, D.R., 1995. Weekday vs. weekend ambient ozone concentrations: discussion and hypotheses with focus on Northern California. Journal of the Air & Waste Management Association, **45**, 967-972.

Atkinson, R., 2000. Atmospheric chemistry of VOCs and NO_x. Atmospheric Environment, **34**, 2063-2101.

Barros, N., Toll, I., Soriano, C., Jiménez, P., Borrego, C., Baldasano, J.M., 2003. Urban Photochemical Pollution in the Iberian Peninsula: the Lisbon and Barcelona Airsheds. Journal of the Air & Waste Management Association, **53**, 347-359.

Byun, D.W., Ching, J.K.S. (Eds.), 1999. Science algorithms of the EPA Models-3 Community Multiscale Air Quality (CMAQ) Modeling System. EPA Report N. EPA-600/R-99/030, Office of Research and Development. U.S. Environmental Protection Agency, Washington, DC.

CARB, 2003. The ozone weekend effect in California. Staff Report. The Planning and Technical Support Division. The Research Division. California Air Resources Board, Sacramento, CA, May 30.

Cleveland, W.S., Graedel, T.E., Kleiner, B., Warner, J.L., 1974. Sunday and workday variations in photochemical air pollutants in New Jersey and New York. Science, **186**, 1037-1038.

Gery, M.W., Whitten, G.Z., Killus, J.P., Dodge, M.C., 1989. A photochemical kinetics mechanism for urban and regional scale computer modeling. Journal of Geophysical Research, **94** (D10), 12925-12956.

Lebron, F., 1975. A comparison of weekend-weekday ozone and hydrocarbon concentrations in the Baltimore-Washington metropolitan area. Atmospheric Environment, **9**, 861-863.

Marr, L.C., Harley, R.A., 2002. Modeling the effect of weekday-weekend differences in motor vehicle emissions of photochemical air pollution in central California. Environmental Science & Technology, **36**, 4099-4106.

MMMD/NCAR, 2001. PSU/NCAR Mesoscale Modeling System Tutorial Class Notes and User's Guide: MM5 Modeling System Version 3.

Parra, R., 2004. Development of the EMICAT2000 model for the estimation of air pollutants emissions in Catalonia and its use in photochemical dispersion models. Ph.D. Dissertation (in Spanish), Polytechnic University of Catalonia (Spain).

Parra R., Baldasano, J.M., 2004. Modeling the on-road traffic emissions from Catalonia (Spain) for photochemical air pollution research. Weekday – weekend differences. In: 12th International Conference on Air Pollution (AP'2004), Rhodes (Greece).

Parra, R., Gassó, S., Baldasano, J.M., 2004. Estimating the biogenic emissions of non-methane volatile organic compounds from the North western Mediterranean vegetation of Catalonia, Spain. The Science of the Total Environment (In Press).

Yarwood, G., Stoeckenius, T.E., Dunker, A.M., 2003. Modeling weekday/ weekend ozone differences in the Los Angeles region for 1997. Journal of the Air & Waste Management Association, **53**, 864-875.

Modelling Weekend Effect in Northeastern Iberia

Speaker: J. M Baldasano

Questioner: D. Steyn

Question: It seems that weekend/weekday changes in emissions are a "natural experiment" we could exploit to understand the effect of mandated emissions changes. Do you think this approach could be useful?

Answer: Yes, because the conditions considered and the detailed emission inventory allow us adopting an episodic methodology to evaluate this type of behavior (difference between weekend/weekday ozone production).

Questioner: R. Bornstein

Question: Could you comment on the Monday effect that you have found?

Answer: The effect observed for Monday ozone responds to the particular behavior for a labor day. The data used for modeling depict a typical pattern for labor days according to the high-resolution emission inventory considered.

Questioner: C. Hogrefe

Question: Are the meteorological conditions over the four days you modeled sufficiently similar to gain confidence that the ozone response is due to the change in emissions rather than fluctuations in meteorology?

Answer: Yes, the conditions for the episode of 13-16 August, 2000 correspond to stagnant meteorological conditions that are dominated by mesoscale processes that are maintained during all the days in the episode.

Questioner: S. T. Rao

Question: Your diurnal profile of emissions indicates that morning time emissions on weekend are injected 2-3 hours later in the morning than on weekdays. Hence, the PBL heights at the time of emissions loading will be quite different for weekday vs. weekend. So, we should expect emissions to be diluted first and allowed to chemically interact on weekends when compared to weekdays. Have you looked into the physical vs. chemical processes' role in dictating the peak O_3 concentration on weekday vs. weekend? Also it is important to consider transport aloft and how O_3 levels on the following day are affected as O_3 and precursors trapped aloft get mixed down as the PBL grows. That is, the multi-day character of O_3 needs to be examined over a larger domain than just looking at O_3 in an urban area in evaluation of the weekday vs. weekend effect. Finally, how can we be sure that emissions alone, not meteorological differences, were the primary reasons for the observed weekday vs. weekend peak O_3 levels.

Answer: The PBL height for the different days of the episode present a similar behavior. In addition, the influence of ozone and precursors trapped aloft has been analyzed through studying the carryover. Results show that the influence of carryover (both on surface and aloft) is negligible for the episode studied, as was also found in other works.

21
Transport and Deposition Patterns of Ozone and Aerosols in the Mediterranean Region

George Kallos[a], Marina Astitha[a], Flora Gofa[a], Michael O'Connor[a], Nikos Mihalopoulos[b], and Zahari Zlatev[c]

1. Introduction

The climatic conditions in the Greater Mediterranean Region (GMR) are known to have significant regional scale characteristics capable of long-range transport. The climatic patterns and the physiographic characteristics of the Mediterranean Region, forces the air quality in the area to exhibit remarkable spatiotemporal variability. In addition, concentrations of various pollutants (primary and/or secondary) are found to be significant in remote locations as well as in multiple-layer structures up to a few kilometers above the surface. For the GMR, besides the production, the term of transport of tropospheric ozone and its precursors should be of great interest, as well as the role of ozone in the production of several other pollutants, such as mercury. During the last years, a great number of studies also focus on the important role of aerosols in the air quality of a specific area, due to the potential impact on human health and ecosystems. Desert dust is one of the crucial components that contribute to the air quality degradation of the GMR. Due to these facts the aerosol concentration and deposition patterns are of great interest, along with ozone and its precursors, especially efforts for predicting air quality degradation episodes.

In the present work an attempt was made to analyze and study the paths and scales of transport and deposition of ozone and aerosols in the Mediterranean Region, due to both natural and anthropogenic influence. Levels of atmospheric gases and aerosol are monitored in ambient air quality networks because of their potential impact on human health, visibility and climate. As it is suggested in

[a] University of Athens, School of Physics, Atmospheric Modeling and Weather Forecasting Group, University Campus, Bldg. PHYS-V, Athens 15784, Greece. Corresponding author: George Kallos, *kallos@mg.uoa.gr*

[b] Environmental Chemical Process Laboratory, Department of Chemistry, University of Crete, P.O Box 1470, 71409, Heraklion, Greece

[c] National Environmental Research Institute, Department for Atmospheric Environment, Roskilde, Denmark

previous studies (Kallos et al, 1995, Millan et al, 1997, Kallos et al, 1999), the synoptic/regional circulation during summer, favors the long-range transport of air pollutants released from Southern and Eastern Europe and Central Mediterranean towards the Eastern Mediterranean, North Africa and Middle East. In order to identify the spatiotemporal patterns and scales of anthropogenic and natural pollutants in the Mediterranean Region, the tools used are advanced modeling systems and quantitative measurements from stations in the region.

Several studies in the past have identified the paths and scales of transport of gases such as ozone and NOx in the atmosphere. A large amount of anthropogenically-generated gases and aerosols, such as sulfur dioxide and sulfates, contribute to the increased concentration of pollutants in the atmosphere and to the reduction of visibility. This is attributed mainly to fine particles, which are capable of long range transport, influencing the air quality in remote locations. In addition to that, natural aerosols, having as main source of origin the Sahara desert, contribute significantly to the increase of particle concentrations in the atmosphere. The Sahara is the largest world desert and Europe is frequently exposed to large amounts of dust generated in intense dust storms. Therefore dust particles affect the air quality of a specific region mainly as episodic phenomena.

A number of studies have focused on the long range transport of air masses over Western and Eastern North Atlantic areas (Prospero et al, 2001), South America and the Mediterranean (di Sarra et al, 2001, Rodriguez et al, 2001). Several studies in Europe and other parts of the world suggest that fine desert particles of the size around 2.5μm are a considerable portion of the entire dust production and can travel thousands of kilometers affecting remote locations (Prospero et al, 2001, Uno et al, 2001).

The work presented here, is a part of a larger effort devoted to contribute to a better understanding of the convoluted effects of both the anthropogenic and naturally produced air pollution and therefore the air quality degradation at various scales. It is focused on the transport and deposition patterns of ozone and aerosols over Southern Europe and Mediterranean. In addition, we examine naturally produced aerosols (Saharan dust) convoluted to the above, with the aid of advanced atmospheric and photochemical models.

2. Model Description

A short description of the modeling systems used for performing simulations is provided below.

The SKIRON/ETA is a modeling system developed at the University of Athens from the Atmospheric Modeling and Weather Forecasting Group (Kallos et al, 1997, Nickovic et al, 2001). It has enhanced capabilities with the unique one to simulate the dust cycle (uptake, transport, deposition).

RAMS (Regional Atmospheric Modeling System) is considered as one of the most advanced atmospheric models (Cotton et al, 2003).

The Comprehensive Air Quality Model with Extensions (CAMx) (Environ, 2003) is an Eulerian photochemical model that allows for integrated assessment of air-pollution over many scales ranging from urban to super-regional (http://www.camx.com). CAMx has also model structures for modeling aerosols, processes that are linked to the CB4 gas phase chemical mechanism. New science modules are introduced for aqueous chemistry (RADM-AQ) inorganic aerosol thermodynamics/partitioning (ISORROPIA) and secondary organic aerosol formation/partitioning (SOAP).

3. Results and Discussion

3.1. Transport and Transformation of Air Pollutants in the Mediterranean Region

The paths and scales of transport and transformation of air pollutants in the Mediterranean Region have been identified in previous work carried out at the framework of various EU projects (Kallos et al, 1997, 1999, Millan et al, 1997). The results showed that the synoptic/regional circulation during summer, favors long-range transport of air pollutants released from Southern and Eastern Europe and Central Mediterranean towards the Eastern Mediterranean, North Africa and Middle East (Figure 1). Aerosols of fine mode behave as gases in the atmosphere (weak gravitational settling). Therefore fine particles exhibit a long range transport pattern similar to that of gaseous pollutants such as O_3 and NOx.

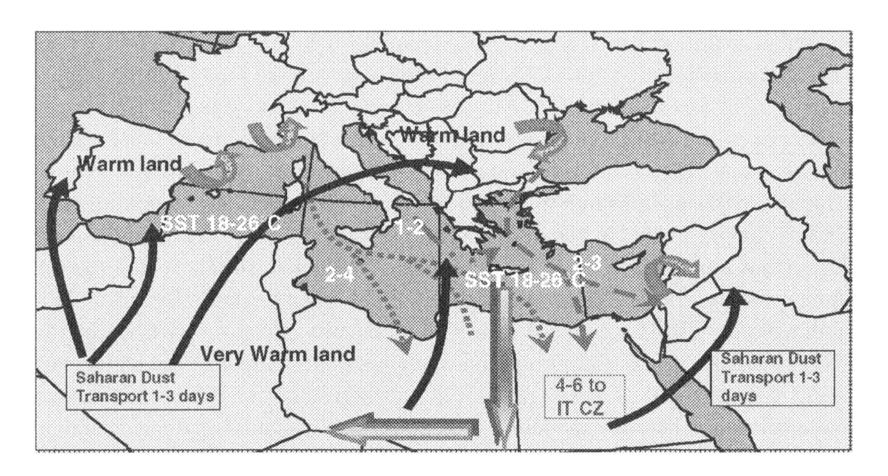

FIGURE 1. Characteristic paths and scales of transport of air masses in the Mediterranean Region.

3.2. Anthropogenic Gases and Aerosols

The atmospheric/photochemical model simulations were performed with the aid of the RAMS and SKIRON/ETA modeling systems in combination with CAMx photochemical model. The most recent version of CAMx (v4.03) was implemented specifically for the purpose of this work, including the newest modules for gas to particle and aqueous phase mechanisms (ISORROPIA).

A number of studies in the past have focused on the identification of the paths and scales of transport and transformation of ozone in the troposphere. The work presented here, focuses on aerosol formation, transport and deposition and the interaction between aerosols and gases in the atmosphere. Emphasis is given to particulate sulfate production and transport. In order to identify the paths and transformation of SO_2 to particulate sulphates, the sulfate ratio was calculated within the code of CAMx model. Sulfate ratio has been used in previous studies (Luria et al, 1996) in order to define the chemical age of air masses, based on measurements of sulfur dioxide and particulate sulfate. Sulfate ratio is characterized as the ratio of sulfate concentration to total sulfur concentration (meaning both SO_2 and particulate sulfate), leading to a dimensionless value from zero to unity. According to Luria (1996), the higher values for sulfate ratio (greater than 0.1) correspond to aged air masses, and the closer the ratio is to unity, the older the air mass and the longer its travel distance. CAMx model code was modified in order to calculate an average sulfate ratio for each hour of the simulation.

The sensitivity tests followed in this work had a similar base case: zero values for every source of SO_2 emissions in the region, zero initial and boundary conditions. Only one point source is implemented in the domain (located near Athens, Greece), emitting SO_2 for only one hour. The preliminary results of this work are presented below.

The sensitivity tests are divided in two categories: a) Injection of SO_2 inside the Marine Boundary Layer (MBL), using a point source in a coastal area with releases near the surface (Figure 2). b) Injection of SO_2 in the free troposphere, using the same location for the point source. The release of SO_2 in the atmosphere, in both cases, occurs for 1h at 00UTC and as a second test, at 12UTC.

The injection of SO_2 emissions (one hour) in the atmosphere of the Eastern Mediterranean Region showed some interesting results. When SO_2 is emitted inside the MBL, as in Figure 2, sulfate production is rather rapid due to the high moisture environment. When SO_2 is emitted in the MBL during night hours (Figure 2a), a major part of SO_2 converts into particulate sulfate within the first 5 hours (that is when sulfate ratio reaches unity). When the emission occurs during day hours (12UTC), a major part of SO_2 converts into sulfate within the first 3 hours (Figure 2b). Both SO_2 and particulate sulfate remain in the atmosphere for 2-3 days, until they have reached the northern coast of Africa and Eastern Mediterranean, where the modeling domain reaches its end. In all the sensitivity tests, the existence of islands in the area contribute significantly in the perturbation and mixing of the air masses. By the time the air masses reach Crete they are considered aged enough, and due to the long atmospheric lifetime of

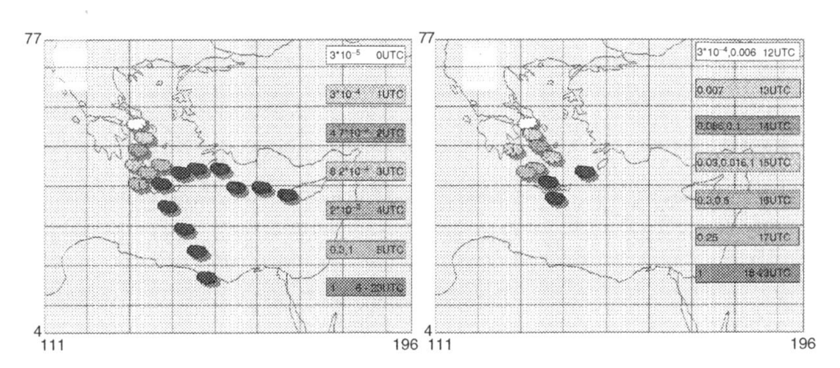

FIGURE 2. Sulfate Ratio for August 1, 2001; Layer 1: 0-50m. Point Source at Lavrio, emissions inside the MBL (a) SO2 emitted for 1h (00-01UTC). (b) SO2 emitted for 1h (12-13UTC).

particulate sulfate the air masses continue their course towards the North African Coast and Middle East, according to the synoptic circulation of this particular time of the year.

After performing the above sensitivity tests, a series of simulations with a state of the art emission inventory for the entire domain, were performed. The emission inventory used is a refined inventory based on EMEP data, with resolution of $16 \times 16 km^2$ (Z. Zlatev, personal contact). The results from this simulation, for the period of 1 to 10 August, 2001 were verified with measurements at Finokalia station, Crete. The inter-comparison of model results with observations showed, in general, good agreement after the second day of the simulation, for both particulate sulfate (Figure 3a) and sulfate ratio (Figure 3b). The differences between observed and modeled values must be analyzed having in mind that the simulation is of regional scale and the comparison between in situ measurements and regional modeling data exhibits certain differences.

An indication of the deposition pattern of particulate sulfate in the greater Mediterranean Region during summer is given in Figure 4, as calculated by the CAMx model (with the detailed emission inventory discussed earlier). The simulations performed showed that deposition patterns exhibit remarkable spatiotemporal variability due to the multi-scale weather circulations and the persistence of certain weather types that dominate in the Region.

3.3. Natural Aerosols

In addition to the anthropogenic produced aerosols, such as sulphates and/or nitrates, desert dust contributes significantly to air quality degradation, due to the episodic character of increased desert dust concentrations. In general, air pollution episodes due to anthropogenic activities occur together with desert dust transport episodes. This is due to the fact that when synoptic conditions favorable to the dust transport (ahead of a trough or behind an anticyclone in the Mediterranean Region), are associated with stable atmospheric conditions and stagnation (transport of warm

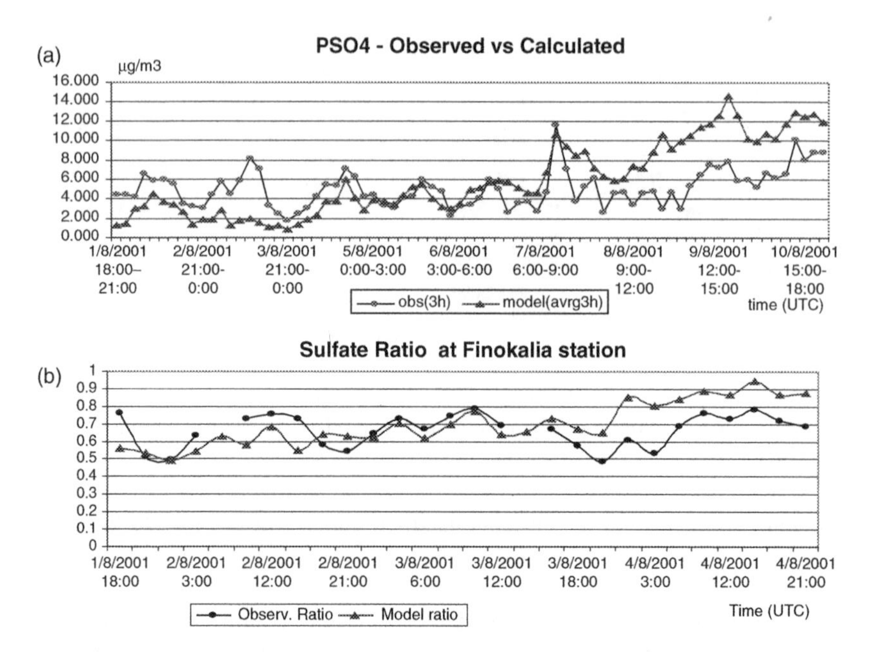

FIGURE 3. Intercomparison between measurements and model calculations of: (a) particulate sulphate, and (b) sulphate ratio, at Finokalia station, Crete, for August 2001. The discrepancies for the first day of the simulation occurred due to the cold start of the photochemical model.

air masses aloft that suppress vertical developments like updraft and convergence zones). The Saharan dust amounts over Mediterranean and Europe on seasonal timescale is examined with the aid of the most recent dust modeling system SKIRON. Since January 2000, the SKIRON/ ETA model runs operationally cover-

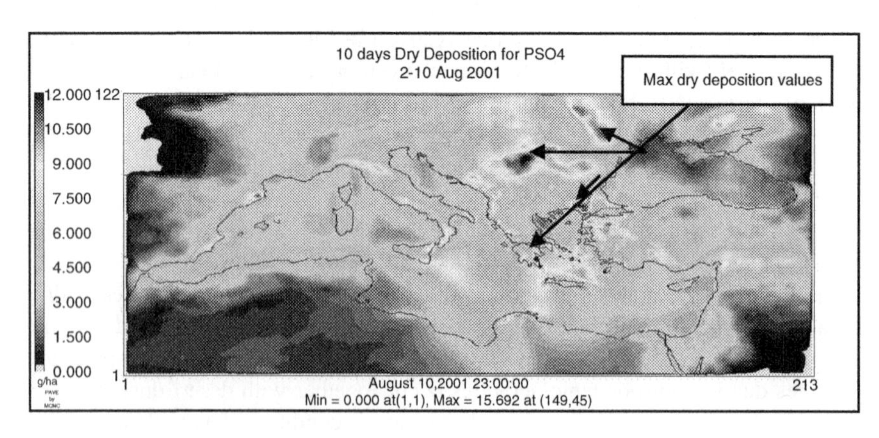

FIGURE 4. Dry deposition for particulate sulphate for a period of 10days, in August 2001, as calculated by CAMx model.

ing the Mediterranean Region, providing 3-day forecasts of dust load and deposition (http://forecast.uoa.gr), among other meteorological parameters. Using the data available, a database of model-derived seasonal amounts of dust deposited on Mediterranean Sea and Europe has been created.

The amount of Saharan dust deposited on the Mediterranean waters or over the European land exhibits significant seasonal and inter-annual variability (Papadopoulos et al, 2003), having a rather episodic character (Figures 5, 6, 7). The strength and the frequency of occurrence of the Saharan dust episodes define the annual deposition amounts and patterns of aerosols to a high degree, alternating

Start date	End date	Saharan dust wet deposition (mg)
10/01/02	14/01/02	0
14/01/02	16/01/02	0
16/01/02	22/01/02	0
22/01/02	30/01/02	0.4
30/01/02	11/02/02	0.6
11/02/02	13/02/02	0
13/02/02	26/02/02	1.3
21/03/02	27/03/02	1.1
27/03/02	03/04/02	0
03/04/02	09/04/02	5.5
09/04/02	***16/04/02***	***179.6***
16/04/02	19/04/02	5.5
	26/05/02	0.4

University of Athens (AM&WFG) SKIRON Forecast
6-h dust deposition (wet) in mg/m^2 (last 6 h) 15.04.02 at 12 UTC

1. 11. 31. 61. 101. 151. 211. 261. 361. 461. 561. 681. 781. 811. 1051. 1201. 1381. 1531. 1711.

FIGURE 5. Desert Dust Episode for April, 2002: (a) as measured at Finokalia station, Crete and (b) as simulated by SKIRON/ETA model. The arrows denote the value of dust deposition in the area of Finokalia station.

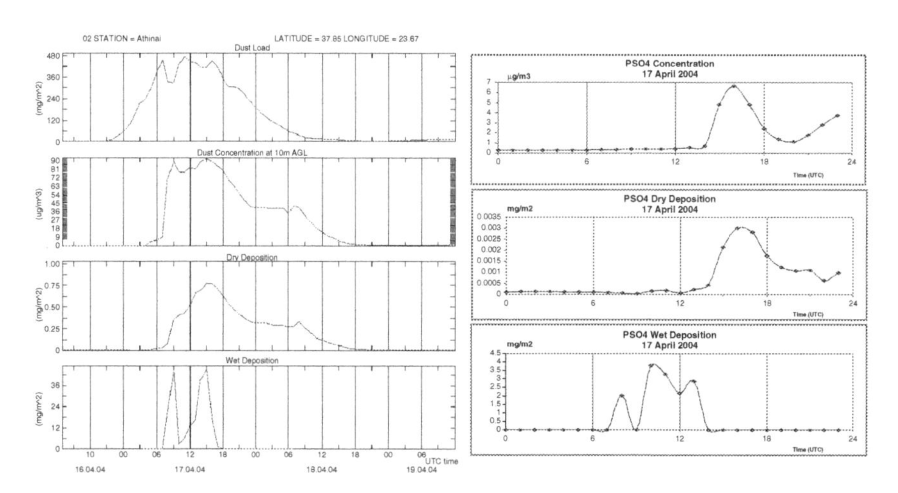

FIGURE 6. (a) (left) Desert Dust episode for April 17, 2004, for the city of Athens, as simulated by SKIRON/ETA model. (b) (right) Particulate sulphate concentration, dry and wet deposition, for April 17, 2004, for the city of Athens, as calculated by CAMx model.

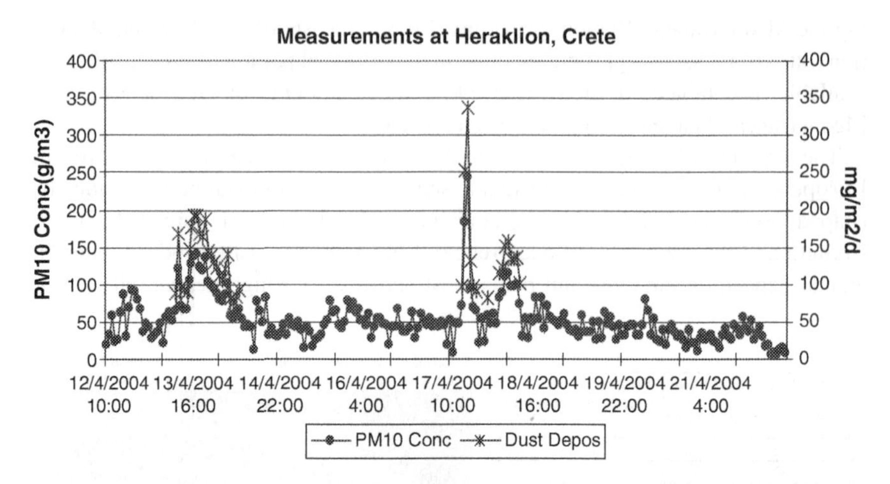

FIGURE 7. Measurements of PM10 concentration and Dust deposition at Heraklion, Crete, during April, 2004. (Dust content in such event is calculated as 80% of the PM10, for this location in the Mediterranean region (Mihalopoulos, personal contact).

the mean annual values. This leads to the fact that long-term modeling and measurement data are essential in understanding the synergetic effects of sulphates and desert dust in the atmosphere of the Mediterranean Region.

4. Conclusions

In this work an attempt was made to identify the transport and deposition patterns of aerosols in the area of the greater Mediterranean Region, focusing on Eastern Mediterranean. This was accomplished with the implication of model simulation in conjunction with measurements in several locations in the area. The remarks concluded from this work are summarized as follows:

The sulfate ratio sensitivity tests performed in this work showed results similar to those of Luria (1996), confirming the long range transport paths of sulfur towards the Middle East coast, during summer. Usually, high concentrations of sulfate, nitrate and other particles of anthropogenic origin, are associated with transport of desert dust due to the formation of stable atmospheric conditions. Deposition patterns over the Mediterranean Region exhibit remarkable spatiotemporal variability. This variability is mainly due to the multi-scale weather circulations that dominate in the Region. The characteristic paths and scales of transport and transformation already discussed define the deposition patterns in the area. To our experience gained from the performance of simulations, it seems that we cannot easily state that over one specific region we have higher amounts of deposition. Deposition amounts for each sub-region depend on source location, paths, mesoscale or regional flow patterns as well as surface characteristics. (e.g The islands in the Aegean Sea act as heating surfaces). There are indications that the multi-scale transport and transformation processes might have significant climatic impacts. More specifically, there can be effects on rain and therefore the water balance in a region where the water budget is critical. This is possible through the increase of the number of CCN and through the direct warming of the lower tropospheric layers (up to about 3 km) without an increase in the specific humidity. Of course, these processes are further more complicated because of the appearance of desert dust particles in the atmosphere which, in a wet environment, should be coated by sulphates and on that way they become very effective CCN.

Acknowledgements This work was supported by the following projects: ADIOS, EU/DG-XII: EVK3-CT-2000-00035, MERCYMS, EU/DG EVK3-2002-00070.

References

Cotton, W.R.; Pielke Sr., R.A.; Walko, R.L.; Liston, G.E.; Tremback, C.J.; Jiang, H.; McAnelly, R.L.; Harrington, J.Y.; Nicholls, M.E.; Carrio, G.G.; McFadden, J.P., 2003, "*RAMS 2001: Current Status and Future Directions, Meteorology and Atmospheric Physics*" (Volume 82 Issue 1-4).

di Sarra, A., T. Di Iorio, M. Cacciani, G. Fiocco, and D. Fuà, 2001. Saharan dust profiles measured by lidar from Lampedusa. J. Geophys. Res., 106, 10, 335-10, 347.

Environ 2003: User's Guide to the Comprehensive Air Quality Model with Extensions (CAMx). Version 4.00. Prepared by ENVIRON International Corporation, Novato, CA.

Kallos G., V. Kotroni, K. Lagouvardos, M. Varinou, A. Papadopoulos, 1995: "Possible mechanisms for long range transport in the eastern Mediterranean". 21st NATO/CCMS Int. Techn. Meeting on Air Pollution Modeling and Its Application, 6-10 Nov., Baltimore, USA. Plenum Press, N.York, Vol 21, pp.99-107.

Kallos, G., V., Kotroni, K., Lagouvardos, M., Varinou, M., Uliasz, A., Papadopoulos 1997: Transport and Transformation of air pollutants from Europe to East Mediterranean Region (T-TRAPEM). Final Report. Athens, Greece, pp.298.

Kallos, G., 1997. The regional weather forecasting system SKIRON: an overview. Proceedings of the symposium on regional weather prediction on parallel computer environments, University of Athens, Greece, pp. 109-122.

Kallos, G., V. Kotroni, K. Lagouvardos, and A. Papadopoulos, 1999: On the transport of air pollutants from Europe to North Africa. *Geophysical Research Letters*. 25, No 5, 619-622.

Luria M., M. Peleg, G. Sharf, D. Siman Tov-Alper, N. Schpitz, Y. Ben Ami, Z. Gawi, B. Lifschitz, A. Yitzchaki, and I. Seter,1996: *Atmospheric Sulphur over the East Mediterranean region*. JGR, 101, (25917).

Millan M.M., Salvador R. Mantilla E. and Kallos G., 1997: "Photo oxidant Dynamics in the Mediterranean Basin in Summer: Results from European Research Projects" JGR - Atmospheres, 102, D7, 8811-8823.

Nickovic, S., G. Kallos, A. Papadopoulos and O. Kakaliagou, 2001: A model for prediction of desert dust cycle in the atmosphere. J. Geophysical Res., Vol. 106, D16, 18113-18129.

Papadopoulos A., P. Katsafados, G. Kallos and S. Nickovic, S. Rodriguez, X. Querol, 2003. Contribution of Desert Dust Transport to Air Quality Degradation of Urban

Environments, Recent Model Developments. 26th NATO/CCMS ITM on Air Pollution Modeling and its Application, Istanbul, Turkey. Proceedings.
Prospero, J. M., I. Olmez and M. Ames. "Al and Fe in PM 2.5 and PM 10 suspended particles in South-Central Florida: The impact of the long range transport of African mineral dust." *Water, Air, and Soil Pollution, 125*, 291-317, 2001.
Rodriguez S., Querol X., Alastuey A., Kallos G. and Kakaliagou O., 2001. Saharan dust inputs to suspended particles time series (PM10 and TSP) in Southern and Eastern Spain. Atm. Env. 35/14, 2433-2447.
Uno I., Amano H., Emori S., Kinoshita K., Matsui I., Sugimoto N., 2001. Trans-Pacific yellow sand transport observed in April 1998: a numerical simulation. JGR, Vol106, D16, 18331-18344.

Transport and Deposition Patterns of Ozone and Aerosols in the Mediterranean Region

Speaker: G. Kallos

Questioner: D. Anfossi
Question: You showed us some figures representing the paths of particles emitted by biomass burning in Africa or in the Indian Ocean. A significant percentage of these particles appear to be at very high levels > 10,000 m. Are you sure that this is realistic or can it be an artifact of the model, which could not be able to stop the rising of particles, due to a possible problem with the turbulence parameterization? at those high levels in the atmosphere?
Answer: It is true that the released particles are relatively high in the troposphere. Although, we must keep in mind that our simulations are during sumer in the equatorial zone where deep convection takes place and the boundary layer depth during day-hours extends up to 4 km as it has been found in previous work. Therefore, the transfer of particles to high layers is something we wanted to show since it is an important mechanism to transfer pollutants at intercontinental scale.

Questioner: Ann-Lise Norman
Question: What size fraction do primary aerosols (eg in Saharan dust) fall into?
Answer: Most of the Saharan dust particles that are transported to long distances (thousands of km) are mainly of the order of PM2.5. Larger particles (of the order of PM10) are transferred also in long distances and fall on the Mediterranean waters. Larger than PM10 particles are falling quickly near their origin.

22
On the Formulation and Implementation of Urban Boundary Conditions for Regional Models

Clemens Mensink*

1. Introduction

"How can we assimilate detailed information from the urban canopy into regional models?" That is the question to be addressed in this contribution. In regional air quality models the exchange of pollutants between the surface and the atmosphere is generally represented by an emission source term and a deposition term. It is common practise that emissions generated at the surface are instantaneously diffused into the grid cells. Many of the dynamics of the urban canopy are not represented in regional models despite the fact that we have a lot of detailed information about these dynamics, e.g. on the canopy structure and roughness elements, on traffic emissions and on street canyon dynamics (Vardoulakis et al., 2003). In particular road transport emissions can be described hourly as a function of the road type, vehicle type, fuel type, traffic volume, vehicle age, trip length distribution and the actual ambient temperature (Mensink et al, 2000).

We present a new approach in which we propose to replace the static emission source terms by a turbulent diffusive boundary flux describing the interactions between the urban canopy and the regional model domain. The flux takes into account the streetwise road transport emissions, the road dimensions, the urban meteorological conditions and the turbulent intermittency in the urban canopy sub-layer. The turbulent diffusive flux is derived from the Prandtl-Taylor hypothesis and describes a vertical exchange of the pollutant over a characteristic length which can be associated with a typical mixing length created by turbulent eddies shedding off at roof level (Mensink et al., 2002).

In section 2, the theoretical formulation of the methodology will be presented. In section 3, the methodology will be illustrated by model applications for benzene and NO_x in the city of Antwerp. The results are presented and discussed in section 4.

* Clemens Mensink, Flemish Institute for Technological Research (Vito), Boeretang 200, B-2400 Mol, Belgium.

2. Methodology

The turbulent diffusive flux describing the interactions between the urban canopy and the regional model domain is derived from the Prandtl-Taylor hypothesis. The vertical exchange of mass and momentum takes place over a characteristic length which is associated with a typical mixing length created by turbulent eddies shedding off at roof level. The street canyon is represented by a box, dimensioned by the length and width of the street and the height of the built up area. Inside the box, only horizontal advection along the street (x-direction) and vertical diffusion processes (z-direction) are considered, together with a continuous source term S. Net contributions of horizontal turbulent fluxes are neglected as well as diffusion in horizontal directions:

$$\frac{\partial \bar{c}}{\partial t} = -\frac{\partial}{\partial x}\left(\overline{v_x \bar{c}}\right) - \frac{\partial}{\partial z}\left(\overline{v'_z\, c'}\right) + D\frac{\partial^2 \bar{c}}{\partial z^2} + S \qquad (1)$$

where the concentration c (μg m^{-3}) has been replaced by a time-smoothed value $\bar{C}$ and a turbulent concentration fluctuation c′. The same is applied to velocity vector $\vec{v}$. The first term on the right hand side in equation (1) represents the advective mass transport, the second term the mass transport due to turbulent fluctuations, and the third term the contribution due to (laminar) diffusion at low wind speeds with coefficient D (m^2 s^{-1}). The vertical turbulent mass flux term in (1) is approximated by applying the eddy diffusivity concept in analogy of Fick's law (Bird et al., 1960):

$$\overline{v'_z\, c'} = -K\frac{\partial \bar{c}}{\partial z} \qquad (2)$$

For a turbulent free stream flow the eddy diffusivity K (m^2 s^{-1}) can be related to a characteristic length scale ℓ (m) and the free stream flow velocity gradient by applying the Prandtl-Taylor hypothesis (Hinze, 1987):

$$\overline{v'_z\, c'} = -\ell^2\left|\frac{dU}{dz}\right|\frac{\partial \bar{c}}{\partial z} \qquad (3)$$

The characteristic length ℓ is associated with a typical mixing length e.g. created by turbulent eddies shedding off at roof level. The velocity gradient over this mixing length is assumed to be constant and equal to the free stream velocity $U_\perp$ above the roof tops in the direction of the eddy shedding, i.e. perpendicular to the street direction, divided by the mixing length ℓ. Thus, conform Prandtl's mixing length theory, the eddy diffusion becomes equal to the product of a mixing length ℓ and some suitable velocity, expressed here by $U_\perp$:

$$K = \ell U_\perp \qquad (4)$$

Substitution of (2) and (4) in equation (1) and reformulation of (1) in terms of a flux balance assuming a steady state approach, i.e. no change in meteorological input, emissions and concentrations during one hour, leads to:

$$C - C_b = \frac{Q}{U_{\equiv} \cdot \left(\frac{H}{L}\right) \cdot W + (D + \ell U_{\perp}) \cdot \left(\frac{W}{H}\right)} \tag{5}$$

where Q is the emission source strength per unit length (μg m^{-1} s^{-1}), C_b the background concentration (μg m^{-3}), H is the height (m), W the width (m) and L the length (m) of the street. In equation (5) the wind speed parallel to the street $U_{\equiv}$ is responsible for the "ventilation" of the street box, whereas wind speed perpendicular to the street $U_{\perp}$ is responsible for the vertical exchange of the pollutant over a characteristic length ℓ. This characteristic length ℓ can be associated with a typical mixing length caused by turbulent eddies shedding off at roof level. D is the diffusion coefficient *at low wind speeds*. Copalle (2001) showed that at low wind speeds this diffusion can play a role. He suggests a value of $D = 1.5$ m^2 s^{-1}. The value of the mixing length could be associated with the varying atmospheric conditions, but for the time being, the value has been tuned and set to $\ell = 1$ m. Model results are not very sensitive with respect to this parameter, as has been shown by Mensink and Lewyckyj (2001). The associated vertical turbulent diffusive flux describing the interactions between the urban canopy and the regional model domain can be expressed as:

$$G = -L \cdot W \cdot (D + \ell U_{\perp}) \frac{C_b - C}{H} \tag{6}$$

Note that this flux can be positive or negative depending on the difference between the concentration C inside the canyon and the concentration C_b above the canyon.

3. Applications

3.1. Calculation of benzene concentrations & fluxes for the entire city of Antwerp

Benzene concentrations and fluxes have been calculated for the city of Antwerp for 4 periods of 5 days in 1998. For these periods diffusive sampler measurements were carried out in 101 streets in Antwerp and at 4 regional background locations (Geyskens et al., 1999). The measurements were carried out in the framework of a European LIFE project MACBETH in which simultaneous benzene measurements were carried out in 6 European cities (Antwerp, Copenhagen, Rouen, Murcia, Padova and Athens). The benzene emissions and concentrations were calculated on an hourly basis, from Monday to Friday for the following periods: 19 – 23 January 1998; 23 – 27 March 1998; 25 – 29 May 1998 and 28 September – 2 October 1998.

For 1963 road segments in Antwerp benzene emissions were calculated by a road transport emission model (Mensink et al, 2000). The concentrations and flux contributions were calculated using equations (5) and (6) using the average measured background concentrations obtained from the four background monitoring locations. The hourly values for wind speed and wind direction were obtained from two meteorological towers located in the city. The hourly temperature was used for the calculation of cold start emission. Wind speed at roof level was calculated from a wind profile described by a power law, with the exponent derived from the wind speed measured at heights of 30 m and 153 m respectively.

3.2. Calculation of NO_x Concentrations & Fluxes for a Single Street Canyon

For the same periods mentioned in section 3.1, NO_x concentrations and fluxes were calculated for a single street canyon in Antwerp, i.e., the "Plantin en Moretuslei". This is a main passage street with moderate traffic volumes of 1200–1300 vehicles passing at peak hours. The emissions were calculated over a length of 436 m. with an average street width of 29 m. and an average building height of 20 m. The street angle is 99° from the north. In the middle of the street there is a green verge with a few trees between the two traffic lanes. The measurement station is located directly at the road side. The monitoring station in the "Plantin en Moretuslei" is located adjacent to the street and approximately 10 m from the nearest traffic lane. The monitoring station is part of the air quality monitoring network in Flanders (Vlaamse Milieumaatschappij, 1999). Background concentrations were obtained from a rural station northeast of the monitoring station. Meteorological data were identical as for the benzene calculations. Hourly NO_x emissions were calculated by the road transport emission model mentioned in 3.1.

4. Results and Discussion

4.1. Results for the Entire City of Antwerp (Benzene)

Figure 1 shows the hourly benzene emissions and concentrations as averaged over the whole city domain (1963 road segments) for the first period. Emissions are relatively low on the first two week days, due to the high daily average temperatures on Monday (5.9°C) and Tuesday (4.4°C) compared to Wednesday (1.0°C), Thursday (–1.8°C) and Friday (0.0°C). Benzene concentrations are relatively low on these first two days due to the relatively high average wind speeds on Monday (8.1 m s^{-1}) and Tuesday (6.2 m s^{-1}) compared to Wednesday (2.5 m s^{-1}), Thursday (3.1 m s^{-1}) and Friday (3.6 m s^{-1}), resulting in stronger ventilation inside the city canyons.

Figure 2 shows the hourly flux as averaged over the whole city domain for the first period. One can observe that the flux is negative for Monday and Tuesday, meaning that there is on average an overall mass transport of benzene from the atmosphere into the city. During daytime and traffic peak hours on Wednesday,

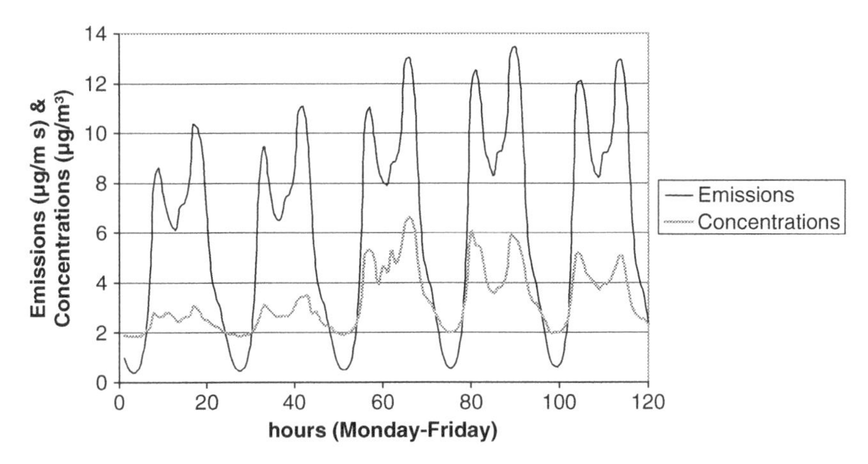

FIGURE 1. Calculated hourly benzene emissions and concentrations as averaged over the Antwerp domain for the period of 19 – 23 January 1998.

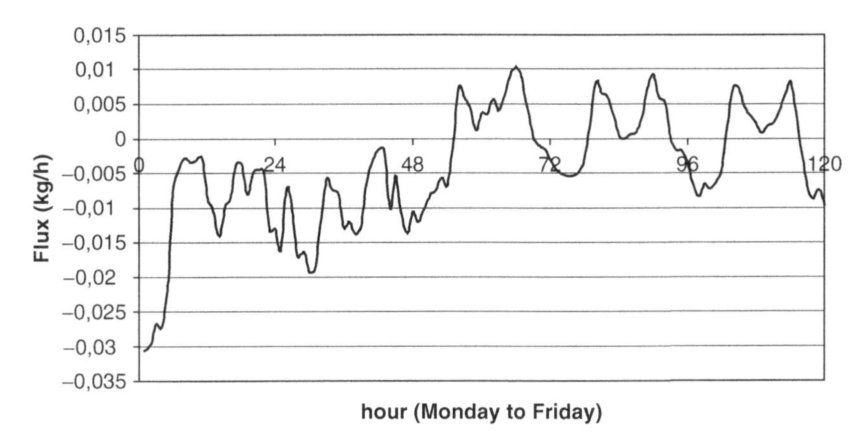

FIGURE 2. Calculated hourly benzene flux as averaged over the Antwerp domain for 19 – 23 January 1998.

Thursday and Friday the flux is positive, indicating that there is a resulting transport of benzene out of the city into the atmosphere. Figure 3 shows a comparison of ensemble averages of measured benzene concentrations (101 locations in Antwerp) with ensemble averages of calculated concentrations in the streets of Antwerp (1963 road segments). The error bar shows the standard deviation for the 101 measurement locations in the city. Figure 3 shows that for the ensemble average the benzene concentrations are well predicted and that the benzene background contribution for the city as a whole varies between 50% and 58% depending on the period considered.

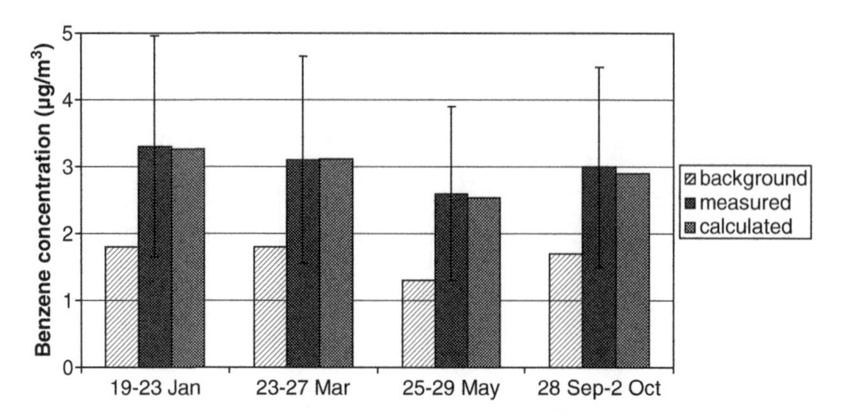

FIGURE 3. Comparison of ensemble averages of measured benzene concentration (101 locations) and background concentrations (4 locations) with ensemble averages of calculated concentrations in the 1963 streets of Antwerp for four measurement periods.

4.2. Results for the Street Canyon (NO_x)

Figure 4 shows the NO_x emissions and NO_x concentrations as calculated and measured for the period in January. Calculated NO_x emissions and concentrations show the same pattern as for the benzene emissions and concentrations, presented in Figure 1, with high temperatures resulting in relatively low emissions on Monday and Tuesday and higher wind speeds resulting in more ventilation and thus lower NO_x concentrations on Monday and Tuesday compared to the rest

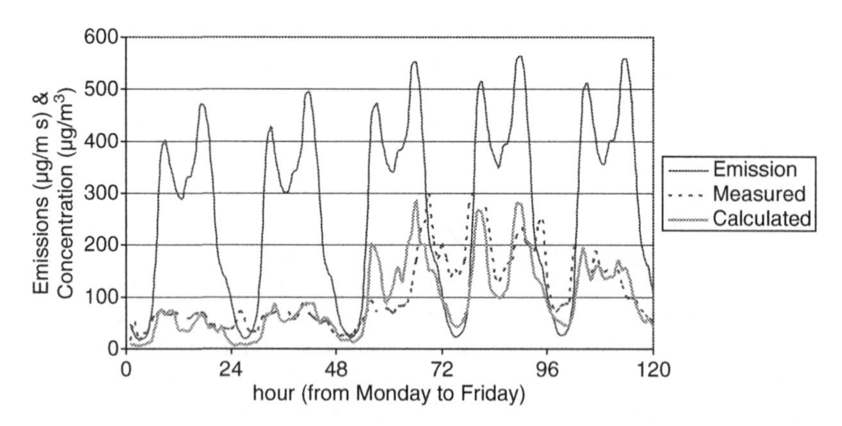

FIGURE 4. Calculated hourly NO_x emissions and calculated and measured NO_x concentrations for a street canyon in the "Plantin en Moretuslei" in Antwerp for the period of 19–23 January 1998.

of the week. This is in very good agreement with the hourly measurements. Figure 5 shows the NO_x emissions and NO_x concentrations as calculated and measured for the autumn period. Also here we can observe a good agreement with the measurements. On the last day of the week the concentrations are lower, despite the fact that the emissions are higher. This is due to the fact that on this Friday the average temperature was lower (8.3°C) compared to the other weekdays (14.7°C–15.9°C), whereas the average wind speeds were relatively high (4.5 m s^{-1}) when compared to Monday (1.6 m s^{-1}), Tuesday (2.7 m s^{-1}), Wednesday (3.8 m s^{-1}) and Thursday (3.0 m s^{-1}). The stronger ventilation of the canyon resulting in lower concentrations is confirmed by the measurements. Thus it seems that the variation in concentrations inside the canyon can be explained quite well by these relatively simple phenomena like horizontal ventilation and vertical transport by turbulent diffusion.

Figure 6 shows the average NO_x concentrations for the four periods. NO_x measurements were not available for the period in March. The NO_x background contribution to the street canyon was 48% for the period in January, 55% for the period in May and 48% for the autumn period, which is somewhat lower than for benzene in the city case. Figure 7 shows the daily variation of the hourly flux as averaged over all four periods. Like in some of the cases for benzene we see a positive flux during daytime and a negative flux during the night. This general picture shows that on average NO_x is transported from the canyon into the atmosphere during daytime, whereas NO_x is transported (and probably deposited) during nighttime.

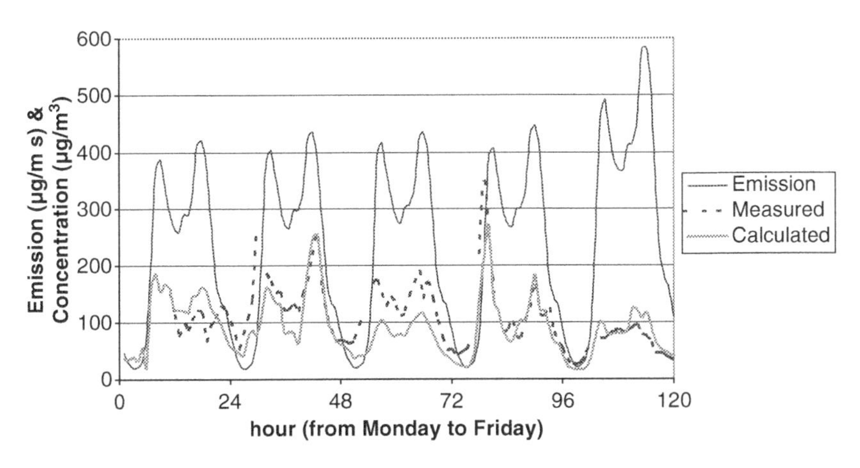

FIGURE 5. Calculated hourly NO_x emissions and calculated and measured NO_x concentrations for a street canyon in the "Plantin en Moretuslei" in Antwerp for the period of 28 September – 2 October 1998.

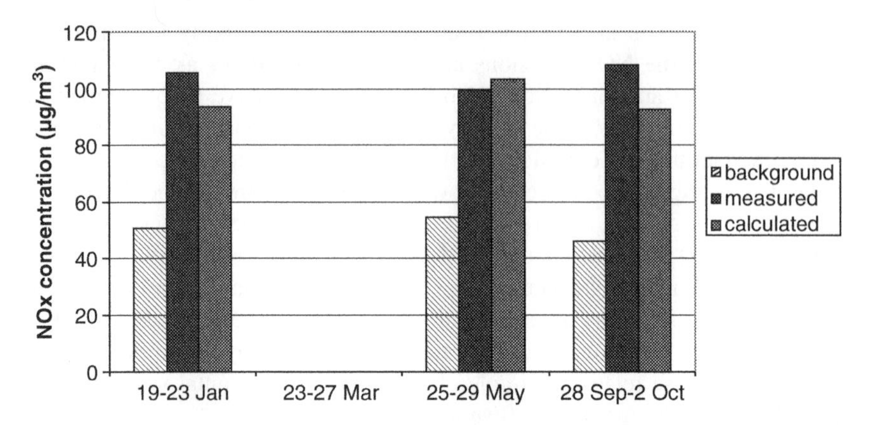

FIGURE 6. Comparison of the measured NO_x concentrations, measured background concentrations and calculated NO_x concentrations inside a street canyon in the "Plantin en Moretuslei" in Antwerp for the four measurement periods.

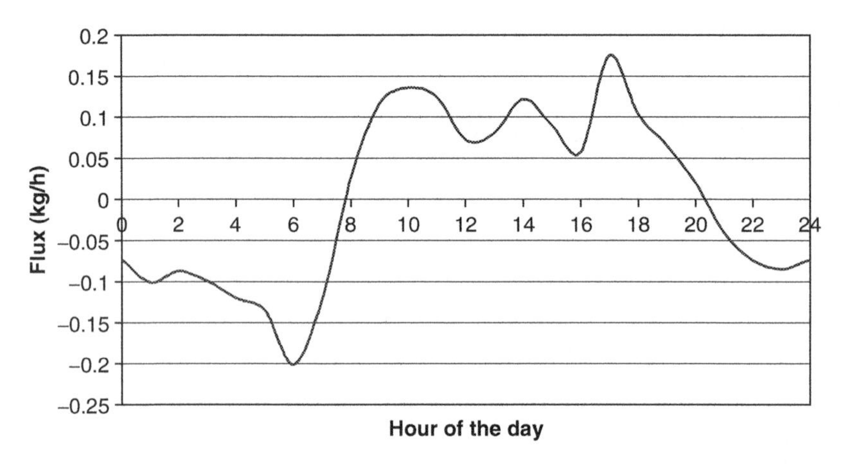

FIGURE 7. Daily pattern of the calculated hourly NO_x flux for the street canyon as averaged over the four measurement periods.

5. Conclusions

It is possible to construct and calculate a turbulent diffusive boundary flux that can replace the static (emission) source & sink terms in the governing equations. This bi-directional flux can take into account some of the dynamics in the urban canopy (traffic, vertical exchange, wind, temperature). It allows a two way interaction between the urban canopy and the regional model domain.

References

Bird, R. B., Stewart, W. E. and Lightfoot, E. N., 1960, *Transport phenomena*, John Wiley & Sons, New York, 780 pp.

Coppalle, A., 2001, A street canyon model for low wind speed conditions, *Int. J. Environment and Pollution* 16, Nos 1-6, pp. 417-424.

Geyskens, F., Bormans, R. Lambrechts, M and Goelen, E., 1999, Measurement campaign in Antwerp for quantification of personal exposure to representative volatile organic compounds, Vito report 1999/DIA/R/048, Mol (in Dutch), 168pp.

Hinze, O., 1987, *Turbulence*, McGraw-Hill, New York, 790 pp.

Mensink, C., De Vlieger, I. and Nys, J., 2000, An urban transport emission model for the Antwerp area. *Atmospheric Environment*, 34, 4594-4602.

Mensink, C. and Lewyckyj, N., 2001, A simple model for the assessment of air quality in streets, *Int. Journal of Vehicle Design* 27, Nos. 1-4, 242-250.

Mensink, C., Lewyckyj, N. and Janssen, L., 2002, A new concept for air quality modelling in street canyons, *Water, Air and Soil Pollution: Focus,* 2 (5-6), 339-349.

Vardoulakis, S., Fisher, B.E. A., Percleous, K. and Gonzalez-Flesca, N., 2003, Modelling air quality in street canyons: a review, *Atmospheric Environment* 37, 155-182.

Vlaamse Milieumaatschappij, 1999, *Air quality in the Flemish region 1998*, Flemish Environmental Agency, Aalst (in Dutch), 178 pp.

On the Formulation and Implementation of Urban Boundary Conditions for Regional Models

Speaker: C. Mensink

Questioner: R. Yamartino

Question: Are the negative NOx fluxes you predict at night consistent with, or the same as, a dry deposition flux?

Answer: The negative fluxes at night are indeed consistent with a dry deposition flux. The units are identical. The flux expressed by equation (6) is bi-directional and acts as an emission flux during daytime and possibly as a deposition flux during nighttime when traffic emissions are low. However in an urban context it is difficult to describe a deposition flux in terms of the resistance analogy, which is commonly used to define dry depositions. How to define the aerodynamic resistance, the quasi-laminar boundary layer resistance and the canopy resistance in the urban canopy? Expression (6) could be an alternative for defining a dry deposition flux in an urban situation.

Questioner: R. San Jose

Question: How important is the background concentration?

Answer: For benzene the contribution from the background varied between 50% and 58% (see also figure 3). For NO_x *this contribution varied between 42% and 55% (see also figure 6).*

Questioner: T. Oke

Question 1st : Comment following previous question : The urban structure (eg. Canyon H/W) is likely to dominate inside the canyon – above roof exchange, together with variability of building height.

Question 2nd : Did you evaluate the net impact of introducing the turbulent diffusive flux vs. using the 'old' source/sink term approach?

Answer: This is indeed the next step in the investigation. The problem of evaluating both approaches is to have a benchmark, i.e. a dataset with measured fluxes in specific urban canopy. These observations could then be compared with the results of the two approaches in order to evaluate them.

Questioner: T. Castelli Silvia

Question: In your formulation you set a value of 1 m for the Prandtl mixing length l : since the characteristic lengths of the street canyon, width and height, are about 30m and 20m respectively, it could be expected that the mixing length also has similar order of magnitude, or anyway be influenced by the characteristic lengths. Could you comment about this aspect?

Answer: The Prandtl mixing length is assumed to be associated with a mixing length created by turbulent eddies shedding off at roof level. So it probably depends on the atmospheric stability characteristics in the urban canopy. For the moment its value is set to 1 m. This is based on experimental results in street canyons where a value of $K = 1.5\ m^2/s$ *has been measured at low wind velocities:* $U_{\perp} \leq 1.5\ m/s$*. A sensitivity study shows that concentrations are not very sensitive to this parameter, although the influence is not linear. Doubling the value of* l *gives a 5% decrease of the concentrations. A value of* $l = 0.5\ m$ *results in a 14% increase of the calculated concentrations.*

23 Computational Model for Transient Pollutants Dispersion in City Intersection and Comparison with Measurements

Jiri Pospisil and Miroslav Jicha*

1. Introduction

Many city intersections are often heavily polluted due to intensive traffic. Dispersion of pollutants originating from traffic is directly connected with the geometry of the urban area and traffic conditions. The urban area is mostly heavily built-up area and buildings and other obstacles that may significantly influence local concentrations. Moving vehicles enhance both micro- and large-scale mixing processes in their surroundings. Not taking into account traffic will lead to neglecting one of the most important phenomena that influences mixing processes in the proximity of traffic paths. The influence of traffic is increasingly important in situations of very low wind speed.

In this study, the authors focus on transient pollutant dispersion due to traffic dynamics in an actual intersection equipped with traffic lights located in the center of the city of Brno. Inclusion of traffic dynamics leads to a better description of dispersion processes. Information about traffic situations were obtained from in-situ measurements. A model based on a Eulerian – Lagrangian approach to moving objects has been developed and integrated into a commercial CFD code StarCD. Results of CFD predictions of NO_x concentrations are compared with measurements acquired from an automatic monitoring station located in the intersection.

2. Mathematical Formulation and Solution Procedure

Dispersion of pollutants can be predicted either for steady or transient traffic situations. The steady approach can be applied where the speed and traffic rates are constant or their changes can be neglected. Typical examples are long street

* Jiri Pospisil & Miroslav Jicha, Brno University of Technology, Faculty of Mechanical Engineering, Technicka 2, 616 69 Brno, Czech Republic

canyons. Transient situations are mostly treated as steady ones by choosing typical meteorological conditions and typical traffic rates. The dynamics of slowing down and speeding up of vehicles is very rarely accounted for.

In this study, the authors focus on traffic dynamics and its impact on pollutant dispersion in two street canyons that form an intersection equipped with traffic lights. A model based on a Eulerian – Lagrangian approach to moving objects[1] has been developed and integrated into a commercial CFD code StarCD. Inclusion of traffic dynamics leads to a better description of the dispersion processes of pollutants in an intersection, where peak values of pollutant concentrations are often located. In the intersection, there are regularly alternating situations with peak production of pollutants when cars drive away and with minimal production in idling situations.

The method includes transient calculation of vehicle movement, emission rates produced by cars and additional sources of turbulence. Quite decisive is the transient approach in the situation of calm or very low wind speed when car movement dominates in the flow field and pollutant dispersion.

2.1. Gas Phase Equations

The set of equations for the conservation of mass, momentum and passive scalar is solved for transient, incompressible turbulent flow. The equation for a general variable ϕ has the form,

$$\frac{\partial(\rho\phi)}{\partial t} + \frac{\partial}{\partial x_i}(\rho u_i \phi) = \frac{\partial}{\partial x_i}\left(\Gamma \frac{\partial \phi}{\partial x_i}\right) + S_\phi + S_\phi^P \tag{1}$$

Variable ϕ stands for velocity components and concentrations. Equation (1) contains an additional source term S_ϕ^P. This term results from the interaction between the continuous phase of ambient air and discrete moving objects. The interaction is treated using a modified Particle-Source-In-Cell (PSIC) technique by Crowe et al.[2]. The additional source term represents the net efflux of mass, momentum and passive scalar transported from the discrete moving objects into the continuous phase. The governing equations for the continuous phase were solved using the commercial CFD code Star-CD based on the finite volume procedure. A quasi-steady model for moving objects and their interaction with the continuous ambient air was integrated into this code.

2.2. Model of Moving Objects and Their Interaction With the Ambient Air

The basis of the quasi-steady approach to moving objects consists in the replacement of the discrete objects with a continuously moving flow of blocks with a specified velocity and mass. The blocks that represent a vehicle, pass through particular control volumes of the solution domain. Each vehicle may be contained in several control volumes of the continuous phase; each of the control volumes contains one block[1].

2.3. Traffic Induced Turbulence

For the continuous phase of air, a non-linear low-Reynolds k-ε model of turbulence was used. An additional source of kinetic energy of turbulence[1] was added along the trajectory the vehicles follow.

3. Mathematical Description of Car Movement in an Intersection

Interaction of a car with ambient air is directly connected with the relative velocity between the car and the air. Car speed is a required parameter for correct description of the interaction. Hence, mathematical formulation of the car speed has been done for accurate description of the behaviour of cars passing through the intersection controlled by traffic lights. The mathematical formulation provides information about the speed of cars in every specified position during all required time steps. The mathematical formulation consists of two parts, namely slowing down and speeding up terms. The slowing down formulation covers gradually varied movements: steady moving of cars, slowing down, stopping and idling of cars in a queue before traffic lights. The slowing-down process begins at the instant traffic lights switch from green to red.

The whole slowing down term can be divided in two major terms, namely transitional and steady slowing down terms.

3.1. Transitional Slowing-Down Term

The transitional slowing down term starts at the instant traffic lights switch from green to red. This instant is marked as t_0. Figure 1 shows car velocity at the instant traffic lights switch and the dashed line indicates a simplified velocity of the first slowing down car.

The first slowing down car starts slowing down at instant t_0, in position X_2,

$$X_2 = X_0 - L_{slow} \tag{2}$$

and stops in position X_0. Cars from interval $(X_2; +\infty)$ go on with their original speed. The transitional slowing down term terminates at instant of stopping of the first slowing down car stops. The following equations are applicable until this term,

$$X_K = X_0 - L_{slow} + v_{steady}(t - t_0) \tag{3}$$

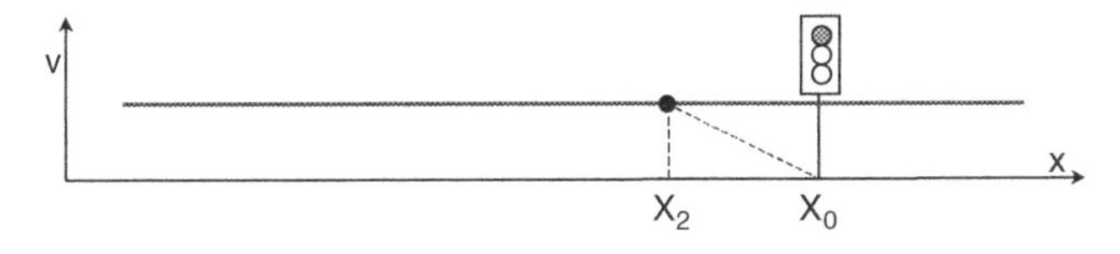

FIGURE 1. Car speed at instant t_0 of over switching from green to red.

and

$$X_2 = X_0 - L_{slow} - n_{za} \cdot \left(L_{car} + L_{gap} \right), \tag{4}$$

where number of cars slowing down cars is calculated as

$$n_{za} = \left(\frac{h_{traffic}}{3600} (t - t_0) \right) \Big/ \left(1 - \frac{L_{car} + L_{gap}}{v_{steady}} \cdot \frac{h_{traffic}}{3600} \right). \tag{5}$$

A description must be supplemented by the immediate position of the first slowing down car

$$X_1 = X_2 + v_{steady} (t - t_0) + \frac{1}{2} a_{slow} (t - t_0)^2. \tag{6}$$

A simplified relation between the position and the speed of cars during the transitional slowing down term is drawn in Figure 2.

Another step of the solution consists in matching the length intervals with the appropriate speed of cars, see Table 1.

3.2. *Steady Slowing-Down Term*

We speak about the steady slowing down term from an instant when the first car stops in front of the traffic lights. Other slowing-down cars stop and form a queue. Spacing of cars is equal to the sum of car length L_{car} and the gap between the cars L_{gap},

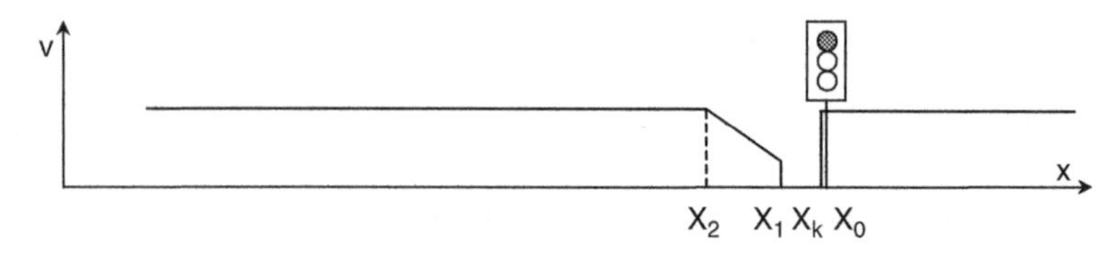

FIGURE 2. Relation between position and speed of cars during the transitional slowing down term.

TABLE 1. Appropriate speed of cars valid for transitional slowing down term.

Position	Speed of cars	
$(-\infty, X_2>$	$v = v_{steady}$	(7)
$(X_1, X_2>$	$v = -\dfrac{X - X_2}{X_1 - X_2}\left(v(X_2) - v(X_1)\right) + v_{steady}$	(8)
$(X_1, X_K>$	N/A	
$(X_K, \infty+>$	$v = v_{steady}$	(9)
X_1	$v = v_{steady} + a_{slow}(t - t_0)$	(10)

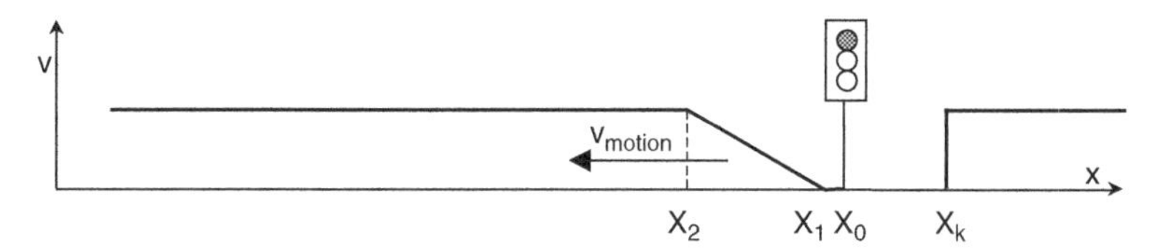

FIGURE 3. Speed of cars as function of position during the steady slowing down term.

$$L_{am} = L_{car} + L_{gap}. \tag{11}$$

The X_1 point is moving against car movement direction due to stopping of the following cars. Speed of motion of the point X_1 is directly connected with traffic density and the distance between cars. Mathematical formulation of the point X_1 position is

$$X_1 = X_0 - L_{am} \cdot \frac{h_{traffic}}{3600} \cdot \left(t - t_0 - t_{X2'X1'}\right). \tag{12}$$

We assume a constant slowing down length for all cars. Therefore, the X_2 point position moves in agreement with the point X_1. Mathematical formulation for the calculation of the position of the point X_2 is in Eq. 3. Figure 3 shows graphically the shift of the slowing-down zone.

Table 2 contains equations for calculation of the car's speed in particular position intervals.

The steady slowing-down term terminates at the instant traffic lights switch from red to green. Then it starts a speeding up term. A similar method was developed for the calculation of speeding up of cars.

4. Results and their Discussion

The intersection studied is formed with two street canyons, namely Kotlarska street and Kounicova street. The street canyons intersect perpendicularly. A sketch of the intersection geometry is drawn in Figure 4. The intersection is located close to the center of city of Brno. Intensive traffic lasts from 6 to 9 pm in both intersecting streets.

TABLE 2. Appropriate speed of cars valid for steady slowing down term.

Position	Speed of cars	
$(-\infty, X_2)$	$v = v_{steady}$	
(X_2, X_1)	$v = -\frac{X - X_2}{X_1 - X_2}\left(v(X_2) - v(X_1)\right) + v_{steady}$	(13)
(X_1, X_0)	idling cars	
(X_0, X_K)	*N/A*	
$(X_K + \infty)$	$v = v_{steady}$	

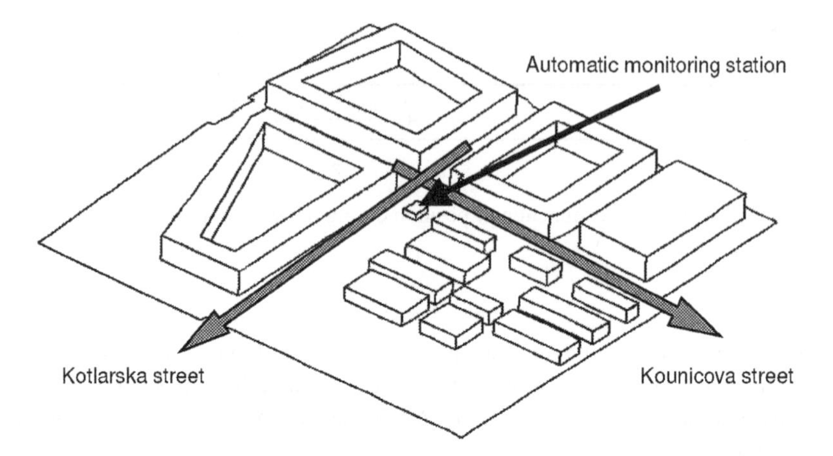

FIGURE 4. Sketch of the intersection.

The intersection is solved in 3D configuration as a finite area with pressure boundary conditions. The aspect ratio (width to the height) of both street canyons is 1.27. Dimensions of the solution domain are $200 \times 200 \times 120$m. In both street canyons two-way traffic was set in four traffic lanes in total.

The actual traffic situation from the intersection was used as input data for the computational modeling. The following data on the traffic were obtained from the in-situ measurements done in July 2001:

- the number of cars that stop in one street canyon on red lights and the number of cars passing through the other street canyon
- the length of car lines waiting on a red light
- slowing-down distance and driving away distance of cars
- traffic light sequence and timing of stop-and-go for traffic lights
- fleet of vehicles

The average traffic rate during the measurement period was set to 360 cars/ hour/ traffic lane, speed of traffic to 50 km/hour, stopping distance 69 m, gap between cars on red light 2.5 m, the average car acceleration when driving away 10 m/s^2, frequency of driving away cars was set to 1 car/s. Identical conditions were assigned in both street canyons. The time intervals on traffic lights are regular and switching from red to green occurs every 60 seconds. During the calculations, the wind velocity profile was assigned at side boundaries of the calculated domain. Wind velocity was set 1.1 m/s at height of 6 m above ground level.

An algorithm for the description of car movement and pollutant production in the intersection was integrated into the solution procedure. As a result, the transient flow field and the ground concentration of a passive scalar were predicted.

Measured NO_x concentrations were obtained from an automatic monitoring station located close to the center of the intersection as shown in Fig. 4. The records of NO_x concentrations acquired during the summer measurements are shown in Fig. 5.

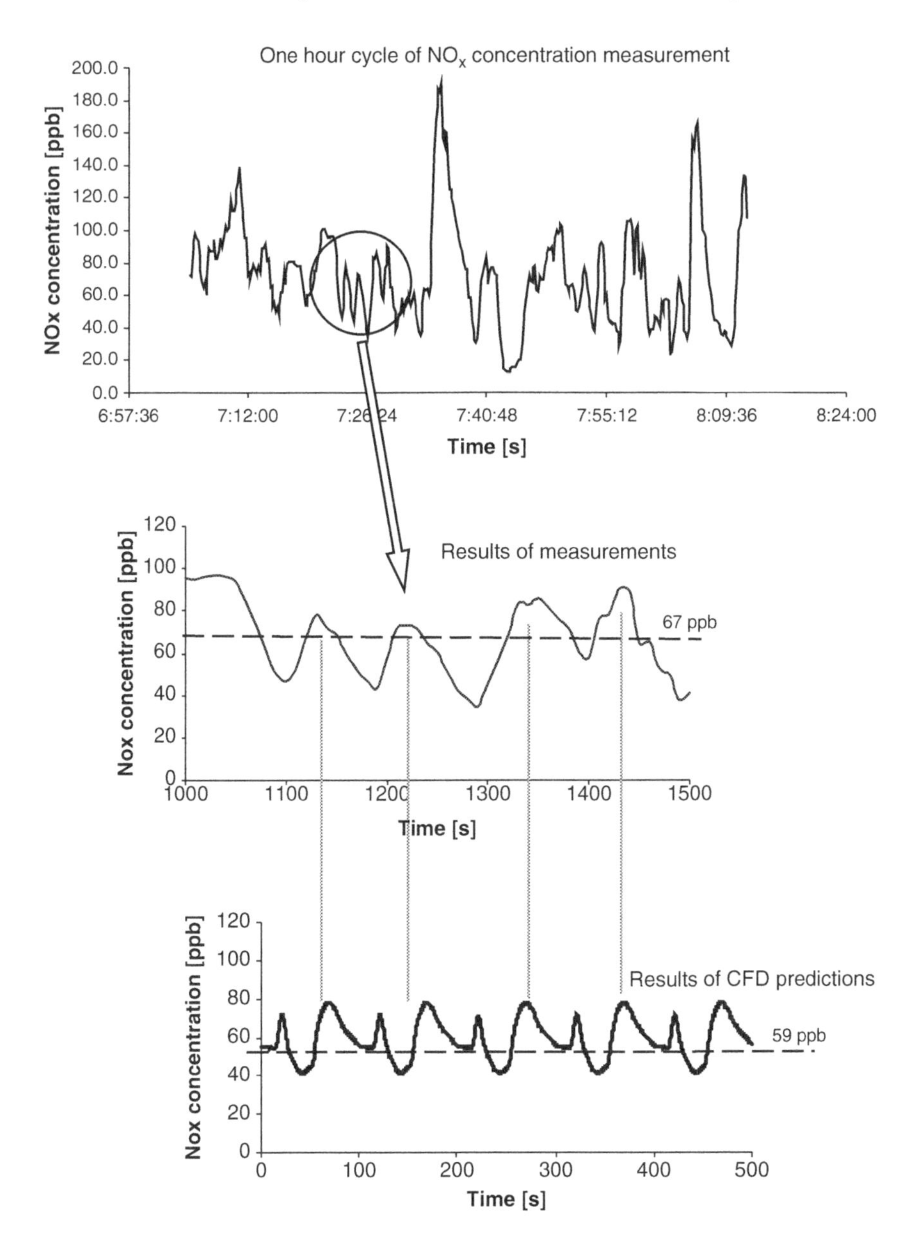

FIGURE 5. Comparasion of measurements and predictions of NO_x concentrations.

We can see an alternating behavior of the concentration. This unsteady character of NO_x concentrations results from a transient velocity field in the intersection and transient pollutant production by cars. The transient velocity field reflects car movement and unsteady meteorological conditions. The highest and the lowest concentration peak values very likely appear at instants of abrupt changes of velocity field in the intersection due to a change of wind direction.

A detail of NOx concentration records is compared with CFD predictions in Fig. 5. The regularly alternating character of concentrations obtained from the CFD prediction results from the set of regularly repetitive traffic cycles. From comparison, corresponding instants of peak values of NOx are clearly seen. Average measured concentrations of NOx are 67 ppb and from CFD prediction 59 ppb during the compared period.

5. Conclusions

An Eulerian-Lagrangian method developed for traffic induced flow (Jicha at al., 2000) was applied to prediction of velocity and concentration in an actual intersection equipped with traffic lights. The method was complemented with a transient speed of cars and transient and locally dependent production of emissions. From the results it can be seen that the traffic dynamics, namely at very low wind speeds, significantly influences the flow field and thus results in a very different field of ground concentrations. The peak values of concentrations are met when the cars start driving away from the intersection. Contrary to this, minimal values are when cars slow down to stop.

Acknowledgment This work is a part of COST Action 715 and Eurotrac-2 subproject SATURN and was financially supported by the Czech Ministry of Education under the grants OC715.80 and OE32/EU1489, respectively and the Brno University of Technology Research Plan No. MSM 262100001.

References

1. Jicha M., Katolicky J., Pospisil J., 2000. Dispersion of pollutants in street canyon under traffic induced flow and turbulence, *J. for Environmental Monitoring and Assessment*, vol. 65, 343-351.
2. Crowe, G.T., Sharma, M.P., Stock, D.E., 1977. The Particle-Source-In-Cell Model for Gas-Droplet Flows, *J. Fluid Eng*., 99, (325-332).
3. Sini J. F., Mestayer P. G., 1997. Traffic-induced urban pollution: A numerical simulation of street dispersion and net production, *22nd NATO/CCMS International Technical Meeting on Air Pollution Modelling and Application*, Clermont-Ferrand, France.

Effects of Climate Change on Air Quality

24 Air Quality in Future Decades – Determining the Relative Impacts of Changes in Climate, Emissions, Global Atmospheric Composition, and Regional Land Use

C. Hogrefe, B. Lynn, B. Solecki, J. Cox, C. Small, K. Knowlton, J. Rosenthal, R. Goldberg, C. Rosenzweig, K. Civerolo, J.-Y. Ku, S. Gaffin, and P. L. Kinney*

1. Introduction

In recent years, there has been a growing realization that regional-scale ozone (O_3) air quality is influenced by processes occurring on global scales, such as the intercontinental transport of pollutants (Jacob et al., 1999; Fiore et al., 2002, 2003; Yienger et al., 2000) and the projected growth in global emissions that alter the chemical composition of the global troposphere (Prather and Ehhalt, 2001; Prather et al., 2003). However, little work has been performed to date to study the potential impacts of regional-scale climate change on near-surface air pollution. Climate change can influence the concentration and distribution of air pollutants through a variety of direct and indirect processes, including the modification of biogenic emissions, the change of chemical reaction rates, changes in mixed-layer heights that affect vertical mixing of pollutants, and modifications of synoptic flow patterns that govern pollutant transport. Another parameter affecting local and regional meteorology and air pollution is land use, and significant land use changes associated with continued urbanization are expected to occur over the same time scales as changes in regional climate.

The results presented in this paper build upon a recent study by Hogrefe et al. (2004a) who presented results of a modeling study aimed at simulating O_3

* C. Hogrefe, ASRC, University at Albany, Albany, NY. K. Civerolo and J.-Y. Ku, New York State Department of Environmental Conservation, Albany, NY.
B. Lynn, C. Rosenzweig and R. Goldberg, NASA-Goddard Institute for Space Studies, New York, NY.
C. Small, S. Gaffin, J. Rosenthal, K. Knowlton, and P.L. Kinney, Columbia University, New York, NY. W.D. Solecki and J. Cox, Hunter College, New York, NY.

concentrations over the eastern United States in three future decades, taking into account the effects of regional climate change for a specific climate change scenario. In this study, we add analysis for a separate climate change scenario. Additionally, further sensitivity simulations were performed to compare the magnitude of O_3 changes due to regional climate change to those that could arise from projected anthropogenic emissions within the modeling domain and changes in chemical boundary conditions. Finally, we present the results of sensitivity simulations in which projected changes in land use were incorporated into the regional climate and air quality modeling.

2. Models and Database

Projections of greenhouse gas and other atmospheric constituents are used as inputs to climate and air quality models to simulate possible future conditions. The Intergovernmental Panel On Climate Change (IPCC) Special Report on Emission Scenarios (SRES) describes various future emissions scenarios for greenhouse gases and ozone precursors based on projections of population, technology change, economic growth, etc. (IPCC, 2000). In this paper, we utilize the emission projections of the SRES A2 and B2 marker scenarios. While the A2 scenario is one of the more pessimistic SRES marker scenarios and is characterized by a large increase of CO_2 emissions, the B2 scenario is relatively optimistic and is characterized by smaller increases in CO_2 emissions (IPCC, 2000).

2.1. Emissions Processing for Air Quality Modeling

As described in Hogrefe et al. (2004a), the county-level U.S. EPA 1996 National Emissions Trends (NET96) inventory is used as the basis for the air quality modeling presented in this study. This emission inventory is processed by the Sparse Matrix Operator Kernel Emissions Modeling System (SMOKE) (Carolina Environmental Programs, 2003) to obtain gridded, hourly, speciated emission inputs for the air quality model. Biogenic emissions are estimated by the Biogenic Emissions Inventory System – Version 2 (BEIS2) that takes into account the effects of temperature and solar radiation on the rates of these emissions (Geron et al., 1994). For sensitivity simulations assessing the relative impact of increased anthropogenic emissions, future year anthropogenic emissions are estimated by multiplying the 1996 base year emission inventory with the regional growth factors for the SRES A2 scenario for 2050 for the so-called OECD90 region that includes many industrialized countries including the U.S. Under this scenario, Emissions of the O_3 precursors NO_x/VOC increase by 125%/60% globally and 29%/8% for the OECD90 region by the 2050s (IPCC, 2000).

2.2. Land Use Change

In this study, we performed sensitivity simulations using a regional land use change scenario consistent with the 2050 A2 SRES scenario that was developed for the

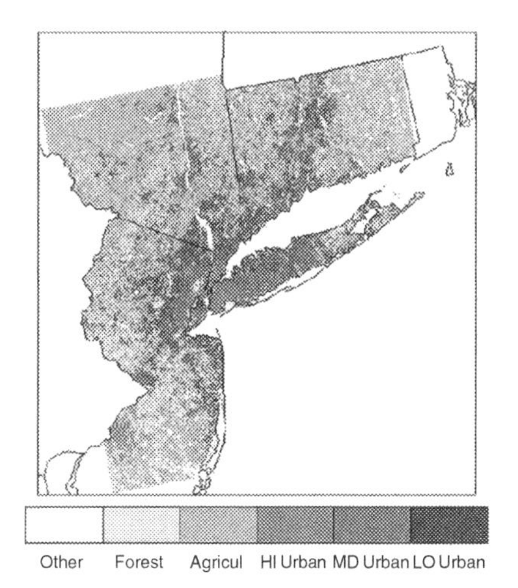

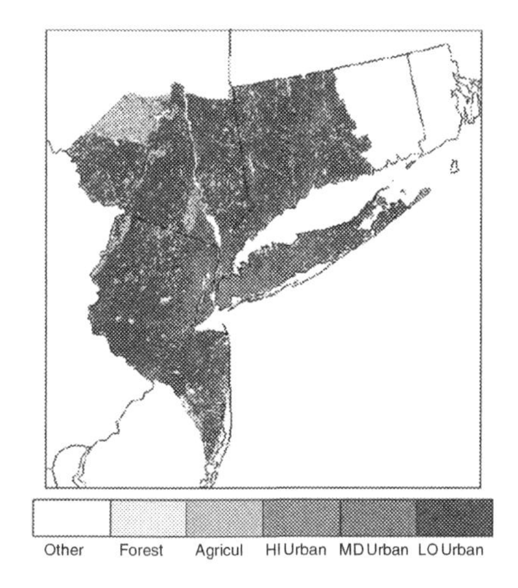

FIGURE 1. Land use categories in the greater New York City metropolitan area as simulated by Solecki and Oliveri (2004) for the 1990s (left) and 2050s A2 scenario (right).

greater New York City metropolitan area by Solecki and Oliveri (2004). Figures 1ab present maps of land use for the 1990s base case and the 2050s A2 scenario. It can be seen that under this land use change scenario, a large number of grid cells in the greater New York City Metropolitan area are predicted to be converted from 'forest' or 'agriculture' to 'low density urban' land use in the 2050s A2 scenario.

2.3. *Global and Regional Climate Modeling*

Current and future year regional climate fields were obtained by coupling the MM5 mesoscale model (Grell et al., 1994) to the Goddard Institute for Space Studies (GISS) $4^{\circ} \times 5^{\circ}$ resolution Global Atmosphere-Ocean Model (GISS-GCM) (Russell et al., 1995) in a one-way mode through initial conditions and lateral boundaries. Simulations were performed for five consecutive summer seasons (June–August) in the 1990s and three future climate scenarios, namely, 2020s A2, 2050s A2, and 2050s B2 at a horizontal resolution of 36 km over the eastern United States. For the sensitivity simulations exploring the effects of land use changes, MM5 simulations at 4 km resolution were performed over the greater New York City metropolitan area for a 20 day period in the model-simulated summer of 1993. The 4 km grid was driven by a 12 km grid which in turn was driven by the original 36 km grid used for the simulations described above. Further details on the setup of this modeling system and results of the future regional climate simulations are described in Lynn et al. (2004) and Hogrefe et al. (2004b).

2.4. *Air Quality Modeling*

Using the SMOKE-processed emissions and the 36 km MM5 regional climate simulations for the five summer seasons in the 1990s, 2020s A2, 2050s A2 and

2050s B2 scenarios, air quality simulations were performed using the Community Multiscale Air Quality (CMAQ) model (Byun and Ching, 1999) at a horizontal resolution of 36 km. The CMAQ modeling domain was slightly smaller than the MM5 modeling domain, and for the analysis presented below, grid cells over the ocean and within a 10-cell radius of the boundary were excluded. The CMAQ evaluation results for simulating O_3 concentrations under present-day climate conditions have been presented in Hogrefe et al. (2004b). For the simulations intended to investigate the role of regional climate change in the absence of changes in emissions and global tropospheric composition, time-invariant climatological profiles for O_3 and its precursors reflecting present day clean-air concentrations were used as boundary conditions (Byun and Ching, 1999). For the sensitivity simulations aimed at investigating the effects of changes in anthropogenic emissions, we utilized the A2-scaled emissions inventory for the 2050s as described in Section 2.1. For the sensitivity simulations aimed at estimating the role of changes in global atmospheric composition, these changes were approximated by changing the CMAQ boundary conditions according to values reported in previous studies. Details of this procedure are described in Hogrefe et al. (2004a). Finally, to investigate the effects of land use changes on CMAQ predicted ozone concentrations, we performed 4 km CMAQ simulations for a 20 day time period in 1993 using the 4 km MM5 fields with and without land use changes.

3. Results and Discussion

3.1. Changes in O_3 due to Regional Climate Change and Changed Biogenic Emissions

Because the U.S. National Ambient Air Quality Standard (NAAQS) for 8-hr O_3 concentrations is set at 84 ppb, model predicted exceedances of this threshold are of particular importance when assessing the effect of climate change on O_3 air quality. To analyze changes in the frequency and duration of extreme O_3 events, the number of days for which the predicted daily maximum 8-hr O_3 concentrations exceeded 84 ppb was determined over all cells (except those grid cells over the ocean and within a 10-cell radius of the boundary), and for each such event the number of consecutive days for which these conditions persisted was tracked. Figure 2 shows the total number of days with daily maximum 8-hr O_3 concentrations exceeding 84 ppb for the 1990s, 2020s A2, 2050s A2, and 2050s B2 simulations, grouped by the number of consecutive days on which this concentration was exceeded. Note that for these simulations, emissions, boundary conditions and land use were held constant. The total number of exceedance days increased from about 47,000 in the 1990s to about 64,000 in the 2020s A2, about 69,000 in the 2050s B2 and about 83,000 in the 2050s A2 simulations, i.e. by roughly 75% in the 2050s A2 scenario. Additional analysis also shows that the persistence of extreme O_3 events increases in the future climate scenarios.

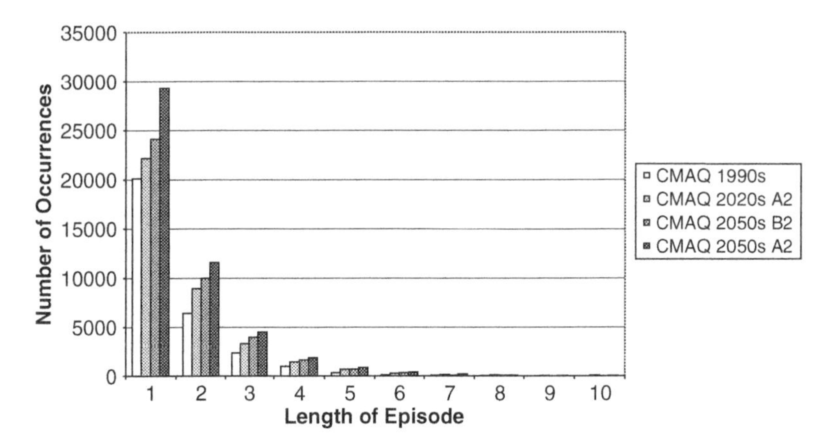

FIGURE 2. Number of days for which prediced daily maximum 8-hr ozone concentrations exceeded 84 ppb over all non-water, non-boundary grid cells for each of the different climate scenarios, grouped by length of episode.

Lynn et al. (2004) showed that the MM5 simulations used as input to the CMAQ simulations performed in this study predict an increase of average summertime daily maximum temperatures of 1.5°C to 3.5°C for most regions in the modeling domain for the 2050s A2 scenario compared to the 1990s. To estimate the first-order contribution of changed biogenic emissions caused by this change in temperatures on the ozone changes shown above, Hogrefe et al. (2004a) performed a series of sensitivity simulations to decouple the increase of biogenic emissions due to rising temperatures from other effects of climate change for the 2050s. Results from these simulations indicate that on a spatial average basis, increased summertime total biogenic emissions and synergistic effects between climate change and increased emissions account for about half of the overall increase in summertime average daily maximum 8-hr O_3 concentrations for the 2050s A2 scenario.

3.2. Relative Impact of Changes in Anthropogenic Emissions vs. Changes in Regional Climate

The analysis presented above focused on determining the effects of climate change on summertime O_3 concentrations over the eastern United States in the absence of changes in anthropogenic emissions within the modeling domain or changes in boundary conditions approximating changes in global atmospheric composition. While a fully coupled, multi-scale dynamics-chemistry model would be necessary to study all interactions between climate and air quality on both regional and global scales, the regional-scale model system described in this study can be used to compare the effects of climate change on air quality presented above to the effects of increases in anthropogenic emissions, and the approximated effects of changes in global atmospheric composition through the specification of altered boundary conditions. To this end, sensitivity simulations were performed for the time periods from 1993-1997 and 2053-2057. To compare the contribution of the three factors to changes in summertime average daily maximum 8-hr O_3 concentrations as well as the 4th highest summertime daily maximum 8-hr O_3 concentration that is of relevance to the NAAQS, these contributions were calculated and averaged over all grid cells except those grid cells over the ocean and within a 10-cell radius of the boundary. Figure 3 shows a bar chart of the contribution of each factor to changes in summertime average daily maximum 8-hr O_3 concentrations (light shading) and to changes in the 4th highest summertime daily maximum 8-hr O_3 concentration in the 2050s A2 scenario simulation (dark shading).

Figure 3 indicates that changed boundary conditions as described in Section 2.3 are the largest contributor (4.8 ppb) to changes in summertime average daily maximum 8-hr O_3 concentration in the 2050s A2 scenario simulation from the 1990s base simulation, followed by the effects of climate change (3.3 ppb) and the effects of increased anthropogenic emissions within the modeling domain

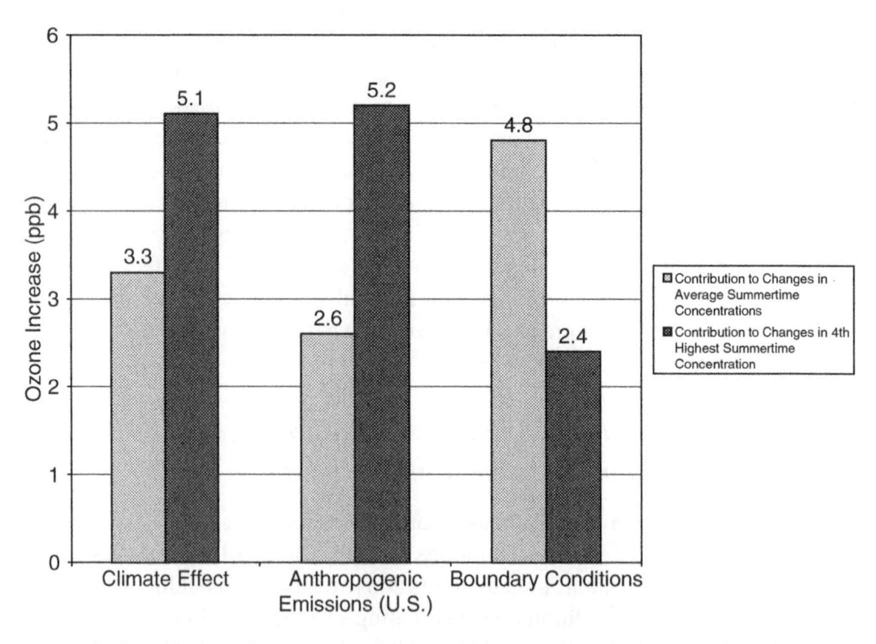

FIGURE 3. Contribution of changes in climate, anthropogenic emissions, and boundary conditions to changes in summertime average daily maximum 8-hr O_3 concentrations (light shading) and to changes in the 4th highest summertime daily maximum 8-hr O_3 concentration in the 2050s A2 scenario simulation (dark shading).

(2.6 ppb). For the 4th highest daily maximum 8-hr O_3 concentration, the effects of climate change and increased anthropogenic emissions account for almost equal increases of 5.1 ppb and 5.2 ppb, respectively, while increased boundary conditions account for an increase of 2.4 ppb. The reduced importance of changed boundary conditions for changes in the 4th highest daily maximum 8-hr O_3 concentration is consistent with the fact that high ozone concentrations in the eastern United States often occur under synoptic conditions characterized by slow-moving high pressure systems, clear skies and lower wind speeds. Such conditions tend to reduce the impacts of inflow from the model boundaries. In summary, while previous studies have pointed out the potentially important contribution of growing global emissions and intercontinental transport to O_3 air quality in the United States for future decades, the results presented above imply that the effects of a changing climate may be of at least equal importance when planning for the future attainment of the NAAQS.

3.3. Effects of Changed Land Use

Figure 4a illustrates that the predicted conversion of many grid cells in the greater New York City metropolitan area to 'low-residential urban' for the 2050s A2 land use scenario shown in Figure 1 leads to increases in episode-maximum temperatures over many grid cells in this area, with temperature increases over 2°C in some areas in Connecticut. When averaged over all non-water cells shown in Figure 4a, the increase in episode-maximum temperature increases is 0.7°C, consistent with the higher heat capacitity and lower moisture availability of urban grid cells compared to agricultural or forest grid cells. As Figure 4b illustrates, the impact of changes in meteorology caused by growing urbanization on episode-maximum 8-hr ozone concentrations is more variable. While most grid cells experienced an increase of ozone and the average increase over

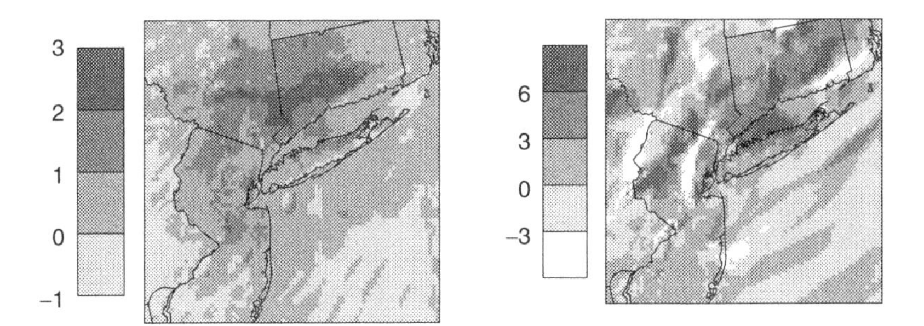

FIGURE 4. Changes in episode-maximum temperature (left) and episode-maximum 8-hr ozone concentration (right) caused by incorporation the changes in land use illustrated in Figure 1 for a 20-day MM5 simulation in 1993 and subsequently using these MM5 fields to perform CMAQ simulations.

all non-water grid cells was 1.5 ppb, some areas experienced decreases of more than 3 ppb in episode-maximum 8-hr ozone concentrations. To investigate this effect further, future analysis will include comparisons of other meteorological variables such as wind fields and mixed layer height between the base and future land use case simulations. The preliminary results presented here suggest that changes in land use can have local impacts of comparable magnitude as the other factors considered in this study, i.e. changes in regional climate, anthropogenic emissions, and chemical boundary conditions.

4. Summary

This paper described the application of a one-way coupled global/regional modeling system to simulate O_3 air quality in future decades over the eastern United States. The CMAQ simulations of O_3 concentrations utilizing the regional climate fields for the A2 and B2 emission scenario for the 2020s and 2050s show an increase in the frequency and duration of extreme O_3 events over the eastern United States in the absence of changes in anthropogenic emissions and boundary conditions. For the 2050s, it was determined that about half of the overall increase in summertime average daily maximum 8-hr O_3 concentrations is related to changes in biogenic emissions. Through additional sensitivity simulations for the 2050s, it was determined that changes in regional climate and increased anthropogenic emissions outweigh the effects of increased boundary conditions when changes in the 4th-highest summertime daily maximum 8-hr O_3 concentration are considered. Thus, while previous studies have pointed out the potentially important contribution of growing global emissions and intercontinental transport to O_3 air quality in the United States for future decades, the results presented in this study imply that the effects of a changing climate may be of at least equal importance when planning for the future attainment of regional-scale air quality standards such as the U.S. NAAQS. Analysis of the effects of land use change on temperature and ozone suggest that such changes can have local impacts of comparable magnitude as the other factors considered in this study, i.e. changes in regional climate, anthropogenic emissions, and chemical boundary conditions.

Acknowledgments This work is supported by the U.S. Environmental Projection Agency under STAR grant R-82873301.

Disclaimer Although the research described in this article has been funded in part by the U.S. Environmental Protection Agency, it has not been subjected to the Agency's required peer and policy review and therefore does not necessarily reflect the views of the Agency and no official endorsement should be inferred.

References

Byun, D.W. and Ching, J.K.S. (eds.), 1999. Science algorithms of the EPA Models-3 Community Multiscale Air Quality Model (CMAQ) modeling system. *EPA/600/R-99/030*, U. S. Environmental Protection Agency, Office of Research and Development, Washington, DC 20460.

Carolina Environmental Programs, 2003: Sparse Matrix Operator Kernel Emission (SMOKE) Modeling System, University of Carolina, Carolina Environmental Programs, Research Triangle Park, NC.

Fiore, A.M., Jacob, D.L., Bey, I., Yantosca, R.M., Field, B.D., Fusco, A.C., and J.G. Wilkinson, 2002: Background ozone over the United States in summer: Origin, trend, and contribution to pollution episodes, 2002: *J. Geophys. Res.*, **107 D(15),** 10.1029/2001JD000982

Fiore, A.M., T.A. Holloway, and M.G. Hastings, 2003: A global perspective on air quality: intercontinental transport and linkages with climate. Environ. Manag., December 2003, 13-22.

Geron, C. D., A. B. Guenther, and T.E. Pierce. 1994. An improved model for estimating emissions of volatile organic compounds from forests in the eastern United States. *J. Geophys. Res.,* **99**, 12,773-12,791.

Grell, G. A., J. Dudhia, and D. Stauffer, 1994: A description of the fifth-generation Penn State/NCAR Mesoscale Model (MM5). *NCAR Technical Note*, 138 pp., TN-398 + STR, National Center for Atmospheric Research, Boulder, CO.

Hogrefe, C., B. Lynn, K. Civerolo, J.-Y. Ku, J. Rosenthal, C. Rosenzweig, R. Goldberg, and P.L. Kinney, 2004a: Simulating changes in regional air pollution due to changes in global and regional climate and emissions, J. Geophys. Res., in press.

Hogrefe, C., J. Biswas, B. Lynn, K. Civerolo, J.-Y. Ku, J. Rosenthal, C. Rosenzweig, R. Goldberg, and P.L. Kinney, 2004b: Simulating regional-scale ozone climatology over the Eastern United States: Model evaluation results, *Atmos. Env.,* **38**, 2627-2638.

Intergovernmental Panel on Climate Change, 2000. Special Report on Emissions Scenarios. Nacenovic, N. and Swart, R. (eds.), Cambridge University Press, Cambridge, United Kingdom, 612 pp.

Intergovernmental Panel on Climate Change, 2001. Climate Change 2001: The Scientific Basis. Houghton J.T., Ding Y., Griggs, D.J., Noguer, M., van der Linden, P.J., Dai, X., Maskel, K., Johnson, C.A. (eds.), Cambridge University Press, Cambridge, United Kingdom, 944 pp.

Jacob, D.J., J.A. Logan, and P.P. Murti, 1999: Effect of rising Asian emissions on surface ozone in the United States, *Geophys. Res. Lett.*, **26**, 2175-2178.

Lynn, B.H., C. Rosenzweig, R. Goldberg, J. Dudhia, C. Hogrefe, D. Rind, L. Druyan, R. Healy, J. Biswas, P. Kinney, and J. Rosenthal, 2004: The GISS-MM5 regional climate modeling system: Part I: sensitivity of simulated current and future climate to model configuration, submitted to *J. Climate*

McCarthy, J.J., O.F. Canziani, N.A. Leary, D.J. Dokken, and K.S. White (eds.), 2001: Climate Change 2001: Impacts, adaptation and vulnerability. Intergovernmental Panel on Climate Change (IPCC). Cambridge University Press, New York, NY.

Prather, M. and D. Ehhalt, 2001: Chapter 4. Atmospheric chemistry and greenhouse gases, in "*Climate change 2001: the scientific basis*", pp. 239-287, edited by J.T. Houghton et al., Cambridge University Press, Cambridge.

Prather, M., et al., 2003: Fresh air in the 21st century? *Geophys. Res. Lett.*, **30**(2), doi:10.1029/2002GL016285

Russell, G.L., J.R. Miller, and D. Rind 1995: A coupled atmosphere-ocean model for transient climate change studies. Atmos.-Ocean 33, 683-730.

Solecki, W.D., and C. Oliveri, 2004: Downscaling climate change scenarios in an urban land use change model *J. Env. Manag.,* **72**, 105-115.

U.S. Environmental Protection Agency, 1994: User's guide to Mobile5 (Mobile source emission factor model), *Rep. EPA/AA/TEB/94/01*, Ann Arbor, MI.

Yienger, J.J., M. Galanter, T.A. Holloway, M.H. Phadnis, S.K. Guttikunda, G.R. Carmichael, W.J. Moxim, and H. Levy II, 2000: The episodic nature of air pollution transport from Asia to North America. *J. Geophys. Res.*, **105**, 26,931-26,946.

25 Calculated Feedback Effects of Climate Change Caused by Anthropogenic Aerosols

Trond Iversen, Jón Egill Kristjánsson, Alf Kirkevåg, and Øyvind Seland*

1. Introduction

Depending on their chemical composition, sizes and shapes, aerosol particles may scatter and absorb solar radiation and act as nuclei for condensation of water vapour and for freezing of water droplets. Availability of cloud condensation (CCN) and ice nuclei (IN) is responsible for the realized water vapour super-saturations in the troposphere.

Human activity inadvertently produces aerosol particles. Production mechanisms include combustion of fossil fuels and biomass, leading to submicron particles containing sulphate, nitrate, black carbon and particulate organic matter. These compounds typically reside up to a week in the troposphere and the mixing ratios have considerable gradients. Depending on their composition as a function of size and shape, particles may scatter and absorb solar radiation and act as CCN. Anthropogenic changes in these properties may directly produce radiative forcing, or indirectly through changes in cloud properties. Considerable attention has been paid to the potential climate influence of anthropogenic particles (e.g. Charlson *et al.*, 1987; Wigley, 1989; Charlson *et al.* 1991; Kiehl and Briegleb, 1993). There is considerable incertitude associated with its quantification, and in particular the indirect effect (Houghton *et al.*, 2001).

First principle calculations of aerosol-climate interactions are computationally impossible due to the complex processes involved. The chemical and physical processes that influence aerosols need to be parameterized in climate models. Climate scenario runs with prescribed aerosol forcing without feedback have proven successful for historical climate periods (Mitchell *et al.*, 1995; Delworth

* Trond Iversen, Jón Egill Kristjánsson, Alf Kirkevåg, and Øyvind Seland, Department of Geosciences, University of Oslo, P.O.Box 1022, Blindern, N-0315 Oslo, Norway. (trond.iversen@geo.uio.no, alf.kirkevag@geo.uio.no, j.e.kristjansson@geo.uio.no, oyvind.seland@geo.uio.no)

and Knutson, 2000), and simplified aerosol forcing has been included in future climate projections (e.g. Roeckner *et al.*, 1999). Several atmospheric GCMs calculate two-way interactions between parameterized aerosol properties and meteorological conditions (e.g. Kiehl *et al.*, 2000; Chin *et al.* 2000; Koch, 2001; Iversen *et al.*, 2001; Kirkevåg and Iversen, 2002, Kristjansson, 2002). However, the response of the climate system also implies changes in geophysical parameters such as sea-surface temperatures (SST) and sea-ice cover. Equilibrium climate simulations with atmospheric GCMs coupled to "slab ocean" models for the upper mixed layer, provide first-order estimates of such changes. Thus the indirect effects of sulphate have been studied in a few papers. Rotstayn *et al.* (1999) used prescribed sulphate concentration (*off-line*) whilst Williams *et al.* (2000) and Rotstayn and Lohman (2002) calculated sulphate as a part of the model (*on-line*). A southward shift of rainfall was found in the Tropics and a strong sea-ice albedo feedback in the Arctic.

Also the present paper discusses the indirect effects studied in an atmospheric GCM (CCM-Oslo) coupled to a slab ocean model. The experiments allow separate discussions of the response in atmospheric dynamics, sea-surface temperature (SST), and sea-ice (*the geophysical feedback*) on one hand, and the response in aerosol processes responsible for indirect forcing (*the chemical feed-back*) on the other.

2. Two Twin Experiments

CCM-Oslo is a well documented extension of the NCAR CCM3.2 global atmospheric model with resolution T42L18. It contains a prognostic cloud scheme (Rasch and Kristjansson, 1998), and calculates aerosol concentrations and interactions with radiation and clouds (Iversen and Seland, 2002, 2003; Kirkevåg and Iversen, 2002; Kristjansson, 2002). Primary marine (sea-salt) and continental (soil dust) aerosols are prescribed. Sulphate and BC are calculated from emissions estimated for the year 2000 (Penner *et al.*, 2001). Aerosols are allocated to production mechanisms, which enables estimated size-distributions and mixing states. Tables for optical parameters and water-activity are used to quantify interactions with radiation and clouds.

A series of 30 year equilibrium runs have been made with CCM-Oslo coupled to a slab ocean. Data from years 11-30 during equilibrium are used for analysis. Two twin experiments are made. *Twin 1* uses monthly aerosols prescribed from the atmospheric model alone; thus the geophysical response is calculated off-line. In the first member of the twin, the anthropogenic contribution of the aerosol emissions is removed whilst the second use all emissions. *Twin 2* uses the same two sets of emissions, but in this case the aerosols are calculated on-line with the geophysical variables. Anthropogenic increments are obtained by taking differences between members 2 and 1 for each twin. Twin 1 only produces geophysical feedbacks whilst twin 2 produces the combined geophysical and chemical feedbacks. Greenhouse gases are kept at the level of year 2000 throughout.

The anthropogenic aerosols are sulphate and black carbon (BC). Of these two components only sulphate has a potential indirect climate effect of any significance.

3. Geophysical Feedback

Table 1 shows global budget numbers for sulphur components in the model. The numbers for Twin 1 are for the prescribed budgets introduced off-line with the geophysical fields in the model. The numbers are similar to many other global models of this type, although the burdens and life-times are on the lower side. This is because the used model version does not include vertical transport in deep convective clouds (Iversen and Seland, 2004). The indirect anthropogenic aerosol forcing of sulphate and black carbon has tentatively been estimated as the sum of short-wave and long-wave radiative cloud forcing taken over the first year of the off-line experiment (Twin 1). In this way a top-of-the-atmosphere "quasi-forcing" by both indirect effects (the droplet radius and cloud life-time effects) is estimated to $-1.55Wm^{-2}$. There is much regional noise that would be reduced by ensemble averaging. The global long-wave forcing is zero.

Fig. 1 shows the 20-year equilibrium climate response of the indirect forcing by anthropogenic sulphate and BC. Shown are average anthropogenic change in 2m temperature, diurnal precipitation, and fractional cloudiness. Surface cooling is widespread in the northern hemisphere with a maximum (more than 3K) in the Arctic. At southern extra-tropical latitudes the cooling amounts to ca. 1K. In the tropics there are patches with a slight warming ($<0.5K$), due to changes in cloudiness. Global cooling matches the negative forcing with a global feedback parameter (top of the atmosphere forcing per 2m temperature response) of $1.34W/(m^2K)$. The feedback is positive at high latitudes due to increased extension of sea-ice and snow cover as the atmosphere cools.

Considerable precipitation changes are seen in the tropics, and the inter-tropical convergence zone is displaced southwards in many regions. This is in agreement with other model results and to some extent also with measured changes during the 20th century (Rotstayn and Lohman, 2002). This response is believed to be linked to the asymmetric cooling of two hemispheres due to more anthropogenic aerosols in the northern hemisphere. The cloud response is closely related to the precipitation, in particular in the Tropics. Although being a small impact, there is a net reduction of total cloudiness due to the indirect effect. The reduction is found in the middle subtropical troposphere; other areas experience increased cloudiness.

4. Chemical Feedback

Differences between corresponding variables calculated in twin 2 and twin 1 are estimates of the chemical feedback of the indirect effects of aerosols. Fig. 2 shows the effect of chemical feedback on the same variables as in Fig. 1, and the

TABLE 1. Global budget parameters for the production of airborne particulate sulphate.

	SO_x [a]	SO_2	SO_4-prod		SO_2		SO_4	SO_4		
	source (TgS/a)	dep. (%)	Aq. (%)	Gas. (%)	burden (TsS)	T [b] (days)	source (TgS/a)	wetdep. (%)	burden (TgS)	T [b] (days)
Twin 1: aerosols off-line										
Natural emissions	22.0	45.2	44.1	10.3	0.07	1.1	12.0	82	0.11	3.3
Total emissions	90.4	40.7	44.8	13.2	0.37	1.5	54.0	85	0.51	3.4
Anthropogenic increment	68.4	39.3	45.0	14.1	0.30	1.6	42.0	86	0.40	3.5
Twin 2: aerosols on-line										
Natural emissions	22.0	45.0	43.5	11.1	0.07	1.2	12.1	82	0.12	3.5
Total emissions	90.4	37.7	47.3	13.4	0.38	1.6	56.5	81	0.60	3.8
Anthropogenic Increment	68.4	35.3	48.5	14.1	0.31	1.7	44.4	81	0.48	3.9
Difference Twin 2 -Twin 1										
Anthropogenic Increment	0.0	–4.0	+3.5	0.0	+0.01	+0.1	+2.4	–5	+0.08	+0.4

[a] SO_x source are emissions of SO_2 and Sulphate plus SO_2 produced by oxidation of DMS.
[b] T are turnover times.

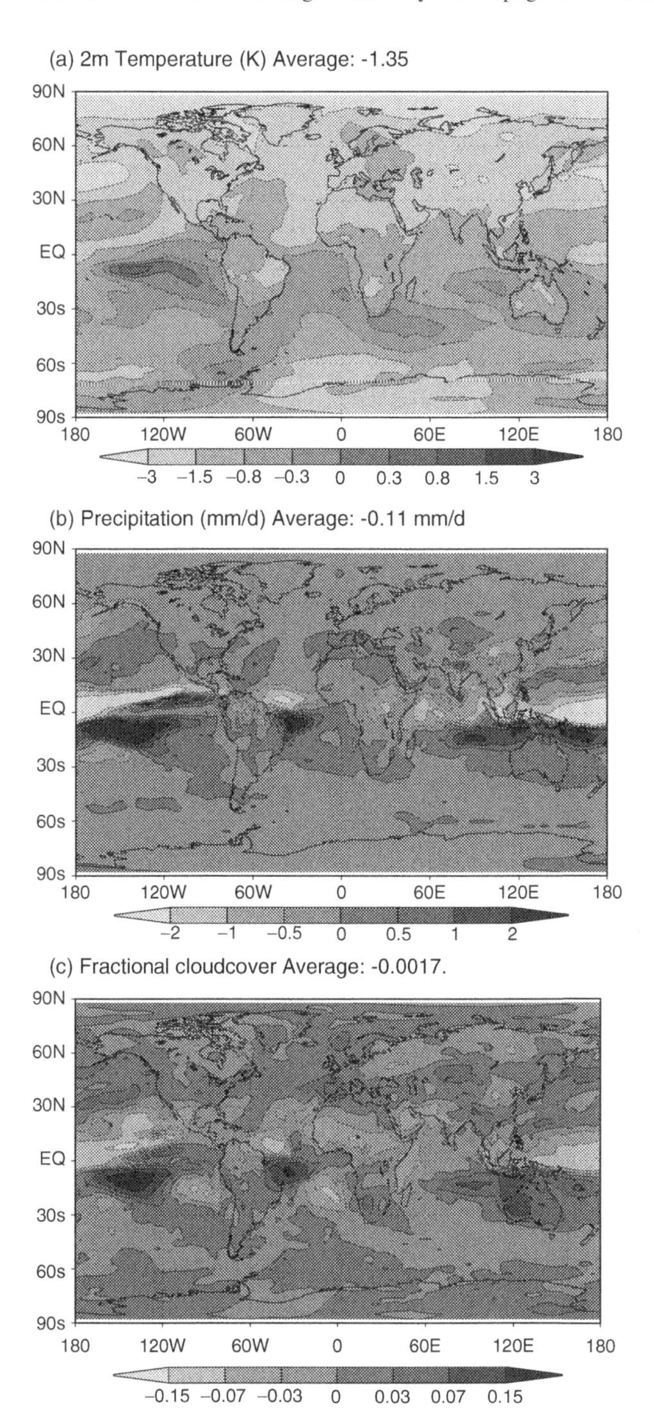

FIGURE 1. Calculated increments in selected equilibrium geophysical variables due to indirect effects of anthropogenic sulphate and black carbon.

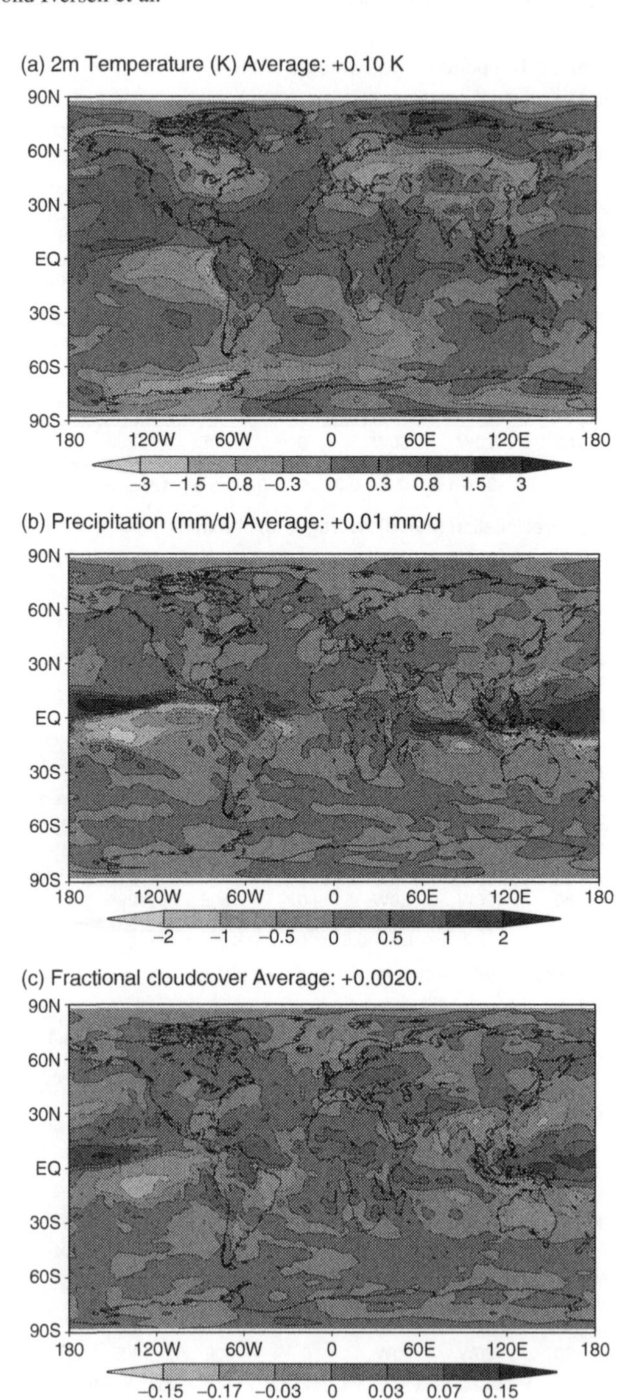

FIGURE 2. Difference between on-line and off-line calculated increments in selected equilibrium variables due to indirect effects of anthropogenic sulphate and black carbon.

patterns in Fig. 2 are to a large extent the opposite of those in Fig. 1. This means that the chemical feedback reduces the indirect climate effects in our calculations: the NH cooling is reduced and the displacement of tropical precipitation and associated clouds are partly reversed. Exceptions are the enhanced cooling in the three major emission regions of the NH and enhanced cloudiness in Europe. The indirect effect is far from cancelled, but the global feedback parameter is increased with 0.11W/(m^{2}K), i.e. 8%. Fig. 3 shows that the chemical feedback increases the amount of sulphate almost everywhere except in tropical regions influenced by the reversed displacement of tropical rainfall, and some oceanic regions where reductions are small. The global burden increases with ca. 17%.

Increased sulphate and reduced indirect effect is a paradox, since sulphate is responsible for the model's indirect effect in the first place. Table 1 shows that the chemical feedback reduces the SO_2 removal and that more SO_2 is transformed to sulphate by oxidation in cloud droplets. Furthermore, sulphate is less efficiently scavenged by precipitation. Reduced scavenging is consistent with the indirect effect causing precipitation decrease. The chemical feedback also produces slightly increased cloudiness globally, but in particular over the emissions in Europe (Fig. 2c).

Why does the indirect effect decrease with increased sulphate burden? First, sulphate produced in cloud droplets will, at most, to a small extent, produce new CCNs. Second, Fig. 4 shows that increased gas-phase sulphate due to chemical feedback is seen in the lower troposphere between 0° and 15°S where cloudiness is negligibly influenced. Figs. 2c and 3 confirm that the sulphate change is anti-correlated with cloudiness change (except in Europe). The main sulphate increase occurs in regions where it is already abundant, whilst the increase is small or

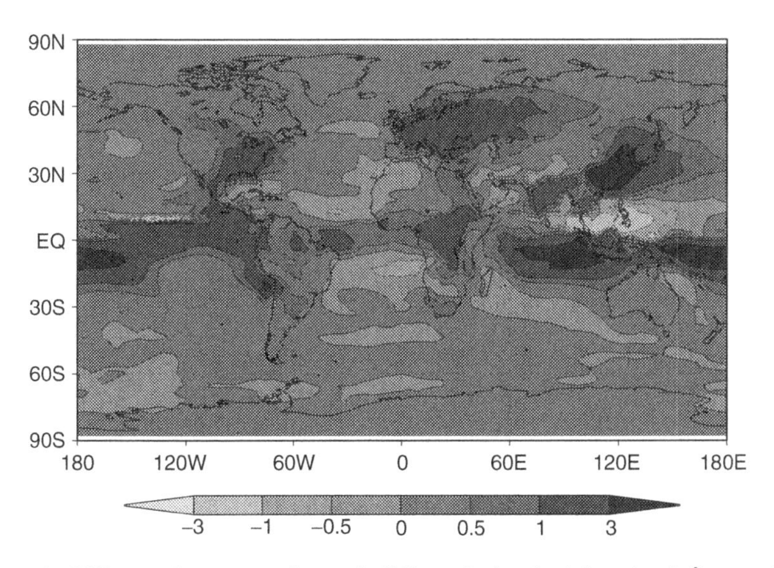

FIGURE 3. Difference between on-line and off-line calculated sulphate (mg/m^2) at equilibrium due to indirect effects of sulphate and black carbon. Average: +0.50 mg/m^2.

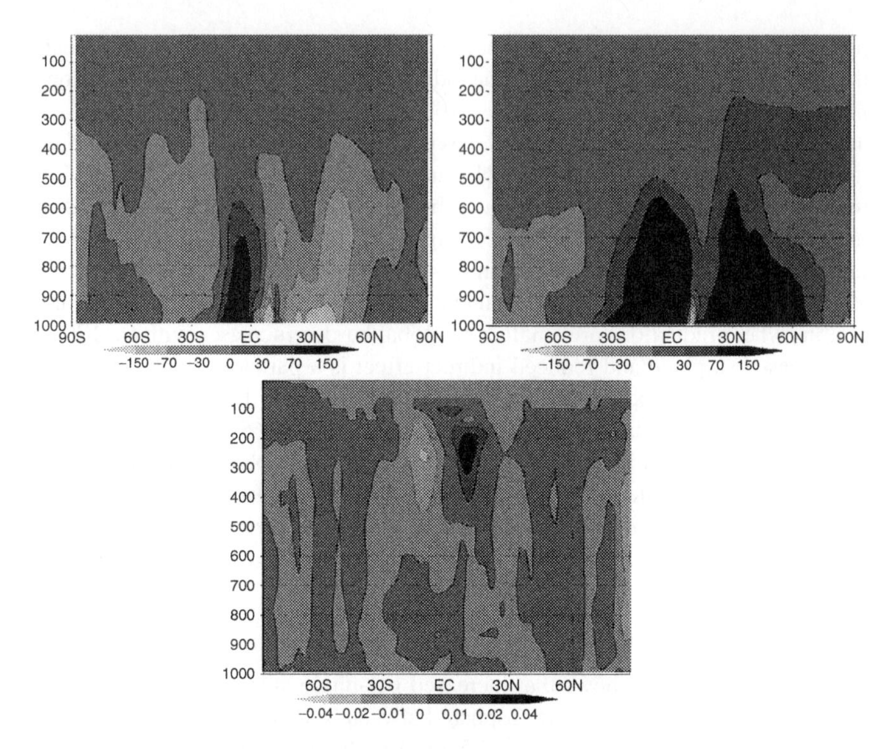

FIGURE 4. Zonal averages of differences between on-line and off-line equilibrium calculations of: **(above left)** sulphate produced in clear air (ng(SO_4)/kg(air)); **(above right)** sulphate produced in cloud droplets (ng(SO_4)/kg(air)); **(below right)** fractional cloudiness.

negative over oceans away from ITCZ. Hence, sulphate increases in regions where clouds are not sensitive to CCN amounts.

The chemical feedback is further strengthened as clouds in the upper tropical troposphere become more abundant when the southward displacement is reversed (Fig. 4). The 20-year averaged global long-wave cloud forcing is +0.21W/m^2 whilst the short-wave is only –0.14 W/m^2. Thus the net is positive (+0.07 W/m^2) despite enhanced sulphate burden. The chemical feedback on the direct effect remains to be estimated. Our result may be sensitive to our treatment of vertical transport in convective clouds.

Acknowledgements This work is partly financed by the Research Council of Norway (RegClim and AerOzClim). Computational costs are covered by a grant from the Research Council's Programme for Supercomputing. IPCC-emissions were provided by J. Penner. We gratefully acknowledge co-operation with P. J. Rasch, and S. Ghan.

References

Charlson, R.J., Lovelock, J.E, Andreae, M.O., Warren, S.W. 1987. Oceanic phytoplankton, atmospheric sulphur, cloud albedo and climate. *Nature*, **326**, 655-661.

Charlson, R.J., Langner, J., Rodhe, H., Leovy, C.B. and Warren, S.G. 1991. Perturbation of the northern hemisphere radiative balance by backscattering from anthropogenic sulphate aerosols. *Tellus*, **43AB**, 152-163.

Chin, M., Rood, R. B., Lin, S.-J., Müller, J-F., and Thompson, A. M. 2000. Atmospheric sulfur cycle simulated in the global model GOCART: Model description and global properties. *J Geophys. Res.* **105**, 24,671-24,687.

Delworth, T.L. and Knutson, T.R. 2000: Simulation of early 20th century global warming. Science, 287, 2246-2250.

Houghton, J.T., Ding,Y., Griggs, D.J., Noguer, M. van der Linden, P.J., Dai, X., Maskell, K., and Johnson, C.A. (eds.), 2001. *Climate Change 2001: The Scientific Basis*. Contribution of Working Group I to the Third Assessment Report of the Intergovernmental Panel of Climate Change. Cambridge University Press, 881 pp.

Iversen, T., Kirkevåg, A., Kristjansson, J.E. and Seland, Ø. 2001: Climate effects of sulphate and black carbon estimated in a global climate model. In: *Air Pollution Modeling and Its Application XIV, Gryning and Schiermeier (Eds.)*. 335-342. Kluwer Academic/Plenum, New York.

Iversen, T., and Seland, Ø., 2002: A scheme for process-tagged SO4 and BC aerosols in NCAR CCM3: Validation and sensitivity to cloud processes. *J. Geophys. Res.*, **107** (D24), 4751, 10.1029/2001JD000885.

Iversen, T., and Seland, Ø., 2003: Correction to "A scheme for process-tagged SO4 and BC aerosols in NCAR CCM3: Validation and sensitivity to cloud processes." *J. Geophys. Res.*, **108** (D16), 4502, 10.1029/2003JD003840.

Iversen, T. and Seland, Ø. 2004. The role of cumulus parameterisation in global and regional sulphur transport. In: *Air Pollution Modeling and Its Application XVI, Borrego and Incecik (Eds.)*. 225-233. Kluwer Academic/Plenum, New York.

Kiehl, J.T. and Briegleb, B.P. 1993. The relative roles of sulfate aerosols and grenhouse gases in climate forcing. *Science*, **260**, 311-314.

Kiehl, J. T., Schneider, T. L., Rasch, P. J., Barth, M. C., and Wong, J. 2000: Radiative forcing due to sulfate aerosols from simulations with the NCAR Community Climate Model (CCM3). *J. Geophys. Res.*, **105**, 1441-1457.

Kirkevåg, A., and Iversen, T., 2002. Global direct radiative forcing by process-parameterized aerosol optical properties. *J. Geophys. Res.*, **107**, (D20) 10.1029/2001D000886.

Kristjánsson, J. E., 2002. Studies of the aerosol indirect effect from sulfate and black carbon aerosols. *J. Geophys. Res.* **107**, 10.1029/2001JD000887.

Koch, D. 2001. Transport and direct radiative forcing of carbonaceous and sulphate aerosols in the GISS GCM. *J Geophys. Res.* **106**, 20,311-20,332.

Mitchell, J.F.B., Johns T.C., Gregory, J.M. and Tett, F.B. 1995: Climate response to increasing levels of greenhouse gases and sulphate aerosols. Nature, 376, 501-504.

Penner, J. E., Andreae, M., Annegarn, H., Barrie, L., Feichter, J., Hegg, D., Jayaraman, A., Leaitch, R., Murphy, D., Nganga, J., and Pitari, G., 2001. Aerosols, their direct and indirect effects. Chapter 5 (pp. 289-348) in: *Climate change 2001: The scientific basis*. Contribution of working group I to the Third Assessment Report of the Intergovernmental Panel on Climate Change. Cambridge University Press.

Rasch, P.J., and Kristjánsson, J.E., 1998: A comparison of the CCM3 model climate using diagnosed and predicted condensate parameterisations. *J. Climate*,**11**,1587-1614.

Roeckner, E, Bengtson, L., Feichter, J. Lelieveld, J., Rodhe, H. 1999: Transient climate change simulations with a coupled atmosphere-ocean GCM including the tropospheric sulfur cycle. *J. Clim.*, **12**, 3004-3032.

Rotstayn, L.D., Ryan, B.F., and Penner, J.E., 2000: Precipitation changes in a GCM resulting from the indirect effects of anthropogenic aerosols. *Geophys.Res.Lett.*, **27**, 3045-3048.

Rotstayn, L. D., and Lohmann, U., 2002. Tropical rainfall trends and the indirect aerosol effect. *J. Climate,* **15**, 2103-2116.

Williams, K. D., Jones, A., Roberts, D. L., Senior, C. A., and Woodage, M. J., 2001. The response of the climate system to the indirect effects of anthropogenic sulfate aerosol. *Climate Dyn.*, **17**, 845-856.

Wigley, T.M.L. 1989. Possible climate change due to SO_2-derived cloud condensation nuclei. *Nature*, **339**, 365-367.

26 Dimethyl Sulphide (DMS) and its Oxidation to Sulphur Dioxide Downwind of an Ocean Iron Fertilization Study, SERIES: A Model for DMS Flux

Ann-Lise Norman[1] and Moire A. Wadleigh

1. Introduction

Aerosol formation in the remote marine atmosphere and its effect on climate has been an area of intense study as a negative feedback to warmer surface oceans. Fine aerosols form cloud condensation nucleii (CCN) and scatter incident radiation back to space. Charlson *et al.*, 1987 proposed that dimethylsulphide (DMS), a gas released from the ocean during turnover of microalgae populations in the ocean's surface, and its oxidation products including sulphur dioxide (SO_2) were crucial pieces to understanding the global climate puzzle.

DMS is produced from cleavage of DMSP (dimethylsulphoproprionate) in the surface ocean, which in turn is released by microalgae during cell lysis. A portion of the DMS in the surface ocean mixed layer is released from the water in response to wave action. Once in the air, it may be oxidized to sulphur dioxide (SO_2), aerosol sulphate and methanesulphonic acid (MSA).

Recent studies have shown biotic productivity increased after deliberate iron additions in regions characterized by high nitrogen, low chlorophyll (HNLC) (e.g. Boyd *et al.*, 2004). The North Pacific, near Ocean Station Papa (OSP 50°N, 145°W) was the subject of a recent iron fertilization experiment in July 2002 called the Subarctic Ecosystem Response to Iron Enrichment Study (SERIES). Iron concentrations of >1 nmol/L were distributed over a 77 km^2 patch of ocean on July 9 and 16. By the end of the experiment the patch size increased to approximately 1000 km^2. Measurements in and above the surface ocean were performed aboard three ships-based and these were complemented by a single land-based

[1] A. L. Norman, Department of Physics and Astronomy, The University of Calgary, Calgary, Alberta, Canada T2N 1N4, annlisen@phas.ucalgary.ca

site on the west coast of Vancouver Island, Canada, to shed light on factors influencing C and S cycling at the water-air interface (Boyd *et al.*, 2004). During the SERIES experiment DMS in air reached mixing ratios of 7600 $pmolmol^{-1}$, 40 times higher than average ocean DMS ratios (Wadleigh *et al.*, 2004).

DMS fluxes over the remote ocean are typically calculated based on surface water DMS concentrations (C_w) and a parameterization based on wind speed, such as that by Wannikhof and McGillis (1999), to calculate ventilation rates (K_{WM}). Flux (F) is calculated using the difference in surface water and atmospheric concentration (in $\mu molL^{-1}$) divided by the dimensionless Henry's Law constant (K). Atmospheric DMS concentrations (C_a) are often too small to impact the calculation of flux. However for the SERIES experiment Wadleigh *et al.*, (2004) showed that DMS fluxes changed by as much as 50 % if C_a was included.

$$F = K_{WM} (C_w - C_a/H)$$

A second method for calculating flux is based on a steady state model that uses atmospheric DMS concentrations, boundary layer height (BLH), and OH plus BrO oxidation as functions of temperature as input parameters (Kouvarakis and Mihalopoulos, 2002, Wadleigh *et al.*, 2004). Entrainment flux (E) is calculated assuming a buffer layer DMS concentration that is 10% that in the boundary layer (Andreae *et al.*, 1993).

$$F_{DMS} = C_a \times BLH \times (K_{OH}[OH] + K_{BrO}[BrO]) - E$$

Mechanisms controlling the oxidation and nucleation of the aerosol precursor gases DMS and SO_2 over the remote ocean have been the subject of numerous modeling exercises (Davis *et al.*, 1999, Sciare *et al.*, 2000, Yvon *et al.*, 1996, von Glasow and Crutzen, 2004). Also Kettle and Andreae (2000) compiled information on discrete shipboard and coastal DMS measurements for use in an oceanic DMS flux database. However, there are considerable temporal and spatial variations in both atmospheric and ocean DMS concentrations. New methods to determine large-scale DMS fluxes present the opportunity for significant improvements in quantifying the magnitude and effects of DMS oxidation. Here we use isotope apportionment techniques to calculate concentrations of DMS derived SO_2 and aerosol sulphate. In addition, DMS fluxes for the North Pacific west of Ucluelet, B.C., Canada, were modeled using SO_2 and sulphate concentrations downwind of the SERIES experiment.

2. Methods

DMS concentrations in the boundary layer and surface waters were made hourly throughout the cruise. Details of measurement methods for atmospheric DMS and ocean DMS concentrations can be found in Wadleigh *et al.* (2004) and Levasseur *et al.* (2004) respectively.

Total aerosol sulphate and SO_2 were collected on board the ship and downwind on the roof of the Coast Guard station at Ucluelet on the west side of Vancouver

Island, British Columbia, Canada using high volume samplers. The two samplers were interfaced to an anemometer situated 10m above ground and were programmed to operate only during ocean winds when relative humidity was less than 99 % and wind speeds were above 1 km/hr. One high volume sampler was used to collect total aerosol sulphate and SO_2. A quartz fibre filter (Whatman 2500) was placed in a filter cassette above a 2nd cellulose acetate filter treated with K_2CO_3 and glycerol to trap SO_2. The second high volume sampler was fitted with a 5-stage cascade impactor (Sierra Series 230) followed by a sixth quartz filter to trap all aerosols less than 0.45 microns in diameter.

Sample volume and frequency was dictated by the minimum amount of sample for isotope analysis, 10 μg of S (minimum 500 minutes). In total, samples were collected during five periods: July 1-6, July 7-9, July 10-15, July 16-28, and July 29 – Aug. 5.

Sample extraction and preparation for isotope and ion analysis is documented elsewhere (Norman *et al.*, 2004). Isotope composition and ion concentrations were blank corrected. Isotope apportionment was based on mass balance calculations assuming Na is a conservative tracer of seawater sulphate (SS).

$$\delta^{34}S = \delta^{34}S_{SS} \times f_{SS} + \delta^{34}S_A \times f_A + \delta^{34}S_B \times f_B$$

$\delta^{34}S$ values express the difference in the ratio of ^{34}S to ^{32}S in a sample relative to an international standard in parts per thousand (‰). This value is assumed to be constant for each of the three sources: sea salt $\delta^{34}S = +21‰$, anthropogenic $\delta^{34}S_A = +2‰$, and biogenic $\delta^{34}S_B = +18‰$. The fractional contribution (f) from each of these sources is then calculated using the ratio of $SO_4/Na = 0.252$ to constrain the sea water contribution.

Measurement uncertainties (1σ) for isotope ratios were ±0.3‰ based on replicate measurements. Uncertainties in concentration values for SO_2 and sulphate were plus/minus 20 %. Replicate measurements of sodium standards were within 10 %.

3. Results and Discussion

SO_2 mixing ratios above the ocean to the west of Ucluelet, Vancouver Island, ranged from 22 to 330 pmolmol^{-1} between July 1 and August 5 with peak values observed during the period July 16-28 after the 2nd iron fertilization at Station Papa. The temporal variation in mixing ratios for total SO_2 and its apportionment between anthropogenic and ocean biogenic sources is shown in Figure 1. Although the mixing ratio for total SO_2 did not exhibit larger values after iron fertilization, SO_2 from DMS oxidation was its dominant source from July 10-15 and was 2.5 times higher than biogenic SO_2 prior to the experiment and returned to initial mixing ratios by August 29.

The proportion of SO_2, non-sea salt aerosol sulphate in total particulate matter and sulphate in particulate matter less than 0.45 μm were compared. Between July 1-5 and 10-15, sulphate in the fine aerosol fraction dominated the total

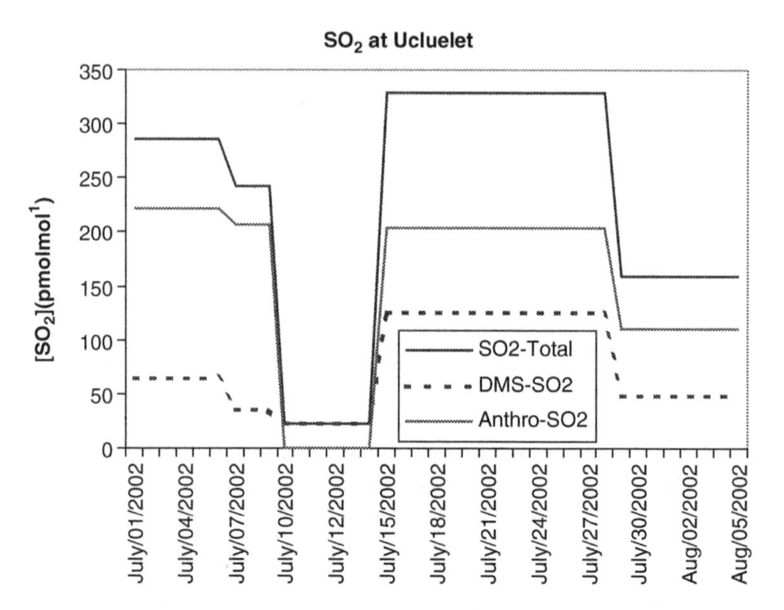

FIGURE 1. SO_2 mixing ratios at Ucluelet and its apportionment between anthropogenic and biogenic sources is shown as a function of time.

aerosol sulphate load. For July 1-5 anthropogenic SO_2 dominated SO_2 mixing ratios (Figure 1) whereas from July 10-15 biogenic SO_2 was its sole source.

Biogenic sulphur at Ucluelet, in the form of SO_2 and total aerosol sulphate increased after the second iron fertilization by a factor of 3 relative to its concentration in air prior to SERIES. Although SO_2 decreased after August 29, biogenic sulphate steadily increased after the first SERIES fertilization as expected if DMS were oxidized to SO_2 and other intermediates prior to sulphate formation. In contrast to increasing atmospheric sulphur concentrations due to biogenic sources, anthropogenic sulphur concentrations remained relatively constant, averaging 0.30 ± 0.15 μgSm^{-3}.

3.1. Model Description

A steady state model was developed to describe SO_2 from DMS oxidation during the SERIES experiment. The model incorporates a production term for DMS oxidation by OH and BrO (P), entrainment (E), and loss due to homogeneous oxidation (Ox), aerosol oxidation (AOx), deposition (D), and in-cloud oxidation (COx).

$$d[SO_2]^{*}BLH/dt = P + E_{SO2} - Ox - AOx - D - COx$$

$$P = R \times F_{DMS}$$

F_{DMS} was calculated from atmospheric DMS concentrations as described in the introduction. DMS fluxes were calculated using values for K_{OH} and K_{BrO} of

1.13×10^{-11} $e^{(-253/T)}$ $cm^3molecule^{-1}s^{-1}$ and 2.54×10^{-14} $e^{(850/T)}$ $cm^3molecule^{-1}s^{-1}$ after DeMore *et al.*, (1997) and Ingham *et al.*, (1999) respectively. Average hydroxyl radical concentrations were assumed to be 9.85×10^5 $moleculescm^{-3}$ which is half the expected value for July at 50° N Lawrence *et al.*, (2001). BrO concentrations were assumed to be 4.03×10^8 $moleculescm^{-3}$, the minimum measured by Leser *et al.*, 2003. R is the branching ratio for SO_2 to other DMS oxidation products. A value of R=0.72 was used here based on the results of Davis *et al.*, (1999).

$$E_{SO2} = 0.1 \times [SO_2] \times v_e$$

Entrainment was calculated from biogenic SO_2 concentrations times the entrainment velocity ($0.35\ cms^{-1}$).

$$Ox = K_{OH} \times [OH] \times [SO_2] \times BLH$$

$K_{OH} = A/((1+A/B) \times 0.5 \times (1/(1+\{LOG10(A/B)\}^2))$ where $A = (3 \times 10^{-31} \times (T/300)^{-3.3})$ and $B = 1.5 \times 10^{-12}$ $moleculescm^{-3}$ from Yin *et al.*, (1990). Average surface temperatures were 11.7°C during SERIES.

$$AOx = 0.5 \times [SO_{4DMS}] \times BLH \times t^{-1}$$

where $[SO_{4DMS}]$ is a measured quantity. Here we assume half the biogenic aerosol sulphate is the result of SO_2 oxidation and derive values for the lifetime of sulphate aerosol during the cruise when DMS fluxes were measured assuming steady state conditions apply for $[SO_2] \times BLH$.

$$D = [SO_{2DMS}] \times C \times v_{av}$$

with constant, $C = 1.22 \times 10^{-3}$ and $v_{av} = 8\ ms^{-1}$ average wind speed (Yvon *et al.*, 1996).

$$COx = 1.275 \times 10^{-6}\ s^{-1} \times [SO_2] \times BLH$$

Here the rate constant for in cloud oxidation is taken from Sciare *et al.*, (2000).

3.2. *Model Results*

The model was applied to the 5 sampling periods. Two periods coincided with shipboard measurements of atmospheric DMS on July 10-15 and July 16-28. These cases were used to constrain the rate constant for in-cloud oxidation. These rate constants were applied to the three sampling periods where no DMS measurements were available to calculate DMS flux. Model output is the DMS flux (Table 1). Derived values for the oxidation rate for aerosols were 9.3 and 17.3 d^{-1} for July 10-15 and 16-28, respectively. These values are not unreasonable; Sciare *et al.*, (2000) used a rate constant of 7.3 d^{-1} for aerosol oxidation over the Atlantic while Aranami and Tsunogai (2004) found that turnover times in the Pacific were on the order of 10 days irrespective of region and season. Modeled DMS fluxes are consistent with values derived from measurements: fluxes on the order of 10 $\mu molem^{-2}d^{-1}$ prior to the SERIES experiment are consistent with DMS fluxes over the Pacific ocean (Aranami and Tsunogai, 2004). After SERIES, DMS fluxes were larger by approximately a factor of four.

TABLE 1. DMS fluxes derived from the model.

Sampling Period Start Date	DMS Flux $t^{-1} = 9.34\ d^{-1}$ (μmolem^{-2} d^{-1})	DMS Flux $t^{-1} = 17.395 d^{-1}$ (μmolem^{-2}d^{-1})
July/01/2002	9.5	15.4
July/07/2002	6.8	11.4
July/29/2002	38.0	69.0

4. Conclusions

Isotope apportionment was used to determine the fraction and concentration of DMS oxidation products and pollutant sulphur downwind of the SERIES iron fertilization experiment in the North Pacific in July 2002. Sulphur concentrations attributable to DMS oxidation were three times higher for westerly winds at Ucluelet, Vancouver Island, Canada after iron fertilization and total aerosol sulphate increased throughout the sampling period (July 1–August 6). Fine DMS aerosol sulphate made up 22 to 99% of total aerosol non-sea salt sulphate. Pollutant concentrations varied, but less so than for DMS oxidation products. SO_2 at Ucluelet averaged 150 ± 90 picomolemole^{-1}, 93 ± 60 ngSm^{-3} for total aerosol sulphate and 24 ± 27 ngSm^{-3} for sulphate in aerosols less than 0.45 micrometers diameter between July 1 and August 5.

DMS fluxes ranging from 9.5 - 69.0 (μmolem^{-2} d^{-1}) were calculated using isotope apportionment techniques on samples collected at Ucluelet on the west coast of Vancouver Island prior to and after the SERIES ocean iron-fertilization experiment. DMS fluxes prior to iron-fertilization (July 1-9) are similar to values found in the literature for the world's oceans. This changed during the experiment. To our knowledge DMS fluxes were 10 times higher during SERIES than any other values reported so far in the literature (Wadleigh *et al.*, 2002). DMS oxidized to sulphur dioxide and aerosol sulphate, was collected downwind during SERIES and their concentrations, derived from isotope apportionment techniques, were approximately doubled. The maximum area of the fertilized patch, determined by an SF_6 tracer, was ~ 1000 km^2 by the time SERIES ended. In contrast, a simplistic estimation to determine the region encompassed by sectored sampling (~120°) and assuming that oxidation mechanisms were not affected by iron-fertilization can be roughly approximated from the average wind speed (8 m/s) and by assuming a conservative transport time of 2 days for SERIES oxidation products. Using these assumptions the region sampled at Ucluelet covered roughly 6×10^6 km^2. It seems unlikely that even a hundred fold increase in DMS fluxes for an area occupying 0.02 % of the region sampled could influence the concentration of oxidation products downwind. Instead, it is more likely that a natural phytoplankton bloom observed just prior to iron-fertilization (Levasseur *et al.*, 2004) caused widespread DMS flux increases over a wide patch of the ocean and that increases attributable to DMS from iron fertilization were superimposed on a coincident and large natural variation. We would like to emphasize however, that if studies by other scientists aboard the SERIES experiment indicate a change in the composition of air above the patch, particularly with respect to halogen chemistry, the statement above (that it is doubtful that DMS from the patch influenced atmospheric SO_2 and sulphate concentrations downwind) needs to be revisited.

Acknowledgements This study was supported by a grant from NSERC and the Canadian Foundation for Climate and Atmospheric Sciences (CFCAS). The authors thank the Canadian Coast Guard at Ucluelet, and Ainslie Campbell and Wing Tang for technical assistance.

References

Aranami, K., and Tsunogai, S., 2004, Seasonal and regional comparison of oceanic and atmospheric dimethylsulfide in the northern North Pacific: Dilution effects on its concentration in winter. *J. of Geophys. Res.* **109** D12303: doi:10.1029/2003JD004288.

Andreae, T.W., Andreae, M.O., Bingemer, H.G., Leck, C., 1993, Measurements of dimethyl sulfide and H_2S over the western North Atlantic and the tropical Atlantic. *J. of Geophys. Res.* **98** (D12):23,389.

Boyd, P.W., Law, C.S., Wong, C.S., Nojiri, Y., Tusda, A., Levasseur, M., Takeda, S., Rivkin, R., Harrison, P.J., Strzepek, T., Gower, J., McKay, R.M., Abraham, E., Arychuk, M., Barwell-Clarke, J., Crawford, W., Crawford, D., Hale, M., Harada, K., Johnson, K, Kiyosawa, H., Kudo, I., Marchetti, A., Miller, W., Needoba, J., Nishioka, J., Ogawa, H., Page, J., Robert, M., Saito, H., Sastri, A., Sherry, N., Soutar, T., Sutherland, N., Taira, Y., Whitney, F., Wong, S.K.E., Yoshimura, T., 2004, The decline and fate of an iron-induced subarctic phytoplankton bloom. *Nature*, **428**:549-553.

Charlson, R.J., Lovelock, J.E., Andraea, M.O., Warren, S.G., 1987, Oceanic plankton, atmospheric sulphur, cloud albedo and climate. *Nature* **326**: 655-661.

Davis, D., Chen, G., Bandy, A., Thornton, D., Eisele, F., Mauldin, L., Tanner, D., Lenschow, D., Fuelberg, H., Huebert, B., Heath, J., Clarke, A., Blake, D. 1999, Dimethyl sulfide oxidation in the equatorial Pacific: Comparison of model simulations with field observations for DMS, SO_2, H_2SO_4(g), MSA(g), MS, and NSS. *J. of Geophys. Res.* **104** (D5): 5765-5784.

DeMore, W.B., Sander, S.P., Golden, D., Hampson, R.F., Kurylo, M.J., Howard, C.J., Ravishankara, A.R., Kolb, C.E., Molina, M.J., 1997, Chemical kinetics and photochemical data for use in stratospheric modeling. Evaluation no.12 , *JPL Publ.* **97**-4.

Ingham, T., Bauer, D., Sander, R., Crutzen, P.J., Crowley, J.N., 1999, Kinetics and products of the reactions BrO+DMS and Br+DMS at 298K., *J. of Phys. Chem., A* **103**: 7199-7209.

Kettle, A.J., and Andreae, M.O., 2000, Flux of dimethylsulfide from the oceans: A comparison of updated data sets and flux models. *J. of Geophys. Res* **105**, D22: 26,793-26,808.

Lawrence, M.G., Jöckel, P., von Kuhlmann, P., 2001, What does the mean global OH concentration tell us? *Atmos. Chem. Phys.* **1**: 37-49.

Leser, H., Honninger, G., Platt, U. 2003, MAX-DOAS measurements of BrO and NO_2 in the marine boundary layer. *Geophys. Res. Lett.*. **30**: 1537, doi: 10.1029/2002GL015811.

Levasseur, M., Scarratt, M., Michaud, S., Merzouk, A., Boyd, P.W., Rivkin, R., Hale, M., Le Clainche, Y., Wong, C.S., Law, C.S., Sherry, N., Tsuda, A., Takeda, S., Matthews, P.,

Harrison, P.J., Miller, W., Kiene, R., Kiyosawa, H., Arychuk, M., Li, W.K.W., Vezina, A., 2004, Iron enrichment decreases DMS production in the subarctic Northeast Pacific. (submitted to Nature).

Kouvarakis, G., Mihalopoulos, N., 2002, Seasonal variation of dimethylsulfide in the gas phase and methanesulfonate and non-sea-salt sulfate in the aerosols phase in the Eastern Mediterranean atmosphere. Atmos. Environ. **36**: 929-938.

Norman, A.L., Belzer, W., and Barrie, L.A., 2004, Insights into the biogenic contribution to total sulphate in aerosol and precipitation in the Fraser Valley afforded by isotopes of sulphur and oxygen. *J. of Geophys. Res* **109**: D05311, doi:10.1029/2002JD003072.

Sciare, J., Baboukas, E., Kanakidou, M., Krischeke, U., Belviso, S., Bardouki, H., Mihalopoulos, N., 2000, Spatial and temporal variability of atmospheric sulfur-containing gases and particles during the Albatross campaign. *J. of Geophys. Res.* **105**, D11: 14,433-14,448.

von Glasow, R., and Crutzen, P.J., 2004., Model study of multiphase DMS oxidation with a focus on halogens. *Atmos. Chem. Phys.* **4**: 589-608.

Wadleigh, M.A., Norman, A.L., Burridge, C., Scarratt, M., Levasseur, M. , Sharma, S. and Wong, C.S. 2004, Trends in atmospheric DMS resulting from the SERIES iron enrichment experiment in the subarctic Pacific Ocean. *Deep Sea Res.* (submitted).

Wanninkhof, R. McGillis, W.R., 1999, A cubic relationship between air-sea CO_2 exchange and wind speed. *Geophys. Res. Lett.* **26**: 1889-1892.

Yin, F., Grosjean, D., and Seinfeld, J.H. 1990, Photooxidation of dimethylsulfide and dimethyldisulfide: Mechanism development. *J. Atmos. Chem.* **11**: 309-364.

Yvon, S.A., and Saltzman, E.S., 1996, Atmospheric sulfur cycling in the tropical Pacific marine boundary layer (12°S, 135°W): A comparison of field data and model results. 2. Sulfur dioxide. *J. of Geophys. Res.* **101** (D3): 6911-6918.

Yvon, S.A., Saltzman, E.S., Cooper, D.J., Bates, T.S., Thompson, A.M., 1996., Atmospheric sulfur cycling in the tropical Pacific marine boundary layer (12°S, 135°W): A comparison of field data and model results. 1. Dimethylsulfide. *J. of Geophys. Res.* **101** (D3): 6899-6909.

DMS and its Oxidation to SO_2 Downwind of an Ocean Iron Fertilization Study: SERIES: A Model for DMS Flux

Speaker: A. L. Norman

Questioner: A. Gross

Question: As I remember Miholopoulos cannot measure BrO at the measurement station at Amsterdam Island, the BrO concentrations are below the detection limit. Which concentration of BrO has been used in your study?

Answer: In the absence of measured BrO concentrations during the SERIES study, the lower concentration reported by Leser et al., 2003 (Geophys. Res. Lett. 30, 2003) of (4×10^8 molecules cm^{-3}) and used by von Glasow and Crutzen in their models of halogen reactions with DMS (Atmos. Chem. Phys. 4, 2004) was used here.

Aerosols as Atmospheric Contaminants

27 Aerosol Modelling with CAMX4 and PMCAMX: A Comparison Study

S. Andreani-Aksoyoglu*, J. Keller, and A. S. H. Prévôt

1. Introduction

Recently there have been significant improvements with respect to aerosol modelling. Models with 1-atmosphere approach are able to simulate both gaseous and particulate pollutants and suitable for episodic and annual calculations (Ackermann et al., 1998; Schell et al., 2001). There are also more complete models including full-science algorithms for aerosol modelling (Griffin et al., 2003; Bessagnet et al., 2004; Zhang et al., 2004). However, such models are more complex and demanding. A model intercomparison study showed that a more complex model approach to the aerosol problem does not automatically lead to better results in a 3-dimensional application (Hass et al., 2003). As long as speciated aerosol measurements with high time resolution are limited, the necessity of using so called full-science aerosol models is questionable. In this study, two air quality models with different approaches were applied to two domains, northern Italy and Switzerland, to investigate the capabilities, strengths and weaknesses.

2. Modelling

2.1. Models

In this study, two CAMx models (The **C**omprehensive **A**ir Quality **M**odel with e**x**tensions) were used. CAMx is an Eulerian photochemical dispersion model that allows for integrated assessment of gaseous and particle air pollution over many scales (Environ, 2004). The first model is the latest version of CAMx (version 4.03) which is called CAMx4. It has a 1-atmosphere approach for gaseous and particulate air pollution modelling. Aqueous sulfate and nitrate formation in cloud water is calculated using RADM aqueous chemistry algorithm

* Laboratory of Atmospheric Chemistry, Paul Scherrer Institute, Villigen PSI 5232 Switzerland

(Chang et al., 1987). Partitioning of condensible organic gases to secondary organic aerosols (SOA) to form a condensed organic solution phase is performed by the semi-volatile equilibrium scheme called SOAP (Strader et al., 1998). There are four SOA classes depending on the gaseous precursors. The first three classes represent SOA formed from anthropogenic precursors and the aerosols in the fourth class are formed from the biogenic precursors. ISORROPIA thermodynamic module is used to calculate the partitioning of inorganic aerosol constituents between gas and particle phases (Nenes et al., 1998). Particle sizes are static. Primary particles are modelled as fine and/or coarse particles whereas secondary species are modelled as fine particles. In this study, the particle size range for these secondary aerosol species was chosen as 0.04 -2.5 μm, as indicated by experimental results.

The second model used in this study is PMCAMx (version 3.01) which contains full science aerosol algorithms and is still under development and testing (Environ, 2003). It is a more complete aerosol model than CAMx4, but it is more demanding as well. Aerosol dynamics and particle size distribution are treated with a sectional approach covering the range of 0.04 μm to 40 μm. PMCAMx contains a more complete aqueous chemistry mechanism (Variable Size Resolution Model) developed by the Carnegie Mellon University. The two models have similar gas phase (CBMIV/SAPRC99), inorganic (ISORROPIA) and organic (SOAP) aerosol treatment. The main difference is in the particle size distribution, aerosol dynamics and aqueous chemistry mechanisms (Table 1).

2.2. *Applications*

2.2.1. Northern Italy

The first application was carried out over a domain covering northern Italy, for the period 12-13 May 1998 (Figure 1). The model domain consists of 47×54 grid cells with a resolution of 3km × 3km. There are 8 layers in a terrain-following coordinate system, the first being 50 m above ground. The model top was set at about 3000 m agl. The same input data from the modelling study using the previous CAMx version 3.10 (Andreani-Aksoyoglu et al., 2004) were used for this application. Simulations started on May 11 at 1200 central European summer time (CEST) and ended on May 13 at 2400 CEST. The first 12 hours were used to initialize the model.

TABLE 1. Comparison of CAMx4 and PMCAMx models.

	CAMx4	PMCAMx
approach	1-atmosphere	full-science PM model
gas-phase mechanism	CBMIV/SAPRC99	CBMIV/SAPRC99
inorganic aerosol module	ISORROPIA	ISORROPIA
organic aerosol module	SOAP	SOAP
aqueous chemistry	RADM	VSRM
particle size	fine/coarse	10-sectional
relative cpu time	1	3

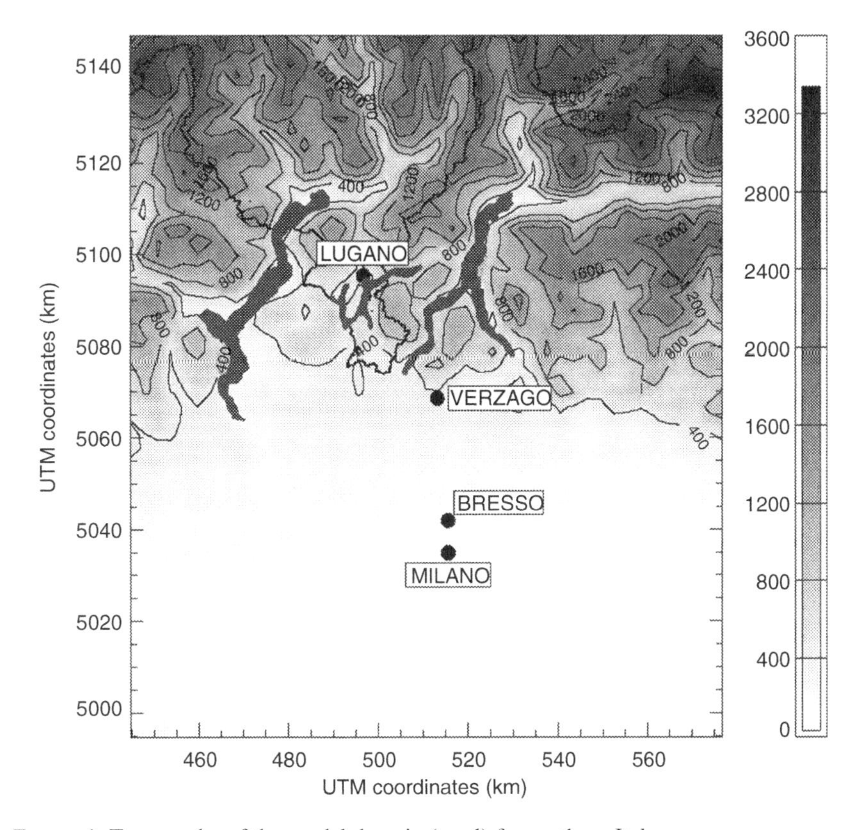

FIGURE 1. Topography of the model domain (masl) for northern Italy.

2.2.2. Switzerland

The second study was carried out in a domain covering Switzerland and some parts of the neighbouring countries (Figure 2) between 4-7 August 2003 which was an exceptionally hot period. The model domain encloses 650 km in the west-east and 450 km in the north-south direction with a horizontal resolution of 9 km. In the vertical direction 10 layers with varying heights were used, lowest being 30 m above ground. The model top is at about 4000 m above ground. For this application, meteorological data were calculated using MM5 meteorological model (PSU/NCAR, 2004). Initial and boundary trace gas conditions were extracted from the output of the European model REM3, provided by the Free University of Berlin. The emission inventory of gaseous species for Europe was based on an inventory provided by the Free University of Berlin. The emissions in Switzerland were compiled from various sources for the year 2000.

CAMx4 results refer to PM2.5 whereas PMCAMx calculates the aerosol concentrations for each of the 10 size bins between 0.04 and 40 μm. The sum of the first 6 sections which corresponds to PM2.5, was compared with CAMx4 results.

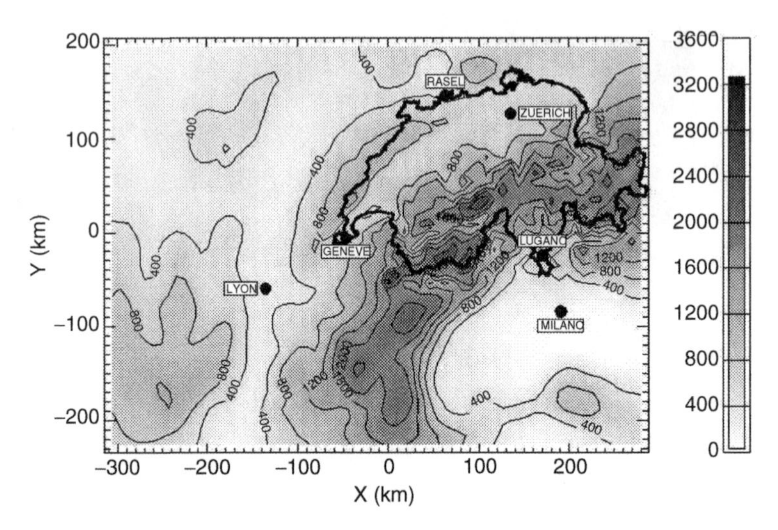

FIGURE 2. Topography of the model domain (masl) for Switzerland.

3. Results

3.1. Northern Italy

Comparison of model results with measurements of secondary inorganic aerosols performed in Verzago, 35 km north of Milan, shows that the particulate NO_3 and NH_4 levels are similar and close to the observations in the afternoon (Figure 3). At night, CAMx4 predicts higher nitrate concentrations than PMCAMx. In case of particulate SO_4, PMCAMx predicts slightly lower concentrations than CAMx4 and both model results are lower than measurements at night. Since both models use the same inorganic aerosol module and the same gas-phase mechanism, similar results are expected. However, it should be kept in mind that the gas-phase

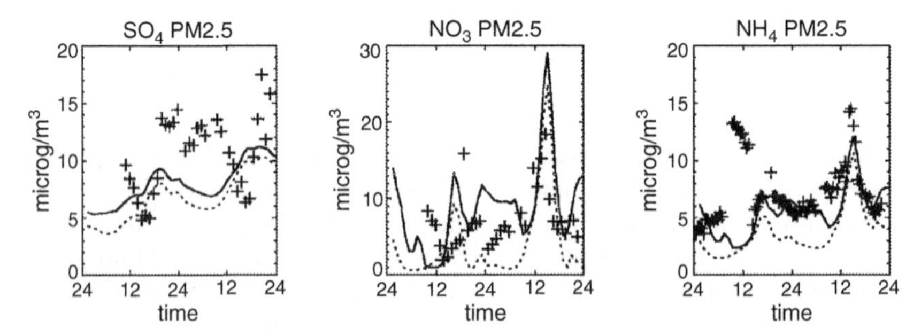

FIGURE 3. Diurnal variation of measured (+) and predicted concentrations (µg m^{-3}) by CAMx4 (solid line) and PMCAMx (dotted line) of particulate $SO_4^{\,2-}$, NO_3^- and NH_4^+ in Verzago.

chemistry of PMCAMx is still based on an earlier version (CAMx 3.01). The updated CBMIV mechanism in CAMx4 may lead to differences in results with respect to sulfate production and nighttime nitrate chemistry. Overestimation of nitrate at night by CAMx4 may also be due in part to assuming all nitrate is fine. The difference in the aqueous chemistry mechanism is not expected to be significant under the clear sky conditions used in this study. Boundary conditions for the particulate species in PMCAMx are not exactly the same as in CAMx4 because of various size sections. This may also lead to differences between the model results.

In general, PMCAMx predicts lower levels at night for all SOA classes (Figure 4). Afternoon levels of SOA1 and SOA3 are the same for both models. On the other hand, model results of SOA2, which is the aerosol product of toluene and xylene look quite different. Since the same aerosol parameters are used in both models, discrepancies are possibly due to differences in CBMIV versions. Biogenic SOA (SOA4) is about 20% of the total predicted SOA. It is difficult to validate the calculated secondary organic aerosol concentrations because there are no direct measurements. Estimations based on black carbon and total organic carbon measurements yielded an SOA range of 4–5 $\mu g/m^3$ for daily average (Andreani-Aksoyoglu et al., 2004). The average of calculated values in Verzago are in the same range (5.4 for CAMx4 and 4.4 for PMCAMx).

3.2. *Switzerland*

The model results over the Swiss domain are compared only with each other due to lack of measurements of aerosol species (Figure 5). There are significant differences in sulfate concentrations calculated by the models. Nitrate concentrations are the same in the afternoon, but night values differ. CAMx4 predicts higher concentrations at night as in the first application.

Some emission sensitivity tests were performed to understand the significance of the difference between the CAMx4 and PMCAMx results. The highest uncertainties in anthropogenic emissions are believed to be in NH_3 and SO_2 emissions, around 20-30% (Hass et al., 2003). Therefore two more simulations with CAMx4 were carried out using 20% reduced SO_2 and NH_3 emissions, respectively. Reduction of SO_2 emissions by 20% did not effect sulfate particle concentrations (Figure 6).

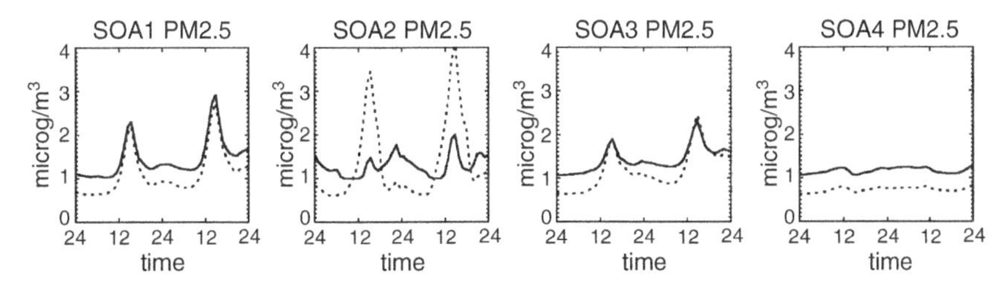

FIGURE 4. Diurnal variation of predicted concentrations ($\mu g\ m^{-3}$) of 4 SOA species by CAMx4 (solid line) and PMCAMx (dotted line) in Verzago.

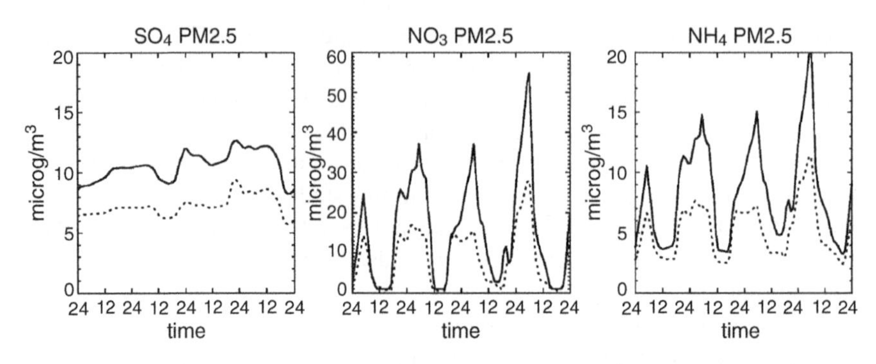

FIGURE 5. Diurnal variation of predicted concentrations (µg m^{-3}) by CAMx4 (solid line) and PMCAMx (dotted line) of particulate SO_4^{2-}, NO_3^- and NH_4^+ in Tänikon.

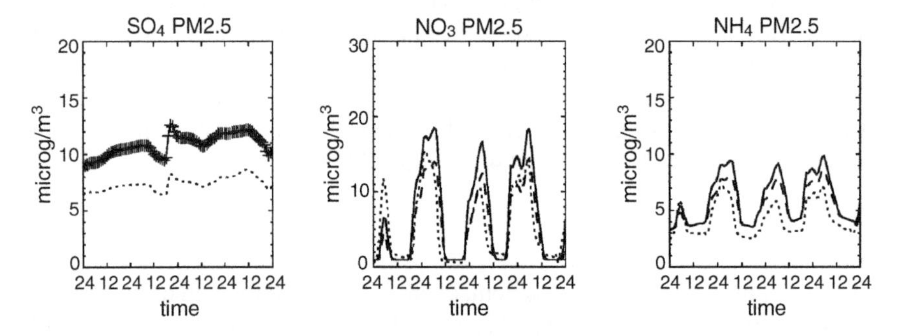

FIGURE 6. Diurnal variation of predicted concentrations (µg m^{-3}) of particulate SO_4^{2-}, NO_3^- and NH_4^+ by PMCAMx (dotted line) and by CAMx4 with 100% emissions (solid line), with 80% SO_2 emissions (+), with 80% NH_3 emissions (dashed line) during 4-7 August 2003 in Zürich.

This means that the difference between CAMx4 and PMCAMx is larger than the uncertainties in SO_2 emissions. On the other hand, reduction in NH_3 emissions caused a significant decrease in particulate nitrate and ammonium, reducing the gap between the two model results. One can say therefore that the difference between the nitrate results of the two models is in the same range as uncertainties in NH_3 emissions.

SOA formation in both models is similar (Figure 7). Anthropogenic secondary aerosol concentrations (SOA1-SOA3) are much lower in northern Switzerland than in the south and in the region of Milan. The biogenic contribution to SOA which is more than 80% in the northern part of Switzerland is therefore much larger than in northern Italy.

After this study was completed, a new version of CAMx has been released (version 4.10s). This version has the option to choose either fine/coarse approach or multi-sectional approach as in PMCAMx. Two additional simulations for the

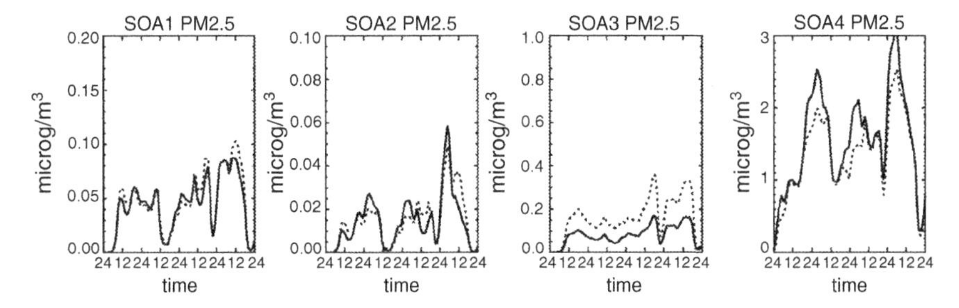

FIGURE 7. Diurnal variation of predicted concentrations (μg m^{-3}) of SOA by CAMx4 (solid line) and PMCAMx (dotted line) during 4-7 August 2003 in Tänikon.

domain in northern Italy were carried out with the new version of CAMx using fine/coarse and multi-sectional approaches, respectively. Both approaches gave the same results in Verzago (Figures 8 and 9). These results indicated that the differences found between the previous CAMx version and PMCAMx were simply due to different model versions.

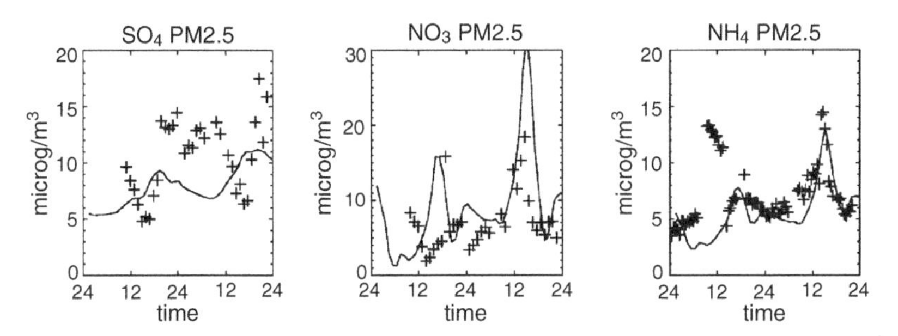

FIGURE 8. Diurnal variation of measured (+) and predicted concentrations (μg m^{-3}) by CAMx4.10s using fine/coarse mode (solid line) and multi-sectional mode (dotted line) in Verzago.

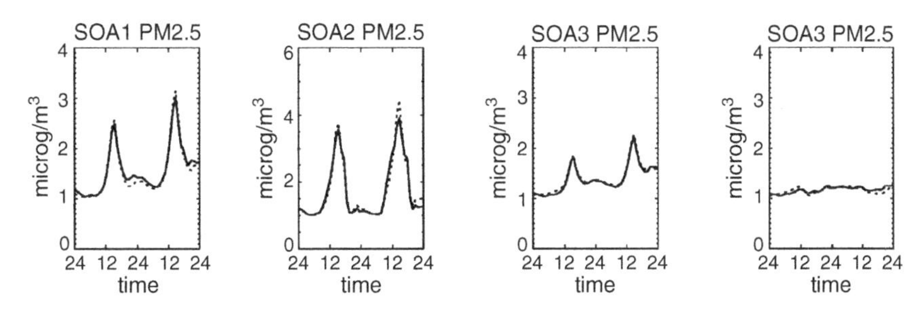

FIGURE 9. Diurnal variation of predicted concentrations (μg m^{-3}) of 4 SOA species by CAMx4.10 using fine/coarse mode (solid line) and using multi-sectional mode (dotted line) in Verzago.

4. Conclusions

On the basis of two applications, the results of two models with different complexity, CAMx4 and PMCAMx, are comparable. Some differences in the gas-phase mechanisms and boundary conditions between the models may lead to discrepancies. Model predictions of inorganic aerosols are rather close to observations for the study in northern Italy. Overestimation of nitrate at night by CAMx4 may be due in part to assuming nitrate only in the fine fraction or due to the different CBMIV versions. The difference in sulfate predictions between two models is larger than the uncertainties in SO_2 emissions. On the other hand, the difference for nitrate is in the same range as the uncertainties in NH_3 emissions. SOA predictions of the two models are almost the same. Both models indicated a lower biogenic contribution to SOA in northern Italy (about 20%) than in northern Switzerland (about 80%). Although SOA predictions cannot be validated by direct measurements, estimates from black carbon and total organic carbon measurements in northern Italy suggest that model results are in the same range. Computer time required by PMCAMx is about 3 times higher than the time needed by CAMx4. Although PMCAMx has the capability of calculating the aerosol species in various size sections, as long as there are limited measurements of speciated aerosols, it is difficult to evaluate the results. Most of the particle measurements in Europe consists of PM10 only. Using more complex models such as PMCAMx requires measurements of aerosol species with smaller size (not only PM10) with higher resolution in time and space. Additional simulations with the new version of CAMx 4.10s which has been released after this study was completed, gave the same results for fine/coarse and multi-sectional approaches. The differences found in the results of previous CAMx version and PMCAMx are therefore, considered to be due to different model versions. Finally, we should consider that the concept of equilibrium of the partitioning between gas phase and aerosol organics is not valid anymore considering studies of Jang et al. (2002) and Kalberer et al. (2004). Due to polymerization reactions in the aerosol, more organics can partition into the aerosols compared to prediction of equilibrium models. Some improvements in the aerosol models are therefore needed in the future.

Acknowledgements We would like to thank the LOOP community for providing us with various data from northern Italy. The output data of European Model REM3 was kindly provided by J. Flemming (FUB) to be used as boundary conditions. We are grateful to R.Stern (FUB), A.Graff (UBA) and M. van Loon (TNO) for European emissions, INFRAS and METEOTEST for the Swiss emission data. We thank Ch. Hüglin (EMPA) for providing the PM2.5 data from NABEL stations. G. Yarwood from ENVIRON is gratefully acknowledged for providing the PMCAMx code.

References

Andreani-Aksoyoglu, S., Prevot, A.S.H., Baltensperger, U., Keller, J., Dommen, J., 2004, Modeling of formation and distribution of secondary aerosols in the Milan area (Italy), *J. Geophys. Res.,* **109:** D05306, doi:10.1029/2003JD004231.

Ackermann, I.J., Hass, H., Memmesheimer, M., Ebel, A., Binkowski, F.S., Shankar, U., 1998, Modal aerosol dynamics model for Europe: Development and first applications, *Atmos. Environ.*, **32**: 2981.

Bessagnet, B., Hodzic A., Vautard R., Beekmann M., Cheinet S., Honore C., Liousse C., Rouil L., 2004, Aerosol modelling with CHIMERE-preliminary evaluation at the continental scale, *Atmos. Environ.*, **38**: 2803.

Chang, J.S., Brost., R.A., Isaksen, I.S.A., Madronich, S., Middleton, P., Stockwell, W.R., Walcek, C.J., 1987, A Three-dimensional Eulerian Acid Deposition Model: Physical Concepts and Formulation, *J. Geophys. Res.*, **92**: 14,681.

Environ, 2003, Final Report CRC project A-30, Development of an advanced photochemical model for particulate matter: PMCAMx, Environ.

Environ, 2004, User's Guide, Comprehensive Air Quality Model with Extensions (CAMx), Version 4.00, Environ International Corporation, California, January 2004.

Griffin, R.J., Nguyen, K., Dabdub, D., Seinfeld, J.H., 2003, A coupled hydrophobic-hydrophilic model for predicting secondary organic aerosol formation, *J. Atmos. Chem.*, **44**: 171.

Hass, H., van Loon, M., Kessler, C., Stern, R., Matthijsen, J., Sauter, F., Zlatev, Z., Langner, J., Foltescu, V., Schaap, M., 2003, Aerosol Modeling: Results and Intercomparison from European Regional-scale Modeling Systems, EUROTRAC 2.

Jang, M., Czoschke, N.M., Lee, S., Kamens, R.M., 2002, Heterogeneous atmospheric aerosol production by acid-catalyzed particle-phase reactions, *Science*, **298**: 814.

Kalberer M., D. Paulsen, M. Sax, M. Steinbacher, J. Dommen, A.S.H. Prevot, R. Fisseha, E. Weingartner, V. Frankevich, R. Zenobi, U. Baltensperger, 2004, Identification of polymers as major components of atmospheric organic aerosols, *Science*, **303:** 1659.

Nenes A., S.N. Pandis, C. Pilinis, 1998, ISORROPIA: A new thermodynamic equilibrium model for multiphase multicomponent inorganic aerosols, *Aquat. Geoch.*, **4**: 123.

PSU, NCAR, 2004, MM5 Version 3 Tutorial Presentations. http://www.mmm.ucar.edu/mm5/mm5v3/tutorial/presentations/tut-presentations.html.

Putaud, J.P. et al., 2004, A European aerosol phenomenology-2:chemical characteristics of particulate matter at kerbside, urban, rural and background sites in Europe, *Atmos. Environ*, **38**: 2579.

Schell B., I.J. Ackermann, H. Hass, F.S. Binkowski, A. Ebel, 2001, Modeling the formation of secondary organic aerosol within a comprehensive air quality modeling system, *J. Geophys. Res.,* **106**: 28275.

Strader R., C. Gurciullo, S.N. Pandis, N. Kumar, F.W. Lurmann, 1998, Development of gas-phase chemistry, secondary organic aerosol, and aqueous-phase chemistry modules for PM modeling. Final Report for CRC Project A21-1 prepared for the Coordinating research Council, Atlanta, GA by Sonoma technology, Inc., Petaluma, CA, STI-97510-1822-FR.

Zhang Y., B. Pun, K. Vijayaraghavan, S.-Y. Wu, C. Seigneur, S.N. Pandis, M. Jacobson, A. Nenes, J.H. Seinfeld, 2004, Development and application of the model of aerosol Dynamics, Reaction, Ionization, and Dissolution (MADRID), *J. Geophys. Res.,* **109**: D01202, doi:10.1029/2003JD003501.

Aerosol Modelling with CAMX4 and PMCAMX: A Comparison Study

Speaker: Andreani-Aksoyoglu

Questioner: P. Bhave
Question: In the multi-section aerosol models present (PMCAMx and CAMx 4.105) what assumptions are made regarding mass transfer between the gas phase and the coarse particle sections.
Answer: Equilibrium approach is used in PMCAMx.

Questioner: G. Yarwood
In reply to Bob Yarmartino
Analysis of difference in PM prediction due to different model algorithms often required detailed investigation due to the couplings present in PM chemistry. e.g. Differences in sulfate lead to differences in available ammonia and thereby differences in nitrate.
In reply to Prakash Bhave
The PMCAMx sectional calculations assumed that mass transport (gas – aerosol) reached equilibrium.

Questioner: M. Jean
Question: Have you done sensitivity analysis to the numerical weather prediction component such as horizontal or vertical resolution physical parameterization options, etc?
Answer: We are using aLMo output to initialize the model and as boundary conditions. Sensitivity analyses are still going on. We are testing various parameters such as resolution, one-way, two-way nesting, time step, nudging etc.

Questioner: R. San Jose
Question: Have you tried to increase the number of vertical layers to be much more in accordance with the number of vertical layers used in the meteorological model?
Answer: In this study we used 8 layers. We are planning to increase a little, maybe up to 10. Increasing the number of layers more than that, is not easy because it will increase the computer time substantially.

28 Source Apportionment of Primary Carbonaceous Aerosol Using the Community Multiscale Air Quality Model

Prakash V. Bhave[*†], George A. Pouliot[*], and Mei Zheng[‡]

1. Introduction

A substantial fraction of fine particulate matter (PM) across the United States is composed of carbon, which may be either emitted in particulate form (i.e., primary) or formed in the atmosphere through gas-to-particle conversion processes (i.e., secondary). Primary carbonaceous aerosol is emitted from numerous sources including motor vehicle exhaust, residential wood combustion, coal combustion, forest fires, agricultural burning, solid waste incineration, food cooking operations, and road dust. Quantifying the primary contributions from each major emission source category is a prerequisite to formulating an effective control strategy for the reduction of carbonaceous aerosol concentrations. A quantitative assessment of secondary carbonaceous aerosol concentrations also is required, but falls outside the scope of the present work.

A common method of primary carbonaceous aerosol source apportionment involves a molecular characterization of emission source effluents and ambient aerosol samples followed by a determination of the linear combination of source signatures that best matches the measured composition of the ambient sample. This method, referred to as an organic tracer-based chemical mass balance (CMB), has been demonstrated using atmospheric aerosol samples collected at a number of receptor sites across the United States (Schauer et al., 1996; Fujita et al., 1998; Zheng et al., 2002; Fine 2002; Fraser et al., 2003). An alternative

[*] Atmospheric Sciences Modeling Division, Air Resources Laboratory, National Oceanic and Atmospheric Administration, Research Triangle Park, NC 27711, U.S.A. On Assignment to the National Exposure Research Laboratory, U.S. Environmental Protection Agency -Office of Research and Development.
[†] Corresponding author. e-mail: bhave.prakash@epa.gov. tel. (919) 541-2194. fax. (919) 541-1379.
[‡] School of Earth and Atmospheric Sciences, Georgia Institute of Technology, Atlanta, GA 30332, U.S.A.

source apportionment methodology makes use of source-specific emission rates and atmospheric transport calculations in a source-oriented air quality modeling framework. The PM emitted from each major source category is tagged at the point of emission and tracked numerically as it is transported through the study region. In this manner, the ambient pollutant concentration increments due to each source of primary carbonaceous aerosol can be estimated at any time and location within the modeling domain. Applications of this method have been limited in large part due to the input requirement of a detailed emission inventory that includes the strengths, temporal distributions, and spatial allocations of each major emission source of carbonaceous PM. Moreover, it is difficult to evaluate results of this method without atmospheric measurements of source-specific chemical tracers. For these reasons, most applications of the source-oriented approach reported to date are for the Los Angeles metropolitan area during intensive field measurement campaigns (Hildemann et al., 1993; Rogge et al., 1996; Fraser et al., 2000). Recently, a global-scale three dimensional model (GEOS-CHEM) was used to track the carbonaceous aerosol contributions from three primary source categories (fossil fuel combustion, biofuel combustion, and biomass burning) across the U.S. in 1998 (Park et al., 2003). That model application was intended to apportion sources at a coarse spatial resolution (2° latitude by 2.5° longitude) for regional visibility calculations and the evaluation was limited by bulk compositional data.

Over the past decade, the U.S. Environmental Protection Agency (EPA) in cooperation with state and local agencies has developed a National Emission Inventory (NEI) for fine PM (EPA, 2001). In addition, the EPA has been developing the Community Multiscale Air Quality (CMAQ) model for the mechanistic prediction of gas and aerosol-phase pollutant concentrations (Byun and Ching, 1999). In the present work, an extension to the CMAQ model is described that allows the user to track the emissions from an arbitrary number of primary aerosol sources as they are transported through the atmosphere. The model is coupled with the NEI to estimate primary carbonaceous aerosol concentration increments contributed by nine major emission categories over the continental U.S. from June 15–August 31, 1999. Model results are evaluated against source-specific molecular measurements collected at eight receptor sites in the southeastern U.S.

2. Emission Inventory and Model Descriptions

Gaseous and particle-phase emissions in the NEI are categorized by geographic region and source classification code (SCC). For typical CMAQ modeling applications, the NEI is processed using the Sparse Matrix Operator Kernel Emissions (SMOKE) model to yield *model-ready* input files that contain chemically, spatially, and temporally resolved pollutant emissions. These gridded emission files include particulate elemental carbon (coded as PEC), organic aerosol (POA), sulfate (PSO4), nitrate (PNO3), and other unspecified fine PM. In the NEI, POA mass is defined implicitly as primary organic carbon times 1.2, to account for the masses of H, O, N, and S atoms that are associated with organic carbon emissions. The temporal resolution of the emission file is hourly and, for the present application, the grid spacing is 32 km. In order to track different sources of carbonaceous aerosol, the fine particle speciation profiles used in the SMOKE model are duplicated to create a source-specific profile for each emission category of interest. In the source-specific profiles, PEC and POA emitted from the first source category are designated respectively as PEC1 and POA1, those from the second source category as PEC2 and POA2, and so forth. Also, the SCC-to-speciation profile reference table used in the SMOKE model is modified to appropriately map each SCC to the newly-created source-specific speciation profiles.

Version 1 of the 1999 NEI serves as the base inventory for the present model application. It was developed by applying growth factors to the 1996 National Emission Trends criteria air pollutant inventory, which is described in detail elsewhere (EPA, 2001). When preparing model-ready emission files for the present study, all NEI estimates of fugitive dust emissions (e.g., from paved and unpaved roads, agricultural tilling, construction activities, etc.) are reduced by a factor of four to account approximately for the dust removal processes that occur within several hundred meters of their sources (Watson and Chow, 2000). Commercial cooking emissions were reported by very few states in the NEI, so emissions from that category are replaced by a more comprehensive 2002 commercial cooking emission inventory (Roe et al., 2004). The resulting fine PM emission inventory is categorized into 2,890 SCCs. To reduce the computational burden that would be associated with tracking each of these sources throughout the modeling domain, emissions associated with each SCC are lumped into nine major source categories plus a tenth miscellaneous category. These nine categories constitute nearly 95% of the total POA and PEC emissions on an annual basis, as shown in Table 1. It should be noted that vegetative detritus, fungal spores, natural windblown dust, and cigarette smoke, are in neither the base nor the model-ready emission inventories.

Version 4.3 of the CMAQ model (2003 public release) is used as the base model configuration for this study. Aerosol components of the CMAQ model are

TABLE 1. Fine particle mass emission totals (tons/yr) in the model-ready inventory.

	EC	OC×1.2	SO_4	NO_3	OTHER	TOTAL
Diesel Exhaust	350000	105000	10600	760	2800	470000
Gasoline Exhaust	18600	100000	2400	600	23000	145000
Biomass Combustion	89000	620000	15500	5700	300000	1030000
Coal Combustion	3000	1770	2800	0	141000	149000
Oil Combustion	5700	6600	860	49	19700	33000
Natural Gas Combustion	0	23000	7400	330	6600	37000
Food Cooking	870	68000	148	14	920	70000
Paved Road Dust	1880	30000	1170	370	135000	168000
Crustal Material	2500	37000	270	750	640000	680000
Other Sources	33000	60000	72000	2400	520000	690000
Grand Total	500000	1050000	113000	11100	1790000	3500000

described in detail elsewhere (Binkowski and Roselle, 2003). In the base configuration, primary carbonaceous aerosols are tracked as four model species to distinguish their size and composition distribution: Aitken mode organic aerosol (AORGPAI), Aitken mode elemental carbon (AECI), accumulation mode organic aerosol (AORGPAJ), and accumulation mode elemental carbon (AECJ). For each carbonaceous aerosol source category tracked in the extended CMAQ model, four species are added. For example, AORGP4I, AEC4I, AORGP4J, and AEC4J, represent primary carbonaceous species originating from coal combustion in the present application. The 40 additional model species are internally mixed within their designated aerosol mode (i.e., Aitken or accumulation) and participate in advection, diffusion, deposition, condensational growth, and coagulation processes in a manner identical to the treatment of AORGPA and AEC species in the base model configuration.

3. Model Results

The extended CMAQ model is used to simulate gaseous and aerosol-phase pollutant concentrations while tracking the contributions from nine major primary PM source categories across the continental U.S. from June 15–August 31, 1999. The modeling domain, meteorological inputs, boundary conditions, and initial spin-up period are identical to those used by Yu et al. (2004). Figure 1 displays model results in the lowest vertical layer averaged over the 78 day simulation period for a subset of source categories. Carbon concentrations are calculated as (AEC + AORGPA/1.2) and summed over the Aitken and accumulation modes.

The spatial pattern of diesel exhaust concentrations resembles closely the U.S. population density distribution, with highest concentrations found over urban areas (see Figure 1a). Concentrations of gasoline exhaust, paved road dust, and food cooking, also exhibit spatial patterns similar to the population density distribution, and therefore, are not displayed in Figure 1. Over most urban areas during the 1999 summer, model results indicate that diesel exhaust makes a larger contribution to primary carbonaceous fine PM than any other source category. The highest seasonal average concentration of diesel exhaust carbon is calculated as 5.2 μg C m^{-3} in the grid cell surrounding New York City. The next highest concentrations are found over northern Ohio, Los Angeles, New Orleans, Phoenix, and Atlanta, ranging from 2.6–3.5 μg C m^{-3}. The highest seasonal average concentrations of gasoline exhaust carbon are 1.0 μg C m^{-3} over the Los Angeles area, 0.7 over New York City, and 0.5 over Chicago, resulting from large volumes of automobile traffic in these cities. Carbon concentrations from paved road dust are roughly twenty times lower than the motor vehicle exhaust contributions. The highest seasonal average primary carbon concentrations from food cooking are 1.4 and 1.2 μg C m^{-3} over New York City and Los Angeles, and 0.5–0.6 μg C m^{-3} over Chicago, San Francisco, and Washington, D.C. It is of interest to note that model calculations of food cooking carbon concentrations exceed those of gasoline exhaust in most urban areas across the U.S., even though total carbon emissions from the latter source category are greater on a national scale (see Table 1).

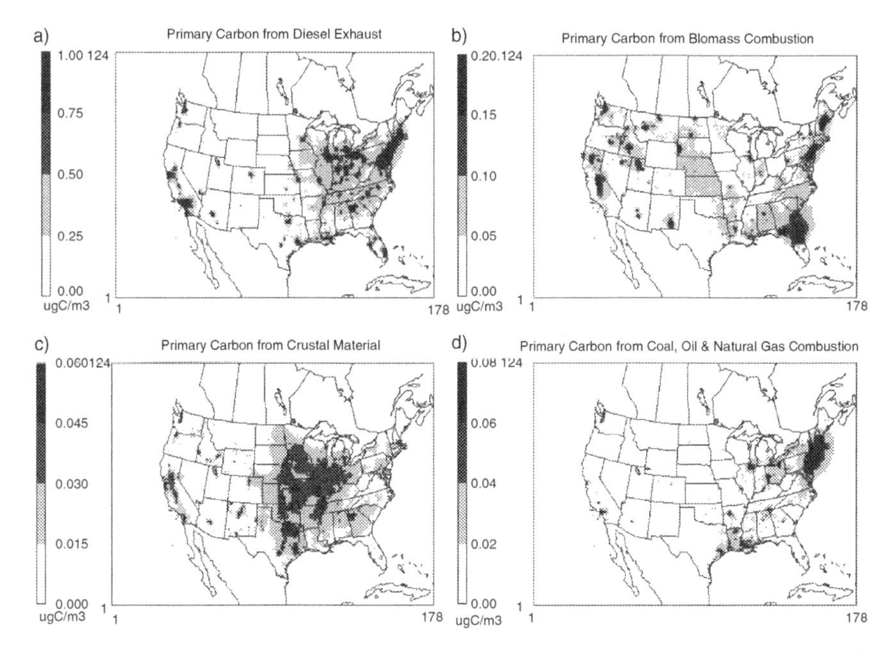

FIGURE 1. Model predictions of fine particle primary carbon concentrations [μg C m^{-3}] from select source categories averaged over the June 15–August 31, 1999 period. Note the differences in scale. Plots are prepared using PAVE by MCNC.

In the model-ready inventory, 90% of biomass combustion carbon emissions during the summer months are from wildfires. Hence, the spatial distribution of primary carbonaceous aerosol concentrations originating from biomass combustion (see Figure 1b) is roughly proportional to the number of acres that burned in 1999. The highest modeled concentrations of biomass combustion carbon are over Florida, Montana, New Mexico, and California. As shown in Figure 1c, carbon concentrations from crustal material are highest over the Midwest and central states. In the inventory, summertime emissions of fine crustal material are dominated by unpaved road dust (51% of total) and agricultural tilling (31%), followed by smaller contributions from construction activities (14%) and beef cattle feedlots (3%). Hence, the spatial patterns of crustal carbon are concentrated over rural and agricultural areas. The inclusion of natural windblown dust emissions (e.g., from desert dust storms) in future inventories would likely increase crustal aerosol concentrations over the arid Southwest.

Although tracked separately in the present model application, the aggregate of coal, oil, and natural gas combustion contributions are displayed in Figure 1d. Coal combustion carbon is highest over the Ohio River valley but exhibits surprisingly low concentrations (max = 40 ng C m^{-3}). The speciation profile for coal combustion emissions designates only 2.7% of the fine particle mass as carbon, based on measurements taken at a Philadelphia power plant over 20 years ago. A number of recent studies estimate the carbonaceous fraction to be over 15%, indicating a need to update this particular speciation profile (Ryan, 2003). Domain-wide maximums from oil and natural gas combustion are found in New Jersey (0.74 and 0.97 μg C m^{-3}, respectively) due to very high emissions from a single utility company. Excluding the New Jersey plumes, the domain-wide maximum concentrations from oil and natural gas combustion are 0.23 and 0.09 μg C m^{-3}, respectively.

4. Model Evaluation

Atmospheric concentrations of about 100 individual organic compounds were measured from fine particle samples collected in 1999 at eight receptor sites across the southeastern U.S. (Zheng et al., 2002). This is the first available set of source-specific carbonaceous concentration data that spans a multi-state geographic region. To obtain model estimates of individual organic compound concentrations at each receptor site, model calculations of source-specific carbon concentrations are multiplied by organic molecular speciation profiles. The source profiles used for diesel exhaust, gasoline exhaust, food cooking, biomass combustion, natural gas combustion, and paved road dust are identical to those described by Zheng et al. (2002). The oil combustion profile is an average of two source tests reported by Rogge et al. (1997). Organic molecular profiles are not available for coal combustion, crustal material, and the numerous miscellaneous sources, so model calculations of carbonaceous aerosol from these source categories are not speciated in the present study.

Figure 2 displays a model evaluation summary comparing the extended CMAQ model results for July 1999 against atmospheric measurements at all eight receptor sites in the Southeast. The ratios of model predictions to observations are displayed along the vertical axis for all cases where the given species was detected above quantifiable limits at the given site. Symbols lying between the two horizontal lines represent cases where model predictions are within a factor of two of the observed concentrations. Shaded symbols represent urban monitoring sites, whereas unfilled symbols correspond to rural (Centreville, Oak Grove, and Yorkville) or suburban (OLF#8) locations. Seventeen organic species and bulk elemental carbon (EC) and organic carbon (OC) are arranged in sections along the horizontal axis, separated by vertical dashed lines that delineate conserved tracers emitted from different source categories. Conserved organic markers unique to paved road dust are unavailable, so it is not possible to directly evaluate model results from that source in the present study.

Daily fine particle EC and OC measurements are obtained from the Southeastern Aerosol Research and Characterization network (Hansen et al., 2003) and averaged over the month of July at each site for comparison against monthly-averaged model predictions. Model calculations of EC fall within a factor of two of observations at more than half of the sampling locations. This level of agreement builds confidence in the transport algorithms used in CMAQ and in the inventory of diesel emissions, because diesel exhaust is the dominant source of atmospheric EC in the U.S. Model predictions of total OC (primary + secondary) are in reasonable agreement with observations in Atlanta, but fall below measurements by a factor of three, on

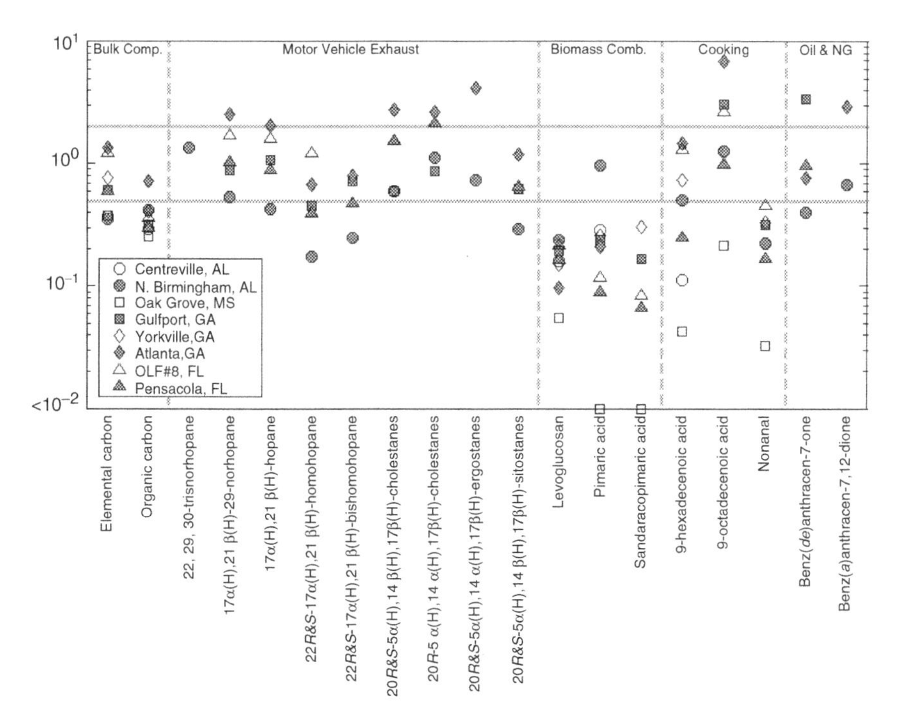

FIGURE 2. Ratios of CMAQ model results to ambient measurements of EC, OC, and individual organic compounds at eight southeastern U.S. sites in July 1999. Horizontal lines bound the region in which model-observation agreement is within a factor of two. Vertical dashed lines distinguish molecular markers specific to different source categories.

average, at the remaining sites. This indicates that total OC is underestimated across the southeastern U.S. From bulk EC and OC measurements alone, it is impossible to determine which source contributions have been underestimated.

Out of 34 quantitative measurements of motor vehicle exhaust tracers, 21 model-observation pairs agree within a factor of two (see Figure 2). Considering that the observed concentrations of these species span a broad range (15 to 500 pg m^{-3}), the displayed level of agreement is quite good and indicates that the contributions of gasoline and diesel exhaust to primary carbon concentrations are captured reasonably well in the model. A systematic model underestimation of biomass combustion throughout the Southeastern U.S. is apparent from comparisons shown in Figure 2. Model estimates of pimaric acid and sandaracopimaric acid at Oak Grove fall short of measurements by more than a factor of 500, indicating that a biomass combustion event near that site is not captured in the model simulation. Measurements of levoglucosan, a well-established tracer for wood smoke, exceed model results by a factor of six on average when the Oak Grove data are excluded (obs. range = 26–80 ng m^{-3}). A close examination of the inventory reveals that none of the annual emissions from agricultural and prescribed forest burning is allocated to the summer months. Emission allocation refinements from these two sources and from wildfires in future inventories will likely increase modeled concentrations of biomass combustion carbon during summer months in the Southeast. Cholesterol is a reliable tracer for meat cooking, but unfortunately, was not quantified in the July 1999 samples due to a limitation of the analytical techniques used in that study. As a substitute for cholesterol, particle-phase nonanal is used as a meat cooking tracer in the present study because it is absent from the remaining source profiles. The nonanal comparisons indicate that emissions of food cooking carbon in the Southeast are underestimated by a factor of 2 to 6 in the present inventory (obs. range = $0.9-2.9$ ng m^{-3}). The other food cooking tracers listed in Figure 2 (9-hexadecenoic acid and 9-octadecenoic acid) are emitted from multiple sources in addition to meat cooking, which may explain much of the scatter shown in Figure 2. Model comparisons against measurements of chemical tracers unique to oil and natural gas combustion are plotted in the rightmost section of Figure 2. The agreement between model calculations and observations for these two species is reasonable, but very few data points are available for comparison.

5. Conclusions

We have extended the CMAQ model to provide a first estimate of the source contributions to fine particle primary carbon concentrations across the United States using a detailed emission inventory and mesoscale meteorological inputs. Spatial distributions of the various source contributions agree qualitatively with our knowledge of emission patterns. Molecular speciation of the source apportioned model results allows the calculation of individual organic compound concentrations at selected receptor sites. Model evaluation against measurements of individual organic compounds reveals that fine particle emission estimates of motor vehicle exhaust and natural gas combustion are reasonably accurate over the southeastern U.S., whereas carbonaceous emissions from biomass combustion and food cooking are biased low by more than a factor of two.

References

Binkowski, F. S., Roselle, S. J., 2003, Models-3 Community Multiscale Air Quality (CMAQ) model aerosol component 1. Model description, *J. Geophys. Res.*, **108**(D6), 4183, doi:10.1029/2001JD001409.

Byun, D. W., Ching, J. K. S., 1999, *Science Algorithms of the EPA Models-3 Community Multiscale Air Quality (CMAQ) Modeling System*, EPA report 600/R-99/030, Washington DC.

EPA, 2001, *Procedures Document for National Emission Inventory, Criteria Air Pollutants 1985-1999*, EPA report 454/R-01-006, Research Triangle Park, NC.

Fine, P. M., 2002, *The Contribution of Biomass Combustion to Ambient Fine Particle Concentrations in the United States*, Ph. D. thesis, Calif. Inst. of Technol., Pasadena, CA.

Fraser, M. P., Kleeman, M. J., Schauer, J. J., Cass, G. R., 2000, Modeling the atmospheric concentrations of individual gas-phase and particle-phase organic compounds, *Environ. Sci. Technol.*, **34**:1302.

Fraser, M. P., Yue, Z. W., Buzcu, B., 2003, Source apportionment of fine particulate matter in Houston, TX, using organic molecular markers, *Atmos. Environ.*, **37**:2117.

Fujita, E. M., Watson, J. G., Chow, J. C., Robinson, N., Richards, L., Kumar, N., 1998, *Northern Front Range Air Quality Study. Volume C: Source Apportionment and Simulation Methods for Evaluation*, Final report to Colorado State University, Fort Collins, CO.

Hansen, D. A., Edgerton, E. S., Hartsell, B. E., Jansen, J. J., Kandasamy, N., Hidy, G. M., Blanchard, C. L., 2003, The southeastern aerosol research and characterization study: Part 1 – Overview, *J. Air & Waste Manage. Assoc.*, **53**:1460.

Hildemann, L. M., Cass, G. R., Mazurek, M. A., Simoneit, B. R. T., 1993, Mathematical modeling of urban organic aerosol: properties measured by high-resolution gas chromatography, *Environ. Sci. Technol.*, **27**:2045.

Park, R. J., Jacob, D. J., Chin, M., Martin, R. V., 2003, Sources of carbonaceous aerosols over the United States and implications for natural visibility, *J. Geophys. Res.*, **108**(D12), 4335, doi:10.1029/2002JD003190.

Roe, S. M., Spivey, M. D., Lindquist, H. C., Hemmer, P., Huntley, R., 2004, National emissions inventory for commercial cooking, *13th Annual Emission Inventory Conference*, Clearwater, FL.

Rogge, W. F., Hildemann, L. M., Mazurek, M. A., Cass, G. R., Simoneit, B. R. T., 1996, Mathematical modeling of atmospheric fine particle-associated primary organic compound concentrations, *J. Geophys. Res.*, **101**:19379.

Rogge, W. F., Hildemann, L. M., Mazurek, M. A., Cass, G. R., Simoneit, B. R. T., 1997, Sources of fine organic aerosol. 8. Boilers burning No. 2 distillate fuel oil, *Environ. Sci. Technol.*, **31**:2731.

Ryan, R., 2003, personal communication.

Schauer, J. J., Rogge, W. F., Hildemann, L. M., Mazurek, M. A., Cass, G. R., Simoneit, B. R. T., 1996, Source apportionment of airborne particulate matter using organic compounds as tracers, *Atmos. Environ.*, **30**:3837.

Watson, J. G., Chow, J. C., 2000, *Reconciling Urban Fugitive Dust Emissions Inventory and Ambient Source Contribution Estimates: Summary of Current Knowledge and Needed Research*, DRI document 6110.4F, Reno, NV.

Yu, S., Dennis, R. L., Bhave, P. V., Eder, B. K., 2004, Primary and secondary organic aerosols over the United States: estimates on the basis of observed organic carbon (OC) and elemental carbon (EC), and air quality modeled primary OC/EC ratios, *Atmos. Environ.*, in press.

Zheng, M., Cass, G. R., Schauer, J. J., Edgerton, E. S., 2002, Source apportionment of PM2.5 in the southeastern United States using solvent-extractable organic compounds as tracers, *Environ. Sci. Technol.*, **36**:2361.

The research presented here was performed under the Memorandum of Understanding between the U.S. Environmental Protection Agency (EPA) and the U.S. Department of Commerce's National Oceanic and Atmospheric Administration (NOAA) and under agreement number DW13921548. *Although it has been reviewed by EPA and NOAA and approved for publication, it does not necessarily reflect their policies or views.*

Source Apportionment of Primary Carbonaceous Aerosol Using the Community Multiscale Air Quality Model

Speaker: P. V. Bhave

Questioner: W. Jiang

Question 1: Where did you get the chemical speciation profiles to split carbon into individual chemical species?

Answer: The speciation profiles are from a series of source characterization experiments conducted and documented by Glen Cass's research group at Caltech.

Question 2: How did you attribute carbon in PM2.5 to the two CMAQ fine modes (i-mode and j-mode)?

Answer: Since we don't have size-resolved emissions data in our current inventory, I used the default assumption in CMAQ for each source category by assigning 99.9% of primary carbon mass emissions to the accumulation mode and 0.1% to the Aitken mode.

Questioner: P. Builtjes

Question: Is food cooking an indoor source, or is it considered to be an external source?

Answer: Although food cooking occurs indoors, the exhaust at commercial facilities is ventilated to the ambient environment.

Questioner: C. Mensink

Question: What are the sources for meat cooking?

Answer: The meat cooking inventory in the U.S. is dominated by commercial cooking sources, such as emissions from Burger King or McDonalds.

Questioner: B. Fisher

Question: The calculated diesel C in cities are representative grid square averages. Presumably diesel C concentrations near to busy roads in cities could be much higher, although most monitoring is made somewhat remote from sources.

Answer: That is correct. Near roadways, the contribution of diesel exhaust to primary carbon would be even larger than I presented here.

Questioner: M. Moran

Question: You used organic tracer measurements from the SEARCH network in the southeastern US and such measurements are also available for the Los Angeles area, but are such measurements also available for other North American locations?

Answer: Organic-tracer measurements have been made by Phil Fine, using semi-annual composites of filters collected in 1995 at IMPROVE sites that span the United States. In a follow-up to her SEARCH analyses, Mei Zheng has begun analyzing seasonal composites of filters collected during 2002 at thirty STN sites spanning the eastern U.S. Also, year-long time series of speciated organic compound data should be available soon from the St. Louis and Pittsburgh supersites.

29
Urban Population Exposure to Particulate Air Pollution Induced by Road Transport

Carlos Borrego, Oxana Tchepel, Ana Margarida Costa, Helena Martins, and Joana Ferreira*

Abstract

In the last years, there has been an increase of scientific studies confirming that long- and short-term exposure to particulate matter pollution leads to adverse health effects.

The determination of accumulated human exposure in urban areas (in the present study focused on Lisbon) is the main objective of the current work combining information on concentrations at different microenvironments and population time-activity pattern data. A link between a mesoscale meteorological model and a local scale model (Computational Fluid Dynamics' based) was developed to define the boundary conditions for the local scale application. The time-activity pattern of the population was derived from statistical information for different sub-population groups and linked to digital city maps. Finally, the hourly PM_{10} concentrations for indoor and outdoor microenvironments were estimated for the Lisbon city centre based on the local scale air quality model application for a chosen day.

The developed methodology is a first approach to estimate population exposure, calculated as the total daily values above the thresholds recommended for long- and short-term health effects. Obtained results reveal that, in fact, in Lisbon city centre a large number of persons are exposed to particulate matter (PM) levels overpassing the legislated limit value. To get more accurate and consistent conclusions, a larger study, including a series of single days, should be performed.

1. Introduction

Air pollution is a major environmental health problem causing approximately 3 million deaths per year in the world, as result of exposure to particulate matter (PM) (WHO, 2001). Portugal, as a European Union Member, should follow the

* Carlos Borrego, Ana Margarida Costa, Helena Martins and Joana Ferreira, Department of Environment and Planning, University of Aveiro, 3810-193 Aveiro, Portugal.
Oxana Tchepel, School of Technology and Management, Polytechnic Institute of Leiria, 2401-951 Leiria, Portugal.

main objectives for management and quality of ambient air, namely those related to particulate matter. One of the current main concerns is the high levels of PM registered, namely PM_{10}, which are responsible for "medium air quality" in large urban areas, like Porto and Lisbon. In fact, during the past four years, the number of exceedances of PM_{10} daily limit value (50 $\mu g.m^{-3}$) overpassed the allowed 35 exceedances per year. Exposure studies can be carried out with the aim of obtaining estimates of exposure to atmospheric pollutants of individuals (personal exposure) or for a large population group (population exposure) and could be based on direct (exposure monitoring) or indirect methods (exposure modelling). Some epidemiological studies need exposure estimates of large population groups for entire cities over long periods of time. Monitoring of personal exposure in such studies should not be applied and exposure modelling techniques are recommended in exposure assessment (WHO, EC, 2002).

Several sources of particulate air pollution should be considered in exposure studies. Indoor sources include cooking, heating appliances and pets and, also, outdoor PM penetrating indoors, usually the most significant source in the absence of smoking (WHO, EC, 2002). Road traffic generally provides the major source of ambient particulate pollution, especially for finer particles. Although atmospheric particles may also be transported over long distances, peak concentrations tend to occur close to roads. Road journeys, whether by car, foot or cycle, thus tend to make up a large proportion of peak exposures (Gulliver and Briggs, 2004).

Since people spend most of their time indoors, and in different indoor places depending on their age and activity, it is essential to evaluate the PM concentrations not only in open air, but also in different indoor locations, called microenvironments.

Although nowadays it is possible to assess air quality in single stations of a monitoring network, the mapping of air pollutants over an area of interest, constitutes a challenging task. For that, one of the possible approaches is the use of numerical models.

The main objective of the current work is the development of a methodology to determine the population exposure in an urban area, combining information on modelled concentrations at different microenvironments and population time-activity pattern data. Two Accumulated Population Exposure Indexes ($APEI_{50}$ and $APEI_{100}$) are proposed in this study in order to estimate the number of persons exposed to PM_{10} concentrations above 50 $\mu g.m^{-3}$ and 100 $\mu g.m^{-3}$ during a day (24 hours). The 50 $\mu g.m^{-3}$ concentration value is the daily limit value settled by the European legislation, while 100 $\mu g.m^{-3}$ is defined by the WHO (WHO, 1987) as the daily-value regularly exceeded in many areas in Europe, especially during winter inversions.

2. Methodology

A methodology was developed to estimate the population exposure to PM_{10} pollutant in the Lisbon city centre, using a link between a mesoscale meteorological model, a CFD model and a population exposure model (Figure 1).

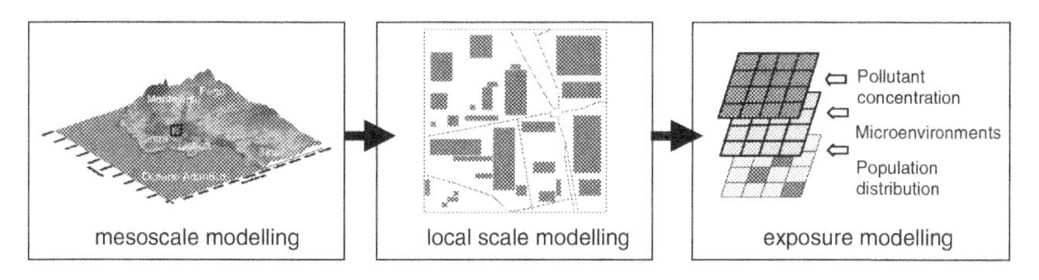

FIGURE 1. Schematic diagram of the applied methodology.

2.1. *Mesoscale Modelling*

MEMO is a three-dimensional Eulerian non-hydrostatic prognostic mesoscale model, which describes the atmospheric boundary layer for unsaturated air. The atmospheric physical phenomena are simulated by numerically approximating a set of equations in terrain-following co-ordinates, including mass continuity and momentum and transport equations for scalar quantities (Flassak and Moussiopoulos, 1987).

The MEMO model input data includes synoptic forcing, topography, land-use and emissions; as output it generates 3D meteorological fields (wind, temperature, turbulent kinetic energy among others) and 3D dispersion fields for a passive pollutant. The model has been successfully applied and verified in various European airsheds, namely Lisbon (Coutinho et al. 1993, Borrego et al., 1999). A PM dispersion and deposition module was included in MEMO, following the approach by Venkatram and Pleim (1999).

Analysis of temporal evolution of the air quality data series and of radiossoundings meteorological data revealed that high levels of PM_{10} (exceeding 300 $\mu g.m^{-3}$) were registered on 28th of February 2000, under the influence of winter inversion conditions, giving good arguments to select this day for the simulation. The MEMO model was applied to a simulation domain of 200km × 200km with a horizontal resolution of 2 km resolution (Figure 1).

Potential temperature and wind speed and direction from radiossoundings taken at 00:00 and 12:00 of February 28th were used to calculate the temperature and wind components necessary to initialise the MEMO model. The PM_{10} emissions needed for model input were based on an average emission factor per capita (5.5 $kg.inhab^{-1}$), recommended by Pulles and Visschedijk (2003) for Europe (this value can be considered representative for Portugal since the way of life is close to European standards).

2.2. *Local Scale Modelling*

VADIS is a CFD model, developed at the University of Aveiro, allowing the evaluation of maximum short-term local concentrations in urban geometries due to traffic road emissions (Martins and Borrego, 2003). Its structure is based in FLOW module for 3D meteorological fields, and DISPER module for 3D pollutant concentration fields (Borrego *et al.*, 2004).

The model has been validated with wind tunnel measurements and more recently, was applied to the Lisbon downtown area, demonstrating a good performance in the calculation of flow and dispersion around obstacles under variable wind conditions when compared with the measured values at the air quality station (Borrego *et al*, 2003).

VADIS study domain is located in the Lisbon city centre, covering an area of 1000m by 1000m, characterized by a set of 29 buildings, with an average height of 12 m, and 8 main roads (Figure 1). Hourly simulations for the estimation of PM_{10} concentrations were conducted for February 28th, 2000. Boundary meteorological conditions and PM_{10} background concentrations were estimated by MEMO.

Hourly PM_{10} traffic emissions were estimated with TREM model using local information (Borrego *et al*, 2003).

2.3. Population Exposure Modelling

The total human exposure to air pollution is the sum of exposures in different locations and times. The approach for modelling population exposure is based on modelled air quality concentrations in different microenvironments of an urban area and on a simple modelling of the number of persons present in the same environments, using Geographical Information Systems (GIS).

The human exposure will depend on the population time-activity pattern and on the concentration levels in the visited microenvironments. A microenvironment is a location that is assumed to have homogeneous pollution concentrations patterns in time and space e.g. at home, at work, in a street.

2.3.1. Study area Characterization

The selected study area, located in Lisbon city centre, corresponding to VADIS domain, is a mainly commercial and administrative zone, with intense road traffic, and where air pollution problems are present. As a working and also residential area, it is highly frequented by people walking on the streets or passing by car or public transports.

For the current work, according to the known characteristics of the region and to some available mapping data, five distinct microenvironments have been distinguished for exposure index quantification purposes: residence, office, school/university, transport and outdoor. Each cell of the study domain defined for local scale dispersion modelling was allocated to a microenvironment (see Figure 4 (a) below).

2.3.2. Population data and time-activity patterns

Data from the year 2001 National Census were gathered, namely for resident population by age groups, employees, number of schools and universities, number of students, for each one of the three sub-municipality regions ("Freguesias") included in the study region. The available data was analysed and used to estimate the hourly average number of persons in each indoor and outdoor microenvironments. Additional data on traffic counting, including buses and private cars, was also required to estimate the number of persons inside vehicles.

Figure 2 presents, for each microenvironment, the daily number of persons, as well as the time-activity pattern of the population, generated with one-hour time resolution. During the day, the total number of persons present in the study area almost duplicates due to the concentration of offices and commercial buildings. Occupation of schools was estimated considering their existence in the study domain and the number of 0-14 and 15-24 year-old residents. Time-activity patterns were estimated based on the lifestyle and habits of a typical Lisbon citizen. The right axis of Figure 2 plot refers to the hourly average number of persons in traffic microenvironment and was calculated taking into consideration the sum of fractions of time that people stay in the area. (e.g. one person is equivalent to 60 persons staying one minute in the area). An occupation of 1.4 persons per car and a variable daily occupation per bus were considered.

2.3.3. Exposure estimation

The exposure modelling is based on the microenvironments concept that considers empirical ratios for outdoor/indoor concentration in different microenvironments and population distribution functions. The prime objective of the model is the quantification of an integrated exposure:

$$APEI50 = \sum_{t}^{24} \sum_{i} (C50_{ti} \times P_{ti}),$$

where *APEI50* is the accumulated exposure population index (μg.m^{-3}.person.hour) for the population exposed above the concentration C=50 μg.m^{-3} during time *t*, for each grid cell *i* and *P* is a number of persons exposed. Therefore, the exposure modelling considers two major steps: (1) estimation of the pollutant concentration for the defined microenvironments, and (2) estimation of the number of persons in each microenvironment using statistical data and time-activity pattern.

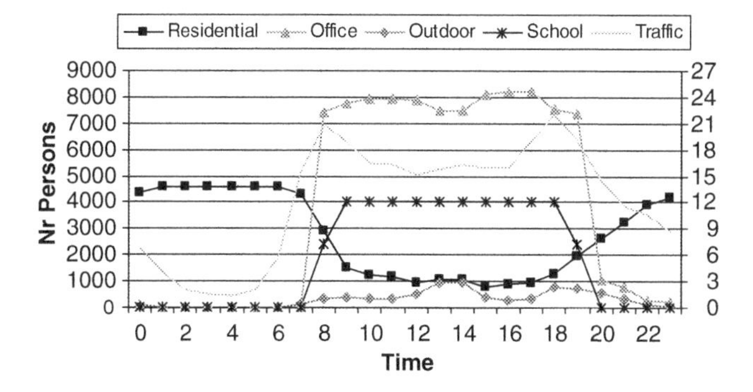

FIGURE 2. Number of persons per microenvironment in study area.

Regressions of indoor on outdoor concentrations proposed in (Burke *et al.*, 2001) were used for residence microenvironment.

$$C_{indoor}(day) = 48 + 0.52C_{outdoor} \quad (1)$$

$$C_{indoor}(night) = 20 + 0.51C_{outdoor} \quad (2)$$

Based on the literature source (Burke et al., 2001), cooking is responsible for 14% of the indoor sources. For this reason, the ratio of outdoor to indoor concentration was recalculated for the office microenvironment in order to exclude this source in the following way:

$$C_{indoor}(day) = 48 \cdot (1 - 0.14) + 0.52C_{outdoor} \quad (3)$$

$$C_{indoor}(night) = 20 \cdot (1 - 0.14) + 0.51C_{outdoor} \quad (4)$$

The contribution of the outdoor pollution to the total indoor concentrations in the car's interior microenvironment is stronger that in the case of residence and office, and can be expressed based on experimental data reported by Gulliver and Briggs (2004) as follows:

$$C_{car} = 13.1 + 0.83C_{outdoor} \quad (5)$$

3. Results and Discussion

MEMO application to the mesoscale simulation domain (see Figure 1) generated as output the boundary conditions to be used by the local scale model: wind velocity and components, temperature, turbulent kinetic energy and PM_{10} concentrations. An example of MEMO results is given in Figure 3 (a) for 9:00. At this hour PM_{10} concentrations reach a maximum of 160 μg.m^{-3} over the city of Lisbon.

Figure 3 (b) represents an example of VADIS application, with wind and dispersion fields estimated for 9:00. As expected, higher PM_{10} concentration values are located in southeast area of the domain, since the main winds blow from Northwest. It is important to notice that all the simulated PM_{10} concentration

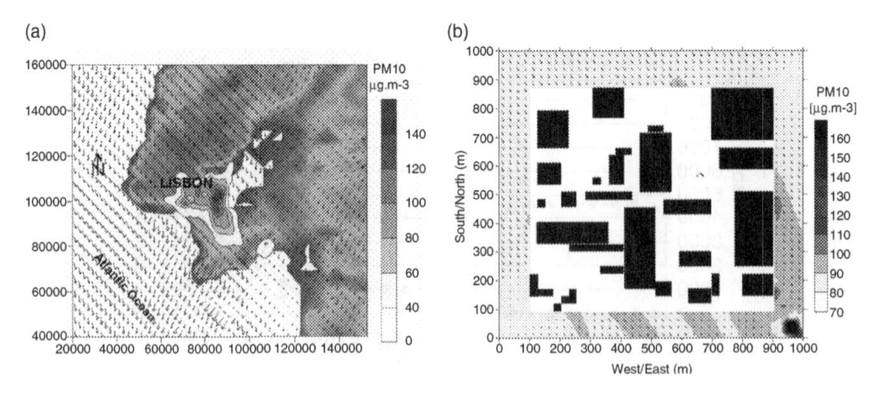

FIGURE 3. MEMO results for PM_{10} dispersion at 9:00 (a); Flow and PM_{10} concentration fields at 9:00 (b).

values inside the domain are above the daily limit value of 50 $\mu g.m^{-3}$ defined by the European Legislation.

As a result of the population exposure modelling performance, Figure 4 (b) presents the APEI50 field obtained for 9:00, which can be analysed together with the microenvironments distribution (Figure 4 (a)). At 9:00 students and working people are already in their respective occupation places, and homes are almost empty. The highest values of APEI50 correspond to schools and universities microenvironment. Relatively high APEI50 values are also verified in some office buildings in the vicinity of roads.

Results of APEI50 and APEI100 temporal evolution for transport microenvironment are shown in Figure 5 (a), revealing that a large number of persons are exposed to high concentrations of PM when going out and coming back home, early in the morning and in late evening. As expected, APEI50 for residence+office and outdoor microenvironment is higher during daytime, as it can be seen in Figure 5 (b).

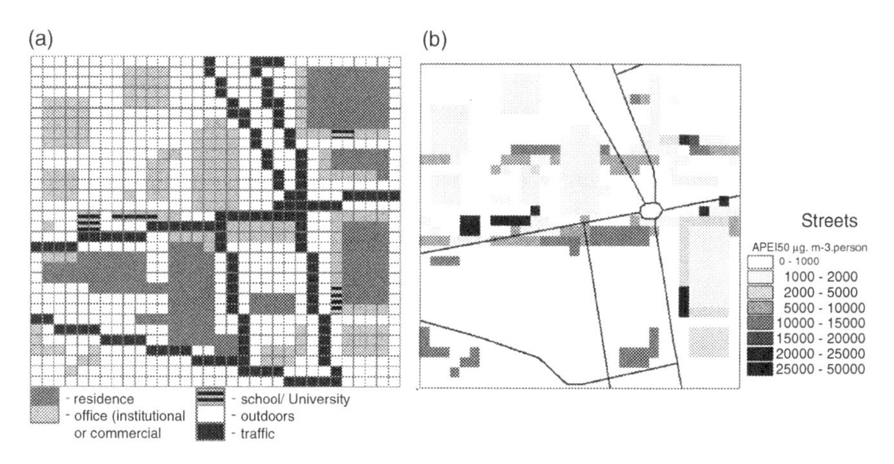

FIGURE 4. Microenvironments distribution in the study area (a) and APEI50 field at 9:00 (b).

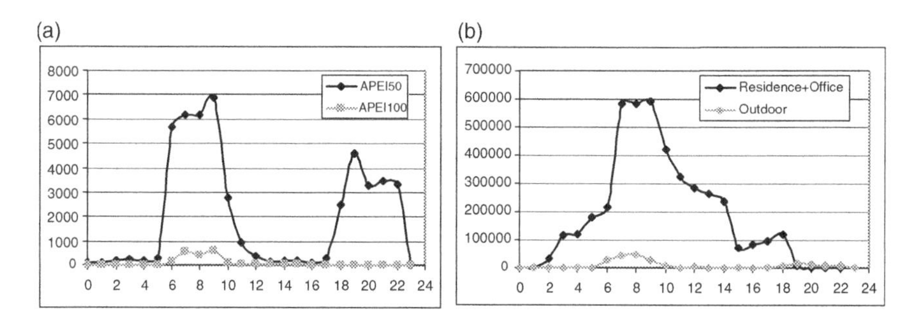

FIGURE 5. Hourly APEI50 and APEI100 for transport microenvironment (a); Hourly APEI50 for Residence+Office and Outdoor microenvironments (b).

The daily APEI50 obtained is $4.6 \times 10^6 \mu g.m^{-3}$.person, meaning that, during the selected day, 92000 persons are exposed to PM_{10} concentrations higher than the allowed limit. To have statistically significant results, a series of individual days has to be studied, in order to obtain more than one value for APEI50.

4. Conclusions

In the last years, there has been an increase of scientific studies confirming that long- and short-term exposure to particulate matter pollution leads to adverse health effects. These studies generally use air pollution measurements from stationary air monitoring sites to determine population exposure levels. However, because of the large local variations in pollution concentrations, these estimates are often associated with high uncertainties. For more accurate exposure estimation, air quality models can be used as important tools with the ability to provide detailed information on pollutants concentration fields.

Results from this study show that the highest values of APEI50 were found in schools and universities, which indicates that a high number of children, more sensitive to air pollution exposure, could be daily exposed to PM_{10} levels above 50 $\mu g.m^{-3}$.

Opposite to the individual exposure that characterises a person's contact with a given type of pollution during a period of time, the population exposure expressed in terms of the accumulated index (APEI50 and APEI100) represents the number of inhabitants exposed to pollutant concentrations above the defined levels (50 $\mu g.m^{-3}$ and 100 $\mu g.m^{-3}$), and could be a good short-term exposure indicator. This type of analysis is valuable for impact assessment on population exposure in connection with traffic management, city planning or emissions reduction strategies.

Acknowledgements The authors wish to thank the financial support of the 3rd EU Framework Program and the Portuguese Ministério da Ciência e do Ensino Superior, for the PhD grants of A.M. Costa (SFRH/ BD/11097/2002), H. Martins (SFRH/BD/13581/2003) and J. Ferreira (SFRH/BD/3347/2000).

References

Borrego, C.; Tchepel, O.; Salmim, L; Amorim, J.H.; Costa, A.M. and Janko, J., 2004, Integrated modelling of road traffic emissions: application to Lisbon air quality management, *Cybern. and Syst.: An Int. Journal* **35** (5-6), pp. 535-548.

Borrego, C.; Tchepel, O.; Costa, A.M.; Amorim, J.H. and Miranda, A.I, 2003, Emission and dispersion modelling of Lisbon air quality at local scale. *Atm. Env*. **37**, pp. 5197-5205.

Borrego, C., Carvalho, A.C., Miranda, A.I., 1999, Numerical simulation of wind field over complex terrain, in: *Measuring and Modelling Investigation of Environmental Processes*, R. San-Jose, ed., WIT Press, U.K.

Burke, J., Zufall, M., Ozkaynak, H., 2001, A population exposure model for particulate matter: case study results for PM2.5 in Philadelphia, PA. *Journal of Exp. Analysis and Env. Epidemiology* **11**, pp. 470-489.

Coutinho, M.; Rocha, A. and Borrego, C., 1993, Numerical simulation of meso-meteorological circulations in the Lisbon region, in: *20th Int. Tech. Meeting of NATO-CCMS on Air Pollution Modelling and its Application*, pp. 31-38.

Flassak, T, Moussiopoulos, N., 1987, Application of an efficient non-hydrostatic mesoscale model. *Boundary Layer Meteorology* **41**, pp.135-147.

Gulliver, J., Briggs, D.J., 2004, Personal exposure to particulate air pollution in transport microenvironments, *Atm. Env*. **38**, pp. 1-8.

Hertel, O., De Leeuw, F., Raaschou-Nielsen, O., Jensen, S., Gee, D., Herbarth, O., Pryor, S., Palmgren, F. and Olsen, E., 2001, Human exposure to outdoor air pollution -IUPAC Technical Report, *Pure Applied Chemistry* **73** (6), pp. 933-958.

Jensen S., 1999, A Geographic Approach to Modelling Human Exposure to Traffic Air Pollution using GIS, PhD Thesis, Ministry of Environment and Energy, National Environmental Research Institute, Denmark.

Kousa, A., Kukkonen, J., Karppinen, A., Aarnio, P. and Koskentalo, T., 2002, A model for evaluating the population exposure to ambient air pollution in an urban area, *Atm. Env.* **36**, pp.2109-2119.

Martins J.M., Borrego C., 1998, Describing the dispersion of pollutants near buildings under low wind speed conditions: real scale and numerical results, in: *Envirosoft 98 – Development and application of computer techniques to environmental studies*, WIT, Las Vegas, pp. 149-158.

Moschandreas, D.J., Saksena, S., 2002, Modelling exposure to particulate matter, *Chemosp.* **49**, pp.1137-1150.

Pulles, T., Visschedijk, A., 2003, Emission estimation methods for particulates: the CEP-MEIP emission factor database, TNO-MEP; *http://www.mep.tno.nl/emissions.*

Venkatram, A. and Pleim, J. (1999) The electrical analogy does not apply to modelling dry deposition of particles, Atmospheric Environment, 33, pp. 3075-3076.

WHO, 2001, WHO Strategy on Air Quality and Health, Occupational and Environmental Health Protection of the Human Environment, World Health Organization, Geneva, 2001.

WHO and EC, 2002, Guidelines for concentration and exposure-response measurement of fine and ultra fine particulate matter for use in epidemiological studies, EUR 20238 EN 2002.

WHO, 1987, Air Quality Guidelines for Europe, WHO Regional Publications, European Series, No.23, Copenhagen.

Urban Pollution Exposure

Speaker: A. M. Costa

Questioner: M. Sofiev

Question: Applying the meso-scale model, you forced it with just one input profile from a single sounding point. It might be better to use data from some NWP model. It may well happen that more accurate meteorological forcing would help in resolving the under-estimation problem, which could originate from incorrect mesoscale transport from neighboring areas with higher PM concentrations.

Answer: Although not presented in the paper, meteorological results from the mesoscale model were validated against measured data, and the model performance was quite reasonable. However, the use of meteorological data fields from a NWP model is a possibility that should be regarded, namely a sensibility study could be conducted to evaluate the mesoscale modeling results following the two input approaches: i) one single sounding point; and ii) meteorological fields from a NWP model. Nevertheless, when trying to explain the under-estimation problem, it is worth mentioning an additional aspect of no less importance: the PM10 emissions needed for model input were based on an average European emission factor per capita, recommended in bibliography. The use of this factor instead of the use of emissions resulting from an integrated emission estimation approach may as well have contributed to the problem in PM modelling.

30 Numerical Simulation of Air Concentration and Deposition of Particulate Metals Emitted from a Copper Smelter and a Coal Fired Power Plant During the 2000 Field Experiments on Characterization of Anthropogenic Plumes

Sreerama M. Daggupaty, Catharine M. Banic, and Philip Cheung*

1. Introduction

In winter and summer of 2000 field experiments were conducted to investigate the physical and chemical evolution of two different plume sources, one is a copper smelter in Quebec and the other is a coal-fired power plant in Ontario. The field experimental data was used to support a modelling study of the transport and deposition of particulate metals from these industrial sources. The purpose of this study is to investigate the particle size dependent concentration and deposition patterns of current emissions of particulate metals and to estimate the proportion subject to long range transport.

2. Methodology

Numerical modelling and simulation experiments were conducted using the BLFMAPS-a system that is a combination of a three dimensional mesoscale meteorological boundary layer forecast model (BLFM, Daggupaty et al 1994) and a set of air pollution transport, dispersion, and deposition (APS) prediction modules. The modelling system was utilized to simulate meteorology and air concentration, dry deposition, and wet deposition of particulate metals emitted

* Meteorological Service of Canada, 4905 Dufferin Street, Toronto, Ontario, Canada. M3H 5T4 Tel: 416-739-4451, Fax: 416-739-4288; e-mail: sam.daggupaty@ec.gc.ca

during research aircraft flight days in January and September 2000 from the power plant site, and during February and July of 2000 from the copper smelter.

The BLFM model utilises twice daily objectively analysed weather data from the Canadian Meteorological Centre (CMC) and predicts meteorological parameters for 12 hours with a five minute time step over a three-dimensional model domain. The model domain has horizontal area of 400 km $\times$ 400 km with grid spacing of 5 km and in the vertical it has 10 un-equally spaced layers with 1.5 m as the lowest level and a top at 3000 m above ground. The predicted meteorological variables, mixed layer depth, and turbulent parameters as function of space and time are used in the air pollution forecast modules. Air pollutant concentration and deposition fluxes are predicted with a 5-minute time step over the model domain. The domain is centered as per the simulation case either over the copper smelter at Rouyn-Noranda in Quebec or over the coal-fired power plant that is located at Nanticoke on the northern shore of Lake Erie in Ontario.

2.1. Emission Rates

An instrumented research aircraft measured meteorological parameters and the chemical and physical properties of the particles in the plume during the field study periods in winter and summer months of 2000 (Banic et al., 2004). In the case of the copper smelter, the industry made in-stack measurements of the emission rate of the metals during seven of the aircraft flights. The fraction of each metal in particles of different sizes (<2 μm, 2 to 8 μm, and >8 μm) was determined by aircraft sampling in the plume (Wong et al., 2004). The emission rate determined in-stack for each metal was prorated by the fraction of the metal in each size bin as observed by the aircraft measurements to get emissions as a function of particle size.

In the case of the power plant in Ontario the stack emissions by particle size (<2.5 μm, 2.5 to 10 μm, and >10 μm) were determined by the industry before the field study. The industry also provided particle emission rates as a function of power production and the actual power produced during the field study periods; we estimated the emission rates for each metal in each size bin by prorating with the actual power production.

For the model simulations presented in this paper for both facilities, the three bins were represented by particles with aerodynamic diameter of 0.25μm, 4μm, and 20μm, for small, medium, and large particles, respectively. In order to readily compare the effect of particle size on the concentration and deposition of metals emitted, we also performed simulations for small and large particles with a common emission rate of 1 g/s.

3. Model Details in Brief

Particulars of the APS portion of BLFMAPS are given briefly in the following. The model equation is the well-known atmospheric transport and dispersion equation and in a terrain-following coordinate (x, y, z*) system it is given as in Eq. (1).

$$\frac{\partial c}{\partial t} = -(u\frac{\partial c}{\partial x} + v\frac{\partial c}{\partial y} + w_*\frac{\partial c}{\partial z_*}) + K_h(\frac{\partial^2 c}{\partial x^2} + \frac{\partial^2 c}{\partial y^2})(\frac{\partial^2 c}{\partial x^2} + \frac{\partial^2 c}{\partial y^2})$$
$$+\frac{\partial}{\partial x_*}(K_z\frac{\partial}{\partial z_*}) + \text{source terms} + \text{skin terms.} \qquad (1)$$

where $z^* = \partial z - h(x,y)$, h is terrain elevation above sea level, c is the pollutant concentration in air and the source term is input emissions and the sink terms include dry and wet deposition processes. Equation (1) is solved numerically in an Eulerian grid by a finite difference approximation and operator splitting scheme. The horizontal advection terms are solved by an efficient modified Bott's scheme. Most of the other numerics are as in BLFM (Daggupaty et al 1994).

3.1. Dry Deposition

Effective dry deposition velocity (V_d^{eff}), which accounts for the influence of subgrid scale heterogeneous land-type effects, is given in Eq. (2) (Ma and Daggupaty, 2000 and Zhang et al 2001). The land use information on 1 km subgrid spacing was used for the effective V_d formulation over the study area.

$$V_d^{eff} = V_g + \left[1/\left(R_a^{eff} + R_d^{eff}\right)\right] \qquad (2)$$

where V_g is the gravitational settling velocity given by Stokes law and it is independent of surface type, but is rather a function of particle size and density. There exists no spatial average of V_g. Ra^{eff} and R_d^{eff} are effective bulk aerodynamic resistance and effective quasi-laminar resistance respectively. These resistances depend upon boundary layer parameters. R_d depends also on collection efficiency of the surface and is determined by various deposition processes (Brownian diffusion, impaction, and interception), surface properties, particle size, and density and atmospheric conditions. We followed Zhang et al., (2001) for the formulation of R_d. The method used for the determination of effective V_d is described in Ma and Daggupaty (2000).

The dry deposition flux (g m^{-2} s^{-1}), is given as $F(x, y)_d = c(x,y,z_{1.5})\ V_d^{eff}(x, y)$; where $c(x,y,z_{1.5})$ is pollutant concentration in air at 1.5 m above ground level.

3.2. Wet Deposition

The wet deposition flux was calculated as product of the vertically integrated concentration, normalised scavenging coefficient and precipitation rate as follows,

$$F(x,y)_w = \Lambda \int_{Zb}^{Zt} I\,c(x,y,z)\,dz$$

where $F(x,y)_w$ is the wet deposition flux (g m^{-2} s^{-1}), Zt, Zb are model top and bottom heights, Λ is the normalised scavenging coefficient ($s^{-1}mm^{-1}$ hr) and I is precipitation intensity (water equivalent in mm hr^{-1}). In summer season Λ is taken as 1.4×10^{-5}, 2.2×10^{-4} and 1.8×10^{-3} for small, medium, and large particles respectively. During winter, the corresponding values for snow are 4.7×10^{-6}, 7.3×10^{-5} and 6.0×10^{-4} respectively (Schwede and Paumier, 1997, Gatz, 1975, Slinn, 1977).

4. Results and Discussion

For the field study periods in each season, February 13 to 28, 2000 in winter and July 22 to August 5, 2000 in summer, for Rouyn-Noranda case and January 16 to 31, 2000 and September 11 to 22, 2000 for the Nanticoke case, the pollution model equations (BLFMAPS) were integrated over the three-dimensional model domain to predict air pollutant concentration and deposition fluxes. The hourly averaged concentration and accumulated deposition fluxes to the surface for the three particle sizes with aerodynamic diameters of 0.25μm, 4μm, and 20μm were evaluated for the periods of all simulations.

4.1. Concentration Distribution

Figure 1 displays the model computed daily averaged surface concentration distribution for small and large particles during the fall period of simulation for power plant case (Nanticoke). For ease of comparison we used an emission rate of 1 g s^{-1} for particles in both the fine (0.25μm) and coarse (20μm) sizes. This simulation clearly shows that the concentration of fine particles exceeds that of the large particles over a broader area by one order of magnitude. This is likely due to lower deposition flux and hence longer residence time for fine particles. Whereas relatively stronger depletion rate for large particles results in lower surface air concentration. Similar deduction was true from the analysis of simulation results for the smelter plume case in Quebec.

The smelter plume was observed to be at a lower height and was dispersing over a shallower depth resulting in higher concentration relative to the power

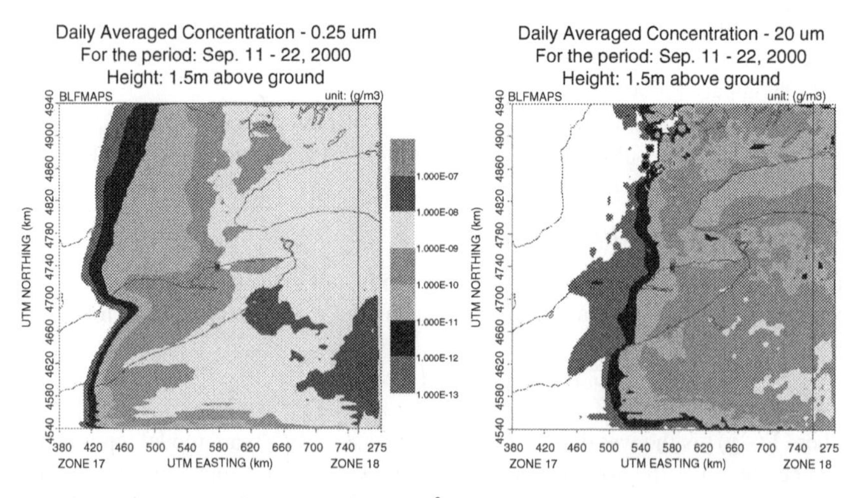

FIGURE 1. Daily averaged concentration (g m^{-3}) for small particles (left) and large particles (right) at 1.5 m agl for the period September 11-22, 2000 with an emission rate of 1g s^{-1}. Nanticoke station is identified with a X in the box on the coast of Lake Erie.

plant plume, which was at a higher height and was dispersing over a deeper mixed layer.

4.2. Comparison of Modelled and Aircraft-Monitored Plume Concentration

Model simulated hourly averaged concentrations with real emission rate for Rouyn-Noranda smelter case for each particulate metal and in each size and for corresponding flight time were compared with the aircraft measured concentration values. The aircraft samples for these flights were collected for particulate metals in three size bins within a short downwind distance of 3 to 18 km from the stack within the plume. The hourly averaged model values at the nearest grid point to the aircraft sampling location and height were used. We should note that the aircraft samples represent in situ conditions in the plume in its actual plume width and vertical depth, whereas the model grid point value represents an average over 25 km^2 area and a depth of about 200 m.

Table 1 presents the concentration ratio of aircraft measured to the model values for particles containing Pb, As, and Cu in large, medium, and small sizes. Almost all the flights in the table show that modelled values are reasonably agreeing with the aircraft values. The ratio is within a factor of three (model is under predicting) and in quite a few cases it is close to one. However in the case of flight 22 (February 21, 18:00 LST), the model values are differing by a factor of 10 to 20. This is a night time winter case with stable stratification during which the plume maintained a "ribbon" like character with little dispersion in the horizontal (1 km wide at 5 km distance downwind) and in vertical (about 200 m deep at 10 km downwind) and exhibiting high concentration. Since the model grid interval is 5 km it can not resolve such fine sub-grid scale gradients in concentration. The model, in distributing the plume through full grid cells, is predicting more dilute concentrations than observed. However, an important observation in Table 1 is that there is no systematic difference in the ratio for the different particle sizes. This indicates that the performance of the model is not biased with respect to particle size.

TABLE 1. Concentration ratio (Aircraft measured/Model simulated) for different metals for Rouyn-Noranda smelter.

Flight # /Date/	Large Particles 20 μm			Medium Particles 4μm			Small Particles		0.25μm
Time (LST)	Pb	As	Cu	Pb	As	Cu	Pb	As	Cu
28/July 26 /11	3.8	1.7	3.8	3.8	1.7	3.8	3.8	1.7	3.8
29/July 26 /15	1.8	n/a	4.5	1.8	n/a	4.4	1.8	1.7	4.5
30/July 27 /13	2.7	1.9	3.3	2.9	1.9	3.5	2.7	1.8	3.3
32/July 29 /11	1.3	0.70	0.99	1.1	0.6	0.90	1.3	0.70	0.99
19/Feb. 18 /15	3.2	2.2	3.2	2.7	n/a	3.0	3.5	2.8	3.2
22/Feb. 21 /18	19	12	12	19	10	10	17	10	11
23/Feb. 22 /16	6.2	n/a	n/a	6.4	4.3	3.3	6.3	4.3	3.3

Aircraft-measured, particle size dependent, particulate-metal-concentration data are being still analysed for the Nanticoke field study, hence similar comparison with the modelled data will be reported in a future study.

4.3. Accumulated Deposition Pattern

In order to study particle size dependent deposition character we simulated concentration and deposition fluxes for the entire period of field experiments in winter and summer with the emission rate of 1g s^{-1} for each size of particulate metal. As a constant particle density of 1 g cm^{-3} was used to correspond with the aerodynamic diameter, the numerical model results are equally applicable for all metals of same particle size. We examined the accumulated dry and wet deposition over applicable periods in different seasons of the field experiments at Rouyn-Noranda and Nanticoke. The effect of particle size is clearly evident in our analysis. The deposition of large particles is significantly larger than that of the small particles (Fig. 2). This can be explained by the higher scavenging coefficient in the wet deposition process and higher deposition velocity in the dry deposition process for coarser particles compared to the fine particles. Thus the larger particles are depleted from the plume at a faster rate leaving the finer particles to be transported over longer distance. The same phenomenon was also noticed in the case of summer season and with Rouyn-Noranda simulations.

4.4. Deposition around Rouyn and Nanticoke Sites

In order to examine the actual impact and total contribution of particulate metals we calculated total deposition flux (i.e., loading) over the period and over different circular areas around Rouyn and Nanticoke. We accumulated the modelled

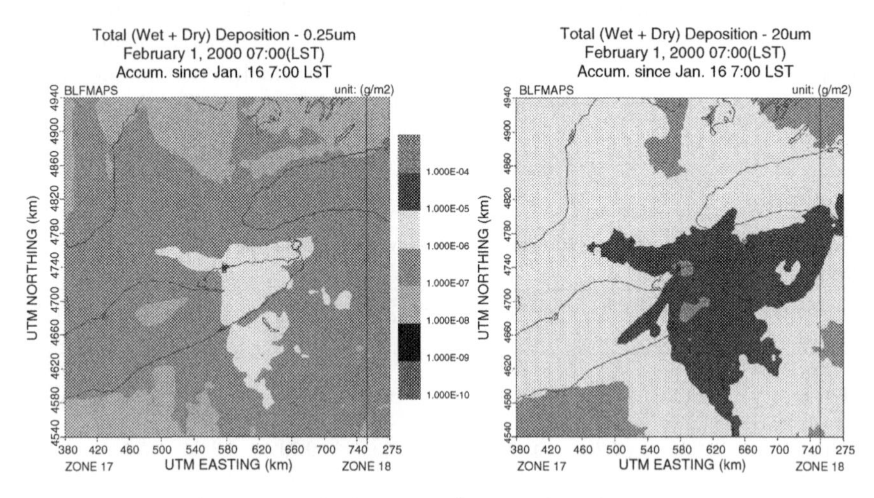

FIGURE 2. Accumulated total deposition (g m^{-2}) for small (left) and large (right) particles for the winter field study period February 13-29, 2000 with emission rate of 1 g/s.

deposition for the field study days in summer, fall, or winter (15, 16 or 12 days as the case may be) over the circular areas ranging from 25 km to 100-km radius from the centre of Rouyn or Nanticoke in our model domain.

Figures 3 and 4 give the total deposition as well as the dry and wet components at Rouyn and Nanticoke sites respectively as a percentage of the emissions over respective periods for large, medium, and small particles. Since these results are expressed as percentage of the total emission over the period, the results are applicable for any particulate metal of same particle size for any emission rate. For example the total deposition of fine particulate Pb within the circular area around the smelter plant with radius of 100 km is 5% of fine particulate Pb emissions over the period of 16 days (February 13–28, 2000). The total deposition (sum of dry and wet deposition) of fine particles is relatively small (5 to 15%) in comparison with coarse particulate (50%) deposition in the case of Rouyn whereas it is 2–7% for fine particles and about 50% in winter and 25% in fall for Nanticoke case. Dry and wet processes are equally contributing to the total flux in the case of large particles, whereas wet deposition exceeds dry deposition for medium particles and dry deposition slightly exceeds wet for fine particles (Fig.3). The primary cause for difference between Rouyn and Nanticoke cases was that the Nanticoke plume was at a much higher level (700 m to 500 m) than of the Rouyn

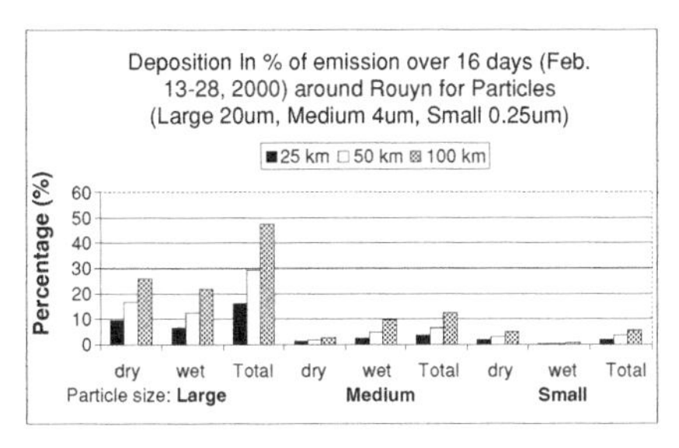

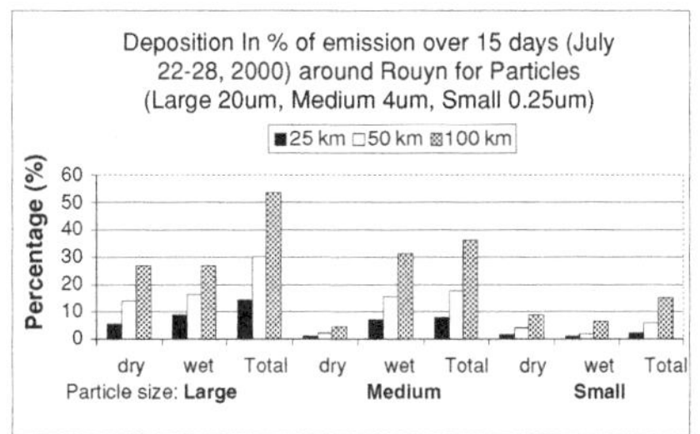

FIGURE 3. Accumulated deposition (% of total emission) around Rouyn over study period in winter (left) and summer study period (right).

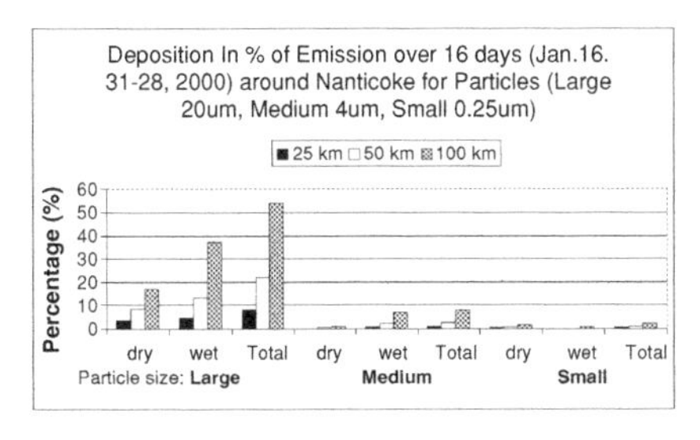

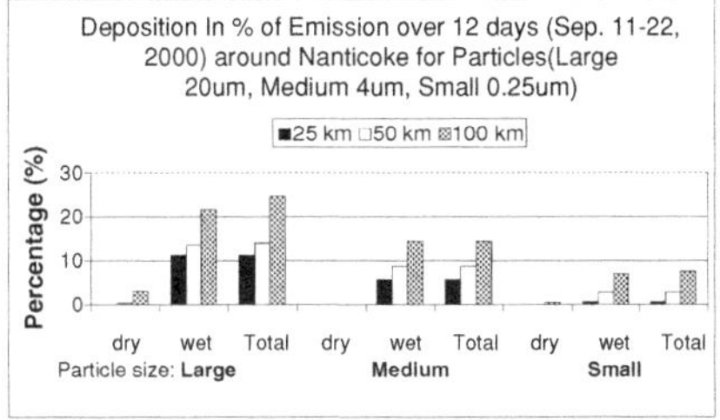

FIGURE 4. Accumulated deposition (% of total emission) around Nanticoke over study period in winter (left) and fall study period (right).

plume (200 m to 300 m), other factors being weather and surface characteristics. During our study period, wet deposition flux is slightly larger during the summer (Fig. 3) than in the winter.

Higher scavenging coefficient explains higher wet deposition process for large and medium particles. Higher V_d and higher scavenging coefficient deplete large particles very efficiently. However one has to note that wet deposition process is effective only in the presence of precipitation and its rate depends upon the intensity and type of precipitation.

5. Conclusions

Model simulated concentration and deposition patterns with particle size dependence showed interesting pattern and phenomenon. Particle size plays a significant role in dry and wet deposition processes. The coarser particles have stronger deposition rate due to higher V_d and higher scavenging rate. Fine particles tend to exhibit higher concentration in the air, and last longer and transport over longer distances. Specifically within the 100-km radius of the smelter plant, fine particle deposition was small, about 5 to 15% of total fine particle emissions in comparison with the coarse particle deposition of 50% of total large particle emissions. Similarly for the power plant at Nanticoke small particle deposition was 2 to 10% and large particle deposition was about 50% in winter and 25% in fall. Model simulated Rouyn plume level air concentrations of Pb, As, and Cu contained large, medium, and fine size particles compare well with the corresponding values measured in situ by the aircraft.

Acknowledgements The study was jointly funded by project #153 of the Toxic Substances Research Initiative which was managed by Health Canada and Environment Canada, the Metals in the Environment Research Network and the Meteorological Service of Canada.

References

Banic, C., W. R. Leaitch, K. Strawbridge, R. Tanabe, H. K. T. Wong, C. Gariépy, A. Simonetti, Z. Nejedly, J. L. Campbell, J. Skeaff, D. Paktunc, J. I. MacPherson, S. M. Daggupaty, H. Geo'nach, A. Chatt, M. Lamoureux., 2004: An aircraft study of the physical and chemical evolution of aerosols in smelter and power plant plumes. Accepted for *Geochemistry: Exploration, Environment, Analysis.*

Daggupaty, S. M., R. S. Tangirala, and H. Sahota, 1994: BLFMESO, A 3 dimensional mesoscale meteorological model for microcomputers. *Boundary-Layer Meteorology*, **71**, 81-107.

Gatz, D. F., 1975: Pollutant aerosol deposition into southern Lake Michigan. *Water, Air, Soil Pollution.*, **5**, 239-251.

Ma, J., and S. M. Daggupaty, 2000: Effective Dry Deposition velocities for Gases and Particles over heterogeneous terrain. *Journal of Applied Meteorology*, **39**, 1379-1390.

Schwede D. B., and J. O. Paumier, 1997. Sensitivity of the industrial source complex model to input deposition parameters, *Journal of Applied Meteorology,* **36**, 1096-1106.
Slinn, W. G. N., 1977: Some approximations of the wet and dry removal of particles and gases from the atmosphere. *Water, Air, Soil Pollution*, **7**, 513-543.
Wong, H. K. T., C. M. Banic, S. Robert, Z. Nejedly and J. L. Campbell. 2004: In-stack and in-plume characterization of particulate metals emitted from a copper smelter. Accepted for *Geochemistry: Exploration, Environment, Analysis.*
Zhang, L., S. Gong, J. Padro, and L. Barrie. 2001: A size-segregated particle dry deposition scheme for an atmospheric aerosol module. *Atmospheric Environment,* **35**, 49-560.

Numerical Simulation of Air Concentration and Deposition of Particulate Metals Emitted from a Copper Smelter and a Coal Fired Power Plant During 2000 Field Experiments on Characterization of Anthropogenic Plumes

Speaker: S. M. Daggupaty

Questioner: R. San Jose
Question: Is there any limit under the Canadian legislation for Pb and other metals?
Answer: Typically ambient air standards are administered by provincial Governments. Here are some limits that are used in the province of Ontario:

Ambient Air quality criteria (24 hour limit) in micrograms per cubic meter for Arsenic is 0.3, for Copper is 50, and for Lead is 2.

31
Aerosol Production in the Marine Boundary Layer Due to Emissions from DMS: Study Based on Theoretical Scenarios Guided by Field Campaign Data

Allan Gross and Alexander Baklanov*

1. Introduction

The main sources of aerosols in the remote Marine Boundary Layer (MBL) are sea salt sources, non sea salt sources, and entrainment of free tropospheric aerosols. Non sea salt aerosols are mainly sulphate derived from the oxidation of gaseous DiMethyl Sulphide (DMS) produced in surface water.[1] DMS is produced by phytoplankton and is estimated to account for approximately 25% of the total global gaseous sulphur released into the atmosphere.[2] DMS can either be dissolved into aqueous-phase aerosols or oxidised to other gas-phase species which can contribute to aerosol formation, e.g. SO_2, H_2SO_4, DiMethyl SulphOxide (DMSO), DiMethyl SulphOne ($DMSO_2$), MethaneSulfInic Acid (MSIA) and Methane Sulphonic Acid (MSA). It has therefore been postulated[3] that DMS emission from the oceans can produce new condensation nuclei and eventually Cloud Condensation Nuclei (CCN). Thus, DMS may have significant influence on the Earth's radiation budget.

In this paper, this DMS-postulate has been investigated for MBL conditions using the Chemistry-Aerosol-Cloud (CAC) model.[4] The CAC model is a box model where the following processes are solved:

$$\begin{aligned} dC_i/dt = {} & \text{chemical production} - \text{chemical loss} + \text{emission} \\ & - \text{dry deposition} - \text{wet deposition} \\ & + \text{entraiment from the free troposphere to the boundary layer} \\ & + \text{aerosol dynamics} \end{aligned}$$

C_i is the concentration of the ith species, which can either be a liquid- or gas-phase species.

---

* Allan Gross, and Alexander Baklanov, Danish Meteorological Institute, Meteorological Research Division, Lyngbyvej 100, DK-2100 Copenhagen Ø, Denmark
Corresponding author: Allan Gross, Phone: +4539157443, Fax: +4539157460, e-mail: agr@dmi.dk.

The aerosol dynamics in the CAC model is base on the modal description of the particle distributions.[5] Analytical solutions are found using the suggestions by Whitby and McMurry[5] and Binkowski.[6] The present aerosol model has three modes: nuclei, accumulation and coarse.

The gas-phase mechanism used is the regional atmospheric chemistry mechanism[7] where a DMS mechanism[4] (21 gas-phase sulphur species and 34 sulphur reactions) is added. For the liquid-phase chemistry we assumed that after a gas-phase species DMS, DMSO, $DMSO_2$, MSA, MSIA, SO_2 and H_2SO_4 has entered the liquid-aerosol it reacts rapidly to form $HSO_4^- + H^+$. Parameters used to calculate the mass transfer of species between the gas- and aqueous-phase are from Refs. 8 and 9.

Depending on the study item different meteorological reanalysed data sets, numerical weather prediction or climate models can be used as the meteorological driver for the CAC model. For the item considered here the 0-dim. version of the CAC model has been used.

In this paper the DMS-postulate has been investigated by performing two series of clean MBL simulations. In Section 2 these scenarios and the simulations of them are presented and discussed. In Section 3 the conclusions for the study are outlined.

2. Modelling Results and Discussion

To investigate the influence DMS can have on the dynamics of non sea salt aerosols in the MBL two types of clean MBL scenarios are setup. The first type is Based On a Literature Research (BLOR) for more details see Gross and Baklonov[4], in the rest of the paper called the BLOR scenarios. The second type is Based On Measurement from three remote Marine Atmospheres (BOMMA). The three remote sites are Cape Grim (Summer and Winter, from 1984-1992) (call CGS and CGW in the paper) and , Amsterdam Island (Summer and Winter, 1984-1991) (call AIS and AIW in the paper), and the EUMELI3 oceanographic cuise in the area south and east of the Canary Islands in 1991, for more details see Capaldo and Pandis.[10] They will in the rest of the paper be call the BOMMA scenarios. In Table 1 the conditions for these two scenario series are outlines. The initial concentrations of the aqueous-species are initialised using the initial aerosol conditions and the initial calculated water mole fraction.[4]

In Figures 1 and 2 the total concentrations of DMS + DMSO + $DMSO_2$ = $DMSO_X$ and inorganic sulphur from the simulations are plotted. We observe that for simulations with low DMS emission rates $DMSO_X$ is slowly increasing during the simulation. For high emission rates the $DMSO_X$ concentrations first increases rapidly followed by a slowly decrease. In general for all six simulations the $DMSO_X$ concentration has reached an equilibrium condition after eight days.

It is observed from field measurements[11,12] that the concentration of DMS is typically around 50 to 600 pptV and the concentrations of DMSO and $DMSO_2$ are much lower than the concentration of DMS. This is also observed for simulations

TABLE 1. Scenarios simulated to study the influence of DMS on the aerosol production in the MBL. The meteorological conditions, initial gas-phase concentrations, initial aerosol conditions and emission of SO_2 are the same for all simulations, only the DMS emissions and temperatured are varied according to the values given in the table.[a,b,c]

Meteorological Conditions:		**Initial Gas-Phase Concentrations:**			
Ground Albedo	0.10	H_2	2 ppmV	→	(2 pptV)
Pressure (mbar)	1013.25	CH_4	1.7 ppmV	→	(1.7 ppmV)
Relative Humidity	90%	CO	0.14 ppmV	→	(0.14 ppmV)
Cloud Frequency	1 d^{-1}	H_2O	3%	→	(3%)
Precipitation Frequency	0.1 d^{-1}	N_2	78%	→	(78%)
		O_2	20%	→	(20%)
Intital Aerosol Conditions:		NO_2	400 pptV	→	(10 pptV)
Nuclei Mode:		H_2O_2	1 pptV	→	(850 pptV)
Number conc.	133 cm^{-3}	HO_2	0 pptV	→	(15 pptV)
log(σ)	0.657	CH_3O_2	0 pptV	→	(14 ppt)
Geo. Mean Dia.	0.8×10^{-6} cm	HNO_3	150 pptV	→	(0 pptV)
		O_3	40 ppbV	→	(18 ppbV)
Accumulation mode:		HCHO	10 pptV	→	(330 pptV)
Number conc.	66.6 cm^{-3}	VOC	5.5 ppbC	→	(0 ppbC)
log(σ)	0.21	SO_2	2 pptV	→	(2 pptV)
Geo. Mean Dia.	$.266 \times 10^{-4}$ cm	DMS	100 pptV	→	(100 pptV)
		MSA	1 pptV	→	(1 pptV)

Emission of SO_2 in pptV/min:	0.014
Emission of DMS in pptV/min:	0.00, 0.06, 0.12, 0.24, 0.36, and 0.48 → ($0.055_{EUMELI3}$, 0.145_{CCS}, 0.025_{CCW}, 0.345_{AIS}, and 0.110_{AIW})
Temperature in K:	288.15, 288.15, 288.15, 288.15, 288.15, and 288.15 → ($296_{EUMELI3}$, 290_{CCS}, 286_{CCW}, 290_{AIS}, and 286_{AIW})

[a] The summertime simulations are started at the first day of the second month of the summer, and the wintertime simulations are started at the first day of the second month of the winter. The EUMELI3 simulation are started at 1 October. The BLOR scenarios are considered as summertime scenarios.
[b] For the initial concentrations, emissions of DMS and the temperatures the parameters given in the parentheses are used in the BOMMA scenarios.
[c] For more details of the used scenarios see Refs. 4 and 10.

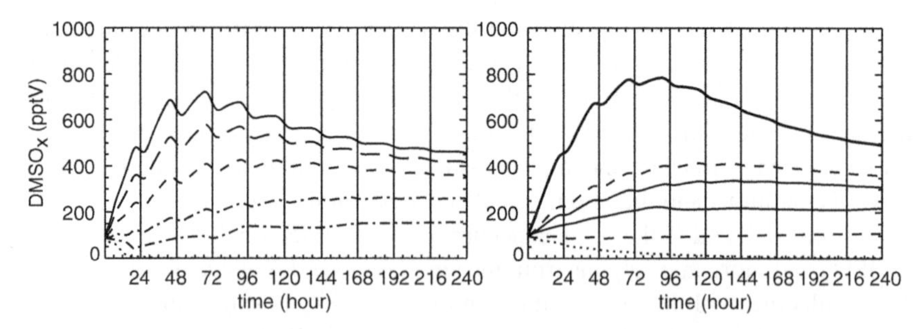

FIGURE 1. Concentration of DMO_X = DMS + DMSO + DMS_2 for the BOLR (left graph) and BOMMA scenarios (right graph) as a function of time. The simulations are started at noon with output every 15 min. The BOLR scenarios: (• • • • •) = 0.00, (– • • • – • • •) = 0.06, (– • – •) = 0.12, (– – –) = 0.24, (— — —) = 0.36, and (——) = 0.48 pptV/min DMS emission, respectively. The BOMMA scenarios: (• • • • •) = 0.00 pptV/min DMS emsision, (– • • • – • • •) = EUMELI3, (– – –) upper curve = CGS, (– – –) lower curve = CGW, (——) upper curve = AIS, and (——) lower curve = AIW, respectively.

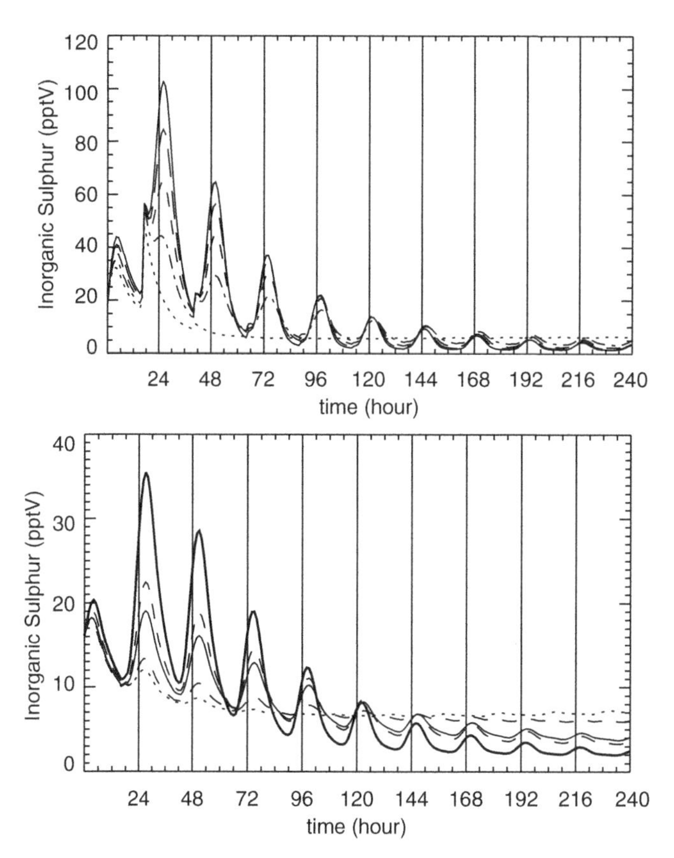

FIGURE 2. Concentration of inorganic sulphur for the BOLR (upper graph) and BOMMA scenarios (lowerr graph) as a function of time. The simulations are started at noon with output every 15 min. The BOLR scenarios: (• • • • •) = 0.00, (– • • • – • • •) = 0.06, (– • – •) = 0.12, (– – –) = 0.24, (— — —) = 0.36, and (——) = 0.48 pptV/min DMS emission, respectively. The BOMMA scenarios: (• • • • •) = 0.00 pptV/min DMS emsision, (– • • • – • • •) = EUMELI3, (– – –) upper curve = CGS, (– – –) lower curve = CGW, (——) upper curve = AIS, and (——) lower curve = AIW, respectively.

performed in this study, i.e. the BOLR and BOMMA scenarios correspond to typical concentrations of DMS in the clean MBL.

The concentration of inorganic sulphur plotted in Figure 2 increases rapidly during the first 24 hours, then it starts to decrease. After six days an equilibrium condition is reached. The equilibrium concentration for all the simulations are approximately 7 pptV. However, for the BLOR scenarios the inorganic sulphur concentration reach a maximum after the first day which are approximately 3 time higher compared to the BOMMA scenarios. The reason for these difference is related to the lower concentrations of NO_2 and O_3 in the BOMMA scenarios compared to the BOLR scenarios.

The influences of DMS on the aerosol production in the MBL are shown in Figure 3. It is observed that the accumulation and nuclei modes develop in a realistic way. The figure shows that the accumulation mode particles for the BOLR scenarios, grow rapidly in the beginning of the simulation followed by a decrease. For the BOMMA scenarios, the accumulation mode particles mainly decrease during the entire simulation. This difference is related to the higher concentration levels of inorganic sulphur in the BOLR scenarios relative to the BOMMA scenarios during the first four days of the simulations.

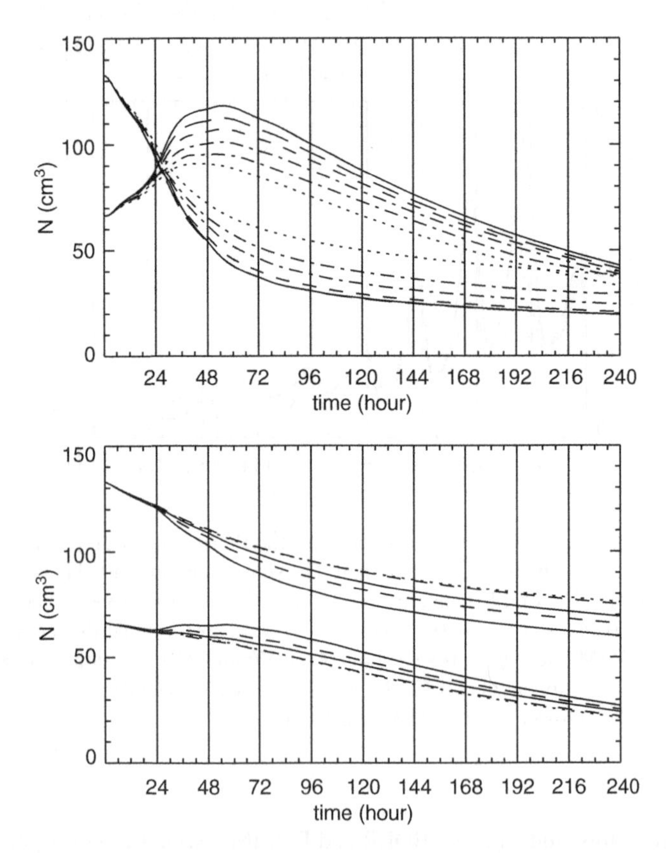

FIGURE 3. Particle number concentration for the BOLR (upper graph) and BOMMA scenarios (lower graph) for the accumulation and nuclei modes as a function of time. The accumulation mode plots are started with 66.6 cm^{-3} and 0.266×10^{-4}cm for N and d_g. The nuclei plots are started with 133 cm^{-3} and 0.8×10^{-6}cm for N and d_g. The simulations are started at noon with output every 15 min. The BOLR scenarios: (• • • • •) = 0.00, (– • • • – • • •) = 0.06, (– • – •) = 0.12, (– – –) = 0.24, (— — —) = 0.36, and (——) = 0.48 pptV/min DMS emission, respectively. The BOMMA scenarios: (• • • • •) = 0.00 pptV/min DMS emsision, (– • • • – • • •) = EUMELI3, (– – –) upper curve = CGS, (– – –) lower curve = CGW, (——) upper curve = AIS, and (——) lower curve = AIW, respectively.

For the nuclei model particles a decrease is observed during the entire simulation for all the scenarios. The development of the nuclei mode particles is realistic, since nuclei particles move to the accumulation mode due to condensation growth.

Even though the accumulation mode particles from the BOLR scenarios in the beginning of the simulation reaches particle number concentrations which are approximately three times higher then those obtained from the BOMMA scenarios, we observe in Figure 3 that after the concentration of $DMSO_X$ and inorganic sulphur compounds have reached its equilibrium condition the particle number concentrations are almost the same. This is an important observation, since it is the accumulation mode particles which are the potential source of new CCN.

The effect of DMS as a possible source of new particles in the MBL, which can have an effect on climate changes, is still uncertain. If we assume that the MBL height is 1 km we get from the simulations that, DMS will, for the BLOR scenarios, contribute approximately from 12.3% to 29.3% to the production of non sea salt accumulation mode particles, see Table 2 DMS % cont. N_{nss}. For the BOMMA scenarios this contribution is from 2.95% (CGW) to 26.8% (AIS), see Table 2.

Typical concentrations of sea salt aerosol in the MBL lie approximately in the interval from 5 to 30 cm^{-3}.[13,14] If this lower and upper boundary is added to the simulated number concentrations of the accumulation mode we can estimate an approximately upper and lower limit of the contribution of DMS to the total production of accumulation mode particles (Table 2 DMA % cont. N_{tot}). These results show that DMS contributes from 6% to 20% of the total production of accumulation mode particles for the BLOR scenarios, and from 1.21% (CGW) to 17.8% (AIS) for the BOMMA scenarios.

The DMS-postulate has lead to many modelling studies with a variety of approaches and different results. The latest studies are performed by Pandis et al.[15], Russel et al.[16], Yoon and Brimblecombe[17] and Pirjola.[18] All these studies and the study presented here use different aerosol physics, chemistry, meteorological conditions and scenarios. These studies can therefore not be compared directly.

TABLE 2. The influence of DMS on particle formation in accumulation mode for the BLOR and BOMMA scenarios. DMS % cont. N_{nss}: DMS contribution in % to accumulation mode non sea salt (nss) aerosols. DMS % cont. N_{tot} upper (lower) limit: the upper (lower) limit of DMS contribution in % to the sea salt plus the non sea salt accumulation mode aerosols.

The BLOR Scenarios	DMS emission in pptV/min				
	0.06	0.12	0.24	0.36	0.48
DMS % cont. N_{nss}	12.3	16.5	21.0	24.5	29.3
DMS % cont. N_{tot} upper limit	9.68	12.6	15.4	17.6	20.3
DMS % cont. N_{tot} lower limit	6.08	7.98	9.92	11.4	13.3
The BOMMA Scenarios	AIS	AIW	CGS	CGW	EUMELI3
DMS % cont. N_{nss}	26.8	13.3	18.3	2.95	12.9
DMS % cont. N_{tot} upper limit	17.8	9.72	12.9	2.33	9.44
DMS % cont. N_{tot} lower limit	10.0	5.24	7.07	1.21	5.08

However, most of the trends observed by Russel et al., Pandis et al. and Yoon and Brimblecombe are also observed in this study, but we get less accumulation mode particles compared to these studies.

An important factor in the CCN formation in the MBL is the sea salt flux from sea surfaces. The CCN from sea salt provides more than 70% of the total CCN, especially for winter seasons, over middle and high latitude regions.[17] In this paper we have not discussed the ability DMS has to form CCN, since it will be done together with a cloud model at the next step. The particle number concentrations in the model are related to the used gas-phase chemistry mechanism. Furthermore, only binary nucleation of H_2SO_4/H_2O is considered at present, and the nucleation of $(NH_4)HSO_4$ might be important.[19] Mereover, it is crucial to understand the chemistry of DMS in liquid- and gas-phase is if the problem investigated in this paper shall be fully understood, and the chemistry of DMS is still highly uncertain.[20] Therefore, many uncertainties remain and more studies are needed.

3. Conclusion

The general conclusion from the results presented in the paper indicates that DMS during the summer period can contribute approximately to 27% of the production of non sea salt sulphate particles in accumulation mode, and from 7% to 18% to the total production of accumulation mode particles in the clean MBL. During the wintertime the influence of DMS is much less, only up to around 13% of the non sea salt production, and between 1% and 10% for the total accumulation mode particle production. The influence of increasing emissions of DMS on the particle number concentration are related to the used gas-phase chemistry mechanism and the binary nucleation of H_2SO_4/H_2O considered in the model.

The chemistry of DMS is still highly uncertain, and many parameters used to describe the mass transport of $DMSO_X$ to the aerosols and the aerosol physics of $DMSO_X$ are almost unknown. Sensitivity studies of the chemistry mechanism and the aerosol parameters are therefore needed in order to achieve better understanding of DMS's influence on the aerosol dynamics in the MBL.

Acknowledgements This study was supported by the European Commission, as a part of the 5FP EU-project: Evaluation of the Climatic Impact of Dimethyl Sulphide (ELCID). The authors are grateful to ELCID-project coordinator Prof. I. Barnes, University of Wuppertal, Germany, for fruitful discussions.

References

1. T. W. Andeae, M.O. Andreae, and G. Schebeska, 1994. Biogenic sulphur emissions and aerosols over the tropical South Atlantic. 1. Dimethylsulfide in seawater and in the atmospheric boundary layer, *J. Geophys. Res* **99**(D11), 22819-22829.

2. R. Van Dingenen, N. R. Jensen, J. Hjorth, and F. Raes, 1994. Peroxynitrate formation during the night-time oxidation of dimethyl sulfide: its role as a reservoir species for aerosol formation, *J. Atm. Chem.* **18**, 211-237.
3. R. J. Charlson, J. E. Lovelock, M. O. Andrea, and S. G. Warren, 1987. Oceanic phytoplankton, atmospheric sulphur, cloud albedo and climate, *Nature*. **326**, 655-661.
4. A. Gross, and A. Baklanov, 2004. Modelling the Influence of Dimethyl Sulphide in the Marine Boundary Layer, *Int. J. Environment and Pollution*. (22 pages, accepted, in press).
5. E.R. Whitby, P.H. 1997. McMurry, Modal aerosol dynamics modelling, *Aerosol Sci. and Technol.* **27**, 673-688.
6. F. Binkowski, 1999. Aerosols in Models-3 CMAQ. In: Science algorithms of the EPA Models-3 community multiscale air quality (CMAQ) modeling system, *EPA/600/R-99/030*, 14.1-14.6.
7. W. R. Stockwell, F. Kirchner, M. Kuhn, and S. Seefeld, 1997. A new mechanism for regional atmospheric chemistry modelling, *J. Geophys. Res.* **102**, 25847-25879.
8. H. Herrmann, B. Ervens, H. W. Jacobi, R. Wolke, P. Nowacki, and R. Zellner, 2000. CAPRAM2.3: A chemical aqueous phase radical mechanism for tropospheric chemistry. *J. Atmos. Chem.* **26**, 231-284.
9. F. Campolongo, A. Saltelli, N. R. Jensen, J. Wilson, and J. Hjorth, 1999. The role of multiphase chemistry in the oxidation of dimethyl sulphide (DMS). A latitude dependent analysis, *J. Atm. Chem.* **32**, 327-356.
10. K. P Capaldo, and S. N. Pandis, 1997. Dimethyl sulfide chemistry in the remote marine atmosphere: evaluation and sensitivity analysis of available mechanisms, *J. Geophys. Res.* **102**, 23251-23267.
11. J. Sciare, M. Kanakidou, and N. Mihalopoulos, 2000. Diurnal and seasonal variation of atmospheric dimethylsulfoxide at Amsterdam Island in the southern Indian Ocean, *J. Geophys. Res.* **105**(D13), 17257-17265.
12. M. Legrand, J. Sciare, B. Jourdain, and C. Genthon, 2001. Sub-daily variations of atmospheric dimethyl sulfide, dimethylsulfoxide, methanesulfonate, and non-sea-salt sulfate aerosols in the atmospheric boundary layer and Dumont d'Urville (coastal Antarctica) during summer, *J. Geophys. Res.* **106** (D13), 14409-14422.
13. D. C. Blanchard, and R. J. Cipriano, 1987. Biological regulation of climate, *Nature* **330**, 526.
14. C. D. O'Dowd, and M. H. Smith, 1993. Physicocchemical properties of aerosols over the Northeast Atlantic: evidence for wind-speed related submicron sea-salt aerosol production. *J. Geophys. Res.* **98**, 1137-1149.
15. S. N. Pandis, L. M. Russel, and J. H. Seinfeld, 1994. The relationship between DMS flux and CCN concentration in remote marine regions, *J. Geophys. Res.* **99**(D8), 16945-16957.
16. L.M. Russel, S.N. Pandis, and J.H. Seinfeld, 1994. Aerosol production and growth in the marine boundary layer, *J. Geophys. Res.* **99**(D10), 20989-21003.
17. Y.J. Yoon, and P. Brimblecombe, 2002. Modelling the contribution of sea salt and dimethyl sulfide derived aerosol to marine CCN, *Atmos. Chem. Phys.* **2**, 17-30.
18. L. Pirjola, C. D. O'Dowd, I. M. Brooks, and M. Kulmala, 2000. Can new particle formation occur in the clean marine boundary layer? *J. Geophys. Res.* **105**, 26531-26546.
19. F. Raes, and R. Van Dingenen, 1995. Comment on "The relationship between DMS flux and CCN concentration in remote marine regions" by S.N. Pandis, L.M. Russell and J.H. Sienfeld, *J. Geophys. Res.* **100**(D7), 14355-14356.
20. A. Gross, I. Barnes, R. M. Sørensen, J. Konsted. and K. V. Mikkelsen, 2004. A Theoretical Study of the Reaction between $CH_3S(OH)CH_3$ and O_2, *J. Phys. Chem. A* (13 pages, accepted, in press).

Aerosol Production in the MBL due to Emissions from DMS

Speaker: Allan Gross

Questioner: M. Moran
Question: Given that two thirds of the rate constants in your DMS gas-phase mechanism are based on "best-guesses" rather than laboratory measurements, have you tried any sensitivity tests where you have substituted different values for those rate constants based on best guesses?
Answer: I have not performed sensitivity tests on the DMS gas-phase mechanisms I presented in the talk. However, I agree such tests should be performed, because it can inform us about which reactions are important and which are of negligible importance. Thus, sensitivity tests can guide scientists to focus on improving the important "best-guessed" DMS rate constants.

32 Modelling the Atmospheric Transport and Environmental Fate of Persistent Organic Pollutants in the Northern Hemisphere using a 3-D Dynamical Model

Kaj M. Hansen, Jesper H. Christensen, Jørgen Brandt,
Lise M. Frohn, and Camilla Geels*

1. Introduction

Persistent organic pollutants (POPs) are a group of chemical compounds with mainly anthropogenic origin; they are semi-volatile, hydrophobic, they bioaccumulate, they have toxic effects on human and wildlife and they display low degradation rates in the environment (Jones and de Voogt, 1999). POPs are emitted to the atmosphere either from industrial production, as by-products from combustion, or intentionally as pesticides used on crops or for insect control. A number of POPs are banned or subject to regulation, e.g. under the UNEP Stockholm convention for POPs and emissions of them have decreased during the last decades (Jones and de Voogt, 1999). However, due to the great persistence large amounts are still cycling in the environment. The volatility of POPs is temperature dependent, which can lead to several consecutive deposition and re-emission events named multi-hop or grasshopper transport (Wania and Mackay, 1996). To contribute to the understanding of these processes several models are developed. The environmental fate of POPs is traditionally studied with box models (e.g. Wania et al., 1999). Recently, atmospheric transport models with high spatiotemporal resolution are also developed to address these issues (e.g. Koziol and Pudykiewicz, 2001; Hansen et al., 2004).

The latest development of the Danish Eulerian Hemispheric Model (DEHM) at the National Environmental Research Institute now includes a description of the air-surface gas-exchange processes of soil, water and snow to study the environmental fate of POPs. The original version of the DEHM model was developed for studying the long-range transport of SO_2 and SO_4 to the Arctic (Christensen, 1997).

* National Environmental Research Institute, Department of Atmospheric Environment, Frederiksborgvej 399, P. O. Box 358, DK-4000, Roskilde, Denmark.

It has been further developed to study transport, transformation and deposition of reactive and elemental mercury (Christensen et al., 2002), concentrations and depositions of various pollutants (Frohn et al., 2002a,b) through the inclusion of an extensive chemistry scheme, and transport and exchange of atmospheric CO_2 (Geels et al., 2001, 2002).

This paper describes the POP version: DEHM-POP with the focus on the air-surface gas-exchange processes that are special for this model version. A general description of the DEHM-POP model system and the included air-surface exchange processes is given here. Thereafter a few examples from the simulations and model evaluation are discussed.

2. The Model System

The DEHM-POP model domain is centred at the North Pole and covers the majority of the Northern Hemisphere with a horizontal resolution of 150 km × 150 km at 60°N. The horizontal model grid extends into the Southern Hemisphere, an expansion of the grid to 135 × 135 cells compared to the 96 × 96 grid cells in previous DEHM model versions (Figure 1). The model is divided into 20 unevenly distributed vertical layers defined on terrain following σ-levels. The finest resolution is within the lowest few kilometres and the uppermost level is at a height of approximately 15 km. The applied numerical schemes have all been carefully

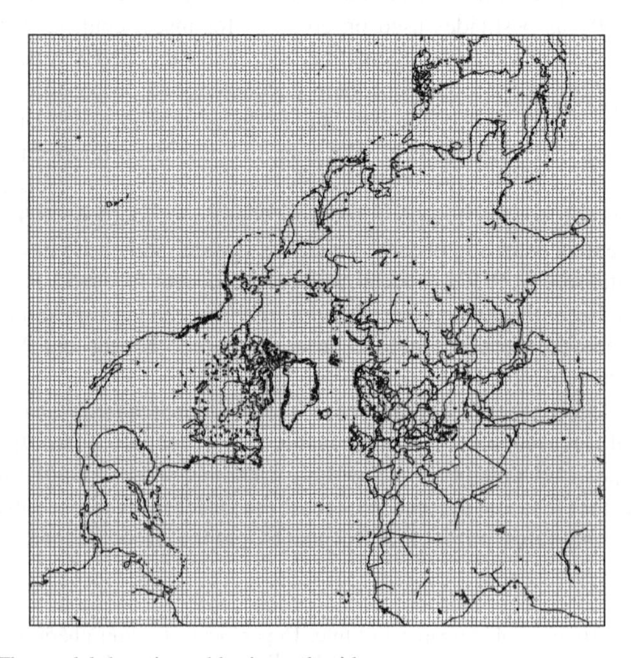

FIGURE 1. The model domain and horizontal grid.

tested for the previous versions of the model (Christensen, 1997; Frohn et al., 2002a). In order to obtain the required high-resolution meteorological data field the MM5 modelling subsystem (Grell et al., 1995), has been implemented as a meteorological driver between the data from ECMWF (2.5°×2.5° spatial and 12 h temporal resolution) and the DEHM model. In this study the domains and resolutions are the same in MM5 and DEHM and the meteorological data are archived at a 3 h interval.

The model is developed using α-hexachlorocyclohexane (α-HCH), the major component of the historically most used insecticide, as a tracer. This is one of the most abundant POPs in air, water and snow. Extensive monitoring data as well as reliable emission estimates exist for this compound. Being one of the most volatile POPs α-HCH is found almost exclusively in the gas-phase, so the particle bound fraction is omitted in this study. Emission input used in this study is adapted from Li et al. (2000). A large amount of α-HCH is found in the world oceans deposited through the historical emissions. This is taken into account by introducing an initial ocean concentration estimated from measurements.

2.1. *Parameterisations of Air-Surface Gas-Exchange Fluxes*

The special characteristics of POPs necessitate a description of air-surface exchange processes, since the compounds not only are deposited to the surface but also can be re-emitted to the air. Therefore the model is expanded with surface compartments representing soil, ocean water, sea ice and snow. The air-soil and air-water exchange processes are adapted from a zonally averaged global multimedia box model (Strand and Hov, 1996). Details of this implementation can be found in Hansen et al. (2004) together with an evaluation of the model including these two surface compartments only. The sea ice is considered to be a passive media without accumulation of contaminants. It acts as a lid on oceans preventing the air-ocean gas exchange. Data on sea-ice extension within the model domain is obtained from the MM5 model together with the meteorological data. The snow module is adapted from a newly developed dynamical model describing the exchange between air and a homogeneous snow pack (Hansen et al., in prep.).

3. Model Evaluation

The model was run for the period 1991–1998. High α-HCH concentrations are found in air over source areas such as India, Southeast Asia, and Mexico, but contaminated air is also found over regions without primary emissions such as the Atlantic and Pacific Oceans, and the Arctic, indicating long-range transport (Figure 2). Observations from several monitoring sites are available for model evaluation. Data used in this paper are from: Dunai Island in northern Russia, Spitzbergen at Svalbard, Alert and Tagish in Northern and Western Canada, respectively, Stórhöfdi in Iceland, Pallas in northern Finland, Lista at the Sothern

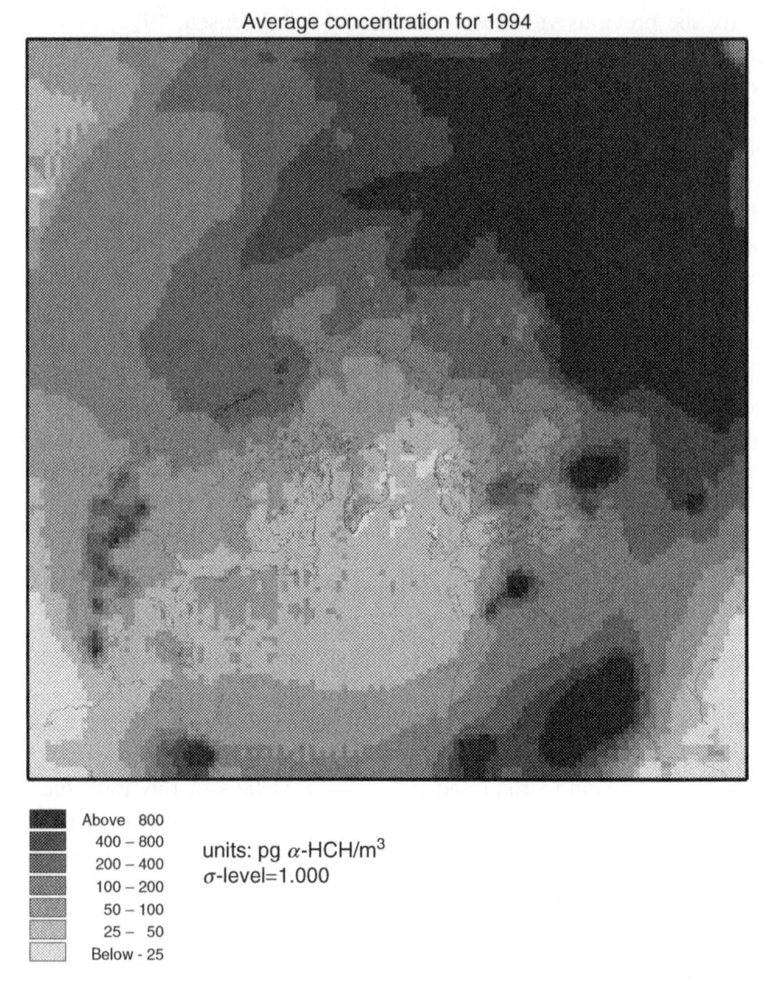

FIGURE 2. Annual average α-HCH concentrations in the lowermost atmospheric model layer for 1994.

tip of Norway and Rörvik and Aspvreten in Sweden (Berg et al., 2001; Aas et al., 2003). The time span and the temporal resolution of the measurements are variable at the sites. As the air concentrations are very low, samples are often integrated over a long time period. The deployment time of the samplers vary from between 1 and 14 days with a frequency of between 1 sample per week to 1 per month. The monitoring stations have been operated between 1 and 8 years within the simulated period. Daily averaged air concentrations from the lowermost atmospheric layer are extracted from the model at each of the monitoring sites. These data are averaged over the deployment time of the individual measurements to enable a direct comparison between measurements and model results.

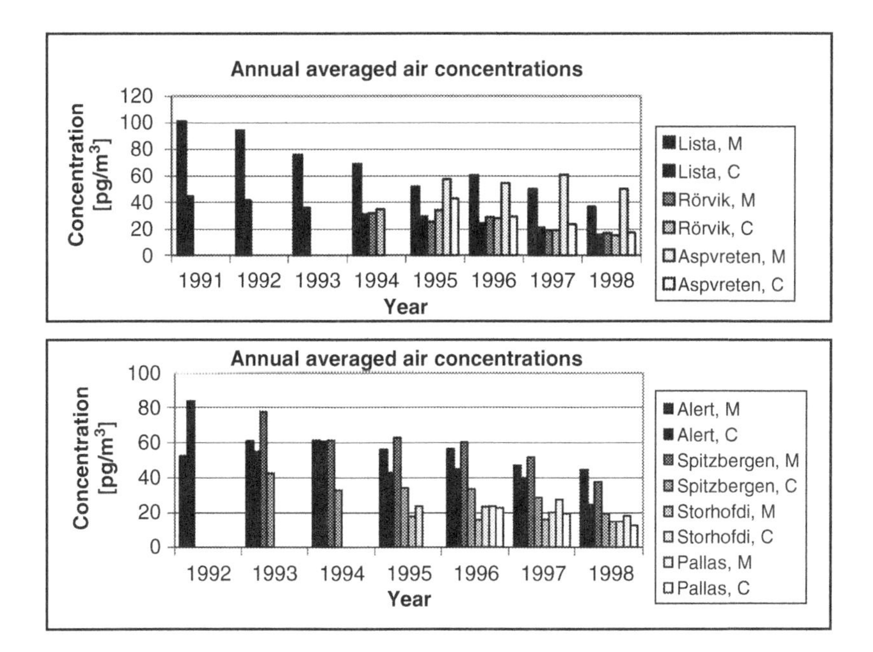

FIGURE 3. Annual averaged measured (M) and calculated (C) air concentrations in pg/m^3 for Lista, Rörvik and Aspvreten (upper figure) and Alert, Spitzbergen, Storhofdi, and Pallas (lower figure).

Annual averages of measured and modelled air concentrations are compared for each of the stations (Figure 3). Air concentrations during the simulated period are generally declining within the model domain reflecting the reduced emissions during the period. Averaged modelled air concentrations are generally within a factor 2 of the averaged measured concentrations. Exceptions are at Lista and Aspvreten where the modelled concentrations are up to a factor 2.5 and 3 lower than measured, respectively, and at Tagish and Dunai Island where the modelled concentrations are up to a factor 6 and 4 higher than measured, respectively (not shown). It is interesting to note that the predicted annual averaged air concentrations at Lista, Rörvik, and Aspvreten, which in the model are separated by only a few grid cells, are very close, whereas the measured concentrations are up to a factor 3 higher at Aspvreten and Lista than at Rörvik (Figure 3). This indicates that local surface exchange processes not described in the model, e.g. between vegetation and air have a great influence on the observed values.

4. Evaluation Against Individual Measurements

To evaluate the short-term average prediction capacity of the model measured α-HCH concentrations are plotted against model calculations in scatter plots. Four examples of the comparison between model results and observations are given in Figure 4. Various statistical parameters are given below the scatter plots.

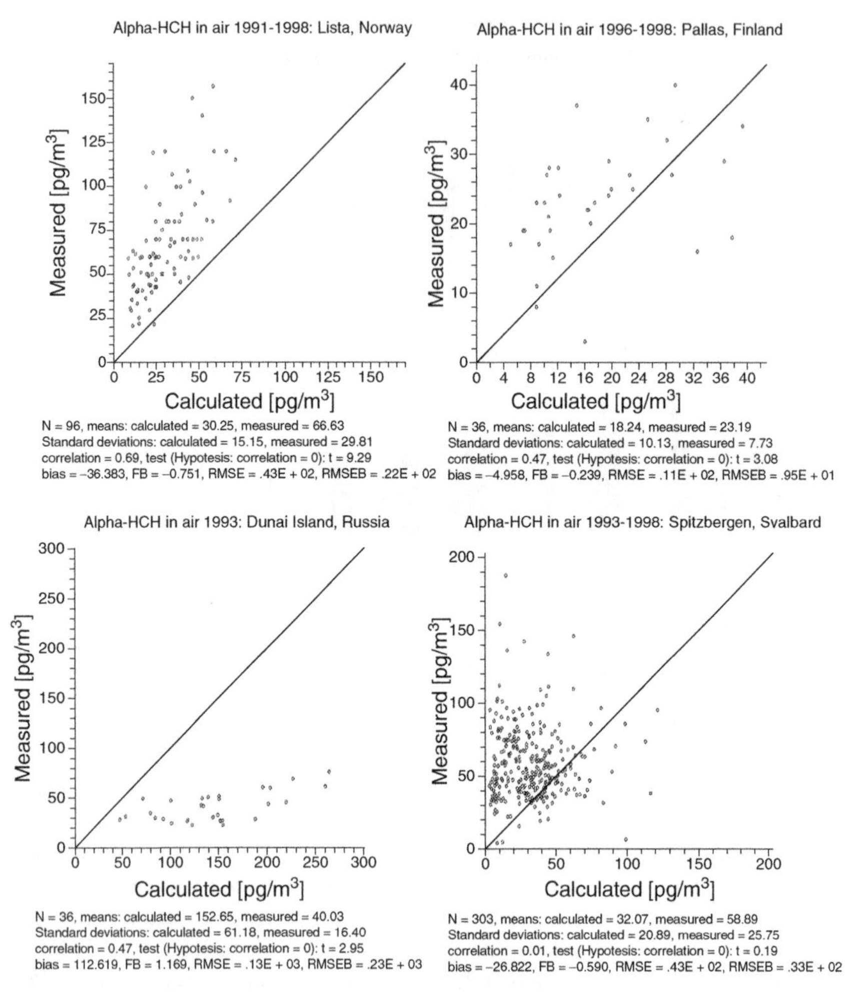

FIGURE 4. Four examples of the measured α-HCH concentrations plotted against calculated concentrations in the lowermost atmospheric model layer: Lista (upper left), Pallas (upper right), Dunai Island (lower left), and Spitzbergen (lower right). The results from the other stations except Rörvik are similar to Spitzbergen.

These include the number of data samples, (N), mean values, standard deviation, correlation coefficients with test for significance, bias, fractional bias (FB), root mean square error (RMSE) and RMSE based on values where the bias has been subtracted (RMSEB). The latter parameters are mostly helpful for comparing results from different model simulations. The correlation found at Lista, Pallas and Dunai Island is significant within a 0.1%, 1% and 1% significance level, respectively (Figure 4). A low correlation (0.24) significant within a 10% significance level is also found for Rörvik (not shown). No correlation is found at the other stations as illustrated by the example from Spitzbergen (Figure 4). The lack of correlation between measured and calculated air concentrations at some

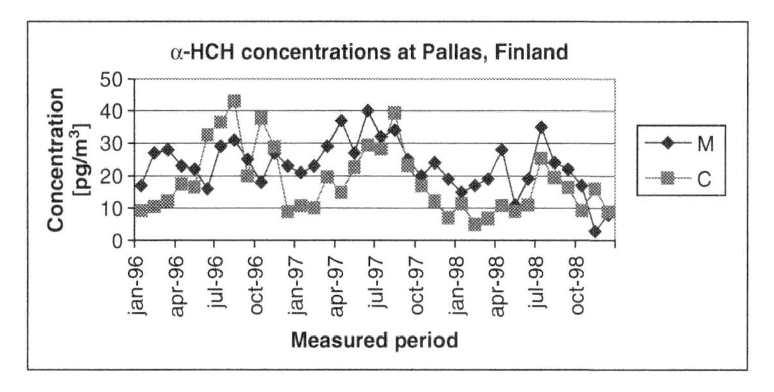

FIGURE 5. Measured (M) and calculated (C) air concentrations in pg/m^3 for Pallas, Finland from 1996 – 1998. The measurements are one-week integrated samples taken once a month.

stations but not at others and the variation in bias from strong negative to strong positive is a further indication that the surface exchange is not well enough parameterised, i.e. there are surface characteristics not described in the model, which possibly influence observed values.

A time series of the measured and calculated concentrations at Pallas, Finland is shown in Figure 5. It is seen that the model at this site captures a large part of the variability in measurements; an apparent annual variation with higher values during summer than during winter is found in both measurements and model results. The variability is not equally well described at all station, as also indicated in Figure 4.

5. Conclusions and Future Work

A hemispheric model for simulations of atmospheric transport and environmental fate of persistent organic pollutants (POPs) has been developed with α-HCH as a model tracer. This model includes modules representing the surface compartments: soil, water, sea ice and snow pack, and the air-surface exchange processes characteristic for POPs are thereby included in the model. Here the model simulations of atmospheric transport of α-HCH for the period 1991–1998 have been described and validated against observations. By comparing the annual averaged model results and measurements it has been shown that the model captures the long-term atmospheric concentrations. However shorter-term averaged concentrations are not well resolved by the model in the whole model domain. This indicates that the air-surface exchange is not well enough characterised by the present surface compartments, where e.g. exchange of air with vegetation and fresh water is not accounted for. A further expansion of the surface compartments is therefore needed. Furthermore, the inclusion of a description of the partitioning between gas-phase and particle-phase compounds is necessary to study POPs in general. These issues are part of the ongoing model development.

Acknowledgments This study is partly funded by the Danish Research Training Council through the Copenhagen Global Change Initiative.

References

Aas, W., Solberg, S., Berg, T., Manø, S., and Yttri, K. E., 2003, Monitoring of long rang transported air pollutants, Annual report for 2002. Kjeller, Norwegian Institute for Air Research, SFT Report 877/2003 NILU OR 23/2003.

Berg, T., Hjellbrekke, A. G., and Larsen R., 2001, Heavy metals and POPs within the EMEP region 1999, EMEP/CCC 9/2001.

Christensen, J. H., 1997, The Danish Eulerian Hemispheric Model – a three-dimensional air pollution model used for the Arctic, *Atmos. Environ.*, 31 (24), pp. 4169-4191.

Christensen, J. H., Brandt, J., Frohn, L. M., and Skov, H., Modelling of mercury with the Danish Eulerian Hemispheric Model. *Atmos. Chem. Phys.* Submitted December 19, 2002.

Frohn, L. M., Christensen, J. H., and Brandt, J., 2002a, Development of a high resolution nested air pollution model – the numerical approach, *J. Comp. Phys.*, **179** (1), pp. 68-94.

Frohn, L. M., Christensen, J. H., and Brandt, J., 2002b, Development and testing of numerical methods for two-way nested air pollution modelling, *Phys. Chem. Earth*, **27** (35), pp. 1487-1494.

Geels, C., Christensen, J. H., Hansen, A. W., Kiilsholm, S., Larsen, N. W., Larsen, S. E., Pedersen, T., and Sørensen, L. L., 2001, Modelling concentrations and fluxes of atmospheric CO_2 in the North East Atlantic region, *Phys. Chem. Earth*, **106** (10), pp. 763-768.

Geels, C., Christensen, J. H., Frohn, L. M., and Brandt, J., 2002, Simulating spatiotemporal variations of atmospheric CO_2 using a nested hemispheric model, *Phys. Chem. Earth*, **27** (35), pp. 1495-1505.

Grell, G. A., Dudhia, J., and Stauffer, D. R., 1995, A description of the Fifth-Generation Penn State/NCAR Mesocale Model (MM5), NCAR/TN-398 + STR, NCAR Technical Note, June 1995, p. 122, Mesoscale and Microscale Meteorology Division, National Center for Atmospheric Research, Boulder, Colorado.

Hansen, K. M., Christense, J. H., Brandt, J., Frohn, L. M., and Geels, C., 2004, Modelling atmospheric transport of α-hexachlorocyclohexane in the Northern Hemisphere with a 3-D dynamical model: DEHM-POP, *Atmos. Chem. Phys.*, **4**, pp. 1125-1137.

Hansen, K. M., Halsall, C. J., and Christensen, J. H., in prep., A dynamic model to study the exchange of gas-phase POPs between air and a seasonal snow pack, in prep. for *Environ. Sci. Technol.*

Jones, K. C. and de Voogt, P. 1999, Persistent organic pollutants (POPs): State of the science, *Environ. Pollut.*, **100**, pp. 209–221.

Koziol, A. S. and Pudykiewicz, J. A., 2001, Global-scale environmental transport of persistent organic pollutants, *Chemosphere*, **45**, pp. 1181–1200.

Li, Y.-F., Scholtz, M. T., and van Heyst, B. J., 2000, Global gridded emission inventories of α-hexachlorocyclohexane, *J. Geophys. Res.*, 1**02**, D5, pp. 6621–6632.

Strand, A. and Hov, Ø., 1996, A model strategy for the simulation of chlorinated hydrocarbon distributions in the global environment, *Water Air Soil Poll.*, **86**, pp. 283–316.

Wania, F. and Mackay, D., 1996, Tracking the distribution of persistent organic pollutants, *Environ. Sci. Technol.*, **30**, 9, pp. 390A–396A.

Wania, F., Mackay, D., Li, Y.-F., Bidleman, T. F., and Strand, A., 1999, Global chemical fate of α-hexachlorocyclohexane. 1. Evaluation of a global distribution model. *Environ. Toxicol. Chem.*, **18**, 7, pp 1390–1399.

33 PM-Measurement Campaign HOVERT: Transport Analysis of Aerosol Components by use of the CTM REM–CALGRID

Andreas Kerschbaumer[1], Matthias Beekmann[2], and Eberhard Reimer[1]

1. Introduction

A one year PM10 and its major components measurement campaign in, and around, Berlin has been carried out from September 2001 to September 2002. The most important challenge was to improve the knowledge about the contribution of anthropogenic urban sources and of long term transports of anthropogenic and natural constituents of air to local concentrations in order to give advice to authorities to elaborate reduction strategies for PM10 concentrations. The validation of the chemical transport model (CTM) REM_Calgrid (Stern et al, 2003) by means of the sampled data was propaedeutic to any further use of it in determining possible sources of air pollution.

By means of a DIGITAL Hi-Vol-Sampler, daily samples of atmospheric PM10-fractions have been collected on quartz-filters from September 2001 until September 2002 at 15 sites in and around the city of Berlin. These filters were conditioned according to usual regulations, in order to be able to determine gravimetrically the total dust content. Subsequently, filter parts were chemically analysed by means of ion chromatography in order to measure main ions (Sulphate, Nitrate, Chloride, Potassium, Magnesium, Calcium, Ammonium) as well as by means of thermo-gravimetry to measure EC (Elemental Carbon) and OC (Organic Carbon).

In order to provide a chemical air mass classification depending on meteorological situations, trajectories have been calculated. Backward trajectories are used to determine source/receptor relationships between the aerosol observations in Berlin and European wide emission maps for PM10. The 3D-trajectories pay attention to the formation of mixing height, clouds and precipitation, so that climatological information about short and long range transports can be deduced.

[1] kerschba@zedat.fu-berlin.de, Institut für Meteorologie, Freie Universität Berlin, 12165 Berlin, Germany
[2] Service d'Aéronomie/Institut Piere Simon Laplace, CNRS, France

REM_CALGRID was run for the measurement period over Europe with a nest over the Berlin-Brandenburg area with a 4 km horizontal resolution. Secondary inorganic aerosol components have been calculated using ISORROPIA (Nenes et al., 1999) while organic PM-constituents have been simulated using SORGAM (Schell *et al.*, 2001).

A systematic budget study for a control volume over the greater Berlin area has been performed for aerosol components using the CTM REM_Calgrid quantifying the contribution of horizontal and vertical advection, mixing, emissions, secondary aerosol formation, and dry and wet deposition to pollution accumulation and loss in the city of Berlin.

2. Aerosol Transport Analysis by Means of Backward Trajectories

Figure 1 shows the location of the monitoring sites in and around Berlin. Measurements at Frankfurter Allee (MP174) and at Beusselstr. (BS) are representative for concentrations near main traffic roads, concentrations in Neukölln (MP42) represent urban background polluted air, Marienfelde (MP27) in the South and Buch (MP77) in the North measure urban influenced air pollution, while the monitoring station at Frohnauer T. (FT) is located at 324 m height at the top of a tower and represents PM-concentrations well above most emission sources. Stations at Paulinenau (PA) and Hasenholz (HH) are representative for rural background air.

Figure 2 shows the annual average concentration of PM10 (left) and of Nitrate, Sulphate, Ammonium, Elemental and Organic Carbon (right) over the whole measurement period. While total PM10 and Elemental Carbon concentrations show a net decrease from traffic sites (30-35 μg/m^3 for PM10 and 4.6 μg/m^3 for EC) toward rural measurement locations (20 μg/m^3 for PM10 and 1.3 μg/m^3 for EC) and toward upper air concentrations (15 μg/m^3 for PM10 and 1 μg/m^3 for EC),

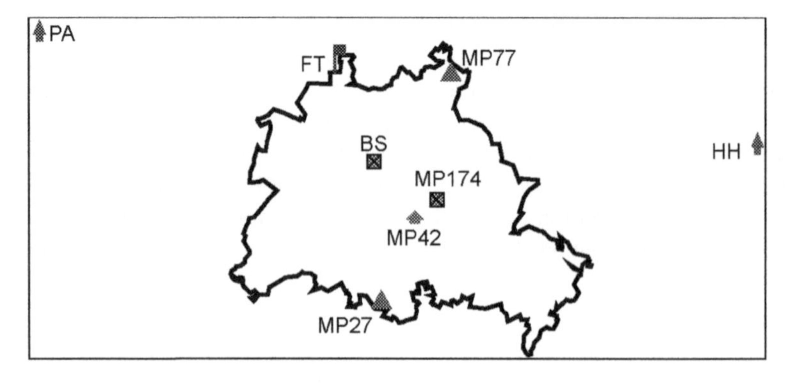

FIGURE 1. Aerosol Monitoring Stations in and around Berlin.

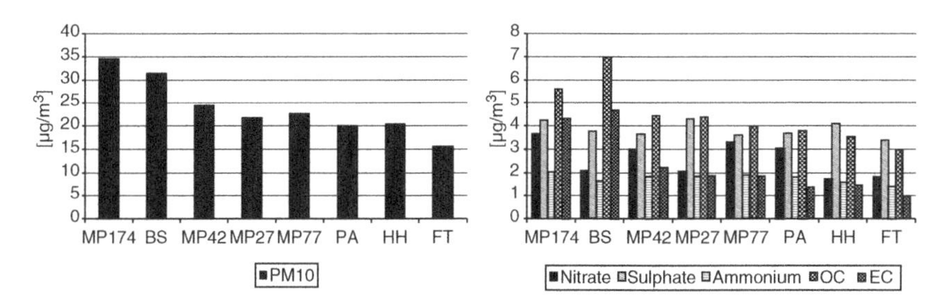

FIGURE 2. Annual mean concentrations of total PM10 (left) and main aerosol components (right) in μg/m³.

secondary inorganic aerosol components exhibit a more homogeneous behaviour: Nitrate ion annual mean concentration at the rural station Paulinenau is 3 μg/m³, which is comparable to the urban background concentration of 3 μg/m³ and to the traffic-related concentration at Frankfurter Allee of 3.6 μg/m³. Sulphate ions show concentrations between 3.2 μg/m³ (upper air station Frohnauer T.) and 4.2 μg/m³ (traffic station Frankfurter Allee) with same values also in the rural background stations (3.6 μg/m³ at Paulinenau and 4 μg/m³ at Hasenholz). Ammonium ion concentrations exhibit the same tendency to be distributed homogeneously over the different measurement sites with values between 1.7 μg/m³ at Paulinenau and 2 μg/m³ at Frankfurter Allee. In conclusion, Organic Carbon concentrations vary from 7 μg/m³ at traffic sites to 3.5 μg/m³ at rural sites.

Figure 3 shows the mean PM composition at different measurement sites. While secondary inorganic aerosol components Sulphate, Nitrate and Ammonium are more or less similar in both regimes, traffic sites exhibit a much stronger Elemental Carbon contribution to PM10 than rural sites.

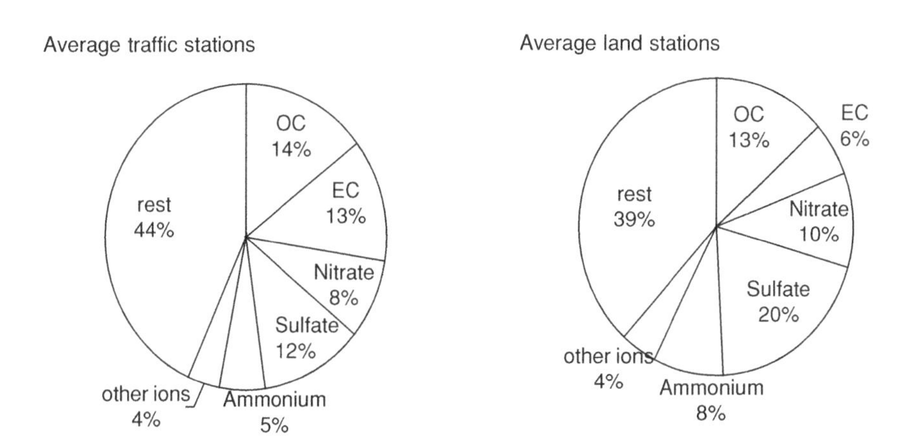

FIGURE 3. PM-composition at traffic sites (left) and at rural stations.

In order to determine the relationship between high concentrations of different species and meteorological air-mass-transport, chemically passive backward trajectories have been calculated.

Meteorological wind-fields are derived from SYNOP-data and interpolated on a Europe-wide grid with a horizontal spacing of 25 km. Vertical analysis is performed on isentropic surfaces, while planetary boundary layer parameters are determined using a bulk method. These regional fields are used as first guess fields for a finer horizontal (2 km grid-spacing) and vertical (1°Celsius) resolution considering additional local wind measurements. Air-mass-trajectories are calculated every 10 minutes using a dynamical-cinematic procedure, starting from the PM-measurement-sites in and around Berlin.

A dense set of 3D backward trajectories is used to analyse the seasonal and episodic relationship between measurements and possible source areas. The surface contacts under 50 m height of the 3D trajectories are integrated over a 1 km^2 grid and weighted by the locally measured concentrations of all species and the given time sections. The resulting concentration matrices for all observational sites are combined and, in result, special source areas or transport paths are determined.

PM10 transport into the Berlin area shows distinctive seasonal and spatial patterns: besides the typical source areas South-East of Berlin down to Katowice, mainly during autumn and winter, during summer there is correlation to the dominantly agricultural areas East to North of Berlin. Different species (ions, Organic-, Elemental Carbon) show distinct transport paths. Sulphate paths are predominantly determined by long range transport from South-East. The relative contribution of transport and of local production of aerosol components in the greater Berlin area is estimated by a direct comparison of influence matrices of suburban and urban observations.

3. REM_Calgrid Validation

REM_CALGRID (Stern et al., 2003) is an Eulerian grid model of medium complexity that can be used on the regional, as well as the urban scale for short-term and long-term simulations of oxidant and aerosol formation. For the study presented here, the model was applied at a modelled resolution of approximately 25 km for an area that covers Central Europe and a nested domain of about 300 km × 300 km around Berlin with a horizontal resolution of about 4 km. The model was run with 4 dynamically changing layers with the reference layer the mixing height, the CBM4 (Gary et al., 1989) photochemistry scheme for the gas-phase chemistry, ISORROPIA for the simulations of secondary inorganic aerosol-components and SORGAM for the secondary organic aerosol-components.

Meteorological input is derived from synoptic meteorological surface data, upper air data and climatological information (Reimer and Scherer, 1992).

In order to obtain a consistent emission database on the different scales, Europe-wide EMEP emissions have been used to scale highly resolved Berlin emissions for all species.

Comparisons with annual means of PM10 at different measurement locations show an underestimation of PM10 at traffic sites, but a good agreement at urban background stations. Rural sites as well as upper air concentrations are underestimated by the CTM, as well. Sulphate fraction concentrations are for all comparison sites underestimated by about 30 percent, while Nitrate concentrations show a good agreement with the annual mean concentration. Ammonium is also well depicted at urban background stations as well as at the measurement site at 324 m height. Elemental and Organic Carbon concentrations are underestimated at traffic sites and overestimated at the urban background location.

The seasonal variations of the PM-compounds are well depicted by the model. REM_Calgrid describes well the monthly urban background patterns of PM10, Ammonium and Organic Carbon, whereas Sulphate is underestimated in winter, Nitrate in spring and Elemental Carbon overestimated in spring. At the city-edges, PM10, Nitrate and Organic Carbon follow well the monthly slope, while Sulphate concentrations are underestimated in winter, Ammonium and Elemental Carbon concentrations are overestimated in autumn.

Correlations between simulated and observed daily averages were especially large for the autumn and winter season (between 0.7 to 0.8 for Sulphate, Ammonium and PM10). Minimum correlations between observed and simulated concentrations were found for Elemental Carbon (0.4) and for Organic Carbon concentrations (0.5). While there is a strong correlation at urban background and traffic stations between Elemental Carbon concentrations and NOx-concentrations in the simulations, this could not be found in the observations. The dependency of EC emissions on traffic seems to be too strong. Therefore the EC concentrations were scaled by NOx concentrations in order to get a better balance between traffic emitted and non traffic emitted EC. Following this procedure, assumed EC traffic emissions seem to be overestimated by a factor of 1.5 to 2.

Correlations of urban and regional tracers, like NOx for urban behaviour and Sulphate for regional behaviour, show that the local influence is overestimated for PM10, Elemental and Organic Carbon concentrations. Differences between urban concentrations and regional background are overestimated by the model.

4. Budget Analysis

A volume over the metropolitan area of Berlin has been chosen to analyse polluted air mass changes due to the single simulated processes within REM_Calgrid (Panitz et al., 1999). Hourly mass differences due to emission-, deposition-, chemical, advective, turbulent and diffusive processes have been summed up in order to determine the importance of every single simulated process in producing or loosing pollution mass within the control-volume of every simulated PM-component (Figure 4).

Primary aerosols are accumulated during the whole year due to emissions, and lost due to horizontal advection through the lateral boundaries of the control volume and due to dry and wet deposition. Coarse PM is accumulated mostly due to emissions and reduced mostly due to horizontal transport. Fine particles show the same behaviour.

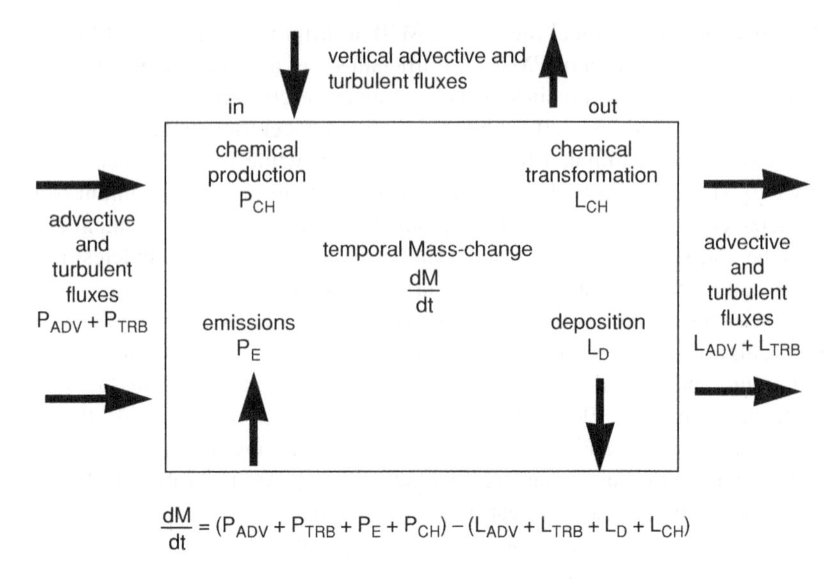

$$\frac{dM}{dt} = (P_{ADV} + P_{TRB} + P_E + P_{CH}) - (L_{ADV} + L_{TRB} + L_D + L_{CH})$$

FIGURE 4. Budget Analysis Method.

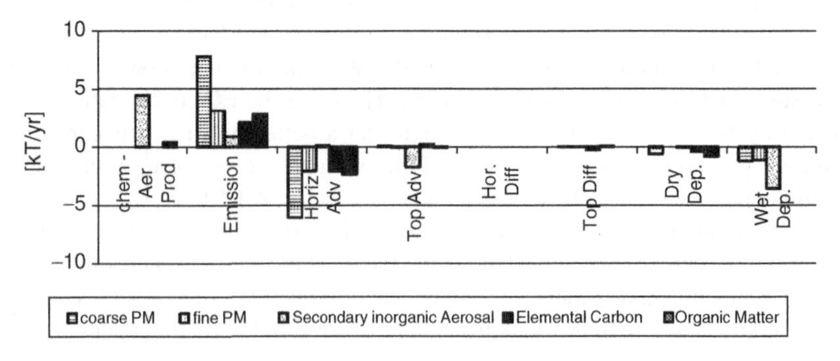

FIGURE 5. Annual mass changes [kT/yr] of PM components due to different simulated processes.

Secondary inorganic aerosol-components, (Sulphate-, Nitrate- and Ammonium ions) are mainly accumulated by chemical production and to a much lesser extent by emissions; loss processes are mainly wet deposition processes and transport through the upper boundary of the control volume. There is also an overall net accumulation of secondary inorganic aerosols due to horizontal advection processes within the control volume over the city of Berlin (Figure 5).

The net annual mass change of PM is almost in equilibrium compared to the single processes of production and loss. This means, that the system is in an overall equilibrium, which, nevertheless, is not zero.

Comparing the main accumulation and loss terms, which are emission and chemistry for production and advection as a loss process, there is a net outflow tendency over the whole year of primary PM-components and Elemental and Organic Carbons from the city toward the surrounding area, whereas secondary inorganic aerosol-components show this to a much lesser extent. Nitrate has a predominant tendency of inflow toward the city; the same is true for Ammonium-ions. Sulphate has an outflow tendency away from the city. Secondary inorganic aerosol-components are also transported upward through the model and control-boundary into the above free atmosphere, whereas Elemental and Organic Carbon show a net inflow tendency from the free atmosphere through the upper boundary toward the city.

Meteorological seasons dependent mass budget analysis for primary PM-components and secondary inorganic components integrated over the whole control volume over Berlin urban area has also been performed. For primary PM particles emission is the only accumulation factor for all seasons and the most important removal processes are horizontal advection and wet deposition. During winter time emissions contribute mostly to accumulation, and advection and deposition cannot totally counterbalance this. (Smaller contributions such as diffusion through the top become important considering the total budget). The net budget for winter presents an accumulation within the city. In summer times emissions contribute often to the accumulation in the city while advection is least compared to the other seasons. Secondary inorganic aerosol-components, on the other hand, show a quite different behaviour. The chemical production process is the most important contribution. Emission is important only for Sulphates, while accumulation or loss due to advection is very dependent on season and on PM-component. For Sulphate, there is a net loss during winter, spring and autumn, but a net accumulation during summer. Nitrate is accumulated in the volume over Berlin for all seasons, except for winter and shows also a loss due to chemical processes during spring and summer. Ammonium is accumulated in the city during spring and summer due to advection. Deposition is most effective for Sulphate during the warm season, and for Nitrate during cold season.

For all secondary inorganic aerosol components, deposition is an important removal process.

Sulphate is accumulated partially by direct emissions in the city and more importantly by chemical production and horizontal advection.

Nitrate is lost through the top and lateral boundaries and to a smaller extent also by chemical reactions. Important accumulation processes are horizontal advection and chemical production.

Ammonium is lost through the top and through the lateral boundaries. The main production processes are horizontal advection and chemical production.

For Elemental Carbon, which is considered a primary aerosol component in the model, horizontal removal is counterbalanced by emissions. Deposition seems to be counterbalanced by accumulation through the top boundary.

Organic Carbon exhibit a balance between horizontal advection and emission. For the secondary organic parts of the organic matter there is a positive accumulation due to chemistry. Some loss is due to advective processes, as well.

5. Summary and Outlooks

A one year measurement campaign has been carried out in and around Berlin sampling PM10 and analysing its main constituents Sulphate-, Nitrate- and Ammonium Ions, Elemental and Organic Carbon and other components of minor fractions.

Observations have shown that PM10 is influenced strongly by long range transport of secondary inorganic aerosol components whose sources may be far away from the measurement sites in Berlin. Pollutant concentration weighted air parcel trajectories give evidence that an important source of secondary inorganic aerosol components is situated to the South-East region of Berlin.

The chemical transport model REM_Calgrid has been applied to simulated the whole measurement campaign period and has been validated with the aerosol samples. An overall good agreement between observations and simulations can be stated; seasonal effects could be simulated as well as daily variations of concentrations. Nevertheless local city contributions on secondary aerosol components are over-estimated and there are episodic problems in long range transport. Furthermore, it is important to analyse special episodes with large errors in modelling, large scale as well as local scale events.

The chemical transport model has been used to understand better the importance of single production and loss processes of aerosols. Distinguished summations of process dependent accumulation and removal of PM-constituents have shown a strong importance of advective processes in the accumulation of secondary inorganic aerosols also within the CTM, which is in agreement with the observations and with the trajectory analysis. Elemental and Organic Carbon exhibit a minor influence of transports to the city which could not be confirmed by the measurements. Some revision of EC-emission data as well as of OC simulations are desirable.

An ongoing better quantification of transported and locally produced parts of aerosol is still needed in order to be of help to local authorities in their decision strategies for reducing air pollution.

Acknowledgements This work has been funded by the BMBF–Germany within the atmospheric research activity programme AFO2000.

References

Gery M.W., G.Z.Whitten, J.P.Killus, and M.C.Dodge, 1989. A Photochemical kinetics mechanism for urban and regional scale computer modeling. J. Geophys. Res., 94, 925-956.

Nenes, A., Pilinis, C., and S.N. Pandis, 1999: Continued development and testing of a new thermodynamic aerosol module for urban and regional air quality models. Atmos. Environ., 33, 1553-1560.

Panitz H.-J., Nester K., Fiedler F.,(1999) Bestimmung der Massenbilanzen chemisch reaktiver Luftschadstoffe in Baden-Württemberg und den Teilregionen Freudenstadt und Stuttgart,

Forschungsbericht FZKA-PEF, PEF – Projekt "Europäisches Forschungszentrum für Maβnahmen zur Luftreinhaltung"

Schell B., I.J. Ackermann, H. Hass, F.S. Binkowski, and A. Ebel, 2001, Modeling the formation of secondary organic aerosol within a comprehensive air quality model system, Journal of Geophysical research, 106, 28275-28293.

Stern, R., Yarmatino, R. und Graff, A., 2003, Dispersion modeling within the European community's air quality directives: long term modeling of O3, PM10 and NO2, in proceedings of 26th ITM on Air Pollution Modelling and its Application. May 26-30, 2003, Istanbul, Turkey.

Reimer, E. and Scherer, B., 1992, An operational meteorological diagnostic system for regional air pollution analysis and long-term modelling, in: Air Pollution Modelling and its Applications IX., van Doop, H.., ed.., Plenum Press.

PM-Measurement Campaign HOVERT: Transport Analysis of Aerosol Components by Use of the CTM REM-Calgrid

Speaker: A. Kerschbaumer

Questioner: P. Bhave

Question: Your measurements indicate the wintertime sulfate concentrations are comparable with the summer concentrations (quite different from the U.S.) and the wintertime sulfate appeared to be one of the greatest areas of model under-predictions. Do you have any thoughts about why the wintertime sulfate is so greatly under-predicted?

Answer: Sulfate measurements show no annual pattern at sites in and around Berlin and magnitudes are very similar in all locations giving a strong indication of long term transport of sulfate and very weak local contribution. The model overestimates the local effects of sulfate in Berlin giving rise to an annual cycle in the simulations. Furthermore, the underestimation of cold season's sulfate concentration is due to a too low chemical production of sulfate during winter.

34
Direct Radiative Forcing due to Anthropogenic Aerosols in East Asia During 21-25 April 2001

Soon-Ung Park* and Lim-Seok Chang*

1. Introduction

Atmospheric aerosols play a major role in the global climate system. Many researchers have conducted studies on the radiative forcing of aerosols for recent years. Aerosol particles are known to cool or warm the atmosphere directly by absorption, scattering and emission of solar and terrestrial radiation and indirectly by changing the albedo and the life time of clouds by acting as cloud condensation nuclei (Charlson et al., 1992). Current estimates suggest that anthropogenic aerosols and biomass burning have enough climate forcing to offset warming caused by greenhouse gases such as carbon dioxide (Kiehl and Briegle, 1993). For example, the present day global mean radiative forcing due to anthropogenic aerosols is estimated to be between –0.3 and –3.5 W m^{-2} which is comparable to the present day greenhouse gases forcing of between 2.0 and 2.8 W m^{-2} (IPCC, 1996).

In spite of great radiative impacts of aerosols, a large uncertainty of key variables makes it impossible to correctly quantify the magnitude of radiative forcing. The uncertainties are due in part to the limited data on aerosol climatology, and in part to the lack of our understanding on the processes responsible for the production, transport, physical and chemical evolution and the removal of aerosols at various spatial and time scales.

To understand radiative forcing of anthropogenic aerosols, we need the spatial distribution of concentrations of these pollutants. Therefore, the objective of this study is to examine the spatial distribution of concentrations of anthropogenic aerosols in East Asia using the gas-phase chemistry of the California Institute of Technology (CIT) airshed model and the aqueous-phase chemistry of the Regional Acid Deposition Model (RADM) and to investigate radiative forcing of anthropogenic aerosols for a period from 21-25 April 2001.

* Soon-Ung Park, Lim-Seok Chang, School of Earth and Environmental Sciences, Seoul National University, Seoul, 151-742, Korea

2. Model Description

The mass conservation of aerosol can be described as the general dynamic equation.

$$\frac{\partial C_i}{\partial t} + \nabla\cdot(uC_i) = \nabla\cdot(K\nabla C_i) + \left(\frac{\partial C_i}{\partial t}\right) con/evap. + \left(\frac{\partial C_i}{\partial t}\right) reac. + \left(\frac{\partial C_i}{\partial t}\right) coag. + \left(\frac{\partial C_i}{\partial t}\right) source/sink \tag{1}$$

where C_i is the mass concentration of species i of particles, K the eddy diffusivity and $\left(\frac{\partial C_i}{\partial t}\right)$ *con/evap.*, $\left(\frac{\partial C_i}{\partial t}\right)$ *reac*, $\left(\frac{\partial C_i}{\partial t}\right)$ *coag.*, $\left(\frac{\partial C_i}{\partial t}\right)$ *source/sink* are respectively the changes of a composition due to condensation/evaporation, heterogeneous reaction on the surface of aerosol, coagulation and source/sink. Aerosol processes include emissions, homogeneous nucleation, condensation, coagulation, transport, sedimentation, hygroscopic growth and dry/wet deposition which are given in Chang and Park (2004) in detail. The gas chemistry of CIT is used to simulate the gaseous species concentrations. The chemical mechanism of CIT (Russell et al., 1988) treats 29 species and 52 reactions including 8 photolytic reactions and is extended to include a SO_2 oxidation (SO_2+OH).

3. Gaseous Pollutants and Carbon Emissions

The gas and aerosol emissions for SO_2, NO_x, NH_3, CO, Non-Methane Volatile Organic Carbon (NMVOC), black and organic carbon are obtained from the most recent and highly resolved emission inventory (Street et al., 2003) (Fig.1). The

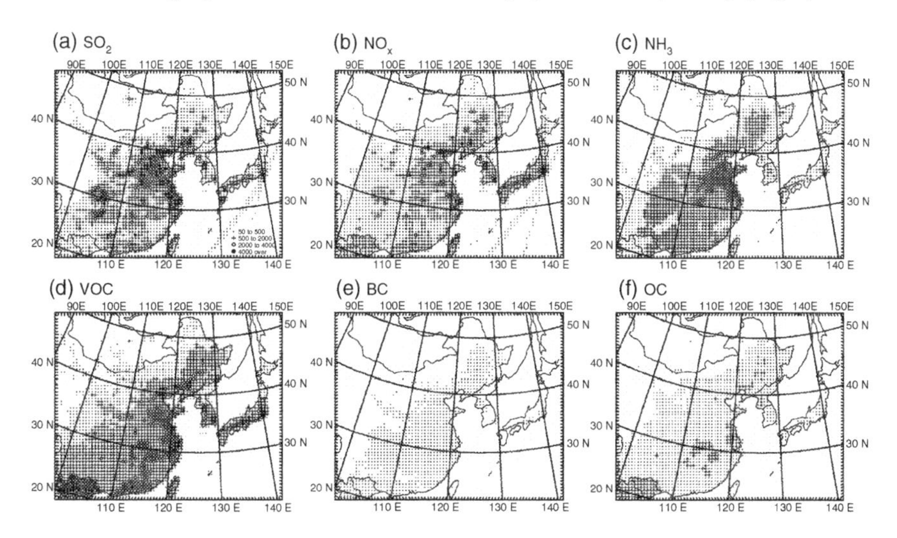

FIGURE 1. Spatial distributions of (a) SO_2, (b) NO_x, (c) NH_3, (d) NMVOC, (e) BC and (f) OC emissions in a grid 60×60 km^2 (Tg gridcell^{-1}) in April over the whole model domain.

sea salt emission are from the empirical relationship from Gong et al. (1997) and Monahan et al. (1986). The spectral BC and OC emissions are assumed to be constant with a diameter less than 1 μm.

4. Results

4.1. Simulated Model Concentrations

Comparisons of modeled and observed hourly surface concentrations of SO_2, NO_2 and O_3 averaged over South Korea for the period from 21-25 April show that the model slightly underestimates observed concentrations but diurnal variations are well simulated (not shown here). The daily mean concentrations of the modeled sulfate, nitrate, ammonium, chloride, black carbon and organic carbon aerosols against those observed at ACE-Asia experimental sites indicate that the model simulates these aerosol concentrations within the error of $\pm 2\ \mu g\ m^{-3}$. However, the organic carbon aerosol is significantly underestimated. This may be due to not taking into account secondary organic aerosols. Fig. 2 shows spatial distributions of the column integrated mean aerosol concentrations on 24 April. All types of aerosols except for the sea salt aerosol that is emitted from the sea have similar spatial distribution patterns with maxima along the 30° N latitude belt and the southern coast line of China. The most abundant aerosol type in East Asia is mixed type aerosol (IOC-BC-OC) and is followed by sea salt which is mainly confined over the sea. The black carbon concentration is relatively low.

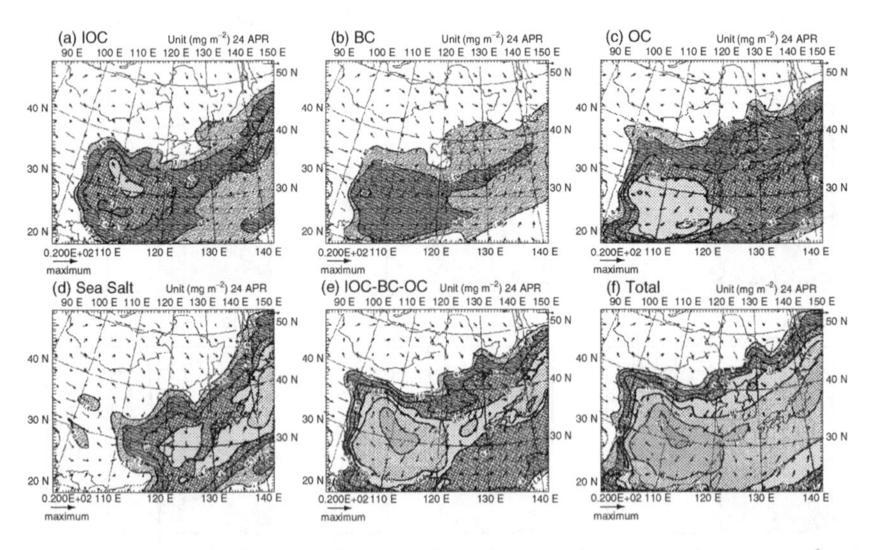

FIGURE 2. Spatial distributions of mean column integrated concentrations (mg m^{-2}) of (a) IOC, (b) BC, (c) OC, (d) Sea Salt, (e) IOC-BC-OC and (f) Total on 24 April 2001. The averaged mean wind vectors at 850 hPa are shown.

4.2. Estimation of Aerosol Direct Radiative Forcing

Vertical profiles of temperature, pressure, H_2O mixing ratio obtained from the MM5 simulation and those of ozone and CO_2 mixing ratios from the Column Radiation Model (CRM) of NCAR are used for the initial profiles in the CRM model. The aerosol direct radiative forcing is obtained as the difference in shortwave net radiative fluxes at TOA and the surface between CRM simulations with and without aerosol mass loading simulated by the present model. Fig. 2 shows the time series of the modeled and observed mean aerosol optical depths (AOD) at the ACE-Asia experimental sites. The observed AOD is averaged one at 0.44 μm and 0.67 μm wavelengths whereas the model is taken at 0.5 μm wavelength. The model simulates quite low values of AOD at all sites compared with observations, especially for the periods of 24-25 April in Korea and 21 to 22 April in China. These periods are associated with the Asian dust event periods. In fact the observed mean TSP concentration in Korea has increased from 60 μg m^{-3} before 24 April to 200 μg m^{-3} on 24 and 25 April. Presently the model accounts only for anthropogenic aerosols but does not take into account dust aerosol for the estimation of radiative forcing. This is why the modeled AOD is lower than the observed one (Fig. 3).

Fig. 4 shows the spatial distribution of daily mean shortwave direct radiative forcing at the surface (SRF) due to each type of aerosol. The spatial distribution pattern of SRF quite resembles that of the column integrated aerosol concentration (Fig. 2). More than –20 Wm^{-2} of SRF occurs along the 30° N latitude over China due to the mixed type aerosol (IOC-BC-OC). However, the direct radiative forcing at the surface contributed by IOC, BC, OC and sea salt aerosols (Figs 4a, b, c and d) is relatively small due to their low concentrations. The radiative forcing at the surface due to all types of anthropogenic aerosols (Fig. 4f) is larger than –50 Wm^{-2} along the 30° N latitude belt in China and the East China sea. This value is the same order of magnitude estimated by Redemann et al. (2000) at the east coast of the US over the Atlantic Ocean.

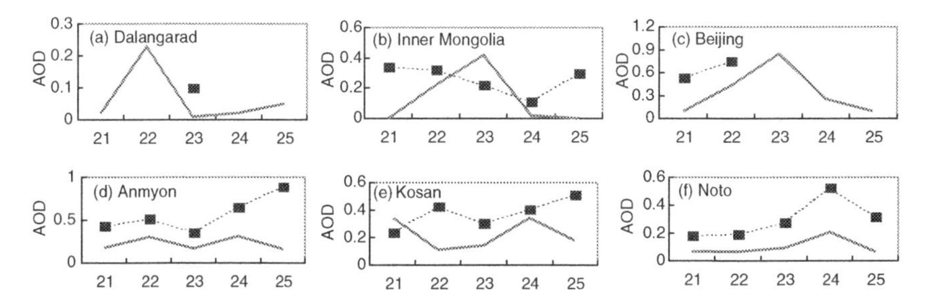

FIGURE 3. Averaged observed (■■) and modeled (—) aerosol optical depths at 0.5 μm wavelength at (a) Dalanzadgad in Mongolia and (b) Inner Mongolia and (c) Beijing in China, (d) Anmyundo and (e) Kosan in Korea, (f) Noto in Japan.

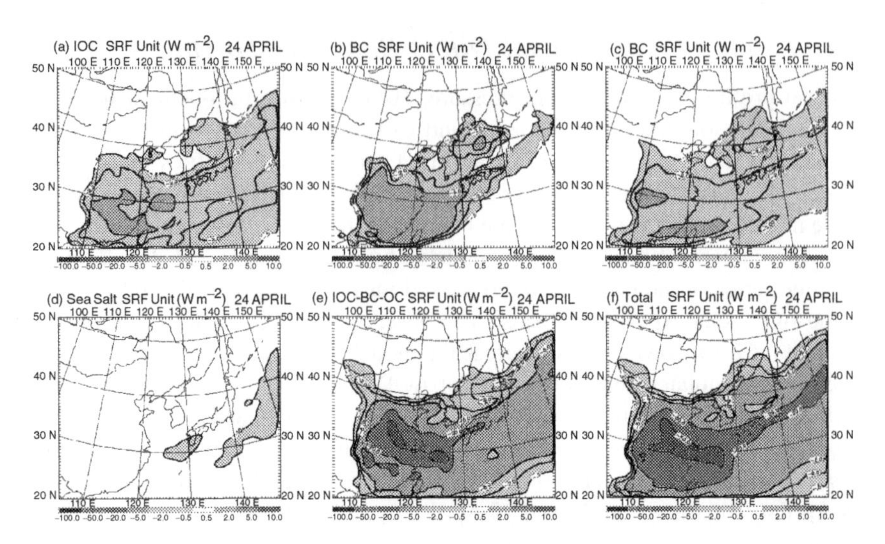

FIGURE 4. Daily mean direct radiative forcing (W m^{-2}) at the surface due to (a) IOC, (b) BC, (c) OC, (d) Sea Salt, (e) IOC-BC-OC and (f) total aerosol on 24 April 2001.

Fig. 5 shows the spatial distributions of the daily mean shortwave aerosol radiative forcing at the top of atmosphere (TOA). The spatial distribution patterns of the direct radiative forcing at TOA due to IOC (Fig. 5a), OC (Fig. 5c) and sea salt (Fig. 5d) aerosols quite resemble those at the surface (Fig. 4) with almost the same magnitude suggesting negligible amounts of atmospheric absorption of

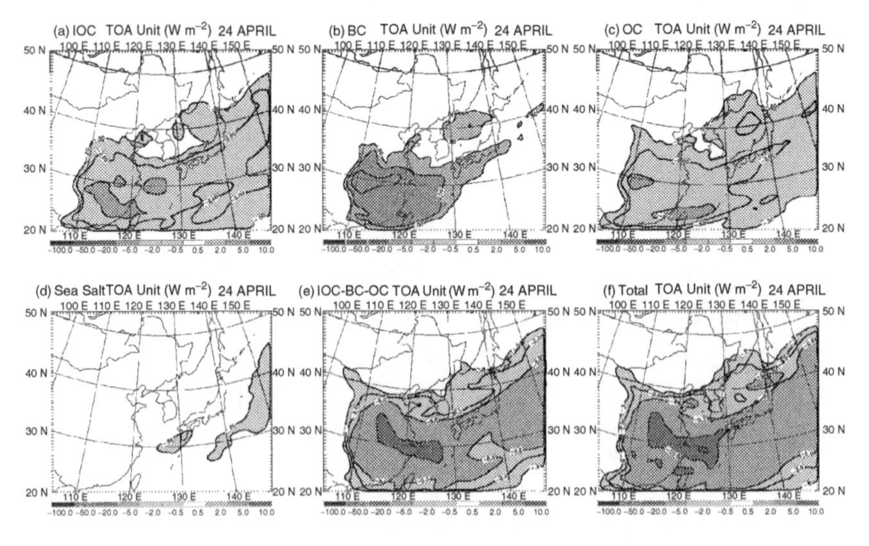

FIGURE 5. The same as in Fig. 4 except for TOA radiative forcing.

shortwave radiation due to these types of aerosols. However, the spatial distribution of the radiative forcing at TOA attributed by BC (Fig. 5b) and the mixed type aerosol (Fig. 5e) is quite similar to that at the surface (Fig. 4). But the radiative forcing at TOA is much reduced and even becomes positive for BC, implying large amounts of atmospheric absorption of shortwave radiation is caused by black carbon (BC) and the mixed type aerosol. The total aerosol radiative forcing at TOA (Fig. 5f) has a similar distribution pattern to that at the surface (Fig. 4f) but its value is much reduced. This is attributed to the highly absorbing aerosols including black carbon and the mixed type aerosol (IOC-BC-OC). The maximum radiative forcing at TOA is about –20 Wm^{-2} that occurs southeastern China. This is largely contributed by the mixed type aerosol.

Averaged total aerosol mass in East Asia is about 22 mg m^{-2}, of which 42% and 21% are, respectively contributed by the mixed type aerosol and sea salt (Fig. 6a). The contribution of BC to the total aerosol mass is the least (about 5%). Mean aerosol radiative forcing at the surface is about –10.4 W m^{-2}, of which 62% and 12% are, respectively contributed by the mixed type aerosol and BC. The contribution of BC to the radiative forcing at the surface is enhanced due to its short wave absorption character (Fig. 6b). On the other hand, mean aerosol radiative forcing at TOA is about –5.9 W m^{-2}. Most of this is contributed by and IOC-BC-OC (57%), IOC (17%) and OC (15%). The contribution of the other types of aerosols to radiative forcing at TOA is small (Fig. 6c), suggesting the dependency of radiative forcing on different types of aerosols. The difference between radiative forcing at TOA and at the surface leads to an estimate of the aerosol

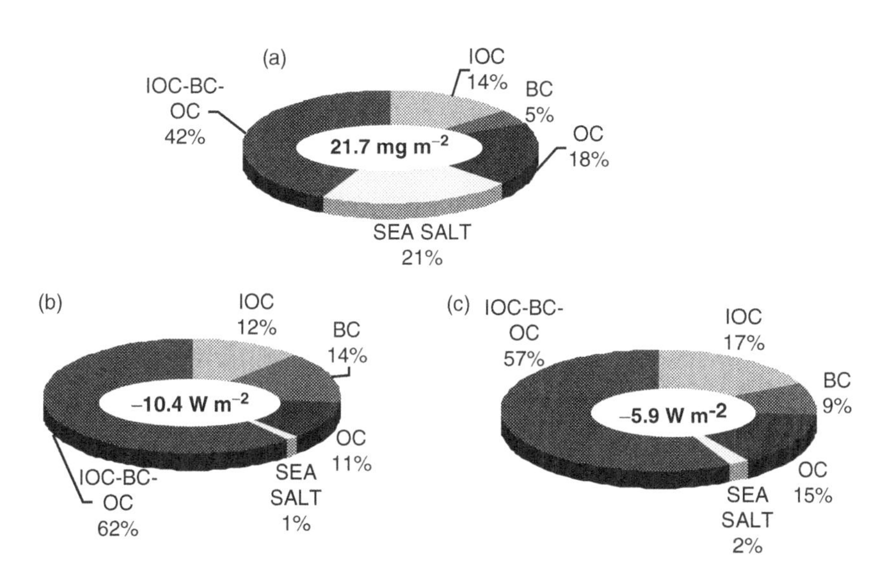

FIGURE 6. The fractional contributions of various aerosols on the (a) total mass (mg m^{-2}), (b) daily mean direct radiative forcing (W m^{-2}) at the surface and (c) at TOA averaged for the whole model period in East Asia.

absorption in the atmosphere. The presently estimated atmospheric aerosol absorption in East Asia is about 4.5 W m^{-2}. This implies that aerosols heat the atmosphere over East Asia due to the absorbing aerosols.

5. Conclusions

An aerosol dynamic model including aerosol processes of nucleation, condensation/evaporation, coagulation, sedimentation, dry and wet deposition and hygroscopic growth coupled with the gas-phase chemistry of the California Institute of Technology (CIT) airshed model and the aqueous-phase chemistry of the Regional Acid Deposition Model (RADM) with the MM5 meteorological model in a grid of 60×60 km^2 has been used to estimate anthropogenic aerosols in East Asia (90-145E, 20-50N) for the period from 21 to 25 April 2001 in the ACE-Asia experimental period. The modeled anthropogenic aerosols are implemented to estimate radiative forcing at the surface and TOA and atmospheric aerosol absorption in East Asia using the NCAR CCM3 CRM model. The results show that the area averaged column integrated anthropogenic aerosol concentration is found to be about 22 mg m^{-2}, of which 42%, 21%, 18%, 14% and 5% are, respectively contributed by the mixed type (IOC-BC-OC), sea -salt, OC, IOC and BC aerosols, implying more than half of the anthropogenic aerosol being contributed by the transformed aerosol in East Asia. The area mean radiative forcing at the surface is found to be about –10.4 W m^{-2}, of which more than 74% is contributed by the mixed type aerosol and IOC. The sea salt and OC aerosols occupy almost 40% of the total anthropogenic aerosol mass in East Asia but their contribution to direct short wave radiative forcing at the surface is less than 12%, suggesting the importance of aerosol types for the estimation of radiative forcing at the surface. The area mean direct radiative forcing at TOA in East Asia is found to be about –5.9 W m^{-2}, of which more than 70% is contributed by the mixed type and IOC aerosol, suggesting the importance of transformed aerosols on radiative forcing at the top of atmosphere. The area mean atmospheric absorption due to aerosol layer in East Asia is estimated to be about 4.6 W m^{-2}. This value is close to the annual mean value estimated by Yu et al. (2001) in the southeastern US. This study confirms the existence of a cooling effect (negative forcing) due to the direct effect of aerosol at the surface and TOA in East Asia. However, the atmosphere of the troposphere above the ground is still heated with the area averaged rate of 4.6 W m^{-2} by an absorbing aerosol layer that consists of BC and the mixed type aerosol.

This study mainly pertains to estimates of short-term direct radiative forcing due to anthropogenic aerosols transformed from various precursors emissions in East Asia. However, in this region, especially in spring, mineral dusts originating from desert areas of inland China are frequently dominated. In fact, a weak Asian dust event has been observed in Korea during the analysis period. The radiative forcing due to anthropogenic aerosol together with dust aerosols is a prerequisite to understand the role of aerosols in climate change in East Asia. Nevertheless, this also requires dust emission, transformation and transport processes that are now at hand.

References

Chang. L.S., Park, S.-U., 2004. Direct radiative forcing due to anthropogenic aerosols in East Asia during April 2001. *Atmospheric Environment* **38**, 4467-4482.

Charlson, R. J., Schwartz, S. E., Hales, J. M., Cess , R. D., Coakley, J.A. JR., Hansen, J. E., Hofmann, D. J., 1992. Climate forcing by anthropogenic aerosols. *Science* **255**, 423-430.

Gong, S., Barrie, L. A., Blanchet, J. P., 1997. Modeling sea salt aerosols in the atmosphere. I Model development. *Journal of Geophysical. Research* **102**, 3805-3818.

Intergovernmental Panel on Climate Change (IPCC), 1996. *Climate Change 1995: The Science of Climate Change,* Cambridge University Press, Cambridge.

Kiehl, J. T., Briegle, B. P., 1993. The relative roles of sulfate aerosols and greenhouse gases in climate forcing. *Science* **260,** 311-314.

Monahan, E. C., Spiel, D.E., Davidson, K. L., 1986. A model of marine aerosol generation via whitecaps and wave disruption, *Oceanic Whitecaps*, E. C. Monahan and G. Mac Nicaill, Eds., D. Reidel, 167-174.

Redemann, J., Turco, R.P., Liou, K.N., Hobbs, P.V., Hartely, W.S., Berstrom, R.W., Browell, E.V., Russel, P.B., 2000. Case studies of the vertical structure of the direct shortwave aerosol radiative forcing during TARFOX. *Journal of Geophysical Research* **105**, 9971-9979

Russell, A., McCue, K. F., Cass, G. R., 1988. Mathematical Modeling of the formation of Nitrogen-containing air pollutants. 1. Evaluation of an Eulerian photochemical model. *Environmental Science & Technology* **22**, 263-271.

Streets, D. G., Bond, T. G., Carmichael, G. R., Fernandes, S. D., Fu, Q., He, D., Klimout, Z., Nelson, S. M., Tsai, N. Y., Wang, M. Q., Woo, J.-H., Yarber, K. F., 2003: An inventory of gaseous and primary aerosol emissions in Asia in the year 2000. Journal of Geophysical Research 108 (D21), 8809, doi:10.1029/2002JD003093.

Yu, S., Zender, C. S., Saxena, V.K., 2001. Direct radiative forcing and atmospheric absorption by boundary layer aerosols in the southeastern US: model estimates on the basis of new observations. *Atmospheric Environment* **35**, 3967-3977.

Direct Radiative Forcing Due to Anthropogenic Aerosols in East Asia During 21-25 April 2001

Speaker: S. U. Park

Questioner: S. Yu

Question 1: What is the value of aerosol single scattering albedo used in your model when you calculate your direct aerosol forcing?

Answer: The single scattering albedo depends upon the size and chemical composition of the aerosol. We have calculated the single scattering albedo using the Mie code with given refractive index. The average single albedo of sulfate and nitrate is about 0.95 and BC is 0.3 and so on.

Question 2: Do you consider cloud effects when you calculate direct aerosol forcing?
Answer: We just only consider the aerosol radiative forcing.

Questioner: A. Chtcherbakov
Question: Why are you using NO_2 in your emission inventory instead of NO_x?
Answer: We have used NO_x rather than NO_2.

35 Modelling Fine Aerosol and Black Carbon over Europe to Address Health and Climate Effects

Martijn Schaap and Peter J. H. Builtjes*

1. Introduction

Atmospheric particulate matter (PM) is a complex mixture of anthropogenic and natural airborne particles. Particulate matter in ambient air has been associated consistently with excess mortality and morbidity in human populations (e.g., Brunekreef, 1997; Hoek et al., 2002). The European air quality standards currently focus on all particles smaller than 10 μm in diameter (PM10), which covers the inhalable size fraction of PM. Mass and composition of PM10 tend to divide into two principal groups: coarse particles, mostly larger than 2.5 μm in aerodynamic diameter, and fine particles, mostly smaller than 2.5 μm in aerodynamic diameter. The fine particles contain secondary aerosols, combustion particles and condensed organic and metal vapours. The larger particles usually contain sea salt, earth crust materials and fugitive dust from roads and industries. Although adverse health effects are associated with elevated levels of both PM10 and PM2.5, these health effects were most strongly and consistently associated with particles derived from fossil fuel combustion (e.g. Hoek et al. 2002), which mostly occur in the PM2.5 size range.

Various components of fine particulate matter (PM2.5) in the atmosphere also have climate-forcing impacts, either contributing to or offsetting the warming effects of greenhouse gases (Kiehl and Briegleb, 1993; Hansen and Sato, 2001). In particular, black carbon (BC) has recently been identified as an important contributor to radiative heating of the atmosphere (Haywood et al., 1997; Myhre et al., 1998). Organic carbon (OC), which is often emitted along with BC, may act to offset some of the global warming impact of BC emissions (Hansen and Sato, 2001). In case of biomass burning aerosol, OC is thought to completely compensate the warming potential of BC (Penner et al., 1998; Grant et al., 1999). However, for fossil fuel derived emissions this is not the case and a net positive forcing remains (Penner et al., 1998; Cooke et al., 1999).

* Martijn Schaap and Peter Builtjes, TNO Institute of Environmental Sciences, Energy and Process Innovation (TNO-MEP), PO Box 342, 7300 AH Apeldoorn, The Netherlands.

Measures to abate climate forcing by reducing BC emissions often would have collateral benefits by reducing emissions of health-related pollutants. A similar but opposite reasoning applies for the abatement of the acidifying secondary inorganic aerosol compounds, which have a cooling effect on climate.

We use a regional model to assess the composition of anthropogenic-induced PM2.5 in Europe with special emphasis on BC. For this purpose we use a recent emission inventory of anthropogenic primary particulate matter in Europe (TNO, 2001), which can be used to derive BC emissions. The burdens of primary emitted particles over Europe are calculated with the LOTOS model, version 5.2 (Schaap et al., 2003; 2004). By combining these results with earlier calculations for the secondary aerosol components sulphate, nitrate and ammonium (Schaap et al., 2003), estimates of PM2.5 levels over Europe are obtained. The consistency of the model calculations is checked with observations at a number of European locations. Finally, a first order estimate of the radiative forcing of BC over Europe is presented.

2. Emission Data

Black Carbon (BC) is mostly released from incomplete combustion of carbonaceous fuels. Black Carbon sources and their distributions over Europe are estimated using the results of a spatially distributed pan-European inventory of anthropogenic PM emissions for the year 1995 (CEPMEIP (TNO, 2001)). CEPMEIP covers nearly all of the relevant sources of primary anthropogenic TSP, PM10, PM4, PM2.5 and PM0.95 emissions, however, without chemical speciation. A European sub-micron BC emission inventory was derived from the CEPMEIP data using the fraction of BC in PM2.5 for each of the source categories. The sub-micron BC fractions were adopted from Streets et al. (2001). The BC fraction of PM2.5 was calculated per major source category (SNAP level 1) for each country separately. Figure 1 gives the BC emission distribution over Europe. The average European PM2.5 and BC emissions are given per major source category in Figure 2. The European total PM2.5 (BC included) emission, excluding the Former Soviet Union (FSU) amounts to 1880 Gg/yr. The corresponding BC emissions are 473 Gg/yr. The largest source contributions are from transport followed by wood burning in households. Other important sources are industrial combustion, energy transformation (especially with liquid fuels) and gas flaring at oil platforms. For a detailed description of the emission data used we refer to Schaap et al. (2004).

3. Model Description

The LOTOS model, version 5.2, has previously been applied for the calculation of secondary aerosol fields over Europe (Schaap et al., 2003). The geographical domain of LOTOS ranges from 10°W to 40°E and from 35°N to 70°N with a spatial resolution of 0.5×0.25°lon-lat. The vertical domain is divided in three

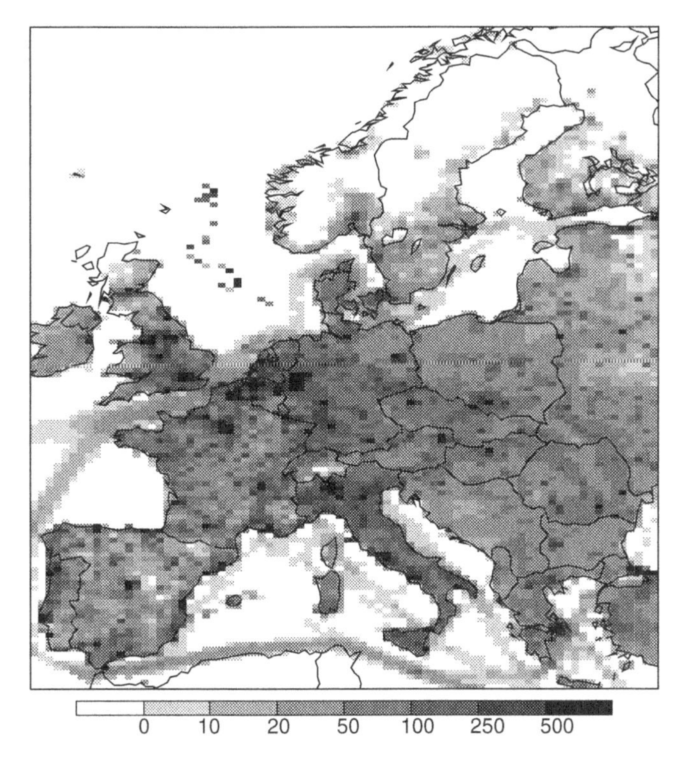

FIGURE 1. Black carbon emissions (Ktonnes/yr) over Europe for 1995.

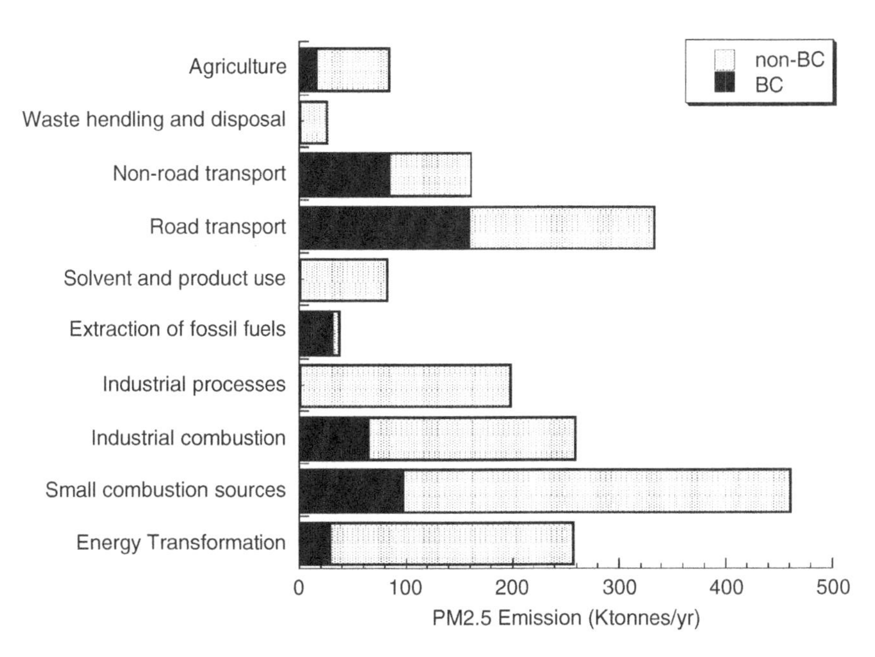

FIGURE 2. Overview of the total European BC and non-BC emissions per source category.

layers and extends to 3.5 km. The lowest layer represents the variable mixing layer on top of which two layers with equal, but variable, depth are located. Every hour, the mixing layer height is updated and the mass in the three layers is redistributed by mass conserving linear interpolation. The vertical extend of the model is considered sufficient, since it has been shown that on average, 80 to 90% of the aerosol burden is located below 3 Km (Banic et al., 1996).

In LOTOS a bulk approach is followed for the description of the aerosols. Hence, our model parameterisations are based upon accumulation mode aerosol since we are interested in mass concentrations. Recent modelling results using explicit aerosol dynamics show that the ageing time of BC is only a few hours (Riemer et al., 2004). Therefore, we assume all BC and primary PM to be hydrophyllic directly after emission. Due to the efficient transfer of small particles, i.e. BC, to the accumulation mode this assumption has probably minor effects on our results. Below cloud scavenging is represented using scavenging coefficients. Due to our limited vertical domain and insufficient data on cloud occurrence, incloud-scavenging is neglected. Dry deposition of the accumulation mode aerosol is parameterised by the surface-atmosphere interaction module DEPAC (Erisman et al., 1994). All boundary conditions for BC and primary PM were assumed to be zero. At the eastern boundary, however, the assumed boundary conditions and, hence, the model results are highly uncertain and hence we choose to present results only west of 30×E. For a more elaborate description of the model, including the SIA formation, we refer to Schaap et al. (2003; 2004).

4. Results

Modelled fields of BC and total primary PM2.5 (including BC) over Europe averaged over 1995 are shown in Figure 3. Modelled BC concentrations are lower than 50 ng/m^3 over remote regions such as northern Scandinavia. In relatively clean areas over Spain and southern Scandinavia BC is about 250 ng/m^3 whereas the calculated BC concentrations exceed 500 ng/m^3 over central Europe and 1000 ng/m^3 in the densely populated areas. The modelled black carbon distribution is very similar to that of the total primary PM2.5 components because the source categories with the highest PM2.5 emissions also have the highest BC fractions. These calculated fields also show that high concentrations of BC occur over densely populated areas, with related health effects.

Comparison with modelled fields for secondary inorganic aerosol, also depicted in Figure 3, shows that the BC mass fraction of PM is relatively low. Annual averaged modelled PM2.5 mass concentration ranges from 2-5 ug/m^3 in the less populated regions to 15 to 25 ug/m^3 in industrialised and densely populated areas in Europe. Hence, on a regional scale BC contributes only 4-10 % of the total fine aerosol mass, which is in agreement with observed data. Near sources, the mass fraction measured is considerably higher since in our model approach the emissions are instantly diluted over the boundary layer and an area of about 25×25 Km^2.

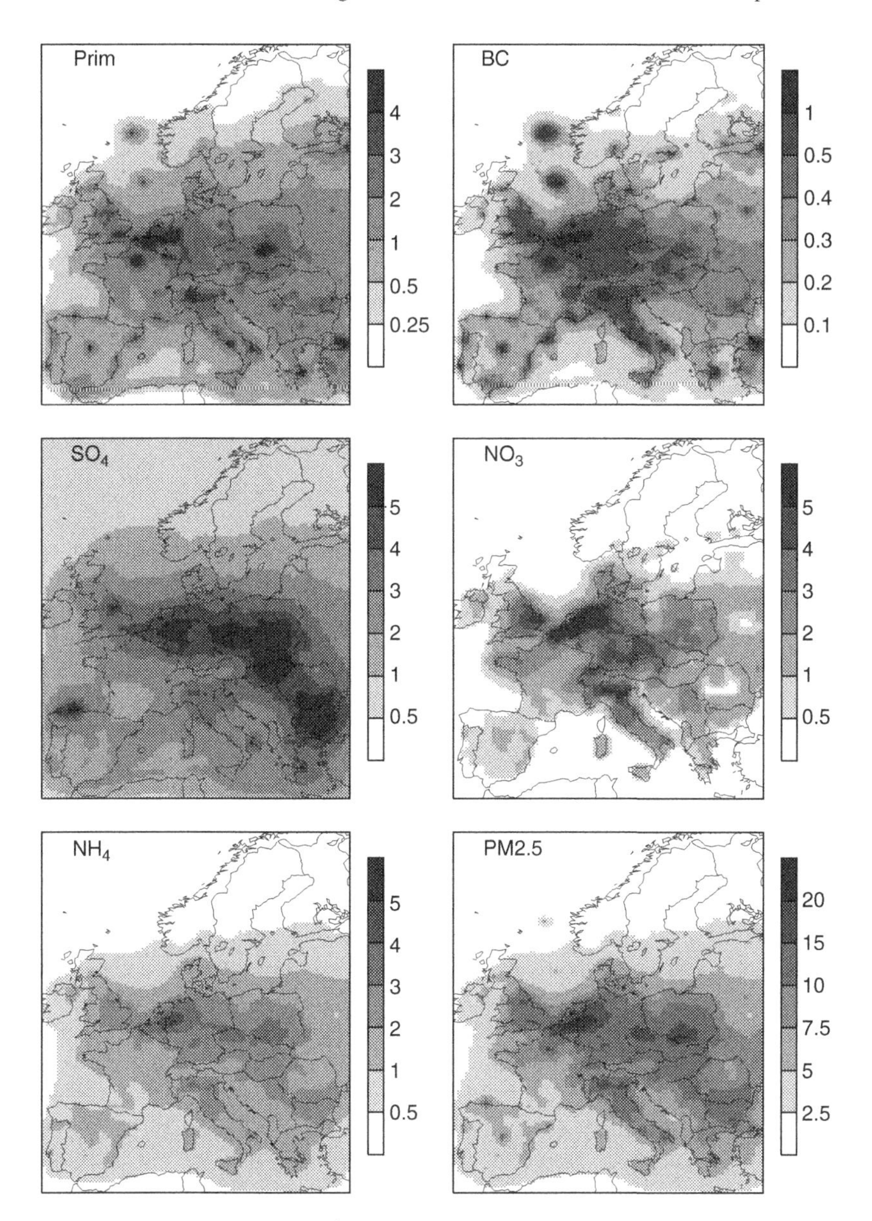

FIGURE 3. Calculated fields (μg/m³) of primary and secondary components over Europe.

Monitoring data on BC for 1995 are not available. Therefore, we have compared our calculated BC concentrations with observations on a number of sites representing a period ranging from the end of the 1980s to 2001 and often obtained in campaigns. The data are plotted as a function of the regional characteristic in Figure 4. The comparison shows that the simulated BC concentrations consistently underestimated those measured by about a factor of 2.

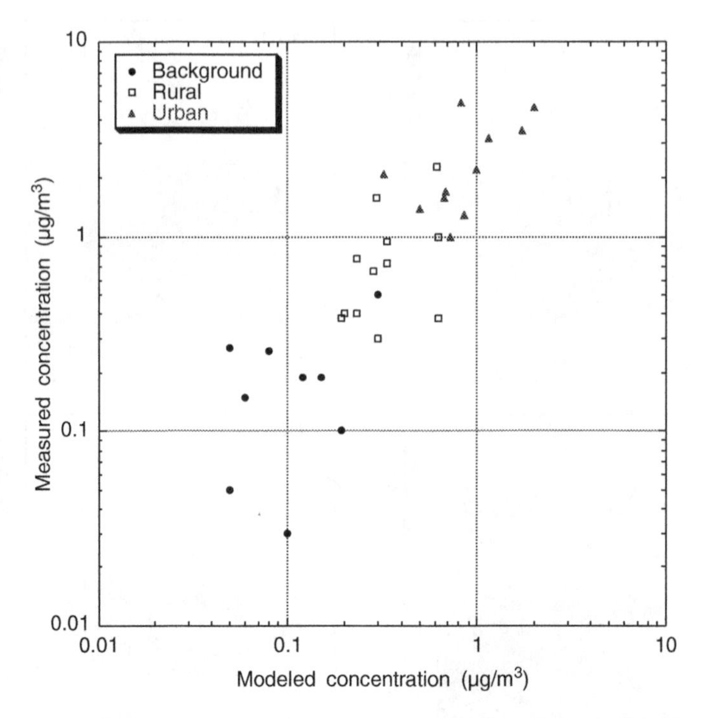

FIGURE 4. Comparison between modelled and measured BC concentrations in Europe.

5. Radiative Forcing due to Black Carbon

Below we present a tentative calculation of the climate forcing by BC over Europe. Since black carbon absorbs solar radiation it does not exert a major forcing over regions with a low surface albedo, e.g. over sea. Maximum forcing by BC occurs over reflective surfaces. We have followed the approach by de Groot et al. (2003) to calculate the forcing of BC. In a first approximation the positive forcing by BC is given by:

$$\Delta F = +2T^2 OD(1 - C) Rs\, S$$

S	Average solar flux at the top of the atmosphere (W/m^2)
T	Fraction of light transmitted through the atmosphere
OD	Optical depth of black Carbon
C	Cloud cover
Rs	Surface albedo

The optical depth (OD) is obtained from the modelled BC burden, assuming a mass absorption coefficient of 10 m^2/g. We neglected the influence of the relative humidity in the absorption coefficient, because no reliable data are available on this subject. Similar approaches to estimate the aerosol optical depth by black

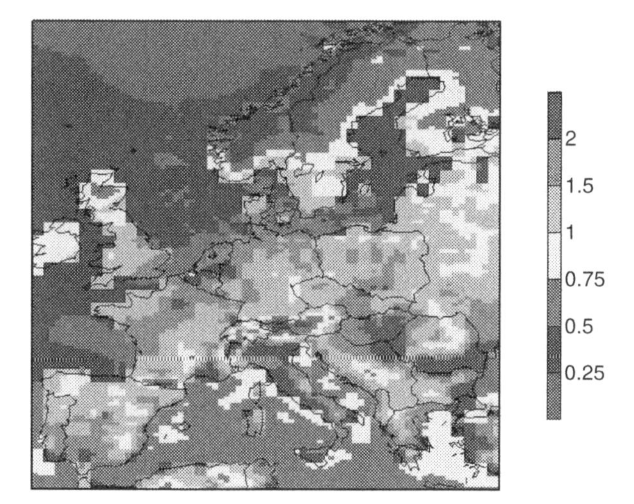

FIGURE 5. Radiative forcing (W/m^2) of BC over Europe.

carbon have been used by other authors, e.g. Tegen et al. (2000). We calculate the clear sky forcing and thus correct for the cloud cover. Seasonal values for the surface albedo are taken from Matthews (1984). The transmission, T, of the atmosphere above the soot-layer was approximated by scaling it with the sine of the solar angle (ten Brink et al., 2001).

The annual average forcing calculated from the BC burden from LOTOS for 1995 is shown in Figure 5. Over the sea, the forcing is mainly lower than 0.5 W/m^2. Above land, typical values for the forcing are about 1 W/m^2, considerably higher than over sea due to the higher surface albedo. Over the most polluted areas such as large cities, the calculated forcing exceeds 2 W/m^2.

The uncertainty in the estimate presented above is large. The BC concentrations in LOTOS underestimate the ambient levels by about a factor of 2, which might indicate that we underestimate the forcing. On the other hand, the uncertainties in measurement data, emissions estimates and atmospheric processes prohibit strong conclusions in this respect. In addition, the forcing efficiency per unit OD used in the literature is most probably more uncertain than a factor of 2. Thus, the uncertainty associated with BC forcing estimates presented here and elsewhere is inevitably very high. However, we feel that we can conclude that the positive forcing of BC is considerable compared to the forcing exerted by greenhouse gases.

6. Synthesis

In this paper we presented an emission data base for BC, a regional study to the distribution of BC and a tentative calculation of its radiative forcing. We estimate the total BC emission to be 0.47 in Europe, excluding the Former Soviet Union.

BC concentrations up to 1-2 ug/m^3 are calculated over densely populated regions, which is only 5-10% of total PM2.5 over Europe. Our model underestimates the BC concentrations over Europe, which hints at an underestimation of the total PM emissions. Comparison with other emission data shows that the uncertainty in the emission estimates is still more than a factor of 2.

Although uncertainties are large, we, among others, show that the climate impact of BC is important over Europe. However, one should bear in mind that along with BC also the cooling OC is emitted. On average the BC to OC ratio is about 1 to 3. To reduce the warming potential of black carbon it is therefore effective to abate sources with a higher BC/OC ratio. These emissions sources are non-road transport (e.g. shipping) and road transport, most notably diesel car emissions. Together with small combustion sources, these are the major emission sources for fine PM and BC within a city, where also the exposure to PM has the highest adverse health impacts. Hence, reducing the BC emissions of these source categories is beneficial for reducing both its adverse health and climate effects.

Acknowledgement We would like to thank Dr. Harry ten Brink for valuable discussions on the study presented here.

References

Banic, C. M., Leaitch, W. R., Isaac, G. A., Couture, M. D., Kleinman, L. I., Springston, S. R., and MacPherson, J. I. 1996. Transport of ozone and sulfur to the North Atlantic atmosphere during the North Atlantic Regional Experiment, J. Geophys. Res., 101 (D22), 29, 091–29,104.

Brunekreef, B., 1997. Air pollution and life expectancy: Is there a relation?, Occup Environ. Med., 54, 781–784.

Cooke, W. F., Liousse, C., Cachier, H., and Feichter, J. 1999. Construction of a I' x I' fossil fuel emission data set for carbonaceous aerosol and implementation and radiative impact in the ECHAM4 model, J. Geophys. Res, 104, 22137–22162.

De Groot, G. J., Builtjes, P. J. H., ten Brink, H. M., Schaap, M., van Loon, M., de Leeuw, G., Vermeulen, A. T. 2003. Climate change in relation to aerosols: Scientific report, ECN report ECN-C—03-030, ECN, Petter, the Netherlands.

Erisman, J. W., van Pul, A., Wyers, P. 1994. Parametrization of surface-resistance for the quantification of atmospheric deposition of acidifying pollutants and ozone, Atmos. Environ., 28, 2595–2607.

Grant, K. E., C.C. Chuang, A. S. Grossman, and J. E. Penner 1999. Modeling the spectral optical properties of ammonium sulfate and biomass aerosols: Parametrization of relative humidity effects and model results, Atmos. Environ., 33, 2603–2620.

Hansen, J. E., and Sato, M. 2001. Trends of measured climate forcing agents, Proceedings of the National Academy of Sciences of the United States of America, 98, 26, 14778–14783.

Haywood, J. M., Roberts, D. L., Slingo, A., Edwards, J. M., and Shine, K. P. 1997. General circulation model calculations of the direct radiative forcing by anthropogenic sulfate and fossil-fuel soot aerosol, J. Clim., 10, 1562–1577.

Hoek, G., Brunekreef, B., Goldbohm, S., Fischer, P., and van den Brandt, P. A. 2002. Association between mortality and indicators of traffic-related air pollution in the Netherlands: a cohort study, The lancet, 360, 1203–1209.

Kiehl, J. T., and B. P. Briegleb 1993. The relative roles of sulfate aerosols and greenhouse gases in climate forcing, Science, 260, 311–314.

Matthews, E. 1984. Vegetation, land-use and seasonal albedo data sets: Documentation and archived data tape, NASA Tech. Memo, 86107.

Myhre, G., Stordal, F., Restad, K., and Isaksen, I. S. A. 1998. Estimation of the direct radiative forcing due to sulfate and soot aerosols, Tellus, 50B, 463–477.

Penner, J. E., Chuang, C. C., and Grant, K. 1998. Climate forcing by carbonaceous and sulfate aerosols, Clim. Dyn., 14, 839–851.

Riemer, N., Vogel, H., and Vogel, N. 2004. A parameterisation of the soot aging for global climate models, Atmospheric Chemistry and Physics Discussions, Vol. 4, pp 2089–2115.

Schaap, M., H. A. C. Denier Van Der Gon, F. J. Dentener, A. J. H. Visschedijk, M. Van Loon, H. M. Ten Brink, J-P Putaud, B. Guillaume, C. Liousse and P.J.H. Builtjes, Anthropogenic Black Carbon and Fine Aerosol Distribution over Europe 2004, J. Geophys. Res., in press.

Schaap, M., van Loon, M., ten Brink, H. M., Dentener, F. D., Builtjes, P. J. H. 2004, Secondary inorganic aerosol simulations for Europe with special attention to nitrate, Atmos. Phys. Chem., 4, 857–874.

Streets, D. G., Gupta, S., Waldhoff, S. T., Wang, M. Q., Bond, T. C., and Yiyun B., 2001. Black carbon emissions in China, Atmospheric environment, vol. 35, 4281–4296.

Tegen, I., Koch, D., Lacis, A. A., amd Sato, M. 2000. Trends in tropospheric aerosols and corresponding impact on direct radiative forcing between 1950 and 1990: A model study, J. Geopys. Res., 105, 26971–26989.

ten Brink, et al. 2001. Aerosol; cycle and influence on the radiation balance = MEMORA: Measurement and Modeling of the reduction of Radiation by Aerosol, ECN report ECN-R—01-003, ECN, Petten, the Netherlands.

TNO 2001, The CEPMEIP database, http://www.air.sk/tno/cepmeip/

Modeling Fine Aerosol and Black Carbon over Europe to Address Health and Climate Effects

Speaker: Peter Builtjes

Questioner: T. Iversen

Question: I liked your multi-model ensemble study, but I would recommend to use the ensemble-members to fit probability-distributions rather than only construct a single concensus by e.g. taking averages. The increase in correlation coefficients for the ensemble mean is a trivial consequence of removing random errors. The full information from the ensemble is better conveyed by fitting probability distributions. Don't you agree?

Answer: I fully agree. This ensemble example I presented is just a first attempt, and a median would already be better than using the mean. The increase in correlation coefficient is indeed trivial. Not trivial is the size of the increase in correlation as such, I would have expected a smaller increase.

Questioner: W. Jiang

Question: What is the complete unit on your emission map? Kton/yr/km2 or kton/yr/grid cell? If the latter what is the grid cell size?

Answer: The correct unit should be tons/yr/grid, with a grid size of 0.25 × 0.5 lat-long, about 30 × 30 km2 on average.

Questioner: C. Mensink

Question: Can you comment on the confidence levels or reliability with respect to the black carbon fraction of the PM2.5 emissions (Figure 2)?

Answer: The uncertainty in the fractions would at least be about 30% on average, but would be a bit better for road transport, may be 20%. The overall uncertainty, so including the total PM 2.5 would be more in the order of 50%. Also the 50% might be too small in view of the underestimation of about a factor of 2 as indicated in fig 4.

Questioner: B. Fisher

Question: In your study you estimate that the annual average forcing from black carbon over Europe is about 1 W/m^2. Is this large enough to matter in global climate models?

Answer: Global average for CO_2+CH_4+N_2O+CFC's is about 2.5 W/m^2, so 1 W/m^2 matters. It should be noted however that this 1W/m2 of BC is a regional scale effect, not a global scale effect.

36 An Approach to Simulation of Long-Range Atmospheric Transport of Natural Allergens: An Example of Birch Pollen

Pilvi Siljamo[*], Mikhail Sofiev, and Hanna Ranta

1. Introduction

Diseases of the respiratory system due to aeroallergens, such as rhinitis and asthma, are major causes of a demand for healthcare, loss of productivity and an increased rate of morbidity. The overall prevalence of seasonal allergic rhinitis (allergic reactions in the upper respiratory system) in Europe is approximately 15%; the asthma rates vary from 2.5%–10%; etc.. Pollenosis accounts for 12–45% of all allergy cases. The sensitisation to pollen allergens is increasing in most European regions.

The adverse health effects of allergens can be reduced by pre-emptive medical measures. However, their planning requires reliable forecasts of the start time of high pollen concentrations in air, as well their levels and durations (Rantio-Lehtimäki, 1994; Rantio-Lehtimäki and Matikainen, 2002). The currently available forecasts are based solely on local observations and do not consider the pollen transport from other regions or countries (Frøsig and Rasmussen, 2003). However, there is a convincing evidence that long-range transport of pollen can significantly modify pollinating seasons (first of all, the start time and duration of high atmospheric pollen concentrations) in many European regions (Corden et al., 2002; Malgorzata et al., 2002; Hjelmroos, 1992). This transport causes unforeseen and sudden increases of concentrations of pollen that can occur up to a month before the start of the local pollen season. The long-range transport can substantially increase the concentrations of allergenic pollen also during the local flowering season.

This is an important problem for Northern Europe and especially for Finland, where the flowering takes place later in spring. The most important pollinating species with respect to long-range transport are the birch, and Finland is neighbored by the Baltic countries, North-Western Russia and Belarus where the

[*] P. Siljamo, M. Sofiev: Finnish Meteorological Institute, PO Box 503, Vuorikatu 24, 00101 Helsinki Finland. E-mail: pilvi.siljamo@fmi.fi
H. Ranta: University of Turku, Aerobiology Unit, Department of Biology, 20014, Turku, Finland, e-mail: hanranta@utu.fi

proportion of birch forests regionally exceeds 40% of forest area (Pisarenko et al., 2000).

The current paper presents an attempt to parameterize the processes of birch pollen production and distribution by means of an atmospheric emission and dispersion model. Below, we outline this integrated model and analyze several recent cases of high pollen concentrations over Finland. An attempt is made to evaluate the possibility of short-term forecasts of such events for the future.

2. Models of the Flowering Seasons

There are two treelike birch species in Europe. Downy birch (*Betula pubescens*) is the most common in the northern part of Europe, while silver birch (*Betula pendula*) is dominating in the southern part of the birch area of distribution. These species resemble each other but respond somewhat differently to the external forcing. In particular, *B. pubescens* is mainly controlled by the accumulated heat during spring time, while *B. pendula* is more light dependent (Luomajoki, 1999). Unfortunately, there is virtually no quantitative information about the geographical distribution of different birches, so we had to describe a "general birch" in the model and neglect the differences between its species.

Several empirical models for predicting the start and duration of flowering seasons can be used. Descriptions of the flowering start time are based on three main principles: (i) climatological averaging of long-term observations of flowering calendar dates (hereinafter referred as *CD*) e.g., (Rötzer and Chmielewski, 2001); (ii) accumulated heat indices (such as the so-called degree-days (*DD*), and period units (*PU*) (Hänninen, 1990, Linkosalo, 2000 a,b, Luomajoki, 1999 and Sarvas, 1972) and (iii) dynamic models (e.g., promoter-inhibitor model of Schaber and Badeck, 2003). The climate-based values are the only ones available over the whole of Europe as harmonized datasets, while the parameterizations (ii) and (iii) are usually based on local or, at best, country-wide observations, and are therefore not representative at the European scale.

Descriptions of other parameters of flowering such as its intensity and the total amount of released pollen also require the use of semi-empirical models that predict the next-year flowering features, based on the conditions of the previous growing season (Masaka and Maguchi, 2001; Herrera et al., 1998, Dahl & Strandhede, 1996, Emberlin et al., 2002).

For the first model application, we used the CD-based climatology as the simplest approach covering the whole of Europe and providing unbiased mean estimates of the flowering timing.

3. Atmospheric Dispersion Model and Supplementary Tools

The first version of the integrated pollen emission and transport model was built on the basis on the emergency modeling system SILAM (Sofiev, 2002; Sofiev and Siljamo, 2003, http://www.fmi.fi/research_air/air_50.html), which is currently used for the operational forecasting of consequences of accidental releases in atmosphere of hazardous substances in the vicinity of Finland. The system is based on a so-called Lagrangian Monte-Carlo random walk dispersion model. The treatment of aerosol is based on a modal representation of the aerosol size spectrum, resistance-based parameterization of dry deposition and scavenging coefficient-based wet deposition description.

Typical birch pollen grain has a size of 20-22 μm. It is fairly light (a full grain filled with protein material has a density of ~ 800 kg m^{-3}) and is approximately spherical. The most efficient removal mechanism from the atmosphere is scavenging by precipitation.

As stated above, the current version of the integrated model uses climatological index *CD* as a first guess for the start of the birch flowering season. Since this index is very crude and, for some years, misplaces the flowering season by up to a couple of weeks, we also utilized the probability forecasting capabilities of SILAM, which can compute an "area of risk" for the specific source, regardless of the actual mass released.

Utilization of more sophisticated heat sum indices or dynamic models is envisaged but it requires further preparatory work in building a unified parameterization suitable for most of Europe, as well as technical developments allowing computations of heat indices over the whole spring (and, in some cases, the whole last autumn and winter) in near-real-time mode.

Another important part of preparation was a delineation of the main source areas of pollen in Europe that affects e.g. Finnish pollinating seasons, which requires inverse type of simulations. This question was approached via adjoint SILAM runs (see below) using the methodology developed by Sofiev (2002) and Sofiev & Siljamo (2004).

4. Inverse Case Studies for Springs 2002-2004

Almost every spring in Finland is characterized by high pollen concentrations recorded at aerobiological monitoring sites approximately one week before the local flowering started. Here we analyze four cases observed during springs 2002-2004. The analysis includes: (i) delineation of potential source areas using the SILAM runs in adjoint mode, (ii) combination of the obtained sensitivity distributions with observed pollen concentrations at the suspected source areas.

Setup of the SILAM adjoint runs used the data of Finnish stations Turku, Kangasala, Kuopio, Oulu and Kevo for 2002, and Turku, Kangasala, Joutseno, Vaasa and Oulu for 2003. In spring 2004, birch pollen grains were observed in Turku, Helsinki, Vaasa, Oulu and Kangasala, but not in Joutseno, Kuopio or Rovaniemi. Corresponding concentrations were approximated with step-wise functions and used as source terms for SILAM. The simulated period covered 4 days preceding the observed peaks. Meteorological data were taken from operational FMI-HIRLAM (HIRLAM 4 for 2002 and HIRLAM 5 for 2003, HIRLAM 6 for 2004) with 3-hour time steps and 20/30/20 km of horizontal resolution, respectively.

In 2002, the flowering in Finland started at the beginning of May, while the birch pollen was registered already since April 22. Late spring of 2003 delayed the flowering, so that the high concentrations were recorded around May 5–6, again well before the local flowering. As it is seen from Figure 1, possible sources of pollen during spring 2002 are located in Baltic countries, Russia and Belarus.

Figure 2 shows quite different pattern for spring 2003 pointing to the south-western sector as a potential source, with maximum pollen age of a couple of days (contrary to 3–5 days in 2002). Baltic countries are important source areas again, but this time the grains can also originate from Sweden and, with less probability, from Poland and Germany. Observations from Riga and Stockholm confirm this hypothesis (Siljamo et al., 2004).

During spring 2004, there were two pollen episodes in Finland: first one from the 15th-17th and the second one from the 19th–22nd of April. The earliest flowering season in South-West of Finland started from the 25th of April, about a week later than the observed concentration peaks. As seen from Figure 3 and Figure 4,

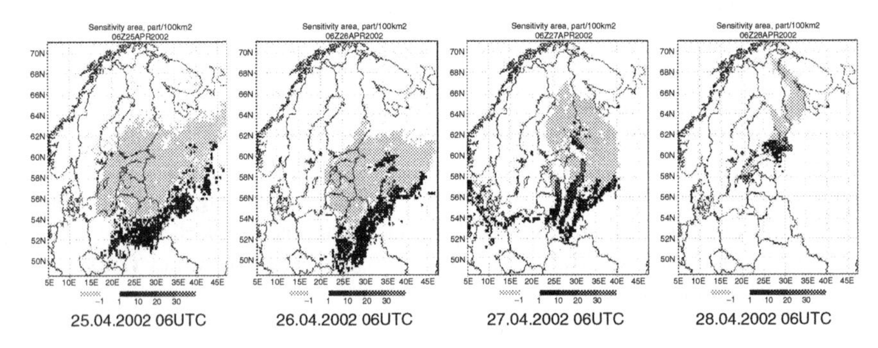

FIGURE 1. Possible source areas of *Betula* pollen in spring 2002, 25th–28th of April. Light grey indicates areas, which could not supply the grains; the dark grey areas could serve as sources. Pollen was observed over Finland during April 27–30, 2002.

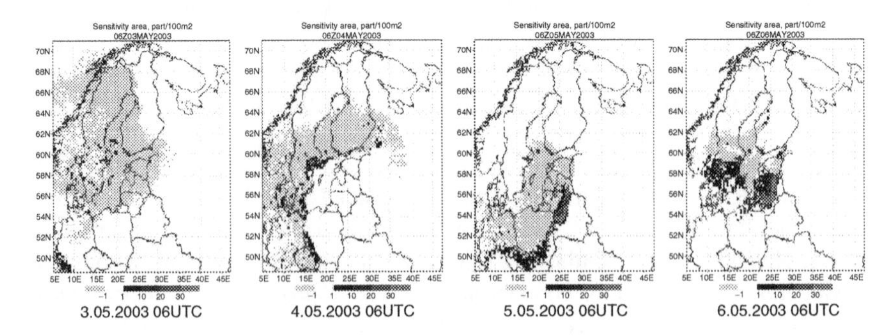

FIGURE 2. Possible source areas of *Betula* pollen in spring 2003, 3rd–6th of May. Light grey indicates areas, which could not supply the grains; the dark grey areas that could serve as sources. Pollen was observed around the 7th of May, 2003.

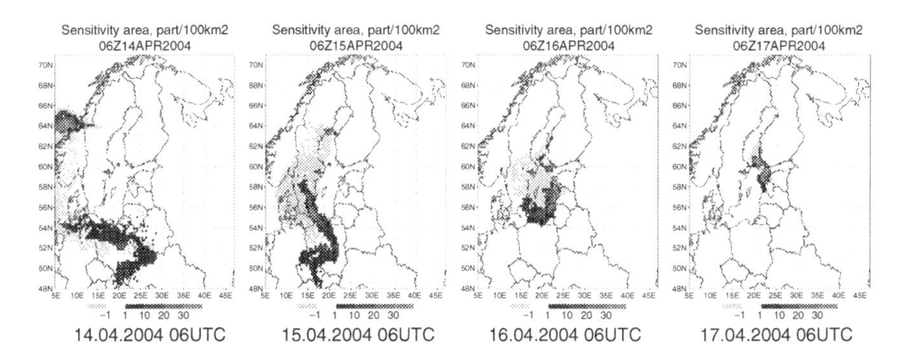

FIGURE 3. Possible source areas of *Betula* pollen in spring 2004, 14th – 17th of April. Light grey indicates areas, which could not supply the grains; the dark grey areas could serve as sources. Pollen was observed on 15th–17th of April, 2004.

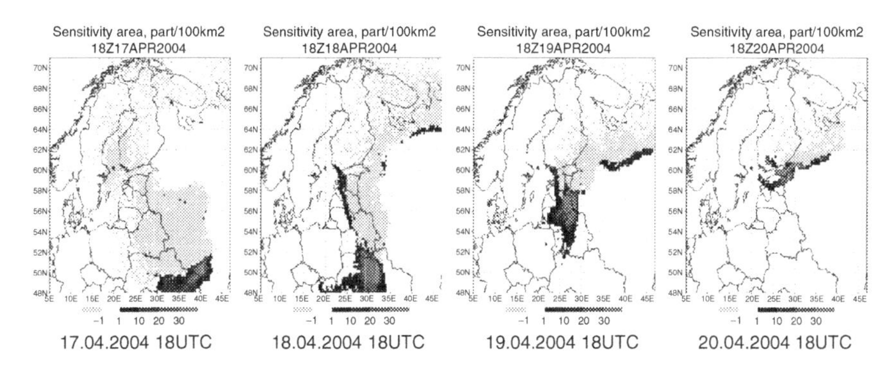

FIGURE 4. Possible source areas of *Betula* pollen in spring 2004, 17th – 20th of April. Light grey indicates areas, which could not supply the grains; the dark grey areas could serve as sources. Pollen was observed on 19th–21st of April, 2004.

these two episodes were created by different pollen sources. Firstly, the pollen was transported from Poland and North-Germany, with the next plume coming from Belarus or Ukraine a couple of days later. According to Figure 4, the Baltic countries could be a source area once again, but the flowering season had not yet started there during the considered time period.

5. Forward Simulation for Spring 2004

Based on the above adjoint simulations, four main source areas were considered for the forward model runs in 2004: birch forests in Finland itself, those in Sweden, Norway, Germany, Poland, Denmark (western sector – see Figure 5, left

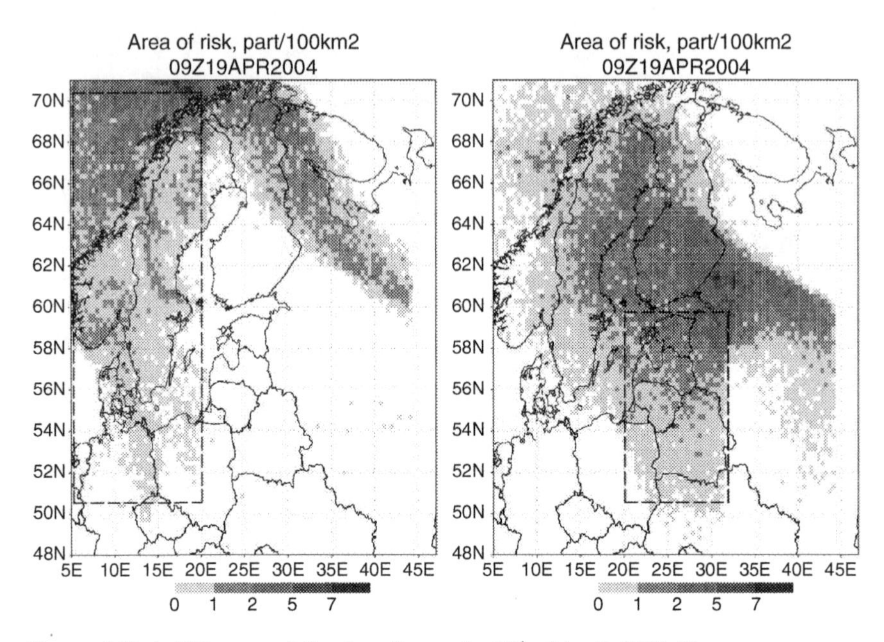

FIGURE 5. Probability area of *Betula* pollen on the 19th of April, 2004. The source areas are outlined by the boxes. Computations start since the 16th of April.

panel), birch in Baltic countries and Belarus (Figure 5, right panel), and the eastern sector (birch forests in Russia).

An application of "classical" dispersion forecast for spring 2004 using the climatological CD index as a flowering trigger (the only available parameterization so far) appeared not very practical because the flowering season started some two weeks earlier than the long-term averages. Therefore, we concentrated on probability forecasts, which show the area of risk for each of the above four sources. The actual pollen concentration could then be computed from these probabilities when the information of flowering in the corresponding areas becomes available.

An example of such probability forecast is shown in Figure 5. The left panel shows the area of risk for the sources in the western sector, while the right panel shows that of the Baltic countries. Start time of computations was the 16th of April, i.e. the pictures demonstrate the 3 days old distribution. From the left panel, it is well seen that the western sector could not create any high-concentration episode in central Finland by that time, which is well in agreement with the inverse case study (Figure 4).

Unlike western sources, southern areas have a high probability to create pollinating episodes. Phenological observations show that flowering in the Baltic countries started 18th-19th of May. It is also possible that pollen grains transported quite a long way from Belarus or Ukraine, which is also confirmed by the adjoint fields in Figure 4.

6. Conclusions

A problem of long-range atmospheric transport of natural allergens is recognized as a potential source of difficult-to-forecast events of high concentrations of allergens in atmosphere of Nordic countries, which can happen weeks before the local flowering starts. The most important species in this context is birch pollen, which features favor a large-scale atmospheric transport of the grains.

The most complicated part of the integrated pollen model is the emission module, which has to provide a very high accuracy of the flowering timing over large territories. A climatological approach that was tested so far has proved to be irrelevant because conditions during the specific years might be significantly different from the long-term averages. Accumulated heat indices seem to be more promising but still exhibit the common weakness of all empirical models – their accuracy is known only for the specific conditions, while any deviations from them can lead to totally incorrect results. In this view, dynamic models of growing and flowering, e.g., promoter-inhibitor models, might provide better accuracy but their implementation is also the most complicated. In particular, computation of such indices would require analysis of long periods of meteorological data.

Analysis of springs of the past three years showed a high variability of sources affecting Finland but also demonstrated that the observed cases so far were conditioned to (i) birch flowering started over some remote area, (ii) meteorological conditions favorable for the transport from this very area towards the receptor territory. The first condition is difficult to simulate but comparably straightforward to observe, while the meteorological patterns are available from NWP forecasts. Therefore, one of the possible ways of making the pollen forecasts is to utilize real-time observations – either from phenological and aerobiological networks or derived from satellites – to describe the source term, with further application of a dispersion model. This method has own complications such as a necessity to apply data assimilation procedures, still barely used in dispersion modeling. Nevertheless, it can well replace the first step in "classical" emission-dispersion-deposition chain of the dispersion models if the above mentioned flowering models appear inaccurate for the short-term forecasting task.

References

Corden, J. M., A. Stach and W. Millington, 2002, A comparison of *Betula* pollen season at two European sites; Derby, United Kingdom and Poznan, Poland (1995-1999). *Aerobiologia*, **18**, pp. 45–53.

Dahl A. and Strandhede S.-O., 1996, Predicting the intensity of the birch pollen season. *Aerobiologia,* **12**, pp. 97–106.

Emberlin J., Detandt M., Gehrig R., Jäger S., Nolard N. and Rantio-Lehtimäki A., 2002, Responses in the start of Betula (birch) pollen seasons to recent changes in spring temperatures across Europe. *Int. J. Biometeorol.* **46**, pp. 159–170.

Frøsig, A. and Rasmussen A., 2003, Pollen-& sporemålinger i Danmark. Sæsonen 2001. Technical report 03–10, Danish Meteorological Institute

Hjelmroos, M., 1992, Long-distance transport of Betula pollen grains and allergic symptoms. *Aerobiologia*, **8**, pp. 231–236.

Hänninen, H., 1990, Modelling bud dormancy release in trees from cool and temperate regions. *Acta forestalia Fennica*, 213, 47p.

Herrera C. M., Jordano P., Guitian J. and Traveset A. 1998, Annual variability in seed production by woody plants and masting concept: reassessment of principles and relationship to pollination and seed dispersal. , *Am.. Nat.* **152**(4): 576–588.

Linkosalo, T. 2000a, Mutual regularity of spring phenology of some boreal tree species: predicting with other species and phenological models. *Canadian Journal of Forest Research,* **30**, pp.667–673

Linkosalo, T. 2000b, *Analyses of the spring phenology of boreal trees and its response to climate change.*, Ph.D. dissertation. University of Helsinki Department of Forest Ecology Publications 22. 55p.

Luomajoki, A., 1999, Differences in the climatic adaptation of silver birch (*Betula pendula*) and downy birch (*Betula pubescens*) in Finland based on male flowering phenology. *Acta Forestalia Fennica*, 263, Finnish Society of Forest Science

Małgorzata, L., M. Mięetus and A. Uruska, 2002, Seasonal variations in the atmospheric *Betula* pollen count in Gdańsk (southern Baltic coast) in relation to meteorological parameters. *Aerobiologia*, **18**, pp. 33–43.

Masaka K., and Maguchi S., 2001, Modelling the masting behaviour of *Betula platyphylla var japonica* using the resource budget model. *Ann. Bot.*, **88**, pp. 1049–1055.

Pisarenko, A. I., Strakhov, V. V., Päivinen, R., Kuusela, K., Dyakun, F. A., and Sdobnova, V. V., 2001, *Development of forest resources in the European part of the Russian Federation*, European Forest Institute Research Report 11, Brill

Rantio-Lehtimäki, A., 1994, Short, medium and long range transported airborne particles in viability and antigenicity analyses. *Aerobiologia* , **10**, pp. 175–181.

Rantio-Lehtimäki, A. and Matikainen, E., 2002, Pollen allergen reports help to understand preseason symptoms, *Aerobiologia*, **18**, pp. 135–140.

Rötzer, T. and Chmielewski, F.-M. , 2001, Phenological maps of Europe. *Clim Res*, **18**, pp. 249–257

Sarvas, R., 1972, *Investigations on the annual cycle of development of forest trees. Active period.* Communicationes Instituti Forestalis Fenniae **76**(3), pp. 1–110.

Schaber, J. and Badeck, F.-W. , 2003, Physiology Based phenology models for forest tree species in Germany. *Int. J. Biometeorol.* **47**, pp 193–201

Siljamo, P., Sofiev, M., Ranta, H., Kalnina, L., and Ekebom, A., 2004, Long range transport of birch pollen. A problem statement and feasibility studies, *Baltic HIRLAM Workshop, St. Petersburgh, 17-20 November, 2003*, pp. 100–103

Sofiev M., 2002, Real time solution of forward and inverse air pollution problems with a numerical dispersion model based on short-term weather forecasts. *HIRLAM Newsletter* **14**, pp.131–138.

Sofiev, M., and Siljamo, P., 2003, Forward and inverse simulations with Finnish emergency model SILAM., In: *Air Pollution Modelling and its Applications XVI*, eds. C.Borrego, S.Incecik, Kluwer Acad. / Plenum Publ. pp.417–425.

An Approach to Simulation of Long-Range Atmospheric Transport of Natural Allergens: An Example of Birch Pollen

Speaker: P. Siljamo

Questioner: P. Bhave
Question: In yesterday's presentation by Dr. Daggupaty, we saw that 50% of the 20 μm particle emissions deposited within 100 km of their source. What is different about the 20 μm birch pollen grains that allow them to be transported 1000 km from their source?
Answer: Birch pollen grain is light. Its density is ~800 kg/m3 while, e.g., density of typical anthropogenic aerosol is 2–4 times more. In addition, we have to trace even 5–10% of the total amount pollen coming from the source area. Concentrations in those areas can be thousands or tens of thousand of pollen grains/m3, while allergic people suffer already from 100 grains/m3 and observation detection limit is 10–100 times lower.

Questioner: S. Galmarini
Question: The location of the source and time of the release is a crucial issue in your study. Have you investigated whether remote sensing can provide indications of the flow or in movement and locations?
Answer: We have not investigated it yet, but we are going to. Satellites can help to specify a source location and timing, but the problem is not easy and straightforward. They can not observe clouds of pollen grains, because it is not dense enough.

Questioner: B. Terliuc
Question: Did you perform a second order inversion of the problem to test the reliability of the model and the inversion process?
Answer: This topic will be discussed tomorrow.

Questioner: M. Jean
Question: Have you considered using other kind of measurements (such as radionuclides) to help in the assessment of the source of pollens?
Answer: No, we have not, but idea is very interesting and worth trying.

37 Cloud Chemistry Modeling: Parcel and 3D Simulations

Aissa-Mounir Sehili, Ralf Wolke, Jürgen Helmert, Martin Simmel, Wolfram Schröder, and Eberhard Renner*

1. Introduction

The interaction of gases and aerosol particles with clouds entails a number of key environmental processes. On the one hand, they directly influence the life cycles of trace constituents and facilitate conversions of these trace constituents. On the other hand, multiphase transformations strongly influence cloud formation. Over a long time, the complexities of the cloud processes involved have discouraged investigators from simultaneously treating all aspects of multiphase chemistry and microphysics with equal rigor. Many recently available models focus either on complex multiphase chemistry only in a few aggregated drop classes (Ervens et al., 2003; Herrmann et al., 2000), or detailed microphysics for strongly simplified chemical mechanisms (Bott, 1999). The last year's efforts invested to develop sophisticated cloud models with complex multiphase chemistry allows more detailed studies on the interaction between microphysical and chemical multiphase processes (Leriche et al., 2000, 2003; Ervens et al., 2004). This is also true of size-resolved parameterizations of cloud chemistry in chemistry-transport codes (Gong et al., 2003).

The paper focuses on the description and the numerical treatment of these interactions in regional chemistry-transport codes. For this purpose, a new coupling scheme between transport processes, microphysics and multiphase chemistry was developed and implemented in the model system LM-MUSCAT (Wolke et al., 2004a,b). This coupling scheme is adapted to the implicit-explicit (IMEX) time integration scheme used in MUSCAT. In contrast to the usual "operator splitting" approach, the mass balance equations for all chemical species and all processes are integrated in a coupled way. The IMEX scheme utilize the special sparse structure of the large system of stiff ordinary differential equations. Evaluation of atmospheric process parameterizations by means of field

* Institute for Tropospheric Research, Permoserstr. 15, D–04303 Leipzig, Germany, *Email: wolke@tropos.de*

measurements is often not possible or very expensive. Therefore, simulation studies under realistic atmospheric conditions which compare the results for these parameterizations with that of detailed parcel models involving an explicit description of the processes is necessary to quantify the error of the parameterization. The size resolved cloud chemistry parcel model SPACCIM ("SPectral Aerosol Cloud Interaction Model") has shown recently its ability to simulate, with an appreciable accuracy, the processing of trace gases and aerosols by local orographic cloud events (Wolke et al., 2005; Tilgner et al., 2005). The model allows a detailed description of the processing of gases and particles shortly before cloud formation, during the cloud life time and shortly after cloud evaporation. The performance of the model is shown for simple chemical mechanisms (with only inorganic chemistry) as well as for very complex mechanisms of the CAPRAM family (Herrmann et al., 2000; Ervens et al., 2003) which contain a detailed description of the organic chemistry. The model is evaluated with data from the FEBUKO[1] field campaign.

The paper presents cloud chemistry results simulated by LM-MUSCAT in comparison to SPACCIM simulations which uses a more detailed process description. The loss in the quality of the process description due to the simplification and its influence on the simulation results is discussed in view of an improvement of cloud chemistry parameterizations in higher dimensional models. Using SPACCIM, the movement of the air parcel follows a predefined trajectory generated by LM-MUSCAT. Entrainment and detrainment processes are considered in a parameterized form. The simulations are performed for one scenarios of the FEBUKO field campaign. For the tests, a more simple inorganic reaction scheme INORG (Sehili et al., 2005) extracted from CAPRAM2.3 by neglecting all organic reactions, is applied. The gas phase chemistry is based on the RACM mechanism (Stockwell et al., 1997). The whole reaction scheme includes 80 gas phase and 20 aqueous phase species. In the liquid phase, 10 dissociations, 4 irreversible reactions and 8 phase transfer equilibria are considered. The reaction system (gas and aqueous phases, phase transfer according to Schwartz) is read from an ASCII data file. Afterwards, all data structures required for the computation of the chemical terms and the corresponding Jacobians are generated. This approach allows large flexibility in the choice of the chemical reacting mechanism.

2. The Parcel Model SPACCIM

Model formulation. Generally in a box model, the chemical multiphase processes are described by the mass balance equations of the species in a size-resolved droplet spectrum. Formally, these system can be written as a system of ordinary differential equations (ODE)

$$\dot{c} = f_{chem}(t, c; m) + f_{henry}(t, c; m) + f_{mphys}(t, c; m) \tag{1}$$

[1] http://projects.tropos.de:8088/afo2000g3/FEBUKO_dateien/febuko.html

where c denotes the vector of mass concentrations related to air volume of the gas phase species and the aqueous species in each particle/drop fraction. The vector m represents the time-dependent microphysical variables which have to be provided simultaneously by a microphysical model. The term f_{chem} stands for the chemical reactions in gas and aqueous phase. Note that the liquid phase chemistry is always performed for ideal solutions. The gas-liquid mass transfer term f_{henry} is parameterized by the approach of Schwartz (1986) under the hypothesis of well mixed droplets. The term

$$f_{mphys} = \sum_{i=1}^{M} \left[T_{ik} c_l^i - T_{ki} c_l^k \right] \tag{2}$$

stands for mass fluxes between different size bins caused by microphysical processes (condensation/evaporation, coagulation/breakup). T_{ik} denotes the time-averaged total mass flux from bin i to bin k. The size-resolved microphysical model allows alternatively a "fixed bin" or "moving bin" discretization of the particle/drop spectrum. Note that droplets having the same size contain the same amount of aerosol.

Coupling scheme. In the SPACCIM approach (Wolke et al., 2005), a multiphase chemistry model is directly coupled with a microphysical model. The two models run separately and exchange information only every coupling time step Δt_{cpl}. Each of the two models uses its own time step control. The coupling scheme provides time-interpolated values of the microphysical variables (temperature, water vapor, liquid water content) and generates time-averaged mass fluxes f_{mphys} over the coupling time interval (Fig. 1). The feedback of changes in the chemical composition to microphysics is also implemented. The SPACCIM approach allows the coupling of a complex multiphase chemistry model with microphysical codes of different types. The exchange of information is organized over well-defined interfaces. The size bin discretisation for the multiphase chemistry is taken from microphysics. For a reduction in the computational costs, the use of coarser resolutions in the multiphase chemistry computations is possible by averaging the microphysical variables.

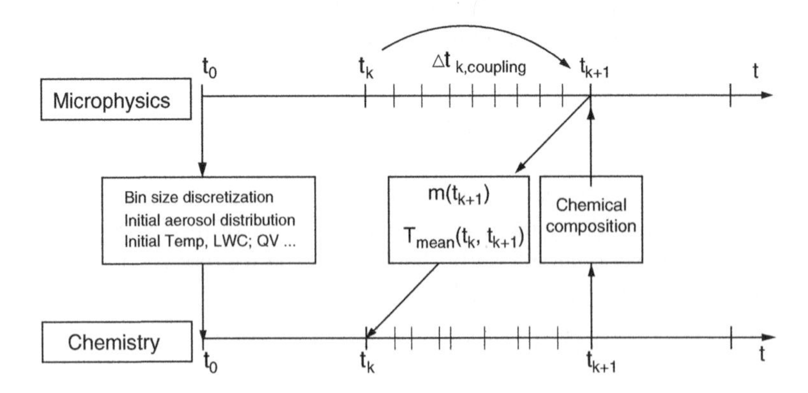

FIGURE 1. Schematic representation of the coupling strategy.

Numerics. The ODE system (1) resulting from multiphase chemical systems is nonlinear, highly coupled and extremely stiff. Because explicit ODE solvers require numerous short time steps in order to maintain stability, only implicit schemes can integrate these systems in an effective way (Sandu et al., 1997). The implicit time integration of (1) is performed by a high order BDF scheme. For this part, a modified version of the popular code LSODE (Hindmarsh, 1983) with an entire replacement of the linear algebra part is used. This direct sparse technique exploits the special structure (sparsity, block structure, different types of coupling) of the corresponding Jacobians (Wolke and Knoth, 2002). In the coupling scheme, the choice of an appropriate coupling time step can cause problems in some cases. The splitting error should be kept small. Otherwise, the number of restarts, which are expensive for higher order implicit solvers, should be as small as possible.

Entrainment and detrainment. The adaptation of the box model to simulate air parcels undergoing forcing along a trajectory delivered by a 3D meteorological model(LM) implies the treatment of new issues that were neglected, or did not show influence, during the simulation of local adiabatic events.

Considering a more general non adiabatic case needs the characterization of the entrainment/detrainment process. In contrast to 1D and higher resolved models where entrainment is parameterized with an entrainment rate depending on boundary conditions or velocities (e.g., linear variation with altitude in 1D models), the parcel resolved model can only involve a constant or quasi-constant rate. In our approach, the entrainment is described by

$$\frac{\mathrm{d}x}{\mathrm{d}t} = \ldots - \mu(x - x_{enu}) \tag{3}$$

x is the value of the variable in the air parcel, x_{enu} the one outside and μ the entrainment rate given in [s^{-1}]. It represents the fraction of the unit volume of the parcel exchanged with the entrained air per unit time. This parameterization includes the exchange of meteorological variables like heat, and moisture as well as physical and chemical ones like aerosols and gas phase species. Recent studies for 1D models (Liu and Seidl, 1998) and preliminary simulations have shown that a magnitude of the order of 10^{-2} to 10^{-4} constitutes a reasonable range for μ. The two boundaries correspond respectively to strong and weak entrainment. In our simulations, two different entrainment rates were taken. A strong entrainment rate of the order of 10^{-2} is used for the meteorological forcing data. The mixing occurs rapidly in order to bring the parcel into the trajectory. To account for aerosol mixing, another approach for the entrainment rate is considered namely $\mu = \mu' \left[\frac{N - N_{ent}}{N_{ent}} \right]$. N represents the total number of particles in the parcel and N_{ent} the total number of entrained particles at a supersaturation of 99,99%. For μ', the value of 10^{-4} is considered. It is clear that in such an approach, the effect of particles entrainment can have a significant influence only when sedimentation is efficient enough. That requires a high liquid water content ratio in the parcel in order to initiate the coagulation process. In the present study, coagulation and consequently deposition plays a negligible role. For the gas phase mixing, several entrainment rates ranging from the weaker value of 10^{-5} to the stronger one of 10^{-2} are tested to figure out the effect of gas phase entrainment.

Further adjustments. Wet deposition and wash-out are considered using the approximate terminal fall velocity. Photolysis frequencies correction due to cloud radiative forcing is also an issue that should be treated during the simulation of air parcels along trajectories delivered by a 3D meteorological model. An option to adjust photolysis rates for the presence of clouds is to use the approach developed for the Regional Acid Deposition Model (RADM) (Chang et al., 1987) given by

$$j = j_{clear}\,[1 + a(F_{cld} - 1)] \tag{4}$$

where j is the corrected photolysis rate and j_{clear} the clear sky one. $0 \le a \le 1$ represent the cloud coverage. F_{cld} is the transmission coefficient depending on the zenith angle and accounting for the reflection effect of the surface and clouds. In this approach, the vertical direction is subdivided into three layers, below cloud, in cloud and above cloud. Accordingly, F_{cld} can take three different formulations

$$F_{cld} = 1 + \alpha_i (1 - t_r) \cos \chi_0 \qquad \textit{above cloud layer}$$

$$F_{cld} = 1.4 + \cos \chi_0 \qquad \textit{in cloud layer}$$

$$F_{cld} = 1.6 t_r + \cos \chi_0 \qquad \textit{below cloud layer}$$

The entities t_r and $(1 - t_r)$ represent respectively the reflection units due to surface and cloud albedo. We take $t_r = 0.5$. χ_0 is the solar zenith angle. The above formulations are valid for $\chi_0 \le 60°$. When $\chi_0 > 60°$ values for F_{cld} evaluated at $\chi_0 = 60°$ are used.

3. LM-MUSCAT

The model system consists of the chemistry-transport model MUSCAT and the online-coupled, non-hydrostatic meteorological model LM (Doms and Schättler, 1999; Schättler and Doms, 1999) which is the operational forecast model of the German weather service DWD. Both codes are parallelized, work on their own predefined fraction of available processors and have their own time step control.

Model features. MUSCAT has a multiblock grid structure and uses the static grid nesting technique (Knoth and Wolke, 1998a). The spatial discretization is performed with a finite-volume method. The time integration consists of a implicit-explicit scheme (Knoth and Wolke, 1998b; Wolke and Knoth, 2000). For the horizontal advection, a second order Runge-Kutta method is used. The rest is integrated by an implicit solver. The gas phase chemistry is described by RACM and aerosols dynamics are treated by a modal technique. Parallelization is performed using MPI with a block distribution among processors. To face the problem of load imbalances between parallel processors, a dynamic load balancing has been implemented. The coupling scheme between LM and MUSCAT simultaneously provides time-averaged wind fields and time-interpolated values of other meteorological fields (vertical exchange coefficient, temperature, humidity, density). Coupling between meteorology and chemistry-transport takes place at each horizontal advection time step only. The wind components from the compressible LM are projected to satisfy the discrete continuity equation via solving an elliptic equation by a preconditioned conjugate gradient method. This is done in parallel on the LM processors. The projected wind fields and the other meteorological data are gathered by one of the LM processors. This processor communicates directly with each of the MUSCAT processors. For a more detailed description of LM-MUSCAT we refer to Wolke et al. (2004a,b).

Implementation of cloud chemistry. For the description of multiphase processes in LM-MUSCAT, the SPACCIM approach is applied to Eulerian grid models. According to the IMEX time integration scheme in MUSCAT, all horizontal mass fluxes and the mass fluxes over cloud boundaries (only for the main aerosol species) are integrated explicitly (Fig. 2). All time-averaged mass fluxes between neighbored vertical cloud grid cells are included in the microphysical fluxes f_{mphys}. Using an appropriate order of variables, the mass fluxes T_{ij} in (2) become tridiagonal matrices. An efficient numerical solution of this extended system requires an adaptation of the sparse linear system solvers.

Clouds are dynamical objects with high spatial and temporal variability. Therefore, the data structures used for the implementation of clouds cells should take into account these dynamics. Furthermore, a dynamical redistribution of blocks to reduces load imbalances caused by cloud inhomogeneities is required to reduce the computing time. The implementation of both features is now being tackled.

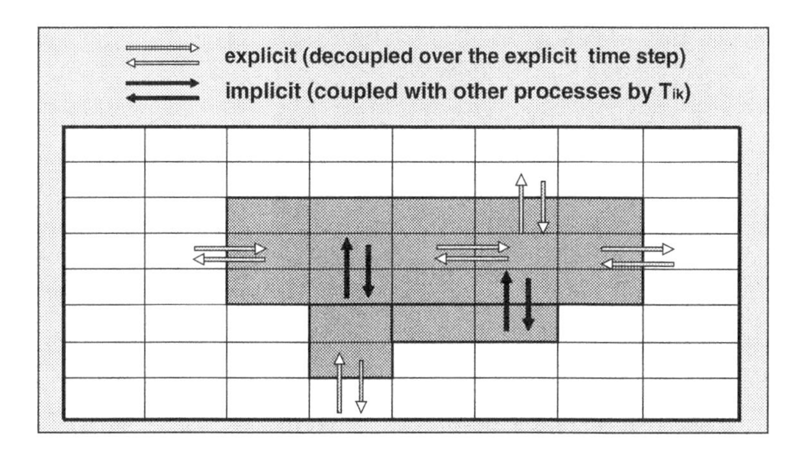

FIGURE 2. Schematic representation of the mass fluxes in the cloud chemistry implementation.

4. Results and Discussions

Description of the scenario. The parcel trajectory is determined inside the domain simulated by the 3D LM-MUSCAT system for a prescribed period. The trajectory is calculated through the definition of a reference point from which backward and forward time resolved interpolated meteorological variables are calculated. That constitutes the whole forcing data of the event. LM-MUSCAT is used on a 100 × 100 × 18 grids cells domain with 2.8 km horizontal grid resolution covering the north east part of Germany (Fig. 3). Cloud chemistry simulations for both models are performed over 5 hours. The air parcel starts traveling at 50.32° N, 11.72° W and 1280 m altitude on 24.10.2001 at 13.00 PM. The background gas phase initial concentrations, as well as initial dry aerosol composition, are delivered by a preliminary LM-MUSCAT run which uses an emission inventory registered by the German environment agencies. The initial dry aerosol number size distribution consists of two lognormal modes, covering the Aitken and the accumulation size range derived from FEBUKO field measurements. The initial pH value is determined through the charge balance and aerosols initial composition and is then computed dynamically throughout the whole simulation time.

Influence of the entrainment rate. Gas phase species evolution shows an important dependence on the gas phase entrainment rate for parcel simulations. A high entrainment factor comparable to the meteorological one makes the concentration of gas phase species strongly depend on the LWC in the parcel.

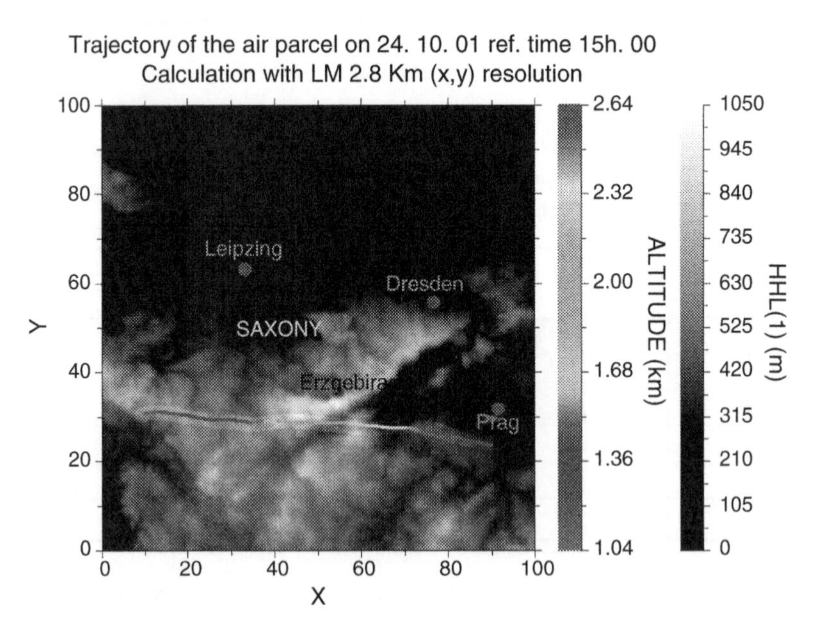

FIGURE 3. Trajectory simulated by LM-MUSCAT for a period of 5 hours (start at 13.00 PM).

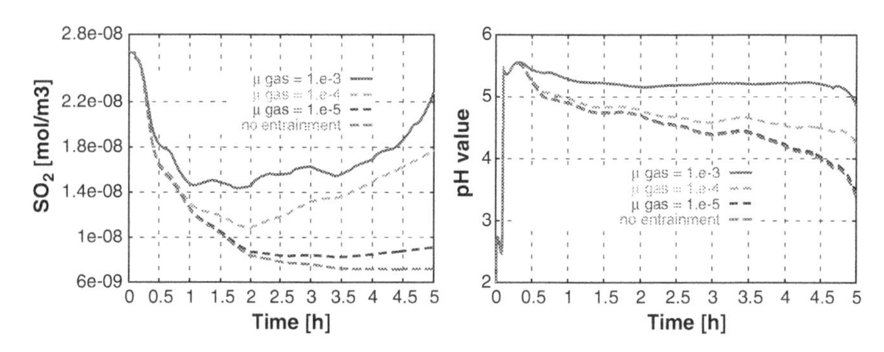

FIGURE 4. Influence of the gas phase entrainment rate on: SO_2 concentration (left), pH value (right) during the simulation with SPACCIM.

The production of gas phase species during evaporation periods is observed (e.g., SO_2, Fig. 4). In this case, the pH value which represents the averaged one over the whole spectrum tends to increase and remain for entrainment rates larger than 10^{-3} over the value of 5. On the other hand, considering smaller or no entrainment leads to a non-realistic description of trace gas processing by clouds. The gas phase moderate entrainment rate of 10^{-4} seems to constitute an appropriate value in order to account for both aspects.

Comparison of parcel and 3D simulations. As shown in Fig. 5 (left), SPACCIM simulates more total liquid water content (LWC) compared to the one given by the forcing data from LM. This deviation reaches its maximum of more than 30% after 2 hours. This is due to the spectral explicit treatment of cloud processes including activation and the description of the water phase transfer feedback on water vapor and air temperature (latent heat release) implemented in SPACCIM. The number size distribution in Fig. 5 (right) illustrates the broadening and the splitting of the spectrum into activated and non activated parts caused by the explicit description of activation after 2 hours corresponding with maximum LWC.

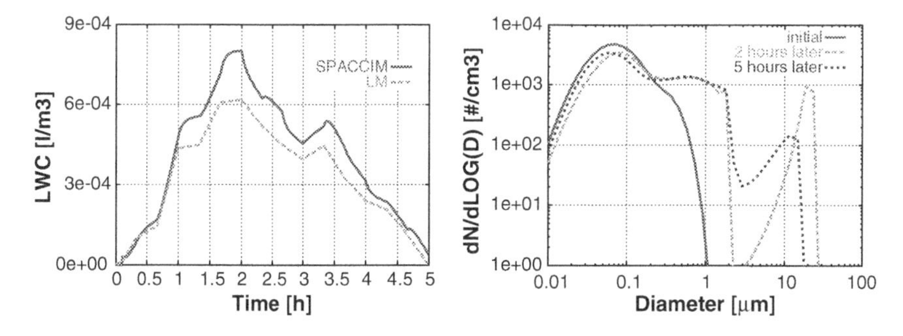

FIGURE 5. Total LWC calculated by SPACCIM compared with the one given by the forcing data from LM (left); Number size distribution evolution for three position of the air parcel simulated with SPACCIM (right).

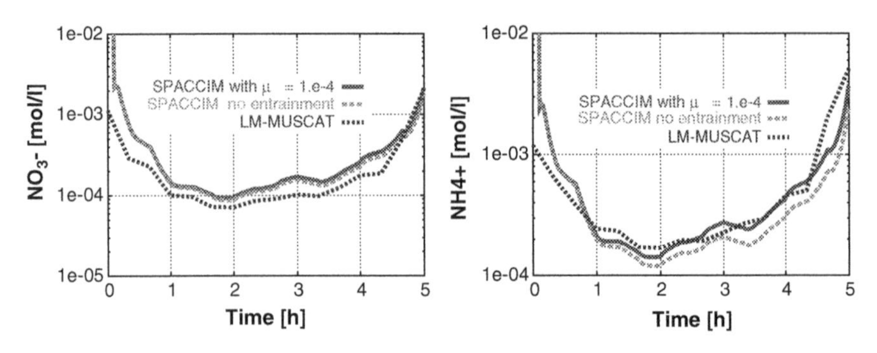

FIGURE 6. Comparison of NO_3^- and NH_4^+ molalities between SPACCIM and LM-MUSCAT during the 5 hours simulations.

Currently, these issues are not implemented in LM-MUSCAT. Furthermore, MUSCAT includes multidimensional cloud chemistry parameterizations involving the same processes described in the parcel model SPACCIM. This leads to a relatively good agreement between the two approaches for aqueous phase species concentration as shown for nitrate and ammonium in Fig. 6. The gas phase entrainment rate of 10^{-4} constitutes an appropriate choice again.

5. Conclusions

In the paper, cloud chemistry modeling in the size-resolved parcel model SPACCIM and the 3D chemistry-transport model LM-MUSCAT are presented. It is shown that the detailed parcel model with necessary adjustments can be used to assess and evaluate the parameterizations applied in the 3D code. The analysis of the entrainment parameterization in SPACCIM indicates that a moderate entrainment rate of 10^{-4} for the gas phase represents a reliable value. On the other hand, entrainment of aerosol becomes significant only in the case of coagulation and sedimentation initiated by higher LWC. The relative good agreement noted between the two approaches for aqueous phase species stresses the seriousness of such comparison studies.

Acknowledgements The work was supported by the BMBF (AFO2000 atmospheric research program, project 07ATF40)and the NIC Jülich. Furthermore, we thank the DWD Offenbach for good cooperation.

References

Bott, A., 1999. A numerical model of the cloud-topped planetary boundary layer: chemistry in marine stratus and the effects on aerosol particles. Atmospheric Environment 33, 1921–1936.

Chang, J. S., Brost, R. A., Isaksen, I. S. A., Madronich, S., Middleton, P., Stockwell, W.R., Walcek, C.J., 1987. A three-dimensional Eulerian acid deposition model: Physical concepts and formulation. Journal of Geophysical Research, 92, 14,681–14,700.

Doms, G., Schättler, U. 1999. The Nonhydrostatic Limited-Area Model *LM* (Lokal Model) of DWD: I. Scientific Documentation (Version *LM*-F90 1.35). German Weather Service, Offenbach.

Ervens, B., George, C., Williams, J. E., Buxton, G. V., Salmon, G. A., Bydder, M., Wilkinson, F., Dentener, F., Mirabel, P., Wolke, R., Herrmann, H., 2002. CAPRAM2.4 (MODAC mechanism): An extended and condensed tropospheric aqueous phase mechanism and its application. Journal of Geophysical Research 108 (D14), 4426, doi:10.1029/2002JD002202.

Ervens, B., Feingold, G., Frost, G. J., Kreidenweis, S. M., 2004. A modeling study of aqueous production of dicarboxylic acids: 1. Chemical pathways and speciated organic mass production. Journal of Geophysical Research 109, D15205, doi:10.1029/2003JD004387.

Gong, W., Dastoor, A. P., Bouchet, V. B., Gong, S., Makar, P. A., Moran, M. D., Pabla, B., 2003. Cloud processing of gases and aerosols in a regional air quality model (AURMAMS) and its evaluation against precipitation-chemistry data. Proceedings of thr Fifth Conference on Atmospheric Chemistry: Gases, Aerosols, and Clouds, 2.3 (CD-ROM), AMS.

Herrmann, H., Ervens, B., Jacobi, H.-W., Wolke, R., Nowacki, P., Zellner, R., 2000. CAPRAM2.3: A chemical aqueous phase radical mechanism for tropospheric chemistry. Journal of Atmospheric Chemistry 36, 231–284.

Chang, A. C., 1987. Scientific Computing. Chapter ODEPACK: A Systematized Collection of ODE Solvers. North-Holland, Amsterdam, pp. 55–74.

Knoth, O., Wolke, R., 1998. An explicit-implicit numerical approach for atmospheric chemistry-transport modeling. Atmospheric Environment, 32, 1785–1797.

Knoth, O., Wolke, R., 1998. Implicit-explicit Runge-Kutta methods for computing atmospheric reactive flow, Applied Numerical Mathematics, 28, 327–341.

Kreidenweis, S. M., Walcek, C. J., Feingold, G., Gong, W., Jacobson, M. Z., KIM, C.-H., Liu, X., Penner, J. E., Nenes, A., Seinfeld, J. H., 2003. Modification of aerosol mass and size distribution due to aqueous-phase SO_2 oxidation in clouds: Comparison of several models. Journal of Geophysical Reasearch, 108(D7), 4213.

Leriche, M., Voisin, D., Chaumerliac, N., Monod, A., Aumont, B., 2000. A model for tropospheric multiphase chemistry: Application to one cloudy event during the CIME experiment. Atmospheric Environment 34 (29/30), 5015–5036.

Leriche, M., Deguillaume, L., Chaumerliac, N., 2003. Modeling study of strong acids formation and partitioning in a polluted cloud during wintertime. Journal of Geophysical Research 108 (D14), 4433, doi:10.1029/2002JD002950.

Liu, X., Seidl, W., 1998. Modelling study of cloud droplet nucleation and in-cloud sulfate production during the Sanitation of the atmosphere (SANA) 2 campaign. Journal of Geophysical Reasearch, 103, D13, 145–158.

Sandu, A., Potra, F. A., Charmichael, G. R., Damian, V., 1996. Efficient implementation of fully implicit methods for atmospheric chemical kinetics. Journal of Computational Physics, 129, 101–110.

Sandu, A., Verwer, J. G., Van Loon, M., Charmichael, G. R., Potra, F. A., Dabdub, D., Seinfeld, J.H., 1997. Benchmarking stiff ODE solvers for atmospheric chemistry problems I: Implicit versus explicit. Atmospheric Environment, 31, 3151–3166.

Schättler, U., Doms, G., 1999. The Nonhydrostatic Limited-Area Model *LM* (Lokal Model) of DWD: II. Implementation Documentation (Version *LM*-F90 1.35). German Weather Service, Offenbach.

Schwartz, S.E., 1986. Mass transport considerations pertinent to aqueous phase reactions of gases in liquid water clouds. NATO ASI Series, Vol. G6, Chemistry of Multiphase Atmospheric Systems. Springer, Berlin, pp. 415–471.

Sehili, A. M., Wolke, R., Knoth, O., Simmel, M., Tilgner, A., Herrmann, H., 2005. Comparison of different model approaches for the simulation of multiphase processes. Atmospheric Environment, submitted.

Stockwell, W. R., Kirchner, F., Kuhn, M., Seefeld, S., 1997. A new mechanism for regional atmospheric chemistry modelling. Journal of Geophysical Reasearch, 102(D22), 25847–25879.

Tilgner, A., Majdik, Z., Sehili, A. M., Simmel, M., Wolke, R., Herrmann, H., 2005. SPACCIM Simulations of Multiphase Chemistry occuring in the FEBUKO Hill-Capped Cloud Experiments. Atmospheric Environment, submitted.

Wolke, R., Knoth, O., 2000. Implicit-explicit Runge-Kutta methods applied to atmospheric chemistry-transport modelling. Environmental Modelling and Software, 15, 711–719.

Wolke, R., Knoth, O., 2002. Time-integration of multiphase chemistry in size-resolved cloud models. Applied Numerical Mathematics, 42, 473–487.

Wolke, R., Hellmuth, O., Knoth, O., Schröder, W., Heinrich, B., Renner, E., 2004 a. The chemistry-transport modeling system LM-MUSCAT: Description and CityDelta applications. In: Air Pollution Modeling and Its Applications XVI, edited by C. Borrego and S. Incecik. Kluver Academic / Plenum Publishers, New York, 427–439.

Wolke, R., Hellmuth, O., Knoth, O., Schröder, W., Renner, E., 2004 b. The parallel model system LM-MUSCAT for chemistry-transport simulations: Coupling scheme, parallelization and application. In: Parallel Computing: Software Technology, Algorithms, Architectures, and Applications (Series: Advanced in Parallel Computing), edited by G. R. Joubert, W. E. Nagel, F.J. Peters, and W.V. Walter. Elsevier, The Netherlands, 363–370.

Wolke, R., Sehili, A. M., Knoth, O., Simmel, M., Madjik, Z., Herrmann, H., 2005. SPACCIM: A parcel model with detailed microphysics and complex multiphase chemistry. Atmospheric Environment, submitted.

38
A Test of Thermodynamic Equilibrium Models and 3-D Air Quality Models for Predictions of Aerosol NO_3^-

Shaocai Yu, Robin Dennis, Shawn Roselle, Athanasios Nenes, John Walker, Brian Eder, Kenneth Schere, Jenisc Swall, and Wayne Robarge[*]

1. Introduction

The inorganic species of sulfate, nitrate and ammonium constitute a major fraction of atmospheric aerosols. The behavior of nitrate is one of the most intriguing aspects of inorganic atmospheric aerosols because particulate nitrate concentrations depend not only on the amount of gas-phase nitric acid, but also on the availability of ammonia and sulfate, together with temperature and relative humidity. Particulate nitrate is produced mainly from the equilibrium reaction between two gas-phase species, HNO_3 and NH_3.

It is a very challenging task to partition the semi-volatile inorganic aerosol components between the gas and aerosol phases correctly. The normalized mean error (NME) for predictions of nitrate is typically three times that for predictions of sulfate for a variety of 3-D air quality models applied to sections of the U.S. (Odman, et al., 2002; Pun, et al, 2004). For an annual average across the entire U.S. the NMEs of the predictions of nitrate from the U.S. EPA Models-3/Community Multiscale Air Quality Model (CMAQ) are two to three times larger than the NMEs for sulfate.

2. Thermodynamic Models and Observational Datasets

Given total (gas + fine particulate phase) concentrations of H_2SO_4, HNO_3, and NH_3, and temperature and RH as inputs, ISORROPIA and AIM (AIM model II is used in this study) predict the partitioning of these inorganic species between

[*] Shaocai Yu[#], Robin Dennis[@], Shawn Roselle[@], Athanasios Nenes[*], John Walker[$], Brian Eder[@], Kenneth Schere[@], Jenise Swall[@], Wayne Robarge[+], NERL, U.S. Environmental Protection Agency, RTP NC, 27711, USA. [#] On Assignment from Science and Technology Corp., Hampton, VA 23666, USA. [@]On assignment from Air Resources Laboratory, National Oceanic and Atmospheric Administration, RTP, NC 27711, USA. [*]Georgia Institute of Technology, Atlanta Georgia 30332-0340. [$]NRMRL, U.S. EPA, RTP NC, 27711, USA. [+]North Carolina State University, Raleigh, NC 27695.

the gas and fine particle ($PM_{2.5}$) phases on the basis of thermodynamic equilibrium. More detailed descriptions of the equilibrium reactions and the solution procedures for AIM and ISORROPIA are given by Wexler and Clegg (2002) and Nenes et al. (1999), respectively.

Three sites were chosen that had high time resolution data to test the equilibrium models. At the Atlanta site (33.78°N, 84.41°W), a total of 325 observational data points were obtained during the SOS/Atlanta '99 Supersite Experiment from August 18 to September 1, 1999, by parsing the 9-minute HNO_3 and 15-minute NH_3 concentrations into 5-minute averages so as to overlap with 5-minute mean concentrations of $PM_{2.5}$ SO_4^{2-}, NO_3^-, and NH_4^+ (Weber et al., 2003) (summer case). At the Pittsburgh Supersite (40.44°N, 79.94°W), Pennsylvania, a total of 313 data points for two-hourly mean concentrations of $PM_{2.5}$ NH_4^+, NO_3^- and SO_4^{2-}, and gas-phase HNO_3 were obtained during the period of January 2 to January 31, 2002 (Wittig et al., 2004) (winter case). At the Clinton Horticultural Crop Research Station (35°01′ N, 78°16′ W), North Carolina, a total of 479 data points for 12-hour (0600-1800 h (EST) day cycle; 1800-0600 h night cycle) mean concentrations of $PM_{2.5}$ NH_4^+, NO_3^- and SO_4^{2-}, and gas-phase NH_3 and HNO_3 were obtained by an annular denuder system from January 20 to November 2, 1999 (Robarge et al., 2002).

3. Results and Discussion

3.1. Test of Thermodynamic Model with Observational Data

Comparisons of observed aerosol NO_3^- and NH_4^+, and gaseous HNO_3 and NH_3 concentrations with those calculated by ISORROPIA and AIM at the three sites are listed in Table 1. At the Atlanta site, 94.8% and 96.0% of the ISORROPIA and AIM predictions of NH_4^+ are within a factor of 1.5 of the observations. ISORROPIA and AIM also predict HNO_3 well, with 86% and 87% of the predictions within a factor of 1.5 of the observations. However, both equilibrium models are unable to replicate a majority of the observed NO_3^- and NH_3 concentrations, see Figure 1 and Table 1. For NO_3^-, only 32% and 48% of the ISORROPIA and AIM predictions are within a factor of 2 of the observations, respectively. For NH_3, ISORROPIA and AIM replicate 25.2% and 51.4% of the observations within a factor of 2, respectively. At the Pittsburgh site, both AIM and ISORROPIA can correctly predict the NO_3^- concentrations to within a factor 2 of the observations for a majority of the data points (>76%) because the TNO_3 concentration is constrained and the aerosol fraction is dominant. On the other hand, both models perform much more poorly on HNO_3, the gas fraction, as compared to the Atlanta situation. At the Clinton site, both models reproduce observed NH_3 concentrations very well (>95% within a factor of 1.5) and reproduce a majority of NH_4^+ concentration data points within a factor of 2 (>92%). Performance of both models for aerosol NO_3^- at the Clinton site is better than at the Atlanta site but worse than at the Pittsburgh site.

TABLE 1. Statistical summaries of the comparison of the modeled (ISORROPIA and AIM) partitioning of total nitrate (gas + aerosol) and total ammonia (gas + aerosol) between gas and aerosol phases with that of observations at the Atlanta Supersite, GA, Pittsburgh Supersite, PA, and Clinton site, NC. The mean concentrations (± standard deviation) of SO_4^{2-}, TNO_3 and TNH_4 ($\mu g\ m^{-3}$), and relative humidity (RH) (%) and temperature (T) (°C) at each site are also listed.

	<C>*			% within a factor of 1.5**		% within a factor of 2**	
Parameters	OBS	ISORROPIA	AIM	ISORROPIA	AIM	ISORROPIA	AIM
Atlanta site (N = 325)							
(SO_4^{2-} = 12.17 ± 6.71, TNH_4 = 4.38 ± 2.39, TNO_3 = 7.57 ± 5.27, RH = 68.9 ± 19.9, T = 25.0 ± 3.3)							
Aerosol NO_3^-	0.53 ± 0.51	0.54 ± 0.92	0.61 ± 0.92	21.8	33.2	31.7	48.3
Gas HNO_3	7.15 ± 4.84	7.13 ± 4.94	7.06 ± 4.92	86.2	87.1	91.7	92.9
Aerosol NH_4^+	3.60 ± 1.77	4.06 ± 2.05	3.85 ± 1.99	94.8	96.0	98.5	98.8
Gas NH_3	0.74 ± 1.06	0.31 ± 0.79	0.50 ± 0.81	16.6	31.4	25.2	51.4
Pittsburgh site (N = 313)							
(SO_4^{2-} = 2.46 ± 1.14, TNH_4 = 1.74 ± 0.77, TNO_3 = 3.08 ± 2.18, RH = 67.1 ± 17.6, T = 3.9 ± 5.9)							
Aerosol NO_3^-	2.09 ± 1.51	2.04 ± 1.74	1.98 ± 1.79	60.8	57.4	77.0	75.7
Gas HNO_3	1.01 ± 0.68	0.96 ± 0.78	1.02 ± 0.74	37.7	39.6	56.5	62.0
Clinton site (N = 479)							
(SO_4^{2-} = 3.64 ± 4.05, TNH_4 = 6.29 ± 5.51, TNO_3 = 0.57 ± 0.51, RH = 79.9 ± 14.2, T = 19.1 ± 7.7)							
Aerosol NO_3^-	0.30 ± 0.26	0.28 ± 0.28	0.24 ± 0.27	58.0	47.2	71.8	62.0
Gas HNO_3	0.27 ± 0.25	0.29 ± 0.28	0.33 ± 0.30	52.4	49.3	78.7	69.5
Aerosol NH_4^+	1.15 ± 1.27	1.44 ± 1.57	1.42 ± 1.54	74.5	76.2	92.5	92.5
Gas NH_3	5.13 ± 4.73	4.86 ± 4.62	4.88 ± 4.63	95.4	96.5	97.5	97.9

* <C> is mean ± standard deviation ($\mu g\ m^{-3}$)

** Percentages (%): are the percentages of the comparison points whose model results are within a factor of 1.5 and 2.0 of the observations. N is number of samples.

There are many possible reasons for the discrepancies between the model predictions and observations in partitioning of TNO_3 for aerosol NO_3^-. To show how the measurement errors in SO_4^{2-} and TNH_4 can contribute to uncertainties in model predictions of aerosol NO_3^-, Gaussian (normally distributed) random errors are added to the input SO_4^{2-} and TNH_4 (base-case concentrations, C_b) to create the sensitivity-case concentrations (C_s) as follows

$$C_s = C_b + \varepsilon_p \tag{1}$$

where ε_p' represents truncated Gaussian random errors with zero mean and standard deviation equal to $15\% \times C_b$. An error of ±15% is used to correspond with the measurement uncertainty for both SO_4^{2-} and TNH_4 that was estimated as part of the U.S. EPA supersite program (Solomon et al., 2003). The errors are truncated so that only values within 2 standard deviations ($2 \times 15\% \times C_b$) are allowed. As shown in Figure 2, the model with the measurement errors in both SO_4^{2-} and TNH_4 can only reproduce 61.3 % of the base-case aerosol NO_3^- within a factor of 2. This indicates that random errors in SO_4^{2-} and TNH_4 measurements can account for most of the discrepancies between the model predictions and observations of aerosol NO_3^- in Figure 1 at the Atlanta site. Similar conclusions

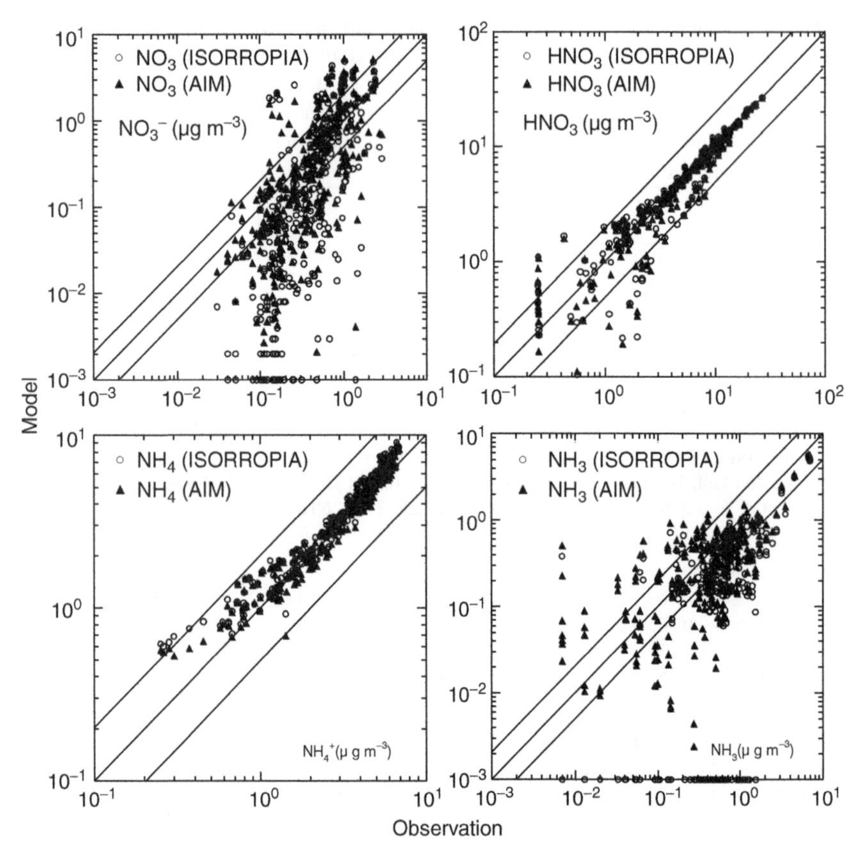

FIGURE 1. Comparison of the modeled (ISORROPIA and AIM) partitioning of total nitrate (gas + aerosol) and total ammonia (gas + aerosol) between gas and aerosol phases with that of observations for aerosol NO_3^-, HNO_3, aerosol NH_4^+ and NH_3 at the Atlanta supersite in summer of 1999. The 1:1, 2:1, and 1:2 lines are shown for reference.

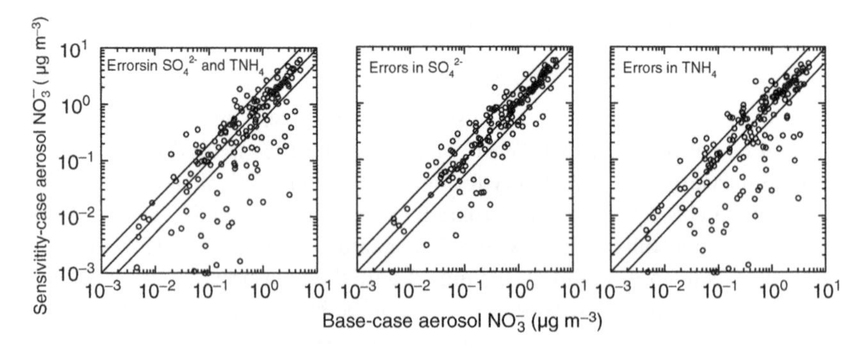

FIGURE 2. Sensitivity-case NO_3^- with assumed Gaussian random errors in observed SO_4^{2-}, TNH_4 vs. the base-case NO_3^- for the ISORROPIA model at the Atlanta site.

can be obtained for the results at the Pittsburgh and Clinton sites and for the AIM model.

3.2. Effects of 3-D Model Prediction Errors in SO_4^{2-}, TNH_4, Temperature and Relative Humidity on Predicting Aerosol NO_3^-

The 3-D CMAQ model can only reproduce 46–79% of SO_4^{2-} and 39–72% of aerosol NH_4^+ within a factor of 1.5 (Yu et al., 2004). This means that the 3-D air quality models are frequently making errors on the order of ± 50% in the simulations of SO_4^{2-} and NH_4^+. To test how much the errors in SO_4^{2-} and TNH_4 associated with predictions from a 3-D air quality model such as CMAQ will affect the predictions of aerosol NO_3^- in the thermodynamic model, sensitivity-case concentrations (C_s) of SO_4^{2-} and TNH_4 are generated by adding independent Gaussian (normally distributed) random errors to their base-case concentrations (C_b) as follows:

$$\ln(C_s) = \ln(C_b) + \epsilon, \epsilon \sim G(0, \sigma = RMSE) \tag{2}$$

where ε represents Gaussian random errors with zero mean and standard deviation equal to the *RMSE*, the root mean square error. The *RMSE* used in this study is obtained from comparisons of the paired 3-D model predictions and observations for each species (Yu et al., 2004). The comparison of predictions of aerosol NO_3^- between the sensitivity-case and the base-case is shown in Figure 3 and summarized in Table 2. The equilibrium models with the 3-D air quality model-derived random errors in SO_4^{2-} and TNH_4 can only predict < 50% and < 62% of aerosol NO_3^- within a factor of 1.5 and 2, respectively, as shown in Table 2, although the modeled means are close to the observations. For ISORROPIA in Table 2, 47% and 60% of the NO_3^- predictions from the sensitivity cases are within a factor of 2 of the base case for Atlanta and Pittsburgh, respectively. This study suggests that a large source of error in predicting aerosol NO_3^- stems from the errors in 3-D model predictions of SO_4^{2-} and TNH_4 for the Eastern U.S. Table 2 and Figure 3 also indicate that errors in TNH_4 are more critical than errors in SO_4^{2-} to prediction of NO_3^- and that the higher the NO_3^- concentration, the less sensitive the predicted NO_3^- concentrations are to the errors in SO_4^{2-} and TNH_4. These results indicate that the ability of 3-D models to simulate aerosol NO_3^- concentrations is limited by uncertainties in predicted SO_4^{2-} and TNH_4.

Additional studies were carried out for the comparison of sensitivity-case NO_3^- for single relative fixed errors of ±10% individually in temperature and RH with those of the base-case in the summer and winter times. In contrast to large effects from the errors in SO_4^{2-} and TNH_4, the responses of the aerosol NO_3^- predictions are less sensitive to errors in temperature and RH. Generally, both models can reproduce a majority of the aerosol NO_3^- data points within a factor of 1.5 if there are only ±10% errors in temperature and RH, especially for the winter times, with somewhat more sensitivity to errors in RH. However, ± 20% errors in both

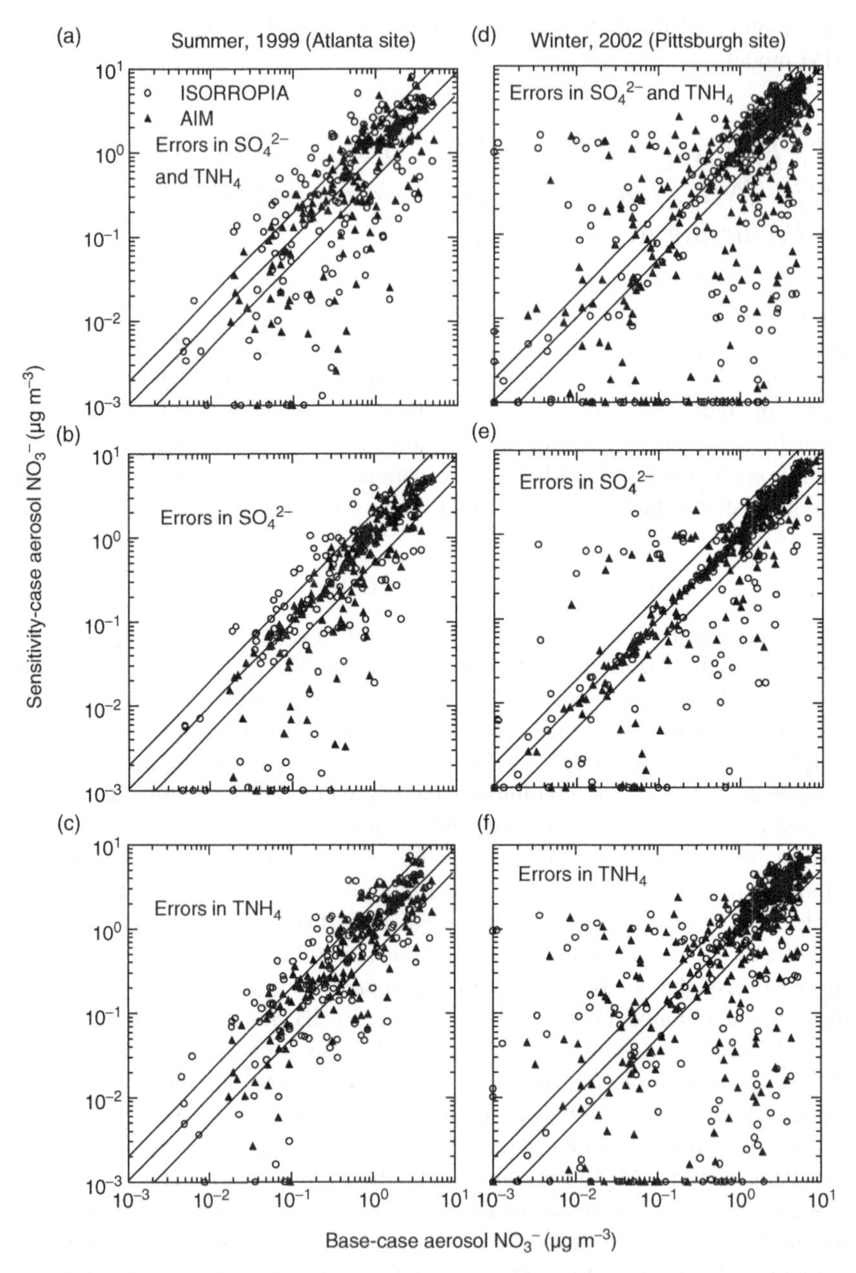

FIGURE 3. Sensitivity-case NO_3^- with assumed Gaussian random errors in SO_4^{2-}, and/or TNH_4 vs. the base-case NO_3^- for AIM and ISORROPIA for summer case ((a), (b), (c)) and winter case ((d), (e), (f)).

TABLE 2. Statistical summaries of the comparisons of the modeled (ISORROPIA and AIM) aerosol NO_3^- for the different sensitivity cases created by the Gaussian random errors (see text explanation) vs. those of the base cases on the basis of observational data at the Atlanta Supersite (summer case) and Pittsburgh Supersite (winter case).

	<C>*			% within a factor of 1.5**		% within a factor of 2**	
Condition	Base-case	ISORROPIA	AIM	ISORROPIA	AIM	ISORROPIA	AIM
Atlanta data (N=163)							
Errors in SO_4^{2-} and TNH_4	0.99±1.12	1.11±1.38	1.11±1.34	30.1	40.5	47.2	62.6
Errors in SO_4^{2-}	0.99±1.12	1.03±1.26	1.05±1.22	43.6	58.9	59.5	71.2
Errors in TNH_4	0.99±1.12	1.10±1.35	1.12±1.30	34.4	42.3	54.6	68.1
Pittsburgh Data (N=312)							
Errors in SO_4^{2-} and TNH_4	2.00±1.72	1.80±1.84	1.80±1.85	48.4	48.1	60.3	60.3
Errors in SO_4^{2-}	2.00±1.72	1.93±1.78	1.91±1.82	70.2	75.6	77.6	84.3
Errors in TNH_4	2.00±1.72	1.81±1.84	1.82±1.86	48.1	46.8	61.2	61.9

* Same as Table 1

temperature and RH can result in neither model being able to reproduce a majority of aerosol NO_3^- data points within a factor of 1.5 (percentage < 42%) (not shown) although both models can still capture 53–69% of aerosol NO_3^- within a factor of 2 in the summer case. For the winter case, the predicted aerosol NO_3^- is much less sensitive to errors in temperature and RH. This may be due to the fact that temperatures in the winter times are very low (3.9 ± 5.9°C), and most of TNO_3 concentration is in the aerosol phase. This is generally in agreement with Takahama et al. (2004), who found that errors in temperature measurements do not contribute significantly to model errors when temperatures are low and most of the nitrate concentration is in the aerosol phase.

4. Summary

The capability of thermodynamic models to reproduce the observed partitioning of TNO_3 and TNH_4 between gas and aerosol phases differed from site to site depending on chemical and meteorological conditions at the site. For example, at the Atlanta site, for NH_4^+ 94% and 96% of ISORROPIA and AIM predictions are within a factor of 1.5 of observations, respectively. For HNO_3, 86 and 87% of ISORROPIA and AIM predictions are within a factor of 1.5 of observations. However, neither model reproduced a majority of observed aerosol NO_3^- and gas NH_3 within a factor of 2 (NO_3^-: < 48% and NH_3: < 51%) at the Atlanta site. At the Pittsburgh site, both models can predict a majority of NO_3^- data points within a factor of 2 (> 76%), especially when NO_3^- concentrations are higher than 1.0 μg m^{-3} (> 89%), whereas both models perform more poorly on HNO_3 than at the Atlanta site. At the Clinton site, both models reproduce observed NH_3 concentrations very well (>95% within a factor of 1.5), and perform a little better on aerosol NO_3^- (47-58% within a factor of 1.5) than at the Atlanta site but worse than at the Pittsburgh site. The different chemical and meteorological conditions at the three sites can explain why both models perform differently in partitioning of TNO_3 and TNH_4. There are many different possible reasons for the discrepancies between the models and observations in partitioning of TNO_3. The sensitivity test indicates that in many cases measurement uncertainties in SO_4^{2-} and TNH_4 can explain a major fraction of the discrepancies between the model predictions and observations in partitioning of TNO_3. Sensitivity tests show that random errors associated with SO_4^{2-} and TNH_4 predictions of the 3-D model can result in the thermodynamic model calculation replicating only 47% and 60% of the base case NO_3^- within a factor of 2 for summer and winter cases, respectively. This suggests that a large source of error in predicting aerosol NO_3^- stems from the errors in 3-D model predictions of SO_4^{2-} and TNH_4 for the Eastern U.S. It was found that errors in TNH_4 are more critical than errors in SO_4^{2-} to prediction of NO_3^- and that the responses of the aerosol NO_3^- predictions are not very sensitive to the errors in temperature and RH under the tested conditions. The ability of 3-D models to simulate aerosol NO_3^- concentrations is limited by uncertainties in predicted SO_4^{2-} and TNH_4. While there is feedback between partitioning and the levels of predicted TNO_3, errors in TNO_3 are much less sensitive to these uncertainties and 3-D models are capable of predicting TNO_3 with accuracy comparable to that of SO_4^{2-} or TNH_4.

References

Byun, D. W. and J. K. S. Ching, Eds.,: 1999. Science algorithms of the EPA Models-3 Community Multi-scale Air Quality (CMAQ) modeling system, EPA/600/R-99/030, Office of Research and Development, U.S. Environmental Protection Agency.

Clegg, S. L., P. Brimblecombe, and A. S. Wexler, 1998. A thermodynamic model of the system H^+-NH_4^+-$SO_4^{2-}$$NO_3^-$$H_2O$ at tropospheric temperatures. *J. Phys. Chem.*, A, 102, 2155-2171.

Nenes, A., C. Pilinis, and S. N. Pandis, 1999. Continued development and testing of a new thermodynamic aerosol module for urban and regional air quality models. *Atmos. Environ.*, 33, 1553-1560.

Odman, M. T., J. W. Boylan, J. G. Wilkinson, A. G. Russell, S. F. Mueller, R. E. Imhoff, K. G. Doty, W. B. Norris, and R. T. McNider, 2002. SAMI Air Quality Modeling, Final Report, Southern Appalachian Mountains Initiative, Asheville, NC.

Pun, B., C. Seigneur, S-.Y. Wu, E. Knipping, and N. Kumar, 2004. Modeling Analysis of the Big Bend Regional Aerosol Visibility Observational (BRAVO) Study, Final Report 1009283, EPRI, Palo Alto, CA.

Robarge, W. P., J. T. Walker, R. B. McCulloch, and G. Murray, 2002. Atmospheric concentrations of ammonia and ammonium at an agricultural site in the southeast United Sates. *Atmos. Environ.*, 36, 1661-1674.

Solomon, P. A., K. Baumann, E. Edgerton, R. Tanner, D. Eatough, W. Modey, H. Marin, D. Savoie, S. Natarajan, M. B. Meyer, and G. Norris. 2003. Comparison of Integrated Samplers for Mass and Composition During the 1999 Atlanta Supersites Project. *J. Geophys. Res.*, 108(D7), 8423, doi:10.1029/2001JD001218.

Takahama, S., B. Wittig, D. V. Vayenas, C. I. Davidson, and S. N. Pandis, 2004. Modeling the diurnal variation of nitrate during the Pittsburgh Air Quality Study, *J. Geophys. Res.*, 109, D16S06, doi:10.1029/2003JD004149.

Wittig, B., S. Takahama, A. Khlystov, S. N. Pandis, S. Hering, B. Kirby, and C. Davidson, 2004. Semi-continuous PM2.5 inorganic composition measurements during the Pittsburgh Air Quality Study. *Atmos. Environ.*, 38, 3201-3213.

Weber, R. J., et al., 2003. Intercomparison of near real-time monitors of $PM_{2.5}$ nitrate and sulfate at the Environmental Protection Agency Atlanta Supersite. *J. Geophys. Res.*, 108, 8421, doi:10.1029/2001JD001220.

Wexler, A. S., and S. L. Clegg, 2002. Atmospheric aerosol models for systems including the ions H^+, NH_4^+, Na^+, SO_4^{2-}, NO_3^-, Cl^-, Br^-, and H_2O, *J. Geophys. Res.*, 107 (D14), doi:10.1029/2001JD000451.

Yu, S. C., R. Dennis, S. Roselle, A. Nenes, J. Walker, B. Eder, K. Schere, J. Swall, W. Robarge, 2004 (in press). An assessment of the ability of 3-D air quality models with current thermodynamic equilibrium models to predict aerosol NO_3^-. *J. Geophys. Res.*

A Test of Thermodynamic Equilibrium Models and 3-D Air Quality Models for Predictions of Aerosol NO_3^-

Speaker: S. Yu

Questioner: P. Bhave

Question: Why is the performance of the equilibrium partitioning modules consistently better in the winter season than in the summer?

Answer: This is because the aerosol NO_3^- concentrations in the winter season are much higher than those in the summer and the aerosol NO_3^- fraction is dominant in the winter season.

Questioner: P. Builtjes

Question: I was surprised by the error of only 15% in the observations of aerosol nitrate. Could you comment on this?

Answer: I guess that Dr. P. Builtjes tried to ask the error of ±15% in the observations of aerosol SO_4^{2-} and TNH_4 (total ammonium). An error of ±15% is used to correspond with the measurement uncertainty for both SO_4^{2-} and TNH_4 that was estimated as part of the U.S. EPA supersite program [Solomon et al., Comparison of Integrated Samplers for Mass and Composition During the 1999 Atlanta Supersites Project. J. Geophys. Res., 108(D7), 8423, doi:10.1029/2001JD001218, 2003].

Questioner: D. Steyn

Question: I am concerned by your use of log-log scatter plots combined with linear measurement of agreement (plus and minus % error). Furthermore I believe if you use linear scatter plots you will decide that both models have zero skill and therefore cannot be ranked by skill.

Answer: The reason for our use of the log-log scale is because there is substantial linear correlation between the 3-D model predictions and observations on a log scale for both SO_4^{2-} and NH_4, i.e., correlation coefficient > 0.67, according to our model evaluation results. The root mean square error (RMSE) between the

3-D model predictions and observations is uniform on a log scale. We use these RMSEs to create a Gaussian random sample set of data that reproduce the pattern of scatter between the 3-D model predictions and observations. Then we transform these data to the linear scale for input into the equilibrium models to assess the effect of the uncertainties in the sensitivity tests.

Disclaimer The research presented here was performed under the Memorandum of Understanding between the U.S. Environmental Protection Agency (EPA) and the U.S. Department of Commerce's National Oceanic and Atmospheric Administration (NOAA) and under agreement number DW13921548. Although it has been reviewed by EPA and NOAA and approved for publication, it does not necessarily reflect their policies or views.

New Developments

39 Comparison of Aggregated and Measured Turbulent Fluxes in an Urban Area

Ekaterina Batchvarova[1,2], Sven-Erik Gryning[3], Mathias W. Rotach[4], and Andreas Christen[5]

1. Introduction

Most of the parameters commonly used to describe the turbulence in the atmospheric boundary layer represent conditions of atmospheric turbulence near the ground and therefore have a small footprint. For a mast of 10 meters above ground the footprint is few hundred meters. A 40-50 meter mast looks over some kilometers in upwind direction. The height of the convective boundary layer is typically 1-2 kilometers in middle latitudes and reflects the conditions several tens of kilometers upwind. The footprint of meteorological characteristics is essentially dependant on atmospheric stability and wind speed (Gryning and Batchvarova, 1999; Kljun et al., 2003). The mixing height determination based on surface data is much more demanding in terms of requirements of homogeneous conditions because it overlooks a large area. It is still commonly accepted practice to use parameterisations for homogeneous terrain, even under inhomogeneous conditions. The results are not always good when compared to measurements.

The sub-grid variability of the mixed-layer height, momentum and sensible heat fluxes is believed to be an important but not yet settled issue for many model applications. Models for long-range transport and dispersion of atmospheric pollutants, Numerical Weather Prediction models, and Climate Models are all characterised by large grid cells that often enclose regions of pronounced non-homogeneities. The estimation of the regional or aggregated momentum and heat fluxes over non-homogeneous surfaces is therefore a central issue when the boundary conditions in models with large grid cells have to be specified.

The size of the grid determines the features that models can resolve. The larger the grid the more of the small scale features in the flow field will not be resolved.

[1] National Institute of Meteorology and Hydrology, Sofia, Bulgaria (permanent);
[2] Hertfordshire University, Hatfield, UK (visiting November 2003–June 2004);
[3] Risø National Laboratory, Roskilde, Denmark;
[4] Swiss Federal Office for Meteorology and Climatology, MeteoSwiss, Zürich, Switzerland;
[5] University of Basel, Institute of Meteorology, Climatology and Remote sensing, Basel, Switzerland.

A multitude of well established models have their roots in parameterisations of surface fluxes that have been developed for homogeneous terrain. The effect of their use in simple aggregation schemes over typical European areas characterized by a patchy structure of villages, towns, various types of agricultural fields and forests is not clear. Consequently the estimation of the regional or aggregated momentum and heat fluxes over non-homogeneous surfaces is a central issue when the boundary conditions in models with large grid cells have to be specified. Here we consider the heterogeneity of an urban area and its surroundings. Therefore the complex structure of the urban boundary layer is sketched in Figure 1.

In this paper we consider the use of boundary layer parameterisations for homogeneous conditions when applied for heterogeneous land cover areas. We consider the actual (measured) convective boundary layer height as formed by the forcing of the surface conditions tens of kilometers upwind. We use a method that is based on the height and the growth rate of the convective layer (Gryning and Batchvarova, 1999) to estimate aggregated surface sensible heat flux values. Further we use in situ turbulence measurements at 40 meters height on a urban meteorological tower to compare the results.

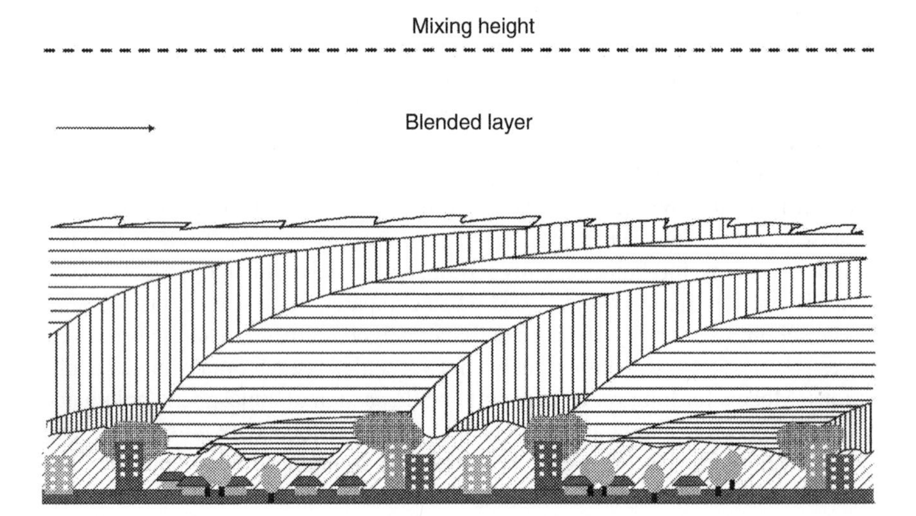

FIGURE 1. Schematic of the boundary layer structure over an urban area. The vertical and horizontal patterns represent the underlying surface of the neighbourhoods of tall and low buildings, respectively. Broad spaced patterns represent the urban internal boundary layers where advection processes are important. Fine spaced patterns show the inertial sub-layers that are in equilibrium with the underlying surface and where Monin-Obukhov scaling applies. The forward slash pattern represents the roughness sub-layer that is highly inhomogeneous both in its vertical and horizontal structure. The dotted pattern represents adjustment zones between neighbourhoods with large accelerations and shear in the flow near the top of the canopy. Above the height where the internal boundary layers are intermixed the effects of the individual neighbourhoods cannot be distinguished any more – the so-called blended layer.

2. The Method

The required information for use of the method to aggregate fluxes can be derived from wind speed and temperature profiles obtained by radio soundings when performed frequently enough to provide a reasonably detailed structure of the mixed-layer development. Alternatively, data from remote sensing techniques like combined wind profiler and radio acoustic sounding systems can be used.

The method was used first over a sub-Arctic area with rather large patches of forest, fields, mires and lakes. The aggregated heat flux was found to be in general agreement with the land-use-weighted average heat flux. Thus, rough aggregation of the heat flux is simple, but it is not known if this can be extended to small patches as in an urban area. Batchvarova et al. (2001) and Gryning and Batchvarova (2002) show that this is not the case for the momentum flux. It is desirable to continue to investigate the blending of fluxes over patchy terrain, and extend the research to include aggregation for the urban area that today is virtually unknown. The urban area is far more complicated that the forest (Figure 1) and research in the formation of the local fluxes and the way they interact and finally forms the blending layer is very relevant for a better understanding of the urban climate (Rotach et al., 2002).

The framework is the mixed-layer growth model by Batchvarova and Gryning (1991) and (1994).

$$\left\{\frac{h^2}{(1+2A)h - 2B\kappa L^{eff}} + \frac{C(u_*^{eff})^2 T}{\gamma g\left[(1+A)h - B\kappa L^{eff}\right]}\right\}\left(\frac{dh}{dt} - w_s\right) = \frac{\left(\overline{w'\theta'}\right)_s^{eff}}{\gamma}. \quad (1)$$

The potential temperature gradient, γ in the free atmosphere, the large scale vertical air velocity, w_s, the height of the boundary layer, h its growth rate, dh/dt and the rate of warming, $\partial\theta/\partial t$ of the free atmosphere above the mixing layer can be extracted from i.e. radio soundings. The warming rate at height z in the free atmosphere is connected to the vertical velocity w_s and the potential temperature gradient at the same height through:

$$\frac{\partial\theta(z)}{\partial t} = -\gamma\, w_s(z). \quad (2)$$

Equation (1) can be solved numerically for the effective vertical turbulent kinematic sensible heat flux at the surface, $\left(\overline{w'\theta'}\right)_s^{eff}$, that is forcing the growth of the boundary layer. The coupled momentum and heat flux solution is discussed in Batchvarova et al. (2001).

A comprehensive discussion on the applicability of this approach is given in Gryning and Batchvarova (1999). Based on footprint analysis, the blending height hypothesis and internal boundary layer theory, the height at which the surface heterogeneities do not influence the flow any more can be determined. For typical values of the meteorological parameters and landscape patchiness, this height is 100-200 meters. The method is applicable when the mixed layer is deeper than the layer where the flow features are blended, the so-called blending height.

3. The Sofia Experiment

Data drawn from a recent urban boundary layer experiment (September/October 2003) in Sofia, Bulgaria, that comprised high resolution boundary layer radiosoundings to determine the mixing height and mixed layer growth and measurements with sonic anemometers at two heights in a sub-urban area of the city of Sofia are used for the analysis. The two sonic anemometers and a fast hygrometer were mounted on the research tower of NIMH at 20 and 40 m height agl (above ground level) and 10 and 30 m above roof, respectively on booms 4 m away of the tower in direction west, Figure 2.

High resolution radiosoundings were performed with Vaisala equipment. Typical convective conditions were chosen for the campaign when 7 soundings per day were performed providing data for the growth of the mixed layer with 2 hours time interval. Five days with such conditions were identified in the period 18 September–8 October 2003.

Following the understanding of aggregation of fluxes over heterogeneous areas, (Gryning and Batchvarova, 1999 and Batchvarova et al., 2001) the observed convective boundary layer was considered forced by the blended thermal and mechanical fluxes over the area. The measuring site (NIMH) is in the south east part of the city and the urban characteristics for it are spread over 3 km to south and east, about 10 km to north and about 20 km to the west. Depending on wind direction, the aggregated fluxes represent different percentages of urban and rural conditions, Figure 3.

4. Results of Modelling

In Figure 4 the comparison of measured and aggregated sensible heat fluxes is presented (upper left panel) for 29 September 2003. The aggregated and measured values are close, suggesting that blended fluxes are representing urban conditions

FIGURE 2. The meteorological tower with sonic anemometers and a Krypton hygrometer (left) and a view from the tower to the west (right).

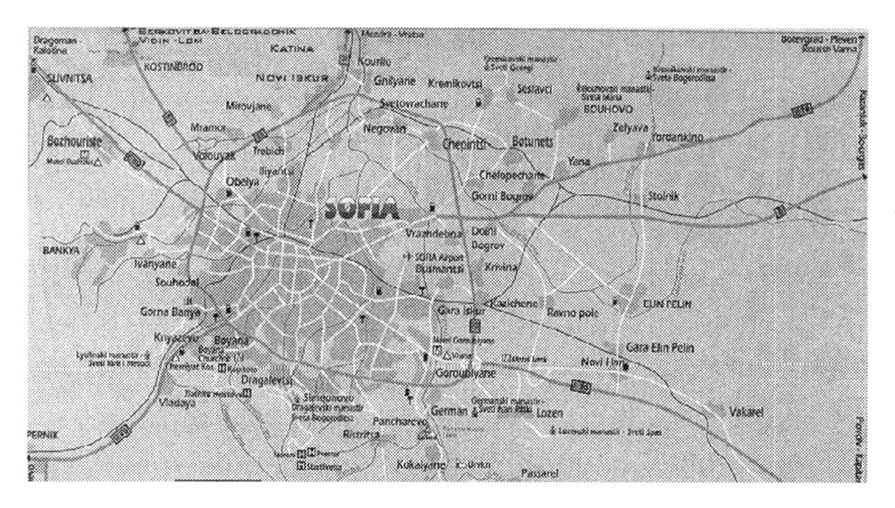

FIGURE 3. Map of Sofia and close rural areas (56 by 28 km approximately). The position of the measurement tower at NIMH is marked.

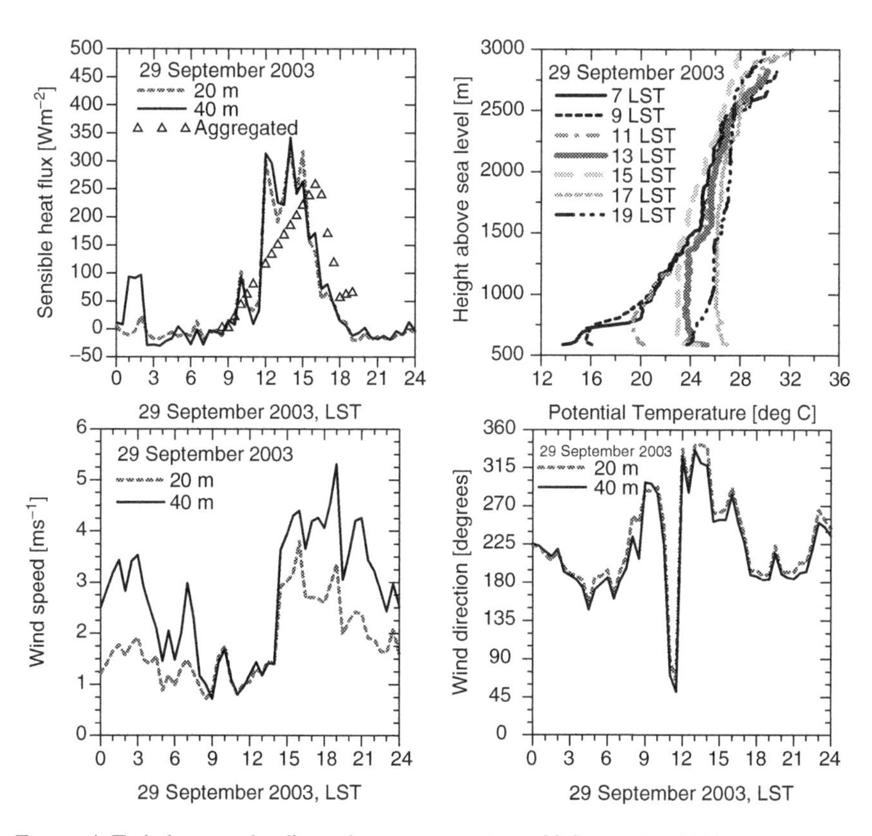

FIGURE 4. Turbulence and radiosonde measurements on 29 September 2003.

for western and northwestern weak winds, Figure 4 (lower panels). At 13.00 local summer time (LST) the mixed layer-height reaches some 1000 meters above the ground (Figure 4, upper right panel).

On 1 October 2003 the aggregated fluxes are smaller than those measured on the tower (Figure 5 upper panel). On this day the wind is easterly (Figure 5, lower panel), thus leading to the conclusion that the footprint of the profile observations contains large rural areas in the upwind direction (Figure 3). These are apparently influencing the blended (i.e., modeled) values as obtained from the soundings. On the other hand, the tower remains within the roughness sub-layer of the urban atmosphere and exhibits larger (urban only) surface heat fluxes. The depth of the convective boundary layer is again about 1000 meters above ground, Figure 5.

It can be seen from the upper left panels of Figures 4 and 5 that the measured heat fluxes at 40 meters are slightly higher as compared to the 20 meters level. This feature is connected to the structure of the urban surface layer, Figure 1, and is discussed on data from several experiments such as BUBBLE (Rotach et al. 2004), Copenhagen (Gryning and Lyck 1984) and Sofia in Batchvarova et al. (2004), Batchvarova and Gryning (2004 & 2005).

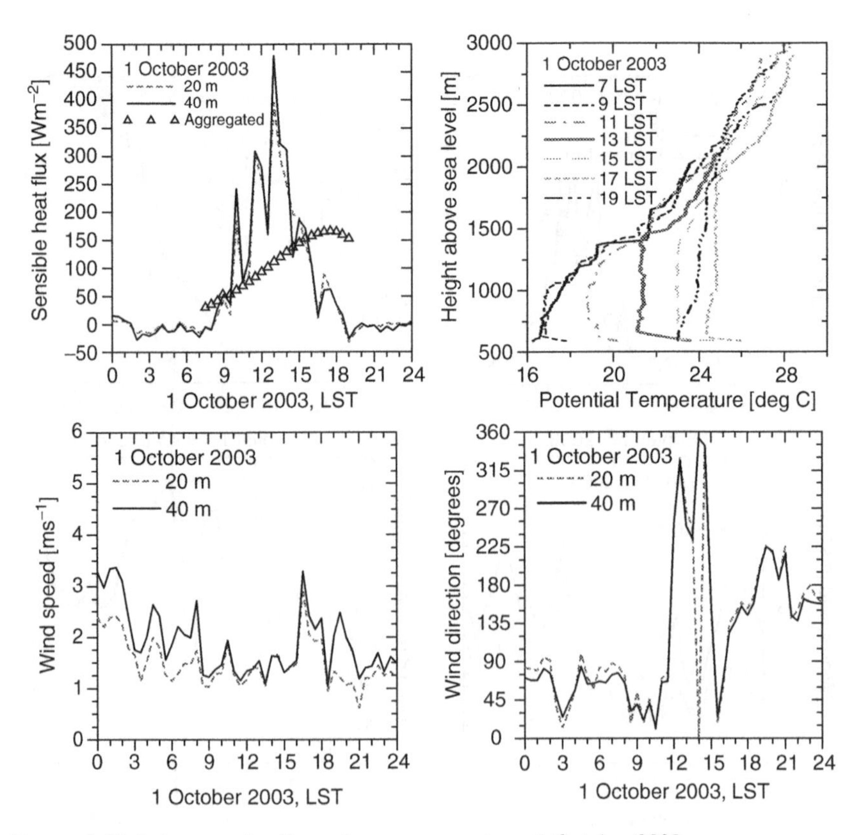

FIGURE 5. Turbulence and radiosonde measurements on 1 October 2003.

5. Discussion

One immediate implication of the results presented here is the caution needed when applying pre-processors based on local measurements in urban areas. In the case of 1st October 2003 the use of the turbulence measurement for estimation of the mixing height with the use of a meteorological pre-processor will give a much deeper mixed layer than is actually observed. Under such circumstances air pollution concentrations could be under-predicted. The representativeness of measuring sites is wind direction dependant. The same considerations are valid if mesoscale model results are compared to measurements. The area over which the modelled parameters are averaged can be different from that represented by measurements. Considering the representativeness of airport stations for urban meteorological and air pollution studies is therefore a complex issue.

Acknowledgements The study reports results from ongoing work that aims further understanding of the complex layer structure and turbulence regime over a city and is a part of Swiss-Bulgarian collaboration project (7IP 065650), COST 715 Working group 1 and a NATO CLG grant (979863). The Sofia experiment was also in the frame of the BULAIR Project EVK2-CT-2002-80024. The input in the experimental work of Ivan Lanzov, Plamen Videnov, Nedialko Valkov and Alexander Gamanov, researchers from NIMH is kindly acknowledged.

References

Batchvarova, E. and Gryning, S.-E., 1991, Applied model for the growth of the daytime mixed layer. Boundary-Layer Meteorol, **56** , 261-274.

Batchvarova, E. and Gryning, S.-E., 1994, An applied model for the height of the daytime mixed layer and the entrainment zone. Boundary-Layer Meteorol. 71, 311-323.

Batchvarova, E., Gryning, S.-E., Hasager, C. B., 2001, Regional fluxes of momentum and sensible heat over a sub-arctic landscape during late winter, Boundary-Layer Meteorol. 99, 489-507.

Batchvarova, E., Gryning, S. E., 2004, Advances in the modelling of meteorology in urban areas for environmental applications. NATO advanced research workshop: Advances in air pollution modelling for environmental security, Borovetz (BG), 8-12 May 2004, Kluwer Academic, In Press.

Batchvarova, E., Gryning, S.-E., Rotach, M. W., Christen, A., 2004, Modelled aggregated turbulent fluxes compared to urban turbulence measurements at different heights. 9th International conference on harmonisation within atmospheric dispersion modelling for regulatory purposes, Garmisch-Partenkirchen (DE), 1-4 Jun 2004. Suppan, P. (ed.), In: Proceedings. Vol. 2. 7-11.

Batchvarova, E. and Gryning, S.-E., 2005, Progress in urban dispersion studies, Theoretical and Applied Climatology. In press.

Gryning, S.-E., Lyck, E., 1984, Atmospheric Dispersion from Elevated Sources in an Urban Area: Comparison between Tracer Experiments and Model Calculations. J. Climate Appl. Meteorol. 23, 651-660.

Gryning S.-E. and Batchvarova E., 1999, Regional heat flux over an inhomogeneous area estimated from the evolution of the mixed-layer, Agricultural and Forest Meteorology, 98-99, 159-168.
Gryning, S.-E. and Batchvarova, E., 2002, Mixing heights in urban areas: will 'rural' parameterizations work?, in Rotach, M.W., B. Fisher and M. Piringer Eds, 2002: Workshop on urban boundary-layer parameterizations, Zurich, Switzerland, 24-25 May 2001, EUR 20355, 99-109.
Kljun, N., Calanca, P., Rotach, M. W. and Schmid, H. P., 2003, A simple parameterisation for flux footprint predictions, Boundary-Layer Meteorology 112, 503- 523.
Rotach, M. W., Fisher, B., and Piringer, M., Eds, 2002, Workshop on urban boundary-layer parameterizations, Zürich, Switzerland, 24-25 May 2001, EUR 20355, 119p.
Rotach M. W., Gryning S.-E., Batchvarova E., Christen A. and Vogt R., 2004, Pollutant dispersion close to an urban surface - BUBBLE Tracer experiment, Journal of Meteorology and Atmospheric Physics, **87**, 39-56.

Comparison of Aggregated and Measured Turbulent Fluxes in an Urban Area

Speaker: E. Batchvarova

Questioner: H. Thielen
Question: What is the right sampling interval for the measurement of friction velocity and turbulent heat flux to be in line with similarity theory?
Answer: The sampling frequency was 40 Hz and the averaging time for the study was 30 minutes. Half-hour averaging is considered appropriate for turbulence measurements in urban environments.

Questioner: T. Newton
Question: How did you calculate the size of the footprints of your sonic anemometers? Do you assume that the footprints are circular?
Answer: Only rough estimates were elaborated here based on detailed analysis in Gryning and Batchvarova 1999 (Gryning S.-E. and E. Batchvarova, 1999: Regional heat flux over an inhomogeneous area estimated from the evolution of the mixed-layer. Agricultural and Forest Meteorology, 98-99, 159-168).

Questioner: T. Oke
Question: Could you please confirm whether the height of measurement is 20 and 40 above ground, or it is it above mean roughness element height (roof-level and tice height)? Obviously this is relevant together with the element spacing to the height of the roughness sub layer.
It would also be helpful to know the direction of the instrument booms to know if the fairly substantial tower has any effect on the fluxes to observed with SE flows.
Answer: The building is 10 m high and the tower is 30 m on top. Yes, in the study we refer to 20 m and 40 m above ground. The booms were stretched 3.5 m away from the tower towards west. The flow from the city is not disturbed therefore. SE and E flows might be disturbed by the tower construction, while NE as well as all Westerly flows should be undisturbed. Specific analysis has not been done so far.

40
Ensemble Dispersion Modeling: "All For One, One For All!"

Stefano Galmarini*

1. Introduction

Almost every country has adopted modeling systems to forecast the consequences of atmospheric dispersion at various scales. In particular long range transport and dispersion (LRTD) models are used to forecast the dispersion of large emissions of harmful pollutants from point sources such as, for example, atmospheric dispersion of radioactive gasses from NPP or other sources.

The comparison of state-of-the-art model results with observations (e.g. Draxler, 1983; Klug *et al.*, 1992; Girardi *et al.*, 1998) has shown unequivocally that among the various approaches to atmospheric dispersion modeling, none is systematically performing better than others. Two are the main sources of uncertainty in the model results: one connected to the atmospheric circulation forecast and one dependent on the way in which atmospheric dispersion has been modeled. Therefore none of the models evaluated can be identified as "the model" to be used. On the other hand, model results are used for the assessment or forecast of conditions that may involve adoption of countermeasures for the protection of the population, for which a high level of accuracy is required.

A way to reduce the uncertainty of a single deterministic prediction and to better exploit dispersion model forecasts for practical applications, is to use the so-called ensemble dispersion modeling (EDM). Ensemble modeling (used in this context in the general sense of combination of several model results), has lately been adopted in different forms also for atmospheric dispersion. EDM has already been applied to air quality modeling (Delle Monache and Stull, 2003), and point source dispersion (Galmarini *et al.*, 2001; Straume, 2001; Warner *et al.*, 2002; Draxler, 2002). Within this context, multi-model ensemble dispersion forecasting is presented as a new approach to the use of LRTD models. The approach consists of the treatment and analysis of several long range dispersion forecasts and on the use of ad-hoc ensemble statistical parameters to determine the level of

* Dr. Stefano Galmarini, Institute for Environment and Sustainability, European Commission, Joint Research Center, I-21020, Italy. e-mail: stefano.galmarini@jrc.it

uncertainty associated with the forecast. An evaluation of the technique is presented that is based on the ETEX-1 tracer measurements as the term of comparison. The concept of *Median Model* will be introduced as the synthesis of the ensemble dispersion methodology. The concept developed for LRTD models can obviously be generalized to other types of dispersion models and atmospheric scales.

2. Ensemble Technique for LRTM

The multi-model ensemble dispersion technique (Galmarini *et al.*, 2001) consists of the analysis of a number of dispersion forecasts produced for the same case study by models that are different from one another and that make use of atmospheric circulation forecasts produced by different Numerical Weather Prediction (NWP) (Table 1). For point dispersion at the continental scale, all models simulate the same release conditions and produce a forecast over a period of time (60 h).

The simultaneous analysis of all the operational dispersion forecasts and the use of specific statistical parameters allow a determination of the level of agreement of the models and the relative uncertainty of the model results. Since the simulations relate to forecast, the uncertainty can only be relative to the model

TABLE 1. List of institutes, dispersion models, dispersion model types and NWP Participating to the ENSEMBLE activity. E = Eulerian, L = Lagrangian, P = Particle

Institute	Dispersion Model	Type	NWP
Canadian Meteorological Centre	CANERM	E	GEM Global
Danish Meteorological Institute	DERMA	L-P	DMI-HIRLAM-E
	DERMA	L-P	DMI-HIRLAM-G
	DERMA	L-P	ECMWF
Deutscher Wetterdienst	GME-LPDM	L	DWD-GME
	LM-LPDM	L	DWD-LM
Finnish Meteorological Institute	SILAM	L	FMI-HIRLAM
KMI (B)	BPaM4D	L	ECMWF
KNMI (NL)	NPK-PUFF	L	ECMWF and HIRLAM
Meteo-France	MEDIA	E	ARPEGE
	MEDIA-nested	E	ARPEGE-ALADIN
Met Office (UK)	NAME	L	UM
RIVM (NL)	NPK-PUFF	L	HIRLAM and ECMWF
Nat. C. Scient. Res. "Demokritos" (G)	DIPCOT	L	ECMWF
	DIPCOT	P	ECMWF
Nat. Inst. Atomic Energy Agency (PO)	RODOS-MATCH	L-E	DMI-HIRLAM
NMHI (BG)	EMAP	E	DWD-GME
Norwegian Meteorol. Institute	SNAP	L	HIRLAM
Risø National Laboratory (DK)	RODOS-LSMC	P	DMI-HIRLAM
	RODOS-MATCH	L-E	DMI-HIRLAM
Savannah River Tech. Center (US)	LPDM	L	RAMS3a
Swedish Meteorol. and Hydrol. Inst.	MATCH	L-E	HIRLAM
ZAMG (A)	TAMOS	L	ECMWF T319L50
	TAMOS	L	ECMWF T319L50

results as no absolute term for comparison exists. In more detail, results of the various models are collected in real-time by means of a web-based system called ENSEMBLE and described in Bianconi *et al.*, 2004. Upon notification of the release characteristics, each modeling group provides a forecast of the following variables: air concentration of a radio nuclide at 0, 200, 500, 1300, 3000 m a.g.l.; time integrated concentration at 0 m a.g.l.; cumulated dry deposition; cumulated wet deposition; precipitation as obtained from NWP. Once a number of model results are available they can be consulted and compared with one another or grouped according to common modeling features in order to determine their level of agreement. An example of ensemble dispersion treatment is given by the so-called Agreement in Threshold Level (ATL). Given an ensemble of M dispersion realizations of the same case and a concentration threshold C_T, ATL is defined as:

$$ATL(x,y,t) = \frac{100}{M}\sum_{k=1}^{M}\delta_k \quad where \quad \begin{cases} \delta_k = 1 & if \quad C_k(x,y,t) \geq C_I \\ \delta_k = 0 & otherwise \end{cases}$$

For a specific concentration value, ATL gives the percentage of models that predict the exceedance of the threshold and its spatial distribution. An example of application of ATL is given in Figure 1a. The figure relates to the simulation of a fictitious release of ^{137}Cs from Dublin (Ireland). The variable analyzed is time integrated concentration at surface 60h after the release. The figure shows the agreement distribution in predicting the exceedance of 10 $Bqhm^{-3}$. Darker colors indicated that several models predict an exceedance in that region.

The figure was produced by combining the results of 14 distinct dispersion models. As can be deduced from the figure, the models' agreement relates to the central part of the distribution whereas there is a portion of the domain where the threshold exceeding is not predicted in all models. ATL provides a clear indication of the area affected by the plume passage that includes the variability of the forecast produced. Moreover it is a quantitative estimate of the models' agreement which, given the different nature of the models and NWP used, is also

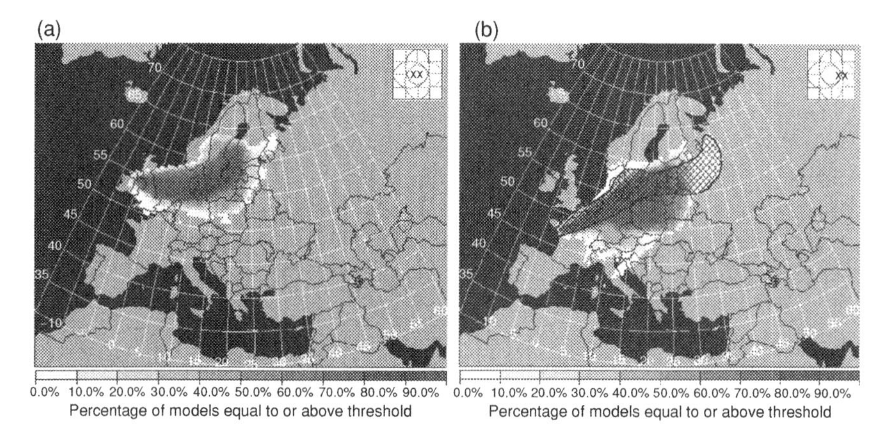

FIGURE 1. ATL distribution for two fictitious releases. See text for details.

an estimate of the forecast reliability. Another example of the relevance of EDM and of the ATL indicator is given in Figure 1b. The figure relates to the ATL calculated for a threshold value of 0.1 $Bqhm^{-3}$ of ^{137}Cs fictitiously released from Nantes (France) and forecasted by 10 models. In spite of a wide area of high model agreement in central Europe, in the Northern part of Italy, Switzerland, Southern Germany and Austria the figure shows an area where a reduced number of models predict an exceedance of the threshold. Furthermore a large area of low agreement is shown in the forefront of the cloud over Russia and the Baltic republics. As shown by the hatched surface, this last region is produced by a single model whose dispersing cloud travels much faster than the one forecasted by the others. Figure 1b is relevant for a series of aspects. First of all, in spite of the wider area of low agreement, any decision based on the ensemble forecast is based on a wide spectrum of scenarios. Secondly, the dispersion forecasts are less consistent with one another than the case of Figure 1a and therefore the reliability of the forecasts could be downgraded. Thirdly, the situation assessment that would be based on the single model producing the hatched surface would lead to a significant underestimation of the threshold exceedance in central Europe and probably to a significant overestimation of the exceedance in the Eastern part of the domain.

3. Evalution with the ETEX-1 Dataset

In order to evaluate the multi-model ensemble dispersion approach, the modeling groups taking part to the ENSEMBLE activities were asked to simulate the ETEX-1 tracer release performed in 1994 (Van Dop H. and K. Nodop, 1998; Girardi et. al., 1999). For the sake of synthesis we shall refer to existing publications for the details on the tracer experiment. The scope of the study is to verify to what extent the ensemble dispersion analysis provides more information compared to the single deterministic forecast and constitutes and improvement of the atmospheric dispersion prediction. In total 16 model results were analyzed in this exercise.

Figure 2a shows a comparison of the ATL obtained for the threshold value of 0.1 $nghm^{-3}$ of time integrated concentration, 24 hours after the release. The grayscale corresponds to the agreement among models in predicting the area of the domain where the threshold was exceeded whereas the hatched surface corresponds to the contour of measured tracer at the threshold level. The figure shows a clear correspondence between the area where the threshold was measured and the area of large model agreement (>60%). This result indicates that the combination of the model results through the ATL seems to be a reliable way to identify the region where the threshold was exceeded and that the single model result can be deceiving in this respect. In fact, the outer fringes of the cloud are predicted by a small number of models and do not overlap with the measured one. Another comparison of ATL with the measured values for a later stage of development of the plume and a higher threshold value (2 $nghm^{-3}$), is given in Figure 2b.

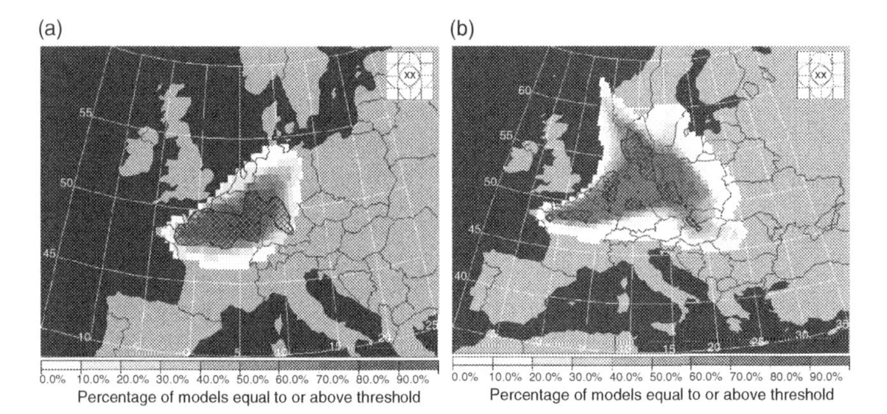

FIGURE 2. ETEX comparison. (a) Grayscale, ATL of time integrated concentration from 14 models, 24 hours after release. Threshold 0.1 nghm^{-3}; hatched contour measurements. (b) Same as (a) 48 hours after the release. Threshold 2 nghm^{-3}.

Provided a large number of model results relating to the same dispersion simulation, one may wonder what is the behavior of the combination of the results according to a specific statistical treatment. In this respect the following analysis has been performed. For the first 20 3-hourly time intervals of the ETEX-1 cloud evolution and at each grid cells of the domain (every 0.5×0.5 deg) the percentiles of the model distribution were calculated from the 15th to the 80th in order to obtain a new set of simulations given by the combination of all the model results. For each of the new model results we calculated the FA2, FA5 (number of model results-measurement couples falling within a factor 2 and 5 respectively), and FOEX (percentage of model results overestimating or underestimating the measurements) with respect to the ETEX-1 measurements. The results of the analysis are given in Figure 3. The figure clearly shows that the optimal results (maximum FA2 and FA5; minimum FOEX) are obtained for the percentiles falling between the 40th and the 50th. This result shows that when a new model result is constructed from the model ensemble using one of these percentiles, the best model performance is obtained. A demonstration of the consequence of using what we will call the *Median Model* (model results obtained by using the 50th percentile value at every grid point in space and time interval) is given in Figure 4. It shows the 0.01 ngm^{-3} contour 48 hours after the ETEX-1 release obtained from the measurements (white), the best performing model of the ensemble (dark gray) and the Median Model (light gray). As one can see the Median Model results are substantially better than those obtained with the best performing model (in this context the best performing model is the one that shows the best values of FA2, FA5 and FOEX in the global analysis).

Since the results of the Median Model are obtained by calculating the 50th percentile of the surface concentration produced by the model ensemble at every grid point and time step, it is important to determine to what extent all the models are

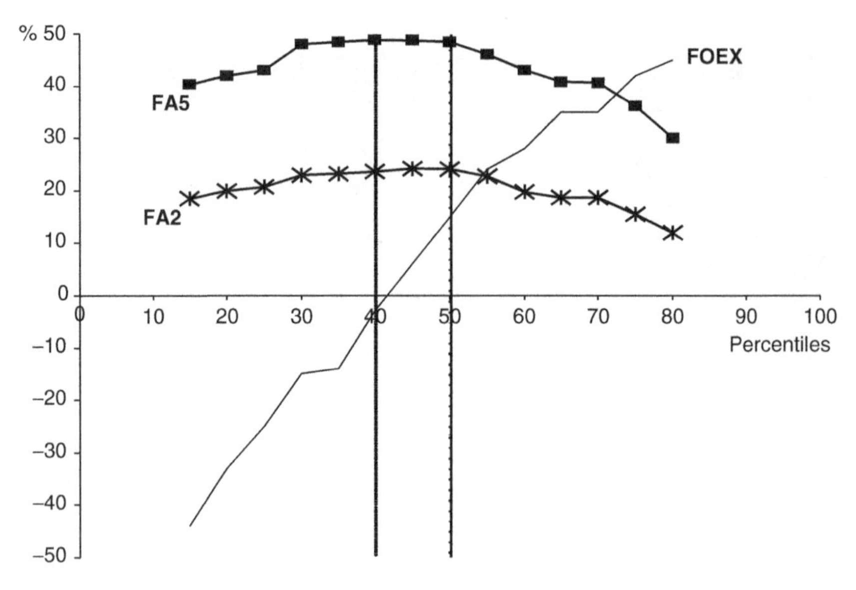

FIGURE 3. FA2, FA5 and FOEX calculated for different percentile values.

contributing in defining the Median Model results and therefore if all the ensemble members are relevant to the improvement of the simulation.

Figure 5 provides the number of times each of the 16 models contributes to the definition of the median value within a specific range of concentration values. With the only exception of m4 all the others do and therefore the Median Model

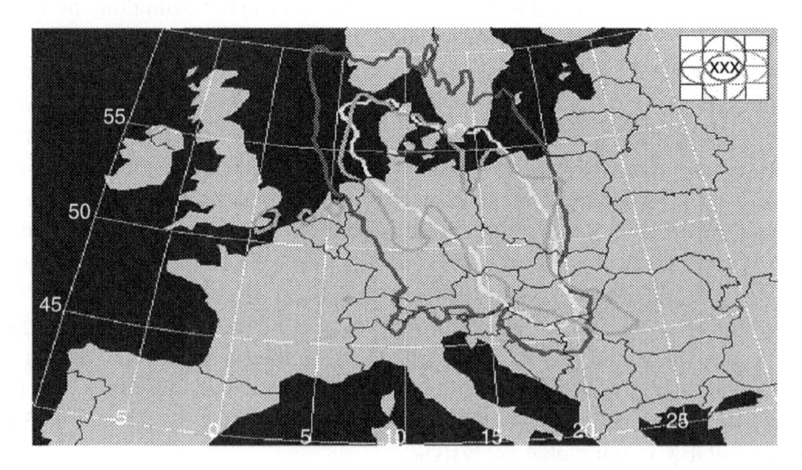

FIGURE 4. 0.01 ngm-3 contour 48 hours after the ETEX-1 release obtained from the measurements (white), the best performing model of the ensemble (dark gray) and the Median Model (light gray).

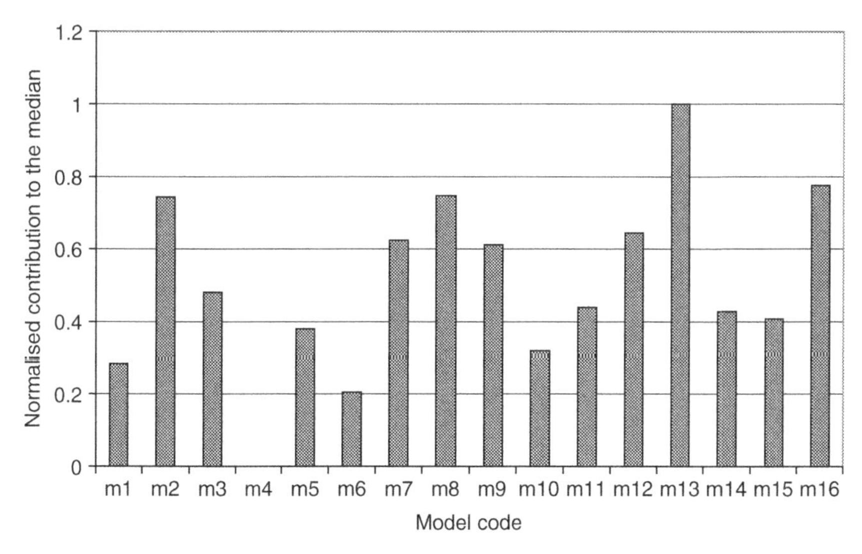

FIGURE 5. Number of times each of the 16 models contribute to the definition of the median value in every point of the domain and for every time interval of the simulation.

results do not coincide with the good performance of one or two models outperforming the others. The result of Figure 5 shows that all model results are relevant to the improved ensemble simulation and also that the model results are in a way complementary in the sense that the combination of the 16 model simulation produces a result that is closer to the measured cloud evolution.

4. Conclusions and Discussion

The paper presented the multi-model ensemble dispersion concept and validation. The analysis of several LRTD models applied to the same case study provides a great deal of information and a better use of the single deterministic model forecast. In particular the simple combination of the results in the so called ATL allows a definition of the level of agreement for the models to be made. Given the fact that the dispersion models are different from one another and use atmospheric circulation forecasts produced by various NWP's, ATL defines the level of reliability of the dispersion predictions.

By means of the measurements collected during the ETEX-1 tracer experiment the methodology was validated. The results obtained with the ATL indicator indeed show that the high agreement region corresponds to the actual location of the measured cloud. The so-called Median Model results were obtained. A comparison with the measurements shows that the it compares very well with the measured cloud and that its performance is superior to that of the best model of the ensemble. It was clearly demonstrated that all model simulations contribute to the definition of the Median Model results. The results obtained show that there is a clear advantage in using an ensemble of model realizations rather than the single model results. The statistical treatment of dispersion model simulation seems to improve considerably the actual result. In the case of dispersion forecast EDM provides a wider spectrum of scenarios and it is of better support to decision making.

What is left to be explained is why the Median Model should improve the results of the single models, whether this is valid for the ETEX case study or could be generalized. The methodology applied here to long range atmospheric dispersion can be extended to other scales or model types.

Acknowledgement This work is based on the results obtained within the ENSEMBLE Consortium (http://ensemble.ei.jrc.it) which is acknowledged. The ENSEMBLE project was supported by the European Commission DEG-RES Nuclear Fission Program.

References

Bianconi R., S. Galmarini and R. Bellasio, 2004, A WWW-based decision support system for the management of accidental releases of radionuclides in the atmosphere. *Environmental Modeling & Software*, 19, 401-411.

Delle Monache L. and R. B. Stull, 2003, An ensemble air-quality forecast over Europe during an ozone episode, Atmos. Environ., 37, 3469-3474.

Draxler, R. R., 1983, Model validation using Kr-85 air concentration data from the 1500 km long-range dispersion experiment. Proceedings, 6th Symp. on Turb. and Diff., Boston, pp. 139-141.

Draxler R.R., 2002, Verification of an Ensemble Dispersion Calculation, *J. App. Meteorol.*, 42, 308-317.

Galmarini S., Bianconi R., Bellasio R and G. Graziani, 2001, Forecasting the consequences of accidental releases of radionuclides in the atmosphere from ensemble dispersion modeling. *Journal of Environmental Radioactivity,* 57, 203-219.

Girardi, F., Graziani, G., van Veltzen, D., Galmarini, S., Mosca, S., Bianconi, R., Bellasio, R., Klug, W. (Eds.), 1998. The ETEX project. EUR Report 181-43 EN. Office for official publications of the European Communities, Luxembourg, 108 pp

Straume A. G., 2001, A more extensive investigation of the use of ensemble forecasts for dispersion model evaluation, *J. Applied Meteor.* 40, 425-445.

Warner T. T., Sheu R. S., Bowers J. F., Sykes R. I., Dodd G. C. and Henn D. S., 2002, Ensemble simulations with coupled atmospheric dynamic and dispersion models: illustrating uncertainties in dosage simulations. *Journal of Applied Meteorology,* 41, 488-504.

Van Dop H. and K. Nodop (eds), 1998, ETEX A European Tracer Experiment, *Atmos. Environ.* 24, 4089-4378.

41
Linking the ETA Model with the Community Multiscale Air Quality (CMAQ) Modeling System: Ozone Boundary Conditions

Pius C. Lee, Jonathan E. Pleim, Rohit Mathur, Jeffery T. McQueen, Marina Tsidulko, Geoff DiMego, Mark Iredell, Tanya L. Otte, George Pouliot, Jeffrey O. Young, David Wong, Daiwen Kang, Mary Hart, and Kenneth L. Schere*

1. Introduction

Until the recent decade, air quality forecasts have been largely based on statistical modeling techniques. There have been significant improvements and innovations made to these statistically based air quality forecast models during past years (Ryan et al., 2000). Forecast fidelity has improved considerably using these methods. Nonetheless, being non-physically-based models, the performance of these models can vary dramatically, both spatially and temporally. Recent strides in computational technology and the increasing speed of supercomputers, combined with scientific improvements in meteorological and air quality models has spurred the development of operational numerical air quality prediction models (e.g., Vaughn et al., 2004, McHenry et al., 2004).

In 2003, NOAA and the U.S. Environmental Protection Agency (EPA) signed a memorandum of agreement to work collaboratively on the development of a national air quality forecast capability. Shortly afterwards, a joint team of scientists from the two agencies developed and evaluated a prototype surface ozone concentration forecast capability for the Eastern U.S. (Davidson et al., 2004). The National Weather Service (NWS) / National Centers for Environmental Prediction (NCEP) ETA model (Black, 1994, Rogers et al., 1996, and Ferrier et al., 2003) with 12-km

* Pius C. Lee, Marina Tsidulko, and Mary Hart, Scientific Applications International Corporation, Camp Springs, MD.
Jonathan E. Pleim, Rohit Mathur, Tanya L. Otte, George Pouliot, Jeffrey O. Young, and Kenneth L. Schere, National Oceanic and Atmospheric Administration, Research Triangle Park, NC, on assignment to the National Exposure Research Laboratory, U.S.E.P.A.
Jeffery T. McQueen, Geoff DiMego, and Mark Iredell, NOAA, NWS / National Centers for Environmental Prediction, Camp Springs, MD.
David Wong, Lockheed Martin Information Technology, Research Triangle Park, NC.
Daiwen Kang, Science and Technology Corporation, Research Triangle Park, NC.

horizontal finite cell size was used to drive the EPA Community Multi-Scale Air Quality (CMAQ) model (Byun et al., 1999) to produce an up to 48 h Ozone (O_3) prediction. McQueen et al. (2004) and Otte et al. (2004) described the challenge of coupling ETA and CMAQ, and running the coupled model on a real time basis.

The general performance of the modeling system is that the system errs on over prediction (Pleim et al., 2003, Ryan et al., 2004). Figure 1 displays a typical time series of computation-domain-wide mean surface O_3 observation (AIRNOW, EPA, 2004), corresponding prediction and bias over 640 monitoring stations in Northeastern U.S.

Throughout this study, these 640 monitoring stations within the computational domain formed the basis of the performance verification (See Figure 2).

A key uncertainty in regional photochemical modeling relates to the specification of Lateral Boundary Conditions (LBC's) for O_3 and its precursor species, within both the boundary layer and the free troposphere. Specification of both temporally and spatially varying boundary conditions is desirable. While surface measurements of O_3 can be used for this purpose to provide some representation of O_3 variations along the lateral boundaries of the surface, this approach alone cannot provide information on variations within the free troposphere. An alternate approach is to use a global scale model to provide vertical variations along

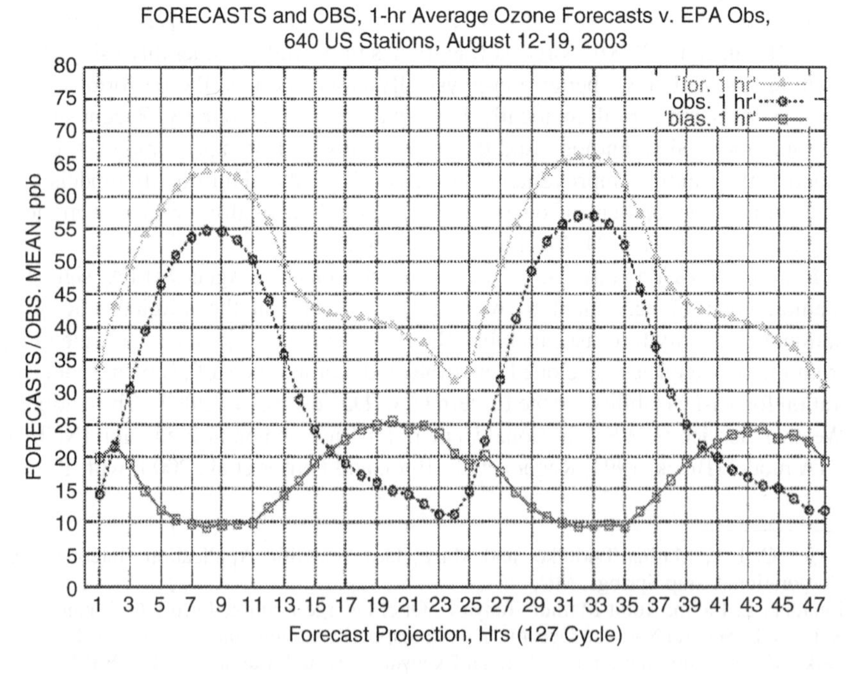

FIGURE 1. Mean predicted CMAQ O_3 concentration (triangle), AIRNOW observed (circle), and bias error (square) by forecast hour for all available 12 UTC cycle CMAQ predictions from August 12th to August 19th, 2003.

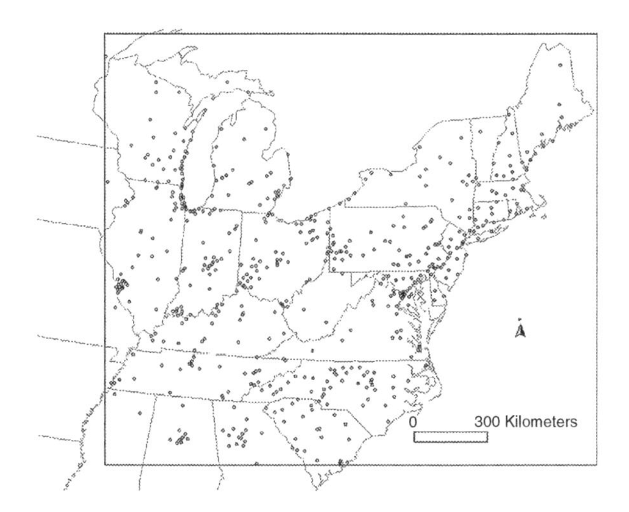

FIGURE 2. EPA/AIRNOW Ozone monitors in the Northeastern U.S. used in the model evaluation. Figure courtesy of the National Weather Service Meteorological Development Laboratory.

the lateral boundary. The current study reports on the issues and impacts associated with such an approach.

2. Lateral O_3 Boundary Condition Schemes

Uncertainty in the O_3 concentration at the boundaries of the computational domain is one of the primary uncertainties. In the 2003 summer runs, a climatologically derived O_3 concentration profile was used. Profiles presented in Figure 3 were prescribed for each of the four boundaries and held static in time.

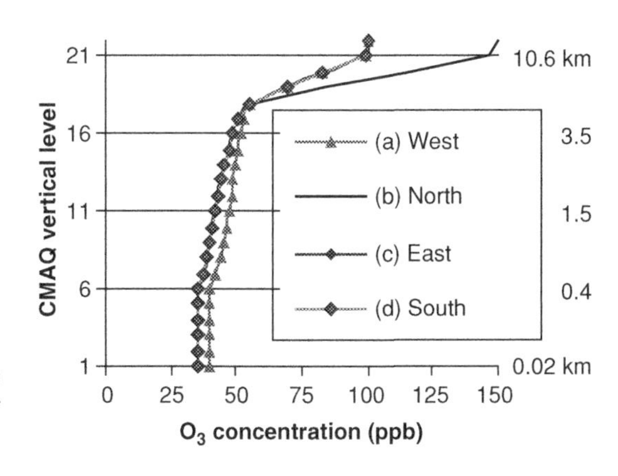

FIGURE 3. Ozone concentration profile used as static LBC's for Case A.

Real-time or near real-time measurements may provide better estimates than climatologic data derived BC's. This study tested the O_3 forecasts that were generated by NCEP's spectral Global Forecast System (GFS) to derive, in part, the CMAQ O_3 LBC's. The GFS model treats O_3 as a 3-D prognostic variable (Moorthi and Iredell, 1998). It is treated as an advection trace species with simple zonally averaged climatological derived production and depletion mechanism (Rood et al., 1991). The GFS ozone is initialized using Solar Backscatter Ultra-Violet-2 (SBUV-2) satellite observations (NCEP, NOAA 2004a). The satellite provides 12 vertical layers of O_3 concentration, with the lowest layer spanning from the surface to 250 mb. The data ingest analysis step within the GFS model system takes the O_3 field from a previous GFS forecast cycle as an initial guess and combines it with the satellite data to generate an updated O_3 field. At NCEP, both the GFS and the ETA-CMAQ model systems are run four times per day at 00, 06, 12 and 18 UTC cycles. Ideally, the GFS analysis O_3 field should be used for all CMAQ cycles. However, the GFS system starts later than the ETA-CMAQ system. There is an hour time lag for the 06 and 18 UTC cycles and a one and a half hour lag for the 00 and 12 UTC cycles between the two systems. The preparation of LBC's for CMAQ starts when ETA has finished 48 forecast hours. At that time, the GFS system is not yet ready to provide its analysis results. Hence, the GFS 6 h forecast of the previous cycle is used by the CMAQ system to derive its LBC's. The GFS system outputs O_3 every three hours on 42 sigma levels over a global 1° resolution grid (NCEP, NOAA, 2004b). The GFS O_3 field is interpolated to the CMAQ 12 km grid spatially and temporally. Another CMAQ input preparation step further extracts time varying O_3 concentration lateral BC's. Figure 4 shows an example of such GFS O_3 derived LBC's. The maxima O_3 lie in the top layer, reaching a magnitude often in excess of hundreds of ppbv.

3. Sensitivity Study

A base case (Case A) is defined as a run of CMAQ with its default static O_3 BC profile in Figure 3. This CMAQ run was started at 12 UTC on May 16, 2004. A 12 UTC May 17 run was initialized using the May 16th 24 hours forecast. The target date for comparison is May 18th, 16 – 40 hours into the May 17th run.

Two sensitivity cases have been devised to investigate the impact of adopting the GFS O_3 for CMAQ's LBC specifications. First, the entire GFS O_3 column is used to derive the O_3 BC profile (Case B), replacing the default profile entirely. Figure 4 presents a sample of such profiles. It is observed that although GFS O_3 in the CMAQ top model layer is from around 300 to around 900 ppbv, in the layers below 6 km (CMAQ layer 18) the O_3 LBC profile based on GFS is usually lower in value than that based on climatologic data, except for the northern boundary. These will be time varying profiles. In Case C, the use of the GFS O_3 profile is limited to above 6 km. This reasoning stems from the fact that SBUV-2 data are not vertically resolved between the surface and 250 mb. The bulk of the total ozone measured is in the stratosphere. Therefore, the GFS has greater confidence in predicting O_3

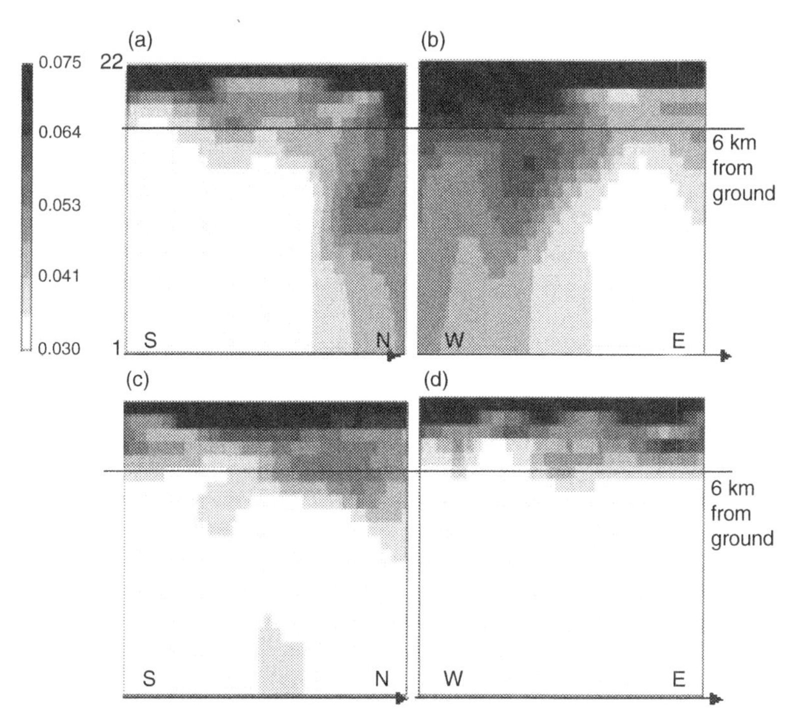

FIGURE 4. Samples of spatially and temporally varying GFS model derived O_3 LBC's at: (a) Western, (b) Northern, (c) Eastern, and (d) Southern boundaries respectively.

concentrations around the tropopause and above, where there are satellite observations, than in those layers below (NCEP, NOAA, 2004c). In the layers below, O_3 prediction depends totally on mechanical advection and a simple seasonally averaged climatologic chemical mechanism within the GFS system. The primary motivation to include O_3 in the GFS was to provide a more accurate estimate of radiative heating in the stratosphere. Therefore, the profile of O_3 below the tropopause is not an intended product of the GFS system. On the other hand, the static LBC O_3 profile of Case A is not very representative in the upper layers. The approach adopted in Case C, attempts to combine the salient features of the two data sets.

4. Meteorological Conditions Around May 18th 2004

On May 16th, a fast moving cold front was migrating southeastward from the central prairies of Canada. In addition, the previous week, strong high pressure system was stationed off the U.S. Eastern Seaboard. This system hampered the eastern movement of the fast moving front. By May 18th, the front became stationary between New England and the Ohio Valley (See Figure 5a). Strong storms formed in advance of the front, and the 24 hour precipitation recorded in the

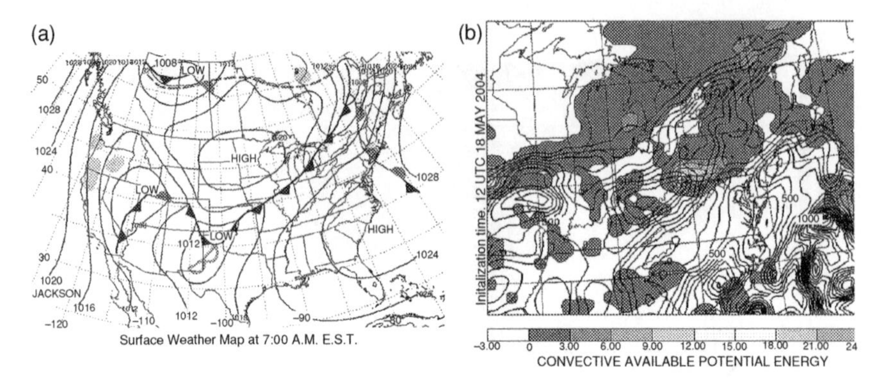

FIGURE 5. Meterological conditions for May 18, 2004: (a) Surface weather map, and (b) ETA's analysis result at 6 UTC for Convective Available Potential Energy (contours at 500 J kg-1 intervals) overlaid with 3 hour accumulated precipitation shaded in 3 mm interval. It indicated a maximum in southwestern PA in excess of 24 mm.

southern part of Ohio registered in excess of 1 inch. In association with this frontal passage, there was rather active cumulus convection. Figure 5b shows a snapshot of the Convection Available Potential Energy (CAPE) at the surface layer as described by the initial analysis of the ETA modeling system. There was also wide-spread precipitation activity over much of the Northeastern U.S., the domain of our prototype system.

5. Tropopause Height Predictions

Both the GFS and the ETA models have refined vertical structure around the tropopause. Two of the objectives in obtaining a good representation of those heights are to improve forecasts of jet level winds and stratospheric-tropospheric exchange. The similarity of the two models in their description of the upper troposphere is important for the assimilation of the GFS O_3 from those layers between the two models. For instance, the activities due to deep convection would influence the sharp gradients of the O_3 concentration near the tropopause. These processes should be described in a compatible manner by the models to preserve these sharp gradients. It is thus of interest to compare the compatibility of their predictions around the tropopause. The predicted tropopause height from the GFS and ETA models on May 18, 2004 agreed well. Typically, the agreement between the two models in tropopause height prediction is rather good during most seasons. This can be partially credited to the similar vertical layer structures of the two models between 300 mb and 20 mb. The GFS model has 13 uneven layers between the heights, spaced at about 20-25 mb intervals. The ETA model has 12 layers between the same heights, spaced at about 20-29 mb intervals. Consistency in tropopause height prediction is a prerequisite to assuming that the two models are compatible in the prediction of tropopause dynamics.

6. Results

Figures 6 a-c present the bias in the 8-hour average maximum O_3 concentrations in the three cases forecasts valid May 18, 2004. Overall, the base Case A has the best performance (See Figure 6a), where sporadic over predictions were clustered around the New England area and western PA. Cases B and C both showed additional clusters of over prediction just behind the cold front. Noticeably, there were additional clusters of high bias between the New England States and the Canadian border, over Lake Erie, and in western Ohio. These regions correspond to areas with prolonged, strong convective activity on that day. Between 03 and 18 UTC, the convective available potential energy at the surface in these 3 regions averaged in excess of 800, 900, and 940 J Kg^{-1}, respectively. Therefore it can be expected that the high BC O_3 concentration in the top layers derived by the GFS will be transported downwards. This is likely to have contributed to the additional high bias of Cases B and C. Namely, the frontal convective movement entrained O_3 from the model top layers to the lower layers through downdrafts associated with clouds.

Case C showed fewer instances of over prediction behind the cold front compared to Case B. For examples, there were fewer additional high biases just north of New York State along its border with Canada and the State of Vermont, and along the mid stretch of Lake Erie. This can be understood from the perspective that the northern boundary condition based on the GFS derived O_3 profile was often higher than that derived from climatologic average, even for the lowest layers below 6 km. Therefore, Case B tends to have the most instances of high bias in these areas close to the northern boundary. However, in the areas close to the western boundary, Case C showed more instances of high bias. In fact, Case B is the only case that had few high bias in western Illinois. This is attributed to the site's proximity to the western boundary, and to the low BC O_3 concentration provided by the GFS forecast in the layers below 6 km.

These results demonstrated three shortcomings in the use of GFS forecast based O_3 BC's. First, the vertical interpolation of the GFS O_3 profile, from the GFS's 42 to CMAQ's 22 sigma levels, is probably too coarse in the layers around the CMAQ model top. Between 300 mb and the model top of 100 mb, there are only 2 vertical layers. This insufficient resolution in CMAQ levels results in a distorted O_3 BC profile, with the peak values extending too far down into the troposphere. Second, the CMAQ vertical resolution is too coarse to describe the O_3 entrainment activity in convective areas. There are only roughly 5 vertical layers from 550 mb and the model top. Third, the dynamics of ETA may not have been adequately represented when the meteorological fields were interpolated from a 60 level step mountain vertical structure into the rather coarse sigma level structure of CMAQ. Therefore, this study suggests further investigation to explore a more

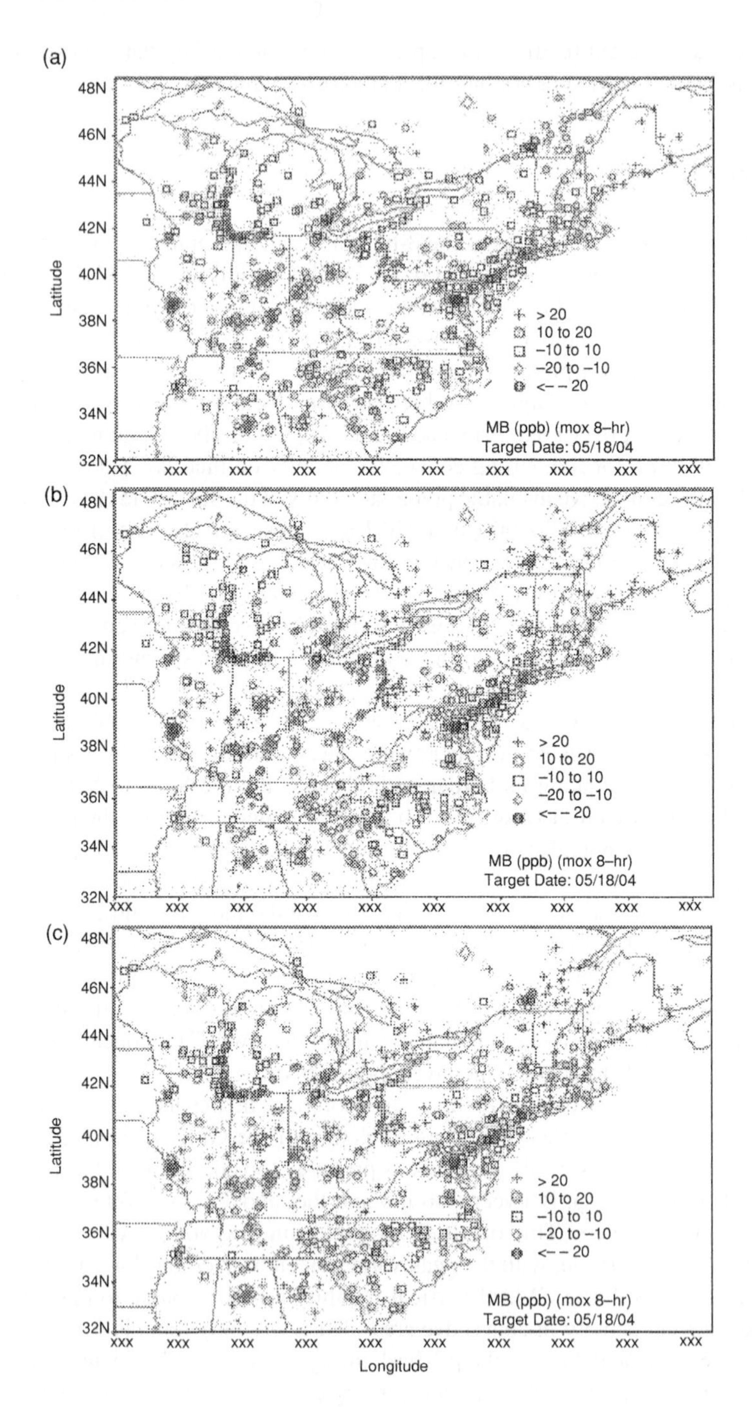

FIGURE 6. Distribution of 8 h average maximum O_3 mean bias for May 18, 2004 for: (a) Case A, (b) Case B, and (c) Case C.

consistent coupling between the various models and increased vertical resolution in CMAQ near the model top.

7. Summary

A prototype ETA-CMAQ operational air quality forecast system had been in use throughout the summer of 2003 – Case A. Model evaluation showed that the system tended to over predict O_3. It has been proposed that the uncertainties associated with the lateral boundary condition for O_3 concentration require investigation. In these summer runs, climatologic data formed the basis for the CMAQ's LBC's, but these data are less reliable in the upper troposphere. Therefore, it was proposed that NCEP's GFS O_3 forecast be used to refine these BC's. Two schemes have been used to ingest GFS O_3 for constructing the CMAQ's LBC profiles. First, the entire GFS O_3 profile has been used to replace the profiles used in the 2003 summer runs – Case B. Second, only the part of the GFS O_3 profiles above 6 km were used – Case C. Results showed that Case A has the least amount of bias when compared to observations. It does not suggest that our confidence in using GFS O_3 profiles to derive air quality model O_3 BC profiles has been decreased. Especially in the upper troposphere, GFS has high confidence of fidelity in predicting O_3 concentrations due to good satellite observations there. It does show that a further investigation of the coupling of ETA and CMAQ in the upper layers is required.

Acknowledgements The views expressed are those of the authors and do not necessarily represent those of the National Weather Service, NOAA or the EPA. The EPA AIRNOW program staff provided the observations necessary for quantitative model evaluation.

Disclaimer The U.S. Environmental Protection Agency, through its Office of Research and Development, partially funded and collaborated in the research described here under Interagency Agreement DW13938634 with NOAA. This document has been subjected to U.S. EPA and NOAA review and approved for publication.

References

AIRNOW, EPA, 2004: Office of Air Quality Planning and Standards, U.S. EPA, AIRNOW Website (http://www.epa.gov/airnow).

Black, T., 1994: The new NMC mesoscale ETA Model: description and forecast examples. *Wea. Forecasting*, **9**, 265-278.

Byun, D. W., J. Young, J, Pleim, M. T. Odman, and K. Alapaty, 1999: Numerical transport algorithms for the Community Multiscale Air Quality (CMAQ) chemical transport model in generalized coordinates. Chapter 7 of *Science algorithms of the EPA Models-3 Community Multiscale Air Quality (CMAQ) Modeling System*. D. W. Byun and J. K. S. Ching, Eds. EPA-600/R-99/030, Office of Research and Development, U.S. Environmental Protection Agency, Washington, D.C. [Available from U.S. EPA, ORD, Washington, D.C. 20460.]

Davidson, P. M., N. Seaman, K. Schere, R. A. Wayland, J. L. Hayes, and K. F. Carey, 2004: National air quality forecasting capability: First steps toward implementation. Preprints, *Sixth Conf. on Atmos. Chem.*, Amer. Met. Soc., Seattle, WA, 12-16 Jan 2004.

Ferrier, B., Y. Lin, D. Parrish, M. Pondeca, E. Rogers, G. Manikin, M. Ek, M. Hart, G. DiMego, K. Mitchell, and H. Chuang, 2003: Changes to the NCEP Meso ETA Analysis and Forecast System: Modified cloud microphysics, assimilation of GOES cloud-top pressure, assimilation of NEXRAD 88D radial wind velocity data. [Available at http://wwwt.emc.ncep.noaa.gov/mmb/tpb.spring03/tpb.htm or from the National Weather Service, Office of Meteorology, 1325 East-West Highway, Silver Spring, MD 20910].

McHenry, J. N., W. F. Ryan, N. L. Seaman, C. J. Coats, Jr., J. Pudykiewicz, S. Arunachalam, and J. M. Vukovich, 2004: A real-time Eulerian photochemical model forecast system: overview and initial ozone forecast performance in the Northeast U.S. corridor. *Bull. Amer. Meteor. Soc.*, **85**, 525-548.

McQueen, J, P. Lee, M. Tsidulko, G. DiMego, T. Otte, J. Pleim, J. Young, G. Pouliot, B. Eder, K. Schere, J. Gorline, M. Schenk, P. Dallavalle, W. Shaffer, N. Seaman, and P. Davidson, 2004: Development and Evaluation of the NOAA/EPA Prototype Air Quality Model Prediction System, Preprints, *Sixth Conf. on Atmos. Chem.*, Amer. Met. Soc., Seattle, WA, 12-16 Jan 2004.

Moorthi, S., and Iredell, M., 1998, Prognostic Ozone: Changes to the 1998 NCEP Operational MRF Model Analysis/Forecast System: The Use of TOVS Level 1-b Radiances and Increased Vertical Diffusion. [Available at http://www.nws.noaa.gov/om/tpb/449.htm from the National Weather Service, Office of Meteorology, 1325 East-West Highway, Silver Spring, MD 20910].

NCEP, NOAA, 2004a: NCEP/GFS total ozone analyses and forecasts Website (http://www.cpc.ncep.noaa.gov/products/stratosphere/strat_a_f/index.html). NOAA Air Resource Laboratory, Silver Spring, MD.

NCEP, NOAA, 2004b: Office note 388 GRIB Website (http:www.nco.ncep.noaa.gov/pmb/docs/on388)

NCEP, NOAA, 2004c: Solar Backscatter UltraViolet Instrument (SBUV/2) Website (http://www.cpc.ncep.noaa.gov/products/stratosphere/sbuv2to/sbuv2to_info.html)

Otte, T. L., and Coauthors, 2004: Linking the ETA Model with the Community Multiscale Air Quality (CMAQ) Modeling System to build a national air quality forecasting system (submitted to *Wea. Forecasting*).

Pleim, J., K. Schere, J. Young, G. Pouliot, T. Otte, 2003: The Models-3 Community Multiscale Air Quality (CMAQ) Model: linking with NWS ETA Model for air quality forecasting, Proceedings, *Air Quality Focus Group Meeting*, Silver Spring, MD, 9-10 Sep 2003, pp 1-23.

Rogers, E., T. Black, D. Deaven, G. DiMego, Q. Zhao, M. Baldwin, N. Junker, and Y. Lin, 1996: Changes to the operational "early" ETA Analysis/Forecast System at the National Centers for Environmental Prediction. *Wea. Forecasting*, **11**, 391-413.

Ryan, W. F., C. A. Petty, and E. D. Luebehusen, 2000: Air quality forecasts in the mid-Atlantic region: current practice and benchmark skill. *Wea. Forecasting*, **15**, 46-60.

Ryan, W. F., P. Davidson, P. Stokols, and K. Carey, 2004: Evaluation of the National Air Quality Forecasting System (NAQFS): Summary of the air quality forecasters focus group workshop. Preprints, *Sixth Conf. on Atmos. Chem.*, Amer. Met. Soc., Seattle, WA, 12-16 Jan 2004.

Rood, R., A. R. Douglas, J. A. Kaye, M. A. Geller, C. Y. Chen, D. J. Allen, E. M. Larsen, E. R. Nash, J. E. Nielsen, 1991: Three-dimensional simulations of wintertime ozone variability in the lower stratosphere. *J. Gephys. Res.*, **96**, # D3, 5055-5071.

Vaughn, J., and Coauthors, 2004: A numerical daily air quality forecast system for the Pacific Northwest. *Bull. Amer. Meteor. Soc.*, **85**, 549-561.

Linking the ETA Model with the CMAQ Modeling System: Ozone Boundary Conditions

Speaker: D. Kang

Questioner: G. Kallos

Question:Do you utilize the reflective transfer profiles from ε for the CMAQ photochemistry or not? If yes, did you find it to be adequate?

Answer: No, CMAQ does not use reflective transfer profiles from ETA. CMAQ derives the radiative fields from the cloud fraction and from the clear sky radiative fields from ETA.

Questioner: R. Yamartino

Question: You mention the difficulty of interpolating the high-resolution ETA model layers to the coarser CMAQ layers. Have you considered using matching of integrated mass fluxes rather than interpolations?

Answer: No. The insightful suggestion brought up is appreciated.

Questioner: R. Bornstein

Question: What needs to be improved in your vertical mixing and/or cumulus-convective schemes to eliminate the high nocturnal transport of tropospheric O_3 to the surface?

Answer: Simulation of large eddy related mixing is an area that needs improvement. A finer vertical and temporal description of the mixing structure would address this problem to a large degree.

Questioner: S. Dorling

Question: Did you choose the May 18th 2004 event in order to emphasize the role vertical transport in the modelers systems and these for test the O_3 performance of the GFS model?

Answer: The model domain of interest experienced a strong cold frontal passage during May 17 and 18, 2004. The GFS model predicted consistently high level of O_3 concentration in the upper layers. The discovery of the importance of the vertical transport and transformation mechanisms for O_3 in the upper layers that resulted in a strong influence of its surface concentration was a discovery from the analysis. Therefore the choice of the dates was really an after-the-fact decision.

Questioner: N. Lin

Question: Just as your presentation mentioned, there are mass sensitive factors that affect the output of the model. Based on your numerical tests, do you have an idea or impression which factor is more sensitive to the output?

Answer: At present the relative importance among mass inconsistency, vertical mixing simulation, photolysis attenuation, and vertical coupling interpolation between the sub-models is an unknown. It is a good topic for further model performance analysis, perhaps with a carefully designed series of sensitivity runs.

42 Mixing in Very Stable Conditions

Larry Mahrt and Dean Vickers

1. Introduction

Recent extensive observations of the nocturnal boundary layer, such as taken in CASES99 (Poulos et al., 2001) and SABLE98 (Cuxart et al., 2000), have revealed important deficiencies in our ability to model the strongly stratified nocturnal boundary layer. Even with more or less continuous and relatively strong turbulence near the surface, the primary source of turbulence can be shear above the surface inversion layer (Mahrt, 1999; Mahrt and Vickers, 2002), often related to a low-level wind maximum (Cuxartet al., 2000; Banta et al., 2002, 2003). Consequently for the CASES99 data, mixing formulations with a $z - less$ limit perform better than traditional parameterizations based on a definable boundary-layer depth (Mahrt and Vickers, 2003).

In an effort to establish some order in what appears to be a wide variety of different nocturnal boundary-layer situations, several investigators have attempted to classify different nocturnal regimes and their relationship to external forcing (e.g., Holtslag and Nieuwstadt, 1986; Mahrt, 1999; Van de Wiel et al., 2003)**. However, a uniform modelling approach is not available for confidently simulating the diverse vertical structures of mixing in the stable boundary layer.

With weak-wind clear-sky conditions, the stability is stronger and the turbulence is weaker, leading to a number of modelling difficulties:

1. In some cases, the boundary layer becomes extremely thin, less than 5 m deep, and not normally resolved. The surface layer may be "squeezed out" due to extension of the influence of boundary-layer depth down to the roughness sublayer, in which case Monin-Obukhov similarity theory would not be valid at any level.
2. For practical reasons, many models do not impose radiative flux divergence, which becomes important near the surface (Sun et al., 2003; Ha and Mahrt, 2003). The radiative flux divergence not only directly affects the stratification but may also affectthe flux-gradient relationship and potentially invalidate formulation of surface fluxes through Monin-Obukhov similarity theory.

3. Very weak turbulence causes a number of modelling problems such as "run away" surface cooling in the model. To prevent these difficulties, models often include conditions on minimum allowable values of wind speed or exchange coefficients or maximum values for the stability function (e.g., Kondo et al., 1978; King et al., 1996).
4. In conditions of very weak turbulence, even small-scale, small-amplitude surface heterogeneity modifies the boundary layer. Horizontal advection of temperature can become relatively large although measurement of such advection is problematic (Sun et al., 2003; Mahrt and Vickers, 2004).
5. With very small downward sensible heat flux to the ground surface, the upward soil heat transfer to the surface becomes one of the main terms in the surface energy balance and may be the primary term balancing the surface net radiative cooling (Mahrt et al., 2001). Then soil characteristics become crucial input information. The thermal conductivity of the soil increases rapidly with increasing soil water content. The model soil and vegetation characteristics exert a strong influence on the entire boundary layer (Van de Wiel et al., 2002).
6. In some cases dew or fog formation strongly influences the heat budgets.
7. With very weak turbulence, transport by gravity waves may become important. Such transport is thought to be more important for momentum than for heat and a possible cause of the observed increase of the eddy Prandtl number with increasing stability (see references in Kim and Mahrt, 1992), although such tendencies are not universally observed.
8. Perhaps the most serious difficulty is untrustworthy observational analysis, leading to poorly constructed similarity theory and transfer relationships for weak turbulence. This study focuses on these analysis problems and their solutions.

2. Improved Turbulent Flux Computations

With very weak turbulence, turbulence may become confined to time scales less than 100 s and sometimes less than 10 s, while mesoscale wave modes can dominate the vertical velocity fluctuations on time scales of a few minutes (e.g., Newsom and Banta, 2003). Therefore, with choice of standard averaging times such as 5 minutes or more to define the perturbations, the perturbations can be dominated by inadvertently capturing mesoscale motions in addition to turbulence. Inadvertent inclusion of mesoscale motions can lead to very large random sampling errors and large scatter for the computed flux due to inadequate sample size of such motions. Discarding flux values of the wrong sign (counter to the gradient) without discarding the other tail of the frequency distribution of the flux, leads to bias; that is, random errors are converted to systematic errors and incorrect similarity relationships.

Vickers and Mahrt (2003) attempt to reduce this problem by specifying a stability-dependent averaging time for defining the perturbations, which shrinks to small values for strong stability. Although, this approach substantially reduces the large random flux errors and frequency of counter-gradient transport, even more dramatic measures may be required for some cases of very weak turbulence above the lowest few metres. To examine the scale dependence of the flux, we decompose the flow into the multi-resolution decomposition (e.g. Howell and Mahrt, 1997; Vickers and Mahrt, 2003), which does not assume periodicity and satisfies Reynolds averaging for individual scales. For weak turbulence, the multi-resolution cospectra show more regular scale dependence compared to the Fourier cospectra and spectral smoothing is not required.

Here, we analyze 4 months of nocturnal eddy-correlation data from Fluxes over Snow-covered Surfaces (FLOSSII) (Mahrt and Vickers, 2004). The following analyses will be posed in terms of heat flux, which can be computed without flux loss due to instrument separation. This loss is potentially greater with very weak turbulence. The turbulent transport of heat may not be a reliable quantitative indicator of the transport coefficients for passive scalars or contaminants.

3. Turbulence and Mesoscale Regimes

To examine the scale dependence of the flux and choose appropriate averaging times to compute the turbulent fluctuations, we composite the cospectra for the set of records with the weakest turbulence (here, arbitrarily 100 one-hour records out of 1294 records). For comparison, we also analyze the 100 one-hour records with the strongest turbulence. The 100 weak-turbulence records represent situations where dispersion would be minimal. These records were selected based on the strength of the fine scale turbulence represented by the vertical velocity variance for 1-s periods averaged over the one-hour record. The strength of the fine-scale turbulence increases systematically with increasing wind speed (not shown).

For the composited cospectra for weak turbulence, the turbulent heat flux on scales less than 10 s is extremely small, orders of magnitude smaller than the composited mesoscale flux. However, the composited turbulence heat flux increases systematically with increasing time scale (well-behaved cospectra), as also occurs for almost all of the individual records. The composited turbulent heat flux is statistically significant (small error bars, Figure 1). The error bars for the weak wind cospectra increase substantially between time scales of 10 and 100 s due to very large between-record variation (very large error bars) and continue to increase for larger time scales (offscale in Figure 1). The sign of the heat flux at these scales frequently switches between records. Compositing the heat flux over different numbers of records indicates that the magnitude of the mesoscale flux on scales greater than 30 s decreases roughly according to the square root of the number of records in the composite, suggesting that the mesoscale flux is largely random error.

Use of averaging times greater than about 30 s to define the turbulent fluctuations leads to serious contamination of the computed turbulence flux by large

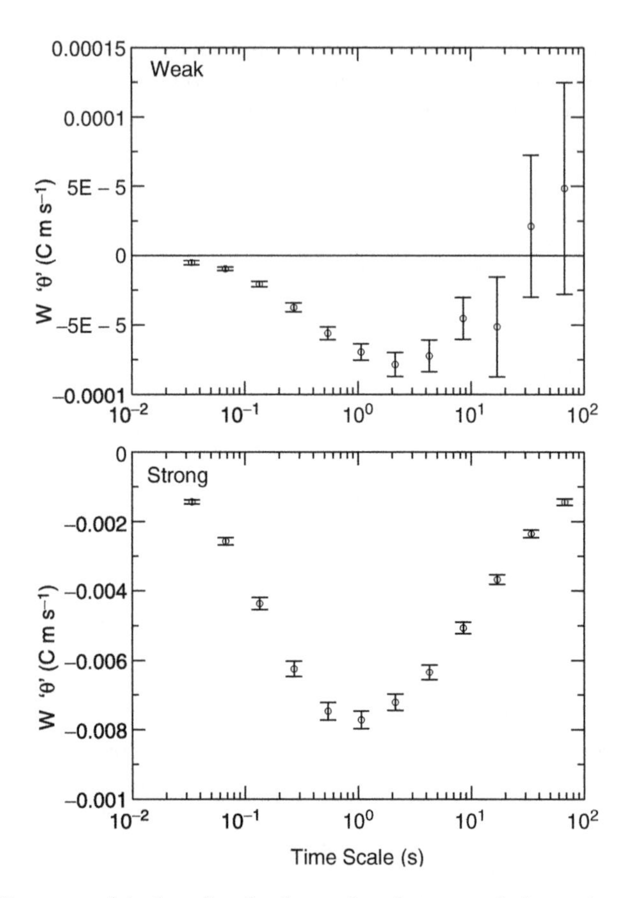

FIGURE 1. Cospectra of the heat flux for the weak and strong turbulence classes. Note that the heat flux for the weak turbulence composite is orders.

random errors associated with the mesoscale flux. For time scales greater than 30 s, the correlation between the vertical velocity fluctuations and the temperature are small compared to turbulent time scales and the vertical velocity fluctuations are orders of magnitude smaller than the horizontal velocity fluctuations. The computed heat flux for such mesoscale motions is large because of very large temperature fluctuations. Note that time scales between 30 s and a few minutes are smaller than normally included in the mesoscale regime.

The composited cospectra for the strong turbulence class exhibits substantially greater turbulence heat flux than mesoscale heat flux (Figure 1). The mesoscale flux for the weak and strong turbulence classes are of similar magnitude. Since the mesoscale flux is not reduced for the weak turbulence class, the estimated turbulence quantities for the weak turbulence class are severely contaminated by inadvertently capturing mesoscale motions, as occurs with standard averaging times. The important mesoscale fluxes can be orders of magnitude larger for individual records.

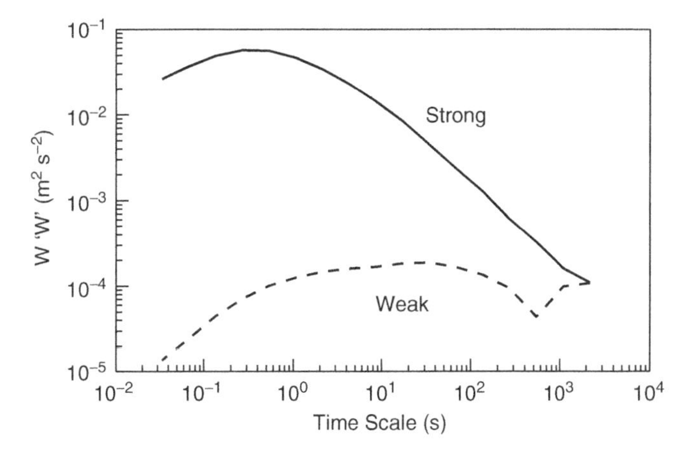

FIGURE 2. Comparison of the spectra for vertical velocity between weak and strong turbulence classes.

For the weak turbulence class, the vertical velocity variance peaks at larger time scales than that for the strong turbulence class and at much larger scales than that for the heat flux (Figure 2). This unexpected result is due to vigorous "small-scale" mesoscale motions at a time scale of about 80 s, which is relatively independent of height and of unknown origin.

Although the fluxes and vertical velocity variances are quite small for the 100 one-hour records of weak turbulence, they can even be several orders of magnitude weaker for subperiods, typically of 20 minutes or shorter duration, where the turbulence is extremely weak. This very weak turbulence would lead to extremely slow dispersion of contaminants. The much stronger mesoscale motion distorts plumes of contaminants through vertical oscillations and horizontal meandering.

4. Similarity Theory

Large random flux errors can lead to misinterpretation of the validity of Monin-Obukhov similarity theory. For stable conditions, the random flux error causes large spurious self-correlation, which is of the same sign as the physical correlation and often larger than the physical correlation (Klipp and Mahrt, 2004). For the 100 weak-turbulence cases, the random error and the scatter of the fluxes are much greater for a 100-s window than for a 10-s window. Ironically, the relationship between ϕ_m and z/L is stronger for fluxes based on the 100-s window. Since the momentum fluxes are more erratic than the heat fluxes, the variation of the Obukhov length is dominated by large spurious variation of u_*^3, leading to a stronger correlation between ϕ_m and z/L due to self-correlation. The self-correlation is proportional to the variance of the common factor, in this case, u_*.

For the weak-turbulence cases with careful flux calculations following Vickers and Mahrt (2003), ϕ_m tends to become independent of z/L for z/L greater than about two (Figure 3). However, the scatter in ϕ_m is still relatively large, suggesting other influences on the flux-gradient relation and/or remaining errors in the flux estimates. Models sometimes specify constant ϕ_m for z/L greater than two to avoid extremely weak turbulence and possible near-surface run-away cooling in the model. For the present data, the asymptotic value of ϕ_m is roughly four. In contrast, the nondimensional temperature gradient continues to increase with increasing stability at a rate similar to that predicted by existing similarity theory. This dependence suggests that the Prandtl number increases with increasing stability and that a maximum value of the stability function for heat flux should not be applied. Monin-Obukhov similarity theory requires revision for very stable

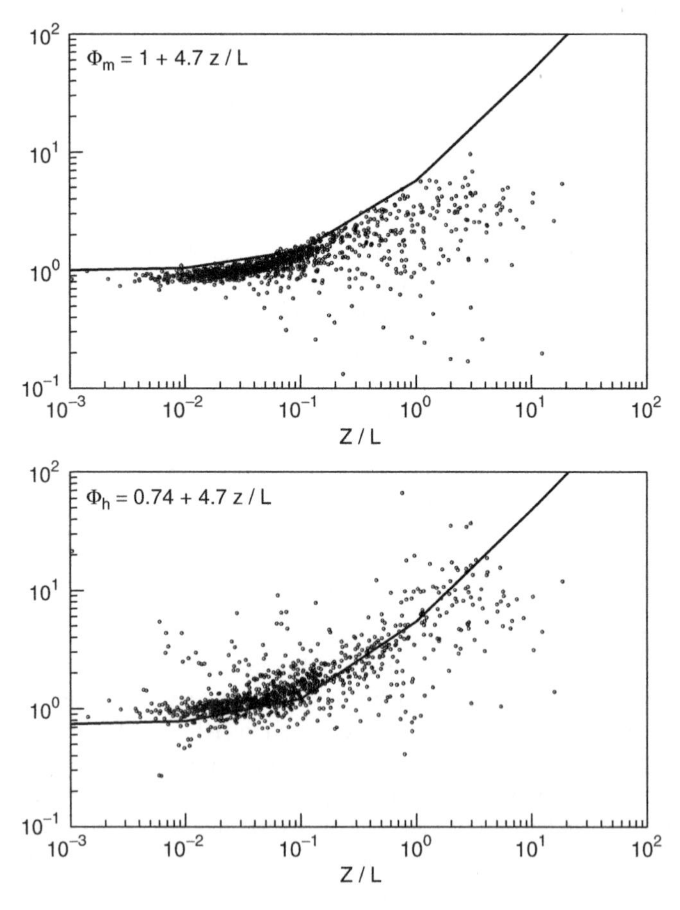

FIGURE 3. Dependence of the nondimensional gradients of momentum and temperature on stability.

conditions. Such revision is currently under construction along with reformulation of dispersion quantities.

Acknowledgments This material is based upon work supported by Grant DAAD19-0210224 from the Army Research Office and Grant 0107617-ATM from the Physical Meteorology Program of the National Sciences Program.

References

Banta, R. M., R. K. Newsom, J. K. Lundquist, Y. L. Pichugina, R. L. Coulter and L. Mahrt, 2002: Nocturnal low-level jet characteristics over Kansas during CASES-99, *Boundary-Layer Meteorol.*, **105**, 221-252.

Banta, R. M., Y. L. Pichugina and R. K. Newsom, 2003: Relationship between low-level jet properties and turbulence kinetic energy in the nocturnal stable boundary layer. *J. Atmos. Sci.*, **60**, 2549-2555.

Cuxart, J. C., Yagüe, G., Morales, E., Terradellas, J., Orbe, J., Calvo, A., Fernandez, M. R., Soler, C., Infante, P., Buenestado, A., Espinalt, H. E., Joergensen, J. M., Rees, J., Vilá, J. M., Redondo, I. L., and Conangla, I. R., Cantalapiedra 2000: Stable atmospheric boundary-layer experiment in Spain (SABLE 98): A report. *Boundary-Layer Meteorol.*, **96**, 337-370.

Ha, K.-J. and L. Mahrt, 2003: Radiative and turbulent fluxes in the nocturnal boundary layer. *Tellus*, **55A**, 317-327.

Holtslag, A. A. M. and Nieuwstadt, F. T. M., 1986: Scaling the Atmospheric Boundary Layer, *Boundary-Layer Meteorol.*, **36**, 201-209.

Howell, J. and L. Mahrt, 1997: Multiresolution flux decomposition. *Boundary-Layer Meteorol.*, **83**, 117-137.

Kim, J., and L. Mahrt, 1992: Simple formulation of turbulent mixing in the stable free atmosphere and nocturnal boundary layer. *Tellus*, **44A**, 381-394.

King, J. C., P. S. Anderson, M. C. Smith and S. D. Mobbs, 1996: The surface energy and mass balance at Halley, Antarctica during winter. *J. Geophys. Res.*, **101**, 19,119-19,128.

Klipp, Cheryl and L. Mahrt, 2004: Flux-gradient relationship, self-correlation and intermittency in the stable boundary layer. To appear in *Quart. J. Roy. Meteorol. Soc.*

Kondo, J., O. Kanechika, and N. Yasuda, 1978: Heat and momentum transfers under strong stability in the atmospheric surface layer. *J. Atmos. Sci.*, **35**, 1012-1021.

Mahrt, L., 1999: Stratified atmospheric boundary layers. *Boundary-Layer Meteorol.*, **90**, 375-396.

Mahrt, L., Vickers, D., Sun, J., Burns, S. and Lenschow, D.: 2001, 'Shallow Drainage Flows'. *Boundary-Layer Meteorol.*, **101**, 243-260.

Mahrt, L. and D. Vickers, 2002: Contrasting vertical structures of nocturnal boundary layers. *Boundary-Layer Meteorol.*, **105**, 351-363.

Mahrt, L. and D. Vickers, 2003: Formulation of turbulent fluxes in the stable boundary layer. *J. Atmos. Sci.*, **60**, 2538-2548.

Mahrt, L. and D. Vickers, 2004: Boundary-layer adjustment over small-scale changes of surface heat flux. To appear in *Boundary-Layer Meteorol.*

Newsom, R. and R. Banta, 2003: Shear-flow instability in the stable nocturnal boundary layer as observed by Doppler lidar during CASES99. *J. Atmos. Sci.*, **60**, 16-33.

Poulos, G., W. Blumen, D. Fritts, J. Lundquist, J. Sun, S. Burns, C. Nappo, R. Banta, R. Newsom, J. Cuxart, E. Terradellas, B. Balsley, M. Jensen, 2001: CASES-99: A

Comprehensive Investigation of the Stable Nocturnal Boundary Layer. *Bull. Amer. Meteorol. Soc.*, **83**, 555-581.

Sun, J., S. Burns, A. Delany, T. Horst, S. Oncley, and D. Lenschow, 2003: Heat balance in nocturnal boundary layers. *J. Appl. Meteorol.*, **42**, 1649-1666.

Van de Wiel, B. J. H., Moene, A., Ronda, R. J., De Bruin, H. A. R., and A. A. M. Holtslag, 2002: Intermittent turbulence in the stable boundary layer over land. Part II: A System Dynamics Approach. *J. Atmos. Sci.*, **59**, 2567-2581.

Vickers, Dean and L. Mahrt, 2003: The cospectral gap and turbulent flux calculations. *J. Atmos. Oceanic Tech.*, **20**, 660-672.

43
Air Quality Ensemble Forecast Over the Lower Fraser Valley, British Columbia

Luca Delle Monache[1], Xingxiu Deng[1], Yongmei Zhou[1], Henryk Modzelewski[1], George Hicks[1], Trina Cannon[1], Roland B. Stull[1], and Colin di Cenzo[2]

1. Introduction

The ensemble-averaging approach is potentially a technique for improving the performance of real-time photochemical air-quality modeling. Ensemble photochemical air-quality forecasts are tested extensively using the Community Multiscale Air Quality (CMAQ) model-system with mesonet observations from the Emergency Weather Net (EmWxNet) and the AQ Data Set over the Lower Fraser Valley (LFV).

The CMAQ model is run daily over a 12 km resolution domain (Figure 1 top) covering southern British Columbia, Washington State, and the northern portion of Oregon State. A 4 km resolution grid (Figure 1 bottom) is nested within the 12km grid, and it covers the southern tip of British Columbia (including Vancouver and the LFV) and the northern part of Washington State (including the Seattle area).

CMAQ is driven by two different meteorological models: the Mesoscale Compressible Community Model (MC2), and the Fifth-Generation NCAR/Penn State Mesoscale Model (MM5).

2. Discussion

Ensemble weather forecasts have been extensively evaluated over the past decade, and have been found to provide better accuracy than any single numerical model run (Wobus and Kalnay, 1995; Molteni et al., 1996; Du et al., 1997;

[1] Luca Delle Monache, Xingxiu Deng, Yongmei Zhou, Henryk Modzelewski, George Hicks, Trina Cannon, Roland B. Stull, University of British Columbia, Vancouver, BC V6T 1Z4 Canada.
[2] Colin di Cenzo, Atmospheric Sciences Section, Environment Canada, Vancouver, BC V6P 6H9 Canada.

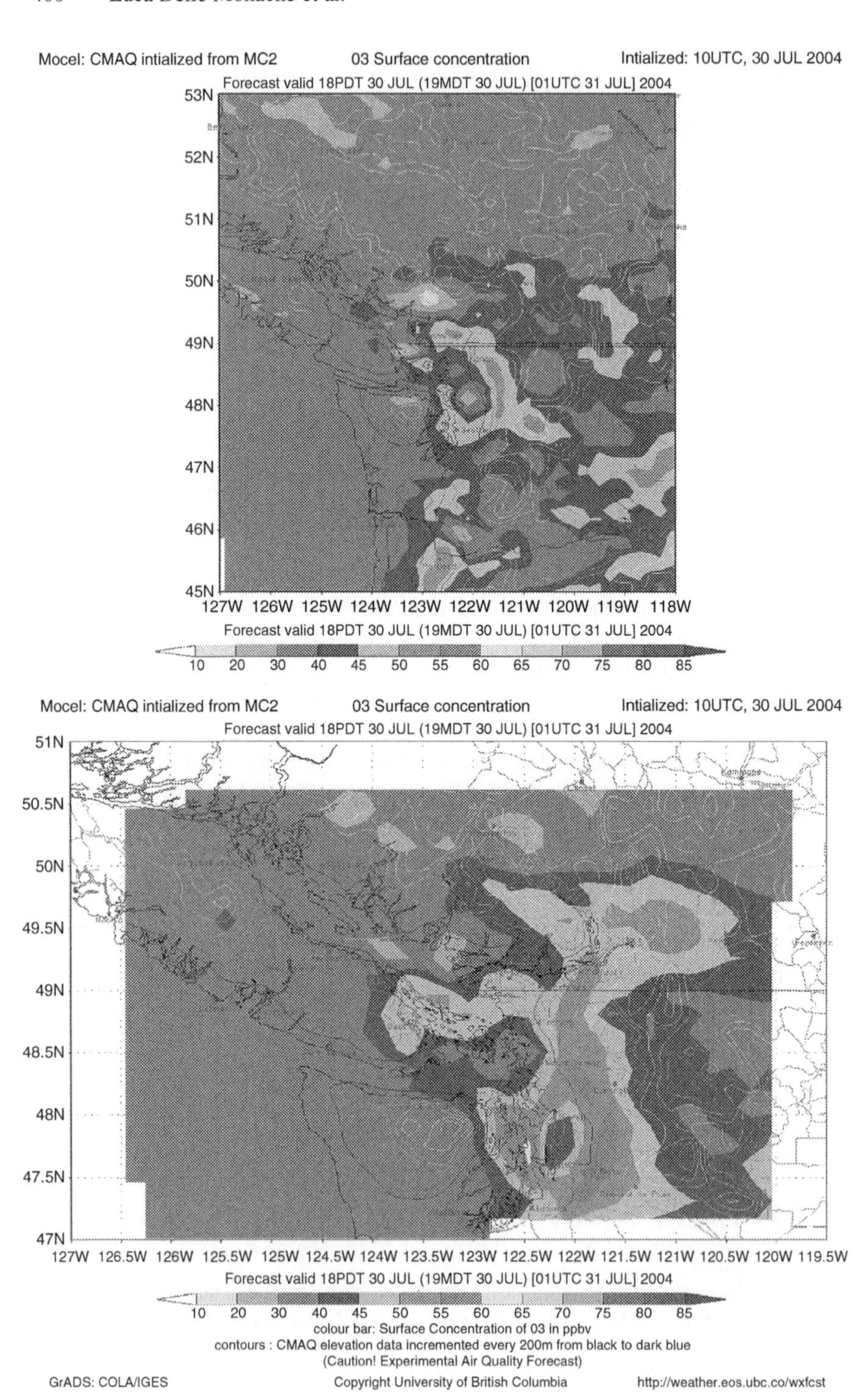

FIGURE 1. Ozone surface concentration for 16 PDT, June 19 2004. Top: 12 km domain. Bottom: 4 km domain.

Hamill and Colucci, 1997; Toth and Kalnay, 1997; Stensrud et al., 1998; Krishnamurti et al., 1999; Evans et al., 2000; Kalnay 2003). Different Numerical Weather Prediction (NWP) models usually perform better for different synoptic situations, and often the best model cannot be anticipated.

Hence, their combination into a multi-model ensemble is usually fruitful. Very clear evidence has been presented by Wandishin et al (2001), Richardson (1999) and the US National Centers for Environmental Prediction (NCEP, 2004: http://www.hpc.ncep.noaa.gov/ensembletraining) that the best short-range forecasts are achieved with multi-model ensembles.

NWP ensembles have been created with different inputs (Toth and Kalnay 1993; Molteni et al. 1996) (initial conditions ICs and/or boundary conditions BCs), different parameterizations within a single model (Stensrud et al. 1998) (physics packages, parameter values), different numerics within a single model (Thomas et al. 2002) (finite difference approximations and solvers, grid resolutions, compiler optimizations), and different models (Hou et al. 2001), trying to take into account different sources of uncertainties.

The ensemble technique can potentially yield similar benefits to air-quality (AQ) modeling, because there are similar code complexities and constraints (Delle Monache and Stull 2003). Different AQ models can be better for different air-pollution episodes, also in ways that cannot always be anticipated. For AQ, the ensemble-mean can be created similarly with different inputs (background concentrations, emissions inventories, meteorology), different parameterizations within a single model (chemistry mechanisms, rate constants, advection and dispersion packages), different numerics within a single model (finite difference approximations and solvers, grid resolutions, compiler optimizations), and different models (Delle Monache and Stull 2003). Given the nonlinear nature of photochemical reactions, the ensemble spread might be useful to account for the uncertainties associated with each component of the modeling process.

Results of an AQ ensemble forecast system are presented. They are validated each day for several locations, for hourly time series of meteorology, ozone, and particulate matter (PM). This data set allows extensive testing, in a wide range of meteorological scenarios and air-pollution episodes.

The ensemble tested in this study has some desirable features. For example, there are differences in the emission data of each ensemble member, partly because the hourly emission values (i.e., biogenic and mobile sources) depend on the meteorology that differs from one mesoscale model to another. These differences can take into account the uncertainty in the emissions estimate, which is often a factor of three or more, and which is the dominant limitation in the photochemical model performance (Russell and Dennis 2000). For the same reason, the different meteorological input fields from MM5 and MC2 allow the ensemble to filter out some of the unpredictable components of the weather. Furthermore, different ensemble members run at different resolutions, which lead to different parcel trajectories, and this allows the ensemble to take into account the uncertainties related to the different but plausible choices of the grid location and resolution.

Acknowledgements The authors thank Todd Plessel of EPA for providing very useful tools to manipulate and visualize Model-3 formatted data.

References

Clappier, A., Perrochet. P., Martilli, A., Muller, F. and Krueger, B. C.:,1996, A new non-hydrostatic mesoscale model using a CVFE (control volume finite element) Discretisation technique, in *Proceedings of EUROTRAC Symposium '96.*, Editors: P. M. Borrell et al., Computational Mechanics Publications, Southampton, pp 527-531.

Delle Monache, L., and R. B. Stull, An ensemble air-quality forecast over Western Europe during an ozone episode. Atmos. Environ., **37**, 3469-3474 (2003).

Du, J., Mullen, S. L., and F. Sanders, Short-range ensemble forecasting of quantitative precipitation. Mon. Wea. Rev., **125**, 2427-2459 (1997).

Evans, R. E., Harrison, M. S. J., and R. Graham, Joint mediumrange ensembles from The Met. Office and ECMWF systems. Mon. Wea. Rev. 128, 3104–3127 (2000).

Hamill, T. M., and S. J. Colucci, Verification of Eta–RSM shortrange ensemble forecasts. Mon. Wea. Rev., **125**, 1322-1327 (1997).

Hou, D., Kalnay, E., and K. K. Droegemeier, Objective verification of the SAMEX '98 ensemble forecasts. Mon. Wea. Rev., **129**, 73-91 (2001).

Jiang, W., Singleton, D. L., Hedley. M., and R. McLaren, Sensitivity of ozone concentrations to VOC and NO_x emissions in the Canadian Lower Fraser Valley. Atmos. Environ., **31**, 627-638 (1997).

Kalnay, E., 2003: Atmospheric Modeling, Data Assimilation and Predictability. Cambridge University Press, New York, 341 pp.

Krishnamurti, T. N., Kishtawal, C. M., LaRow, T. E., Bachiochi, D. R., Zhang, Z., Willford, C. E., Gadgil, S., and S. Surendran, Improved weather and seasonal climate forecast from multimodel superensemble. Science, **285**, 1548-1550 (1999).

Molteni, F., Buizza, R., Palmer, T. N., and T. Petroliagis, The new ECMWF ensemble prediction system: methodology and validation. Quart. J. Roy. Meteor. Soc., **122**, 73-119 (1996).

Richardson, D. S., 1999: Ensembles using multiple models and analyses: report to ECMWF Scientific Advisory Committee.

Russell, A., and R. Dennis, NARSTO critical review of photochemical models and modeling. Atmos. Environ., **34**, 2283-2324 (2000).

Stensrud, D. J., Bao, J.-W., and T. T. Warner, Ensemble forecasting of mesoscale convective systems. 12th AMS Conference on Numerical Weather Prediction, Phoenix, AZ, 265-268 (1998).

Thomas, S. J., Hacker, J. P., Desgagné, M., and R. B. Stull, An ensemble analysis of forecast errors related to floating point performance. Wea. Forecasting, **17**, 898-906 (2002).

Toth, Z., and E. Kalnay, 1993: Ensemble forecasting at NMC: the generation of perturbations. Bull. Amer. Meteor. Soc., **74**, 2317-2330 (1993).

Toth, Z., and E. Kalnay, 1997: Ensemble forecasting at NCEP: the breeding method. Mon. Wea. Rev., 125, 3297-3318 (1997).

Wandishin, M. S., S. L. Mullen, D. J. Stensrud, and H. E. Brooks: Evaluation of a short-range multimodel ensemble system, **129**, 729-747 (2001).

Wobus, R., and E. Kalnay, 1995: Three years of operational prediction of forecast skill. Mon. Wea. Rev., **123**, 2132-2148 (1995).

44 Developments and Results from a Global Multiscale Air Quality Model (GEM-AQ)

Lori Neary, Jacek. W. Kaminski, Alexandru Lupu, and John C. McConnell

1. Introduction

The GEM-AQ model (Global Environmental Multiscale model with Air Quality processes) is based on Canada's operational weather prediction model developed by the Meteorological Services of Canada (MSC). The chemical module is included in the host meteorological model "online", so that the chemical species are advected at each dynamical timestep and the meteorological information can be used in the chemical process calculations. The processes include over 100 gas phase chemical reactions, anthropogenic and biogenic emissions, transport due to vertical diffusion and convection, dry deposition and wet scavenging. We have recently introduced size-resolved aerosols which undergo processes such as coagulation, nucleation, dry and wet scavenging. GEM-AQ is capable of running on a global uniform or global variable resolution domain. The variable resolution capability allows a high resolution simulation over an area of interest without the computational overhead of a global high resolution domain, as well as eliminating the need for boundary conditions in the case of limited area modeling. The capabilities of this modelling system give us a tool to help examine local air quality issues while keeping the global picture in mind. The design philosophy behind GEM-AQ allows for the integration of different physics and chemistry modules into a single computational platform. The major advantage of this approach allows for the development and improvement of parameterizations which can easily be included in the system. In addition, the use of GEM-AQ as the core of the modelling system will permit the incorporation of data assimilation techniques into the model validation and application studies on processes by researchers participating in the Multiscale Air Quality Modelling Network (MAQNet).

Department of Earth and Space Science, York University, Toronto, Ontario, Canada, M3J 1P3 lori@yorku.ca

2. MAQNet Overview

The Multiscale Air Quality Network (MAQNet) is a collaboration of Canadian universities and research institutions whose objective is to develop a multiscale three-dimensional tropospheric air quality modelling and chemical data assimilation system. This network is funded by the Canadian Foundation for Climate and Atmospheric Studies (CFCAS).

List of partners and their contributions:

- Dalhousie University
  - Improvements to aerosol module
- Université de Sherbrooke
  - Evaluation of aerosol module
- McGill University
  - Chemical data assimilation
- Université du Québec à Montréal
  - Gas phase chemistry evaluation, transport improvements
- York University
  - Model maintenance/development, wet chemistry improvements
- University of Toronto
  - Aerosol module, addition of POPs
- Meteorological Services of Canada
  - Wet chemistry improvements
- University of British Columbia
  - MC2-AQ evaluation
- Warsaw University of Technology
  - MC2-AQ evaluation

The GEM-AQ model provides the base for a unified modelling environment that the participating members of the network can use to study a wide variety of air quality issues, taking advantage of the model's multiscale capabilities. More information about MAQNet and the network projects can be found at http://www.maqnet.ca.

3. Model Description

The base of the GEM-AQ model is the Meteorological Services of Canada's operational weather prediction and data assimilation system, the Global Environmental Multiscale model (Côté et al., 2002). The GEM-AQ model is used to produce five and ten day forecasts on a global uniform resolution grid of 0.9×0.9 degrees and 48 hour regional forecasts over North America using a rotated variable resolution grid. The current operational grid has a uniform high resolution window with a grid square size of about 15km. From this central core, the grid squares increase in size. Figure 1 shows an example of a variable resolution grid. The high resolution window can be placed over any region of interest.

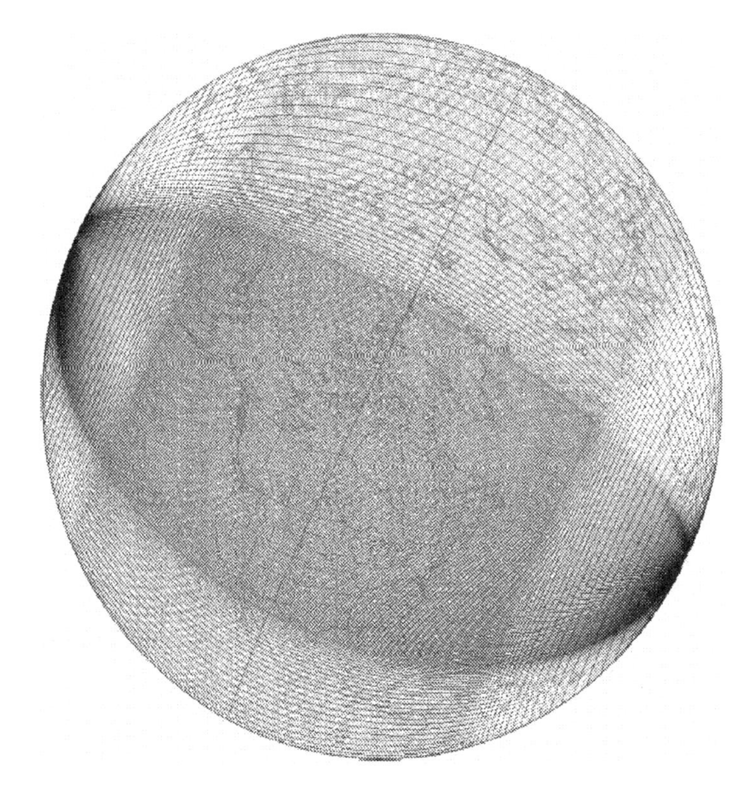

FIGURE 1. An example of a GEM-AQ variable resolution grid, 255×289 total grid points, 0.33° resolution over the central window (180×235 grid points).

The ability to run on a variable resolution grid makes GEM-AQ ideal for multi-scale air quality modelling. The overall computational cost of a simulation can be reduced by eliminating the need to run a larger domain to provide meteorological and chemical boundary conditions, as would be the case in a limited area model. Running a simulation on a global domain also helps to include the effects of long-range transport on regional air quality.

Processes which have been traditionally treated by a separate chemical transport model have been included within the GEM-AQ framework. The chemical tracers are calculated on the same grid and use the same advection algorithm as is used by the meteorological model, and the physical parameterizations affecting the chemical tracers have, to the greatest extent possible, been harmonized with those used in the meteorological calculations.

The gas phase chemical mechanism included is based on ADOM IIB (Lurmann *et al.*, 1986). It includes 49 gas phase chemical species and over 100 reactions. Other processes included are dry and wet deposition, transport due to convection and vertical diffusion.

GEM-AQ also has an aerosol physics package, CAM – Canadian Aerosol Module (Gong *et al.*, 2003). This includes 5 size-resolved aerosol types (12 size bins each), which undergo processes such as coagulation, nucleation, dry and wet scavenging. The aerosol types included at present are sulphate, sea-salt, organic carbon, black carbon and soil dust.

The same chemical package has been incorporated into MSC's non-hydrostatic limited area model, the Mesoscale Compressible Community Model (MC2) as well. A full description of the meteorological model MC2 is given in Tanguay *et al.* (1990), Benoit *et al.* (1997), and Mailhot *et al.* (1998). A more complete description of the air quality module may be found in Plummer *et al.* (2001) and Kaminski *et al.* (2002).

Recent developments to the chemical package include a more advanced dry deposition scheme from A. Robichaud at MSC as well as the inclusion of biogenic emissions calculated at each timestep using model parameters. The scheme is based on the model by Guenther *et al.*, 1995, 1999.

4. Model Results

Both GEM-AQ and MC2-AQ are currently being evaluated for a variety of air quality scenarios. On the global scale, work is being done assimilating MOPITT carbon monoxide and GOME tropospheric ozone. On the regional scale, model ozone output comparisons are being done with US-EPA AIRNOW sites across North America. Some work is being done to examine the model performance over Brazil for comparisons to measurements from the TROCCINOX campaign.

Urban scale modelling scenarios are being carried out with MC2-AQ for the ESCOMPTE measurement campaign in Marseilles, France and for Pacific 2001, Lower Fraser Valley, British Columbia. Results and description of the ESCOMPTE simulations can be found in these proceedings by Kaminski et al.

4.1. Quebec 2002 Forest Fires

The extremely warm and dry summer of 2002 has created conditions for widespread wildfire activity across central and eastern Canada. On July 2 and 3, thunderstorms sparked more than 85 fires in central and western Quebec, however, cloudiness and weak tropospheric winds prevented these fires from intensifying until July 5 and 6. By July 8, about 250,000 acres of forest had been burnt. The smoke from the fanned fires was transported rapidly and with little dispersion into southern Quebec, central and eastern Ontario, and north-eastern United States by a low-pressure system over Nova Scotia. The plume stretched as far south as Washington, DC.

GEM-AQ has been run for the period July 5-10 on a 95×88 rotated variable resolution grid having a 40×40 core window of uniform 0.2° resolution centred on 50°N, 75°W, and 28 hybrid vertical levels. Figure 2 shows the high resolution portion of the domain. High resolution (0.2°) objective analysis data have been used for updating the meteorological fields every 12 hours.

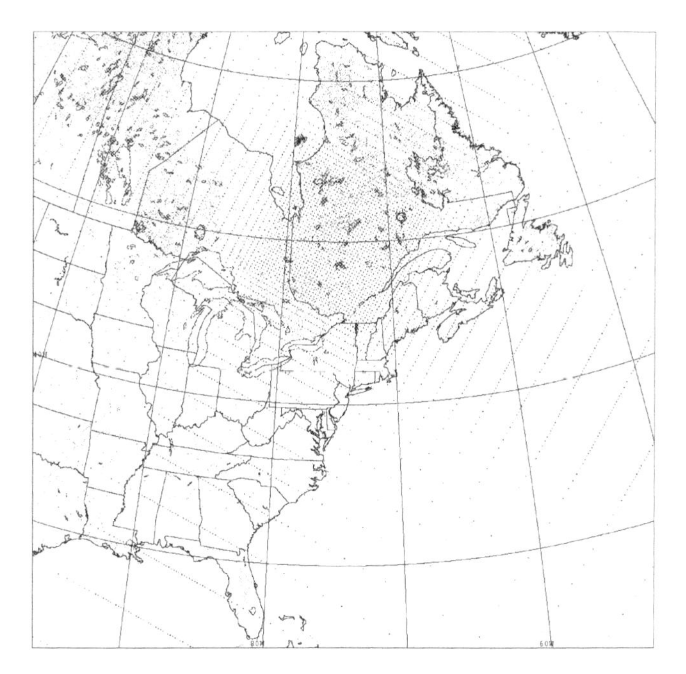

FIGURE 2. A portion of the GEM-AQ global grid used for the Quebec 2002 forest fire simulation.

The emission fluxes of smoke aerosol particles have been generated from the GOES Wildfire Automated Biomass Burning Algorithm (WF_ABBA) products. The smoke has been uniformly injected into grid columns between altitudes of 1 and 5 km, while the source fluxes have been updated every 30 minutes (emission fluxes provided by Jeffery S. Reid of Marine Meteorology Division, Naval Research Laboratory, Monterey, CA). The smoke particles are assumed to be make of 6% black carbon and 94% organic carbon, and are distributed into 12 size bins spaced logarithmically in radius between 0.005 and 20.48 μm. The initial distribution is unimodal and lognormal with a volume radius of 0.145 μm and a standard deviation of 2. The particles grow by coagulation (Brownian, turbulent, and gravitational) only.

The output of the model has been compared with cloud screened aerosol optical depths (AOD), Angström exponents, and effective particle radii measured at 10 AERONET (Aerosol Robotic Network)/AEROCAN (Canadian Sun-Photometer Network) stations across eastern Canada and north-eastern United States. Figure 4 shows the locations of the 10 stations used for this comparison.

The model is generally able to reproduce the evolution of the smoke plume (Figure 3) and its optical properties. The calculated radius of the distribution shows good correlation with the measurement, though the values themselves are systematically lower, most probably because of the artificial dilution at the source introduced by the initial distribution of the smoke mass over the model grid (Figure 5).

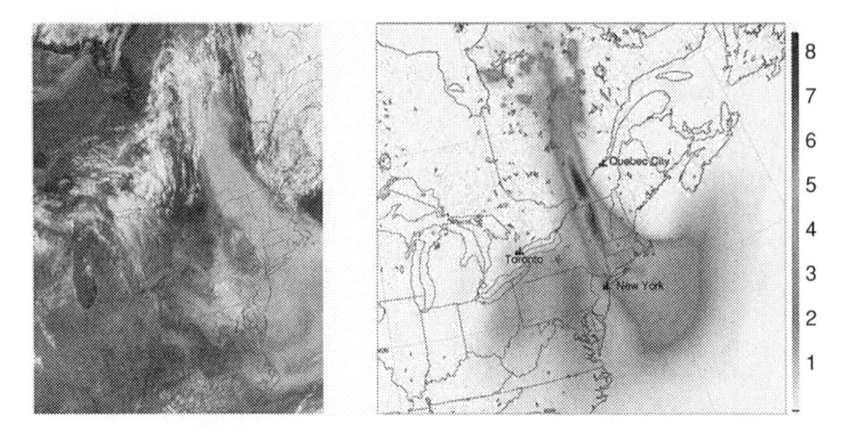

FIGURE 3. MODIS visible image of eastern Canada and north-eastern United States for July 7, 2002 (left) and modelled aerosol optical depth (right). Active wildfires are shown as red squares on the satellite photograph. The smoke plume blankets southern Québec and extends southwards over the Great Lakes and eastern United States into the Atlantic Ocean.

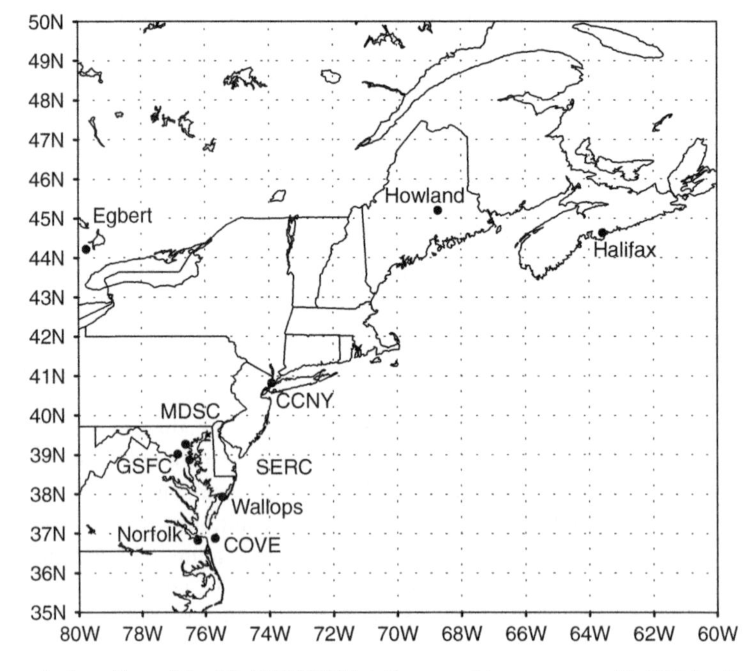

FIGURE 4. Location of the 10 AERONET stations used to compare with GEM-AQ.

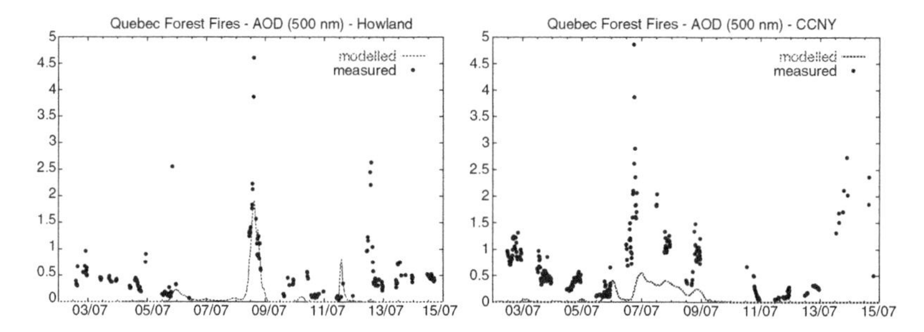

FIGURE 5. 340 nm AOD measurements and model results for Howland, Maine (left) and New York (right).

Acknowledgement The authors acknowledge support from the Canadian Foundation for Climate and Atmospheric Sciences and from the Meteorological Service of Canada, Environment Canada.

References

Benoit, R., Desgagne, M., Pellerin, P., Pellerin, S., Chartier, Y., 1997, The Canadian MC2: a semi-Lagrangian, semi-implicit wideband atmospheric model suited for fine scale process studies and simulation. *Mon. Wea Rev.* **125**: 2382-2414.

Côté, J., S. Gravel, A. Méthot, A. Patoine, M. Roch, A. Staniforth, 2002, The operational CMC-MRB Global Environmental Multiscale (GEM) Model: Part I – Design considerations and formulation, *Mon. Wea. Rev.,* **126**, 1373-1395.

Gong, S. L., L. A. Barrie, J.-P. Blanchet, K. von Salzen, U. Lohmann, G. Lesins, L. Spacek, L. M. Zhang, E. Girard, H. Lin, R. Leaitch, H. Leighton, P. Chylek, P. Huang, 2003, Canadian Aerosol Module: A size-segregated simulation of atmospheric aerosol processes for climate and air quality models, 1. Module development, *J. Geophys. Res.*, **108**(D1), 4007, doi:10.1029/2001JD002002.

Kaminski, J.W., Plummer, D.A., Neary, L., McConnell, J.C., Struzewska J. and Lobocki, L., 2002, First Application of MC2-AQ to Multiscale Air Quality Modelling over Europe, *Phys. Chem. Earth*, **27**: 1517-1524.

Guenther, A., B. Baugh, G. Brasseur, J. Geenberg, P. Harley, L. Klinger, D. Serca, L. Vierling, 1999, Isoprene emission estimates and uncertainties for the Central African EXPRESSO study domain, *J. Geophys. Res.*, **104(D23)**: 30625-30639.

Guenther, A., C.N. Hewitt, D. Erickson, R. Fall, C. Geron, T. Graedel, P. Harley, L. Klinger,. M. Lerdau, W.A. McKay, T. Pierce, R. Scholes, R. Steinbrecher, R. Tallamraju, J. Taylor and P. Zimmerman, 1995, A global model of natural volatile organic compound emissions, *J. Geophys. Res.*, **100(D5)**: 8873-8891.

Lurmann, F.W., A.C. Lloyd and R. Atkinson, 1986, A chemical mechanism for use in long-range transport/acid deposition computer modeling, *J. Geophys. Res.,* **91**: 10905-10936.

Mailhot, J., Belair, S., Benoit, R., Bilodeau, B., Delage, Y., Fillion, L., Garand, L., Girard C. and Tremblay, A. 1998, *Scientific description of the RPN physics library – Version 3.6.* Recherché en Prevision Numerique, Atmospheric Environment Service, Dorval, Quebec, 188 pp.

Plummer, D. A., McConnell, J. C., Neary, L., Kaminski, J., Benoit, R., Drummond, J., Narayan, J., Young, V., and Hastie, D.R., 2001: Assessment of emissions data for the Toronto region using aircraft-based measurements and the air quality model. *Atmos. Env.* **35**: 6453-6463.

Tanguay, M., Robert, A. and Laprise, R., 1990, A semi-implicit semi-Lagrangian fully compressible regional forecast model, *Mon. Wea. Rev.*, **118**, 1970-1980

Developments and Results from a Global Multiscale Air Quality Model (GEM-AQ)

Speaker: L. Neary

Questioner: J. H. Baldasano

Question: Do you obtain better results with the emission inventory of 1 km versus 4 km?

Answer: MC2-AQ was run at 1km and 3km, with the 3km simulation giving results more closely matching observations. We have not determined whether this is due to the inventory or the input climatological and geophysical fields for MC2. Model results are very sensitive to soil moisture, etc. Information about injection height for some of the emissions might also improve MC2-AQ results. We are continuing to adjust the surface parameters to better evaluate the model and inventory.

Questioner: W. Jiang

Question: What are the 5 aerosol types in GEM-AQ? Without nitrate and ammonium, it would be hard to get sulfate/nitrate/ammonium together in the aerosol module?

Answer: The 5 bin-resolved aerosol types are sulphate, sea-salt, organic carbon, black carbon and soil dust. There actually is bulk nitrate and ammonium included but nitrates are not size-resolved at this time.

45
A Variable Time-Step Algorithm for Air Quality Models

M. Talat Odman and Yongtao Hu*

1. Introduction

Air quality models (AQMs) are based on the atmospheric transport and chemistry equation:

$$\frac{\partial c}{\partial t} + \nabla \cdot (\mathbf{u}c) = \nabla \cdot (\mathbf{K}\nabla c) + R(c). \tag{1}$$

Here c is a vector of pollutant concentrations, $\mathbf{u}$ is the wind field, $\mathbf{K}$ represents parameterized atmospheric turbulence and R represents chemical production (or loss) of c. The dependence of the variables on the spatial variable x and time t is not shown here for simplicity. Both $\mathbf{u}$ and $\mathbf{K}$ are given (usually provided by a prognostic meteorological model) so that the problem is linear with respect to the transport part. Characteristic times differ from one process to another. In particular, the range of characteristic times for chemical reactions in R spans several orders of magnitude.

After spatial discretization of Eq. (1) a semi-discrete system of the form

$$\dot{w} = F(w) \tag{2}$$

is obtained. The new variable w consists of c and some other parameters and F is a vector function. The computational power required to solve this system is enormous due to the stiffness caused by the wide range of characteristic times. In addition, various numerical difficulties associated with special requirements of each transport and chemistry process must be dealt with. Therefore, in AQMs, Eq. 2 is divided into smaller pieces. A common approach is process splitting:

$$\dot{w} = F(w) \equiv F_A(w) + F_D(w) + F_R(w) \tag{3}$$

where F is split into functions representing different processes (Blom and Verwer, 2000). The function F_A contains horizontal and vertical advection terms. Advection is the dominant horizontal transport process; vertical advection is usually less

* Talat Odman, Georgia Institute of Technology, Atlanta, Georgia, 30332-0512, talat.odman @ce.gatech.edu, Telephone: 404-894-2783, Fax: 404-894-8266.
Yongtao Hu, Georgia Institute of Technology, Atlanta, Georgia, 30332-0512.

important. Some models, e.g., the Urban-to-Regional Multiscale (URM) model (Boylan et al., 2002), combine horizontal diffusion with advection while others keep them separate. F_D and F_R contain the vertical diffusion (and horizontal diffusion, if not already included in F_A) and chemical reaction terms, respectively. Note that not all processes modeled in current AQMs are shown in Eq. (3). A difficult choice is whether to make emissions part of F_D or F_R. While there are valid arguments for both choices, the URM model solves the problem by combining F_D and F_R into a single diffusion-chemistry function.

Splitting drastically reduces the computational resources required by Eq. (2). It also allows using custom-built numerical solvers for each piece or process. It is much easier to deal with process-specific problems individually rather than trying to develop a general solver. For example, the advection operator is usually made nonlinear to achieve positivity either through filtering or flux-limiting. The only disadvantage of splitting is that it introduces an error into the solution unless F_A; F_D and F_R commute with each other (Lanser and Verwer, 1999). The conditions for this are that **u**, **K**, and R do not vary spatially, and that R is linear in c. Since these conditions are not satisfied in any realistic AQM, the question is not whether there would be splitting errors but how large they are.

Splitting methods are classified as first or second order based on the order of the splitting error they introduce. Most splitting methods are first order. Strang splitting (Strang, 1968), which is believed to be second order, became very popular in AQMs after being adopted by McRae et al. (1982). It advances the solution in time by the following sequence of operators:

$$c^{n+1} = \Phi_A\left(t^{n+\frac{1}{2}}; \frac{\Delta t}{2}\right) \Phi_D\left(t^{n+\frac{1}{2}}; \frac{\Delta t}{2}\right) \Phi_R(t^n; \Delta t)\, \Phi_D\left(t^n; \frac{\Delta t}{2}\right) \Phi_A\left(t^n; \frac{\Delta t}{2}\right) c^n. \quad (4)$$

Note that the transport operators Φ_A and Φ_D are applied for one half of the splitting time step, Δt, symmetrically around the chemistry operator Φ_R. Recently, Sportisse (2000) argued that, for R linear in c, Strang splitting is only first order unless the stiff operator Φ_R is applied last.

The splitting time step determines the frequency by which processes of differing characteristic times are "synchronized" or "coupled." In general, the characteristic time for advection, which is equal to the grid size divided by the wind speed, is selected as the splitting time step. Current AQMs use a global (splitting is dropped from hereon) time step, which is determined by the maximum wind speed in the entire domain. A global time step may be inefficient when the characteristic times for a large number of grid cells are larger than Δt. This is often the case when the modeling domain includes a body of water or extends to altitudes above the boundary layer. High wind speeds over water or aloft dictate small time steps over the entire domain even if the local wind speeds over a large fraction of the domain may be very low. If the grid is non-uniform such as in the Adaptive Grid Model (Odman et al., 2002), where the grid sizes differ by two orders of magnitude, the inefficiency of having a single time step is even more significant.

In this paper, a local time-step algorithm for use in AQMs is described. Its accuracy and computation-time benefits are evaluated via comparisons with a model that uses a global time step. Potential speedups for typical domains and wind fields are discussed.

2. Methodology

We developed the variable time-step algorithm (VARTSTEP) to be able to use local time steps in AQMs. VARTSTEP allows each grid cell to advance by its own time step. The version of the algorithm described here is two-dimensional since it uses the same time step for an entire vertical column. To our knowledge, local time steps have only been used in the vertical direction therefore this is the first attempt at using local time steps in the horizontal domain of an AQM. Extension of the algorithm to third dimension is straightforward. We implemented VARTSTEP in the Community Multiscale Air Quality (CMAQ) model (Byun and Ching, 1999). CMAQ is a uniform grid AQM therefore the maximum wind speed determines the global time step. Version 4.3 of CMAQ introduced a limited local time-step capability in the vertical direction. The user is allowed to specify an altitude up to which wind speeds are considered in determining the global time step, and stronger winds aloft are ignored. We preserved this capability in the VARTSTEP version of CMAQ, which is also referred to as VARTSTEP-CMAQ.

VARTSTEP assigns every vertical column its own time step, Δt_i, which satisfies two conditions. The first condition is

$$\frac{u_i^{\max} \Delta t_i}{\Delta x_i} \leq 1 \tag{5}$$

where $u_i^{\max}$ is the maximum wind speed in vertical column i (up to the user-specified altitude) and Δx_i is the horizontal grid size for that vertical column (subscript i is used considering that grid size may be non-uniform, as in the adaptive grid model). The second condition is that Δt_i be an integer multiple of the global time step and an integer divisor of the output time step. CMAQ generates outputs typically once every hour. Therefore, if the output time step is 60 min and if the global time step is 5 min, the local time step can be 5, 10, 15, 20, 30, or 60 min. However, it cannot be 25 min, for example, since 25 min does not divide 60 min evenly.

The model clock is advanced in increments equal to the global time step. Operators are applied to the grid cell concentrations of vertical column i for the duration of Δt_i, the local time step, (i.e., concentrations are advanced by Δt_i) only if the clock time is an integer multiple of Δt_i (i.e., $t = \mathrm{N} \times \Delta t_i$). Note that since there is a single time step for each vertical column, there are no difficulties involved in doing this with the vertical diffusion operator, Φ_D. The chemical reaction operator, Φ_R, does not pose any problems either since it is applied to one grid cell at a time. However, the horizontal transport operator, Φ_A, requires special attention since neighboring grid cells in the horizontal may have different time steps.

Concentrations in grid cells with shorter local time steps are updated as usual by adding the horizontal fluxes (advective and diffusive) coming from cells with longer time steps. This situation is shown by the arrow marked "Pass" in Fig. 1. For simplicity, let us assume that there are no other cells but Cell 1 and Cell 2. If

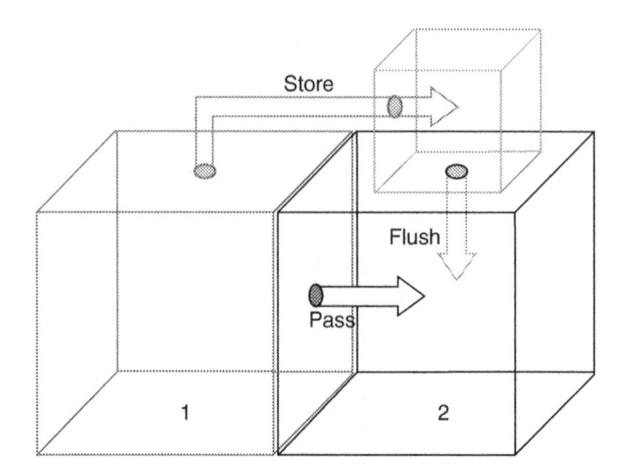

FIGURE 1. Concept of a reservoir used in VARTSTEP algorithm. The flux from Cell 1 is either passed directly to Cell 2 or stored in a reservoir which is flushed when it is time to update Cell 2 concentrations.

$\Delta t_2 < \Delta t_1$, any flux from Cell 1 must be passed to Cell 2 when it's time to update Cell 2 concentrations (i.e., when $t = N \times \Delta t_2$). Concentrations in cells with longer time steps remain constant until the time of update. In other words, if $\Delta t_2 > \Delta t_1$ and if $t \neq N \times \Delta t_2$ any possible horizontal flux from Cell 1 to Cell 2 is directed to a reservoir. This is shown in Fig.1 by the arrow marked "Store". When the time comes for updating Cell 2 concentrations (i.e., when $t = N \times \Delta t_2$) not only any possible flux from Cell 1 is passed to Cell 2, but the mass accumulated in the reservoir is also added to Cell 2. This is represented by the arrow marked "Flush" in Fig. 1. Since all fluxes are eventually added to the appropriate cell, the algorithm is strictly mass conservative.

Above, we ignored any other neighboring cells to simplify the description of the algorithm. In practice, since there are always other neighboring cells, some with shorter local time steps, there may always be some mass accumulated in the reservoir. The actual implementation of VARTSTEP accounts for this possibility and always flushes the reservoir when it is updating grid cell concentrations. Finally, note that VARTSTEP increases memory requirements, due to its need to store fluxes, by an amount equal to the size of the concentration array.

3. Results and Discussion

Results presented here are from an air quality simulation in southeastern U.S. during 1-9 January 2002. Version 4.3 of CMAQ was used for calculating the ambient fine particulate matter ($PM_{2.5}$) concentrations. The grid resolution was 12-km over the entire domain. CMAQ uses a global time step which varied between the user-specified minimum of 5 min and the Courant-number-limited maximum of 7 min and 30 s. In addition to this "benchmark" simulation, a second simulation was conducted with the VARTSTEP version of CMAQ. The computation time for the simulation with VARTSTEP-CMAQ was 35% less than the benchmark CMAQ simulation. The times spent in the aerosol, chemistry and vertical diffusion modules were reduced approximately by 66%, 25%, and 50%, respectively. Considering that the algorithm is only partially implemented in the model (for example, horizontal advection still uses a global time step) this is a significant level of speedup. When the algorithm is fully implemented a factor-of-two speedup might be achieved.

$PM_{2.5}$ concentrations for 5:00 UTC, 5 January 2002 resulting from the two simulations are shown in Fig. 2. This is the time when domain average $PM_{2.5}$ concentrations are highest. Both simulations produced a similar $PM_{2.5}$ distribution over the southeastern U.S. with a peak in Long Island, New York. The value of the peak is 66.4 $\mu g/m^3$ according to the benchmark simulation and 65.6 $\mu g/m^3$ according to the VARTSTEP-CMAQ simulation.

Daily average observations of $PM_{2.5}$ and its composition are available through the Interagency Monitoring of Protected Visual Environments (IMPROVE) network. The network covers areas designated as "Class-1" for visibility protection such as national parks and wilderness areas and there are 22 such observation sites in the region shown in Fig. 2. Eight of these sites are clustered along the Southern Appalachian Mountains, six sites are distributed along the coastline and the remaining eight sites are scattered throughout the domain. Here we will not get into the comparison of modeled concentrations with observations however we will perform a detailed comparison of the VARTSTEP-CMAQ results with benchmark CMAQ results at these 22 sites where performance is of utmost concern. Fig. 3 compares the two sets of model results for daily average $PM_{2.5}$ concentrations at these 22 sites during the January 1-9, 2002 period.

There is strong correlation ($R^2 = 0.98$) between the benchmark and the results obtained with the VARTSTEP version of CMAQ. However, with the exception of a few points, VARTSTEP-CMAQ results are lower than the benchmark, on average, by about 8%. There are two main reasons for this bias. First, we did not

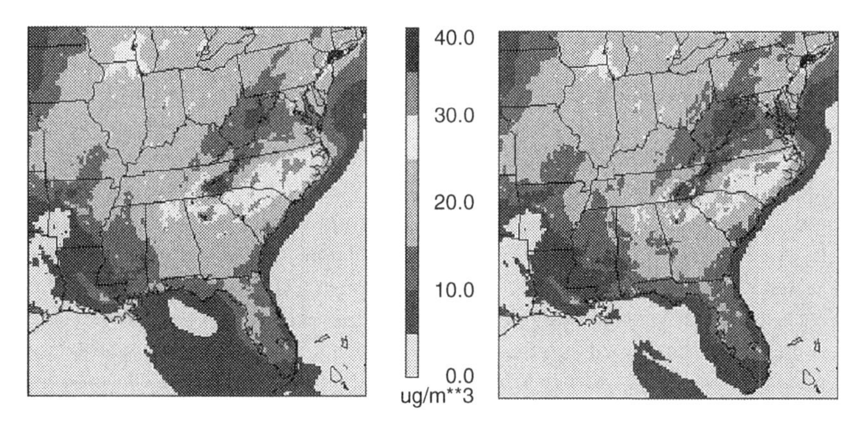

FIGURE 2. Hourly average $PM_{2.5}$ concentrations over southeastern U.S. at 5:00 UTC, 5 January 2002 as calculated by the benchmark CMAQ simulation (left) and the VARTSTEP-CMAQ simulation (right).

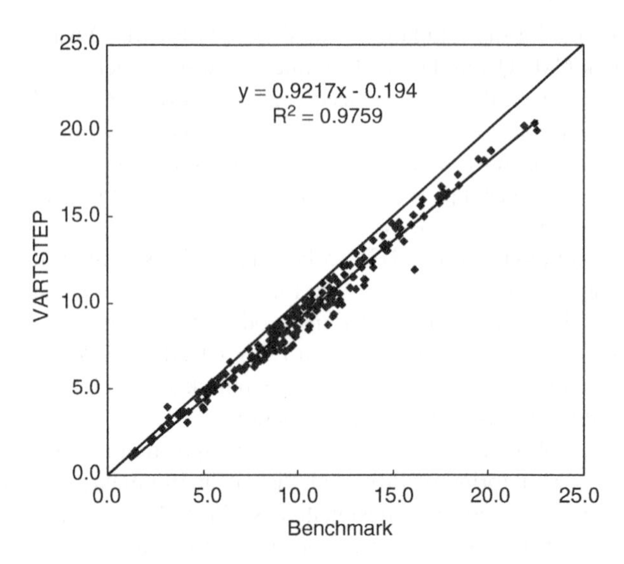

FIGURE 3. Daily average $PM_{2.5}$ concentrations at 22 Class-1 areas in southeastern U.S. during 1-9 January 2002: Comparison of the results obtained from the variable time-step version of CMAQ (y-axis) with the benchmark obtained from Version 4.3 of CMAQ (x-axis).

undertake some straight-forward but time consuming programming tasks at the time this paper was written. This resulted in some differences between CMAQ and VARTSTEP-CMAQ. For example, while meteorological variables used in CMAQ are evaluated at the middle of the global time step, those used in VARTSTEP-CMAQ are not necessarily evaluated at the middle of the local time step. Also, the emission rates are fixed during each hour (step function) in VARTSTEP-CMAQ while they have a linear profile in CMAQ. The second and probably more important reason is a coding error recently discovered in the vertical advection module of CMAQ Version 4.3. The code assumed uniform grid spacing despite the non-uniform vertical layer structure in CMAQ. This may result in more mass being assigned to the surface layer. On the other hand, VARTSTEP uses a different vertical advection module that adjusts vertical velocities for strict mass conservation (Odman and Russell, 2000) therefore it was not affected by this coding error.

Since visibility is the primary concern in this study, it is important to accurately model not just the total $PM_{2.5}$ but also its components. It is well known that different components of $PM_{2.5}$ affect light extinction differently. Contributions to light extinction of sulfate and nitrate particles can be substantial in the presence of water vapor. Elemental carbon can also be an important visibility degradation agent while other particles are relatively less important (Malm et al., 2000). Daily average concentrations of sulfate, nitrate, ammonium, soil, elemental carbon and organic carbon components of $PM_{2.5}$ were also compared at the 22 Class-1 areas during the January 1-9, 2002 period. The scatter plots in Fig. 4 have the concentrations calculated by VARTSTEP-CMAQ on the y-axis and those calculated by CMAQ Version 4.3 on the x-axis.

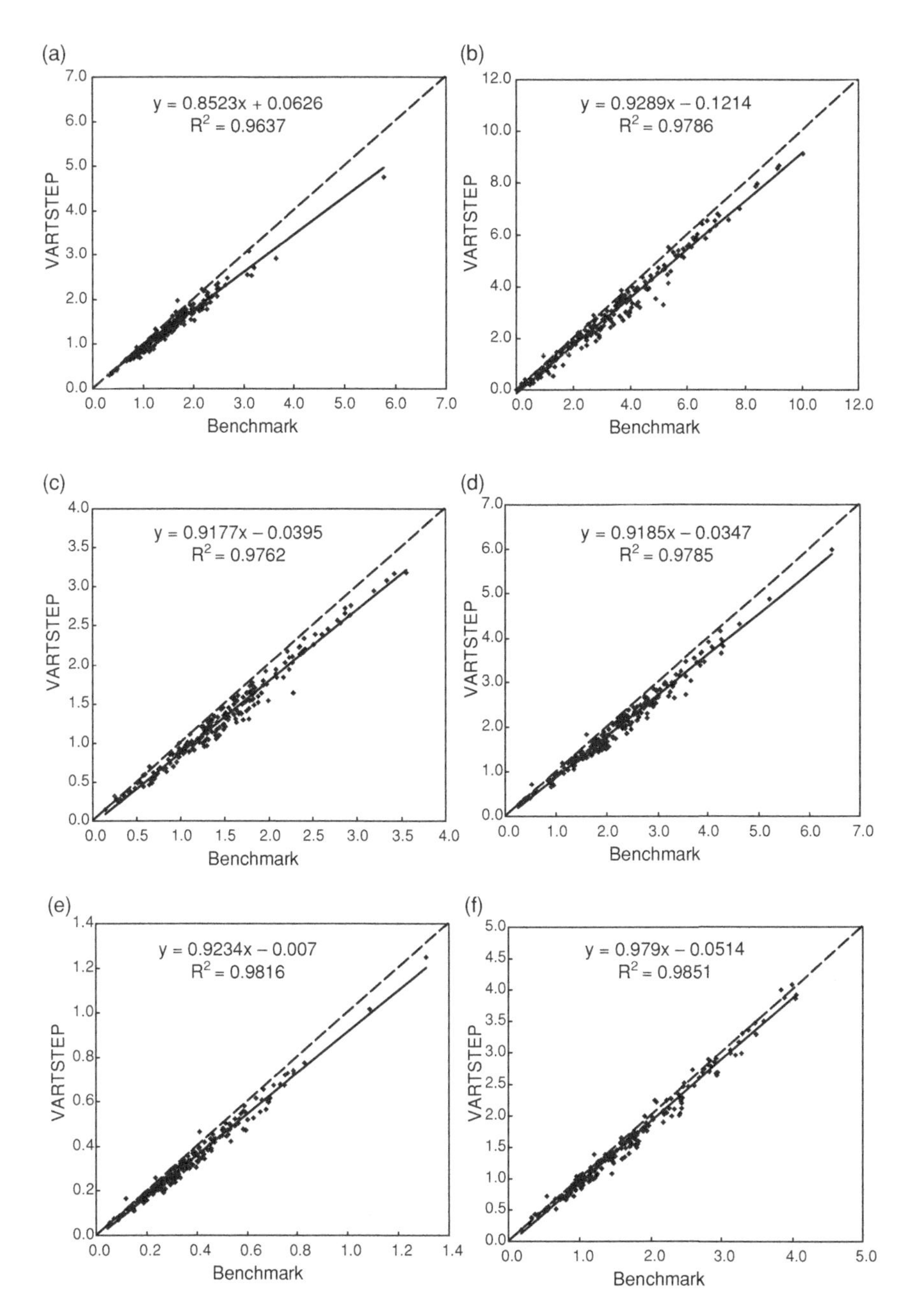

FIGURE 4. Comparison of daily average concentrations calculated by VARTSTEP-CMAQ (y-axis) with the benchmark (x-axis) for various components of $PM_{2.5}$ at 22 Class-1 areas in southeastern U.S. during 1-9 January 2002: (a) Sulfate, (b) Nitrate, (c) Ammonium, (d) Soil, (e) Elemental Carbon, (f) Organic Carbon.

The two model results are highly correlated ($R^2 > 0.96$) but the concentrations of all $PM_{2.5}$ components calculated by VARTSTEP are lower than the benchmark. On the average, sulfate concentrations are about 15% lower but it should be noted that sulfate is only the third largest component of $PM_{2.5}$. In addition, relative humidity is also low in winter therefore, compared to summer time, sulfate has a much smaller contribution to light extinction. Nitrate, the largest component of $PM_{2.5}$, is 7% lower. Ammonium, soil (second largest component) and elemental carbon are each 8% lower. Organic carbon which is the fourth largest component is only 2% lower. Based on these results, one may expect VARTSTEP-CMAQ light extinction to be about 10% lower than the benchmark.

4. Conclusions

The use of a global time step is a source of inefficiency in current AQMs. The variable time-step algorithm, VARTSTEP, was developed to allow the use of local time steps. A two-dimensional version of the algorithm was implemented in CMAQ; it reduced the computation time by 35%. A factor-of-two speedup is likely when the algorithm is fully implemented. The differences between $PM_{2.5}$ concentrations of the VARTSTEP-CMAQ and benchmark CMAQ are small and explainable in terms of known modeling issues. These differences should not significantly affect model performance nor should they change the conclusions drawn from the model. The Adaptive Grid Model (Odman et al., 2002) will greatly benefit from VARTSTEP: a factor-of-100 speedup is expected.

Acknowledgements This research was supported by the Visibility Improvement State and Tribal Association of the Southeast (VISTAS) under contract number V-2003-12 to Georgia Institute of Technology.

References

G. J. McRae, W. R. Goodin, J. H. Seinfeld, 1982. Numerical solution of the atmospheric diffusion equation for chemically reacting flows, *J. Comput. Phys.* **45**, 1-42.

Blom, J. G. and Verwer, J. G, 2000. A comparison of integration methods for atmospheric transport-chemistry problems, *J. Comput. Appl. Math.* **126**, 381-396.

Boylan, J. W., Odman, M. T., Wilkinson, J. G., Russell, A. G., Doty, K. G., Norris, W. B., and McNider, R. T., 2002. Development of a comprehensive multiscale one atmosphere modeling system: application to the southern Appalachian mountains, *Atmos. Environ.* **36**, 3721-3734.

Byun, D. W. and Ching, J. K. S., 1999. *Science Algorithms of the EPA Models-3 Community Multiscale Air Quality (CMAQ) Modeling System*, U.S. Environmental Protection Agency, Washington, DC.

Lanser, D., Verwer, J. G., 1999. Analysis of operator splitting for advection-diffusion-reaction problems from air pollution modeling, *J. Comput. Appl. Math.* **111,** 201-216.

Malm, W. C.; Pitchford M. L.; Scruggs M.; Sisler J. F.; Ames R.; Copeland S.; Gebhart K. A.; Day D. E. 2000. *Spatial and Seasonal Patterns and Temporal Variability of Haze and its Constituents in the United States, Report III*; ISSN: 0737-5352-47; Colorado State University: Fort Collins, CO.

Odman, M. T., Khan, M. N., Srivastava, R. K. and McRae, D. S., 2002. Initial application of the adaptive grid air pollution model, in: *Air Pollution Modeling and its Application XV*, C. Borrego and G. Schayes, eds., Kluwer Academic/Plenum Publishers, New York, pp. 319-328.

Odman, M. T. and Russell, A. G., 2000. Mass conservative coupling of non-hydrostatic meteorological models with air quality models, in: *Air Pollution Modeling and its Application XIII*, S.-E. Gryning and E. Batchvarova, eds., Kluwer Academic/Plenum Publishers, New York, pp. 651-660.

Sportisse, B., 2000. An Analysis of Operator Splitting Techniques in the Stiff Case, *J. Comput. Phys.* **161,** 140-168.

Strang, G., 1968. On the construction and comparison of difference schemes, *SIAM J. Numer. Anal.* **5**, 506-517.

A Variable Time-Step Algorithm for Air Quality Models

Speaker: T. Odman

Questioner: A. L. Norman

Question: What were the criteria used to determine the shape of the grid cells?

Answer: I suppose the question is related to the Adaptive Grid Model. So far we used several criteria for adapting the grid. The grid adaptation algorithm (adaptor) uses a user-defined weight function that can be built as a linear combination of various variables. In ozone simulations, we used surface-layer NO concentrations, surface-layer NO_x *concentrations, vertical column* NO_x*, ozone production rate, and combinations of these variables as weight functions. Actually, we used the curvatures of these variables. The weight function is used in moving the grid nodes and this movement determines the shape of the grid cells. Grid nodes are pulled towards the higher weight, which should be an indication of the need for finer grid resolution. For example, a large curvature in* NO_x *concentrations is generally associated with a rapid change in* NO_x *concentrations, which is likely due to ozone formation. The adaptor would pull the grid nodes towards such areas. As a result, fine grid cells would be generated around these areas while grid cells elsewhere would get coarser.*

Questioner: C. Mensink

Question: What is the reason for the systemic under prediction of the new time stepping algorithm compared to the benchmark?

Answer: Recently, a coding error was discovered in the vertical advection module of CMAQ Version 4.3, which is the version used to generate the benchmark results. The code assumed uniform grid spacing even though the vertical layers are not uniformly spaced in CMAQ. This would lead to larger concentrations in thinner layers and smaller concentrations in thicker layers. Since the surface-layer is the thinnest layer, benchmark surface-layer concentrations are over predictions compared to the corrected Version 4.3 of CMAQ. On the other hand, VARTSTEP-CMAQ uses a different vertical advection module, one that adjusts vertical velocities for strict mass conservation (Odman and Russell, 2000), therefore it is not affected by this coding error. Consequently, VARTSTEP-CMAQ results look like under predictions while in fact the benchmark results are over predictions.

46
Temporal Signatures of Observations and Model Outputs: Do Time Series Decomposition Methods Capture Relevant Time Scales?

P. S. Porter, J. Swall, R. Gilliam, E. L. Gego, C. Hogrefe, A. Gilliland, J. S. Irwin, and S. T. Rao*[1]

1. Introduction

Time series decomposition methods were applied to meteorological and air quality data and their numerical model estimates. Decomposition techniques express a time series as the sum of a small number of independent modes which hypothetically represent identifiable forcings, thereby helping to untangle complex processes. Mode-to-mode comparison of observed and modeled data provides a mechanism for model evaluation.

The decomposition methods included empirical orthogonal functions (EOF), empirical mode decomposition (EMD), and wavelet filters (WF). EOF, a linear method designed for stationary time series, is principal component analysis (PCA) applied to time-lagged copies of a given time series. EMD is a relatively new nonlinear method that operates locally in time and is suitable for nonstationary and nonlinear processes; it is not, in theory, band-width limited, and the number of modes is automatically determined. Wavelet filters are band-width guided with the number of modes set by the analyst.

The purpose of this paper is to compare the performance of decomposition techniques in characterizing time scales in meteorological and air quality variables. The time series for this study, modeled and observed temperature and PM2.5, were chosen because they represent relatively easy and difficult tests,

[1]* P.S. Porter, University of Idaho, Idaho Falls, ID, U.S.A.
C. Hogrefe, ASRC, University at Albany, Albany, NY, U.S.A.
A. Gilliland, R. Gilliam, J. Swall, J. R. Irwin, and Associates, S.T. Rao, NOAA Atmospheric Sciences Modeling Division, Research Triangle Park, NC, U.S.A.
E. Gego, University Corporation for Atmospheric Research, Idaho Falls, ID, U.S.A.

respectively, for decomposition methods. Modeled estimates of temperature are forced to closely track observations from a dense observation network; temporal modes of observations and model temperature time series, estimates should therefore be in close agreement. Comparison of modeled and observed PM2.5, on the other hand, is a more difficult test for decomposition techniques.

Aiding this comparison is an analysis of simulated time series that have features in common with observations. These features include a smooth wave with a period that slowly changes from 40 to 60 days, a cosine wave with a period of one week, and additive red noise. Use of a 40 to 60 day simulated wave was motivated by the Julian-Madden effect observed in some temperature time series, while the question of a one week wave has been the object of many air quality studies as it is considered evidence of anthropogenic forcing.

2. Methods and Time Series

2.1. *Time series*

Time series used in this study include temperature observed at station KLGB (Long Beach, CA, latitude 33.82, longitude -118.15, altitude 10 m), temperature from an annual meteorology-chemistry interface program (MCIP) output of an MM5 simulation (see McNally (2003) and Hogrefe et al (2004) for details), model output for the grid cell containing KLGB (1.5 meter temperature), fine particulate matter (PM2.5) observed at AIRS station 516500004 (Virginia), and CMAQ and REMSAD PM2.5 model output for the grid cell containing the Virginia site. Simulated time series include an oscillating wave with a period varying between 40 and 60 days, the oscillating wave plus noise, a cosine wave with a period of one week, and the weekly cosine wave plus noise. The added noise was a synthetic fractionally integrated autoregressive (AR) process. All time series spanned one year at an hourly sampling rate.

2.2. *Time domain Empirical Orthogonal Functions (EOF)*

Principal component analysis, or PCA, simplifies multivariate problems by reducing a large number of correlated variables to a smaller number that are independent. EOF modes are in essence a collection of AR models whose coefficients are the eigenvectors of the autocovariance matrix of a given time series; as such, EOF modes are not necessarily band-width limited. The number of temporal modes created by EOF, a choice of the analyst that influences the outcome, equals the dimension of the autocovariance matrix, though, in practice, a much smaller number of modes accounts for most process variability. Simulation studies (not shown) indicated that a lag time of at least 289 hours was needed to capture time series features of interest (for example, variation with a period of at least 7 days), hence this value was chosen. (A maximum value of one third of the sample size has been recommended by Vautard et al (1992). EOF, like all the methods used here, are unreliable at time series endpoints. However, there is no general agreement on how much of the end to trim, so here they are left in. Detailed information about EOF and related techniques can be found in Vautard et al (1992).

2.3. *Empirical Mode Decomposition (EMD)*

Empirical mode decomposition (EMD), a temporal decomposition technique relatively new to air quality data analysis, is a nonlinear method that appears to be superior to alternative techniques when applied to nonlinear and or nonstationary processes (Huang et al 1998). Briefly, local minima are connected with a smooth line (in this case, cubic splines) as are local maxima. Original observations minus the average of the smoothed minima and maxima equal the 1st EMD mode. The process is repeated until there are fewer than two local minima or maxima. Criteria for empirical mode formation include: (1) the number of zero crossings = number of extrema (maxima + minima), and (2) modes are roughly symmetric about 0 (local mean = mean of nearby maxima and minima). The Matlab code developed by Rilling et al (2002) was used to compute the EMD modes presented here.

2.4. *Wavelet filters (KZFT)*

While a periodogram is a graph of average energy vs. frequency for the duration of a time series, a wavelet diagram is a local (in time) spectrum revealing changes as a process evolves. Complex features of observations and model outputs may be compared in a single contour image. Wavelet analysis may also reveal signals with slowly changing amplitude and/or phase. Wavelet diagrams may illuminate nonlinear processes for which forcings lead to responses across a range of frequencies. Here we use the KZFT mother wavelet, which is based on the iterated moving average (Zurbenko and Porter, 1998; Hogrefe et al, 2003), and can operate in a missing data environment (see also Torrence and Compo (1998), a very accessible guide to wavelet analysis).

3. Results and Discussion

3.1. *Observed and Modeled Temperature*

Wavelet images of observed and MCIP temperature (Figures 1A and 1B) show several areas of agreement. For example, high amplitude centered roughly at day 30 and a period of 15 days appears in both. Examination of synoptic charts indicates that energy at the 15 day period on day 30 results from a shift in the synoptic pattern from large ridge across the west to an abnormally deep trough. The PNA (Pacific North Atlantic Oscillation) index for that period also has 15 day energy at roughly the same time (not shown) that could be classified as energy from the upper part of the synoptic range, although it is typically less than 15 days over most other parts of the US in the winter. In addition, a ridge that may represent a signal

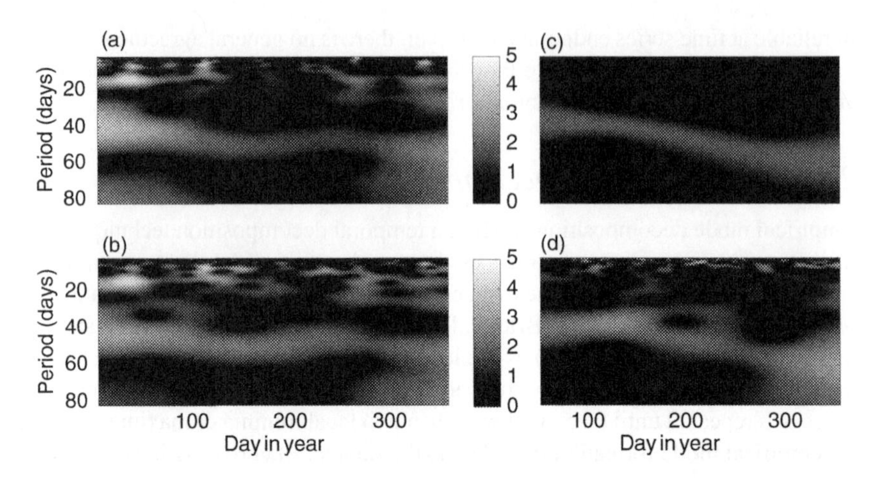

FIGURE 1. Wavelet images: (a) Observed temperature, (b) MCIP temperature, (c) Noise-free wave, (d) Noisy wave.

with a slowly changing period (from roughly 40 to 60 days) appears in both modeled and observed temperature. The point-by-point correlation coefficient for Figures 1A and 1B is 0.44 (R). Figures 1C (the 40-60 days oscillatory wave) and 1D (Figure 1C plus noise) illustrate the effect of noise on a wavelet image. The noise used here has energy across the entire spectrum and leads to features throughout the wavelet spectrum covered by Figure 1. The ridge is distorted as well, with uneven amplitude that nearly disappears at roughly day 260.

Smoothed periodograms (Figure 2) indicate strong agreement between observed and modeled temperature, with a correlation coefficient between Figures 2A and 2B of 0.77 (R for the log scale raw (unsmoothed) periodograms).

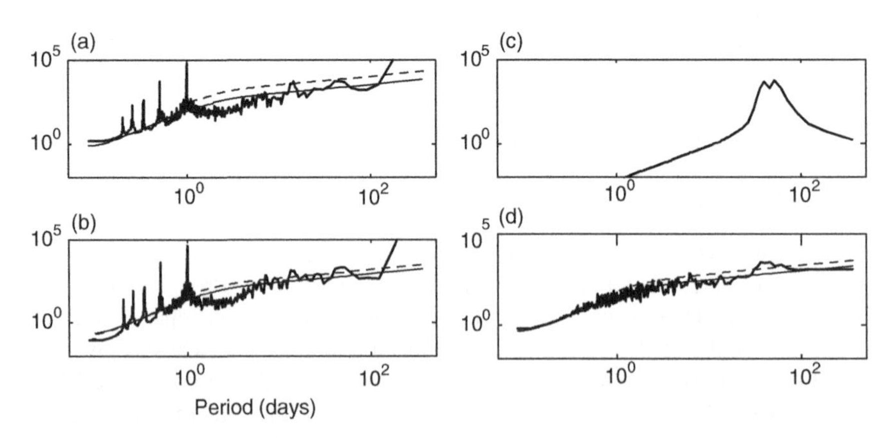

FIGURE 2. Smoothed periodograms: (a) Observed temperature, (b) MCIP temperature, (c) Noise-free wave, (d) Noisy wave. Lines: periodogram = heavy; noise floor = thin; 95% upper bound on noise floor = dashed.

Spikes at 24 hours and harmonics appear in both 2A and 2B, but many fine details at periods greater than about 2 days are present in both as well (R = 0.79 for log scale raw (unsmoothed) peridograms at periods > 2 days)

Focusing on the 40-60 day ridges, they appear on all the periodograms as broadband noise, and were it not for the wavelet image, would probably be identified as such. If the broadband energy in the observations and the model output were coherent signals and not artifacts of the wavelet method, one might expect the various decomposition methods to produce similar results, not only for a given time series, but also for the observations and model estimates.

Would the decomposition methods being addressed here, when applied to white-noise driven processes with energy in the 40-60 day band, produce modes with similar phase and amplitude? Figure 3A shows extraction of a noise-free wave using EOF, EMD, and wavelets. The true wave has constant amplitude of 2 and is nearly indistinguishable from the EMD and wavelet extractions. The EOF extraction (1st mode) has smaller amplitude but identical phase. The first two EOF modes contained 94.0% and 5.9% of total variance, respectively; EMD produced 4 modes, with the 1st having greater than 99.9% of the total variance. Added noise leads to amplitude distortion in all three methods, but their phases are in close agreement (Figure 3B), with EOF appearing to have more amplitude distortion than EMD and wavelets. Added noise also increases the number of significant modes identified by EOF and EMD. The first EOF mode for the noisy wave explains 56.7% of the total variance, and EMD now has 12 modes; mode 9, shown in Figure 3B, explains 23% of the total variance (the true value being 17.8%). Physical explanations for this particular signal, not to mention analyses of climatological time series exhibiting similar features, might lend it credence.

Signals extracted from the observed and MCIP data appear similar to the simulated wave plus noise case (Figures 3C and 3D). The three methods produce waves

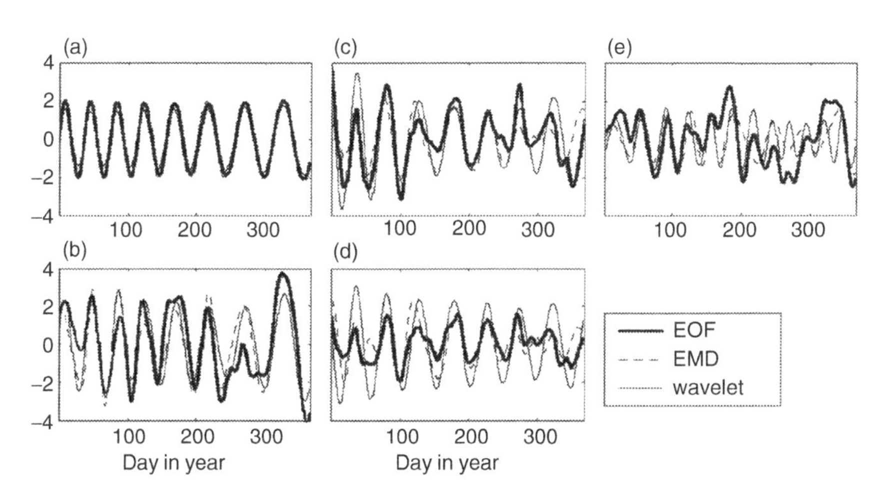

FIGURE 3. Extracted waves: (a) Pure wave, (b) Noisy wave, (c) Observed, (d) MCIP, (e) Noise.

with matching phases and differing amplitudes (the EOF modes in the Figures 3C and 3D were produced by subtracting a smoothed version of the time series from the first EOF mode). The EOF waves shown represent 2.3% and 2.1% of the observed and MCIP temperatures, respectively. EMD produced 12 and 11 modes, respectively, for observations and MCIP estimates; the EMD waves shown in Figures 3C and 3D represent 2.9% and 2.5% of observed and MCIP variance, respectively. Hilbert spectra for the observed and MCIP EMD modes of Figure 3B (not shown) indicate that noise also causes frequency distortion.

Strong agreement among EOF, EMD and wavelet modes suggests that the extracted signals are not part of the noise background, but before we get too excited, it is worth looking at waves extracted by these methods from pure noise. Noise only subjected to the same treatment as the other time series, produced a weak ridge with period varying between 25 and 40 days (not shown). Waves extracted from this time series by the decomposition methods appear to have matching phases for the 1st half of the year but not the 2nd (Figure 3E).

Periodograms of EOF and EMD modes extracted from the synthetic, observed and modeled time series support the result of Flandrin et al (2003), who showed that EMD acts as a bank of 'overlapping band-pass filters' when applied to noise (Figure 4). EOF and EMD applied to pure waves (Figures 4A and 4B) are simple to interpret: a single mode explains the vast majority of the variability of the noise-free wave (notice the log scale). Added noise, on the other hand, results in both methods operating as filter banks (Figures 4C-4D). Applied to climate time series, these methods lead to many (the actual number being proportional to the log of the sample size) band-width limited modes, not just a few modes attributable to clearly definable forcings (Figures 4E -4H). Diurnal modes and harmonics are clearly identified as such, but instead of a single diurnal mode, as one might expect, there are several.

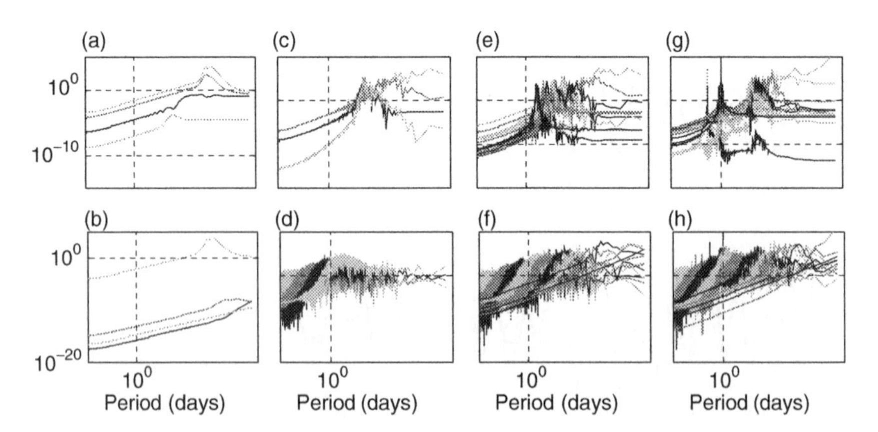

FIGURE 4. Periodograms of temporal components. Noise-free wave: (a) EOF, (b) EMD; Noise: (c) EOF, (d) EMD; Observed temperature: (e) EOF, (f) EMD; MCIP temperature: (g) EOF, (h) EMD.

3.2. Observed and Modeled PM2.5

Smoothed periodograms of observed and modeled PM2.5 have small spikes at a period of seven days (Figures 5A-5C). In considering whether these peaks represent periodic variation, we analyzed a cosine wave with a seven day period (amplitude = 1) buried in the noise used in the earlier example (Figure 5D). Wavelet images of these time series (Figure 6) are inconclusive; though all the images have energy at a period of 7 days, there is no distinct band that would characterize a strictly periodic process. EMD resulted in 13, 12, 12, and 12 modes for observed, CMAQ, REMSAD and simulated processes, respectively.

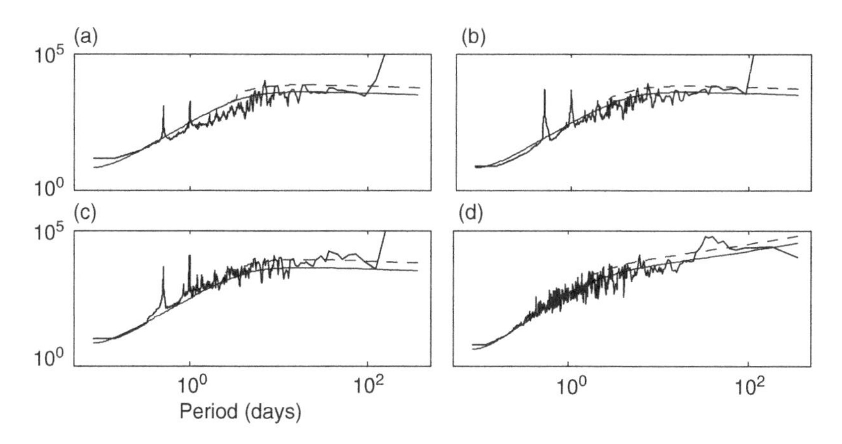

FIGURE 5. Smoothed periodograms of PM2.5: (a) Observed, (b) CMAQ; (c) REMSAD, (d) Noisy wave. Lines: periodogram = heavy; noise floor = thin; 95% upper bound on noise floor = dashed.

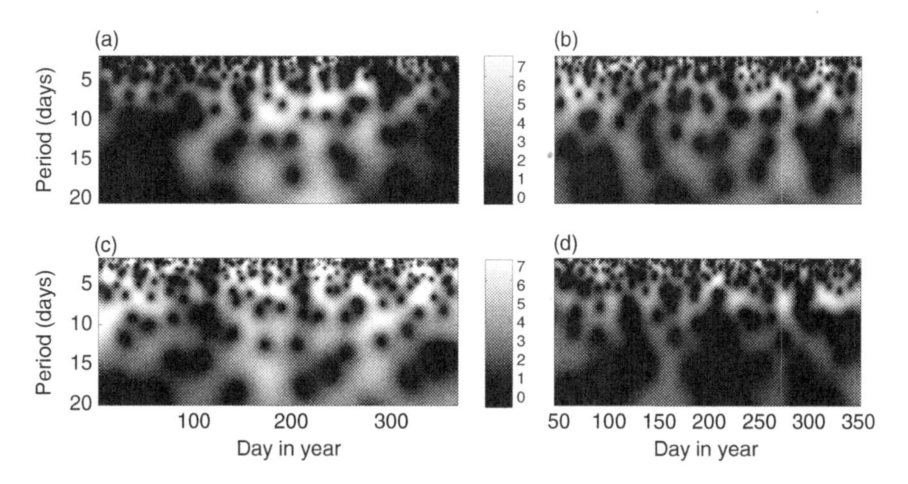

FIGURE 6. Wavelet images of PM2.5: (a) Observed, (b) CMAQ, (c) REMSAD, (d) Simulation.

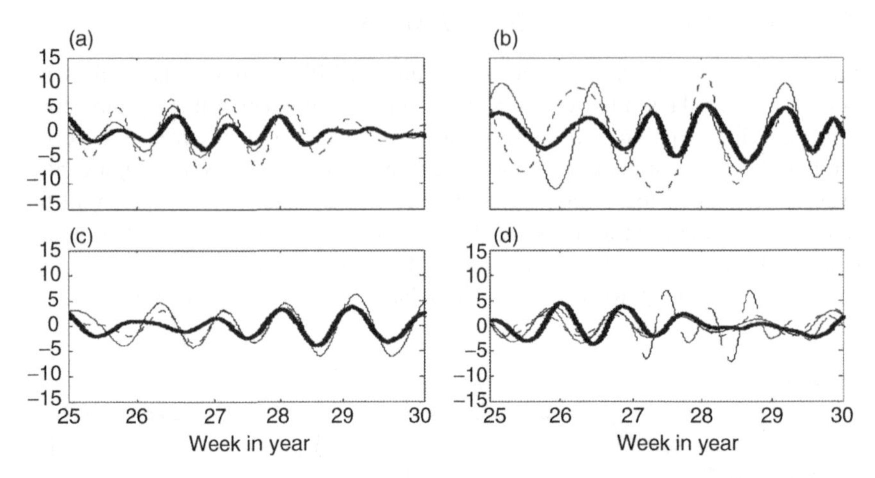

FIGURE 7. Extracted weekly waves: (a) EOF, (b) EMD, (c) Wavelets, (d) All applied to noisy wave. Lines for A, B, and C: observed = thin solid, CMAQ = heavy, REMSAD = dashed; Lines for D: EOF = heavy, EMD = thin, wavelet = dashed, pure wave = dotted.

The periodograms of the EOF and EMD modes of these time series resemble those produced for the temperature time series in that they display what would be expected from overlapping band-pass filters (not shown). Modes with energy nearest seven days plus the wavelet with a period of seven days were compared (Figure 7). EOF and wavelet signals have differing and varying amplitudes but are approximately in phase (Figures 7A and 7C). The EMD signals extracted from observed and CMAQ time series are similar in phase, but the REMSAD EMD mode is not in phase with the others (Figure 7B). Waves extracted from the noisy seven day cosine wave are not in phase either, which is a little surprising. The ratio of the signal to the total variance for the simulation was 4.4% (much smaller than the 17.4% of the 40-60 day simulated wave). Signal to total variance for the other time series shown in Figure 7 (shown on the figure), range from 4.5 to 23.4

4. Summary

The simulation portion of this study indicates that all the methods we tested have difficulty detecting signals embedded in noise. The wavelet method appeared to have less amplitude and frequency distortion, but, like all the methods, detected a phantom wave (Figure 3E), casting doubt on the idea that several methods finding similar phenomena confirms a particular result. More intensive study of the effects of noise on these methods is warranted. Previous studies of this nature used much larger signal/noise ratios than encountered in the hourly PM2.5 observations we analyzed. One should also attempt to identify forcings by analyzing nearby observations and climatological time series affected by similar forcings.

The aims of observation and model output comparisons involving temporal modes are slightly less ambitious than those of signal detection. For example, metrics can be created from modal amplitudes and periodograms without speculation about possible forcings. We plan to continue these analyses using observed and modeled ozone.

Acknowledgments and Disclaimer The research presented here was performed under the Memorandum of Understanding between the U.S. Environmental Protection Agency (EPA) and the U.S. Department of Commerce's National Oceanic and Atmospheric Administration (NOAA) and under agreement number DW13921548. The Department of Commerce partially funded the research described here under contracts with Dr. E. Gego (EA133R-03-SE-0710), with the University of Idaho to Dr. P. S. Porter (EA133R-03-SE-0372), and with the State University of New York to Dr. C. Hogrefe (EA133R-03-SE-0650).

Although this manuscript has been peer reviewed by EPA and NOAA and approved for publication, it does not necessarily represent their views or policies.

References

Flandrin, P, Rilling, G. and Goncalves, P, 2003, Empirical mode decomposition as a filter bank, *IEEE Signal Processing Letters*. 11: 112-114.

Hogrefe, C., Vempaty, S., Rao, S. T., Porter, P. S., 2003, A comparison of four techniques for separating different time scales in atmospheric variables, *Atmospheric Environment*. 37: 313-325.

Huang, N. E., Zheng, S., Long, S. R., Manli, C. W., Shih, H. H., Zheng, Q, Yen, N., Tung, C. C., and Liu, H. H., 1998, The empirical mode decomposition and the Hilbert spectrum for nonlinear and non-stationary time series analysis, *Proc. R. Soc. Lond. A*. 454: 903-995.

Rilling, G., Flandrin, P., and Gonçalves, P., 2002, Matlab code for emd, http://perso.ens-lyon.fr/patrick.flandrin/emd.html

Vautard, R. R. Yiou, and Ghil, M., 1992, Single spectrum analysis: A toolkit for short, noisy chaotic signals, *Phys. D Amsterdam*. 58: 95-126.

Zurbenko I. and Porter, P. S., 1998, Construction of high resolution wavelets, *IEEE Journal of Signal Processing*. 65: 163-173.

47 Wind Tunnel Study of the Exchange Between a Street Canyon and the External Flow

P. Salizzoni[1], N. Grosjean, P. Méjean, R. J. Perkins, L. Soulhac, and R. Vanliefferinge[2]

1. Introduction

In the last three decades, in order to model pollutant dispersion inside a street canyon, several models have been proposed to describe mean concentration and retention time of pollutant inside the canyon, in function of the flow dynamics of the external flow (Berkowicz, 2000, Soulhac, 2000, De Paul and Sheih, 1986, Caton et al., 2003). Some of these models take account just of the mean velocity at the roof height U_h, some other do also consider the turbulence intensity of the incoming flow.

The aim of this study is to evaluate how different conditions of the external flow induce different velocity field inside the canyon, and to find an appropriate velocity and length scale to characterize the mass exchange between the recirculating region and the external flow.

2. Experimental Setup and Measurement Techniques

The study has been carried in a recirculating wind tunnel in the Laboratoire de Mécanique des Fluides et d'Acoustique at the Ecole Centrale de Lyon.

The goal of this study is to investigate the structure of the mean and the fluctuating velocity field inside a street canyon with a fixed aspect ratio between height and width $h/b = 1$, by varying the condition of the external flow. In order to change the upwind boundary conditions, three different incoming wind profiles have been reproduced, with different turbulent intensities and mean velocity characteristics.

[1] Politecnico di Torino, DIASP, Italy
[2] Laboratoire de Mécanique des Fluides et d'Acoustique, CNRS UMR 5509 Ecole Centrale de Lyon, France

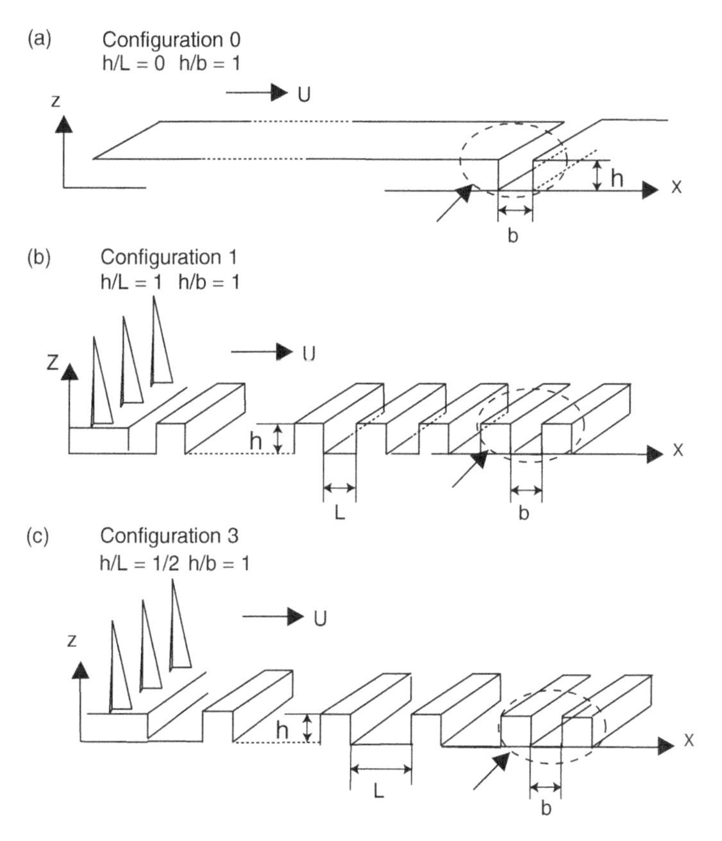

FIGURE 1. Experimental set up.

The first wind profile (configuration 0, figure 1a) is a turbulent boundary layer over a smooth wall. The other two boundary layers have been obtained combining the effect of three spires located at the entrance of the test section – with a lateral spacing equal to half the spire height (Irwin, 1981) – and the effect of wall roughness, due to an array of 2D parallel canyons, made of a set of square section bars (0.06m × 0.06m) placed normal to the wind. The aspect ratio of the 2D parallel canyons varied from h/L = 1 (configuration 1, figure 1b), to reproduce a typical skimming flow regime, to h/L = 1/2 (configuration 3, figure 1c), corresponding to a wake interference flow, according with Oke's classification (Oke, 1988). The experiments have been carried keeping a constant external velocity $U_{max} = 6.8$ m/s.

Measurements of external velocities were carried out by means of hot-wire anemometry in configuration 1 and 3, for external velocities in configuration 0 a PIV system was used instead. Within the cavity a PIV system was used for all configurations.

In Figure 2 are shown the mean velocity, vertical fluctuation velocity and turbulence intensity profiles corresponding to the three configurations.

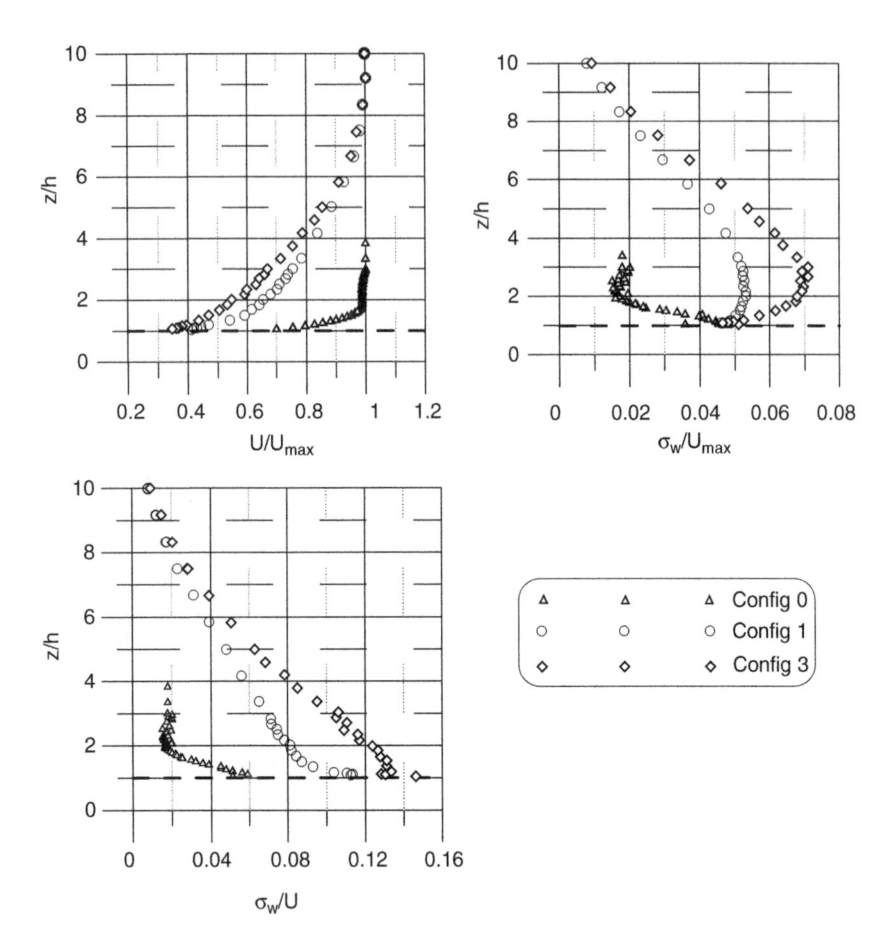

FIGURE 2. External flow profiles.

The mass exchange between the canyon and the external flow has been estimated by measuring the time for the pollutant to be washed out from the cavity, only for the configuration 1. The tracer gas concentrations were measured using a Flame Ionisation Detector. Ethane was used as the passive tracer, since its molecular weight is nearly the same as that of air.

3. Results

3.1. *Velocity Field Inside the Canyon*

In Figure 3 are shown the profiles of the mean horizontal velocity U, as function of z at the mid-length of the cavity ($x = 0$), and of the mean vertical velocity W, as a function of x, at the cavity mid-height. Those profiles have been at first made dimensionless by the velocity at roof height U_h, which is used as reference velocity in many dispersion models (Berkowicz, 2000, Soulhac, 2000, De Paul and Sheih, 1986) within canyons. A more appropriate choice of the velocity scale, however, seems to be U_{int}, the wind velocity inside the canyon just below the mixing layer at the top of the cavity. A closer regroupment of experimental results, using U_{int} as velocity scale, can be seen in Figure 3.

In Figure 4 are shown the turbulent kinetic energy fields corresponding to the three configurations studied. At first glance we can observe two important features:

- the shear layer is clearly confined in a small portion of the domain at the top of the cavity, where the turbulent exchange takes place. This allows an easy evaluation of a typical length scale of the turbulent exchange process, that could be estimated, for the three configurations, of the order of $l_{turb} \approx 10^{-1}b$ (b is the canyon width)

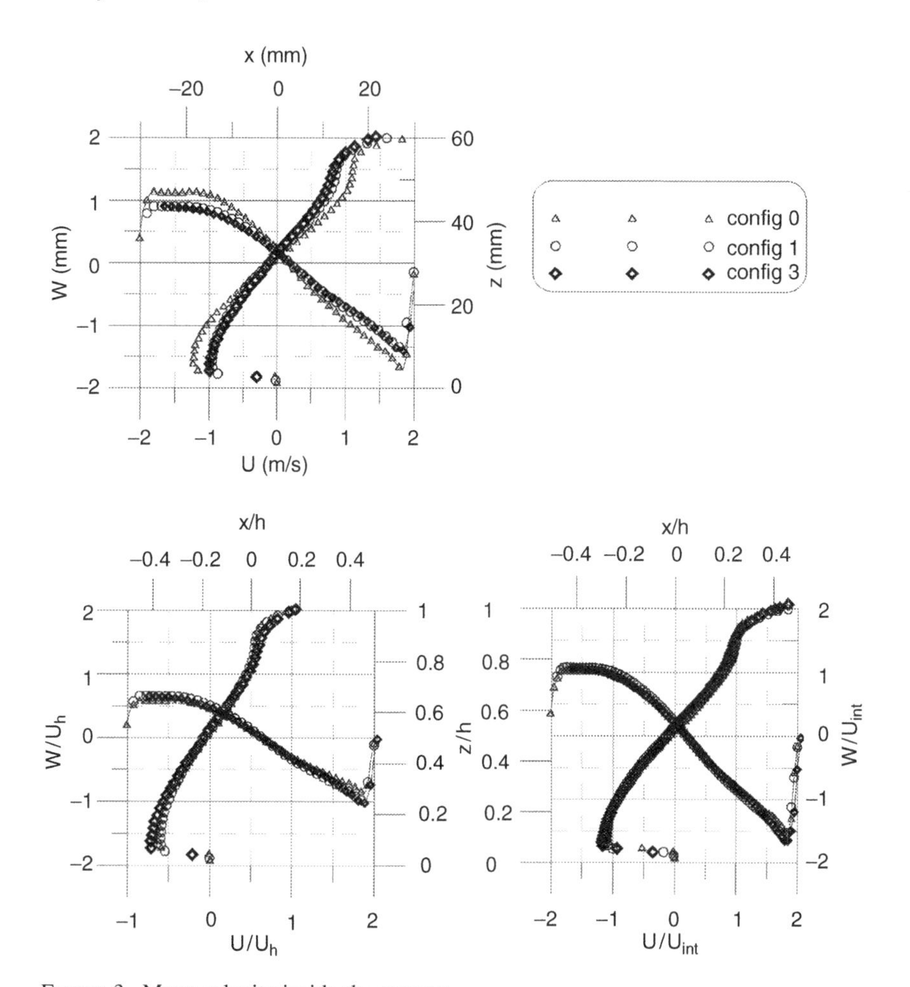

FIGURE 3. Mean velocity inside the canyon.

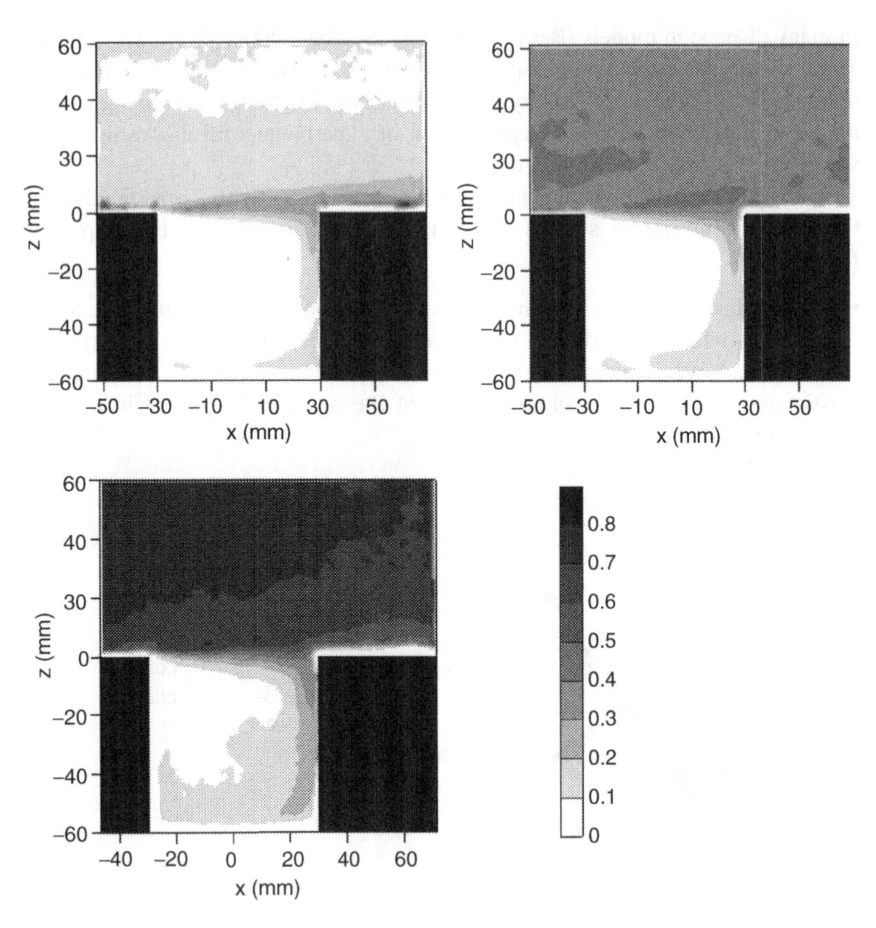

FIGURE 4. Dimensionless turbulent kinetic energy inside the canyon $(1/2q^2)/U_{int}^2$.

- the different external conditions of the turbulent kinetic energy do not seem to be reflected inside the canyon, where the turbulent kinetic energy is very low. The shear layer seems to act as a filter for the external turbulence which does not penetrate to a large amount inside the canyon.

3.2. *Transfer at the Top of the Cavity*

In order to evaluate the typical time scale for mass transfer between the recirculating region and the external flow, we measured the temporal evolution of ethane concentration in the cavity as it empties (Caton et al., 2003). The concentration was measured close to the up-wind wall of the cavity, and the experiment was repeated 30 times, for each configuration, to allow an "ensemble" average for the signals.

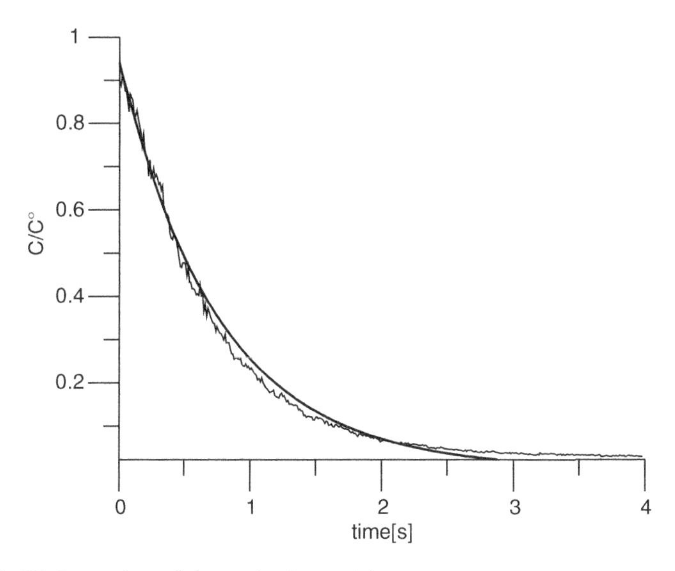

FIGURE 5. Wash out time of the cavity (seconds).

The mass exchange between the canyon and the external flow can be simply described by a mass balance:

$$\frac{d(VC)}{dt} = -w_{exchange} SC$$

where $w_{exchange}$, is the velocity scale of the global mass exchange between the canyon and the external flow, $V = Sh$ is the cavity volume, S the exchange surface, h is the canyon height and C is the mean concentration in the canyon. Substituting and integrating we obtain

$$\frac{C}{C_0} = \exp(-t/T)$$

where T is the retention time scale of pollutants inside the cavity. This is a simple model, describing the canyon as a box with uniform pollutant concentration and a discontinuity surface at the top, where the mass exchange takes place. Fitting the experimental wash-out diagram by mean of an exponential law, an experimental retention time T_s has been estimated,

$$T_s = 0.77s$$

If we consider the velocity field of the cavity as made up of two different regions – the shear layer at the top and the bulk of the fluid inside the cavity – we can estimate the time T_{turb} required for a polluted particle to cross the shear-layer from inside the canyon to the external flow as

$$T_{turb} = \frac{l_{turb}}{\sigma_w} \simeq \frac{5^* 10^{-3}}{5^* 10^{-1}} \simeq 10^{-2} s << T_s$$

where $l_{turb} \simeq 10^{-1\,b} \simeq 5^{*}\ 10^{-3}\,m$ is the maximum width of the shear layer and σ_w is the standard deviation of the vertical velocity fluctuation, whose value at the interface is about 0.5 m/s (Figure 2). The time T_{turb} is by far smaller than T_s, the retention time of the pollutant in the cavity. The turbulent transport through the shear layer can be considered almost instantaneous in the scale T_s.

Therefore, the wash-out time of the cavity depends mainly on the velocity field inside it. By taking an ensemble of trajectories along which a generic particle is brought to the shear layer, starting from a generic position inside the cavity, we can calculate the average number <n> of revolutions required. By estimating the length of any revolution approximately as that of a circular path with a diameter h, we obtain a single revolution time

$$T_r = \frac{\pi^* h}{U_{\text{int}}} \simeq 0.188s$$

and an averaged number of revolutions for a particle before being captured in the shear-layer dynamics as

$$<n> = T_s / T_r \simeq 4$$

which seems to be a reasonable value.

This simple wash-out time analysis shows that, in a square cavity, the pollutant transport outside the canyon is controlled over by the advective transport in the core of it. This means that enhancing the vertical fluctuations at the top of the cavity is not likely to reduce the wash-out time, if not by influencing the velocity field in the body of the cavity.

4. Conclusions

The influence of varying upwind profiles on the flow inside a squared street canyon has been studied, using wind tunnel experiments. The results show that the turbulence level inside the cavity is not very sensitive to the external turbulence condition, since the shear layer at the interface acts as a filter for the incoming turbulent structures. The suitable velocity scale for the recirculation region within the cavity is not U_h (the mean velocity at roof height); a more appropriate scale is U_{int}, the velocity at the top of the cavity just below the mixing layer.

An analysis of the wash-out time shows that the pollutant escape is limited by the advective transfer inside the canyon, and not by the turbulent exchange processes at the interface. Once a velocity inside the canyon in chosen as the correct scale, no more turbulence intensity effect should be taken into account in evaluating the retention time of pollutant inside the canyon.

References

Berkowicz, R. 2000. Operational street pollution model – a parametrised street pollution model. *Environmental Monitoring and Assessment* **65** (1-2), 323.

Caton, F., Britter, R. E. & Dalziel S. 2003. Dispersion Mechanism in a Street Canyon. *Atmospheric Environment* **37**, 693-702.

DePaul, F. & Sheih, C., 1986. Measurment of wind velocities in a street canyon. *Atmospheric Environment* **20** (3), 455-459.
Irwin, H.P.A.H. 1981. *The Design of Spires for Wind Simulation.* J. Wind Eng. And Industrial Aerodynamics, Amsterdam, pp. 361-366.
Oke, T. R. 1988. Street design and urban canopy layer climate. *Energy and Buildings* **11**, 103-113.
Soulhac, L. 2000. Modélisation de la dispersion atmosphérique à l'interieur de la canopée urbaine. *Phd Thesis*, Ecole Centrale de Lyon.

Wind Tunnel Study of the Exchange Between a Street Canyon and the External Flow

Speaker: P. Salizzoni

Questioner: R. Bornstein

Question: What is the effect of canyon aspect ratio and solar heating of canyon walls on your conclusions?

Answer: In this paper we focus our attention on a square cavity, with an aspect ratio $H/W = 1$. We can observe that the flow in the cavity don't seem to be very sensitive to the turbulence intensity conditions in the external flow. We could observe similar flow characteristics in the case of a narrow cavity, with an increased aspect ratio $H/W = 2$. This seems to be a property of flows within cavities in a skimming flow regime, according to Oke's classification. As long as the canyon aspect ratio decreases, in a wake interference regime, we can observe two important features:

- *the turbulent structures of the shear layer have more space to grow and seem to have a stronger interaction with the recirculation flow inside the cavity*
- *the flow in the cavity is more sensitive to the external turbulence.*

On the base of an analysis of the order of magnitude of the advective velocity U, as compared to the convective velocity $u_c = \sqrt{gL\frac{\Delta\rho}{\rho}}$, I consider that the buoyancy effects would not be important on the dynamics of the cavity inner flow (quite an other matter if we consider the larger scale external flow). I would be looking for the numerical experiments I have been told about during the congress, which seem to show a different conclusion.

48 An Example of Application of Data Assimilation Technique and Adjoint Modelling to an Inverse Dispersion Problem Based on the ETEX Experiment

Mikhail Sofiev* and Evgeniy Atlaskin

1. Introduction

Inverse problems is one of comparably new and quickly developing areas. The problems themselves were known for ages but their exact solutions in many cases were either non-existing or requiring a forbidding amount of computations. It is a development of both mathematical methods (first of all, regularization techniques and statistical optimal filters) and new generations of computers that made some of the inverse problems approachable. In the field of atmospheric dispersion, an example of inverse problem is a so-called "compliance regime control" problem introduced by the Convention on Long-Range Transboundary Air Pollution (LRTAP). The task is to evaluate the true emission of primary acid pollutants in Europe from long-term observations and their comparison with model simulations. Host (1996) has shown that such a problem can be approached at a qualitative level only, which has been confirmed by another attempt of Sofiev & Sofieva (2000). In both papers, it is shown that without a heavy use of a-priory assumptions on emission values the uncertainties by far exceed the values themselves, while utilization of a-priory information destroys the signal from the observations together with the noise.

Current paper approaches another type of inverse dispersion problem: determining a source of an accidental release. The main difference from the above long-term inverse problem is that the time scale is about a week, the averaging times for both measured and modeled values do not exceed a few hours, and the emission fluxes are highly irregular both in space and time. Another significant

* M.Sofiev Finnish Meteorological Institute, Sahaajankatu 20 E, 00880 Helsinki Finland; *mikhail.sofiev@fmi.fi*
E.Atlaskin Russian State Hydrometeorological University, St. Petersburg, Russia.

simplification is that the measurement accuracy is of little concern because in cases of accidental releases the typical values of concentrations within the plume are very high, while the background level can be neglected. Finally, one can assume that the source size is small compared to the dispersion domain.

As shown below, these simplifications allow an approximate solution of the inverse problem. The developed methodology is applied to the inverse problem constructed on the basis of the European Tracer Experiment ETEX dataset.

2. Methodology Overview

The methodology consists of three main parts. The first one is based on forward and adjoint dispersion equations as presented by Sofiev (2002) and Sofiev & Siljamo (2003). The second one is a 4-dimensional variational data assimilation 4D-VAR, which description can be found in e.g. (Dimet & Talagrand 1986), with modifications presented below. More information on adjoint formalism and variational assimilation methods can be found in a review of Courtier *et al.* (1993). The third part, also outlined below, is based on a probabilistic interpretation of the adjoint simulation results.

A typical data assimilation problem in meteorology, after (Dimet & Talagrand 1986), consists of a set of differential equations, symbolically written as

$$F(\Phi) = 0 \tag{1}$$

where Φ denotes meteorological fields under considerations (in case of dispersion problem – concentration fields). These fields should verify (1) over a spatial and temporal domain Ω with a border $\partial\Omega$. In dispersion applications, the set (1) describes a dynamic evolution of the concentration fields.

Should one has a set of measurements $\tilde{\Phi}$ of the field Φ, the assimilation of these measurements can be formulated as follows: find a solution of (1) that minimizes a functional:

$$J = \int_{\Omega} \left\| \Phi - \tilde{\Phi} \right\| d\Omega \tag{2}$$

where $\|\bullet\|$ denotes an appropriate norm.

If the solution of the constraint (1) is uniquely defined by some conditions V, one can resolve the constraint and reformulate the problem with unknown V instead of Φ. In many meteorological problems V is defined on $\partial\Omega$ (actually, it is just a set of initial and boundary conditions for Φ), which reduces the overall dimension of the problem (a so-called method of reduction of control variable widely used in control theory).

For the needs of the current study, the above assimilation problem has to be modified because the control variable V contains not only initial and boundary conditions as in meteorological problem but also an emission intensity function $f(\vec{x},t)$. As a partial compensation of this extra complexity, the accidental release assumption allows putting initial concentrations to zero and assuming transparent the down-wind border and a zero concentration along the up-wind one.

Repeating the considerations of Dimet & Talagrand (1986) for the modified problem, one can come to the following formulations: find a solution of the dispersion problem

$$L\varphi = f; \varphi(t=t_0)=0; \begin{bmatrix} \varphi(\vec{x}=\partial\Omega)=0-upwind \\ \nabla\varphi(\vec{x}=\partial\Omega)=0-downwind \end{bmatrix} \tag{3}$$

which minimizes the cost function J from (2). Here L is a differential operator in the dispersion equation, u is a concentration, the emission flux f is the only free parameter that can be varied in searching for the minimum.

It is possible to show that, having a start guess f_0 and performing a standard cycle for 4D-VAR technique: "forward dispersion simulations → model-measurement comparison → adjoint simulations with the model-measurement difference as an input", one obtains the gradient of the cost function:

$$\nabla J = 2\sum_{n=1}^{N} R_n^*[\varphi_n - \tilde{\varphi}_n] = 2\sum_{n=1}^{N} R_n^*[R_n[f_0] - \tilde{\varphi}_n] \tag{4}$$

In the equation (4), it is assumed that there are N observations, and operators R_n, R_n^* provide solutions of the forward and adjoint dispersion equations for the specific n^{th} measurement point, respectively. Derivation of (4) requires linearity of the problem, which is therefore assumed hereinafter. This is not a limitation for the current case because the non-linearities are connected with chemistry and deposition processes, which are not important for the ETEX experiment.

Illustrating the equation (4), the data assimilation cycle can schematically be presented in the following form.

In Figure 1, Δf is an increment of the emission field as a result of the iteration, cost function J is taken as a sum of squares of model-measurement deviations, so that <> denotes an appropriate scalar product between the modelled and

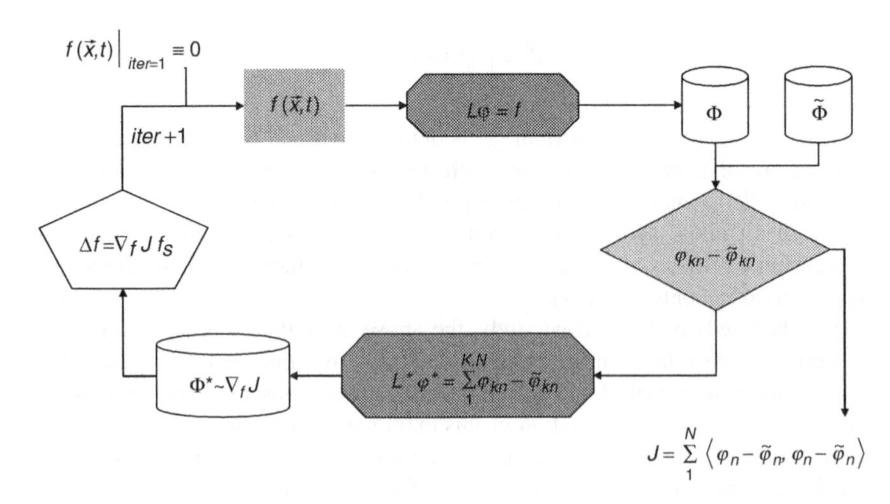

FIGURE 1. A scheme of one data assimilation iteration for inverse dispersion problem.

measured concentration vectors. Computation of the emission increment is straightforward since the gradient of the cost function with regard to the emission is known explicitly, so virtually any numerical optimization method suits to this task. For the below example, the steepest gradient algorithm was applied.

Numerical solutions of both forward and adjoint dispersion problems were performed with the Finnish Emergency Modelling System SILAM (Sofiev & Siljamo, 2003). The kernel of the system consists of the Lagrangian advection routine with Monte-Carlo random-walk diffusion representation. The system is capable of solving both forward and adjoint dispersion problems in exactly the same manner and with very small disturbances, which ensures consistency of results of the iteration cycle.

One more specifics of the considered problem comes out of the nature of accidental releases. The above equations are formulated for non-zero observation values only. Indeed, due to assumed linearity of the dispersion equation, zero emission flux evidently leads to zero concentration level. The same is true for adjoint part: simultaneous modelled and measured zeroes lead to zero φ^*. However, this leads to a large loss of observational information: in most accidental release cases, the bulk of observations report just zeroes, while a small fraction of them manifests concentrations significantly above the background. This was true also for the ETEX experiment. Therefore, an attempt was made to utilize this part of the dataset.

There is no mathematically strict methodology for utilization of these observations so far, it requires further investigations. However, a potentially high value of such "all-zero" information is quite evident: this is a message that corresponding areas are not affected by any emission source. One can interpret the results of adjoint simulations φ^* for non-zero measurements as a density of probability distribution to find the source in the corresponding area (strictly speaking, it is proportional to such a density but this coefficient is constant and does not affect gradient of the cost function). Analogously, a corresponding distribution from the zero point would be a probability NOT to find the source in the corresponding area. Because both simulations include diffusive term, and the meteorological data, dispersion model and measurements themselves contain inevitable uncertainties, both probability densities will be disturbed, blurred and, possibly, overlap due to such distortions. It is then quite natural to use the φ^* distribution from "all-zero" points to correct the φ^* from non-zeroes preventing its expansion to the areas where no sources can be located.

The simplest way of utilizing such information is to subtract the distributions with some weighting coefficient α:

$$\varphi^*_{final} = \varphi^*_{non-zero} - \alpha\varphi^*_{all-zero} \tag{5}$$

The value of the scaling coefficient can be selected so that φ^*_{final} could never be positive near the peaks of distribution from "all-zeroes". For that, it is enough to take:

$$\alpha = (2 \div 5)\frac{\max(\varphi^*_{non-zero})}{\max(\varphi^*_{all-zero})} \tag{6}$$

Since the emission values cannot be negative, the rule (5)-(6) ensures that emission is always zero in areas heavily affecting the stations that reported zero concentrations.

It should be stressed again that the above "trick" is not yet a correct solution of the problem of utilization of zero observations. For example, the resulting distribution cannot be interpreted as probability because it takes negative values in the areas affecting the zero-observation points. However, from physical "common sense" such or similar procedure is reasonable because it makes use of otherwise abandoned information, which amount might be bigger than the size of dataset with non-zero observations. Certainly, if some iteration results in non-zero emission affecting the zero observation, the next cycle will start to reduce its value, finally ending to a certain equilibrium emission value, which, however, will be non-zero. This would mean a blurring of the emission field and require extra iterations for setting the equilibrium. Therefore, the above treatment should both save computations and provide less blurred emission image.

3. Application to the ETEX Experiment

The above methodology was applied for solving a real inverse dispersion problem built on the basis of the first European Tracer Experiment ETEX (Graziani *et al.*, 1998). The experiment was conducted on 23-26.10.1994, with a point source of an inert non-deposible tracer located in Western France. The release conditions were fully controlled and thus are known. The tracer could was followed by ~150 stations during nearly 3 days over Europe, which comprised the observational dataset for the inverse problem.

3.1. First Guess Generation and Assimilation Cycles

There were no a-priory assumptions about the source spatial or temporal features, except for a "fixed location" requirement: the source place and spatial dimensions were considered unknown but not changing during the release. As a result, the first guess for emission is a zero-valued field. The time-integrated results of the adjoint runs are presented in Figure 2 (distributions for non-zero and zero measurements are subtracted as in (5), with scaling factor in (6) taken equal to 3). As seen from the picture, the first-guess emission is still blurred and exhibits several local peaks related to specific observations.

After six iterations the situation changes dramatically (Figure 3). There is a strong maximum dominating the whole field, while the local peaks became an evident high-frequency spatial noise. The low-level "background" still exists and represents an effect of blurring as well as a lack of information in the corresponding areas.

The effect of blurring leads to a significant distortion of spatially-integrated emission flux (Figure 3, right panel). It is seen that a false second peak builds up from the small distributed sources and the background and attracts more

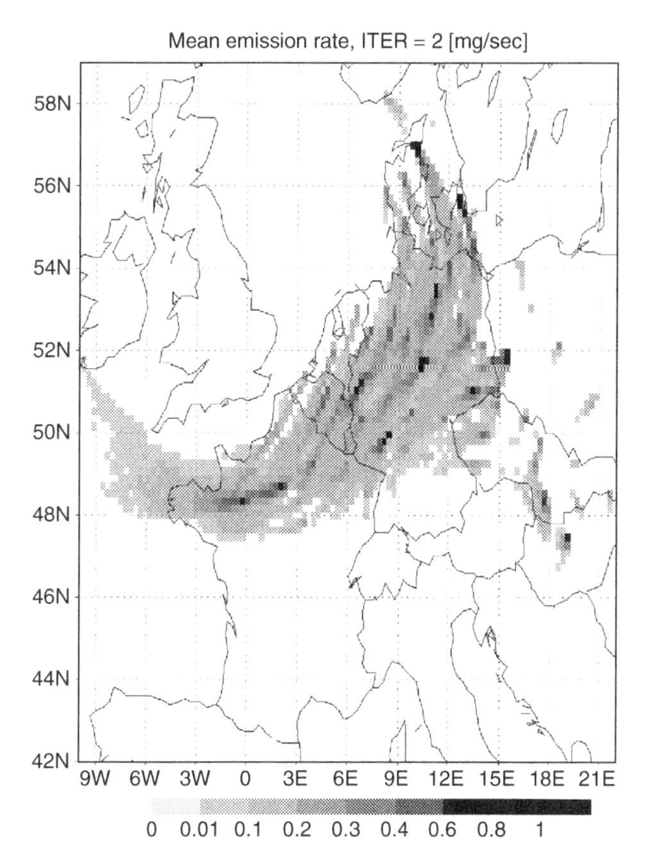

FIGURE 2. Time-integrated emission after the first-guess cycle. [mg/sec]

than half of the total emitted mass. Dealing with this problem requires certain post-processing of the assimilation results. The most straightforward way is a low-pass filtration followed by a cutting-out of the background. This leads to certain distortions of the field and, in particular, results in largely under-estimated absolute emission level. However, the gain is evident (see Figure 4): the source location becomes completely clear and the time variation becomes much closer to an actual shape.

It is worth mentioning that the high-emission area in Figure 4 almost exactly corresponds to the location of the true source.

Certain questions are arising from shifted peak of the source strength and almost 10-fold under-estimation of the flux (Figure 4, right panel). The later one is not very surprising because most of the emission was cut out, so this is a payment for the non-ideal methodology. It can be corrected either by a simple scaling so that modeled and measured mean values correspond to each other (which is evidently a very crude operation), or by the next set of assimilation iterations performed with fixed source location.

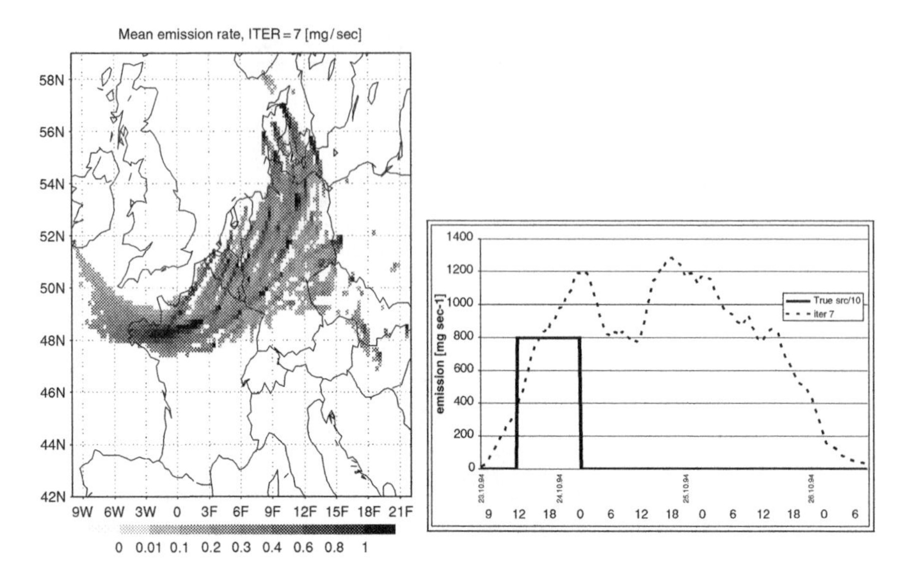

FIGURE 3. Time-integrated emission map after the sixth assimilation cycle and space-integrated temporal emission variation for all iterations. [mg/sec]

The shift is less trivial. It arises either from slightly over-stated wind velocity in the meteorological data or from too strong vertical mixing that brings the plumes to higher altitudes where the wind velocity is stronger. Since SILAM vertical diffusion module is indeed quite simple, the latter cause seems to be possible.

Considering the results it should be mentioned that time correlation coefficient even after six iterations does not exceed 0.7 (Figure 5), which means that the sets of data – observation and meteorology – are not fully compatible with each other. At least using the SILAM model for dispersion assessments, it appeared impossible to create the emission field (the system was totally free to select the field of any complexity in space and time) exactly meeting the observations. The shape of the closest field shows up after the first iteration but after that the correlation coefficient practically does not change, so that the main gain in the cost function value is obtained due to enrichment of the main maximum and overall growth of the emission rates.

Such seemingly inconsistent behavior is not surprising because it reflects the uncertainties in observed values and errors in the meteorological fields of the HIRLAM model. In fact, the level of 0.7 can be considered as a quantitative measure of these uncertainties. Roughly speaking, almost 30% of observed concentration variations cannot be resolved with SILAM model using the HIRLAM meteorological field – whatever emission field is used.

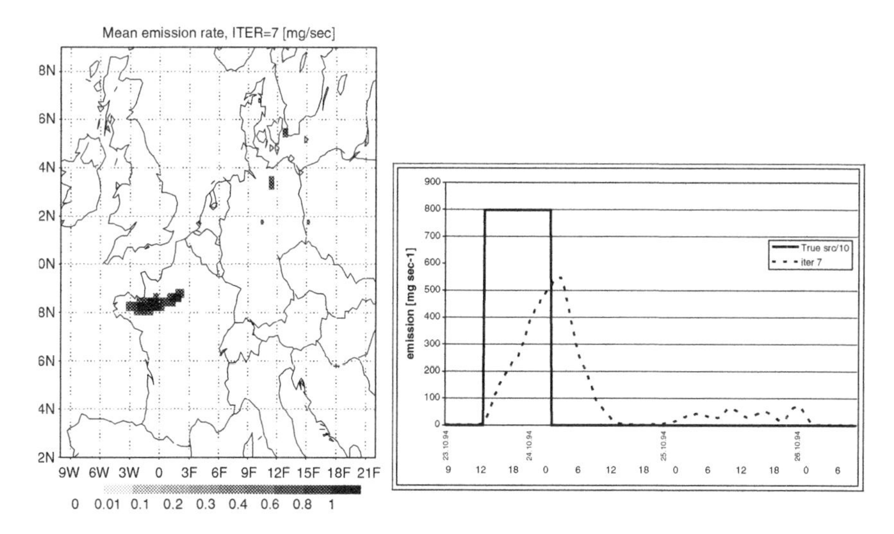

FIGURE 4. Time-integrated emission map after the sixth assimilation cycle and space-integrated temporal emission variation for all iterations. [mg/sec]

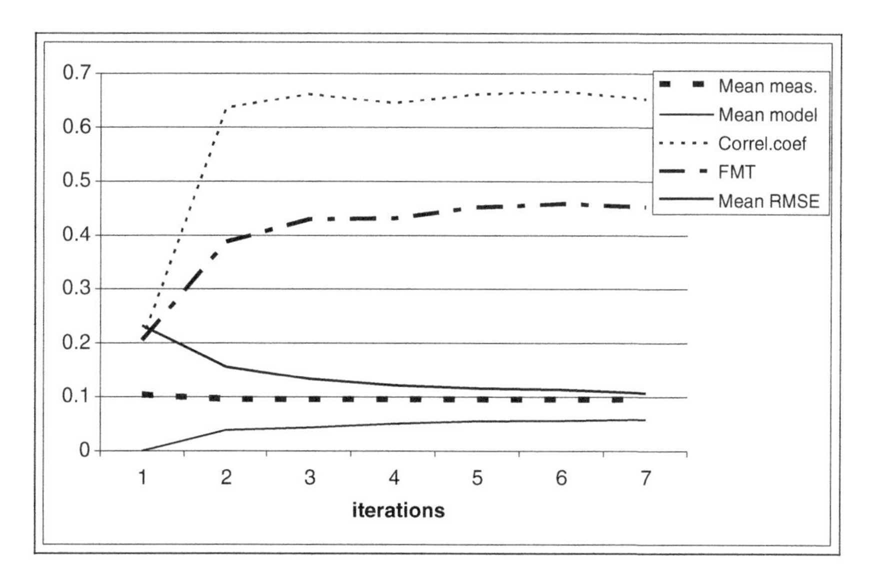

FIGURE 5. Measures of the model-measurement agreement for the assimilation iterations. Legend: Mean meas – average over all observed values [ng m^{-3}]; Mean model – average of modeled values corresponding to observations [ng m^{-3}]; Correl coef – mean correlation coefficient between modeled and observed time series; FMT – Figure of Merit in Time after Mosca *et al.* (1998); Mean RMSE – mean over all stations Root Mean Square Error [ng m^{-3}], also reflecting the value of the cost function.

4. Conclusions

It is shown that the combination of adjoint dispersion formalism, deterministic and probabilistic interpretation of its results and variational data assimilation technique allows approaching full-scale inverse dispersion problems with satisfactory results.

The developed methodology was applied to the real case of European Tracer EXperiment without any a-priory knowledge about the source location or time variation. The delineated area where the source is located is very close to the actual place of release. Time variation of emission intensity is also quite close to the original pattern, except for the absolute level destroyed by the applied filtrations. Its re-stating requires an extra assimilation cycle to be performed after the source location is delineated.

Acknowledgements The authors would like to thank the SILAM development team: Mikko Ilvonen, Pilvi Siljamo, Ilkka Valkama for the co-operation and Eugene Genikhovich for long and fruitful discussions and valuable comments.

References

Courtier, P., Derber, J., Errico, R., Louis, J.-F., Vukichevich, T., 1993, Important literature on the use of adjoint, variational methods and the Kalman filter in meterology. *Tellus*, **45A**, 342-357.

Dimet, F.-X., Talagrand, O., 1986, Variational algorithms for analysis and assimilation of meteorological observations: theoretical aspects. *Tellus*, **38A**, 97-110.

Graziani, G., Klug, W., Mosca, S. (1998) Real-time long-range dispersion model evaluation of the ETEX first release. *Office for official publications of the European Communities*, L-2985, Luxembourg, 215 p.

Host, G., 1996, A statistical method for estimation of European sulphur emissions using EMEP monitoring data. EMEP/MSC-W Note 3/1996, Olso, Norway.

Mosca, S., Graziani, G., Klug, W., Bellasio, R., Biaconi, R. (1998) A statistical methodology for the evaluation of long-range dispersion models: an application to the ETEX exercise. *Atmos. Environ.*, **32**, 24, 4307-4324

Sofiev, M., Sofieva, V. (2000) Methodology for emission estimation of the atmospheric pollution on the basis of mathematical modelling and measurement data. *J. of Mathematical modelling and Computer Experiment*, Russian Academy of Sciences, **v.12**, N 4, pp. 20-32. Russian edition.

Sofiev M., 2002, Real time solution of forward and inverse air pollution problems with a numerical dispersion model based on short-term weather forecasts. *HIRLAM Newsletter*, **14**, pp. 131-138.

Sofiev, M., and Siljamo, P., 2003, Forward and inverse simulations with Finnish emergency model SILAM., In: *Air Pollution Modelling and its Applications XVI*, eds. C.Borrego, S.Incecik, Kluwer Acad. / Plenum Publ. pp.417-425.

An Example of Application of Data Assimilation Technique and Adjoint Modelling to an Inverse Dispersion Problem Based on the ETEX Experiment

Speaker: M. Sofiev

Questioner: P. Builtjes
Question: You are using adjoint approach. Did you consider to use extended Kalman Filtering?
Answer: Yes, I did. FMI uses the Kalman filtration for assimilation of stratospheric observations into our stratospheric model. However, I preferred variational algorithm for the following reasons: (i) SILAM possesses the adjoint formulations of the dispersion equation, which are the key pre-requisite; (ii) having the adjoint model ready, it is straightforward to implement the variational assimilation, which would then require less work than Kalman filter; (iii) I have a feeling (not exact estimates, however) that the variational assimilation works at least as fast as the Kalman filter – or faster.

Questioner: S. Galmarini
Question: The abundancy of monitoring is a crucial parameter of your methodology, how to define a minimum abundancy level?
Answer: Formally speaking, the stations must be located so close to each other that their spatial structure functions overlap or, at least, do not leave "non-covered" area in-between. The next question is how to define these structure functions – and this is a heavy problem, also in meteorological assimilation. Strength of the current approach is that it seems to work even when the density of stations is clearly insufficient. Indeed, if you look at the non-processed output of the iterations, you see a lot of small peaks, which are isolated both in space and in time. In fact, these are just the responses of the system to individual observations, which stay "on its own" because there were no neighboring stations to interfere with. However, as we saw, this stochastic pattern can be eliminated, while the "true" peak remains stable and much more pronounced. It might be interesting to see how the results will respond to reduction of the set of stations, but this is a Monte-Carlo type experiment, which would require tens if not hundreds of assimilation cycles – and the resources I do not have at the moment.

Questioner: M. Jean
Question 1: Comment. In this kind of work we will mainly obtained few positive measurements and many "zero" measurements. The "zeroes" are as important as the "positive" measurements since they will contribute to bound the problem. In emergency situation, it is important that the zero measurements be reported.

Question 2: From a measurement point of view, ETEX had both the coverage and temporal resolution. It is the ideal situation. Are you considering more realistic event such as the Algecvias incident?
Answer: Thanks for your comment. Indeed, the utilization of the zero measurements is one of main goals of this work. Concerning the cases with fewer observations – yes, we are preparing now the input datasets for Algeciras and Chernobyl catastrophes.

49 Micro-Swift-Spray (MSS): A New Modelling System for the Simulation of Dispersion at Microscale. General Description and Validation

G. Tinarelli[1], G. Brusasca, O. Oldrini, D. Anfossi, S. Trini Castelli, and J. Moussafir

1. Introduction

Dispersion at microscale should be simulated using 3D codes, since they are able to account for very complex situations. Targets of these models can be road traffic simulations, such as the reconstruction of the high pollution episodes into street canyons or car parks or, more generally, the reconstruction of dispersion patterns generated by the presence of obstacles. These problems are traditionally solved using CFD models adapted to the atmospheric PBL, suitable only for short term simulations in a limited number of cases, due to their high CPU demand. Here we present a different approach to reproduce the microscale scenarios with a lower demand of computational time. It potentially allows a wider range of applications, such as emergency responses, statistical calculations of impacts due to different conditions and climatological or long term impacts. This method allows an exact representation of buildings directly generated by a GIS (as .shp files). A first guess of the 3D mean flow is computed using all available and relevant meteorological data (inside or outside the target domain). This field is then modified using analytical corrections due to the obstacles (*Kaplan and Dinar*, 1996). At last, the mass consistent field is determined considering the topography and the filled cells representing the buildings. Turbulence is diagnostically estimated considering the distance to the nearest obstacle as a mixing length. This micro-scale quick solution to the flow problem is called Micro SWIFT. Dispersion is simulated through a modified version of SPRAY (Micro SPRAY), a Monte-Carlo Lagrangian dispersion code (*Tinarelli et al.*, 1994), where particles are reflected by the ground and by obstacles, and discretisation time steps are set up in order to better follow the large

[1] Arianet S.r.l., via Gilino 9, 20128 Milano, Italy

gradients of meteorological fields induced by obstacles. The model system Micro SWIFT and Micro SPRAY is named MSS. Different comparisons both with other approaches, such as CFD and wind tunnel, and field data have been obtained. Some street canyon applications regarding typical situation in Italy, in terms of both geometry and traffic emissions, are under test. Preliminary results will be briefly shown here.

2. Model Description

2.1. *Micro SWIFT*

SWIFT is an analytically modified mass consistent interpolator over complex terrain. Given a topography, meteorological data and buildings, a mass consistent 3D wind field is generated by following the steps below:

- according to meteorological data, a first guess of the mean flow is computed through interpolation
- this first guess is then modified to take into account obstacles by creating analytical empirical zones where the flow is modified according to buildings, these being isolated or not (see *Röckle R.*, 1990 or *Kaplan H. and Dinar N.*, 1996). An example of zones attached to a rectangular obstacle is shown on Figure 1.
- finally, the flow is adjusted in order to satisfy the continuity equation and impermeability on the ground and on the building walls

Figure 1 illustrates two examples of the typical structure of flow zones generated by the SWIFT code.

Micro SWIFT is also able to derive a diagnostic turbulence (namely the Turbulent Kinetic Energy (TKE) and its dissipation rate) to be considered by Micro SPRAY inside the flow zones modified by obstacles. To this aim, a local deformation tensor is computed from wind fields, and a mixing length is derived as a function of the minimum distance to buildings and ground. The TKE field is then calculated supposing an equilibrium between production and dissipation terms.

2.2. *Micro SPRAY*

Micro SPRAY is a Lagrangian particle dispersion model directly derived from SPRAY code (*Tinarelli et al.*, 1994, 1998), able to take into account the presence of obstacles. The dispersion of an airborne pollutant is simulated following the motion of a large number of fictitious particles, each representing a part of the emitted mass from sources of general shapes. Particles movement is obtained applying a simple equation of motion where the particle velocity is split into two components: a mean one, or "transport-component" $\overline{U(t)}$, which is defined by the local wind reconstructed by Micro SWIFT, and a stochastic one, simulating the dispersion and reproducing the atmospheric turbulence. The stochastic

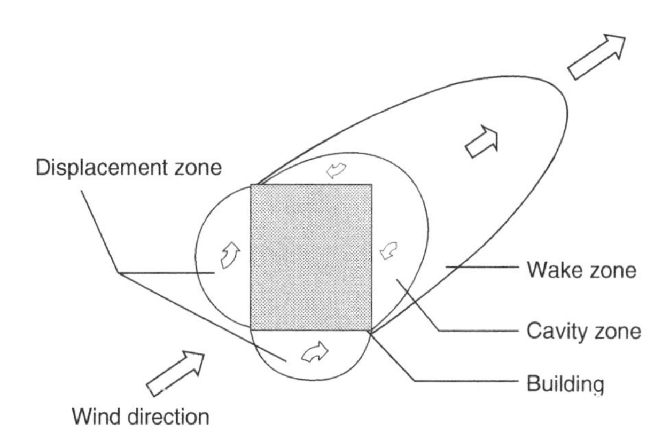

FIGURE 1. Top view of an isolated (left) and non-isolated buildings (right) and their flow zones structure.

component of the particle motion is obtained by solving a 3D form of the Langevin equation for the random velocity (*Thomson*, 1987).

$$dU_p(t) = a(X, U)\,dt + \sqrt{B_0(X)\,dt}\,d\mu_p \tag{1a}$$

and it is coupled to the equation for the displacement:

$$\frac{dX_p(t)}{dt} = \overline{U(t)} + U_p(t) \tag{1b}$$

where **Xp(t)** and **Up(t)** represent the 3D particle position and velocity vectors defined on a fixed Cartesian reference frame; a and B_0 are determined so to fulfill the 'Well-mixed' condition; dμ is a stochastic standardized Gaussian term (zero mean and unit variance). Model input Lagrangian time scales and wind velocity variances are derived from TKE and dissipation rate ε provided by Micro SWIFT (following Rodean, 1996).

Obstacles or buildings are taken into account by setting as impermeable some of the cells of the terrain following grid where meteorological fields are defined.

TABLE 1. Ratio between the computed (cn) and measured (ce) concentration maximum 40m downwind the source for ABC, MISKAM and MSS models.

cn/ce	ABC	MISKAM	MSS
[1/10,1/5]	1	0	1
[1/5,1/2]	5	8	1
[1/2,2]	4	4	9
[2,5]	2	0	0
[5,10]	0	0	1
number of values	12	12	12

3. Preliminary Validation Cases

Preliminary validation has been obtained for different sets of data, ranging from CFD and wind tunnel to real field experiments.

3.1. U-shape Building

The data of the U-shape isolated building are wind tunnel data obtained at the Institute of Hydrology and Water Resources (University of Karlsruhe, Germany). Runs were also performed using the German CFD code MISKAM and a model similar to the MSS approach, called ABC. A tracer gas was released from 3 different locations and for different wind directions. Concentration measurements were made at a plane 40m (20cm in the wind tunnel) downwind the source.

Figure 2 shows the structure of the experiment and a comparison of maximum concentrations obtained for different wind directions in this plane with a source in B. Table 1 shows the ratio between measured/computed maximum.

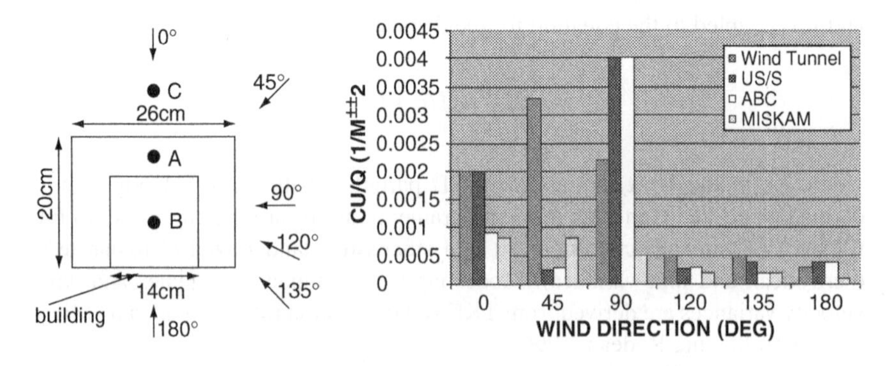

FIGURE 2. Wind directions and release point locations for the U-shape experiment (left) and comparison of maximum concentrations, source in B position.

3.2. *EMU Wind Tunnel L-Shape Building*

Project EMU is concerned with quantifying the uncertainty in CFD predictions of near-field dispersion of toxic and hazardous gas releases at complex industrial sites. This project aims at quantifying these uncertainties through a programme of CFD simulations and wind-tunnel experiments, performed at the University of Surrey. A dispersion experiment taking into account the presence of a single L-shape building has been performed. Figure 3 shows the location of several concentration sensors downwind the obstacle. An 'open door' source located at the arrow position was considered.

Figure 4 shows the comparison between the computed and measured crosswind concentration profiles close to the ground. The agreement is quite noticeable.

4. URBAN 2000

URBAN 2000 is a meteorological and tracer field experiment to study the urban environment and ultimately its effect on atmospheric dispersion that was conducted in Salt Lake City in October 2000. Experiments were performed to investigate transport and diffusion around a single downtown building, within and through the downtown area and into the greater Salt Lake City urban area. For our purposes, an urban domain of about 400m × 400m around the release point was considered (see Figure 5). We simulated 30min out of a 1h SF6 release and compared the results to sonic anemometers measurements for the wind and sample bags for the concentration.

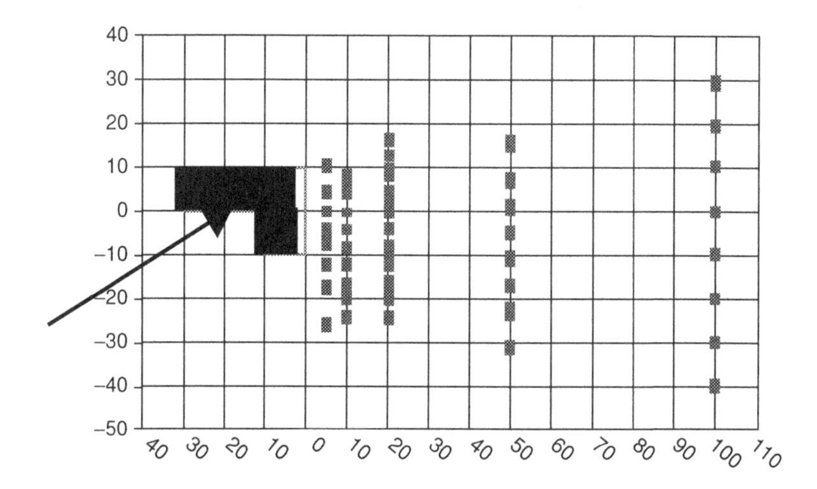

FIGURE 3. Plan view of EMU L-shaped building with sensor locations (grey squares). Black arrow indicates the source position. Wind blows from left to right.

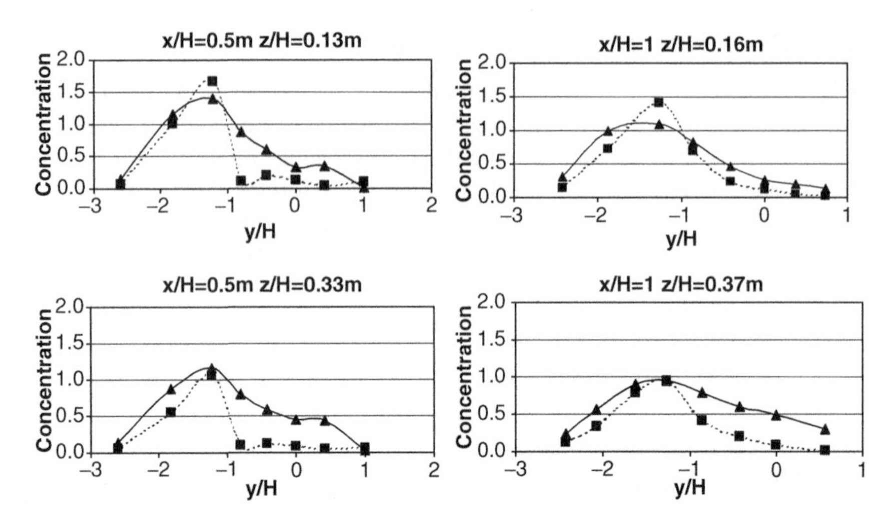

FIGURE 4. Crosswind profiles measured (solid line and triangles) and computed (dotted lines and squares) concentrations at different heights and different y positions depicted in Figure 3. H = 10 is the height of the L-shape obstacle.

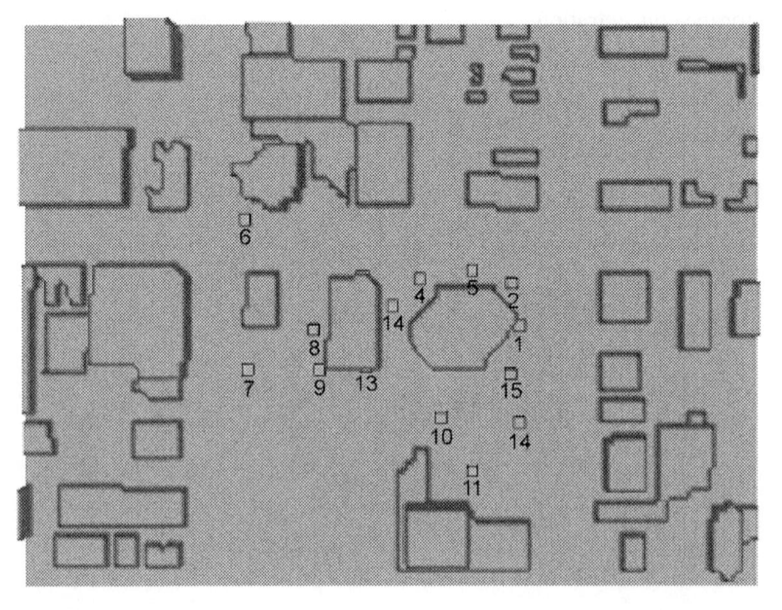

FIGURE 5. Plan view of the domain considered, numbers represent sampler positions for glc measurements.

In figure 6 flow and concentration fields close to ground level are represented. It is worth noting that the solution creates realistic recirculations which generate upwind dispersion of the material. Figure 7 shows a comparison between measured and computed concentrations at different points at different times. The model tends to correctly capture the order of magnitude of concentrations and the position of maximum, even if not all the details are well reproduced.

FIGURE 6. Streamlines & glc field generated 30min after release. Source = white dot.

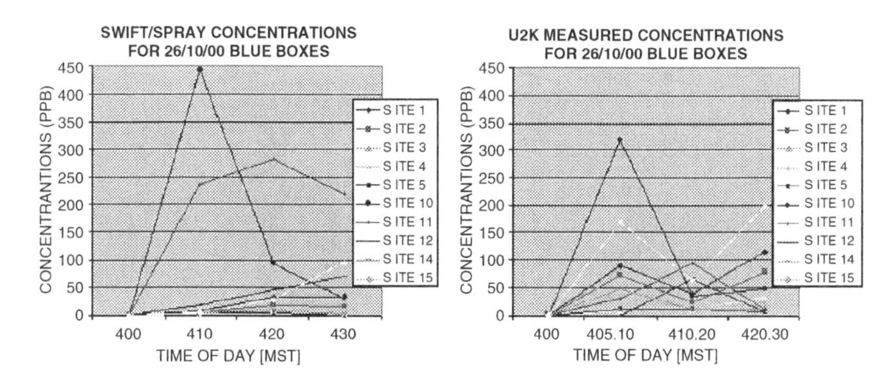

FIGURE 7. Comparison between computed (left) and measured (right) ground level concentrations at the points indicated in figure 5.

4.1. Bologna Experiment

A field experiment has been conducted in the center of Bologna (historical middle age town located in northern Italy) in order to collect information about meteorology and road emissions to reconstruct the ground pollution levels, in particular for PM10 and PM2.5. The region under consideration covers an area of about 500×500 m^2 (Piazza Maggiore) around the main Cathedral, and it is characterized by a composite urban environment, with deep canyons and a complex network of both narrow and wide roads. Normal vehicular traffic is possible only on a part of these roads while in some of the others a limited traffic level is admitted or totally forbidden. Figure 8 shows a plan view of the zone taken into account in this experiment. Standard meteorological (cup anemometers and thermometers) and more sophisticated (sonic anemometers) measurements were collected in order to reconstruct both the mean flow and the turbulence. Concentration levels of many pollutant and traffic levels, characterized in terms of both number and type of vehicles, are measured at points where different traffic regimes are admitted. MSS is applied in these conditions in order to reconstruct both the local flow and the pollution levels on the entire considered domain.

This work is still in progress. Preliminary simulations with MSS have been produced in order to verify the complexity of the wind and concentration fields produced. The horizontal grid step adopted in this case is 4m, while the vertical extension of the computational domain is 200m. Figures 9 show a detail, in the area around the Cathedral, of the mean wind and concentration fields generated at 08:00 in the morning. The complexity of the generated wind fields is quite noticeable, generating many low wind speed sub-regions where the direction is almost random, alternated by regions where channelling effects tend to reinforce the flow. The simulated concentration field shows a general tendency of the code to limit the main impact inside the confined region where the linear

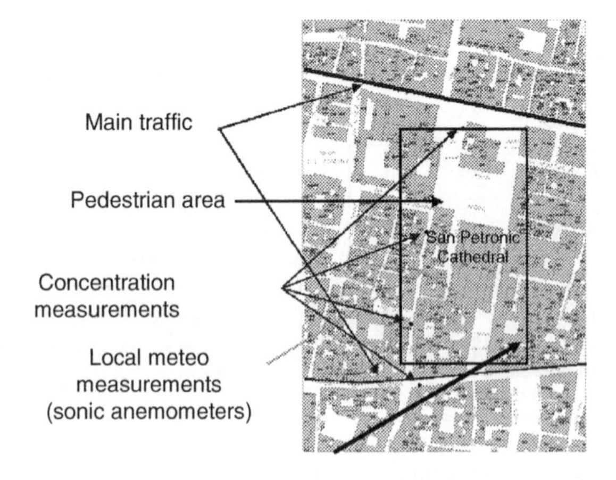

FIGURE 8. Plan view of the central area of Bologna taken into account for the experiment.

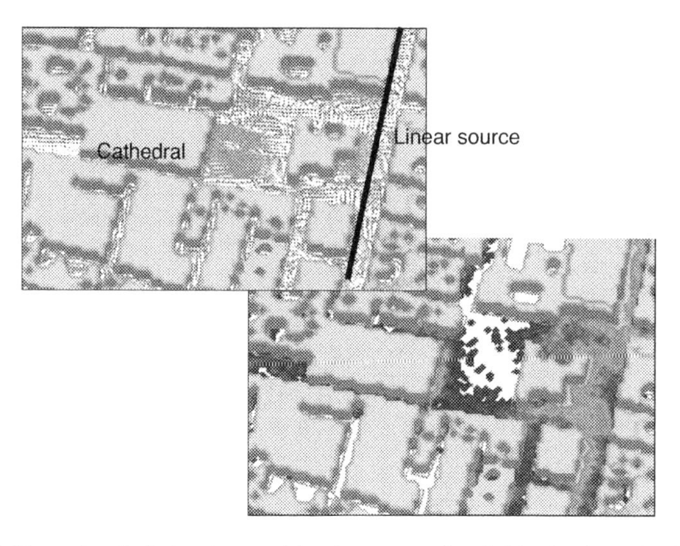

FIGURE 9. Example of wind and ground level concentration fields simulation in the centre of Bologna.

emission is present. It also predicts an intrusion of pollutant in the closer canyons.

5. Conclusion

The validations here presented show a good agreement between measured and computed data, especially when considering the CPU cost of a typical MSS computation (the 30min long urban 2000 simulation taking less than 5min on a PIV laptop). Some other validations need to be performed in real cases (such as the one under test at Bologna) in order to assess the quality of the method in a real environment. However, the encouraging preliminary results obtained in more controlled situation, permit to be optimistic.

References

Kaplan H. and Dinard N., **1996:** A Lagrangian dispersion model for calculating concentration distribution within a built-up domain. *Atmospheric Environment,* **30** **(24)**, 4197–4207.

Röckle R., **1990:** Bestimmung der Strömungsverhältnisse im Bereich komplexer Bebauungsstrukturen. *PhD. thesis,* Darmstadt, Germany.

Rodean H. C., **1996:** Stochastic Lagrangian models of turbulent diffusion. *American Meteorological Society Meteorological Monographs,* **26** **(48)**.

Thomson D. J. **1987:** Criteria for the selection of stochastic models of particle trajectories in turbulent flows. *J. Fluid Mech.*, **180**, 529–556.

Tinarelli G., Anfossi D., Brusasca G., Ferrero E., Giostra U., Morselli M.G., Moussafir J., Tampieri F., Trombetti F., **1994:** Lagrangian particle simulation of tracer dispersion in the lee of a schematic two-dimensional hill", *Journal of Applied Meteorology*, **33**, **N. 6**, 744–756.

Tinarelli G., Anfossi D., Bider M., Ferrero E., Trini Castelli S, **1998:** A new high performance version of the Lagrangian particle dispersion model SPRAY, some case studies, Proceedings of the 23rd CCMS-NATO meeting, Varna, 28 September–2 October 1998, 499–507, Kluwer Academic.

50 New Developments on RAMS-Hg Model

Antigoni Voudouri* and George Kallos*

1. Introduction

In this work recent developments in the Regional Atmospheric Modelling System (RAMS) version 4.3 (Cotton et al., 2003) coupled with modules describing the atmospheric mercury cycle are presented. The mechanisms that describe reemission of Hg^0 from water and soil as well as the wet and dry deposition of mercury species have been implemented in RAMS. A detailed chemistry module has also been tested. The model performance has been examined for a representative summer week during which observations of the wet deposited mercury were available. Model calculations have been compared with observations made during the 14 to 26 August 1997 simulation period in NE USA. An inter-comparison was also performed with the previous version of the model (Voudouri et al., 2004) used over the same domain, and simulation period.

2. Model Description

The various atmospheric and surface processes of mercury species implemented on RAMS model are illustrated in Figure 1 and briefly described below:

***Anthropogenic Emissions*:** The model can use both emission inventories prepared for past and ongoing projects, MAMCS and MERCYMS respectively, for Europe and the 'Modeling of the Mercury Processes over NE USA' project supported by NYSERDA for USA. The atmospheric emissions of mercury from anthropogenic sources in Europe (Pacyna et al., 2001) had been compiled during the MAMCS project (Pirrone et al., 2003) and updated at the framework of MERCYMS in the Mercury Emission Inventory (MEI). The New York State Department of Environmental Conservation provided the emission data used in

* University of Athens, School of Physics, AM & WF Group, University Campus, Bldg PHYS-5, 15784 Athens, Greece. Corresponding author: George Kallos, kallos@mg.uoa.gr

this study. Walcek et al., (2003) performed the spatial distribution of total gaseous and particulate emission rates in eastern North America. Mercury emissions from the Global Mercury Emission Inventory can also be utilized for other domains. The various sources (e.g. power plants, waste incinerators, coal combustion) are automatically allocated within the model domain according to the geographic co-ordinates, type and characteristics of sources.

***Natural Emissions-Reemissions-Atmosphere-surface exchange*:** Approximately 25% of mercury emitted into the atmosphere is related to natural sources of the pollutant (Fitzgerald, 1997). The earth mantle, volcanoes as well as water surfaces and soil are the main natural sources of mercury. The atmospheric mercury from both anthropogenic and natural sources is deposited by wet or dry deposition to the surface. The soluble oxidised forms of mercury previously deposited to the surface are reduced to highly volatile species able to be re-emitted back to the atmosphere (Stein et al., 1996). Both natural emissions and re-emissions were considered in the developed model and treated accordingly. Fluxes of mercury from soil and water had been considered constant at the previous version of the model. Mercury fluxes from soil are calculated at the presented version as a function of soil temperature (Capri and Lindberg 1998; Xu et al. 1999). For air-water exchange of mercury, wind speed at 10m above surface, whitecap coverage, friction velocity and Hg^0 concentration in air and water were considered (Mackay and Yeun 1983, Shannon and Voldner 1995, Xu et al. 1999).

***Chemistry module*:** The modified chemistry module deals not only with the gas and aqueous phase chemistry reactions of mercury species with other reactants but also with photochemical, bimolecular and termolecular reactions that form these reactants. The photochemical reactions of ozone (O_3) and hydrogen peroxide (H_2O_2) both in aqueous and gaseous phase are treated within the chemistry module using the Fast-J scheme proposed by Wild et al. (2000). Other reactions include the bimolecular reactions of SO_x, CO and CO_2 with O_2, H_2O, OH and H_2O_2. The gas and liquid phase reactions of mercury considered in the chemistry module are those with O_3, H_2O_2, chlorines and sulphite (Munthe et al., 1991, Munthe 1992). One of the benefits of the chemistry module incorporated in RAMS is its flexibility, the ability to calculate on line the rate constants of the reactions for various temperatures, pressures and water content as well as the simplicity to add new reactions to the database.

***Dry Deposition module*:** Dry deposition is the transport of the gaseous and particulate species of a pollutant from the atmosphere onto the surfaces in the absence of precipitation. In most deposition models (Wesely and Hicks, 2000) the deposited quantity over a given surface is the product of the pollutant's concentration, at the first model level and the deposition velocity. In the dry deposition process the velocity is calculated using the resistance method. Using this method, deposition is calculated as the sum of various resistances for the gaseous species (Hicks, 1985) and the settling velocity for particles. The values of the resistances depend upon meteorological conditions as well as on the properties of the surface. The deposition velocity of Hg associated with particles, (Hg^p), was calculated by distributing its mass according to a lognormal particle size distribution. The geometric mass mean diameter and the geometric standard deviation were chosen to be 0.4 μm and 1.5 μm respectively. The whole particle size distribution is subdivided into 15 size intervals and the deposition velocity is calculated for each interval. Thus the deposition velocity of Hg^p is obtained as a weighted average of the previous velocities.

***Wet Deposition module*:** The wet removal process concerns the soluble chemical species (Hg^2 and its compounds), and also the particulate matter scavenged only from below the precipitating clouds. Wet scavenging of the divalent mercury (Hg^2) is assumed to occur in and below clouds. As Hg^2 has similar aqueous solubility with HNO_3 (Xu et al., 2000), it is assumed to be an irreversibly soluble gas and its scavenging coefficient is calculated accordingly (Seinfeld, 1986). In cloud, Hg^2 can be removed by interstitial cloud air by dissolution into cloud drops. The local rate of removal of the irreversibly soluble gas with a concentration depends

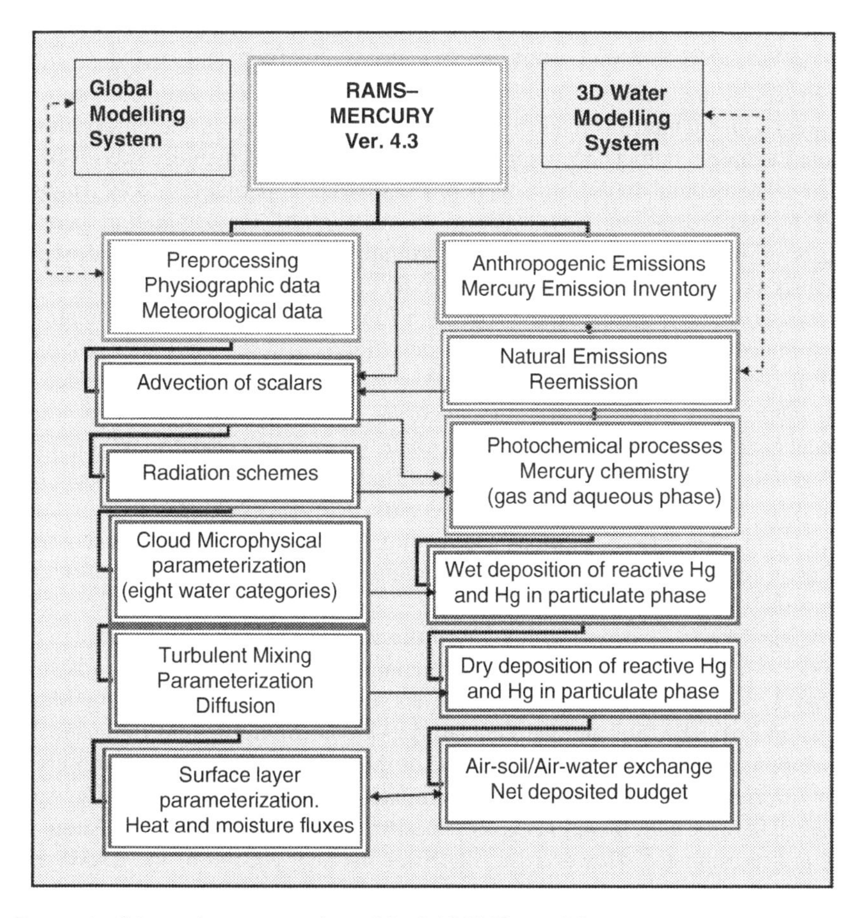

FIGURE 1. Schematic representation of the RAMS-Hg model structure.

on the scavenging coefficient of the gas in the cloud and on the concentration of the pollutant.

3. Results and Discussion

In this study, simulations performed during 14 to 26 August 1997 with both versions of RAMS – namely RAMS ver. 4.3 and RAMS ver. 3b – have been compared with available deposition measurements from several locations of the NE part of US. The simulation domain covers the area of US East of the Rocky Mountains. More specifically, the Mercury Deposition Network (MDN) provided wet deposition measurements at sites upwind and downwind of NY State. MDN deposition observations at selected sites within the MDN – namely Allegheny Portage at Pennsylvania, Dorset and St. Andrews at Canada, Bridgton, Acadia and Greenville at Maine – have been compared with the accumulated wet deposition of mercury from both model versions. The available observations for these stations represent the weekly measured wet deposition of all mercury species, for the periods 12 to 19 August 1997 and 19 to 26 August 1997. Only two deposition observations are available for each station during the model simulation period. However, an attempt was made to inter-compare model outputs and observations. From the model outputs accumulated wet deposition of all mercury species have been calculated for the first five days of the simulation period (14 to 19 August) and for the remaining days (19 to 26 August) of the simulation period. Observations have also been treated accordingly, in order to achieve greater consistency between the observations and model calculations.

A comparison has been made between the observations and model calculations performed with both versions of the model. Figure 2 presents a scatter plot produced using these data. A regression line with zero intercept is also presented in this Figure. The wet deposited amounts of Hg using the previous version of the model seem to be more widely spread than the ones calculated using the new RAMS-Hg version presented in this study. Both models tend to overestimate the deposited amounts of mercury. The overestimated deposited quantities of mercury from both model versions are within the acceptable limits, taken into account the observation errors, the uncertainties of the observation network and the weekly-type measurements.

A quantitative approach to the above discussed differences between the two versions of the model is made through a comparison of statistics summarised in Table 1. The statistical variables used are the Bias (BIAS), the Root Mean Squared Error (RMSE) and the Standard Deviation (SD). The bias estimates the correspondence between the mean value of the modelled value and the observation. If $bias < 0$ (> 0), the model underestimates (overestimates) the specific variables. BIAS for both model versions is positive indicating that wet deposition of Hg is overestimated. However the BIAS for the new version of the RAMS-Hg model is equal to $0.1915 \mu g/m^2$ rather than $0.4349 \mu g/m^2$ calculated for the previous version of the model. The implementation of the developed mercury modules

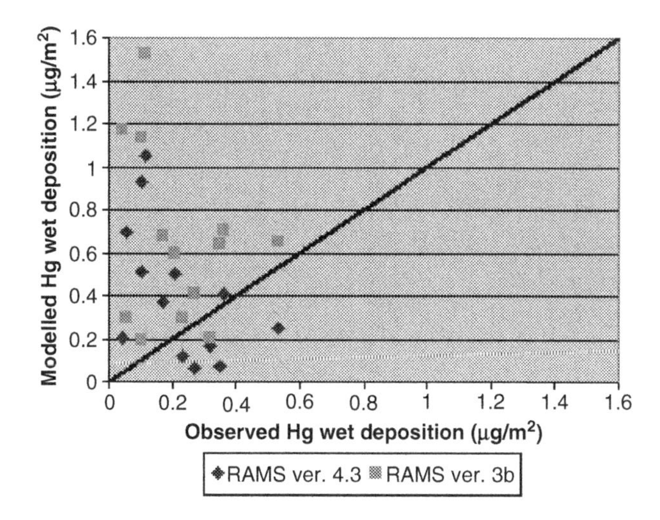

FIGURE 2. Scatter plot of modelled – using RAMS ver. 4.3 and RAMS ver. 3b – versus measured wet deposition of Hg (μg/m²). A forced regression line (y = x) is also shown.

TABLE 1. Statistics comparison between RAMS4.3 and RAMS3b

Model	BIAS (μg/m²)	SD (μg/m²)	RMSE (μg/m²)
RAMS4.3	0.1915	0.3183	0.1936
RAMS3b	0.4349	0.4073	0.3898

on the atmospheric model in addition to the new developments on the RAMS physics (Cotton et al., 2003), seem to have improved the performance of RAMS-Hg. This is also pronounced by the RMSE and the SD of both model versions.

This RMSE is considered to be one of the most popular measures in the estimation of the modelled value accuracy (Wilks, 1995). It is mostly used in grid-point fields. This value is not dimensionless but it has the same units as the validated field. It is an important measure as it provides a quantitative measure of the model performance. The RMSE of the previous model version is 0.3898 μg/m²., almost two times the value calculated for the new RAMS-Hg model (0.1936 μg/m²).

The SD is one of the several indices of variability used to describe the dispersion among the measures in a given population. Numerically, the standard deviation is the square root of the variance. The standard deviation can be readily conceptualized as a distance along the scale of measurement. The SD is higher for the previous version of the model indicating that the new developments decreased the spread of the modelled values leading to more reliable results.

Figure 3 illustrates the total (wet and dry) deposited Hg^2 (in μg/m²) at 0000 UTC on 26 August 1997 after 12 days of simulation, estimated using RAMS ver. 3b and RAMS 4.3. Differences on the deposition pattern of Hg^2 are mainly

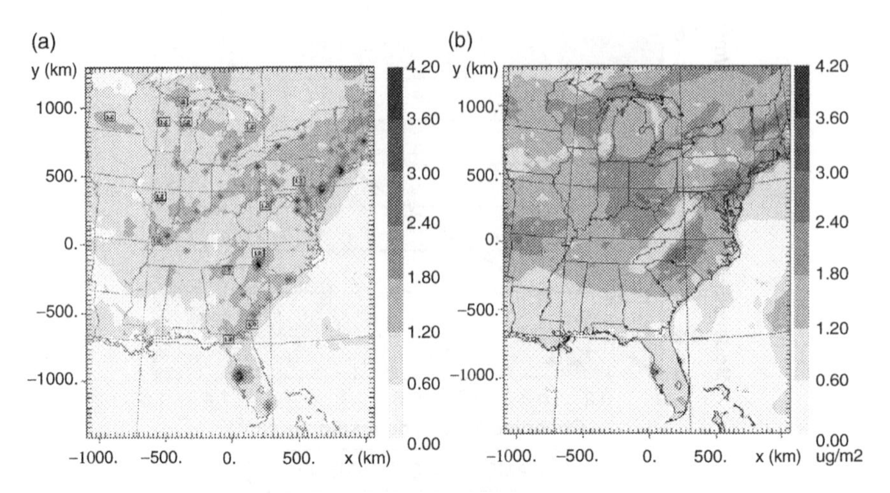

FIGURE 3. Total (wet and dry) deposited Hg^2 (in μg/m^2) at 0000 UTC on 26 August 1997 after 12 days of simulation, estimated (a) from RAMS model ver. 3b and (b) RAMS 4.3.

allocated near the location of the sources. This can be attributed to differences in the microphysics of both versions of the original atmospheric model and to the updated wet and dry deposition modules implemented on the new RAMS-Hg model. The total (wet and dry) deposited Hg^P (in μg/m^2) at 0000 UTC on 26 August 1997 after 12 days of simulation, estimated using RAMS ver.3b and RAMS 4.3 is also presented in Figure 4. The calculated deposited quantities of Hg^P using the new RAMS – Hg are much higher than the ones calculated using the

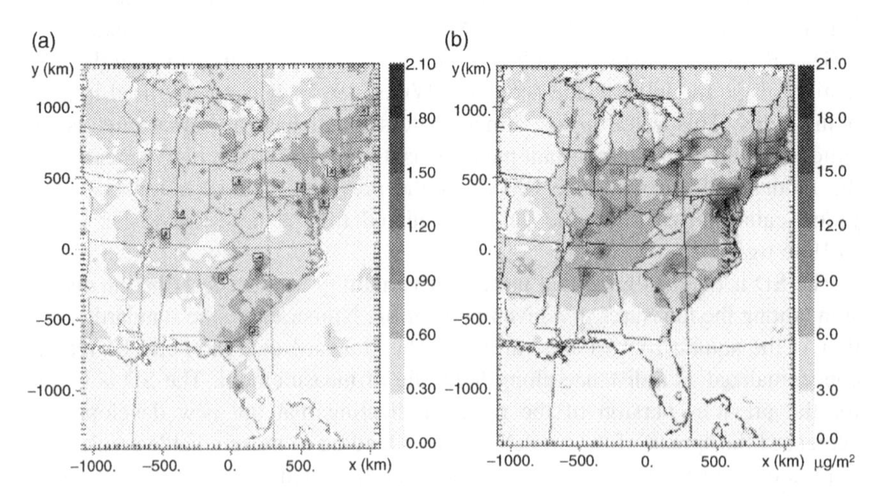

FIGURE 4. Total (wet and dry) deposited Hg^P (in μg/m^2) at 0000 UTC on 26 August 1997 after 12 days of simulation, estimated (a) from RAMS model ver. 3b and (b) RAMS 4.3.

previous version (one order of magnitude). This is mainly attributed on the new dry deposition module used to calculate the dry deposited quantities of Hg^P. The dry deposition scheme proposed by Sehmel (1980) was used at the RAMS ver.3b while the scheme proposed by Wesely and Hicks, (2000) was adopted at the RAMS ver. 4.3.

An attempt was also made to estimate the net Hg flux using the new developed RAMS-Hg model over the domain during the simulation period. Net Hg flux is calculated by subtracting the modelled deposited from the emitted Hg. The net Hg flux (in $\mu g/m^2$) and the total accumulated precipitation (in mm) are illustrated in Figure 5. The net Hg flux is positive over land and especially over areas where the precipitated quantity is high. Over the sea, where there are no sources, negative fluxes dominate indicating that reemission from the water surface is the controlling factor. It should be noted that the budget seems to be in equilibrium, over the soil surfaces at the western USA and for the simulation period.

4. Conclusions

This study presents the main features of the updated version of the RAMS atmospheric model coupled with mercury modules. The latest system is based on the basic concepts adapted at the previous version. New modules for describing the physical and chemical processes of mercury have been implemented. The lack of systematic and detailed mercury measurements remains a problem for a detailed model validation. Based on the limited deposition measurements from a previous study an attempt was made to evaluate the model performance. The wet deposited amount of Hg calculated using the new modelling system was found to be in a closer agreement with observations compared to the previous one. Further and

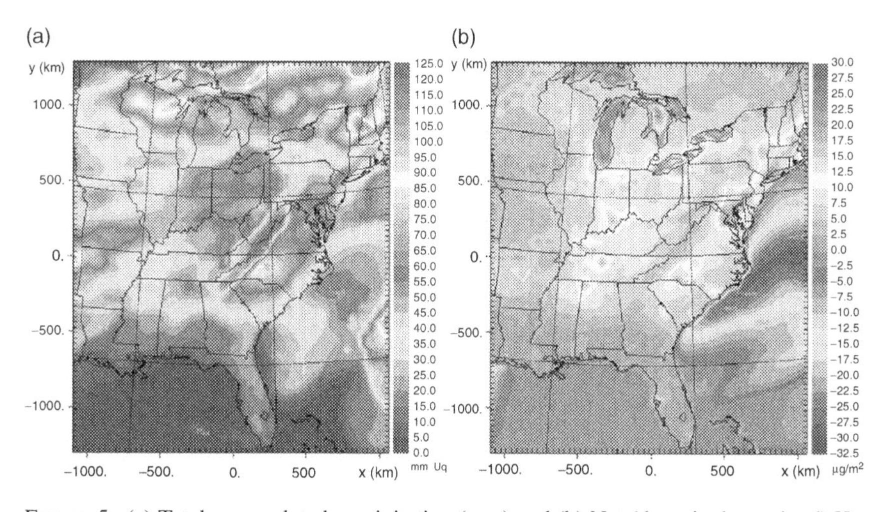

FIGURE 5. (a) Total accumulated precipitation (mm) and (b) Net (deposited – emitted) Hg flux ($\mu g/m^2$) during 14 to 26 August 1997. From RAMS4.3 – Hg model.

more detailed validation is required. This detailed validation is scheduled to perform at the framework of the ongoing EU project MERCYMS.

Acknowledgements The present work was supported by MERCYMS (EVK3-2002-00070) project of the DG-Research of EU and by NYSERDA project #6488 (NYSERDA RF PTAEO#1015266-20439-1).

References

Capri A., Lindberg S. E., 1998. *Application of a Teflon dynamic flux chamber for quantifying soil mercury flux: tests and results over background soil.* Atm. Env. 32, 873-882.

Cotton W. R., R. A. Pielke, R. L. Walko, G. E. Liston, C. J. Tremback, H. Jiang, R. L. McAnelly, J. Y. Harrington, M. E. Nicholls, G. G. Carrio and J. P. McFadden, 2003. *RAMS 2001: Current status and future directions.* Met. Atm. Phys. 82, 5-29.

Fitzgerald, W. F., Engstrom, D. R., Mason, R. P., and Nater, E. A., 1997. *The case for atmospheric mercury contamination in remote areas.* Env. Sci. Techn., 32, 1-7.

Mackay D. and Yeun A. T. K. 1983. *Mass transfer coefficient correlations for volatilization of organic solutes from water.* Env. Science and Technology 17, 211-217.

Munthe, J. Z. F. Xiao, and O. Lindqvist 1991. *The Aqueous Reduction of Divalent Mercury by Sulfite.* Water Air Soil Pollut. 56: pp. 621-630.

Munthe J, 1992. *The Aqueous Oxidation of Elemental Mercury by Ozone.* Atm. En.v Part A-General Topics. 26(8): pp. 1461-1468.

Pacyna E. G., J. M. Pacyna, N. Pirrone, 2001. *European emissions of atmospheric mercury from anthropogenic sources in 1995.* Atm. Env. 35, 2987-1996.

Pirrone, N., Ferrara, R., Hedgecock, I. M., Kallos. G., Mamane, Y., Munthe, J., Pacyna, J. M., Pytharoulis, I., Sprovieri, F., Voudouri, A., Wangberg, I., 2003. *Dynamic Processes of Mercury Over the Mediterranean Region:Summary of Results from the MAMCS Project.* Atm. Env. 37. S21-S37.

Sehmel, G. A., 1980: Particle and gas dry deposition: A review. Atm. Env., **14**, 983-1011.

Seinfeld J. H. 1986. *Atmospheric Chemistry and Physics of Air Pollution.* John Wiley & Sons, New York, N.Y.

Shannon J. D., and E. C. Voldner, 1995: *Modeling atmospheric concentrations of mercury and deposition to the Great Lakes.* Atm. Env., 29, 1649-1661.

Stein E. D., Cohen Y., Winer A. M., 1996. *Environmental distribution and transformation of mercury compounds.* Environmental Science and Technology 26(1), 1-43.

Voudouri A., I. Pytharoulis and G. Kallos, 2004: *Mercury budget estimates for the State of New York.* Env. Fluid Mechanics, pp. 1-21 (in press).

Walcek, C., S. De Santis, T. Gentile, 2003, *Preparation of mercury emissions inventory for eastern North America.* Environmental Pollution, 123, 375-381.

Wesely M. L. and B. B. Hicks, 2000. *A review of the current status of knowledge on dry deposition.* Atm. Env. 34, 2261-2282.

Wild O., X. Zhu and M. Prather, 2000: *Fast-J: Accurate Simulation of In- and Below-Cloud Photolysis in the Tropospheric Chemical Models.* J. Atm. Chem. 37 245-282.

Wilks, D. S., 1995: *Statistical Methods in the Atmospheric Sciences*, Academic Press NY, p. 467.

Xu, X., Yang., X., Miller, D. R., Helble, J. J., and Carley R. J., 1999: *Formulation of bi-directional atmosphere-surface exchanges of elemental mercury*. Atm. Env., 33, 4345-4355.

Xu X., X. Yang, D. R. Miller, J. J. Helble and R. J. Carley, 2000. *A regional scale modeling study of atmospheric transport and transformation of mercury. I. Model development and evaluation*. Atm. Env., 34, Issue 28, 4933-4944.

New Developments on RAMS-Hg Model

Speaker: Kallos

Questioner: S. T. Rao

Question: Regarding the differences in the simulated total deposition of Hg, have you analyzed whether the meteorological fields in RAMS3 and RAMS4 or the chemical mechanisms in the two models are contributing to these differences? In others, how important are meteorology vs. chemistry in dictating the deposition fields?

Answer: According to what we have seen in our analysis, the differences can be attributed to both: meteorology and chemistry. Since the two RAMS versions have significant differences in the microphysics, separation is not possible. According to our experience, meteorology and especially precipitation is the key element to look when we study mercury processes in the atmosphere. Direct coupling of mercury processes with the atmospheric model is considered as necessary despite the fact that such coupling is computationally expensive.

51
Adaptation of Analytic Diffusivity Formulations to Eulerian Grid Model Layers of Finite Thickness

R. J. Yamartino[1], J. Flemming[2], and R. M. Stern[3]

1. Introduction

A Kalman Filter based sensitivity study of the influence of boundary layer parameters (e.g., vertical turbulent exchange coefficient and the mixing layer height) and emission data was conducted on the predicted-observed concentration errors of the REM-CALGRID (RCG) model. To accomplish this, operational ground-based O_3 and NO_2 were assimilated into the RCG model during a simulation covering summer and winter conditions. Analysis of the noise values indicated that a systematic error in the vertical exchange process was periodically impacting deposition fluxes and ground level concentrations. While the Kalman filtering analysis didn't tell us, in detail, the correct way to proceed, it did identify the problematic module and the statistical-based findings were the starting point for revisiting the vertical flux formulation within the model.

A variety of analytic vertical diffusivity formulations are available in the literature. Such formulations of the diffusivity, $K_z(z)$, are directly applicable within analytic plume models, but use of, or adaptation of, these analytic, vertical diffusivity formulations to the finite thickness layers found within Eulerian grid models is generally left unspecified. Some models simply evaluate the value of the diffusivity function at the height of the layer interface, while still others consider integral-averages of the $K_z(z)$ function over the finite thickness of each layer. Both of these approaches can now be shown to be incorrect, simply because if K_z were to drop to zero at any particular point, then all diffusion through this point would be blocked.

[1] Integrals Unlimited, 509 Chandler's Wharf, Portland, ME 04101 USA; rjy@maine.rr.com
[2] E.C.M.W.F, Shinfield Park, Reading, RG2 9AX, UK; Johannes.Flemming@ecmwf.int
[3] Institut für Meteorologie, Carl-Heinrich-Becker Weg 6-10, D-12165 Berlin; rstern@zedat.fu-berlin.de

The RCG model initially used the K_z values evaluated at the z-height point location of the layer interfaces, and the relevant interfacial flux was computed using the product of this diffusivity and the concentration gradient at the interface. This concentration gradient was at first taken as the "bulk" concentration gradient, computed as the difference of the concentrations in the adjacent layers divided by the distance, DZ, between the grid points of these adjacent layers. For the interface at the top of the lowest or surface layer, this "traditional" separation is just $DZ_{1T} = (\Delta h_1 + \Delta h_2)/2$, where $\Delta h_1 = h_1 - h_0$ and $\Delta h_2 = h_2 - h_1$ are the depths of the lowest (i.e., surface) and second layers, respectively, and $h_0, h_1, h_2, \ldots$ are the grid cell face heights with $h_0 = 0$. This "traditional" determination of the "derivative length scale", DZ_{1T}, caused the interfacial flux at the top of the fixed-depth (i.e., $h_1 = 20$m) surface layer to drop dramatically during daytime conditions, when the dynamic layer above it became quite thick. Given that there is little reason to believe that the flux-relevant, concentration gradient at the cell interface equals this "bulk" gradient, we first embarked on an empirical search for a revised "derivative length scale", DZ_1. One attempt involved choosing the smaller of the adjacent layer depths. This left the gradients nearly unchanged for stable conditions (i.e., when the second layer was shallow and approximately the depth of the surface layer anyway), but caused gradients during daytime conditions to increase dramatically over their value computed assuming the traditional DZ_{1T} grid point separation. In addition, choosing $DZ_1 = \Delta h_1$ causes the depth of the surface layer to cancel out in the flux expressions, when assuming a typical near-surface diffusivity profile proportional to z. Other attempts involved developing a "blended" length scale,

$DZ_{1B} = 1/[\alpha/\Delta h_1 + (1-\alpha)/DZ_{1T}]$, combining reciprocal weightings of the depth of the surface layer and the traditional measure DZ_{1T} using a weighting factor of $\alpha \approx 0.5$. Though this definition led to improved surface concentration predictions, it was rather unsatisfying from an intellectual point of view, as it is little more than "tuning". This then inspired a rethinking of the diffusion transport mechanism and led to a revised formulation, not of the $K_z(z)$ formulae, but of how they are utilized with layers of finite thickness.

In this paper, we first show that an effective diffusivity, K_{12}, between any two discrete points $z = z_1$ and $z = z_2$ can be computed via the resistance analogue as:

$$\frac{1}{K_{12}} = \frac{1}{(z_2 - z_1)} \int_{z1}^{z2} dz/Kz(z)$$

Further consideration of the actual flux through the bounding interface leads to a further double-integral averaging over z_1 and z_2, with z_1 varying between layer-1's lower and upper z-bounds of $z = h_0$ and $z = h_1$, and with z_2 varying between layer-2's z-bounds of h_1 and h_2, respectively.

Explicit formulae for the single- and triple-integral formulations are presented for neutral, stable, and unstable diffusivity regimes, along with comparisons of the new results against the simpler "point" diffusivity calculation evaluated at the $z = h_1$ interface.

2. Diffusivity Formalisms for Finite Thickness Layers

The use of rather thick (or deep layers) in photochemical grid models, such as REM-CALGRID (RCG) (Stern and Yamartino, 2002; Stern, 1994) creates two uncertainties, including:

1. selection of the appropriate method to compute an average numerical diffusivity, K_{za}, suitable for use at the interface of two thick layers; and,
2. choice of an ideal length scale, ∂z, that should be used when calculating the mixing ratio derivative, $\partial C/\partial z$, to be used in the flux expression,

$$F = -K_{za} \cdot \rho \cdot \partial C/\partial z. \tag{1}$$

These issues may, in fact, not be separate given that the flux is only sensitive to the product of the diffusivity times the concentration gradient.

The CMAQ report of Byun et al (1999) simply defines the layer-average diffusivity, K_{za}, as $K_{za} = \int_{h0}^{h1} dz \cdot Kz\,(z)$ for the diffusivity appropriate at the top of layer 1, which spans the z interval from $z = h_0$ to the level top at $z = h_1$.

This sort of definition immediately encounters a problem when one considers layer 1 as being made up of a number of sublayers. Suppose that there are 10 sublayers and that in nine of these sublayers $K_z = 1.1$ and $K_z = 0.1$ in one of these sublayers. The average diffusivity is simply $K_{za} = 1.0\ m^2/s$, independent of where the low diffusivity sublayer is located. We know that if this low diffusivity sublayer is located at the bottom of layer 1, far from the top interface, it will have little impact, but what if it is located directly next to the top interface? In this case, we know that it will act to significantly block transport to (or from) the layer 2 above. Thus, it is clear that a simple average cannot give the correct result.

Instead, suppose that we take the “resistance” approach used in dry deposition modeling (e.g., McRae et al., 1982), where an atmospheric resistance term is computed to correct the deposition velocity (i.e., nominally computed at a reference height of z = 1 m) for the finite thickness of a model's lowest layer.

In dry deposition modeling, the flux, F, through an interface is computed as

$$F = -V_d \cdot \rho \cdot (C_1 - C_0), \tag{2}$$

with the canopy layer concentration, C_0, usually taken to be zero.

The resistance R, defined as $R = 1/V_d$, proves to be convenient, as its constituent contributions are based on the physical transport pathways involved and one may establish a rather firm connection between this resistance and its electrical resistance analog. This analog provides the additional advantage that the rules for adding resistances in series and parallel are well established.

Comparing Eqs. (1), (2) and the resistance definition suggests that the resistance R will be of the form $R \approx \Delta z / K$ and have the expected units of seconds/ meter.

Now consider first the resistance to be anticipated as a molecule migrates from a position within layer 1 up to the interface H_1 bounding level 1 and level 2 above. One reasonable definition of that resistance is:

$R_1 = \int_{z1}^{h1} dz/Kz(z)$ for the integral resistance from $z = z_1$ to $z = h_1$ defining the top of layer 1, which spans the z interval from $z = z_1$ within level 1 up to the level top at $z_1 = h_1$. Similarly, one may define the resistance from this interface to a point within layer 2 as:

$R_2 = \int_{h1}^{z2} dz/Kz(z)$ for the integral resistance from $z = h_1$ to $z = z_2$ within layer 2.

Thus, the resistance encountered in propagating from anywhere within level 1 to anywhere within level 2 is simply the series resistance, R_{12}, given as: $R_{12} = R_1 + R_2$, or

$$R_{12} = \int_{z1}^{z2} dz / Kz(z) \text{ for the integral resistance from } z = z_1 = \text{to } z = z_2. \quad (3)$$

Following the dry deposition resistance analog, where only the single integral resistance from the grid point center of level 1 down to a nominal height of z = 1m is ever considered, Eq. (3) would provide the equivalent resistance by choosing the integration endpoints as the grid point locations, but to obtain a full understanding of the mean flux between layers 1 and 2, one should consider the double-integral average of z_1 ranging from $z_1 = h_0$ to $z_1 = h_1$ and of z_2 ranging from $z_2 = h_1$ to $z_2 = h_2$ to yield:

$$<R_{12}> \equiv \frac{1}{[(h_1 - h_0) \cdot (h_2 - h_1)]} \int_{h_0}^{h_1} dz_1 \int_{h_1}^{h_2} dz_2 \cdot R_{12} \quad (4)$$

3. Examples

3.1. Constant Diffusivity

As a sanity check on the development in the previous section, we consider the case of a constant diffusivity profile $K_z(z) = K$. In this case, straightforward substitution into Eq. (3) provides the point-to-point resistance as: $R_{12} = (z_2 - z_1) / K$. Assuming that the two endpoints of interest are the gridpoints sited at the center of each layer, such that $z_1 = (h_1 + h_0)/2$ and $z_2 = (h_2 + h_1)/2$, the resistance is just $R_{12} = 1/2 \cdot (h_2 - h_0) / K$ or $R_{12} = (z_2 - z_1) / K$, as expected.

Evaluation of the Eq. (4), double-integral-averaged resistance also yields: $<R_{12}> = 1/2 \cdot (h_2 - h_0) / K$. While not surprising that a uniform K leads to $R_{12} = <R_{12}>$, it is somewhat surprising that these resistances, and thus the initial transport fluxes, are independent of h_1, and thus independent of the partitioning of this constant diffusivity region into layers of vastly different thickness. The other result of interest is that for the case of two layers of equal thickness $\Delta z = 1/2 \cdot (h_2 - h_0)$, we arrive at the expected flux result of $F = -K \cdot \rho \cdot (C_2 - C_1) / \Delta z$, where ρ is evaluated at the interface and Δz also corresponds to the grid-point separation for equal thickness layers.

3.2. Neutral Diffusivity Profile

The case of assuming a neutral diffusivity profile of $K_z(z) = k' \cdot u_* \cdot z$, where $k' = k/0.74$, k is the Von Karman constant, 0.74 is the constant of van Ulden, and u^* is the friction velocity, also leads to some interesting results. Use of Eq. (3) leads to a resistance $R_{12} = [\log(z_2) - \log(z_1)] / (k' \cdot u_*)$, or $R_{12} = \log[(h_2 + h_1) / (h_1 + h_0)] / (k' \cdot u_*)$ assuming each level's grid points are located at the center of that level.

Let us now compare the flux one would obtain at the interface at $z = h_1$ with the two alternative formulations of flux. In the Eq. (1), gradient approach used in RCG and many other present day models, the flux would be computed as: $F_1 = -K_1 \cdot \rho_1 \cdot (C_2 - C_1) / \lambda$, where λ is the derivative "length scale", often cited as:

$$\lambda = (z_2 - z_1) = \tfrac{1}{2} \cdot (h_2 - h_0), \tag{5}$$

and where K_1 and ρ_1 are the diffusivity and density, respectively, at the interface $z = h_1$.

Now with the resistance approach, utilizing the R_{12} quantity, one computes a flux as $F_1 = -(1/R_{12}) \cdot \rho_1 \cdot (C_2 - C_1)$. Equating these two expressions for F_1, one obtains

$$\text{the simple expression:} \quad \lambda = K_1 \cdot R_{12}, \tag{6a}$$

$$\text{or alternatively,} \quad \lambda = K(h_1) \cdot <R_{12}>. \tag{6b}$$

Substituting $K_1 = k' \cdot u_* \cdot h_1$ into this relation yields $\lambda = h_1 \cdot \log(z_2/z_1)$ as the appropriate value of the "length scale". Further defining z_1 and z_2 as the mid-level grid point values and setting $h_0 = 0$ (i.e., ground level), one obtains a clear definition of the derivative "length scale" λ as:

$$\lambda = h_1 \cdot \log(1 + h_2/h_1) \quad \text{or} \quad \lambda/h_1 = \log(1 + h_2/h_1) \tag{7}$$

under conditions appropriate to the neutral diffusivity profile.

What is interesting about Eq. (7) is that the frequently invoked assumption of $\lambda = (z_2 - z_1) = h_2/2$ (by also assuming $h_0 = 0$ in Eq. (5)) shows a strong linear dependence of λ on the depth, h_2, of the level above the surface layer, whereas Eq. (7) clearly shows a much weaker logarithmic dependence on h_2.

The foregoing discussion assumes that the "dry deposition" analog of using the resistance between grid points is appropriate; however, if one chooses to use the double-integral averaged $<R_{12}>$, rather that this point value difference, R_{12}, invoked above, one finds that

$$<R_{12}> \equiv \frac{1/(k' \cdot u_*) \cdot z_2 \cdot z_1}{[(h_1 - h_0) \cdot (h_2 - h_1)]} \cdot [\log z_2 - \log z_1] \Big|_{z_2 = h_1}^{z_2 = h_2} \Big|_{z_1 = h_0}^{z_1 = h_1}$$

or

$$<R_{12}> = [1/(k' u_*)] \cdot [(h_1 - h_0) \cdot h_2 \cdot \log(h_2) - (h_2 - h_0) \cdot h_1 \cdot \log(h_1) + (h_2 - h_1) \cdot h_0 \cdot \log(h_0)] / [(h_1 - h_0) \cdot (h_2 - h_1)].$$

In the usual case of $h_0 = 0$ for layer 1 as the surface layer, this result simplifies to $<R_{12}> = [1/(k' \cdot u*)] \cdot h_2 \cdot \log(h_2/h_1) / (h_2 - h_1)$, so that the equivalent to Eq. (7) becomes:

$$\lambda / h_1 = \log(h_2 / h_1) / (1 - h_1 / h_2). \tag{8}$$

Equations (5, 7-8) now provide three separate estimates for the length scale λ. In the table below one sees how striking the contrast is between the rapid, linear variation of λ with h_2/h_1 for the traditional Eq. (5) estimate and the slower log dependence variations obtained from Eqs. (7-8).

h_2/h_1	λ/h_1 Eq. (5)	λ/h_1 Eq. (7)	λ/h_1 Eq. (8)
2	1.0	1.0986	1.3863
3	1.5	1.3863	1.6479
4	2.0	1.6094	1.8484
5	2.5	1.7918	2.0118
10	5.0	2.3979	2.5584
20	10.0	3.0445	3.1534
50	25.0	3.2581	3.2846

It is also interesting that for the equal layer thickness case of $h_2/h_1 = 2$, the effective length scale from Eqs. (7-8) exceeds the actual gridpoint spacing of $1.0 \cdot h_1$. This result persists with Eq. (8) past the ratio of $h_2/h_1 = 3$, and may at first appear surprising; however, Eq. (6) blends the dissimilar quantities of a K_z at $z = h_1$ with an integral resistance between midpoint levels. Perhaps, it is more remarkable that, for the equal layer case, the shift in the length scale from the Eq. (5) grid-point separation value is rather modest (i.e., only 10% for Eq. (7) and 39% for Eq. (8)). In any case, the real differences with Eq. (5) emerge for $h_2/h_1 \gg 1$, and here, fortunately, the results for the more complex Eq. (8) converges toward the simpler Eq. (7) results.

Given that Eq. (7) parallels the resistance modeling approach used in dry deposition modeling and that the results for the Eq. (8) evaluation of the double-integral-average do not differ significantly from Eq. (7) results for larger h_2/h_1, it hardly seems worth the effort (i.e., at the present time) to compute Eq. (8), especially as the computational difficulty of evaluating Eq. (8) could increase significantly for the non-neutral stability cases.

3.3. Stable and Unstable Diffusivity Profiles

The LASAT-based formulae for the stable and unstable diffusivity profiles contain more complex z depenences than the simple neutral profile of $K_z(z) = k' \cdot u* \cdot z$, where $k' = k/0.74$, k is the Von Karman constant, 0.74 is the constant of van Ulden, and u* is the friction velocity. For the stable atmosphere, we presently assume:

$K_z(z) = k' \cdot u* \cdot z \cdot (1 - z/H)^{3/2} / (1 + 6.3 \cdot z/L_{MO})$, where H is the mixing depth and L_{MO} is the Monin-Obukhov length scale with $L_{MO} > 0$.

Inserting this expression for $K_z(z)$ into Eq. (3) and performing the integration, one obtains:

$$R_{12} \cdot (k' \cdot u_*) = \log(P) + 2 \cdot (1 + 6.3 \cdot H/L_{MO}) \cdot [(1 - z_2/H)^{-1/2} - (1 - z_1/H)^{-1/2}], \quad (9a)$$

where $P = [1 - (1 - z_2/H)^{1/2}] \cdot [1 + (1 - z_1/H)^{1/2}] / \{[1 - (1 - z_1/H)^{1/2}] \cdot [1 + (1 - z_2/H)^{1/2}]\}$, and where $z_1 = (h_1 + h_0)/2$ and $z_2 = (h_2 + h_1)/2$ corresponds to the mid-point definition.

Recalling that the flux equivalency demand leading to Eq. (6), folds all of the complexities into the length scale, λ, one obtains the result:

$$\lambda/h_1 = (1 - h_1/H)^{3/2} \cdot \{\log(P) + 2 \cdot (1 + 6.3 \cdot H/L_{MO}) . [(1 - z_2/H)^{-1/2} - (1 - z_1/H)^{-1/2}]\} / (1 + 6.3 \cdot h_1/L_{MO}), \quad (9b)$$

where, once again, mid-point level values for z_1 and z_2 are assumed.

The convective diffusivity profile: $K_z(z) = k' \cdot u_* \cdot z \cdot (1 + \alpha \cdot z)^{1/2} \cdot (1 - z/H)$, where $\alpha \equiv 9/|L_{MO}|$ for $L_{MO} < 0$, is only slightly more difficult to deal with inside the resistance integral. The result of the integration in Eq. (3) is:

$$R_{12} \cdot (k' \cdot u_*) = 2 \cdot [\tanh^{-1}(Q_2/q) - \tanh^{-1}(Q_1/q)]/q - 2 . [\tanh^{-1}(Q_2) - \tanh^{-1}(Q_1)], \quad (10a)$$

where $Q_2 = (1 + \alpha \cdot z_2)^{1/2}$, $Q_2 = (1 + \alpha \cdot z_2)^{1/2}$, $q = (1 + \alpha \cdot H)^{1/2}$, and $\tanh^{-1}(x)$ is the hyperbolic arctangent, having the more familiar equivalence of $\tanh^{-1}(x) = 1/2 \cdot \log[(1+x)/(1-x)]$ for $x^2 < 1$.

Again utilizing Eq. (6), one obtains the result: $\lambda/h_1 = 2 \cdot (1 - h_1/H) \cdot (1 + \alpha \cdot h_1)^{1/2} . \{[\tanh^{-1}(Q_2/q) - \tanh^{-1}(Q_1/q)]/q - [\tanh^{-1}(Q_2) - \tanh^{-1}(Q_1)]\}$, which is more simply expressed as:

$$\lambda/h_1 = (1 - h_1/H) \cdot (1 + \alpha \cdot h_1)^{1/2} \cdot \{\log[(q + Q_2) \cdot (q - Q_1)/((q + Q_1) \cdot (q - Q_2))]/ q - \log[(1 + Q_2) \cdot (Q_1 - 1)/((1 + Q_1) \cdot (Q_2 - 1))]\} \quad (10b)$$

where, once again, mid-point level values for z_1 and z_2 are assumed.

4. Results and Conclusions

Continuing with the previous approach of adjusting the length scale, we consider a very stable ($L_{MO} = 40$m, $H = 100$m) application of Eq. (9) and a very unstable ($L_{MO} = -10$m, $H = 1600$m) application of Eq. (10). The results are presented below along with the previous neutral results.

h_2/h_1	λ/h_1 Eq. (5)	λ/h_1 Eq. (7)	λ/h_1 Eq. (9b)	λ/h_1 Eq. (10b)
2	1.0	1.0986	1.0250	1.1851
3	1.5	1.3863	1.6397	1.4084
4	2.0	1.6094	2.4032	1.5624
5	2.5	1.7918	3.4142	1.6770
10	5.0	2.3979	-	1.9974
20	10.0	3.0445	-	2.2607
50	25.0	3.2581	-	-

In the above examples, it is clear that any simple estimate for the derivative length scale λ in terms of the depth, h_1, of the lowest layer (e.g., a constant $\lambda/h_1 = 1$ as assumed in the original versions of RCG) is inadequate. On the other hand, computing a K_z at the interface $z = h_1$ coupled with the traditionally assumed $\lambda = (z_2 - z_1) = 1/2 \cdot h_2$ expressed by Eq. (5), can also lead one far from the appropriate result, particularly under convective conditions with thick layers and large associated h_2/h_1.

Also, it is clear that if one is going to the computational expense of computing λ/h_1, as expressed by Eqs. (9b,10b), one might as well simply compute a modified K'_z, given as:

$$K'_z \equiv K'_z(z=h_1)=(z_2-z_1)/R_{12} \tag{11}$$

and use the traditional $\lambda = (z_2 - z_1)$. That is, instead of computing a new quantity λ, simply compute the "average" K based on the reciprocal of the integrated resistance.

5. Summary

This paper shows how to take "point" diffusivity profiles, defined as some function of z, and apply them to grid models having layers of finite thickness. Rather than use the diffusivity value defined at the level interfaces, an integrated resistance is computed via Eq. (3) and a modified interface diffusivity computed via Eq. (11). A more complex Eq. (4), involving double-integral averaging of the Eq. (3) resistance, may also be employed together with Eq. (11), but most of the improvement over the point diffusivity result is realized using the simpler Eq. (3) plus Eq. (11) combination. Explicit results are presented for typical neutral, stable, and convective diffusivity profiles. This method of a combining varying diffusivities via averaging or integral smoothing of the reciprocal of the diffusivities is also consistent with analytic solutions of the 1-d diffusion equation through strata of varying diffusivity.

Acknowledgements This work has been initiated and funded by the German Federal Environmental Agency (Umweltbundesamt) within the R&D-projects 298 41 252, 299 43 246, and most recently by R&D contract 202 43 270 to the Meteorological Institute of the Free University of Berlin.

References

Byun, D. W., J. Young, J. Pleim, M. T. Odman, and K. Alapaty, 1999. Numerical Transport Algorithms for the Community Multiscale Air Quality (CMAQ) Chemical Transport Model in Generalized Coordinates, EPA/600/R-99/030, Chapter 7.

Flemming, J., van Loon, M., Stern, R., Data Assimilation for CTM based on Optimum Interpolation and Kalman Filter. Proceedings of *26th ITM on Air Pollution Modelling and its Application. May 26-30, 2003, Istanbul, Turkey.*

Hass, H., Builtjes, P. J. H., Simpson, D., and Stern, R., Comparison of model results obtained with several European regional air quality models, *Atmos. Env. 31 (1997), 3259-3279.*

McRae, G. J., W. R. Goodin and J. H. Seinfeld, 1982. Numerical solution of the atmospheric diffusion equation for chemically reacting flows. *J. Comp. Phys.*, **45**, 1-42.

Stern, R. and R. J. Yamartino, 2002. Development and first results of a new photochemical model for simulating ozone and PM-10 over extended periods. Proceedings of the 12th AMS/AWMA Conf. on the Applications of Air Pollution Meteorology, Norfolk, Virginia, May 20-24, AMS, Boston, MA.

Stern, R., 1994. Development and application of a three-dimensional photochemical dispersion model using different chemical mechanisms. Meteorologische Abhandlungen, Serie A Vol. 8, Dietrich Reimer, Berlin (in German).

Yamartino, R. J., J. Scire, G. R. Carmichael, and Y. S. Chang; The CALGRID mesoscale photochemical grid model-I. Model formulation, *Atmos. Environ.*, 26A (1992), 1493-1512.

Adaptation of Analytic Diffusivity Formulations to Eulerian Grid Model Layers of Finite Thickness

Speaker: R. Yamartino

Questioner: R. Bornstein

Comment:The early AM O_3 peak is connective and somewhat random in time and space, and thus they would average out when the average O_3 from 74 sites (like done here) are presented on one graph.

Questioner: S. T. Rao

Question: Do you have any explanation for the double peak structure in the ozone concentrations? Is the morning time peak related to entrainment of ozone trapped aloft as the PBL is growing and the later peak is related primarily to chemical production? In other words, have you looked into the relative roles of physical and chemical mechanisms dictating the diurnal profile of ozone concentrations in the rural areas?

Answer: Both comment and question relate to the plot of summertime hourly ozone (i.e., shown below, but not in the original reprint). All three curves represent an average over all summer days and over 74 rural stations in Germany, so much of the random space/time structure mentioned above would be erased. No "double-peak" structure is seen in the observed data, and the only semblance of a double peak is seen in the "base-case" model predictions using the original specification of Kz at the current dynamic-layer interface height. This erroneous structure is seen to vanish with the implementation of the resistance-based, integral Kz algorithm described in this paper. I also note that the model still (i) underpredicts O3 at night and (ii) predicts a too rapid rise in O3 during morning hours, but these attributes are presently thought to arise respectively from (i) the presumption of an overly stable, nighttime surface layer and (ii) neglect of canopy damping in the meteorological model, which then allows mixing heights to climb too quickly in the morning. A recent model version with process analysis features will enable us to look into the relative roles of the physical and chemical mechanisms dictating this average diurnal profile.

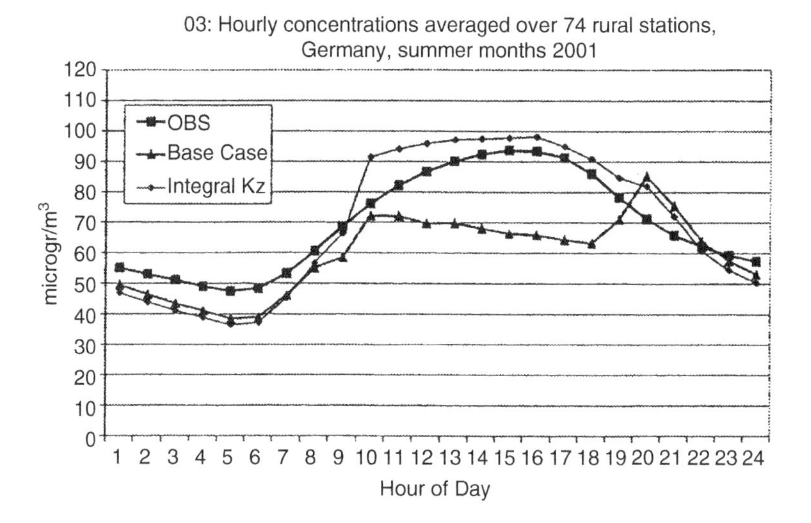

52 Particulate Matter Source Apportionment Technology (PSAT) in the CAMx Photochemical Grid Model

Greg Yarwood, Ralph E. Morris, and Gary M. Wilson*

1. Introduction

Airborne particulate matter (PM) is important because it causes health problems and environmental degradation and accordingly many countries implement programs to control PM pollution (e.g., EPA, 1996). In recent years the emphasis on controlling PM pollution has shifted toward problems associated with fine PM ($PM_{2.5}$ with particle diameter less than 2.5 μm) because it is more strongly associated with serious health effects than coarse PM. Knowing what sources contribute to fine $PM_{2.5}$ is essential for developing effective control strategies. Many components of $PM_{2.5}$ are secondary pollutants and so photochemical models are important tools for PM air quality planning. The Comprehensive Air quality Model with extensions CAMx; (ENVIRON, 2003) is one of the photochemical grid models being used to understand PM pollution and visibility impairment in the US and Europe. The Particulate Matter Source Apportionment Technology (PSAT) has been developed for CAMx to provide geographic and source category specific PM source apportionment. PM source apportionment information from PSAT is useful for:

(1) Understanding model performance and thereby improving model inputs/formulation.
(2) Performing culpability assessments to identify sources that contribute significantly to PM pollution.
(3) Designing the most effective and cost-effective PM control strategies.

Source apportionment for primary PM is relatively simple to obtain from any air pollution model because source-receptor relationships are essentially linear for primary pollutants. Gaussian steady-state models and Lagrangian puff models

* ENVIRON International Corporation, 101 Rowland Way, Suite 220, Novato, CA 94945-5010. e-mail: gyarwood@environcorp.com

have been used extensively to model primary PM pollution from specific sources, which provides source apportionment. The Gaussian and Lagrangian approaches work for primary PM because the models can assume that emissions from separate sources do not interact. This assumption breaks down for secondary PM pollutants (e.g., sulfate, nitrate, ammonium, secondary organic aerosol) and so puff models may dramatically simplify the chemistry (to eliminate interactions between sources) so that they can be applied to secondary PM. Eulerian photochemical grid models are better suited to model secondary pollutants because they account for chemical interactions between sources. Grid models do not naturally provide source apportionment because the impact of all sources has been combined in the total pollutant concentration. PSAT has been developed to retain the advantage of using a grid model to describe the chemistry of secondary PM formation and also provide source apportionment.

This paper starts by reviewing several approaches to PM source attribution in grid models that were considered when PSAT was designed. Next, the PSAT algorithms are explained. Finally, example PSAT results are presented and compared to results from other methods.

2. Approaches to PM Source Attribution in Grid Models

The need for PM source apportionment (PSAT) in CAMx was discussed in the introduction. Several potential approaches were considered at the PSAT design stage and these are discussed below. The approaches considered fall into two general categories, which we have called sensitivity analysis and reactive tracers. The reactive tracer approaches also could be called tagged species approaches. This section concludes with a discussion of an important fundamental difference between sensitivity analysis and source apportionment that explains why these two concepts should not be confused when dealing with secondary pollutants.

2.1. Sensitivity Analysis Methods

Sensitivity analysis methods measure the model output response to an input change, e.g., the change in sulfate concentration due to a change in SOx emissions. In general, sensitivity methods will not provide source apportionment if the relationship between model input and output is non-linear. For example, if sulfate formation is non-linearly relates to SOx emissions, the sum of sulfate (SO4) sensitivities over all SOx sources will not equal the model total sulfate concentration. This concept is discussed further in section 2.3.

Brute Force or Direct Method. The brute force method estimates first-order sensitivity coefficients (e.g., dSO4/dSOx) by making a small input change (dSOx) and measuring the change in model output (dSO4). This method is simple and can be applied to any model, but is inefficient because a complete model run is required for each sensitivity coefficient to be determined. Accuracy also may be an issue for the brute force method because ideally the input change (dSOx) should be vanishingly small but for small input changes the output change may be contaminated by numerical precision or model "noise." Higher-order sensitivity coefficients also can be estimated by the brute force method.

Zero-out modeling. The zero-out method differs from the brute force method in that a specific emissions input is set to zero and the change in output measured. Zero-out modeling can be used with any model but is inefficient because a complete model run is required for each source. Accuracy also may be an issue if the zero-out method is applied to a small emissions source. This method has been used extensively for source attribution because it seems intuitively obvious that removing a source should reveal the source's impact. However, because zero-out modeling is a sensitivity method it does not provide source apportionment for non-linear systems because the sum of zero-out impacts over all sources will not equal the total concentration, as discussed further in section 2.3.

Decoupled direct method (DDM). The DDM provides the same type of sensitivity information as the brute force method but using a computational method that is directly implemented in the host model (Dunker, 1981). The DDM has potential advantages of greater efficiency and accuracy relative to the brute force method (Dunker at al., 2002a) and the DDM implementation in CAMx (Dunker at al., 2002a,b) is currently being extended to PM species. Drawbacks of DDM are that the implementation is technically challenging, that using DDM for many sensitivities simultaneously requires large computer memory, and because DDM is a sensitivity method it does not provide source apportionment for non-linear systems (section 2.3.)

Other sensitivity methods. There are other sensitivity methods that provide similar information to DDM such as adjoint methods (e.g., Menut at al., 2000; Elbern and Schmidt, 1999) and automatic differentiation in FORTRAN (ADIFOR; e.g., Sandhu, 1997). They have similar advantages and disadvantages as DDM but may be less computationally efficient (Dunker et al., 2002a).

2.2. Reactive Tracer Methods

Reactive tracers (or tagged species) are extra species added to a grid model to track pollutants from specific sources. For example, a standard grid model calculates concentrations for a species X that has many sources and so the concentration of X is the total concentration due to all sources. A reactive tracer (x_i) is assigned to for each source (i) with the intention that the sum of the reactive tracers will equal total concentration $\left(X = \sum x_i\right)$. The challenge is to develop numerical algorithms for solving the reactive tracer concentrations that ensure that this equality is maintained. Depending upon the formulation of the tracer algorithms, it may be possible to model tracers for a single source of interest and omit tracers for all other sources, or it may be necessary to include tracers for all sources (as is the case for PSAT). Reactive tracers can potentially provide true source apportionment $\left(X = \sum x_i\right)$, however the numerical value of the source apportionment will depend upon assumptions within the reactive tracer formulation. In particular, for any process that is non-linear in species concentrations (e.g., chemistry) there is no unique way to assign the total concentration change to the reactive tracers. This issue is discussed further in section 2.3.

Source Oriented External Mixture (SOEM). Kleeman and Cass (2001) developed an approach called SOEM that tracks primary PM from different source categories/regions using that tagged species that are considered to represent seed particles. Reactive tracers are added to track secondary PM and related gases from different source categories/regions and source apportioned secondary PM condenses onto the seed particles. Chemical change for secondary PM and related gases is accounted for by expanding the chemical mechanism to treat different source regions/categories as separate precursor and product species. This requires thousands of chemical reactions and hundreds to thousands of chemical species, depending upon the number of source regions/categories. The main advantage of the SOME method is that it is potentially accurate, and the main disadvantage is computational demand.

PSAT and TSSA. The PM Source Apportionment Technology (PSAT) uses reactive tracers to apportion primary PM, secondary PM and gaseous precursors to secondary PM among different source categories and source regions. The PSAT methodology is described in section 4. PSAT was developed from the related ozone source apportionment method (OSAT) already implemented in CAMx (Dunker at al., 2002b). Tonnesen and Wang (2004) are independently developing a method very similar to PSAT called Tagged Species Source Apportionment (TSSA) in a different photochemical grid model. Advantages of PSAT and TSSA are expected to be high efficiency and flexibility to study different source categories/regions. The accuracy of the PSAT (and TSSA) source apportionment results must be evaluated, as for any other method.

2.3. Source Apportionment and Source Sensitivity

Consider a chemical species X that has two sources A and B (so $X = x_A + x_B$) and which undergoes a second order self-reaction with rate constant k. The rate of chemical change is:

$$dX/dt = -kX^2$$

$$dX/dt = -k(x_A + x_B)^2$$

$$dX/dt = -kx_A^2 - kx_B^2 - 2kx_Ax_B$$

The homogeneous rate terms kx_A^2 and kx_B^2 clearly describe chemical change for pollutants from sources A and B (x_A and x_B), but the inhomogeneous rate term $2kx_Ax_B$ is not uniquely associated with either source A or B. A reactive tracer (or tagged species) source apportionment method must deal with this inhomogeneous rate term either by developing rules to apportion the inhomogeneous term to sources A and B or modifying the chemistry to eliminate the inhomogeneous term. For a sensitivity method, the homogeneous quadratic rate terms generate second-order homogeneous sensitivity coefficients (s_{AA} and s_{BB})

and the inhomogeneous rate term generates a second-order inhomogeneous sensitivity coefficient (s_{AB}). Consequently, the total concentration of X is incompletely described by the first-order sensitivity coefficients (s_A and s_B) that resemble source apportionments.

The example presented above is a simple case of a non-linear chemical system that illustrates why source apportionment and source sensitivity are not the same thing for nonlinear systems. The implications are:

(1) There is no unique source apportionment for chemical species that depend upon nonlinear reactions, such as secondary PM species and ozone. Nonetheless, reasonable source apportionment schemes can be developed and are useful tools for achieving the objectives listed in the introduction.
(2) Sensitivity coefficients are not source apportionments for chemical species that depend upon nonlinear reactions, such as secondary PM species and ozone. Sensitivity coefficients are most applicable for predicting the model response to an input change (e.g., control strategy) and are very useful for this purpose (within the range of linear model response), but sensitivity coefficients should be used with care for source apportionment or culpability assessment. Likewise, source apportionments should be used with care for predicting model response to input changes.

3. PSAT Methodology

PSAT is designed to source apportion the following PM species modeled in CAMx:

- Sulfate
- Particulate nitrate
- Ammonium
- Particulate mercury
- Secondary organic aerosol (SOA)
- Six categories of primary PM

The PSAT "reactive tracers" that are added for each source category/region (i) are described below. In general, a single tracer can track primary PM species whereas secondary PM species require several tracers to track the relationship between gaseous precursors and the resulting PM. Nitrate and secondary organic PM are the most complex aerosol species to apportion because the emitted gases (NO, VOCs) are several steps removed from the resulting PM (nitrate, SOA). The reactive tracers used by PSAT are listed below for each class of PM.

<u>Sulfur</u>

$SO2_i$ Primary SO_2 emissions

$PS4_i$ Particulate sulfate ion from primary emissions plus secondarily formed sulfate

Nitrogen

RGN_i Reactive gaseous nitrogen including primary NOx ($NO + NO_2$) emissions plus nitrate radical (NO_3), nitrous acid (HONO) and dinitrogen pentoxide (N_2O_5).
TPN_i Gaseous peroxyl acetyl nitrate (PAN) plus peroxy nitric acid (PNA)
NTR_i Organic nitrates (RNO_3)
$HN3_i$ Gaseous nitric acid (HNO_3)
$PN3_i$ Particulate nitrate ion from primary emissions plus secondarily formed nitrate

Ammonia/Ammonium

$NH3_i$ Gaseous ammonia (NH_3)
$PN4_i$ Particulate ammonium (NH_4)

Secondary Organic

ALK_i Alkane/Paraffin secondary organic aerosol precursors
ARO_i Aromatic (toluene and xylene) secondary organic aerosol precursors
CRE_i Cresol Secondary secondary organic aerosol precursors
TRP_i Biogenic olefin (terpene) secondary organic aerosol precursors
$CG1_i$ Condensable gases from toluene and xylene reactions (low volatility)
$CG2_i$ Condensable gases from toluene and xylene reactions (high volatility)
$CG3_i$ Condensable gases from alkane reactions
$CG4_i$ Condensable gases from terpene reactions
$CG5_i$ Condensable gases from cresol reactions
$PS1_i$ Secondary organic aerosol associated with CG1
$PS2_i$ Secondary organic aerosol associated with CG2
$PS3_i$ Secondary organic aerosol associated with CG3
$PS4_i$ Secondary organic aerosol associated with CG4
$PS5_i$ Secondary organic aerosol associated with CG5

Mercury

$HG0_i$ Elemental Mercury vapor
$HG2_i$ Reactive gaseous Mercury vapor
HGP_i Particulate Mercury

Primary Particulate

PEC_i Primary Elemental Carbon
POA_i Primary Organic Aerosol
PFC_i Fine Crustal PM
PFN_i Other Fine Particulate
PCC_i Coarse Crustal PM
PCS_i Other Coarse Particulate

PSAT includes a total of 32 tracers for each source group (i) if source apportionment is applied to all types of PM. Since source apportionment may not always be needed for all species, the PSAT implementation is flexible and allows source apportionment of selected chemical classes in each CAMx simulation. For example, source apportionment for sulfate/nitrate/ammonium requires just 9 tracers per source group.

The PSAT approach to source apportionment is described below. Consider two model species A and B that are apportioned by reactive tracers a_i and b_i, respectively. Reactive tracers must be included for all sources of A and B including emissions, initial conditions and boundary conditions so that complete source apportionment is obtained, i.e., $A = \Sigma a_i$ and $B = \Sigma b_i$.

The general approach to modeling change over a model time step Δt is illustrated for a chemical reaction:

$$A \rightarrow B$$

The general equation for species destruction is:

$$a_i(t + \Delta t) = a_i(t) + \Delta A \times [a_i / \Sigma a_i]$$

Here the relative apportionment of A is preserved as the total amount changes. This equation applies to chemical removal of A and also physical removal by processes such as deposition.

The general equation for species production (e.g, chemical production) is:

$$b_i(t+\Delta t) = b_i(t) + \Delta B \times [a_i / \Sigma a_i]$$

Here the product B inherits the apportionment of the precursor A.

In some cases, source category specific weighting factors (w_i) must be added to the equation for species destruction:

$$a_i(t+\Delta t) = a_i(t) + \Delta A \times [w_i a_i / \Sigma w_i a_i]$$

An example is chemical decay of the aromatic VOC tracers (ARO), which must be weighted by the average OH rate constant of each ARO_i. ARO tracers for different source groups have different average VOC reactivities because the relative amounts of toluenes and xylenes differ between source categories.

In some cases, source category specific weighting factors (w_i) must be added to the equation for species production:

$$b_i(t+\Delta t) = b_i(t) + \Delta B \times [w_i a_i / \Sigma w_i a_i]$$

An example is chemical production of condensable gases (CG1 or CG2) from aromatic VOC tracers, which must be weighted by aerosol yield weighting factors. The aerosol yield weighting factors depend upon the relative amounts of toluenes and xylenes in each source group.

Several aerosol reactions are treated as equilibria:

$$A \leftrightarrow B$$

If A and B reach equilibrium at each time step, it follows that their source apportionments also reach equilibrium:

$$a_i(t+\Delta t) = [a_i(t) + b_i(t)] \times [A / (A+B)]$$

$$b_i(t+\Delta t) = [a_i(t) + b_i(t)] \times [B / (A+B)]$$

Examples are the equilibrium between gas phase nitric acid and aerosol nitrate, gas phase ammonium and aerosol ammonium, and condensable organic gases (CG) and secondary organic aerosols (SOA).

4. Testing PSAT

The initial PSAT testing was for primary PM species (e.g., fine crustal material) because these species are the most straightforward to implement and test (because there is no chemistry). The results for primary PM species are not shown here but the tests were successful in that the sum of PSAT tracers over all source groups remained identical to the total model concentration ($A = \Sigma a_i$) for all grid cells and hours. The primary PM tests confirmed the correct implementation of PSAT for all CAMx algorithms apart from chemistry.

The next stage of PSAT testing included the chemistry algorithms and results are shown here for two different types of testing for sulfate and nitrate. Sulfate was tested by comparing PSAT to zero out results in full 3-D CAMx simulations. As discussed above, zero out results are not expected to agree perfectly with PSAT because zero out does not give true source apportionment in non-linear systems. However, non-linearity in sulfate formation chemistry is expected to be less important than for other secondary PM species, making sulfate the best candidate for zero out testing.

The most complex PSAT chemistry algorithm is for nitrate. Zero out tests were not used for nitrate because the relationship between NO emissions and nitric acid may be highly non-linear. Therefore, conduct 1-D (box model) tests were used to evaluate PSAT results for nitrate by comparing against the method used by Kleeman and Cass (2001) to track nitrate apportionment in the Source-Oriented External Mixture (SOEM) method, discussed above.

4.1. PSAT Sulfate Testing

The PSAT performance for sulfate was tested using a CAMx database developed by the Midwest RPO (MPRO) for PM and visibility modeling of the Eastern US. The modeling was from June 18 to July 21, 2001 and used a 36-km modeling grid with meteorology developed using the mesoscale model version 5 (MM5). The modeling domain was sub-divided to geographic areas according to regional planning organizations (RPOs) responsible for developing regional visibility and PM control strategies in the U.S. The sub division of the modeling domain to RPOs is shown in Figure 1 and the RPOs are labeled by their respective acronyms (MRPO, MANE-VU, VISTAS, CERAP and WRAP). The state of Illinois (IL) was split out from the Midwest RPO (MPRO) to test the ability of PSAT to apportion the contribution from a single state. Four hypothetical point sources were added near the middle of the MRPO, MANE-VU, VISTAS and CENRAP areas (shown by the + symbols in Figure 1) to test the ability of PSAT to track contributions from single sources. The hypothetical point sources were chosen to be

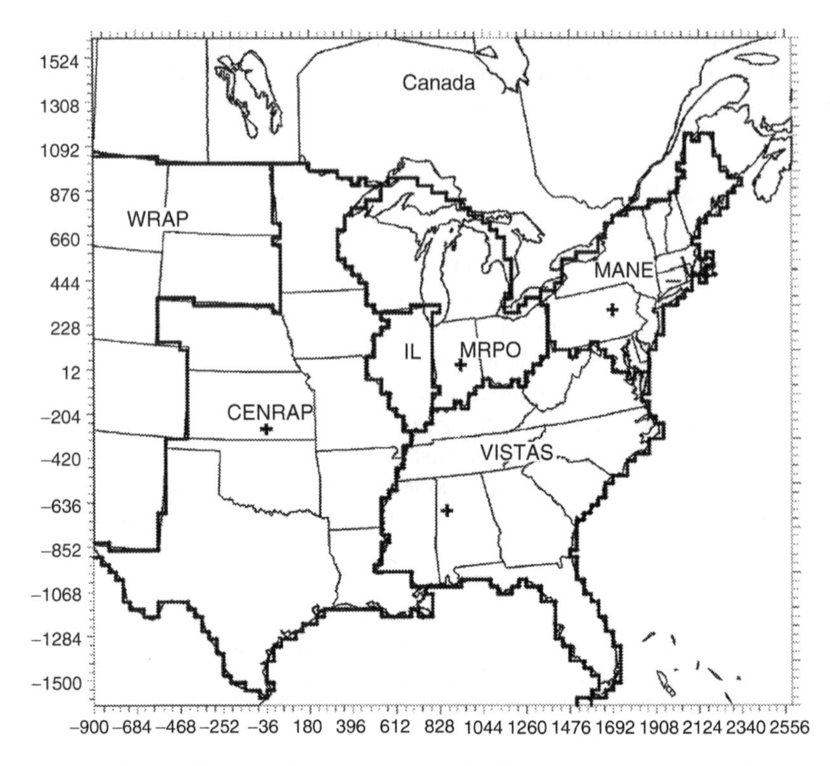

FIGURE 1. The CAMx modeling domain for PSAT testing showing sub-division to geographic areas and locations of four hypothetical point sources (+ symbols).

generally representative of a large coal-fired utility source but do not represent actual sources. In total, sulfate was apportioned to 11 source groups including a remainder group for Canada, Mexico and over water areas.

The sulfate impacts from the hypothetical MRPO point source are compared in Figure 2 at a single hour (hour 15) on 28 June 2001. The spatial distribution of sulfate impacts is very similar in the PSAT and zero out results as shown by the edge of the impacts plume (0.1 $\mu g/m^3$ level). There are differences in the areas of larger impacts (e.g. the 1 and 2 $\mu g/m^3$ levels) and these are due to the effects of non-linear chemistry in the zero out test. As discussed above, sensitivity methods such as zero out do not provide accurate source apportionments for non-linear processes. Sulfate formation can be limited by the availability of oxidants, especially hydrogen peroxide, which will tend to depress the maximum impact levels in zero out runs as well as shift impacts further downwind (to where oxidant availability is no longer limiting). The oxidant limiting effect on zero out sulfate impacts is most easily seen from the 2 $\mu g/m^3$ level extending further downwind over Lake Michigan in the zero out result than the PSAT result.

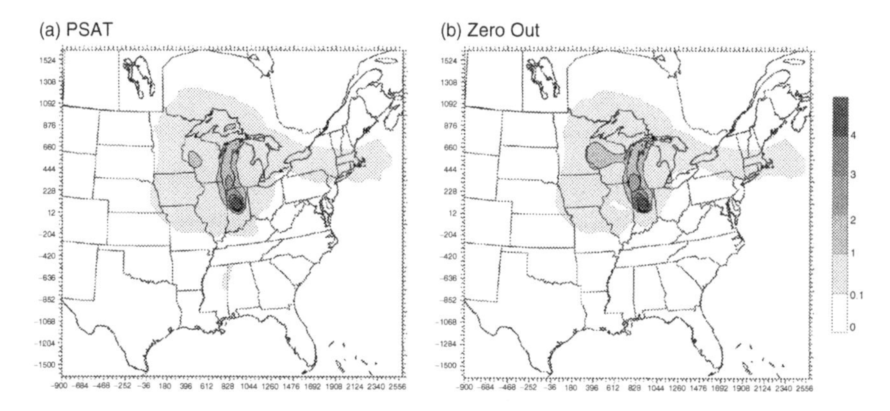

FIGURE 2. Comparison of sulfate impacts (μg/m³) from the hypothetical MRPO point source on 28 June 2001 at hour 15: (a) PSAT result; (b) Zero out result.

The episode average sulfate impacts from the hypothetical MRPO point source are compared in Figure 3 for the entire 28 June to July 21, 2001 modeling period. The spatial distribution of sulfate impacts is very similar in the PSAT and zero out results. The maximum impact occurs very close to the source and is higher in the PSAT result (2.2 μg/m³) than the zero out result (1.8 μg/m³) due to the effect of oxidant limitation on sulfate impacts in the zero out result.

The PSAT sulfate tests provide a comparison of the efficiency of the PSAT method compared to zero out modeling. Zero out modeling requires a new model run for each source contribution determined, so the incremental time for each "apportionment" is the same as for the model base case. In contrast, the marginal cost for each PSAT source apportionment was about 2% of the time required for

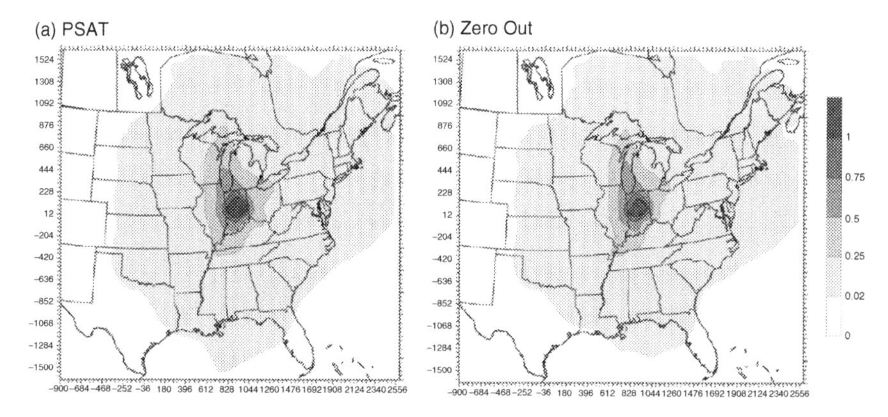

FIGURE 3. Comparison of episode average (June 18 to July 21, 2001) sulfate impacts (μg/m³) from the hypothetical MRPO point source: (a) PSAT result; (b) Zero out result.

the base case, so PSAT is about 50 times more efficient than zero out modeling for sulfate apportionment.

4.2. PSAT Nitrate Testing

The PSAT nitrate algorithms were tested using CAMx for a 1-D (box model) problem in order to focus upon the ability of the PSAT chemical algorithms to track nitrate apportionment. The box model problem was for summer conditions and PSAT was used to apportion nitrate between 20 ppb of initial NOx and 100 ppb of NOx emissions injected continuously through the 24 hour run. The PSAT results were compared to the SOEM algorithm of Kleeman and Cass (2001) where the reactive nitrogen (NOy) reactions in the chemical mechanism are duplicated to provide two different types of NOy (think of them as "red" NOy and "blue" NOy). There was no ammonia in the box model so that nitric acid remained in the gas phase rather than forming PM nitrate.

The PSAT apportionment of NOy to initial conditions (ICs) and emissions during the 24 hour box model simulation is shown in Figure 4. The total NOy apportioned to ICs remains constant at 20 ppb throughout the simulation but the apportionment changes over time from NOx (RGN-IC) at the start to PAN (TPN-IC), organic nitrates (NTR-IC) and nitric acid (HN3-IC) later in the simulation. The NOy apportioned to emissions increased linearly throughout the simulation and the apportionment also evolved as the NOx emissions (RGN-E) reacted. At the start of the simulation RGN (NOx) is dominated by ICs

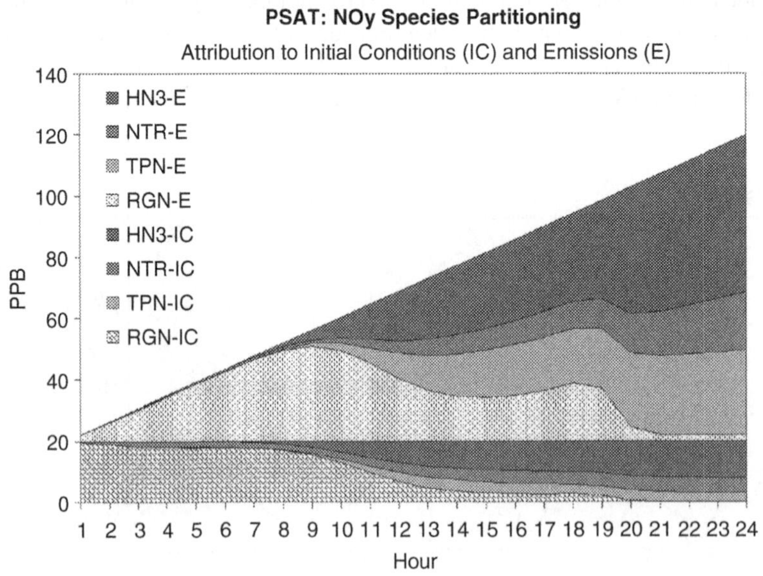

FIGURE 4. PSAT apportionment of reactive nitrogen species to initial conditions and emissions during a 24 hour box model simulation.

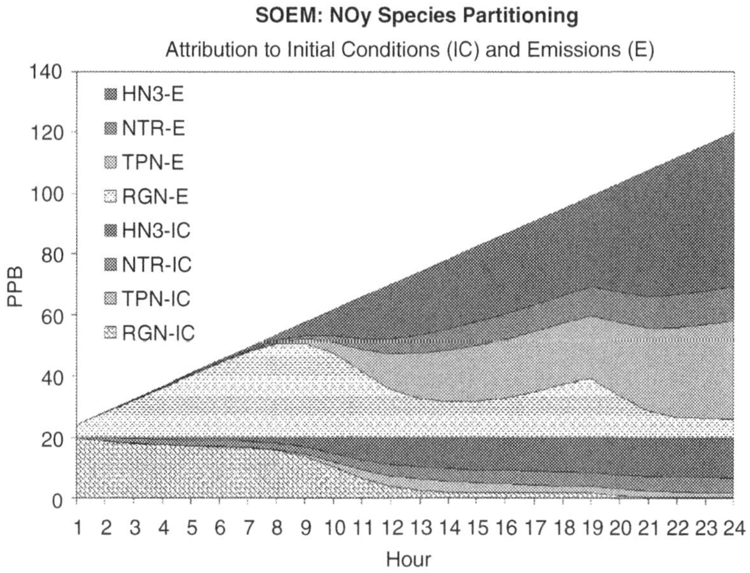

FIGURE 5. Source-Oriented External Mixture (SOEM) apportionment of reactive nitrogen species to initial conditions and emissions during a 24-hour box model simulation.

whereas late in the day (hour 18) RGN is dominated by emissions. The source apportionments shown in Figure 4 are reasonable and show necessary attributes (such as conserving the total 20 ppb of ICs) for a source apportionment scheme.

The SOEM apportionment of NOy to ICs and emissions during the 24-hour box model simulation is shown in Figure 5. The time evolution of the source apportionments for the ICs and emissions is very similar for SOEM (Figure 5) and PSAT (Figure 4). It is not clear that either method is more "correct" or indeed that a correct source apportionment result exists for this test, but the consistency between the PSAT and the SOEM results is encouraging.

5. Summary

Several approaches to PM source apportionment were considered and a reactive tracer (or tagged species) method has been developed for the PM Source Apportionment Technology (PSAT) implemented in CAMx. Initial tests show that PSAT results are reasonable and that PSAT is much more efficient than sensitivity methods such as zero out modeling. Theoretical considerations suggest that PSAT may be a better approach to source apportionment than sensitivity methods because it is better able to deal with non-linear chemistry. Further testing is needed using real world applications. The PSAT algorithms will become publicly available in a future CAMx release.

Acknowledgment The Midwest Regional Planning Organization (MRPO) and the Lake Michigan Air Directors Consortium (LADCO) sponsored the PSAT development. The authors acknowledge the helpful discussions with Mike Koerber and Kirk Baker of LADCO.

References

Dunker A. M., 1981, Efficient calculations of sensitivity coefficients for complex atmospheric models, Atmos. Environ. 1981, 15, 1155-1161.

Dunker A. M., G. Yarwood, J. P. Ortmann, G. M. Wilson, 2002a, The decoupled direct method for sensitivity analysis in a three-dimensional air quality model – implementation, accuracy and efficiency, *Environ. Sci. Technol.*, 36:2965-2976.

Dunker A. M., G. Yarwood, J. P. Ortmann, G. M. Wilson, 2002b, Comparison of source apportionment and source sensitivity of ozone in a three-dimensional air quality model, *Environ. Sci. Technol.*, 36:2953-2964.

Elbern, H., Schmidt, H. A four-dimensional variational chemistry data assimilation scheme for eulerian chemistry transport modeling, *J. Geophys.Res.* 1999, *104*, 18583-18598.1999

ENVIRON, 2003, User's guide to the Comprehensive Air Quality model with extensions (CAMx) version 4.00 (January, 2003); http://www.camx.com

EPA, 1996, Air quality criteria for particulate matter, Environmental Protection Agency report EPA/600/P-95/001aF-cF.3v. Research Triangle Park, NC.

EPA, 2003, Air quality criteria for particulate matter, Environmental Protection Agency draft report EPA/600/P-99/002aB. Research Triangle Park, NC

Kleeman M. J., G. R. Cass, 2001, A 3D Eulerian source-oriented model for an externally mixed aerosol, *Environmental Science and Technology*, 35: 4834-4848.

Menut L., R. Vautard, M. Beekmann, C.Honoré, 2000, Sensitivity of Photochemical Pollution using the Adjoint of a Simplified Chemistry-Transport Model, *Journal of Geophysical Research – Atmospheres,* 105, D12: 15,379-15,402.

Sandhu A. 1997, Sensitivity analysis of ODE via automatic differentiation, MS Thesis, University of Iowa, IA.

Tonnesen G., B. Wang, 2004, CMAQ Tagged Species Source Apportionment, July 22, 2004; http://www.wrapair.org/forums/aoh/meetings/040722/UCR_tssa_tracer_v2.ppt

Particulate Matter Source Apportionment Technology (PSAT) in the CAMx Photochemical Grid Model

Speaker: G. Yarwood

Questioner: S. Andreani Aksoyoglu

Question: Can PSAT be used also with area emissions?

PSAT can be used with area emissions and some of the examples in my presentation are for area source emission categories such as mobile sources. PSAT apportions the contributions from all pollution sources in CAMx, namely, area emissions, point emissions, boundary conditions and initial condition. The emissions contributions can be broken out by source category and/or geographic area down to an individual source.

Questioner: D. Cohan

Question: Dunker et al. 2002 demonstrated substantial discrepancies between ozone source apportionment technology and the decoupled direct method in the attributing ozone. Do you expect to find more consistency between source apportionment technology and DDM in the case of particulates?

Answer: In my presentation and paper I illustrate how source sensitivity differs from source apportionment in non-linear systems. Therefore, differences were to be expected between source rankings from DDM and ozone source apportionment (OSAT) and will be largest when ozone chemistry is most non-linear. A good example is where ozone chemistry is inhibited by high NO_2 *concentrations. For this situation, the DDM calculates a negative first-order sensitivity of ozone to NOx emissions, which is difficult to interpret as a source apportionment because a negative source contribution does not fit within the common understanding of source apportionment (i.e., a pie chart of contributions). For the same situation, OSAT apportions zero (or small) ozone production to NOx. So, the results from source sensitivity and source apportionment methods are related, but not equal. Care may be needed when using first-order sensitivity information as the basis for control-strategy design because, for example, negative sensitivity (e.g., negative ozone sensitivity to NOx) should not lead to a decision to increase emissions in order to reduce pollution. Care may be needed when using source apportionment information to estimate the level of emissions reduction required to meet an air quality objective because the response of pollution to emissions changes may be non-linear: Specifically, more control may be needed than is suggested by a linear interpretation of the source apportionment results (i.e., linear roll-back).*

Questioner: D. Cohan

Question: The sensitivity, zero out, and PSAT techniques all give interchangeable results in linear situations but as you well pointed out they have different results and meanings in nonlinear contexts. As you noted for the case of oxidant-limited sulfate formation, PSAT may properly attribute substantial amount of sulfate to SO_2 ***from a source, but reducing*** SO_2 ***from that source may not actually change sulfate concentrations because of substitution from other sources. How do you think these three techniques should be applied in regulatory applications and how could the subtle differences between their meanings be explained to regulators?***

Answer: In my opinion, both sensitivity (e.g., DDM) and source apportionment (e.g., PSAT) methods are valuable in regulatory settings because they can provide detailed understanding of which sources cause pollution problems. The traditional zero-out method can provide the same type of information, but in practice zero-out is so burdensome to apply in detail (many simulations are required) that usually only a few sources are examined. Differences between sensitivity and apportionment results can be understood and anticipated, as illustrated by the previous question. Methods like DDM and PSAT should be used to provide a big picture of the source contributions – large versus small contributions – to guide the selection of control strategies. The evaluation of control strategies impacts should be done in the traditional way of making a new model run with emissions reduced. If emissions are reduced substantially by a control strategy, it may be wise to re-run PSAT or DDM at the reduced emission level before selecting the next strategy.

Questioner: R. Yamartino

Question 1: Why can't one simply assign the source specific "color" to the S, N, and C atoms from this source and carry through operators, rather than form these separate nitrogen and organic families?

Question 2: As CAMx contains both PSAT and DDM, can you apply the DDM on top of the PSAT such that questions such as determining the impact of a 50% reduction of a specific source's emissions can be addressed?

Answer 1: I think the result is the same regardless of whether apportionment is framed in terms of S, N, and C atoms or SO4, NO3 and organic carbon.

Answer 2: Predicting the response of a nonlinear system to a specific input change (e.g. control strategy) requires sensitivity rather than apportionment information. The response to a small input change can be accurately described with first-order sensitivities but the response to a large change will require higher-order sensitivities. For your example, predicting the response to a 50% reduction of a specific source's emissions could be done using DDM but would likely require first- and second-order sensitivities.

Questioner: B. Fisher

Question: For application to control strategies, one may wish to reduce one source category by one half, say, and another by one quarter, etc. Can the PSAT source attribution method be adapted to this situation?

Since PSAT provides source attribution for all sources that contribute to a specific pollutant there is no problem in obtaining simultaneously the contributions of two (or more) sources. The response of pollution to emission reductions may be non-linear, as discussed in questions above.

Questioner: R. San Jose

Question: What is the computational cost of source apportionment technique compared with the "classical" approach?

Answer: The PSAT source apportionment approach is much more efficient than the "classical" approach of brute force sensitivity or zero-out modeling. The efficiency of PSAT (and the related OSAT method) is seen both in reduced CPU time and less disk storage. PSAT requires more memory, which may be a limiting consideration. For CPU time, the efficiency varies according to the pollutant being apportioned: we see efficiencies of ~50 for sulfate and ~10 for ozone. An efficiency of 50 means that PSAT can obtain apportionments for 50 sources in the CPU time needed to obtain apportionment for one source using zero-out. The corresponding saving in disk space may be more than 100 depending upon model configuration. The efficiency of PSAT improves as source apportionment is carried out in greater detail, i.e., more sources are apportioned separately.

Model Assessment and Verification

53
Testing Physics and Chemistry Sensitivities in the U.S. EPA Community Multiscale Air Quality Modeling System (CMAQ)

J. R. Arnold and Robin L. Dennis*

1. Introduction

Uncertainties in key elements of emissions and meteorology inputs to air quality models (AQMs) can range from 50 to 100% with some areas of emissions uncertainty even higher (Russell and Dennis, 2000). Uncertainties in the chemical mechanisms are thought to be smaller (Russell and Dennis, 2000) but can range to 30% or more as new techniques are applied to re-measure reaction rate constants and yields. Single perturbation sensitivity analyses have traditionally been used with AQMs to characterize effects of these uncertainties on peak predicted ozone concentration ($[O_3]$).

However, confidence in AQM applications depends on understanding the physical and chemical model dynamics in full emissions cases and in cases with proposed controls on oxides of nitrogen ($NO + NO_2 = NO_X$) and/or volatile organic compounds (VOC). With a sensitivity analysis, emissions control runs depict the true photochemical system change which we define as the O_3 control response, $\delta O_3/\delta E_{NOx}$ and $\delta O_3/\delta E_{VOC}$, where δE_{NOx} or δE_{VOC} is the type and amount of control required to reduce O_3 to acceptable levels. The model's most important sensitivity, correspondingly, is how its control response is changed by uncertain inputs and parameters, not simply how the resultant $[O_3]$ might be changed by the uncertainties in a full emissions case. We define that latter quantity as ΔO_3, or the change in $[O_3]$ due only to the uncertainty perturbation with full emissions.

* J. R. Arnold and Robin L. Dennis, Atmospheric Sciences Modeling Division, Air Resources Laboratory, U.S. National Oceanic and Atmospheric Administration; on assignment to the National Exposure Research Laboratory, U.S. Environmental Protection Agency, Research Triangle Park, NC, USA 27711.

2. Experimental Design and Model Configuration

The Community Multiscale Air Quality (CMAQ) modeling system (Byun and Ching, 1999) v4.2 was used in a matrix of sensitivity simulations that varied either the model's vertical mixing solution or its chemical mechanism. For the vertical mixing sensitivity CMAQ's solution was varied away from K-theory with a lower limit at the regional average value of 1.0 m^2/sec (Kz1) to 0.02 m^2/sec (Kz002), a value twice the minimum observed in two other AQMs, and then again from K-theory to the asymmetric convective mixing (ACM) solution of Pleim and Chang (1992). For the chemistry sensitivity either of two older mechanisms, CB4 and RADM2, were substituted for the larger and newer SAPRC99. Each perturbation ensemble was run with a full emissions case and with a 50% NO_X control (NO_X50) and a 50% VOC+CO (VOC50) control case.

CMAQ was run with horizontal grid spacings of 32 km for the continental U.S. and with one-way nested progeny domains of 8 km for the southeast U.S. and two 2 km domains centered on Atlanta, GA, and on Nashville, TN for the period 4–14 July 1999. The meteorological driver was MM5v3.5-PX (Xiu and Pleim, 2000) configured with 30 sigma levels and a nominal 38 m surface layer. For the chemical transport model these were reduced to 21 vertical levels with 11 layers in the lowest 1000 m. Emissions were processed from the 1999 National Emissions Inventory with the Sparse Matrix Operator Kernel Emissions modeling system v1.4 (SMOKE, 2002). All physics perturbation runs were made with SAPRC99; all chemistry perturbation runs were made using Kz1. The modified Euler backward iterative (MEBI) solver was used for all simulations.

Full emissions runs are compared to observations at special chemistry sites in each 2 km domains: the SouthEastern Aerosol Research and CHaracterization Atlanta urban core site at Jefferson Street (JST) (SEARCH, 1999); and the 1999 Southern Oxidants Study site in the Nashville suburbs at Cornelia Fort Airpark (CFA) (SOS, 1999).

Owing to limited space, results from the chemistry ensembles are shown here for JST only and from the physics ensembles for CFA only; however, companion results at each site for the complementary sensitivity not shown here are very similar in all cases.

3. Results

3.1. Chemistry Ensembles for Atlanta

Figures 1a-c show the $[O_3]$ time series at JST for the three ensemble members SAPRC99, CB4, and RADM2 for all days in the analysis at 32 km (1a), 8 km (1b), and 2 km (1c). (Note the data gap from 6–11 July when lightning-induced power disruptions at this site took most instrumentation off-line.) Relative to CB4 and RADM2, SAPRC99 predicts more $[O_3]$ most every day, often increasing CMAQ's peak overprediction bias. SAPRC99's increased peak prediction holds

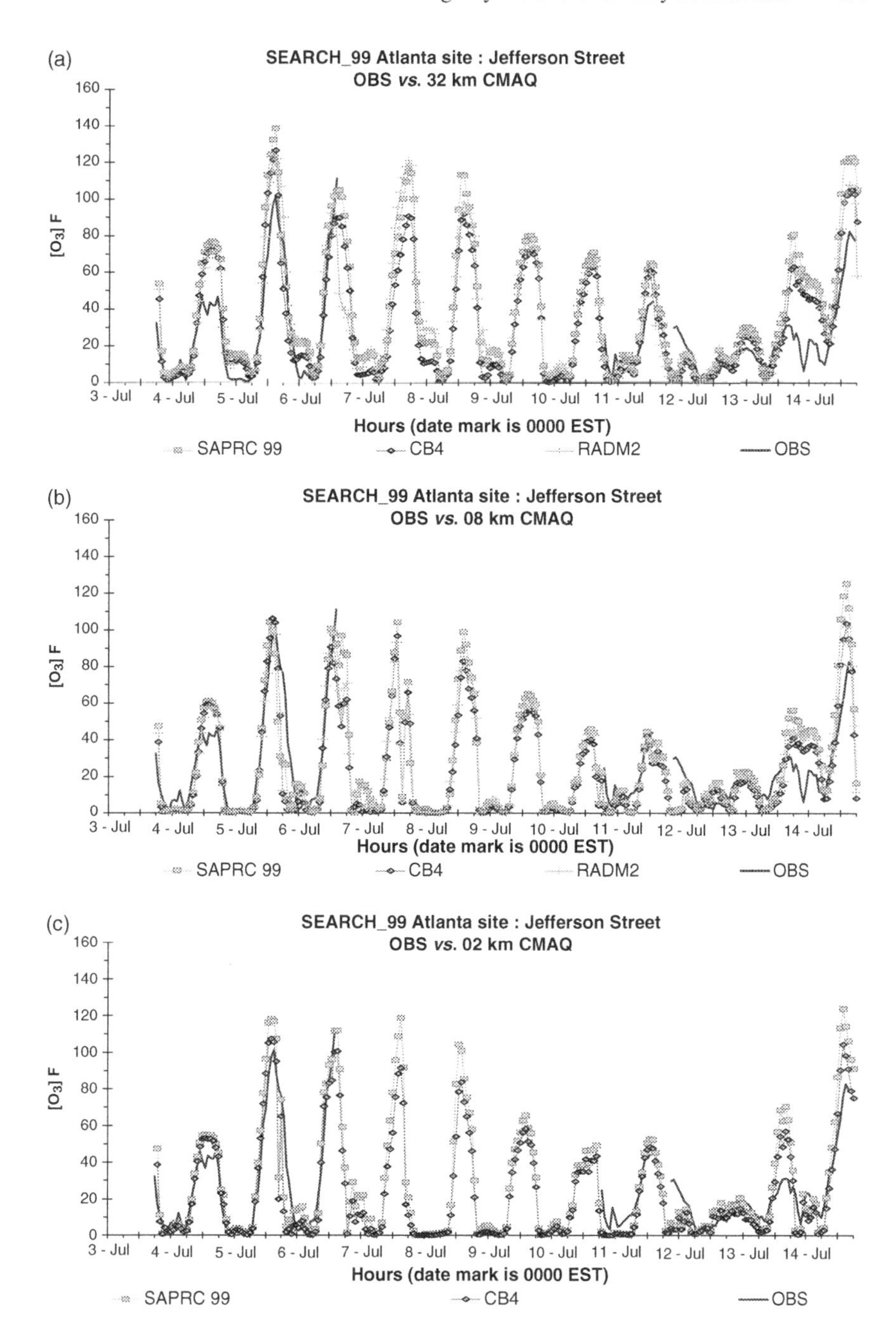

FIGURE 1. O_3 time series of CMAQ chemistry ensembles: 32 km (a); 8 km (b); 2 km (c).

even as the general overprediction recedes at the two finer grid spacings. At the finer grids, 8 km models with all chemistries perform better than the 32 km: the 1-hour normalized bias for peak $[O_3]$ with SAPRC99 falls to 33.2% at 8 km from 63.0% at 32km; to 17.9% at 8 km from 51.5% at 32km with CB4; and to 24.5% from 56.2% with RADM2. The 2 km solutions, however, demonstrate no improvement in 1-hour peak $[O_3]$ bias over the 8 km with either SAPRC99 or CB4: 43.3% bias with SAPRC99 at 2 km and 23.5% bias with CB4 at 2 km. (The RADM2 2 km model was not run for Atlanta.) In addition, the 8 km models perform best for the lowest daytime O_3 peaks and for low overnight $[O_3]$.

Figures 2a-b display the NO_X50 $\delta O_3/\delta E_{NOx}$ response for CB4 (2a) and RADM2 (2b) with time limited to 1000 – 1700 h EST. JST shows hours of NO_X limitation – when NO_X reductions reduce $[O_3]$ – and of NO_X superabundance or radical limitation – when NO_X reductions increase $[O_3]$. These benefits and disbenefits are predicted with all chemistries at all grid spacings though there are differences: CB4 and RADM2 predict less O_3 benefit and more disbenefit than SAPRC99, as well as disbenefits in some hours when SAPRC99 predicts benefits. Work to understand more precisely the causes of these distinctions across the chemistries is continuing.

Figures 3a-b combine the ΔO_3 response from Figure 1 with the $\delta O_3/\delta E_{NOx}$ response in Figure 2. The x-axis for either CB4 (3a) or RADM2 (3b) scales SAPRC99's O_3 difference relative to the other chemistries in the full emissions case expressed here as per cent. The y-axes then scale the per cent residual difference between the predicted change in $[O_3]$ by CB4 and RADM2 relative to the change predicted by SAPRC99 for NO_X50, *i.e.*, the differential sensitivity to each mechanism in the NO_X50 case. That differential sensitivity varies by more than ±15% with CB4 and by more than ±10% with RADM2, and, importantly, cannot be predicted reliably from the model ΔO_3 response. Moreover, for most hours at 32 km and 8 km, and for nearly all hours with the 2 km CB4, CB4 and RADM2 are less sensitive to the NO_X50 control than is SAPRC99, shown as a positive predicted residual value on the y-axes. The relations among chemical mechanisms shown here for NO_X50 generally hold for VOC50 at JST and for both controls at CFA, too, where $\delta O_3/\delta E_{NOx}$ and $\delta O_3/\delta E_{VOC}$ cannot be reliably predicted from the ΔO_3 either.

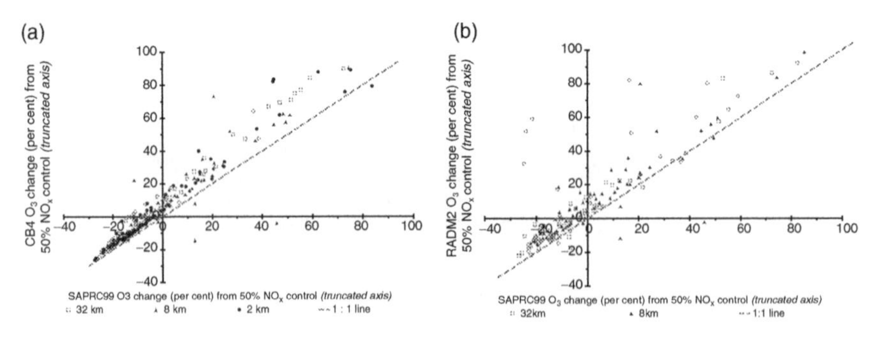

FIGURE 2. O_3 response (per cent) to the 50% NO_X control at Jefferson Street. SAPRC99 *vs.* CB4 (a) and RADM2 (b).

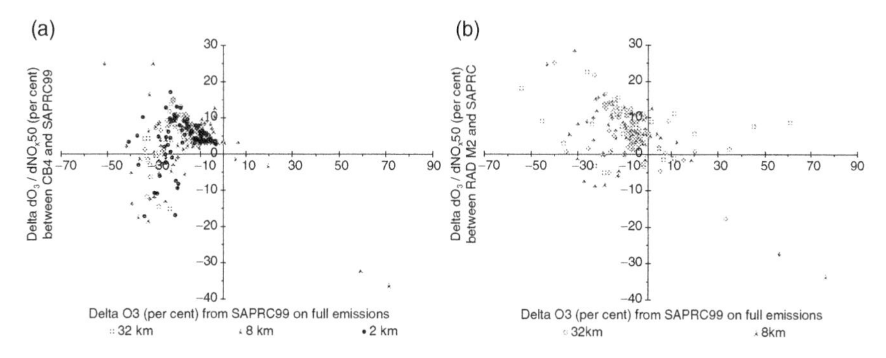

FIGURE 3. Combined O_3 response (per cent) at Jefferson Street. Full emissions difference (x-axis) *vs.* 50% NO_X control difference (y-axis). SAPRC99 *vs.* CB4 (a) and RADM2 (b). CB4 and RADM2 O_3 response normalized to SAPRC99 full emissions and control response.

3.2. Physics Ensembles for Nashville

Figures 4a-c show the $[O_3]$ time series at CFA for the three ensemble members Kz1, ACM, and Kz002, for all days in the analysis at 32 km (4a), 8 km (4b), and 2 km (4c). ACM physics increases CMAQ's O_3 overprediction bias most every hour at all three grid spacings; *e.g.*, from a 1-hour peak normalized bias of ~21% with Kz1 and Kz002 at 2 km to ~35% with ACM. Grid effects at CFA differ from those at JST even when models are configured with Kz1 and SAPRC99 for both sites: at CFA 8 km and 2 km results for Kz1 are not better than the 32 km, with a 1-hour peak bias increasing from 16% at 32 km to 19.5% at 8 km to 21.5% at 2 km. Ozone peak bias values at 2 km and 8 km are statistically indistinguishable at CFA in ACM and Kz002: ~36% ACM and ~21% Kz002.

Kz002's effect lowering overnight O_3 minima is as expected. Although only Kz002 reproduces these lowest values, this comes with substantial overprediction of NO_Y (see Figures 5a-c) and other conserved species, and without better fits to the daytime peak O_3. Moreover, while not reaching the very low overnight levels attained with Kz002, finer grid spacing does lower O_3 minima as the NO_X in each smaller cell is further constrained, and is achieved without introducing the very large errors of the super-stable Kz002.

Figures 6a-b show the O_3 control response, $\delta O_3/\delta E_{VOC}$, for 1000–1700 h EST at all grid spacings for the VOC50 and substituting ACM (6a) or Kz002 (6b) for Kz1. ACM and Kz002 are less responsive than Kz1 to the VOC control here, nearly uniformly so at 32 km and by as much as 50%.

Figures 7a-b combine the ΔO_3 response with the $\delta O_3/\delta E_{VOC}$ response. Reading along the x-axis in each plot reveals the consistent increase in O_3, often more than 20%, when ACM is substituted for Kz1 (7a), and the smaller and more variable response when using Kz002 (7b). As at JST with the chemistry ensemble, here the differential sensitivity to the vertical mixing solution cannot be reliably predicted from the ΔO_3 response, even though the $\delta O_3/\delta E_{VOC}$ response is smaller

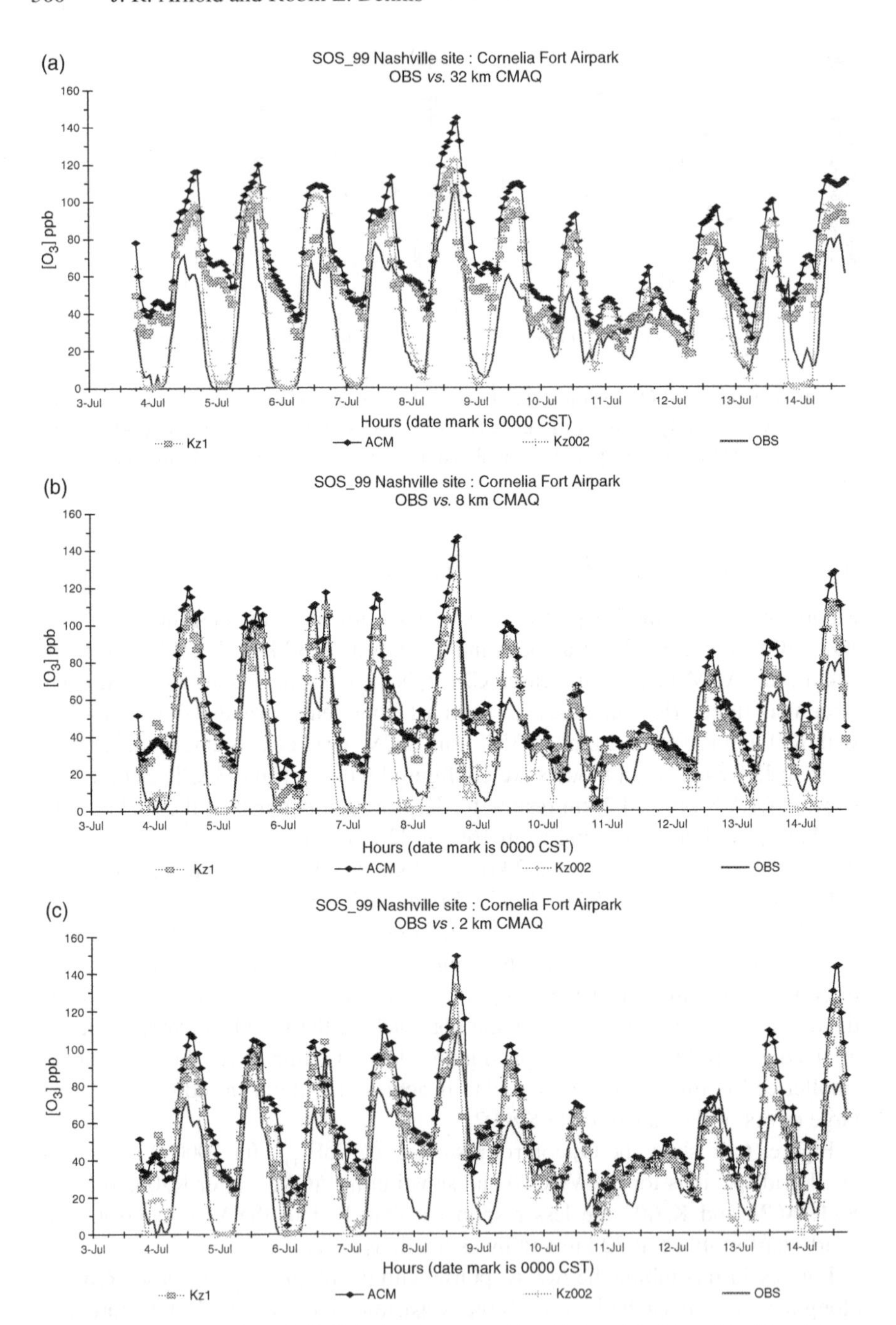

FIGURE 4. O_3 time series of CMAQ physics ensembles: 32 km (a); 8 km (b); 2 km (c).

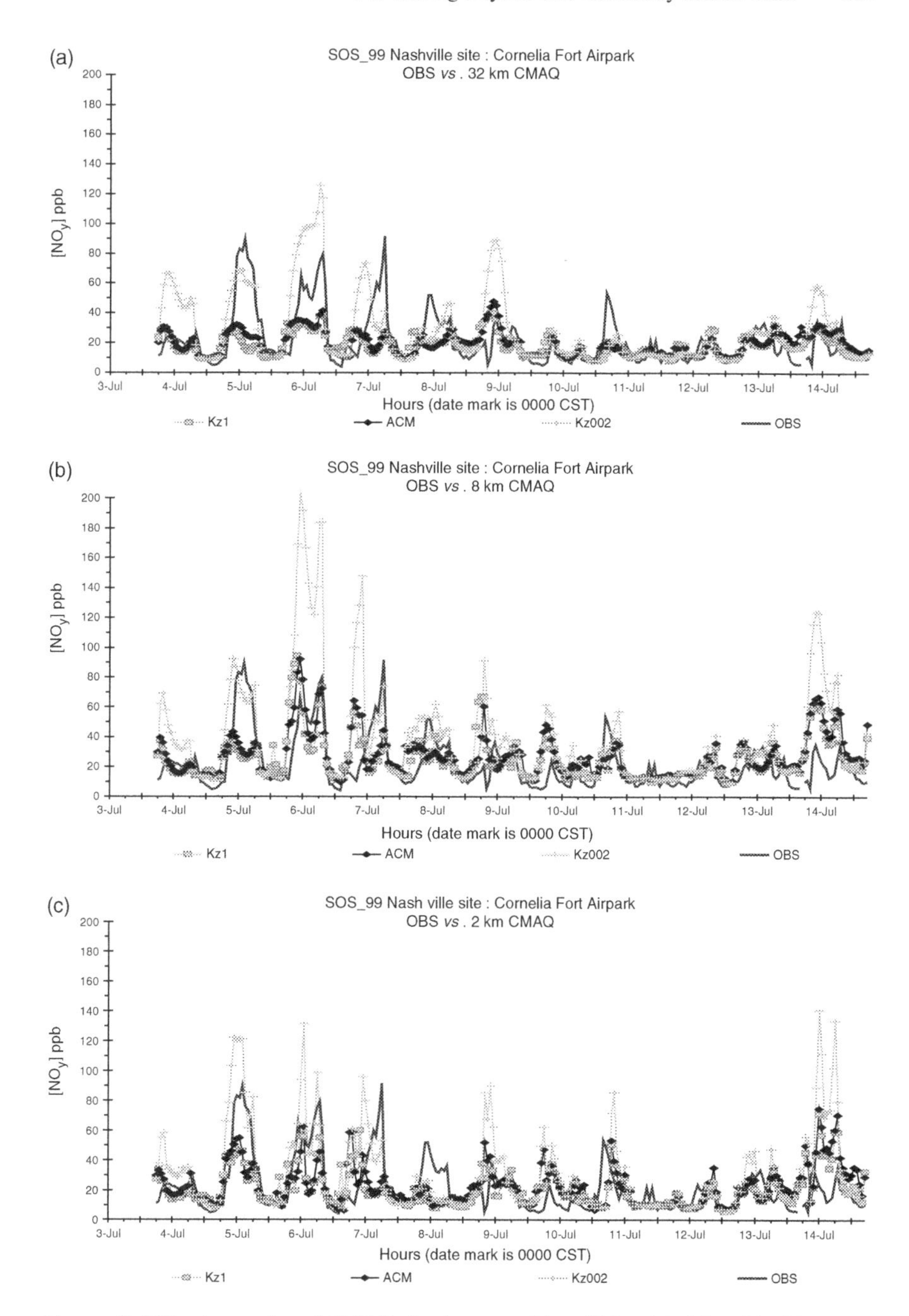

FIGURE 5. NO_Y time series of CMAQ physics ensembles: 32 km (a); 8 km (b); 2 km (c).

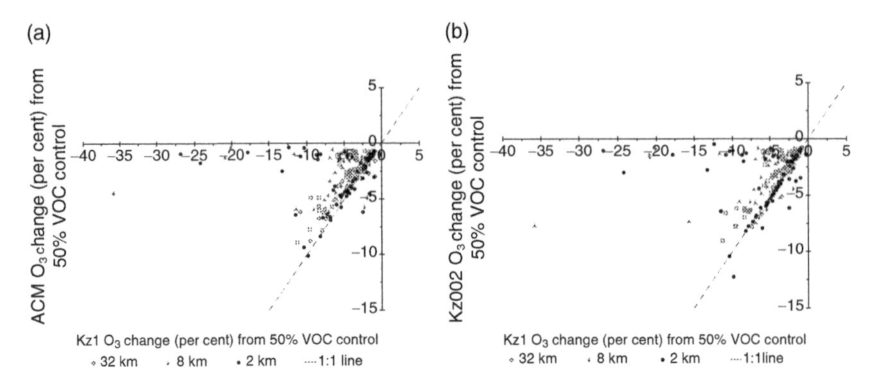

FIGURE 6. O_3 response to 50% VOC control at Cornelia Fort Airpark; Kz1 *vs.* ACM (a) and Kz002 (b).

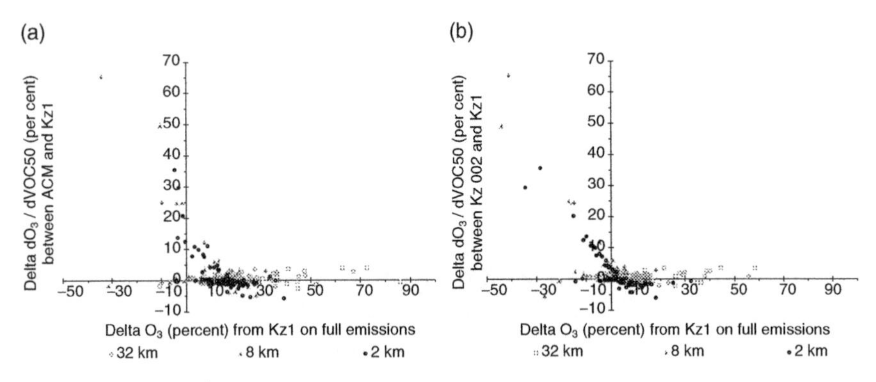

FIGURE 7. Combined O_3 response (per cent) at Cornelia Fort Airpark. Full emissions difference (x-axis) *vs.* 50% VOC control difference (y-axis). Kz1 *vs.* ACM (a) and Kz002 (b). ACM and Kz002 O_3 response normalized to Kz1 full emissions and control response.

than the $\delta O_3/\delta E_{NOx}$ response at JST, most often less than 5% here and with no apparent bias for Kz1 over ACM or Kz002.

4. Conclusions

Choice of chemical mechanism or vertical mixing solution can alter CMAQ's relative O_3 control response by more than 15%.

From the chemistry ensemble: SAPRC99 is generally more responsive than the older mechanisms CB4 and RADM2 to the 50% NO_X and 50% VOC controls tested here even though SAPRC99 is also generally more efficient, producing more O_3 from the same NO_X as CB4 and RADM2 at the two sites.

From the vertical mixing ensemble: ACM increases CMAQ's daytime O_3 overprediction over Kz1 and does not improve fits to nighttime minima. Kz002 does not improve fits to daytime O_3 but nighttime minima are improved over Kz1 and ACM. However, Kz002 introduces consistent and large overpredictions of conserved species including, importantly, NO_Y. Both ACM and Kz002 are more responsive than Kz1 to the NO_X50 control at CFA (not show here) but display no bias with the VOC50 control.

The differential sensitivities to uncertainties in chemistry and physics illustrated here can be substantial yet neither their magnitude nor their sign can be predicted reliably from the model's ΔO_3 response. Hence, uncertain inputs and model components should be tested in a series of emissions control simulations to evaluate the $\delta O_3/\delta E_{NOx}$ or $\delta O_3/\delta E_{VOC}$ differential control response directly to ensure confidence when modeling future case control applications.

References

Byun, D.-W., and Ching, J. K. S. (eds.), 1999, *Science Algorithms of the EPA Models-3 Community Multiscale Air Quality (CMAQ) Modeling System.* US EPA Report No. EPA/600/Region-99/030, Office of Research and Development, Washington, DC; and http://www.epa.gov/asmdnerl/models-3

Pleim, J. E., and Chang, J. S., 1992, A nonlocal closure model for vertical mixing in the convective boundary layer, *Atmospheric Environment.* **26A**:965.

Russell, A. G., and Dennis, R. L., 2000, NARSTO critical review of photochemical models and modeling, *Atmospheric Environment.* **34**, 12-14:2283.

SEARCH, 1999; http://atmospheric-research.com/studies/SEARCH/

SMOKE, 2002; http://www.cep.unc.edu/empd/products/smoke/index.shtml

SOS, 1999; http://www.al.noaa.gov/WWWHD/pubdocs/SOS/sos99.groundsites.html

Xiu, A., and Pleim, J. E., 2000, Development of a land surface model. Part I: Application in a mesoscale meteorology model. *Journal of Applied Meteorology.* **40**:192.

This work was performed under a Memorandum of Understanding between the U.S. Environmental Protection Agency (EPA) and the National Oceanic and Atmospheric Administration (NOAA) of the U.S. Department of Commerce and under agreement number DW13921548. Although it has been reviewed by EPA and NOAA and approved for publication, it does not necessarily reflect policies or views of either agency.

Testing Physics & Chemistry Sensitivities in the US EPA CMAQ Modeling System

Speaker: J. Arnold

Questioner: W. Jiang

Question 1: Why are VOC and CO controlled together in your control case?

Question 2: How much are the O_3 concentration changes in the case are due to the VOC control and CO control, respectively?

Answer: Thanks, Weimin, for your question. We control CO with VOC in our modeled emissions control cases because real-world anthropogenic VOC

controls will limit CO emissions as well. With CMAQ's process analysis and integrated reaction rate tools it would be possible to determine the fraction of the O_3 differential sensitivity from each mechanism due to the changed VOC and CO emissions, but because of the very large number of simulations in our model sensitivity experiment we ran without these options turned-on. Even without the model result, though, I think we could guess that most of the differential sensitivity in the O_3 response in our results here are due to the VOC control and less to the controled CO. Although the mass of CO is much larger, and so could contribute a large effect over longer temporal extents, its reactivity on the urban scale is significantly slower than the VOC and so contributes a smaller effect there.

54
Real-Time Regional Air Quality Modelling in Support of the ICARTT 2004 Campaign

V. S. Bouchet[1*], S. Ménard[1], S. Gaudreault[1], S. Cousineau[1], R. Moffet[1], L.-P. Crevier[1], W. Gong[2], P. A. Makar[2], M. D. Moran[2], and B. Pabla[2]

1. Introduction

During the summer of 2004, the International Consortium for Atmospheric Research on Transport and Transformation (ICARTT) coordinated the largest field study related to air quality so far. Over 500 scientists from 5 different countries, 11 planes, 1 ship, 2 satellites and numerous ground sites were involved in characterizing the emissions of particulate matter (PM), ozone and their precursors, their chemical transformations, and their removal during transport to and over the North Atlantic and into Europe. The study was organized around three main objectives: characterizing the regional air quality in eastern North America; assessing the export of smog related air pollution from eastern North America and its evolution as it travels towards Europe; and assessing the possible effects of the PM associated with the smog on our climate.

For one of the first times, if not the first, flight planning for the 11 planes was being conducted with the assistance of air quality forecasters in the field. The R. H. Brown, cruising in the golf of Maine, also had a dedicated forecaster for the length of the study. With objectives ranging from the regional to the hemispheric scale, multiple modelling teams were involved, providing regional and/or global forecasts to their respective scientific crews on the planes or ship. The 2004 ICARTT campaign provided a great opportunity for the modeling teams to test their systems in real-time and compare their forecasts. A formal intercomparison of all the regional scale forecasts was organized in parallel of

[1] Véronique S. Bouchet, Sylvain Ménard, Stéphane Gaudreault, Sophie Cousineau, Richard Moffet and Louis-Philippe Crevier, Meteorological Service of Canada, Canadian Meteorological Centre, 2121 Route TransCanadienne, Dorval, Québec, H9P 1J3, Canada.
[2] Wanmin Gong, Paul A. Makar, Michael D. Moran and Balbir Pabla, Meteorological Service of Canada, Air Quality Research Branch, 4905 Dufferin Street, Downsview, Ontario, M3H 5T4, Canada.
[*] Veronique.Bouchet@ec.gc.ca

the measurement campaign, using the real-time AIRMAP ozone observations as reference data.

The following paper will describe the modelling effort that supported the Canadian Convair 580, using two models, the Canadian Hemispheric and Regional Ozone and NO_x System (CHRONOS) and A Unified Regional Air quality Modelling System (AURAMS).

2. Description of the Canadian Campaign

The objectives of the Canadian-ICARTT measurement campaign, lead by scientists from the Meteorological Service of Canada, were established in collaboration with the modelling team. The aim was to provide the data needed to evaluate and improve the weakest components of Canadian forecast models. The Convair 580 missions were divided between two sub-studies, and the decision to perform one type of flight or the other based on the weather and air quality forecasts in the field. The Canadian campaign spanned from July 18th to August 20th, 2004 (Leaitch et al., 2004).

The first flight type was dedicated to the study of chemical transformations and transport by clouds (CTC flights). For that part of the experiment, the plane operated out of Cleveland, Ohio. The sampling was preferentially performed in non-precipitating stratocumulus and fair weather cumulus contained in the planetary boundary layer. The Convair 580 carried an Aerosol Mass Spectrometer (AMS), a Particle In Liquid Sampler (PILS), a nephelometer, particles and droplet impactors and analyzers for O_3, NO, NO_2, NO_y, SO_2, NH_3, H_2O_2, HNO_3, HCHO and CO to address questions related to: the importance and mechanisms of cloud activation in shaping the PM size distribution and the formation of particulate $SO_4^=$; the chemical pathways of in-cloud production of particulate NO_3^- during daytime and nighttime; the presumed absence of effect of non-precipitating clouds on the organic PM component; and the presumed preferential vertical transport of VOCs and CO relative to SO_x and NO_x in towering cumulus and cumulonimbus.

The second experiment focused on the transport of regional air pollution into the Atlantic Provinces of Canada (TIMs flights). The plane was then based in Bangor, Maine, where it joined the other American planes participating in ICARTT. With the same instrumentation as for the CTC flights, the Convair sampled the smog transported to the Canadian Maritimes, and provided detailed information on the PM size and composition within that air mass.

In addition to the main aircraft study, scientists from the Meteorological Service of Canada participated in the NASA-INTEX Ozonesonde Network Study (IONS) project, also under the ICARTT umbrella. Ozonesonde releases were increased at 11 ground sites across North America, including 3 Canadian sites. Launches from the R. H. Brown research vessel in the golf of Maine were also performed. During July and August 2004, releases were performed from all sites on Mondays, Wednesdays and Fridays, and daily for a small subset of locations.

As part of a coordinated effort with the Convair 580, additional ozonesonde releases were added when the plane was flying in the Canadian Maritimes. The vertical ozone profiles provided by the sondes are instrumental to evaluate the vertical structure of the forecast models as well as to corroborate profiles performed from the aircraft.

3. Description of the Canadian Modelling Support

Forecast support for the Canadian team, in the field, relied on a series of weather and air quality products developed and/or customized for this experiment. The forecasting team had access to real-time weather forecasts, data and imagery from the Canadian Meteorological Centre and the Ontario Storm Prediction Centre via a secure internet access. In addition to the global operational weather forecasts at 100 km and 15 km spatial resolution, a limited area version of the Global Environmental Multiscale (GEM) model (Côté et al., 1998) was also run at 2.5 km spatial resolution, on a daily basis and, for a domain encompassing southern Ontario and Northern Ohio.

The air quality products can be divided in 3 categories. Forward and backward trajectories based on the long and medium range GEM forecasts were available twice a day, for 23 locations hand picked by the scientists involved in the study. The length of the trajectories varied between 48 and 120 hours. Forecasters in the field were also able to calculate trajectories from the field from a precompiled list of locations of interest. Figure 1 presents a sample back-trajectory for Yarmouth, N.S., issued on July 21st and valid for July 22nd, 2004. It clearly highlights a situation favorable for a TIMs flight on July 22nd.

The second set of air quality information came from two Canadian three-dimensional air quality models, CHRONOS (Pudykiewicz et al., 1997) and AURAMS (Moran et al., 1998, Bouchet et al., 2004, Makar et al., 2004), run in real time at the Canadian Meteorological Centre (CMC) of the Meteorological Service of Canada. CHRONOS is an off line regional photo-oxidant model driven by GEM, which includes a simplified bisectional representation of particulate matter formation. Secondary organic formation and heterogeneous reactions for the sulphate-nitrate-ammonium-water system are described in CHRONOS, but in-cloud aqueous processes and aerosol microphysics are not represented. The model uses emission information based on the 1990 Canadian and U.S. national emission inventories and upscaled to the 1995/96 Canadian and U.S. levels. CHRONOS is run at a regional scale (21 km) but on a continent-wide, North American, domain. Simulations are performed in two different configurations: with the operational version described above and with an experimental version which assimilates surface ozone data. Real time observations from AIRNow (USEPA, 2004) covering the entire domain are used here, but the assimilation is restricted to 3h and limited to the 1200 UTC simulation.

AURAMS is a regional photo-oxidant model, initially based on CHRONOS, which includes a sophisticated PM parameterization. The dynamic of internally

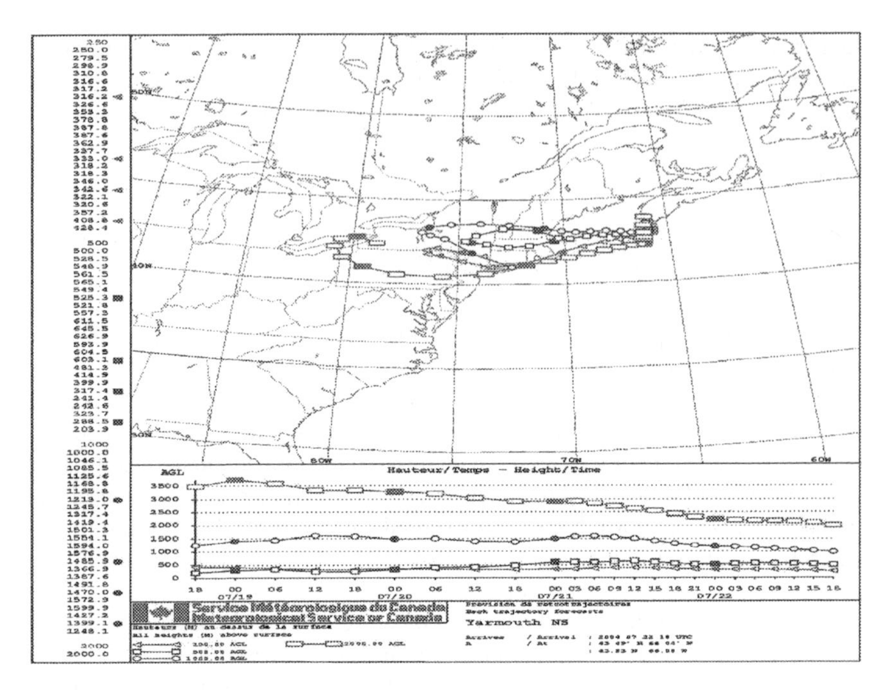

FIGURE 1. Back trajectories arriving at Yarmouth, Nova Scotia. Issued on July 21st and valid for July 22nd, 2004 at 2 pm EDT.

mixed aerosol is parameterized with the Canadian Aerosol Module (CAM), which uses a sectional approach (Gong et al., 2003). The simulated aerosols are composed of up to 8 chemical species (sulphate, nitrate, ammonium, elemental carbon, organic carbon, sea-salt, crustal material and aerosol bound water) and divided into 12 bins ranging in size from 0.01 to 40.96 μm. Interactions with photo-oxidants, in-cloud processes-including aerosol activation, aqueous phase chemistry, cloud to rain conversion and evaporation – and heterogeneous chemical reactions are also represented. AURAMS uses the same emission data as CHRONOS, with an additional scheme which decomposes the bulk PM emissions, on the fly, for each source stream, as a function of size and composition. Due to time constraints, the heterogeneous and aqueous modules currently operate in bulk mode with a subsequent mass redistribution to each individual bins. Instruments such as the AMS and the PILS provide measurements that will allow evaluation of the detailed PM module in AURAMS. For the ICARTT experiment, AURAMS is run on an eastern North American domain at 42 km spatial resolution. No data assimilation is performed in AURAMS.

From the field, the forecasting team had access to results from the 00 UTC CHRONOS operational run (publicly available), a 00 UTC and a 1200 UTC run from the experimental CHRONOS version which includes data assimilation of surface ozone, and a 00 UTC AURAMS simulation. A typical 48 h forecast on

8 processors of CMC supercomputer (IBM pSeries 690) took approximately 2.5 h with either version of CHRONOS using the current 1h timestep, and 4 h with AURAMS using a timestep of 900s. However due to additional meteorological parameters required to drive AURAMS, a second GEM run was required adding 3 h to the total execution time. Output from AURAMS and CHRONOS were processed at CMC and the following products were sent to the forecasters: animated surface and 500m maps of SO_2, $PM_{2.5}$ $SO_4^{=}$, $PM_{2.5}$ NO_3^-, H_2O_2, NO_x and O_3 for the entire 48 h forecast as illustrated in Figure 2; animated cross sections for pre-selected areas around Lake Erie and in the Maritimes for the same species as illustrated in Figure 3; and animated maps of gas-phase and in-cloud production of $PM_{2.5}$ $SO_4^{=}$ (AURAMS only) as illustrated in Figure 4. These products were equally available to all ICARTT participants. When warranted, the forecasting team could also download the entire suite of model output containing the detailed size information for PM sulphate, nitrate and ammonium.

The last component of the air quality forecast products was an objective analysis, based on the optimal interpolation theory, of the real-time O_3 observations gathered by AIR*Now*. An increment analysis is performed on the measurements and used to correct CHRONOS output on the continental domain. The objective analysis maps were generated every hour between 06 and 24 UTC with a 1 h time delay. A sample objective analysis map is presented in Figure 5.

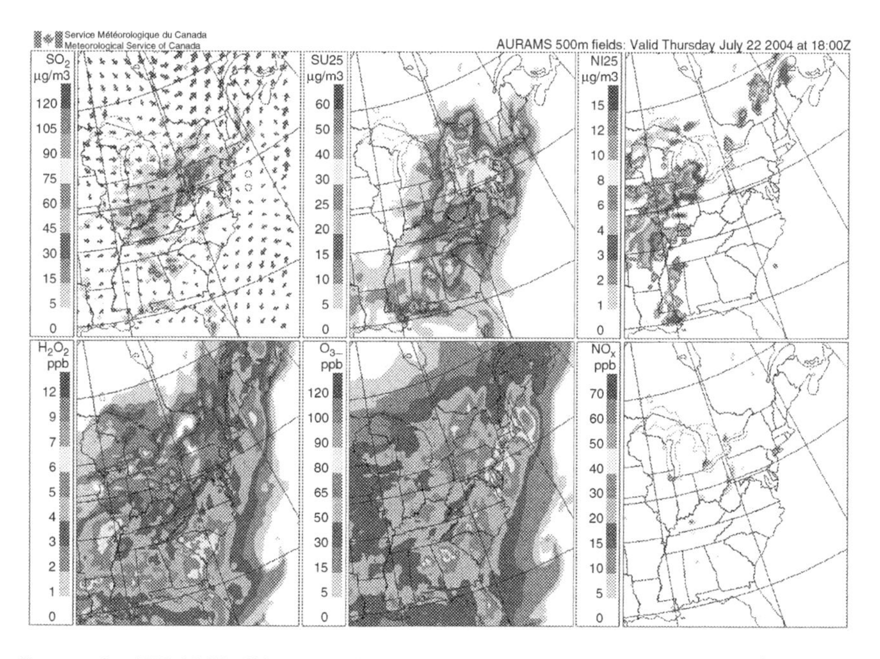

FIGURE 2. AURAMS 500m map issued July 21st, and valid for July 22nd, 2004 at 1800UTC. Top row: SO_2 (left), $PM_{2.5}$ $SO_4^{=}$ (middle), $PM_{2.5}$ NO_3^- (right); Bottom row: H_2O_2 (left), O_3 (middle) and NO_x (right).

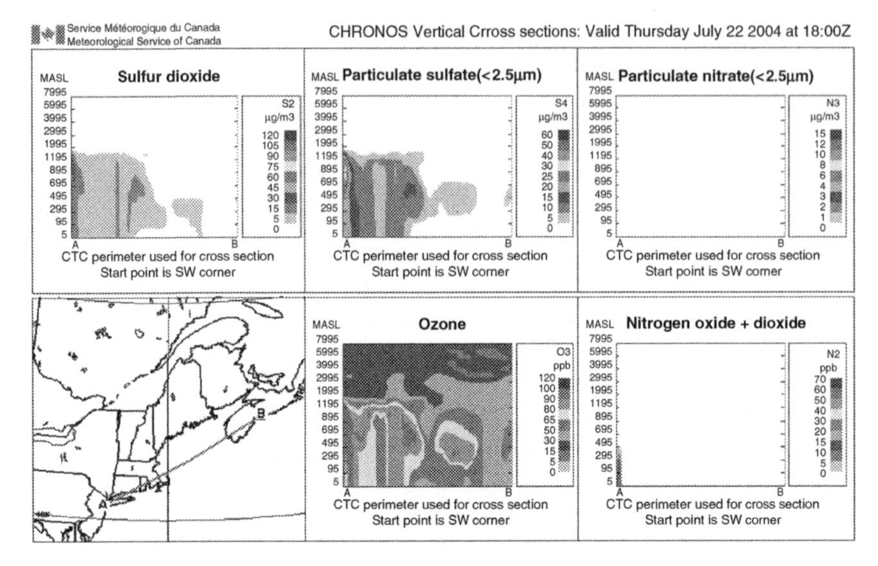

FIGURE 3. CHRONOS vertical cross-section along the red A-B axis on bottom left panel. Issued July 21st, and valid for July 22nd, 2004 at 1800UTC. Top row: SO_2 (left, ppb), $PM_{2.5}$ $SO_4^=$ (middle, µg/m^3), $PM_{2.5}$ NO_3^- (right, µg/m^3); Bottom row: O_3 (middle, ppb) and NO_x (right, ppb).

4. A Sample Air Quality Forecast

Forecasters in the field provided one extended briefing per day to the aircraft scientists to prepare flights for the following two days as well as a weather and air quality update a few hours before take-off. Figures 1, 2, 3 and 4 were used as part of the air quality briefing on the morning of July 21st, 2004. Looking at backwards trajectories from Yarmouth (Fig. 1 and similar plots not shown) and other locations in New Brunswick and Nova Scotia, the forecasters determined that by the evening of July 21st air from Hamilton would cross over northern New Brunswick, followed by air masses from Sarnia, Detroit/Windsor, Toledo and Cleveland during the afternoon of July 22nd. A shift in wind pattern was forecasted to bring stagnant air from New-York city that had very slowly traveled over the Gulf of Maine to move over southern New Brunswick in the eve of July 22nd. Ozone surface concentrations, as deduced from AURAMS and CHRONOS output (map not shown), were not expected to exceed 50 and 80 ppb respectively, but Figure 3 clearly shows that most of the pollution was predicted to stay aloft between 250 and 1200m. This is in agreement with the backward trajectory in Figure 1 showing that the air mass arriving at Yarmouth at 2 pm EDT on July 22, would have been contained in a layer around 500m agl for most of the previous 4 days. Figures 2 and 3 highlight that the event was expected to be largely associated with ozone pollution, with high levels of PM mostly forecasted in Pennsylvania and Ontario. Interestingly, output from the 1200UTC CHRONOS

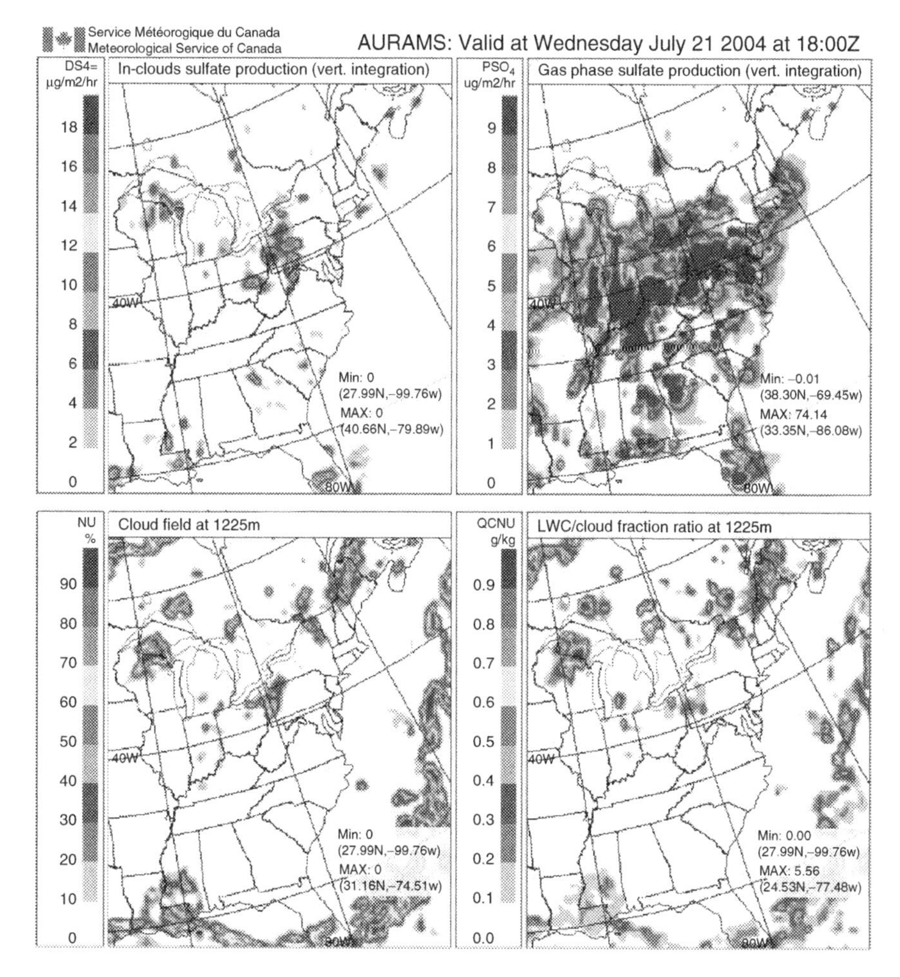

FIGURE 4. Gas-phase (top left) and in-cloud (top right) production of $PM_{2.5}$ $SO_4^=$ (vertically integrated) as forecasted by AURAMS on July 21st and valid for July 22nd, 2004. Bottom left: cloud fraction at 1225m used by AURAMS; bottom right: liquid water content (> 0.01 g/kg) over cloud fraction at 1225m.

run, which included assimilation of surface ozone data, showed a much more polluted air mass, compared to AURAMS, moving further in land, compared to AURAMS, into New Brunswick and Nova Scotia. From Figure 6, illustrating the 500m ozone field from the 1200UTC CHRONOS simulation, ozone concentrations were forecasted to reach over 100 ppb at the tip of Nova Scotia while AURAMS only forecasted up to 65 ppb in that area. AURAMS had similarly high ozone concentrations aloft but forecasted it to stay south of Nova Scotia and New Brunswick (Figure 2).

As seen in Figure 2, AURAMS forecasted some interesting PM activity developing south of Lake Erie on both July 21st and 22nd. Figure 4 outlined the

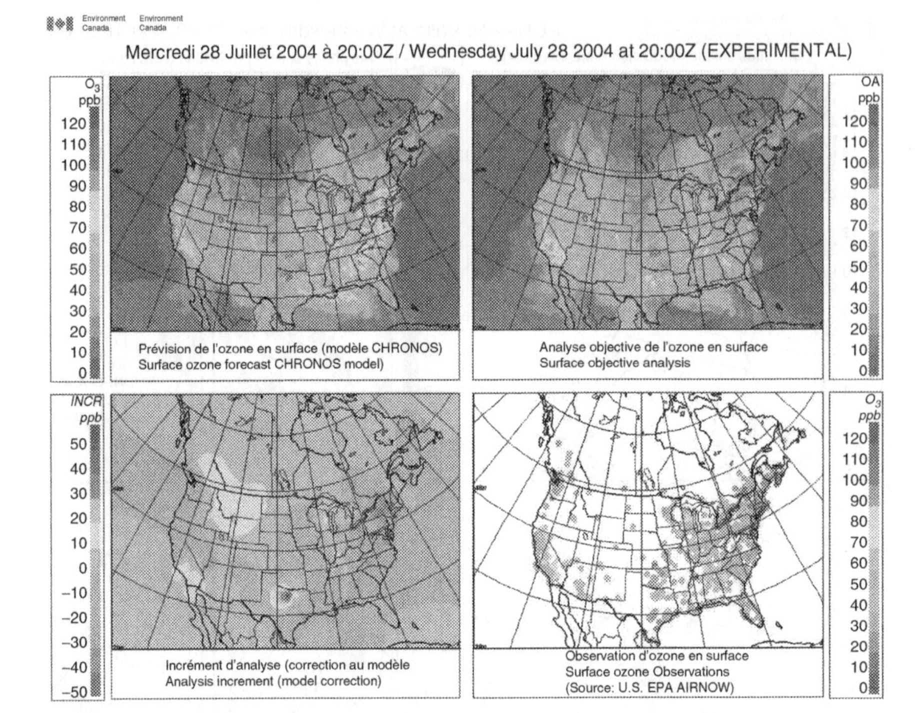

FIGURE 5. Ozone objective analysis, valid for July 28, 2004 at 2000 UTC.

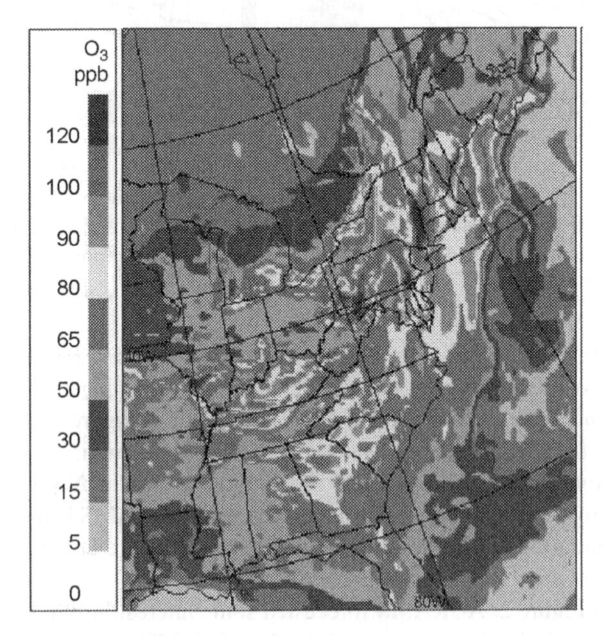

FIGURE 6. CHRONOS 500m ozone field issued July 21st and valid for July 22nd, 2004 at 2200UTC.

locations of the clouds of interest with relatively large amounts of sulphate gas-phase production occurring from Illinois to New Jersey, but cloud activity with in-cloud sulphate production limited to the south to south west of Lake Erie. Based on these elements, the recommendation and decision were to have a CTC flight late morning or early afternoon on July 21st, then travel to Bangor in the evening to prepare for 2 TIMs flights on July 22nd. The first one was to happen in northern New Brunswick in the morning and the second in southern New Brunswick in the afternoon.

Accordingly, the plane went for a CTC mission on July 21st, and sampled over the Maine-New-Brunswick border and the Bay of Fundy the next day. They encountered high level of pollution, with the highest levels of ozone (over 100 ppb) and PM measured above the marine boundary layer. In the afternoon, they also observed the predicted narrowing of the plume, with high levels of pollution on the western part of the flight path, but very clean levels to the east from air masses coming from the middle of the Atlantic.

5. Perspectives

The close collaboration, in the field, between the two teams of scientists has proven very profitable both in maximizing the allocated flight hours for the project and providing real-time evaluation for the air quality models. Building on that strength, real-time automatic evaluation tools are being developed during the campaign to complete the forecasting suite. They will provide up to the hour statistics on the model performance versus AIR*Now* surface observations and ozonesonde profiles when available.

As part of the ICARTT campaign, forecast output from AURAMS and CHRONOS were also provide to NOAA to participate in the real-time evaluation of regional models. Results from the comparisons, as well as an ozone ensemble forecast, were posted at http://www.etl.noaa.gov/programs/2004/neaqs/verification/ during the length of the study. A detailed analysis of the intercomparison results will be undertaken in the fall of 2005 to derive information on the strengths and weaknesses of AURAMS and CHRONOS.

References

Bouchet, V. S., M.. Moran, L.-P. Crevier, A. Dastoor, S. Gong, W. Gong, P. Makar, S. Ménard, B. Pabla and L. Zhang, 2004. Wintertime and summertime evaluation of the regional PM air quality model AURAMS. *Air Poll. Modelling & Applications XVI, Borrego and Incecik, Kluwer Academic/Plenum Publishers*, pp. 97-104.

Côté, J., J.-G. Desmarais, S. Gravel, A. Méthot, A. Patoine, M. Roch and A. Staniforth, 1998. The operational CMC/MRB Global Environmental Multiscale (GEM) model. Part I: Design considerations and formulation. *Mon. Wea. Rev.*, 126, 1373-1395.

Gong S. L. et al., 2003. Canadian aerosol module: a size-segregated simulation of atmospheric aerosol processes for climate and air quality models 1. Module development. *J.Geophys. Res.*, 108, 4007-4033.

Leaitch R., et al., 2004. Airborne Measurements for Study of the Processing of Atmospheric Sulphur, Nitrogen and Organic Species by Cloud. *3rd Int. conf. fog, fog collection & dew*, Cape Town, S. Africa, Oct. 11-15.

Makar, P. A., V. Bouchet, L. P. Crevier, S. Gong, W. Gong, S. Menard, M. Moran, B. Pabla, S. Venkatesh, 2004. AURAMS runs during the Pacific2001 time period – a Model/Measurement Comparison., *Air Poll. Modelling & Applications XVI, Borrego and Incecik, Kluwer Academic/Plenum Publishers,* pp. 153-160.

Moran, M. D., A. P. Dastoor, S.-L. Gong, W. Gong and P. A. Makar, 1998. Conceptual design for the AES unified regional air quality modelling system. Air Quality Research Branch, Meteorological Service of Canada, Downsview, Ontario M3H 5T4, Canada.

Pudykiewicz, J.A., A. Kallaur and P.K. Smolarkiewicz, 1997. Semi-Lagrangian modelling of tropospheric ozone, *Tellus*, 49B, 231-248.

U.S. EPA, 2004. U.S. Environmental Protection Agency, Office of Air Quality Planning and Sstandards, AirNow. www.epa.gov/airnow

Real-Time Regional Air Quality Modeling in Support of the ICARTT 2004 Campaign

Speaker: V. S. Bouchet

Questioner: S. T. Rao

Question 1: Did all the models in ICARTT 2004 use the same lateral boundary conditions?

Question 2: Are there any plans for harmonizing the input variables (emissions, IC/BC, meteorology, grid spacing etc.) for your model intercomparison study?

Answer: All regional forecast models were run in their native mode during the ICARTT experiment. Although there was some recommendation made as far as what emission fields to use as input, groups were free to chose their individual set-ups. So boundary conditions were not identical from team to team.

There is no plan at this time to rerun the ICARTT period with harmonized input or boundary conditions. However the results are being used to investigate how the forecasts could benefit from an ensemble calculation.

Questioner: W. Jiang

Question: You mentioned that the model forecast was not so accurate. Could you give us some quantitative feeling on the meaning of "not so accurate"?

Answer: Generally speaking, the two Canadian models were able to locate the major areas of pollution as well as an estimate of the altitude in the regions of interest. However the amount of pollution, especially for some individual PM chemical fraction such as nitrate, were often overestimated, leading to disappointing results on flights that were dedicated to measure nitrate formation.

55
High Time-Resolved Comparisons for In-Depth Probing of CMAQ Fine-Particle and Gas Predictions

Robin L. Dennis, Shawn J. Roselle, Rob Gilliam, and Jeff Arnold*

1. Introduction

In this paper, two major sources of bias in the Community Multi-scale Air Quality Model (CMAQ), one physical and one chemical process, are examined. The examination is conducted with hourly gas and particle data for the inorganic system of sulfate, total ammonia, also called NH_X, (gaseous ammonia, NH_3 plus aerosol ammonium, NH_4^+) and total nitrate (gaseous nitric acid, HNO_3 plus aerosol nitrate, NO_3^-) and with hourly gas and particle data for inert or conservative species. The physical source of bias stems from the meteorological inputs related to mixing, in particular the behavior of the simulated mixed layer in the evening. The chemical source of bias stems from the nighttime heterogeneous production of HNO_3 from N_2O_5. The analyses are carried out for a summer and a winter period to examine the seasonal dependence of the biases.

2. General Model and Data Description

2.1. CMAQ

CMAQ is an Eulerian model that simulates the atmospheric transport, transformation, and deposition of photochemical oxidants (ozone), particulate matter, airborne toxics and acidic and nutrient species (Byun and Ching, 1999). The 2004 release version of CMAQ was used for these simulations. The meteorological fields were

* Robin L. Dennis, Shawn J. Roselle, Rob Gilliam@, and Jeff Arnold#, NERL, U.S. Environmental Protection Agency, RTP NC, 27711, USA. # On Assignment from University Corporation for Atmospheric Research, CO 80303, USA. @On assignment from Air Resources Laboratory, National Oceanic and Atmospheric Administration, RTP, NC 27711, USA.

derived from MM5, the Fifth-Generation Pennsylvania State University/National Center for Atmospheric Research Mesoscale Model (Grell et al., 1994), with data assimilation and use of the Pleim-Xiu land-surface model (PX) option (Pleim and Xiu, 1995). The modeling domain covered the contiguous U.S. with a 36-km horizontal grid dimension. A 24-layer vertical structure was used that reached to the top of the free troposphere. The simulations were performed with the SAPRC99 gas-phase chemical mechanism, with the U.S. EPA 2001 National Emissions Inventory and biogenic emissions from BEIS 3.12. Two periods corresponding to EPA Supersite Program intensives were simulated: July 2001 and January 2002.

2.2. *Observational Data*

The data used are high time resolution gas and particle data from July 2001 and January 2002 taken at two supersites in the EPA Supersite Program: Jefferson Street, Atlanta, a Southeastern Aerosol Research and Characterization Study (SEARCH) site (web address in references) and Schenley Park, Pittsburgh (Wittig et al., 2004; Takahama et al., 2004). In addition, companion (SEARCH) sites in the Southeastern US. were used.

3. Results

3.1. *Bias Stemming from Physical Process*

Previous comparisons of CMAQ predictions of conservative species against data taken in 1995 in Nashville, TN (as part of the Southern Oxidant Study (SOS)) indicated that there was a systematic nighttime over-prediction of conservative species in the model. Comparison of simulations of mixing heights from MM5 (using the PX land-surface model option) and from radar profilers around Nashville indicated: (1) that the mixed layer heights were in good agreement during the mid-day and (2) that the mixed layer in MM5 was collapsing too soon in the late afternoon. It was hypothesized that the premature collapse of the boundary layer was contributing significantly to the nighttime over-predictions.

Comparisons against aircraft spirals over surface sites also indicated that during the day the atmosphere is well mixed and the surface concentrations are representative of the overall column concentration levels. Because pollutants in CMAQ are very well mixed in the vertical and the mixed layer heights appears to be in reasonable agreement when using the PX option with MM5, we expect the mid-day predicted and measured concentrations to be in good agreement when daytime emissions are reasonably correct.

If there is a premature collapse of the mixed layer, we expect a very rapid rise in predicted surface air concentrations leading to an over-prediction or positive bias in the late afternoon across many inert (e.g., CO and elemental carbon (EC) or quasi-inert species (e.g., NO_Y and NH_X) that are emitted late in the day and at night. The key here is the rate of increase from mid-day levels. We expect little to no effect for species such as sulfate whose gas-phase formation shuts down as the sun goes down.

We start with EC as the example inert tracer. The average diurnal pattern of EC predictions and measurements at the Atlanta supersite location at Jefferson Street is shown in Figure 1a for July 2001 (summer) and 1b for January 2002 (winter). To better illustrate the relative afternoon rate of increase, the scales are adjusted so that the mid-day levels for model and measurements match. Indeed, we find a very rapid rise in surface air concentrations and a rapid increase in bias in the late afternoon. The bias is most pronounced in summer and least in winter. We will explain why later in this section. The Atlanta diurnal patterns for CO and NO_Y (not shown) are very similar in each season to that in Figure 1 for EC. There appear to be nighttime sources of CO, EC and NO_X, consistent with diesel emissions, close to this site that are important in the winter and affect the nighttime comparisons. Diesel emissions are implicated because the EC to NO_Y relationship of the nighttime emissions is the same as for the AM and PM drive peaks. These sources are either not in the emissions inventory or not handled well by the chemical transport model. Figure 2 shows the diurnal pattern of EC comparisons in Pittsburgh for summer to show the similarity in another urban area (without the nighttime local source issue). Figures 3 shows the Pittsburgh diurnal patterns for SO_4^{2-} for summer, indicating that, as expected, sulfate concentrations are not seriously affected, because the SO_4^{2-} gas-phase production shuts down as the OH levels go to near zero in the evening. The winter diurnal comparison is very similar but with a slight sulfate over-prediction.

Figure 4 shows the monthly-averaged diurnal temperature bias for MM5 for July 2001 for Atlanta, Pittsburgh and the continental area of the Eastern U.S. excluding Florida. Given the regular diurnal pattern in the summer of the rise and collapse of the boundary layer and the subsequent decrease in wind speed, we expect the temperature bias to provide a good indication of what is happening in the boundary layer. For Atlanta and Pittsburgh, the temperature bias goes positive at mid-day and then steadily falls to negative values shortly after 1600 EST and

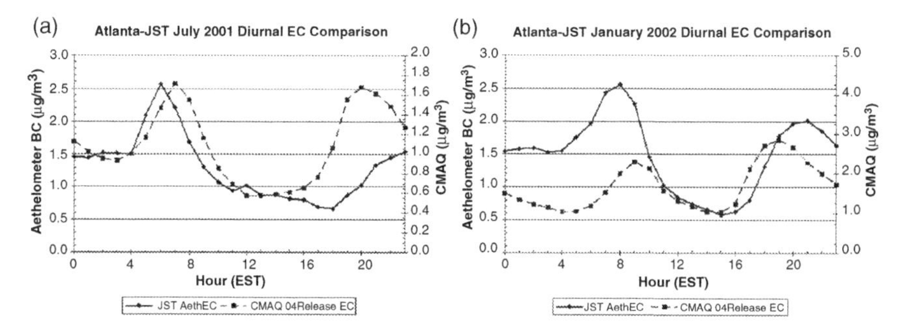

FIGURE 1. Comparison of the observed Athelometer Black Carbon and predicted Elemental Carbon hourly diurnal pattern based on a monthly average of each hour at Atlanta for (a) July 2001 and (b) January 2002.

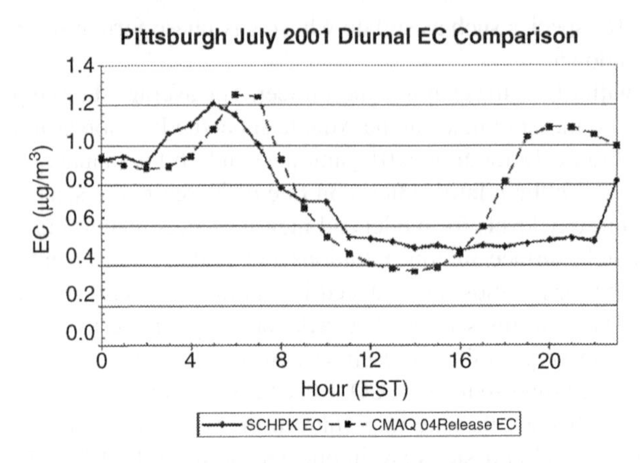

FIGURE 2. Comparison of diurnal pattern of EC at Pittsburgh for July 2001.

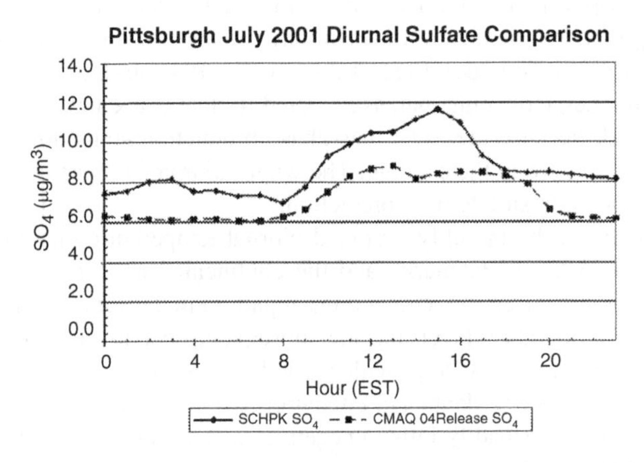

FIGURE 3. Comparison of diurnal pattern of sulfate at Pittsburgh for July 2001.

continues declining until 1900 EST and then reverses direction. We associate this rapid over-cooling in the MM5 with the premature collapse of the mixed layer. Interestingly, the pattern of cold bias maximum in the morning and afternoon is seen to be a systematic feature across the entire Eastern US. So we expect this issue of evening over-prediction to be present across the entire model domain. We expect Pittsburgh and Atlanta, with their larger swings in temperature bias, to be representative of urban areas; thus, we expect this issue to have a strong effect in urban areas. In winter there is a flat, smooth temperature bias with only a monotonic rise towards no bias during the day and then a monotonic fall again to a constant evening bias level. In winter there is more competition between winds (mechanical turbulence) and density stratification and the atmosphere is less stationary, hence, the atmosphere does not become stable as often (or reaches that

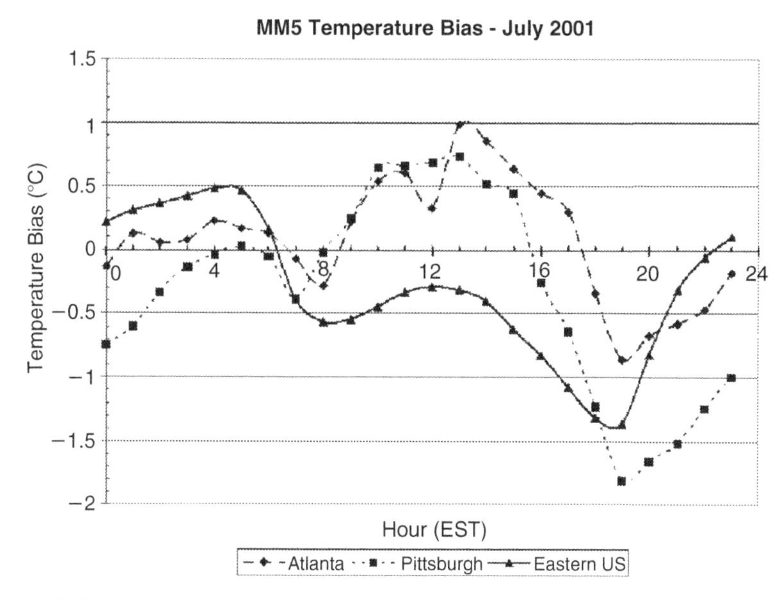

FIGURE 4. Plot of the monthly averaged hourly diurnal pattern of MM5 temperature bias for July 2001 at Atlanta, Pittsburgh and across the Eastern U.S.

state much later in the evening). Thus, we expect less of an impact on the mixed layer and, hence, the impact on surface concentrations is expected to be less in winter than in summer.

Figure 5a presents the July 2001 diurnal pattern of NH_X in Pittsburgh, where there is a strong diurnal swing in the CMAQ predictions with under-prediction during the day and over-prediction at night, whereas there is little to no diurnal variation in the measurements. Figure 5b shows that the agreement between the 24-hour average predictions and observations looks fairly good; however, the daily averages have covered up the fact that there are compensating errors involved, possibly raising questions about emissions. The 24-hour averages also cover up the fact that the diurnally varying NH_X errors will significantly affect the partitioning of total-nitrate to aerosol nitrate, creating a companion bias for particulate nitrate. The exact same behavior of compensating errors was seen with the August 1999 Atlanta supersite comparisons for EC, raising questions about the accuracy of the EC emissions-questions that would be missed when only looking at daily averaged concentrations. Figure 5c shows that there are also larger diurnal swings in the January 2002 predictions of NH_X for Pittsburgh compared to the data, with peaks in the morning and afternoon. Although not as dramatic as the summer situation, these diurnal swings will affect the total-nitrate partitioning as well.

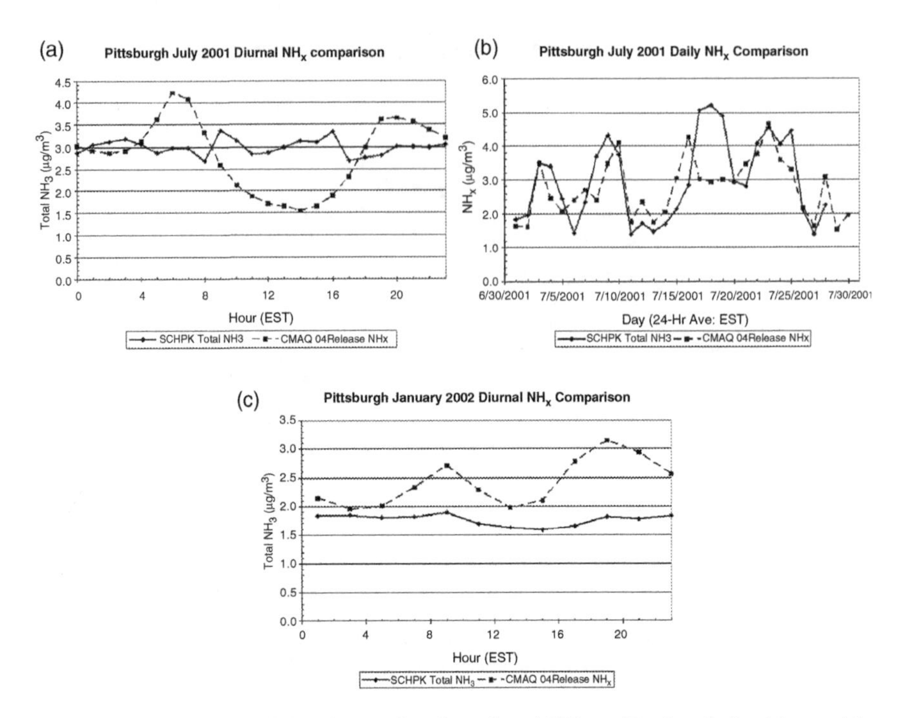

FIGURE 5. Comparison of the observed and predicted NH_X at Pittsburgh for (a) monthly averaged hourly diurnal pattern in July 2001, (b) daily 24-hour average in July 2001, and (c) monthly averaged hourly diurnal pattern in January 2002.

3.2. Bias Stemming from Chemical Process

Previous comparisons of simulation predictions from the CMAQ 2002 public release version showed a very large over-prediction of fine particle nitrate (less than 2.5 microns in size), to the point of being unacceptable. Comparisons against hourly data for January 2002 at the special sites of Atlanta and Pittsburgh showed huge over-predictions of nitric acid and/or total nitrate, especially at night. CMAQ includes the nighttime heterogeneous production of HNO_3 from N_2O_5 on wetted particles. The over-predictions peaked at night, suggesting an issue with this nighttime heterogeneous production of nitric acid.

The heterogeneous reaction probabilities being used in the 2002 version of CMAQ were based on Dentener and Crutzen (1993). Recent estimates of the N_2O_5 hydrolysis reaction probability are however two to three orders of magnitude smaller than those suggested by Dentener and Crutzen (1993). The parameterization in CMAQ was updated to reflect the latest research (Riemer et al., 2003, based on experiments reported in Mentel et al., 1999), which suggested the reaction probability is much smaller, in the range of 0.02 and would be inhibited by the presence of nitrate on the aerosols, reducing it to 0.002 when nitrate is a dominant component of the mixed aerosol. In addition, the gas-phase reaction of the nitrate radical with water that produces N_2O_5 has been turned off in CMAQ. The argument is that this reaction is highly uncertain and could be a chamber wall artifact that belongs in the chamber wall model and not in the chemical mechanism. The SAPRC mechanism developer did not object to this decision, noting there was a significant degree of uncertainty about the gas-phase reaction (Carter, 2003).

A test of these new literature values for the heterogeneous reaction probabilities showed a dramatic improvement in the predictions of CMAQ for nitric acid and aerosol nitrate, although CMAQ potentially is still over-predicting nitric acid at night. Sensitivity tests with CMAQ were also conducted to further explore the degree of over-prediction in which the nighttime production of HNO_3 from N_2O_5 was turned off completely.

Figure 6 shows the monthly-averaged diurnal concentration patterns for CMAQ-predicted HNO_3 at Atlanta for (a) summer and (b) winter for the base CMAQ that includes the heterogeneous production of HNO_3 compared to CMAQ with the heterogeneous pathway completely turned off. Both model versions are compared to monthly-averaged diurnal measurements. The differences at Pittsburgh for winter (not shown) are very similar for the two CMAQ model versions compared to measurements.

The comparisons of the base case and sensitivity study case to the measurements show that eliminating altogether the nighttime heterogeneous production of nitric acid in CMAQ brings its predictions much more in line with the nighttime levels of nitric acid at both special sites, although now the predicted HNO_3 levels can be below the measurements. The comparisons of these CMAQ sensitivity runs show that the nighttime chemistry is most important to the overall HNO_3 budget in winter. Figure 6(b) also shows that there can be a noticeable over-prediction of nitric acid occurring in daylight hours in the winter. That is, daytime photochemical mechanisms for ozone production (either CB4 or SAPRC99) are also contributing to the HNO_3 or total-nitrate over-prediction during the winter.

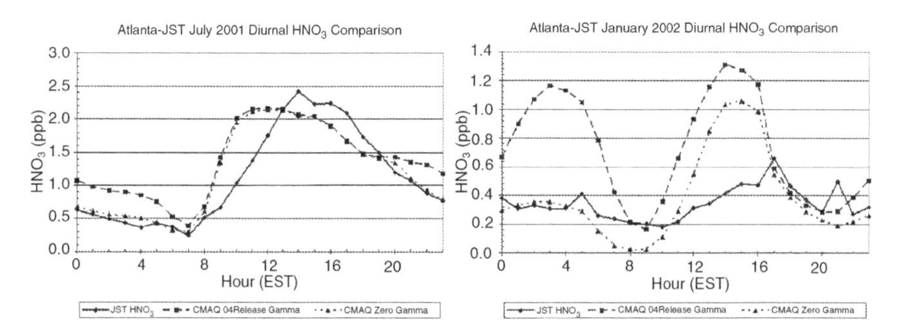

FIGURE 6. Comparison of the observed and predicted (base CMAQ and CMAQ with heterogeneous chemistry turned off) monthly averaged hourly diurnal pattern of HNO_3 at Atlanta for (a) July 2001 and (b) January 2002.

These high time resolution evaluations indicate that nighttime conversion of N_2O_5 to HNO_3 is the source of the majority of HNO_3 in winter and daytime photochemistry is the source of a majority of HNO_3 in summer. These time-resolved resolution evaluations also indicate that the heterogeneous reaction probability is still too high; thus, the "nitrate problem" has been greatly ameliorated, but not eliminated. There are suggestions in the laboratory research community and in recent ambient observations that a variety of factors exist that further inhibit these nighttime reactions, including the presence of organic aerosols or mixtures that include organics, but none are published and available for use by the CMAQ developers at this time. In addition, a wintertime issue with the photochemical production of HNO_3 has been suggested.

4. Summary

There is a systematic over-prediction in the late afternoon for species that are either emitted or produced at the surface in the late afternoon and early evening. The most consistent explanation at this time is that this over-prediction is due to a premature collapse of the boundary layer in the model. For example, the late afternoon and early morning cold bias of MM5 is consistent with this explanation. The over-prediction appears to be relatively larger in urban areas and smaller in the rural areas, lending further support to the hypothesis that an important source of the problem could be the inability of the meteorological models to adequately account for the urban heat island. The degree of over-prediction also appears to be larger in summer than in winter.

Nighttime over-predictions can create compensating errors for some of the pollutants, such as EC in Atlanta (masking errors), and create systematic biases for others, such as CO, NO_Y, and summer NH_X in Atlanta (giving an incorrect sense of error). These biases also can amplify the tendency of CMAQ to over-predict fine-particle nitrate. This analysis shows that comparisons against 24-hour averages are unable to discern whether the model is getting the right answer for the right reason or for compensating wrong reason and that these comparisons form a necessary but not sufficient component of model evaluation.

The diurnal analyses together with a variety of sensitivity analyses show that the influence of the nighttime heterogeneous reactions on overall HNO_3 production is much larger in winter than in summer. In summer, the daytime, photochemical production of HNO_3 is dominant. The analysis also indicates that there can be five sources of error affecting the levels and diurnal pattern of HNO_3 concentrations: (1) error in the nighttime heterogeneous production, (2) error in the daytime photochemical production, (3) error in the NO_X emissions, (4) error in the pbl height and mixing, and (5) error in the NH_3 emissions (affecting the partitioning of total-nitrate). High time-resolution analyses of several species or species combinations besides HNO_3, including NO_Y, EC, NH_X, NH_3, SO_4^{2-}, and total-nitrate, are needed to sort out the possible sources or error and to check for compensating errors.

A serious chemical problem, relating to the heterogeneous production of HNO_3 at night in CMAQ, was, to an acceptable degree, fixed in the 2003 public release version of CMAQ with the help of these high time-resolution evaluations. They indicate, however, that the nighttime production of nitric acid in the 2003 and 2004 versions of CMAQ most likely is still too high across much of the U.S., leading to a systematic over-prediction of total nitrate and, hence, particulate nitrate. This is corroborated by comparisons against the CASTNet data (Eder and Yu, 2004). In a relative sense the degree of over-prediction of particulate nitrate is larger in the winter, when it is more easily formed due to the lower temperatures, than in the summer. Interestingly, the two sources of bias (physical and chemical process) have roughly opposite seasonal dependencies. Thus, the seasonal balancing of surface NH_X and total-nitrate concentrations will vary across the seasons. In certain areas of the country CMAQ might be expected to predict ammonia limitation more often than it should in the colder months. Thus, CMAQ might be expected to over-emphasize the nitrate replacement (in absolute concentration terms) that can potentially offset part of the reduction in sulfate that will accompany reductions of SO_2 emissions. The hypothesis that this bias is affecting CMAQ's predicted changes of fine particles associated with emissions reductions is now being tested with further high time-resolution data coupled with model sensitivity analyses. We are continuing to investigate further both sources of bias with the intent of improving the predictions of CMAQ.

Disclaimer The research presented here was performed under the Memorandum of Understanding between the U.S. Environmental Protection Agency (EPA) and the U.S. Department of Commerce's National Oceanic and Atmospheric Administration (NOAA) under agreement number DW13921548. Although it has been reviewed by EPA and NOAA and approved for publication, it does not necessarily reflect their policies or views.

References

Byun, D. W. and J. K. S. Ching, Eds., Science algorithms of the EPA Models-3 Community Multi-scale Air Quality (CMAQ) modeling system, EPA Report No. EPA/600/R-99/030, Office of Research and Development, U.S. Environmental Protection Agency, Research Triangle Park, NC, 1999.

Carter, W., 2003, personal communication.

Dentener F. J. and Crutzen P. J. 1993. Reaction of N_2O_5 on the tropospheric aerosols: impact on the global distribution of NO_X, O_3, and OH. *J. Geophys. Res.*, 98, 7149-7163.

Eder, B. and Yu, S, A Performance Evaluation of the 2004. Release of Models-3 CMAQ. Proceedings of the 27th NATO CCMS ITM, 25-29 October 204, Banff Center, Canada, 2004.

Grell, G. A., Dudhia, J., Stauffer, D. R., 1994. A description of the Fifth-Generation Penn State/NCAR Mesoscale Model (MM5), Report NCAR/TN-389+STR, 138 pp., National Center for Atmospheric Research, Boulder, Colorado.

Mentel T. F, Sohn M. and Wahner A. Nitrate effect in the heterogeneous hydrolysis of dinitrogen pentoxide on aqueous aerosols. *Phys. Chem. Chem. Phys.*, 1, 5451-5457, 1999.

Pleim, J. E. and A. Xiu. Development and testing of a surface flux and planetary boundary layer model for application in mesoscale models, *J. Applied Meteor.*, 34, 16-32, 1995.

Riemer N., Vogel H., Vogel B., Schell B., Ackermann I., Kessler, C. and Hass H. Impact of the heterogeneous hydrolysis of N_2O_5 on chemistry and nitrate aerosol formation in the lower troposphere under photosmog conditions. *J. Geophys. Res.*, Vol. 108, No. D4, 4144, doi:10.1029/202JD002436, 2003.

SEARCH. Data can be accessed at http://www.atmospheric-research.com

Takahama, S., B. Wittig, D. V. Vayenas, C. I. Davidson, and S. N. Pandis, Modeling the diurnal variation of nitrate during the Pittsburgh Air Quality Study, *J. Geophys. Res.*, 2004 (in press).

Wittig, B., S. Takahama, A. Khlystov, S. N. Pandis, S. Hering, B. Kirby, and C. Davidson, Semi-continuous PM2.5 inorganic composition measurements during the Pittsburgh Air Quality Study, *Atmos. Environ.*, 38, 3201-3213, 2004.

56 Sensitivity Analysis of the EUROS Model for the 2003 Summer Smog Episode in Belgium

Felix Deutsch, Stefan Adriaensen, Filip Lefebre, and Clemens Mensink*

1. Introduction

Like in other Western European countries, the summer of 2003 was characterized by exceptional warm weather conditions and two strong ozone episodes from 14-20 July and in the first half of August. The August episode counted 12 consecutive days of ozone exceedances, which never happened before. The total number of days of ozone exceedances in 2003 was 65, which is the highest number ever registered. At the European level, long-term objectives for the reduction of ozone concentrations have been defined in the Framework Directive 96/62/EC. According to the ozone daughter directive (2002/3/EC), the target values should be attained by the member states by the year 2010. In order to reach these objectives, most of the member states will have to reduce drastically the emission of pollutants responsible for ozone formation, i.e. nitrogen oxides (NO_x) and non-methane volatile organic compounds (VOC). The emission reductions are prescribed for each of the EU member states, by means of national emission ceilings under the Gothenburg Protocol (1999) and the EU directive on National Emission Ceilings (2001/81/EC).

Several modelling tools have been developed and implemented for the assessment of ozone abatement strategies in Belgium (Mensink *et al.*, 2002). One of these tools is the EUROS model, an Eulerian air quality model that simulates tropospheric ozone over Europe on a long-term base. EUROS was originally developed at RIVM and, coupled with a state-of-the-art user interface, it is currently operational as an on-line service tool for policy support at the Interregional Cell for the Environment (IRCEL) in Brussels (Delobbe *et al.*, 2002[a]).

When comparing the results of the EUROS model with measurements for the August episode, it was found that the agreement was poor as can be seen in figure 1 for an urban monitoring station (Sint-Agatha-Berchem) and for a rural monitoring station (Dessel). Results show an underestimation of the ozone peak values for both stations.

* Flemish Institute for Technological Research (Vito), Boeretang 200, B-2400 Mol, Belgium.

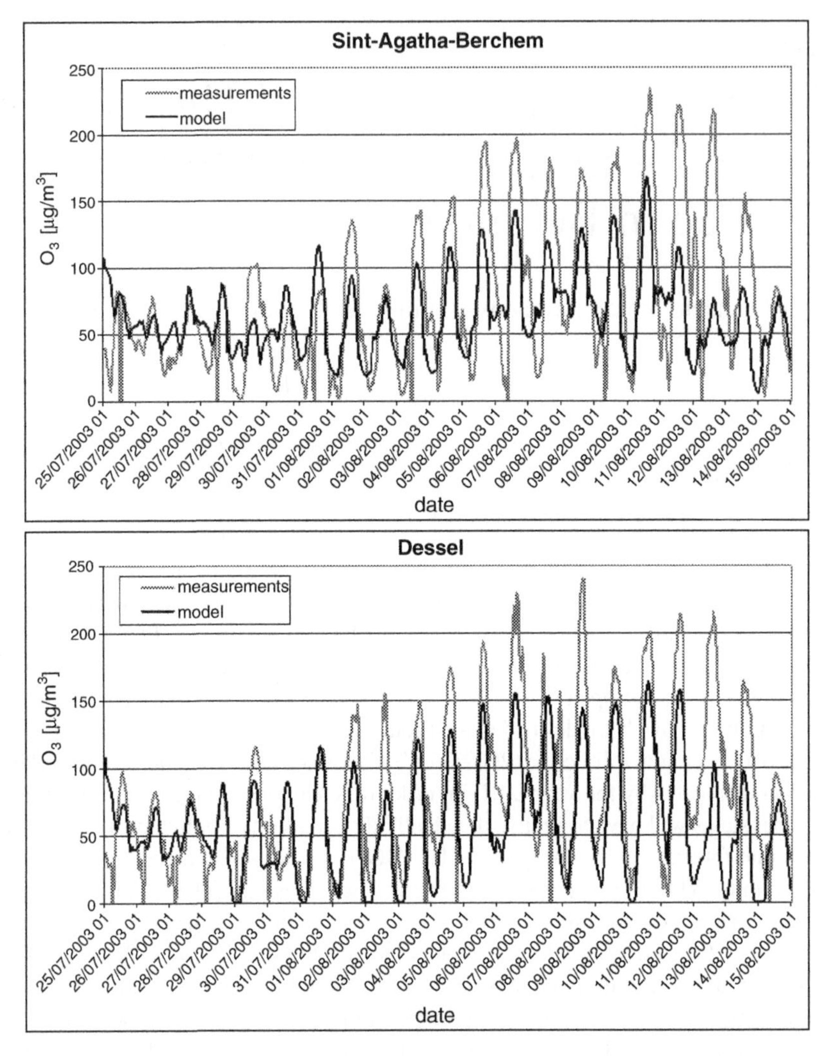

FIGURE 1. Comparison of measured and modelled ozone-concentrations for the urban monitoring station Sint-Agatha-Berchem (upper) and for the rural monitoring station Dessel (lower).

2. Methodology

The EUROS model is a 4-layer atmospheric model that simulates hourly tropospheric ozone concentrations over Europe on a long-term basis with a standard resolution of 60 km. A grid refinement procedure allows refining the spatial resolution in certain areas of the model domain, for example Belgium. A detailed emission module describes the emission of three pollutant categories (NO_x, VOC, SO_2) and for 6 different emission sectors (traffic, space heating, refinery, solvents use, combustion, industry). Both point sources and area sources are included. As far as the meteorology is concerned, a new three-dimensional input data set for EUROS has been generated from the ECMWF meteorological data (European Centre for Medium-Range Weather Forecasts, Reading, UK). Moreover, an important atmospheric parameter in EUROS is the mixing height, i.e. the height of the atmospheric layer adjacent to the ground, where the pollutants are well mixed through the action of turbulence and convection processes. Several methods have been explored to estimate the mixing height from the meteorological data set (Delobbe *et al.*, 2002[b]). The results have been compared with observational data and with the results of detailed model simulations. Based on these results, a new method has been proposed for the determination of the mixing height in air quality models (Delobbe *et al.*, 2001).

In a more detailed analysis we evaluated the model sensitivity towards the following parameters and model options:

- Influence of grid refinement and grid resolution (60 km versus 15 km);
- Influence of the advection scheme (4-layer versus 10-layer);
- Influence of the method to determine the mixing height (modified Bulk Richardson method versus empirical method);
- Influence of the difference between ECMWF and observed surface temperature;
- Influence of the time step in the operational splitting (30 min. versus 6 min.);
- Influence of the reservoir layer height parameterisation.

3. Results and Discussion

3.1. Influence of Grid Refinement and Grid Resolution

Figure 2 shows the effect of grid refinement on the results in the urban and the rural area. Local grid refinement was applied to a zone of 500 × 300 km, covering Belgium and the main regions of influence. For this zone the grid resolution was increased from 60 km (MAXLEV=1) to 15 km (MAXLEV=3), whereas the grid resolution stayed the same (60 km) outside this domain. As can be seen in the figures, local grid refinement helps to improve the results, especially at night time, i.e. for the minimum values. The effect is strongest for the urban station. This has to do with the titration effect. Sources associated with NO emissions (traffic, large point sources) are diffused into a smaller grid cell with a resolution of 15 × 15 km instead of a 60 × 60 km grid cell. This leads to higher initial concentrations of NO and thus in a stronger titration effect, leading to lower night time ozone concentrations.

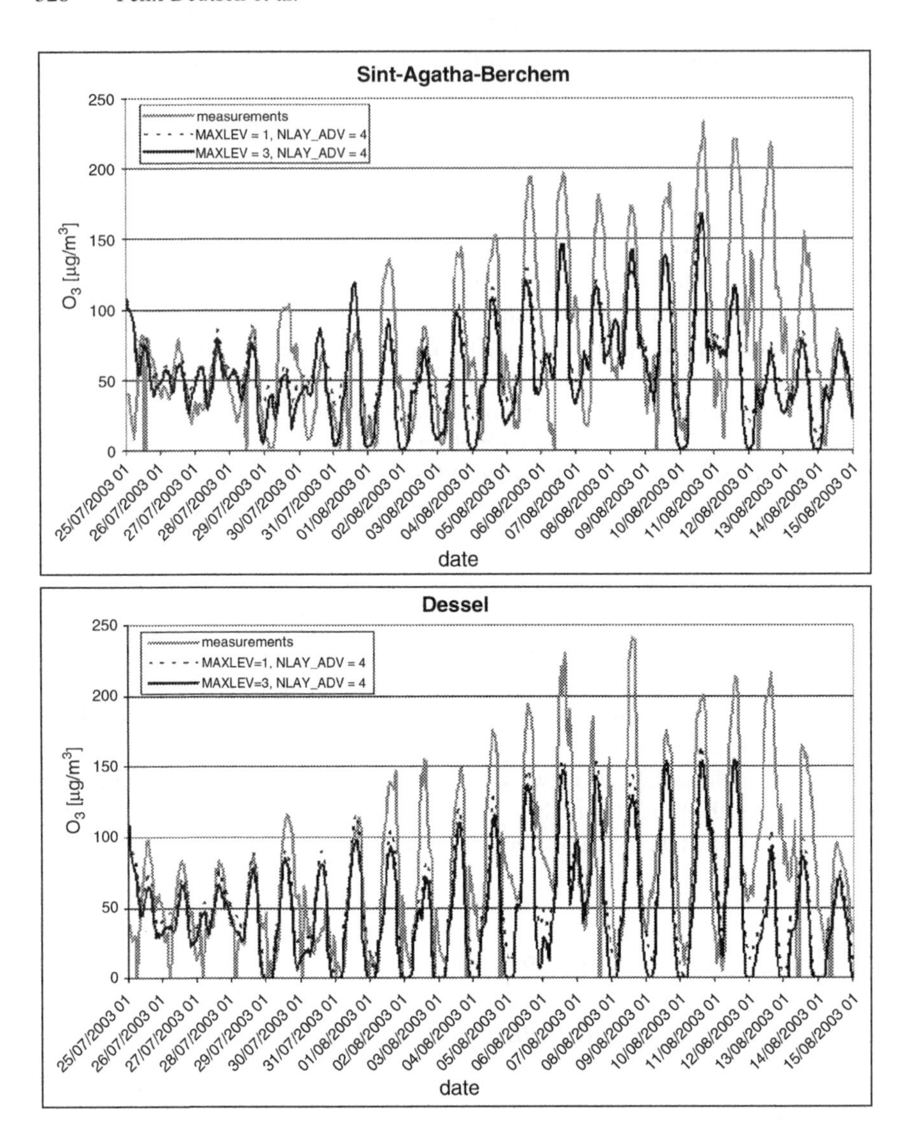

FIGURE 2. Influence of the grid refinement on model results for the urban monitoring station Sint-Agatha-Berchem (upper) and for the rural monitoring station Dessel (lower).

3.2. Influence of the Advection Scheme

Figure 3 shows the effect of changing the number of advection layers from 4 (NLAY_ADV = 4) to 10 (NLAY_ADV = 10) for the transport phase. The model results are only improving when an increase of the number of advection layers is combined with a simultaneous refinement of the grid. Results are similar for the urban and the rural station, but only shown here for the urban station (Sint-Agatha-Berchem).

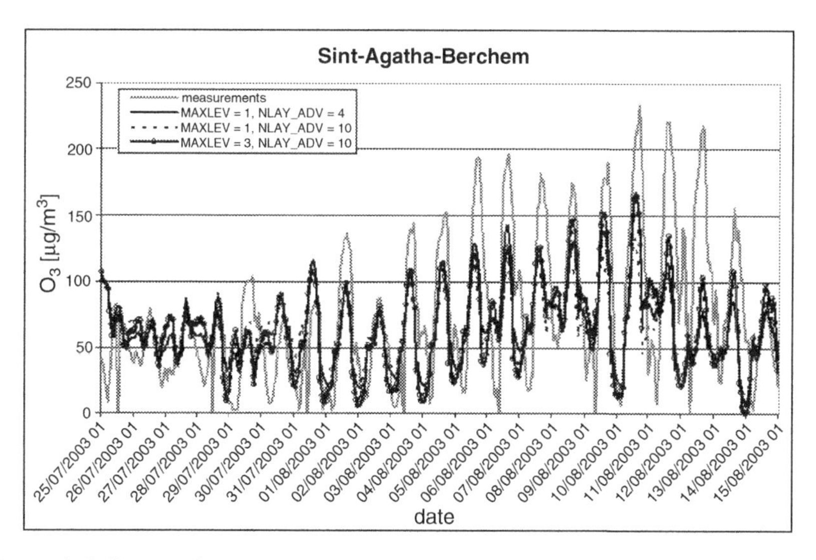

FIGURE 3. Influence of the advection scheme in combination with grid refinement for the urban monitoring station Sint-Agatha-Berchem.

3.3. Influence of the Method to Determine the Mixing Height

Figure 4 shows ozone concentrations obtained with two different methods to determine the mixing height: an empirical method with a daily rising and collapsing mixing layer, independent of the actual meteorological situation and, secondly, a method based on the Bulk Richardson Number (BRN), using ECMWF meteorological parameters. No large differences can be observed between both methods.

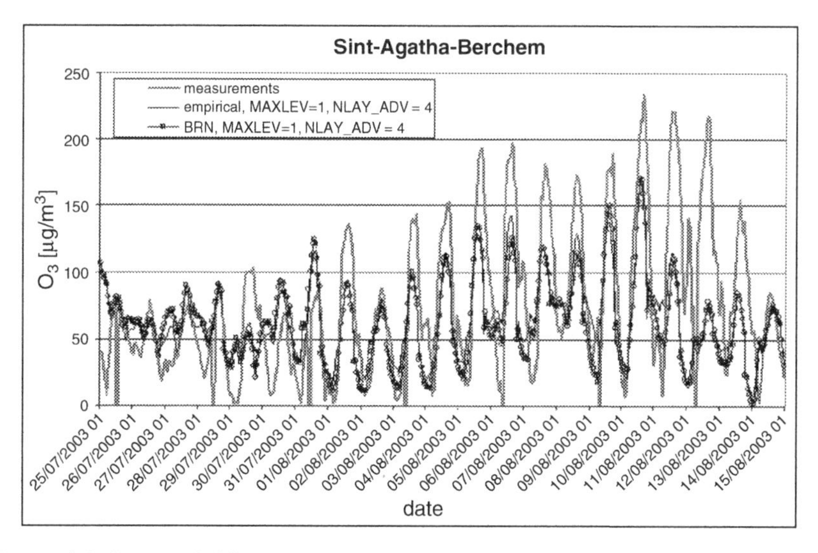

FIGURE 4. Influence of different methods to calculate the mixing height in EUROS for the urban monitoring station Sint-Agatha-Berchem.

3.4. Influence of the Difference Between ECMWF and Observed Surface Temperature

EUROS uses 6-hourly surface temperatures from the ECMWF-database and a linear interpolation to obtain hourly temperature values. However, figure 5 shows that this linear interpolation results in the "missing" of the daily temperature maximum in the early afternoon, resulting in a serious underestimation of the surface temperatures. However, this underestimation does not explain the low ozone-concentrations calculated by the model. Figure 6 shows only slightly higher ozone concentrations obtained in model-runs with an increased temperature by 3 or even 10 degrees.

3.5. Influence of the Time Step in the Operational Splitting

Model runs with a time step in the chemistry-routines of 6 minutes showed somewhat better results than runs carried out with a time step of 30 minutes when also the grid refinement procedure was applied (Figure 7).

3.6. Influence of the Reservoir Layer Height Parameterisation

Figure 8 shows the application of two different reservoir layer height parameterisations. The reservoir layer height can be set to stay "constant" at a certain height (here 1200 m), only rising if the mixing layer exceeds this height, but

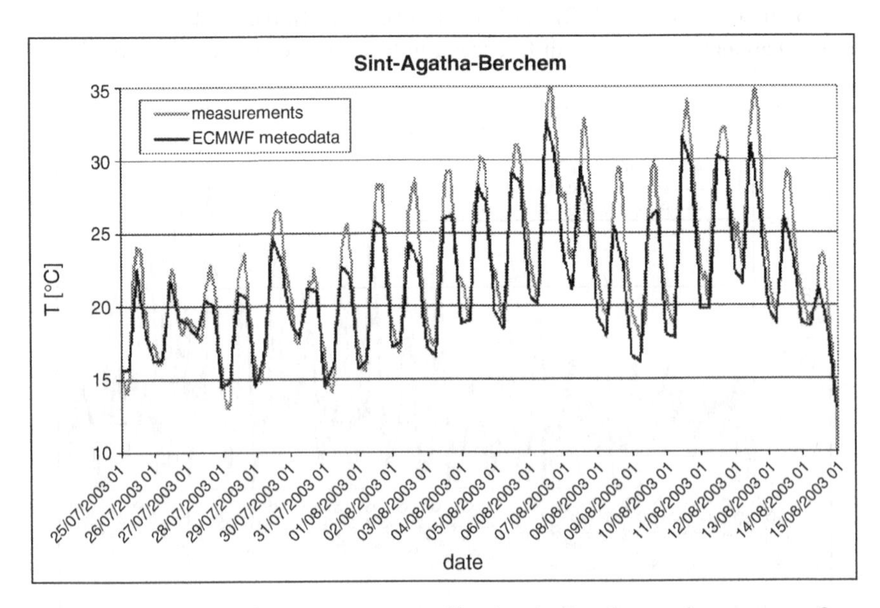

FIGURE 5. Measured surface temperature at Sint-Agatha-Berchem and temperature from ECMWF database.

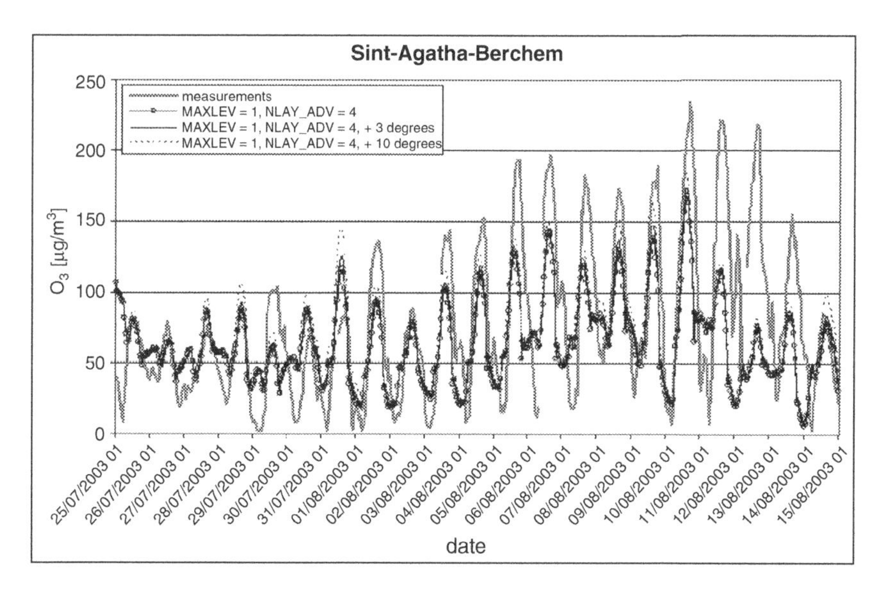

FIGURE 6. Influence of increased temperature on modelled ozone-concentrations for the urban monitoring station Sint-Agatha-Berchem.

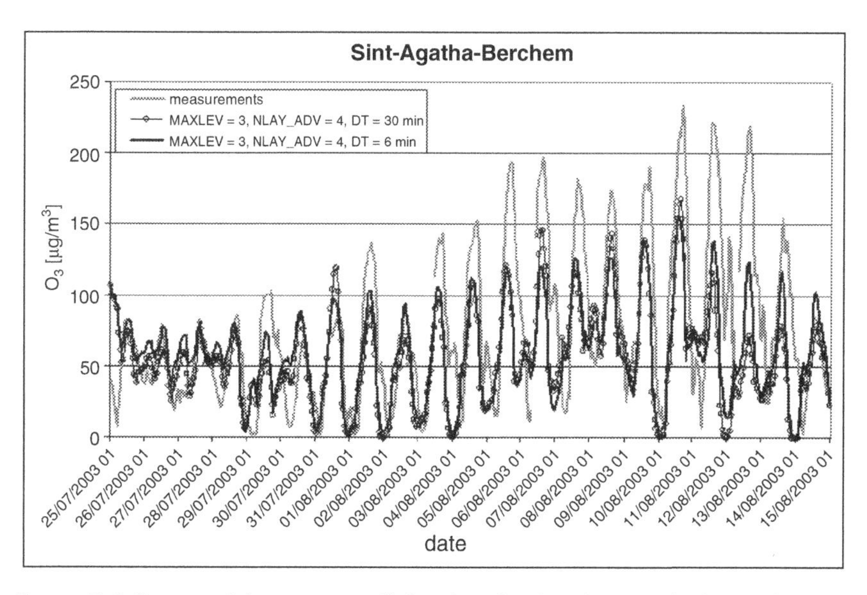

FIGURE 7. Influence of the operator splitting time for the urban monitoring station Sint-Agatha-Berchem.

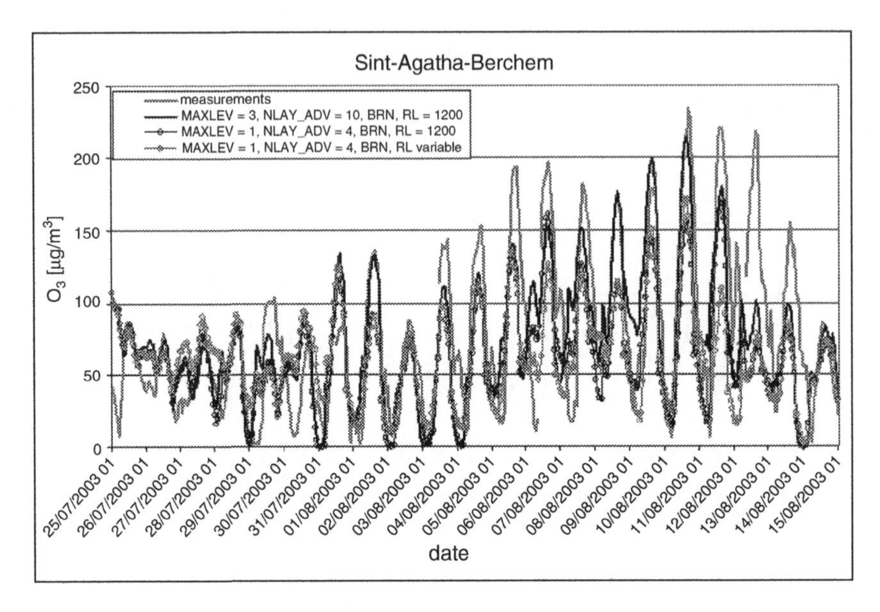

FIGURE 8. Influence of the parameterisation of the reservoir layer height for the urban monitoring station Sint-Agatha-Berchem.

never falling below this prescribed value. Alternatively (this is the "default" in EUROS), the reservoir layer height can fall to lower values when the mixing layer collapses in the afternoon. It can clearly be seen that the application of a constant reservoir layer height leads to much better results. Especially when using also the grid refinement, the employment of a constant reservoir layer height gives a great improvement of the modelled ozone-concentrations. This can be explained by a marked reduction of the loss of ozone and precursor compounds to the top layer of the model when using a constant height for the reservoir layer. Strong movements of the reservoir layer, especially towards lower heights, counteract its function as reservoir for ozone and precursor compounds for the next day.

4. Conclusions

It was found that the combination of grid refinement, constant reservoir layer height and reduction of the operational splitting time step greatly improved the model results for the nightly minimum ozone concentrations. These results can be attributed to improved mixing properties of NO. Also the modelling of ozone peak concentrations improved, especially for the urban station. The results also revealed that the temperatures during the ozone peak concentrations are underestimated systematically by 2°C – 6°C. This is caused by the linear interpolation technique that is used to interpolate temperatures between 12 h and 18 h UTC from the ECMWF data.

References

Delobbe, L., Mensink, C., Schayes, G., Passelecq, Quinet, A., Ch., Dumont, G., Demuth C., Lith, D. van, Matthijsen, J., 2002a, BelEUROS: Implementation and extension of the EUROS model for policy support in Belgium, *Global Change and Sustainable Development*, Federal Science Policy Office, Brussels, pp. 199-205.

Delobbe, L., Brasseur, O., Matthijsen, J., Mensink, C., Sauter, F. L., Schayes, G. and Swart, D. P. J., 2002b, Evaluation of mixing height parameterizations for air pollution models, *Air Pollution Modelling and Its Applications XV*, edited by C. Borrego and G. Schayes, Kluwer Academic/Plenum Publishers, pp. 425-433.

Delobbe, L., Brasseur, O. and Mensink, C., 2001, Determination of the Mixing Height from ECMWF Data for Use in the Regional Photo-chemical Smog model EUROS, in: Midgley, P. M., Reuther, M. J., Williams, M. (Eds.): *Proceedings of the EUROTRAC Symposium 2000*, Springer Verlag, Berlin, Heidelberg, pp. 822-826.

Mensink, C., Delobbe, L. and Colles, A., 2002, A policy oriented model system for the assessment of long-term effects of emission reductions on ozone, *Air Pollution Modelling and Its Applications XV*, edited by C. Borrego and G. Schayes, Kluwer Academic/Plenum Publishers, pp. 3-11.

Sensitivity Analysis of the EUROS Model for the 2003 Summer Smog Episode in Belgium

Speaker: Mensink

Questioner: P. Builtjes

Question: Which VOC-speciation for anthropogenic emissions did you use, and could this explain the underestimation of peak ozone?

Answer: The speciation of the anthropogenic VOC emissions is based on a pre-defined yearly average sector dependent distribution provided by EMEP. Only biogenic VOC emissions are temperature dependent in the model. Since Belgium is NO_x*-limited, VOC emissions do have a strong effect on peak ozone concentrations. Increasing the VOC emission in the model to levels of 1990 did increase the peak ozone concentrations considerably. However there is no evidence that the VOC speciation was different during the summer 2003 or that there were more anthropogenic VOC emissions during this period.*

Questioner: R. Bornstein

Question: Making the daytime reservoir layer heights higher over the urban areas (than the rural areas) might help your concentration results.

Answer: Since the results are very sensitive to the height of the reservoir layer, this might indeed be helpful. It would need a spatial and temporal variation of the reservoir layer height in the individual grid cells, which gives the model a more three dimensional character and abandons the simplified four layer concept.

57
A Performance Evaluation of the 2004 Release of Models-3 CMAQ

Brian K. Eder and Shaocai Yu*

1. Introduction

The Clean Air Act and its Amendments require that the U.S. Environmental Protection Agency (EPA) establish National Ambient Air Quality Standards for O_3 and particulate matter and to assess current and future air quality regulations, designed to protect human health and welfare. Air quality models, such as EPA's Models-3 Community Multi-scale Air Quality (CMAQ) model, provide one of the most reliable tools for performing such assessments. CMAQ simulates air concentrations and deposition of numerous pollutants on a myriad of spatial and temporal scales to support both regulatory assessment as well as scientific studies conducted by research institutions. In order characterize its performance and to build confidence in the air quality regulatory community, CMAQ, like any model, needs to be evaluated using observational data. Accordingly, this evaluation compares concentrations of various species (SO_4, NO_3, $PM_{2.5}$, NH_4, EC, OC, and O_3 (not available at press time)), simulated by CMAQ with data collected by the Interagency Monitoring of PROtected Visual Environments (IMPROVE) network, the Clean Air Status and Trends Network (CASTNet) and the Speciated Trends Network (STN).

2. CMAQ General Description

CMAQ is an Eulerian model that simulates the atmospheric and surface processes affecting the transport, transformation and deposition of air pollutants and their precursors [*Byun and Ching*, 1999]. CMAQ follows first principles and employs a "one atmosphere" philosophy that tackles the complex interactions among

* Brian Eder@ and Shaocai Yu#, NERL, U.S. Environmental Protection Agency, RTP NC, 27711, USA. # On Assignment from Science and Technology Corp., VA 23666, USA. @On assignment from Air Resources Laboratory, National Oceanic and Atmospheric Administration, RTP, NC 27711, USA.

multiple atmospheric pollutants and between regional and urban scales. Pollutants considered within CMAQ include tropospheric ozone, particulate matter and airborne toxics, as well as acidic and nutrient species. The model also calculates visibility parameters.

2.1. CMAQ Simulation Attributes

This evaluation focused on a full, one year simulation of 2001 using the 2004 release of CMAQ. The modeling domain covered the contiguous U.S. using a 36 km grid resolution and a 14-layer vertical resolution (set on a sigma coordinate). The simulation used the CB-IV gas-phase chemistry mechanism. The meteorological fields were derived from MM5, the Fifth-Generation Pennsylvania State University/National Center for Atmospheric Research (NCAR) Mesoscale Model. Emissions of gas-phase SO_2, CO, NO, NO_2, NH_3, and VOC were based on EPA's 2001 National Emissions Inventory. Primary anthropogenic $PM_{2.5}$ emissions were separated into different species including particle SO_4, NO_3, OC, EC. Emissions of HC, CO, NO_x, and particulate matter from cars, trucks, and motorcycles are based on MOBILE6, while biogenic emissions were obtained from BEIS 3.12.

3. Observational Data

3.1. Improve

IMPROVE is a collaborative monitoring effort governed by a steering committee comprised of Federal, Regional and State organizations designed to: (1) establish current visibility and aerosol conditions; (2) identify the chemical species and emission sources responsible for visibility degradation; and (3) document long-term visibility trends at over 100 remote locations nationwide. A majority of the sites located in the western United States (Figure 1). IMPROVE monitors collect 24-hr integrated samples every third day (midnight to midnight LST). Given CMAQ's one year simulation and IMPROVE's sampling schedule, a total of 115 days were available for comparison. IMPROVE species used in this evaluation include $PM_{2.5}$, SO_4, NO_3, EC and OC.

3.2. CASTNet

The Clean Air Status and Trends Network evolved from EPA's National Dry Deposition Network (NDDN) in 1990. The concentration data are collected at predominately rural sites, the majority of which are in the eastern United States, using an open-faced, 3-stage filter pack. The filter packs, which are exposed for 1-week intervals (i.e., Tuesday to Tuesday) at a flow rate of 1.5 liters per minute (3.0 liters per minute for western sites), utilize a Teflon filter for collection of the particulate species. Again, given CMAQ's one year simulation period and

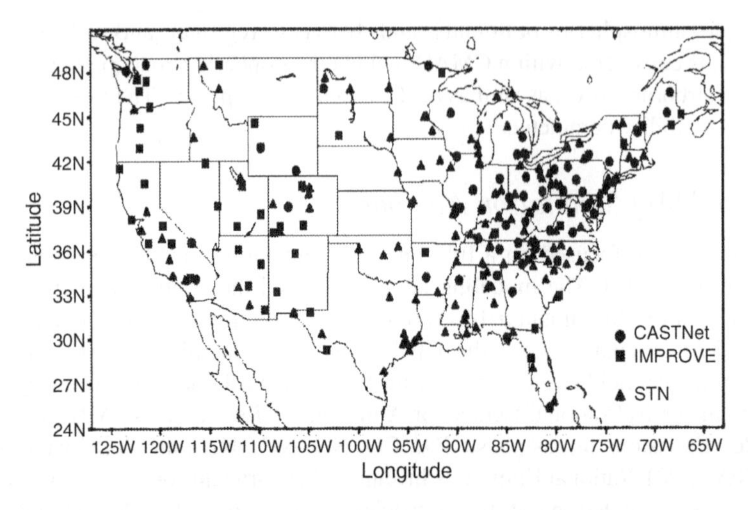

FIGURE 1. Stations location map of the networks used in the evaluation.

CASTNet's weekly sampling schedule, a total of 51 weekly observations were available from a total of 73 sites. CASTNet species used in this evaluation include: SO_4, NO_3, and NH_4.

3.3. *STN*

The more recently established Speciated Trends Network, developed by EPA, follows the protocol of the IMPROVE network (i.e. every third day collection) with the exception that most of the sites are found in urban areas. The main objectives of the STN are to: provide annual and seasonal spatial characterization of aerosols; provide air quality trends analysis; and track the progress of control programs. The number of STN sites available during 2001 varied as the new network was being deployed. STN species used in this evaluation include: SO_4, NO_3, NH_4 and $PM_{2.5}$.

4. Statistics

Because of the noted differences in sampling protocols, evaluation statistics were calculated separately for each network. Monitored values were assigned to the CMAQ grid cells without interpolation. On the rare occasion when more than one monitor was located within a 36 km grid cell, the average of the monitors would be used to represent that grid cell. In addition to general summary statistics (not shown), two measures of model bias: the Mean Bias (MB) and the Normalized Mean Bias (NMB) and two measures of model error: the Root Mean Square Error (RMSE) and Normalized Mean Error (NME) were calculated as seen below:

$$MB = \frac{1}{N}\sum_{1}^{N}(\text{Model} - \text{Obs}) \qquad NMB = \frac{\sum_{1}^{N}(\text{Model} - \text{Obs})}{\sum_{1}^{N}(\text{Obs})} \cdot 100\%$$

$$RMSE = \left(\frac{1}{N}\sum_{1}^{N}(\text{Model} - \text{Obs})^2\right)^{0.5} \qquad NME = \frac{\sum_{1}^{N}\left|(\text{Model} - \text{Obs})\right|}{\sum_{1}^{N}(\text{Obs})} \cdot 100\%$$

Scatter plots of monthly aggregated concentrations of CMAQ and observations are also provided for each network and specie, with two-to-one reference lines (Figures 2-6).

5. Results

Examination of the scatterplots and tables reveals that CMAQ varies in its ability to accurately simulate the various species. Simulations of $\mathbf{SO_4}$ are by far the best (Table 1). Correlation coefficients (Pearson's) associated with each data set are high, ranging from 0.78 (STN) to 0.92 (CASTNet) and the vast majority of the aggregated monthly simulations are within a factor of two of the observations (Fig. 2). The bias is small, with the NMBs ranging from –4% (CASTNet) to 6% (STN). The errors are relatively small as well, with NMEs ranging from

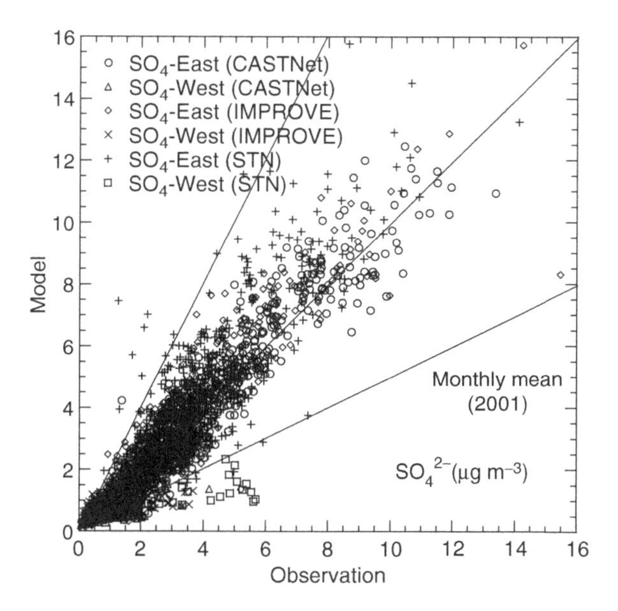

FIGURE 2. Scatterplot of SO_4 concentrations (with 2:1, 1:1, 0.5:1 ratios lines).

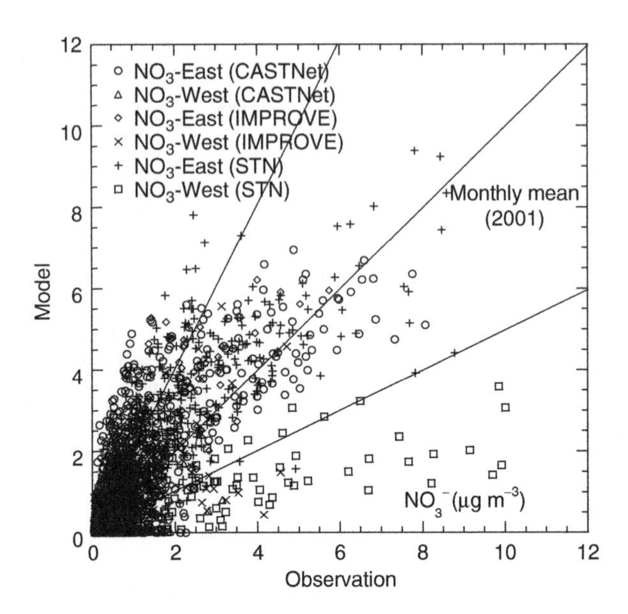

FIGURE 3. Scatterplot of NO_3 concentrations (with 2:1, 1:1, 0.5:1 ratios lines).

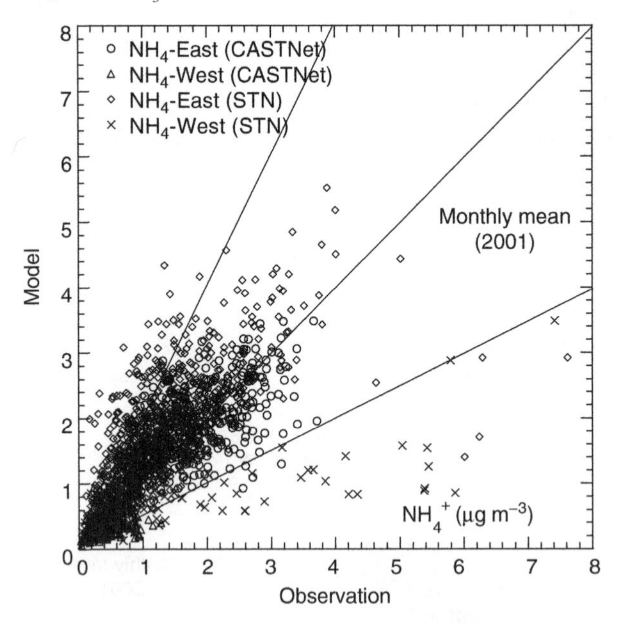

FIGURE 4. Scatterplot of NH_4 concentrations (with 2:1, 1:1, 0.5:1 ratios lines).

24% (CASTNet) to 43% (STN). The model generally performs better in eastern locations as opposed to western locations (not shown) – likely a result of greater experience inherent in CMAQ and its predecessors in simulating eastern locations.

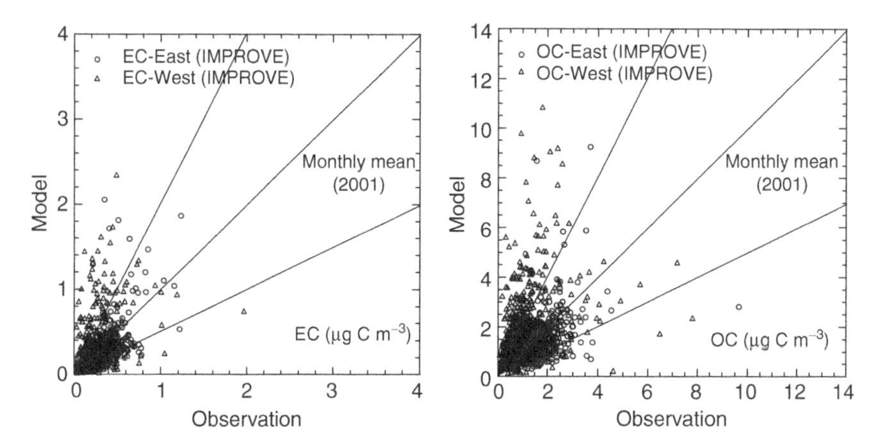

FIGURE 5. Scatterplot of EC (left panel) and OC (right panel) concentrations (with 2:1, 1:1, 0.5:1 ratios lines).

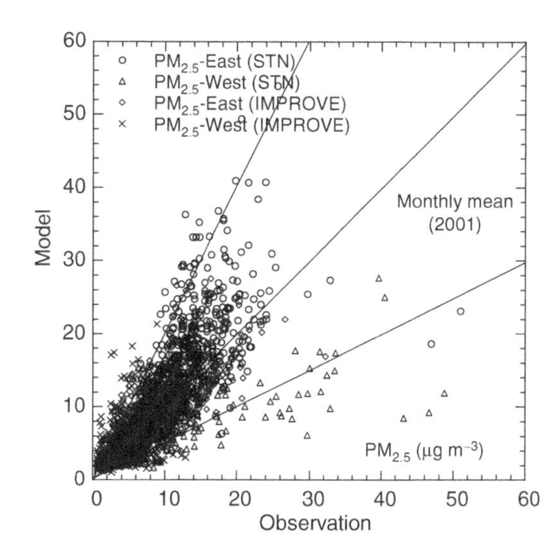

FIGURE 6. Scatterplot of $PM_{2.5}$ concentrations (with 2:1, 1:1, 0.5:1 ratios lines).

TABLE 1. SO_4 statistics.

	CASTNet	IMPROVE	STN
n	3,737	13,447	6,970
r	0.92	0.86	0.78
MB (ppb)	−0.12	0.04	0.22
NMB (%)	−4.0	2.0	6.0
RMSE (ppb)	1.12	1.29	2.32
NME (%)	24.0	40.0	43.0

CMAQ simulations of NO_3 are not nearly as good as those for SO_4 (Table 2). Correlations are lower, ranging from 0.42 (STN) to 0.73 (CASTNet) and the NMBs are larger, ranging from –5% (STN) to 27% (IMPROVE)). The NMEs are much larger, ranging from 76% (CASTNet) to 102% (IMPROVE). When examined over space, the NMEs exhibit little if any difference from one area of the United States to the next. This is not the case for the NMB, however, as CMAQ tends to over predict in the eastern U.S. and under predict in the western United States (with a few exceptions).

The quality of NH_4 simulations is similar to, but not quite as good as that of SO_4 (Table 3). Most aggregated monthly simulations are within a factor of two of the observations and the correlations range from 0.58 (STN) to 0.82 (CASTNet). The NMBs are positive and small (7% for CASTNet and 25% for STN), with the majority of over prediction occurring in the eastern U.S.. The NMEs range from 34% (CASTNet) to 66% (STN) with the error, for the most part, being equally distributed over the U.S.

The quality of OC and EC simulations are similar and fairly poor, with correlations of 0.46 for EC and 0.34 for OC (Table 4 & 5 respectively). Note that many monthly aggregated simulations fall outside the factor of two lines (especially for OC and especially for western stations). For OC the NMB and NME are 34% and 82%, respectively, while for EC they are –2% and 62%, respectively. This relatively poor performance is not surprising, given the crude physical representation of organics within the CMAQ aerosol component and the uncertain emission inventories of organics. There is a marked spatial difference with both species in that CMAQ tends to perform considerably better in the eastern U.S. (most NMEs < 50%) when compared to the western U.S. (most NMEs > 50%). OC NMBs are also considerably larger in the western U.S.

TABLE 2. NO_3 statistics.

	CASTNet	IMPROVE	STN
n	3,763	13,398	6,130
r	0.73	0.59	0.42
MB (ppb)	0.26	0.13	–0.08
NMB (%)	26.0	27.0	–5.0
RMSE (ppb)	1.17	1.05	2.88
NME (%)	76.0	102.0	80.0

TABLE 3. NH_4 statistics.

	CASTNet	STN
n	3,737	6,970
r	0.82	0.58
MB (ppb)	0.08	0.32
NMB (%)	7.0	25.0
RMSE (ppb)	0.56	1.29
NME (%)	34.0	66.0

TABLE 4. EC statistics.

	IMPROVE
n	13,441
r	0.46
MB (ppb)	–0.01
NMB (%)	–2.4
RMSE (ppb)	0.28
NME (%)	61.6

TABLE 5. OC statistics.

	IMPROVE
n	13,427
r	0.34
MB (ppb)	0.38
NMB (%)	34.5
RMSE (ppb)	1.77
NME (%)	82.6

TABLE 6. $PM_{2.5}$ statistics.

	IMPROVE	STN
n	13,217	6,419
r	0.68	0.52
MB (ppb)	0.54	0.97
NMB (%)	9.0	7.0
RMSE (ppb)	4.45	9.21
NME (%)	50.0	47.0

Simulations of $PM_{2.5}$ concentrations (which are composites of the other species), are fairly good as the majority of the simulations lie within a factor of two of the observations. The correlations range from 0.52 (STN) to 0.68 (IMPROVE). The NMBs are small and similar (7-9%) and the NMEs range from 47-50%. As with the other species, the model does somewhat better in the eastern U.S. as opposed to the western U.S. (Table 6).

6. Summary

This performance evaluation of the 2004 release of CMAQ reveals that the model's ability to accurately simulate the various species continues to improve, especially for NO_3 concentrations, which have improved markedly since an earlier evaluation (Mebust et al., 2003). Both SO_4 and NH_4 continue to be well simulated by the model, as does $PM_{2.5}$. Although simulations of the carbon species are somewhat deficient, improvements in both OC and EC simulations are expected with future releases of CMAQ as the scientific community's understanding of these species matures. Potential areas of research into the sources of the deficiencies identified in this evaluation include uncertainties in emissions inventories, imperfect representation of the meteorological fields, a as well as an incomplete understanding of aerosol dynamics in the CMAQ aerosol component.

Acknowledgements The authors would like to thank members of NERL's MEARB branch, especially Alfreida Torian and Steven Howard for the processing of the model and observation datasets.

References

Byun, D.W. and J.K.S. Ching, *Science algorithms of the EPA Models-3 Community Multiscale Air Quality (CMAQ) modeling system*, EPA-600/R-99/030, US EPA, US Government Printing Office, Washington D.C., 1999.

Mebust, M., Eder, B., Binkowski, B. and Roselle, S., Model-3 CMAQ model aerosol component, 2. Model evaluation. JGR Vol. 108, No. D6, 2003.

A Performance Evaluation of the 2004 Release of Models-3 CMAQ

Speaker: B. K. Eder

Questioner: W. Jiang

Question: CMAQ give mass concentrations of 6 organic species, while measurements give organic carbon concentrations. There are not directly comparable quantities. How did you compare them in your evaluation?

Questioner: D. Cohan

Question: Since in certain applications it is important to capture day-to-day variability of concentrations and in other applications the spatial pattern averaged over longer periods may be of more importance, have you looked at computing evolution statistics over a variety of temporal scales?

Questioner: L. Delle Monache

Question: There is any difference between the value of K-min in CMAQ v4.4 and CMAQ v4.3?

Disclaimer The research presented here was performed under the Memorandum of Understanding between the U.S. Environmental Protection Agency (EPA) and the U.S. Department of Commerce's National Oceanic and Atmospheric Administration (NOAA) and under agreement number DW13921548. Although it has been reviewed by EPA and NOAA and approved for publication, it does not necessarily reflect their policies or views.

58 Objective Reduction of the Space-Time Domain Dimensionality for Evaluating Model Performance

E. Gégo, P. S. Porter, C. Hogrefe, R. Gilliam, A. Gilliland, J. Swall, J. Irwin, and S. T. Rao*

1. Introduction

In the United States, photochemical air quality models are the principal tools used by governmental agencies to develop emission reduction strategies aimed at achieving National Ambient Air Quality Standards (NAAQS). Before they can be applied with confidence in a regulatory setting, models' ability to simulate key features embedded in the air quality observations at an acceptable level must be assessed. With this concern in mind, the U.S. Environmental Protection Agency (EPA) has recently completed several runs of the Community Multiscale Air Quality model (CMAQ) and the Regional Modeling System for Aerosols and Deposition model (REMSAD) to simulate air quality over the contiguous United States during the year 2001 with a horizontal cell size of 36 km×36 km. The meteorological model MM5 and the emission processor SMOKE were used to generate the input fields necessary for CMAQ and REMSAD. See Hogrefe et al (2004[a]) and Eder and Yu (2004) for more information about model settings.

Since these annual model simulations generate a huge amount of information, failure to properly organize the results may lead to confusion and hamper the evaluation procedure. The challenge is therefore to identify a technique that would make use of all pertinent observations over a large region and clearly indicate which spatial and temporal features are reproduced by the model.

To address this challenge, we propose a procedure to objectively condense the spatial and temporal observational domain into a limited number of homogeneous

* E. Gégo, Idaho Falls, ID, U.S.A. (e.gego@onewest.net)
P. S. Porter, University of Idaho, Idaho Falls, ID, U.S.A.
C. Hogrefe, ASRC, University at Albany, Albany, NY, U.S.A. R.
J. Irwin, Irwin and Assoc., Raleigh, NC, U.S.A.
R. Gilliam, A. Gilliland, J. Swall and S.T. Rao, NOAA Atmospheric Sciences Modeling Division, On Assignment to the U.S. Environmental Protection Agency, Research Triangle Park, NC, U.S.AE.

categories. This procedure follows naturally from two previous studies by Gego et al. (2004) and Hogrefe et al. (2004b). Gego et al (2004) showed that rotated principal component analysis (RPCA) could be used to delineate homogenous zones in terms of fluctuations in ammonium and sulfate aerosol concentrations within the United States (U.S.). We believe that the same technique can be applied to other variables and that model performance may differ between these regions. Hogrefe et al. (2004b) showed that different synoptic regimes, identified from daily maps of sea-level barometric pressure, led to contrasted spatial distributions of ozone and temperature during the summer months. Extrapolating from the findings of Hogrefe et al. (2004b), we speculate that model performance may vary by synoptic regime. Therefore, in this study, we intend to evaluate model performance on the basis of synoptic regime as well as on the basis of season, the latter being commonly utilized. As illustrated in section 2.3, synoptic typing can also be achieved using RPCA.

Following this categorization of observations in the spatial and the temporal domain, the evaluation procedure can be carried out by separately inspecting the observations and the corresponding model predictions in these spatial and temporal classes. It is the intent of this paper to investigate whether organizing the evaluation procedure in such a fashion can provide clearer indications about model performance, by specifying which specific spatial or temporal classes, if any, are noticeably better or worse simulated. We illustrate the use of the proposed technique for evaluating model predictions of 10 m wind speed fields in the eastern United States (east of the Rocky mountains). The meteorological model simulation, which is described in Section 2, is used as input to the annual CMAQ and REMSAD runs mentioned above. Wind speed was chosen to illustrate the proposed technique because of the impact that errors in the characterization of flow fields can have on the concentrations estimated by the photochemical models.

2. Brief Description of Model Setting and Database

The meteorological model simulations evaluated here were generated using the Penn State/NCAR mesoscale model MM5 (Grell et al., 1994), and were processed by the Meteorological-Chemistry Interface Processor (MCIP version 2.2) prior to analysis. Details of the MM5 settings can be found in McNally (2003). The observed 10 m wind speeds used to assess model performance were retrieved from the Techniques Development Laboratory's (TDL) data set maintained by the Data Support Section at the National Center for Atmospheric Research (NCAR-DSS). Because of computational constraints only 233 sites, fairly uniformly distributed throughout the eastern U.S., were selected from the TDL database.

Synoptic typing was performed on the sea level pressure fields resulting from Eta analysis and available from NCAR. Eta analysis fields are produced by the Eta Data Assimilation System from surface observations soundings and remote sensing data (see Black (1994) and Rogers et al. (1995) for more details). Eta analysis fields were used because sea level pressure maps are not among the

MCIP outputs. For this application, 1200 UTC Eta pressure fields were retrieved for each day during the year 2001 and were regridded from their original 40 km grid size to the 36 km grid size utilized by MCIP/MM5. Note that Eta pressure fields are only used to determine the synoptic classes, they are not directly used in the evaluation.

3. Methods

3.1. Rotated Principal Component Analysis (RPCA) for Division of the Spatial and Temporal Domain into Homogeneous Subsets

Originally designed to facilitate interpretation of large data sets involving numerous mutually dependent variables, PCA allows determination of the 'true' dimensionality of a data set through systematic quantification of correlations among all variables (i.e., identifying the redundancies). The starting element of a PCA is therefore the correlation matrix (or covariance matrix) of the variables, of which the eigenvalues and eigenvectors are first determined. The principal components (PCs) are then calculated. They are mutually independent linear combinations of all the variables, the coefficients of which are the eigenvectors of the correlation (or covariance) matrix. The variances of successive PCs correspond to the successive eigenvalues. Also important in a PCA are the loadings, which are obtained by multiplying the elements of each eigenvector by the square root of the associated eigenvalue. Loadings reveal the strength of the correlation between each PC and the original variables.

As higher order eigenvalues become progressively smaller, successive PCs explain less and less of the variance of the original data, leading to the conclusion that the information in the original data set can be reasonably described by a limited number of PCs. In this study, that number was determined with the Scree test (Cattell, 1966) according to which the number of PCs retained should correspond to the order at which an abrupt variation (elbow) is observed in a plot of ranked eigenvalues (from first to last). The PCs retained are then orthogonally rotated using the varimax procedure (Kaiser, 1958) to better segregate variables with similar characteristics. Variables with similar characteristics are more strongly correlated (larger loadings) with a given rotated PC than with the others. In our application, variables with similar characteristics (similar loadings) are said to belong to a 'homogeneous' group. Often, the number of groups formed does not exceed the number of PCs retained. However, a single PC may be at the origin of two separate 'opposite' groups, respectively formed around large positive and large negative loadings with that PC. Successive execution of PCA and varimax, simply referred to as RPCA, is often performed on the standardized scores (mean = 0; variance = 1) of the variables and not on the variables themselves in order to limit the impact of heteroscedasticity (inequality of variances) and confer on each variable the same importance (same weight) in the analysis.

Applying RPCA to time series of wind speed observed at all monitoring sites (each site is a variable) allows grouping of sites into a limited number of categories, each of which corresponds to a specific temporal evolution or 'mode of variation' (Eder, 1989). When the modes of variation correspond to unified geographical areas, RPCA is a simple and objective way to divide the domain into temporally homogenous regions; i.e., areas where wind speed fluctuations are synchronous. When applied to daily sea level pressure maps (each map is a variable), RPCA allows identification of the number of distinct synoptic patterns observed intermittently throughout the course of 2001 and permits assignment of each day to a particular pattern.

3.2. Evaluation of Model Results

The root mean squared error (RMSE) and the mean gross error (MGE) of the daily average wind speed were calculated for each combination of synoptic class and region by incorporating all pairs of relevant observations and model values. To compare the synoptic-based with a more classical calendar-based classification, the evaluation statistics were also calculated for each of the four quarters of 2001.

4. Results and Discussion

4.1. Synoptic Typing Analysis

Application of the Scree test to the standardized daily sea level pressure maps revealed that retaining the first five PCs would be appropriate since 76% of data variability is explained. Submitting the five retained PCs to the varimax procedure led to identification of ten groups, i.e., ten different synoptic classes. Several pairs of classes are geographic opposites, with low pressure areas in one group corresponding to high pressure areas in the other. Opposite pattern pairs include 1 and 8 and 3 and 4. Figure 1 shows the synoptic regimes corresponding to the

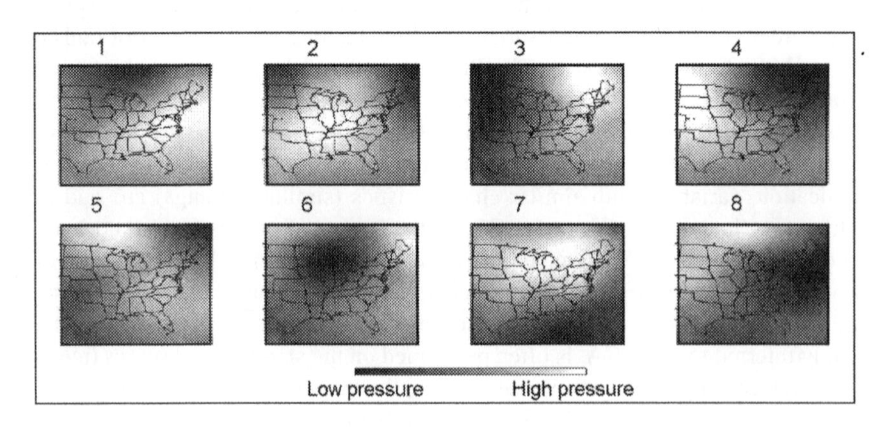

FIGURE 1. Synoptic patterns identified with RPCA for the eastern U.S. during 2001.

eight most prominent classes. Each map is calculated as the average of all daily maps included in the corresponding class. Two synoptic patterns are neither displayed nor considered for subsequent analyses because they are too scarce. A total of 358 of the 365 days are represented in the eight patterns retained.

As detailed in Table 1, the frequency of occurrence of each pattern varies from 89 days for synoptic pattern 1 to 16 days for synoptic pattern 8. The temporal distributions of each pattern throughout the year are distinct as well. Pattern 1 (often referred to as 'the Bermuda High') and Pattern 3, for instance, are much rarer during winter months as compared to other seasons. Pattern 2 and 4, on the other hand, are more common during fall and winter.

4.2. Delineation of Temporally Homogenous Regions

RPCA was applied to the daily mean wind speed data (365 days) from the 233 TDL observation sites. The Scree test indicated that retaining the first nine PCs should suffice to adequately summarize the data set (69% of the variance explained by nine components). Submitting the nine retained components to varimax led to 8 large groups made of at least 10 sites and two smaller groups corresponding to one and two sites only. These two smaller groups were discarded for further evaluation. Figure 2 shows the regions defined by the eight prominent groups. These regions constitute the base of our spatial classification.

TABLE 1. Seasonal distribution of the 8 synoptic patterns.

	Synoptic pattern							
Season	1	2	3	4	5	6	7	8
Winter	8	22	5	17	10	4	5	10
Spring	24	12	16	7	16	11	1	1
Summer	30	11	17	9	12	0	8	4
Fall	27	20	12	17	4	3	3	1
Total	89	65	50	50	52	18	17	16

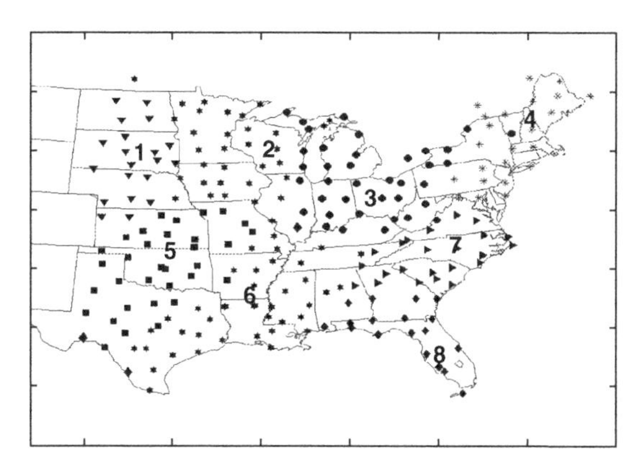

FIGURE 2. Homogenous zones for wind speed fluctuations.

4.3. Evaluation of Model Results

4.3.1. Evaluation by synoptic class

Table 2 presents the root mean square error (RMSE) and mean gross error (MGE) of daily mean wind speed predictions for each region and synoptic pattern, as well as the statistics for each region averaged over all synoptic patterns (last column); and for each synoptic pattern averaged over all sites (last row). Assuming all conditions are met for its meaningful application, the Friedman's ANOVA (non-parametric test) procedure indicated that there are no differences in RMSE among the synoptic patterns. However, not all MGE errors are equal ($p=0.014$). Looking further, a test of multiple comparisons indicates that the MGEs of synoptic groups can be classified according to: (2, 7) > (3, 5, 6), with the other groups falling between.

With respect to geographic regions the Friedman's ANOVA indicates significant regional effects for both RMSE ($p=10^{-4}$) and MGE ($p=10^{-8}$). Multiple comparisons of RMSE indicate the following classifications: 4 > (1, 2, 3, 5, 6, 8); 7 > (6, 8); 5 > 8. Wind speed is generally underestimated (negative bias) in the western portion of the domain (Regions 1 and 5), overestimated along the Atlantic Ocean shores (Regions 4, 7 and 8), and apparently adequately simulated in the center of the domain (Regions 2, 3 and 6). These observations are supported by multiple comparison classes for MGE, which are: 7 > (1, 2, 3, 5, 6); 4 > (1, 2, 3, 5); (6, 8) > (1, 5); (2, 3) > 1.

TABLE 2. Evaluation of daily mean wind speed (m/s) by region and synoptic pattern.

	Synoptic pattern								
Region	1	2	3	4	5	6	7	8	Average
1 RMSE	1.14	1.14	1.25	1.07	1.20	1.33	1.00	0.97	1.16
MGE	–0.55	–0.49	–0.74	–0.46	–0.71	–0.77	–0.34	–0.30	–0.57
2 RMSE	1.01	1.05	1.05	1.21	0.93	1.28	1.06	1.07	1.07
MGE	0.03	0.25	–0.07	0.00	0.18	–0.16	0.34	0.35	0.09
3 RMSE	1.04	1.08	0.98	1.30	0.97	1.25	0.95	1.21	1.09
MGE	0.14	0.18	0.15	–0.10	0.17	–0.21	0.42	0.05	0.09
4 RMSE	1.32	1.54	1.25	1.50	1.41	1.38	1.42	1.42	1.41
MGE	0.61	0.62	0.63	0.30	0.43	0.67	0.74	0.55	0.54
5 RMSE	1.22	1.05	1.34	1.03	1.30	1.33	1.09	1.01	1.18
MGE	–0.45	–0.18	–0.65	–0.06	–0.62	–0.26	–0.20	0.05	–0.33
6 RMSE	0.96	1.08	1.01	1.02	1.00	1.10	1.07	1.04	1.02
MGE	0.28	0.44	0.02	0.40	0.09	–0.03	0.45	0.25	0.25
7 RMSE	1.15	1.49	1.15	1.29	1.11	1.05	1.50	1.31	1.26
MGE	0.67	0.81	0.57	0.57	0.60	0.44	0.74	0.50	0.63
8 RMSE	0.97	1.00	0.91	0.89	0.89	0.90	1.16	1.10	0.96
MGE	0.29	0.40	0.26	0.35	0.25	0.32	0.17	0.24	0.31
Average RMSE	1.11	1.18	1.13	1.19	1.11	1.22	1.15	1.15	
MGE	0.11	0.24	–0.01	0.12	0.03	–0.04	0.29	0.21	

4.3.2. Evaluation by season

Table 3 presents the same evaluation statistics (RMSE and MGE) organized by season instead of synoptic pattern. As observed with the synoptic classification, wind speed is underestimated along the western border of the domain (Regions 1 and 5) and overestimated along the Atlantic coast (Regions 4, 7 and 8), during all four seasons.

Friedman's ANOVA performed on seasons indicates that not all RMSEs can be considered equal ($p = 10^{-5}$), while all MGEs are ($p = 0.94$). This apparent incongruity between the two evaluation statistics underlines the necessity to calculate different evaluation statistics before judging model quality. Multiple comparisons of seasonal RMSE shows: (fall, winter) > (spring, summer). With respect to geographic region, Friedman's ANOVA indicates that neither RMSEs nor MGEs can be considered equal ($p = 10^{-3}$ for both). Multiple comparisons of geographic RMSEs show: 4 > (2, 3, 6, 8); 7 > (2, 6, 8); 5 > (6, 8); 1 > 8. Similarly, multiple comparisons of MGE values indicate: (7, 4) > (1, 2, 3, 5); (6, 8) > 1.

Sorting model results by season is a typical process and judging model performance organized in such a fashion is straightforward. Understanding the effects of different synoptic scenarios, on the other hand, is more complex; while a season represents a block of about 90 successive days, the synoptic patterns are intermittently observed with different frequencies in all 4 seasons throughout the year. Yet, considering the seasonal evaluation results of Table 3 and the seasonal distribution of each synoptic class of Table 1 helps one to understand the arithmetical origin of the synoptic evaluation results and may provide some insights about their meteorological meaning. Synoptic pattern 2, for instance, is characterized by a higher RMSE value than the other patterns (arithmetically higher

TABLE 3. Evaluation of daily mean wind speed (m/s), by region and season.

Region	Winter	Spring	Summer	Fall	Average
1 RMSE	1.21	1.13	0.98	1.30	1.16
MGE	–0.54	–0.54	–0.52	–0.66	–0.57
2 RMSE	1.17	1.02	0.92	1.15	1.06
MGE	0.11	0.41	0.30	–0.20	0.09
3 RMSE	1.19	1.02	0.92	1.21	1.08
MGE	–0.03	0.24	0.23	–0.07	0.09
4 RMSE	1.60	1.29	1.20	1.50	1.40
MGE	0.45	0.55	0.53	0.54	0.54
5 RMSE	1.18	1.25	1.02	1.26	1.18
MGE	–0.21	–0.45	–0.34	–0.34	–0.33
6 RMSE	1.08	1.01	0.87	1.11	1.02
MGE	0.18	0.17	0.26	0.38	0.25
7 RMSE	1.50	1.08	1.07	1.34	1.25
MGE	0.66	0.55	0.53	0.79	0.63
8 RMSE	1.01	0.94	0.86	1.04	0.96
MGE	0.38	0.26	0.17	0.43	0.31
Average RMSE	1.24	1.10	0.99	1.24	
MGE	0.11	0.11	0.14	0.08	

value, not in terms of statistical significance) because it occurs mostly during fall and winter, seasons for which the highest RMSE values were also calculated. Conversely, patterns 1 and 3 display some of the smallest RMSE values because they occur mostly during seasons characterized by smaller RMSEs.

4.3.3. Discussion

The division by PCA of the eastern U.S. territory into distinct homogenous zones has the advantage of relying on the characteristics of the observational data rather than on artificial limits imposed by political boundaries (e.g., state borders) to group model results for evaluation. The value of dividing the eastern United States into homogenous zones in terms of wind speed fluctuations for the purpose of model evaluation is evident in the differences in model performance between adjacent regions. It appears that, for the 2001 simulation considered in this study, neither the seasonal nor the synoptic based temporal classification allowed detection of time periods with systematic weaknesses in wind speed estimates. While different synoptic regimes are characterized by different spatial wind field patterns, they do not seem to lead to distinguishable model skills. At the origin of this absence of contrast may be the data assimilation scheme used to 'nudge' MM5 (and therefore MCIP) fields towards existing observations. Our intent for future work is to apply the same technique to CMAQ and REMSAD concentration estimates, which, unlike the meteorological fields, are not subjected to any assimilation with observations. By doing so, we will be able to provide insights on the value of temporal classification (synoptic typing or seasonal segregation) for the purpose of model evaluation.

Acknowledgments The research presented here was performed under the Memorandum of Understanding between the U.S. Environmental Protection Agency (EPA) and the U.S. Department of Commerce's National Oceanic and Atmospheric Administration (NOAA) and under agreement number DW13921548. This paper has been reviewed in accordance with the EPA's peer and administrative review policies and approved for presentation and publication. This research was partially funded by the Department of Commerce through contracts with Dr. E. Gego (EA133R-03-SE-0710), with the University of Idaho to Dr. P. S. Porter (EA133R-03-SE-0372), and with the State University of New York to Dr. C. Hogrefe (EA133R-03-SE-0650).

References

Black, T. L., 1994: The new NMC mesoscale Eta model: Description and forecast examples. Wea. Forecasting, 9, 265-278.

Cattell, R. B, 1966. The Scree test for the number of factors. Multivariate Behavioral Res, 1, 245-276.

Eder, B. K., 1989. A principal component analysis of SO_4^{2-} precipitation concentrations over the eastern United States, Atmos. Environ. 23 (12), 2739-1750.

Eder, B. K. and S. Yu, 2004: A performance evaluation of the 2004 release of MODELS-3 CMAQ. 27th NATO/CCMS, Banff, Alb. Canada, Oct. 25-29, 2004.
Gego, E., P. S. Porter, J. S. Irwin, C. Hogrefe, and S. T. Rao, 2004: Assessing the comparability of ammonium, nitrate and sulfate concentrations measured by three air quality monitoring networks, Pure Appl. Geoph., in press.
Grell, G. A., J. Dudhia, and D. Stauffer, 1994: A description of the fifth-generation Penn State/NCAR Mesoscale Model (MM5). *NCAR* Technical Note, 138 pp., TN-398 + STR, National Center for Atmospheric Research, Boulder, CO.
Hogrefe, C., J. Biswas, B. Lynn, K. Civerolo, J.-Y. Ku, J. Rosenthal, C. Rosenzweig, R. Goldberg, and P.L. Kinney, 2004[b]: Simulating regional-scale ozone climatology over the Eastern United States: Model evaluation results, Atmos. Env., 38, 2627-2638.
Hogrefe C., J. M. Jones, A. Gilliland, P. S. Porter, E. Gego, R. Gilliam, J. Swall, J. Irwin, and S.T. Rao, 2004[a] Evaluation of an annual simulation of ozone and fine particulate matter over the continental United States – Which temporal features are captured 27th NATO/CCMS, Banff, Alb. Canada, Oct. 25-29, 2004.
Kaiser H. F., 1958. The varimax criterion for analytical rotation in factor analysis, Psychometrika, 23, 187-201.
McNally, D., 2003: Annual application of MM5 for calendar year 2001, prepared for U.S. EPA by D. McNally, Alpine Geophysics, Arvada, CO, March 2003, 179 pp.
Rogers, E., T. L. Black, and G. J. DiMego, 1995: The regional analysis system for the operational "early" Eta model: Original 80 km configuration and recent changes. Wea. Forecasting, 10, 810-825.

Objective Reduction of the Space-Time Domain Dimensionality for Evaluating Model Performance

Speaker: E. Gego

Questioner: S. Dorling

Question: I would like to have seen you use an existing synoptic classification based on a longer data record (e.g. 10 years plus). Also, by not undertaking a seasonal synoptic typing, you may have very different conditions (e.g. of temp/cloud cover) within each synoptic type.

Answer: Since our objective was to evaluate model results for year 2001 only, we don't think that basing our synoptic classification on a longer data record was necessary. We just needed to characterize year 2001 to the best of our ability.

It is true that by not undertaking a seasonal synoptic typing, we group different conditions (e. g. of temp/cloud cover) within each synoptic type. However, differentiating 4 seasonal groups for each of the 8 synoptic patterns observed during 2001 would lead to 32 temporal classes (4 seasons x 8 synoptic patterns) to look at, situation we wanted to avoid since our objective was to introduce a synthetic classification technique for presentation of the evaluation results. The situation would be different for a model diagnostic effort, for which I would definitely use a seasonal synoptic typing.

Questioner: M. Sofiev

Question: A comment continuing the previous question regarding the correlation coefficient:

The clustering methods often rely on other norms than corr. coeff. because of possible misleading and lack of robustness of their metric. Therefore, it might be useful to consider the cluster analysis on the basis of some appropriate norm – the results might appear somewhat different.

Answer: Classical clustering methods are indeed often based on the notion of 'distances' between observations and not on correlation. However, we do not use any traditional clustering method. Our technique uses RPCA to group observations into 'homogenous groups'. The beauty of RPCA and the reason we chose to utilize it is that it analyses 'similarities' between entire time series of observations. The number of 'groups' decided upon is based on the eigen values of the correlation matrix submitted to PCA and not on any inappropriate norm such as a threshold value of the correlation coefficient.

59 Cloud Processing of Gases and Aerosols in a Regional Air Quality Model (AURAMS): Evaluation Against Aircraft Data

Wanmin Gong, Véronique S. Bouchet, Paul A. Makar, Michael D. Moran, Sunling Gong, and W. Richard Leaitch*

1. Introduction

Clouds play an active role in the processing and cycling of chemicals in the atmosphere. Gases and aerosols can enter cloud droplets through absorption/condensation (of soluble gases) and activation and impact scavenging (of aerosol particles). Once inside the cloud droplets these tracers can dissolve, dissociate, and undergo chemical reactions. It is believed that aqueous phase chemistry in cloud is the largest contributor to sulphate aerosol production. Some of the aqueous-phase tracers will be removed from the atmosphere when precipitation forms and reaches the ground. However, the majority of clouds are non-precipitating, and upon cloud dissipation and evaporation, the tracers, physically and chemically altered, will be released back to the atmosphere. Updrafts and downdrafts in convective clouds are also efficient ways of redistributing atmospheric tracers in the vertical. It is therefore important to represent these cloud-related physical and chemical processes when modelling the transport and transformation of atmospheric chemical tracers, particularly aerosols.

A new multiple-pollutant (unified) regional air-quality modelling system, AURAMS, with size- and chemical-composition-resolved aerosols is being developed at the Meteorological Service of Canada. In the current version of AURAMS, many of the cloud processes mentioned above are represented. Evaluation of the model performance is underway for several selected periods (e.g., Makar et al., 2004a,b; Bouchet et al., 2004). In this paper we will focus on a comparison against aircraft measurements conducted during the EMEFS-1 campaign in the summer of

* Wanmin Gong, Paul A. Makar, and Michael D. Moran, Sunling Gong, and W. Richard Leaitch, AQRB, Meteorological Service of Canada, 4905 Dufferin Street, Downsview, Ontario, M3H 5T4, Canada.
Véronique Bouchet, CMC, Meteorological Service of Canada, 2121, Voie de Service Nord No. 404 Route Transcanadienne, Dorval, Quebec, H3P 1J3, Canada.

1988. We will examine the impact of cloud processes on gases and aerosols in two selected cases, one cumulus and the other stratocumulus.

2. Model and Simulation

2.1. AURAMS and Its Cloud Processes

AURAMS consists of these primary components: an emissions processor; a meteorological driver model; and a chemical transport model closely coupled with a size-resolved (sectional) and chemically speciated aerosol module. Processes represented in the chemical transport model include emission, advection and diffusion, gas-phase chemistry, gas-to-particle conversion (including secondary organic-matter aerosol formation and multi-phase thermodynamics of the sulphate-nitrate-ammonia-water system), aerosol microphysics, cloud processing of gases and aerosols, wet and dry deposition, and sedimentation. In the following, we will briefly describe the representation of the various cloud processes in current AURAMS. For a detailed description of the model and the various components see Moran et al. (1998), Gong, S. et al. (2003), Zhang et al. (2002), Makar et al. (2003), Gong, W. et al. (2003).

Cloud processing of gases and aerosols in AURAMS currently includes aerosol activation (or nucleation scavenging), aqueous-phase chemistry, tracer scavenging and removal due to precipitation. The current representation of aerosol activation, inherited from the Canadian Aerosol Module, CAM (Gong, S. et al., 2003), is based on a simple empirical relationship between the aerosol number density and the number density of cloud droplets formed on the activated aerosols (CCN) as described in Jones et al. (1994). The aqueous-phase chemistry mechanism in the present study is adapted from ADOM (Acid Deposition and Oxidant Model: Venkatram et al., 1988). It is focused on sulfur chemistry and includes non-equilibrium mass transfer of SO_2, O_3, H_2O_2, ROOH, HNO_3, NH_3 and CO_2 and oxidation of S(IV) to S(VI) by dissolved ozone, hydrogen peroxide, organic peroxides and oxygen (in the presence of trace metals). The aqueous-phase chemistry is coupled with the size-resolved aerosol chemical components in AURAMS. Sulphate, nitrate, and ammonium are directly affected by the aqueous processes while the remaining aerosol components are indirectly modified due to the internally mixed aerosol assumption. Both aerosol mass and size distribution are modified by aqueous-phase chemistry. Wet deposition in AURAMS considers the processes of tracer scavenging and transport by precipitation. The scavenging process includes both tracer transfer from cloud droplets due to the auto-conversion process and the direct impact scavenging of aerosol particles and soluble gases by falling hydrometeors (liquid or solid). Tracers captured in precipitation are removed from the atmosphere when the precipitation reaches the ground. However, part or all of the precipitation may evaporate before reaching the ground, in which case the tracers will remain in the atmosphere but will be vertically redistributed.

2.2. *Model Simulation*

A simulation of regional oxidants and particulate matter (PM) over eastern Canada and the north eastern U.S. using AURAMS was conducted for the period of August 1-6, 1988. The simulation period was chosen to make use of the relatively extensive measurement data collected during the EMEFS campaign (Eulerian Model Evaluation Field Study, Hansen et al, 1991), which included monitoring data collected from various participating networks and detailed chemistry measurements at several enhanced surface sites as well as aircraft measurements.

The actual simulation started from July 30, 1988, to allow for a 2-day model "spin-up". The model was run on an 85 × 105 × 28 grid with 42 km spacing in the horizontal. The 28 unevenly spaced vertical levels range from the surface to 25 km. AURAMS is driven, off-line at 900 s time interval, by the Canadian operational numerical weather forecast model (GEM) (Côté et al., 1998), at 24 km resolution. Emissions used for the simulation are based on the 1990 Canadian and U.S National Inventories. Bulk primary PM emissions included in these inventories are chemically and size segregated in AURAMS by primary emission source stream. Twelve size bins (logarithmically evenly distributed between 0.01 and 40 μm) are used and 8 aerosol composition categories (sulfate, nitrate, ammonium, organic carbon, elemental carbon, crustal material, sea salt and aerosol-bound water) are currently represented in AURAMS.

The 6-day study period corresponds to a regional oxidant event over the eastern North America. A relatively dry stagnant high pressure system dominated the first half of the period, while a low pressure system moved into the region during the last two days accompanied by stronger flow, more clouds, and precipitation. High concentrations of $PM_{2.5}$, over 100 $\mu g\ m^{-3}$ in places, is simulated over the Ohio Valley extending to the St. Lawrence River Valley for this period, with sulfate being the dominant component. Under the stagnant condition high concentrations of particulate sulfate are found mostly close to the sources (e.g., large power plants over the Ohio valley), while the impacts of transport and aqueous-phase oxidation are evident under more transient condition.

3. Airborne Observations

During the first intensive measurement campaign of EMEFS (July and August 1988), the National Research Council of Canada DHC-6 Twin Otter aircraft was deployed to collect both air and cloud samples for microphysical and chemistry analysis over south-central Ontario, mostly over and between Egbert (44.23°N, 79.78°W) and Dorset (45.23°N, 78.93°W), where intensive surface chemistry sites were located. The measurements included SO_2, H_2O_2, aerosol and cloud droplet size distribution, cloud liquid water content, particulate sulfate, nitrate and ammonium, HNO_3, NH_3, as well as cloud-water $SO_4^=$, NO_3^-, NH_4^+, Cl^-, Na^+, K^+, and H^+. Some of the cloud water samples were also analysed for aqueous-phase H_2O_2. Liu et al. (1993), Macdonald et al. (1995), and Leaitch (1996) gave detailed descriptions of the instrumentation, the collection methods, and the flights conducted during the study.

Within the present model simulation period (August 1-6, 1988), cloud water samples were collected from 3 flights: flight 20 on August 4, flight 24 on August 5, and flight 25 on August 6. Both flight 20 and 24 were conducted under south-westerly flow of warm moist air during the latter part of the regional pollution episode. The clouds sampled during these two flights were of cumulus and towering cumulus, some precipitating on the 4th (flight 20). In contrast, the clouds sampled during flight 25 were stratocumulus behind a cold front that had advanced through the study region. The cumulus case of August 5 (flight 24) has been discussed considerably in Leaitch et al. (1991), Leaitch (1996), and Leighton et al. (1996). In this study we will focus on the latter two flights, i.e. flight 24 and flight 25.

4. Evaluation Against Aircraft Observations

Aircraft data collected by the NRC Twin Otter during the 1988 EMEFS campaign has been used in two previous model evaluation studies: Macdonald et al. (1993) compared aircraft clear-air measurements with ADOM simulation results in a diagnostic evaluation effort; Leighton et al. (1996) used the aircraft data to evaluate a 3-D cloud chemistry model based on three convective cloud cases. In the study of Leighton et al. (1996) clear-air and/or surface measurements were used to initialize the cloud chemistry model while model results were evaluated against cloud microphysics and chemistry measurements. Here we attempt to evaluate the model cloud processes in AURAMS with the aircraft data (both cloud-water and clear-air). In this case the model results are taken out of the 8-day (including 2-day start-up) model simulation which is not constrained in any way by the measurements. Also note that currently the model does not advect cloud tracers explicitly-clouds are generated at the beginning of, according to the microphysical parameters from the meteorological driver model, and artificially evaporated at the end of each advection time step.

4.1. Flight 24 (1734–1941 Z, August 5, 1988)

Flight 24 was conducted during the early-to-mid afternoon on August 5, 1988. It consisted a clear-air profiling over Dorset to 4150 m (ASL) between 1745 and 1817 Z and cloud water collections, also over Dorset, at three levels (4.6, 3.1, and 2.2 km) between 1817 and 1849 Z, followed by below-cloud filter sample for VOC from Dorset to Egbert. The cloud selected for sampling was a growing cumulus with base at 1.1 km and top at around 4.6 km (Leaitch, 1996).

Figure 1 shows the vertical profiles of cloud water $SO4^{=}$, $NO3^{-}$, $NH4^{+}$ and cloud liquid water content from the model. They are taken from the grid where Dorset is located and averaged between 17 and 18 Z (when the cloud water samples were collected). Also shown are profiles averaged over the 9 grids (centred

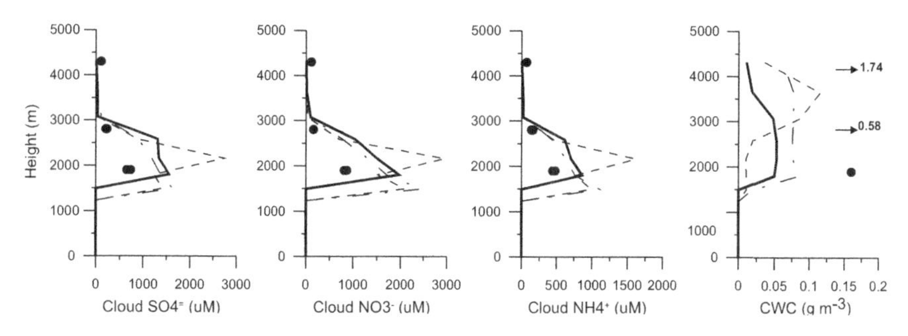

FIGURE 1. Vertical profiles of cloud SO4=, NO3−, NH4+, and liquid water content at grid (42, 62) – Dorset grid. Thick solid lines are averaged between 17 and 18Z on August 5; dashed lines are 9-grid average at 18 z and dash-dot lines are 9-grid average at 17 Z, both centred at the Dorset grid. Observations are indicated by filled circles.

at the Dorset grid) for 17 Z and 18 Z. In the case of cloud ions (in μM) this is the weighted average by cloud liquid water content of each grids. The modelled cloud base is between 1.2 and 1.5 km and the cloud top extends beyond 4 km, which is consistent with cumulus type of cloud. However the liquid water content (< 0.1 g m^{-3}) is significantly smaller than those observed by the aircraft, which were 0.16, 0.58 and 1.74 g m^{-3} at 1.9, 2.8, and 4.3 km (AGL) respectively. The modelled liquid water content has been corrected for cloud fraction although at 42 km resolution this could represent an average of several cumulus cells. The modelled cloud ion concentrations compared with observation reasonably well at mid and upper levels but are over predicted at lower levels. This would at least be partly due to the significantly lower liquid water content in the model.

Vertical profiles of SO_2, H_2O_2, O_3 and pSO_4-2.5 are shown in Figure 2. These are to be compared to observed profiles in Liu et al. (1993, Figure 8) and Leaitch (1996, Figure 3). However note that the observed profiles were taken in clear air,

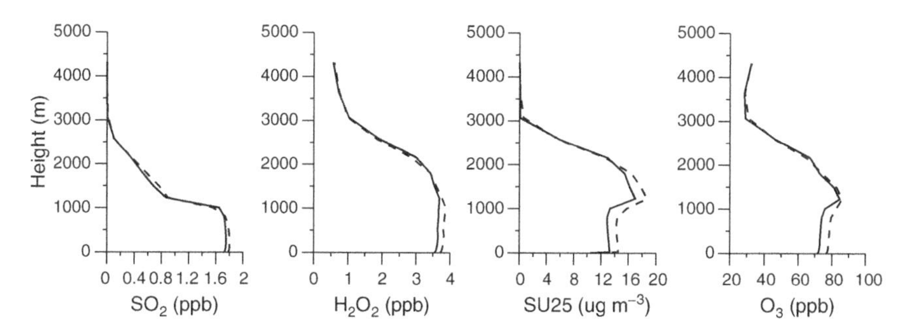

FIGURE 2. Vertical profiles of SO_2, H_2O_2, pSO_4-2.5μm, and O_3 at grid (42, 62) – Dorset grid at 18Z solid lines. Dashed lines are 9-grid average centred at the Dorset grid.

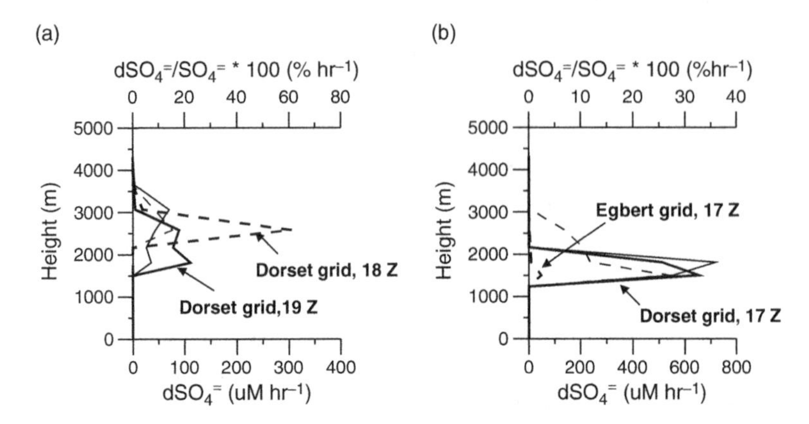

FIGURE 3. Modelled sulfate production rate ($dSO_4^=$) due to aqueous-phase oxidation: (a). August 5 case (flight 24); (b). August 6 case. (Thin lines are relative production rates (% hr-1).

while the modelled profiles reflect the state after the clouds are temporarily evaporated (at the end of an advection time step) in the model. Similar to observation, SO_2 is constant at 1.8 ppb (though lower than observed 4 ppb), below 1 km and decrease sharply with height above that level. H_2O_2 and O_3 also compared well with the measurement: H_2O_2 at about 3.5 ppb below cloud (compared to observed 2.5 ppb) and O_3 at 70-80 ppb (same as observed) for lowest 1.5 km. The observed O_3 profile shows the evidence of convective transport of below cloud air being brought up to the detrainment level near the cloud top (Leaitch, 1996). This is not seen from the model simulation, which is not too surprising since the model currently does not have the representation of sub-grid scale convective tracer mixing. The modelled particulate sulfate (2.5 μm) is at 14 μg m^{-3} at lowest levels which is somewhat lower than the observed 30 μg m^{-3} (Leighton et al., 1996). There is a marked increase in sulfate around 1.5 km which could be due to either, or a combination of, transport or/and in-cloud oxidation. Figure 3a shows the modelled sulfate production rate due to in-cloud oxidation (hourly averaged). The in-cloud production in this case is relative small, between 100 and 300 μM per hour on average, due to very low liquid water content in the model. This amounts to a 10-20% (per hr.) production of sulfate at lower part of the cloud. The production is relatively more significant higher up in the cloud where cloud-water sulfate concentration is lower. Liu et al., (1993) estimated that between 20 and 50% of cloud-water sulfate can be attributed to in-cloud due to oxidation based on their analysis of the data, which would imply a much greater production rate given the average life-time of a cumulus cell being much less than 1 hour.

4.2. Flight 25 (1551–1827 Z, August 6, 1988)

Flight 25 was conducted during the early afternoon on August 6, 1988. It consisted a clear-air profiling over Egbert to 2750 m (ASL) between 1621 and 1636 Z, cloud water collections at a altitude of 1980 m (ASL) on transit from Egbert to Dorset between 1643 and 1717 Z, and another clear-air profiling over Dorset between 1731 and 1813 Z. The cloud sampled in this flight is stratocumulus deck topped by a strong temperature inversion varying between 1800 and 2100 m (Liu et al., 1993).

Model predicted cloud water ion concentrations are shown in Figure 4 along with observations. The agreement between model and observation is good (at the level where measurements are taken). The model predicts higher cloud water ion concentrations at lower levels, more significantly over and near Dorset. There is a considerable variation in model predicted cloud water content between Egbert and Dorset which is not seen from the measurements. Modelled clouds are thinner than the August-5 case, with lower base. These are consistent with the stratocumulus type. The clear-air measurements indicate much cleaner air behind the cold front. This, however, is not quite reflected by the modelled profiles shown in Figure 5. SO_2 is lower than previous day but significantly higher than the observed levels

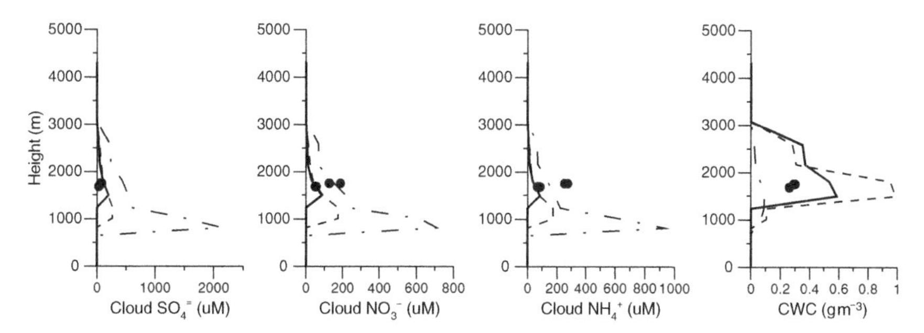

FIGURE 4. Vertical profiles of cloud $SO_4^{=}$, NO_3^-, NH_4^+, and liquid water content: solid lines – average between Egbert (42, 59) and Dorset (42, 62) at 17 Z on August 6; dashed lines are 9-grid average centred at the Egbert grid and dash-dot lines are 9-grid average centred at the Dorset grid, both at 17 Z. Observations are indicated by filled circles.

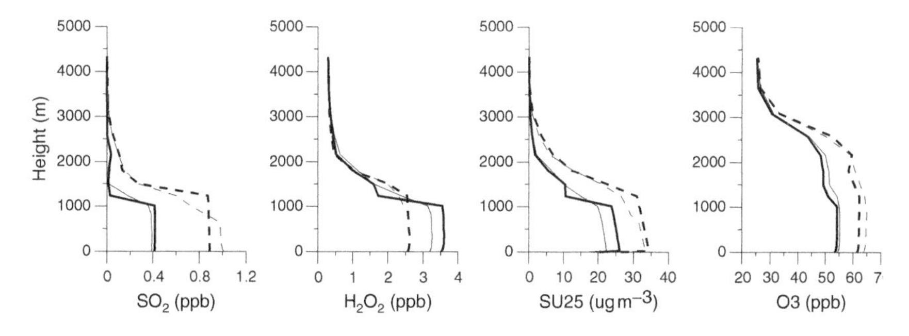

FIGURE 5. Vertical profiles of SO_2, H_2O_2, pSO_4-2.5μm, and O_3 at Egbert grid (dashed lines), between 16 and 17 Z, and Dorset grid (solid lines), at 18Z. (Thinner lines, solid and dashed, are for 9-grid averages).

(0.2-0.3 ppb near the surface). Model predicted ozone is much higher than observed at lower levels (~30 ppb from the observation) although the aircraft measurements did show much higher ozone above 2 km over Dorset. Modelled pSO_4 also seems to be too high. One possible reason for the model over-prediction in this case could be insufficient wet scavenging in the model due to significantly lower precipitation predicted by the meteorological model compared to observation (Gong et al., 2003). In-cloud sulfate production is much more significant in this case shown in Figure 3b.

5. Summary and Conclusions

In this study, we have presented a preliminary evaluation of cloud processes in a regional air quality model (AURAMS) against aircraft observations for two selected cases of different cloud type. A number of issues are involved in such an evaluation. One key issue is how model clouds compare with observations. For the convective (cumulus) case, as expected, the meteorological model, with the regional configuration used in the simulation, has more difficulties in getting the clouds at the right place and time. As seen the modelled cloud amount is significantly lower than the observations. The comparison is better in the case of the stratocumulus clouds behind a cold front (the second case) in terms of cloud water content. With regard to in-cloud production of sulfate, the model seems to under predict sulfate production in cloud in the cumulus cloud case compared to the estimate based on measurements, which may at least be partly attributed to the low cloud water amount predicted by the model. In contrast, the net impact of in-cloud production on regional ambient particulate sulfate may have been overestimated by the model due to insufficient wet removal. In this study we have focused on the mass issue. The impact of cloud processing on aerosol size spectrum is also an issue to be looked into in terms of model evaluation, which is not addressed in this study but should be in future studies. Our understanding of cloud processing of gases and aerosols can also benefit from more detailed measurements and cloud scale modelling, which will lead to improved representations of these processes in regional models.

Reference

Bouchet, V. S., M. D. Moran, L-P. Crevier, A. P. Dastoor, S. Gong,, W. Gong, P. A. Makar, S. Menard, B. Pabla and L. Zhang, 2004. Wintertime and summertime evaluation of the regional PM air quality model AURAMS, in "Air pollution modeling and its application XVI", Edit. C. Borrego and S. Incecik, Kluwer/Plenum publishers, 97-104.

Côté, J., Desmarais, J.-G., Gravel, S., Méthot, A., Patoine, A., Roch, M., and Staniforth, A., 1998a: The operational CMC-MRB Global Environment Multiscale (GEM) model. Part I: Design consideratios and formulation. *Monthly Weather Rev.*, 126, 1373-1395.

Gong, S. L., Barrie, L. A., Blanchet, J.-P., von Salzen, K., Lohmann, U., Lesins, G., Spacek, L., Zhang, L. M., Girard, E., Lin, H., Leaitch, R., Leighton, H., Chylek, P., and Huang, P., 2003: Canadian Aerosol Module: A size segregated simulation of atmospheric aerosol processes for climate and air quality models. 1. Module Development, *J. Geophys. Res.*, 108. D1, 4007.

Gong, W., A. P. Dastoor, V. B. Bouchét, S. Gong, P. A. Makar, M. D. Moran and B. Pabla, 2003: Cloud processing of gases and aerosols in a regional air quality model (AURAMS) and its evaluation against precipitation-chemistry data, Proceedings of the Fifth Conference on Atmospheric Chemistry: Gases, Aerosols, and Clouds, 2.3 (CD-ROM), American Meteorological Society, Boston.

Hansen, D. A., K. J. Puckett, J. J. Jansen, M. Lusis, and J. S. Vickery, 1991: The Eulerian model evaluation field study (EMEFS), in Proceedings of 7th Joint Conference on Applications of Air Pollution Meteorology with AWMA, pp. 58-62, Am. Meteorol. Soc., Boston, Mass.

Jones, A., D. L. Roberts, and J. Slingo, 1994: A climate model study of indirect radiative forcing by anthropogenic sulphate aerosols, *Nature*, 370, 450-453.

Leaitch, W.R., 1996. Observations pertaining to the effect of chemical transformation in cloud on the anthropogenic aerosol size distribution. *Aerosol Science Technology*, Vol. 25, 157-173.

Leaitch, W. R. G. A. Issac, J. W. Strapp, K. G. Anlauf, H. A. Wiebe, and J. I. MacPherson, 1991. Chemical and microphysical case studies of towering cumuli in Ontario, 1988. In the Proceedings of the AMS 7th Joint Conference of Applications of Air Pollution Meteorology with AWMA, 232-237, Jan. 14-18, New Orleans, Louisiana.

Leighton, H. G., L. Lauzon, and W. R. Leaitch, 1996. Evaluation of a three-dimensional cloud chemistry model. *Atmospheric Environment*, Vol. 30, no. 21, 3651-3665.

Liu, P. S. K., W. R. Leaitch, A. M. Macdonald, G. A. Isaac, J.W. Strapp, and H.A. Wiebe, 1993. Sulphate production in summer cloud over Ontario, Canada. *Tellus*, 45B, 368-389.

Macdonald, A. M., K. G. Anlauf, C. M. Banic, W. R. Leaitch, and H. A. Wiebe, 1995. Airborne measurements of aqueous and gaseous hydrogen peroxide during spring and summer in Ontario, Canada, *J. Geophys. Res.*, Vol. 100, No. D4, 7253-7262.

Macdonald, A. M., C. M. Banic, W. R. Leaitch, and K.J.Puckett, 1993. Evaluation of the eulerian acid deposition and oxidant model (ADOM) with summer 1988 aircraft data. Atmos. Environ., Vol. 27A, No. 6, 1019-1034.

Makar, P. A., V. S. Bouchet, and A. Nenes, 2003 : Inorganic chemistry calculations using HETV – a vectorized solver for the SO42–NO3–NH4+ system based on ISOROPIA algorithms, *Atmos. Environment*, 37, 2279-2294.

Makar, P. A., V. Bouchet, L. P. Crevier, S. Gong, W. Gong, S. Menard, M. Moran, B. Pabla, S. Venkatesh, 2004a. AURAMS runs during the Pacific2001 time period–a model/measurement comparison., in "Air pollution modeling and its application XVI", Edit. C. Borrego and S. Incecik, Kluwer/Plenum publishers, 153-160.

Makar, P. A., V. S. Bouchet, W. Gong, M. D. Moran, S. Gong, A.P. Dastoor, K. Hayden, H. Boudries, J. Brook, K. Strawbridge, K. Anlauf, S.M. Li, 2004b. AURAMS/Pacific2001 measurement intensive comparison (these proceedings).

Moran, M. D., A. P. Dastoor, S.-L. Gong, W. Gong and P. A. Makar, 1998. Conceptual design for the AES unified regional air quality modelling system. Air Quality Research Branch, Meteorological Service of Canada, Downsview, Ontario M3H 5T4, Canada.

Venkatram, A., P. K. Karamchandani and P. K. Misra, 1988: Testing a comprehensive acid deposition model. *Atmos. Environment*, 22, 737-747.

Zhang, L., M. D. Moran, P. A. Makar, J. R. Brook and S. Gong, 2002: Modelling gaseous dry deposition in AURAMS – A Unified Regional Air-quality Modelling System. *Atmos. Environment*, 36, 537-560.

60 Evaluation of an Annual Simulation of Ozone and Fine Particulate Matter over the Continental United States – Which Temporal Features are Captured?

C. Hogrefe, J. M. Jones, A. Gilliland, P. S. Porter, E. Gego, R. Gilliam, J. Swall, J. Irwin, and S. T. Rao*

1. Introduction

Motivated by growing concerns about the detrimental effects of fine particulate matter ($PM_{2.5}$) on human health, the U.S. Environmental Protection Agency (EPA) recently promulgated a National Ambient Air Quality Standard (NAAQS) for $PM_{2.5}$. The $PM_{2.5}$ standard includes a 24-hour limit (65 $\mu g/m^3$ for the 98th percentile) and annual (15 $\mu g/m^3$) limit. Except for a few cases, the annual standard will be the primary concern for attainment issues. Over the next several years, grid-based photochemical models such as the Community Multiscale Air Quality (CMAQ) model (Byun and Ching, 1999) will be used by regulatory agencies to design emission control strategies aimed at meeting and maintaining the NAAQS for O_3 and $PM_{2.5}$. The evaluation of these models for a simulation of current conditions is a necessary prerequisite for using them to simulate future conditions. The evaluation presented in this study focuses on determining the temporal patterns in all components of the modeling system (meteorology, emissions and air quality) and comparing them against available observations. Furthermore, we briefly investigated the weekday/weekend differences in the observed and predicted pollutant concentrations and outlined steps for future research. Since anthropogenic emissions are known to have a distinct weekly cycle, such analyses would help us in evaluating the modeling system's ability to accurately reproduce the observed response to emission changes.

* C. Hogrefe and J.M. Jones, ASRC, University at Albany, Albany, NY, U.S.A.
A. Gilliland, R. Gilliam, J. Swall, J. Irwin, and S. T. Rao, NOAA Atmospheric Sciences Modeling Division, On Assignment to the U.S. Environmental Protection Agency, Research Triangle Park, NC, U.S.A.
P.S. Porter, University of Idaho, Idaho Falls, ID, U.S.A. E. Gego, University Corporation for Atmospheric Research, Idaho Falls, ID, U.S.A.

2. Models and Database

Meteorological fields for the photochemical simulations were prepared by the MM5 model (Grell et al., 1994) version 3.6.1 over the continental United States at a horizontal resolution of 36 km for the time period from January 1 – December 31, 2001 (McNally, 2003). The MM5 fields were then processed by the Meteorology-Chemistry Interface Program (MCIP) version 2.2. Emissions were processed by the SMOKE processor (Carolina Environmental Programs, 2003) which incorporated the MOBILE6 module (U.S. EPA, 2003) for mobile source emissions and the BEIS3.12 model for biogenic emissions (http://www.epa.gov/asmdnerl/biogen.html). The emission inventory was based on the USEPA National Emissions Inventory for 2001, which relies on state reported values. The seasonality of the ammonia emissions, an important consideration for prediction of $PM_{2.5}$, was estimated based on seasonal information from Gilliland et al. (2003) and Pinder et al. (2004). These meteorological and emission fields were then provided as input to two photochemical models, namely CMAQ (February 2004 version) and REMSAD version 7.061 (ICF Consulting, 2002), both run at a resolution of 36 km over the continental United States. Chemical boundary conditions for both models were prepared from a global simulation with the GEOS-CHEM model (Bey et al., 2001).

This study utilizes a variety of observations from different networks. Observations of surface temperature and wind speed were retrieved from the TDL data set maintained by the Data Support Section at the National Center for Atmospheric Research (NCAR-DSS). Hourly surface ozone observations, hourly $PM_{2.5}$ concentrations measured by tapered element oscillation microbalance (TEOM) monitors (http://www.rpco.com/products/ambprod/amb1400/), and 24-hr average $PM_{2.5}$ concentrations measured at monitors following the Federal Reference Method (FRM) protocol were retrieved from EPA's Air Quality System (AQS) database (http://www.epa.gov/air/data/aqsdb.html). Speciated $PM_{2.5}$ measurements were obtained from the Interagency Monitoring to Protect Visual Environments (IMPROVE) network, the Clean Air Status and Trends Network (CASTNet), and Speciated Trends Network (STN). Because of differences in measurement techniques and instrumentation, sampling frequencies, and site location criteria, model performance was calculated on a species-by-species and network-by-network basis. The analysis presented in this paper focuses on the Eastern United States. Monitoring sites were only included in the analysis if at least 70% of the data were available.

Following the approach outlined in Rao et al. (1997), Hogrefe et al. (2000) and Hogrefe et al. (2001), a spectral decomposition technique was applied to compare temporal variations in observed and predicted time series. To this end, time series of meteorological variables and pollutant concentrations were spectrally decomposed into fluctuations occurring on the intraday (time period less than 12 hours), diurnal (12-48 hours), synoptic (2-21 days) and baseline (greater than 21 days) time scales using the Kolmogorov-Zurbenko (KZ) filter as described in Hogrefe et al. (2000). Note that the intraday and diurnal components could only be estimated for variables measured hourly, while the synoptic and baseline components could be estimated for variables measured hourly, daily or weekly. All analyses presented in this paper were performed over the entire annual cycle from January 1–December 31, 2001, with the exception of the ozone weekday/weekend analysis which was performed for June 1–August 31, 2001.

3. Results and Discussion

3.1. Correlations on Different Time Scales

Correlations between different temporal components embedded in time series of the observed and predicted variables were computed for temperature, wind speed and ozone as well as total and speciated $PM_{2.5}$ from the different measurement networks (Tables 1a-b). The correlations were computed at each site for a given variable/network/model combination, and Tables 1a-b list the median value of the correlation across all sites for a given variable/network/model combination. For the meteorological variables (temperature and wind speed), correlations increase

TABLE 1(a). Correlations between different temporal components embedded in hourly time series of observed and predicted temperature, wind speed, ozone and total $PM_{2.5}$. Median values are shown for each network/variable.

	#Sites	Intra-day	Diurnal	Synoptic	Baseline
Temperature TDL/MM5	738	0.18	0.90	0.95	0.99
Wind Speed TDL/MM5	735	0.02	0.60	0.84	0.90
O_3 AQS/CMAQ	193	0.07	0.70	0.64	0.87
$PM_{2.5}$ TEOM	67				
CMAQ REMSAD		0.01 0.03	0.25 0.25	0.70 0.63	0.04 0.10

TABLE 1(b). Correlations between different synoptic and baseline components embedded in time series of observed and predicted $PM_{2.5}$ from different networks. Median values are shown for each network/variable.

		Synoptic		Baseline	
	#Sites	CMAQ	REMSAD	CMAQ	REMSAD
$PM_{2.5}$ FRM (daily)	938	0.68	0.65	0.60	0.51
$PM_{2.5}$ STN (daily)	25	0.60	0.63	0.38	0.35
SO_4 Improve (daily)	44	0.77	0.70	0.89	0.77
SO_4 CASTnet (weekly)	48	0.85	0.72	0.94	0.88
SO_4 STN (daily)	23	0.72	0.70	0.85	0.74
NO_3 Improve (daily)	44	0.46	0.54	0.88	0.78
NO_3 CASTnet (weekly)	48	0.51	0.46	0.89	0.83
NO_3 STN (daily)	23	0.39	0.42	0.83	0.66
NH_4 CASTnet (weekly)	48	0.71	0.72	0.55	0.45
NH_4 STN (daily)	23	0.63	0.66	0.52	0.37
EC STN (daily)	23	0.41	0.39	0.15	0.32
OC STN (daily)	22	0.48	0.55	0.24	0.28
Crustal STN (daily)	23	0.34	0.29	–0.35	–0.39

with increasing time scale, i.e. correlations are lowest for the intra-day component ($r < 0.2$) and highest for the baseline component ($r > 0.9$).

While correlations are relatively high for the diurnal component ($r > 0.6$), part of this correlation is due to the inherent cyclical nature of this component, and correlations are lower when the time series of the diurnal amplitudes are considered (not shown). It is not surprising that the correlation is highest on the synoptic and baseline time scale since MM5 model predictions were nudged towards analysis fields using 4-Dimensional Data Assimilation techniques. For ozone, correlations follow a similar pattern as temperature and wind speed, with correlations on the intraday time scale being less than 0.1 and correlations on the baseline time scale being 0.87. The results presented here are consistent with those presented in Hogrefe et al. (2001) who analyzed ozone from a three-months summertime simulation over the Eastern United States.

Except for hourly measurements of total $PM_{2.5}$ by TEOM instruments retrieved from EPA's AQS, all $PM_{2.5}$ measurements analyzed in this study are based on filter samples of either 24-hr average or 7-day average concentrations. Consequently, the intra-day and diurnal components could only be estimated for the comparison of CMAQ and REMSAD model predictions with total $PM_{2.5}$ measurements by TEOM instruments. It is striking that the correlations between TEOM observations and model predictions are poor on the diurnal and baseline components for both CMAQ and REMSAD. Figures 1 a and b show the average observed and predicted diurnal cycles and the time series of the observed and predicted baseline components averaged over all TEOM monitors and corresponding model grid cells. Although the time of occurrence of maxima are simulated well, there is a large difference in the amplitude of the diurnal forcing; both models

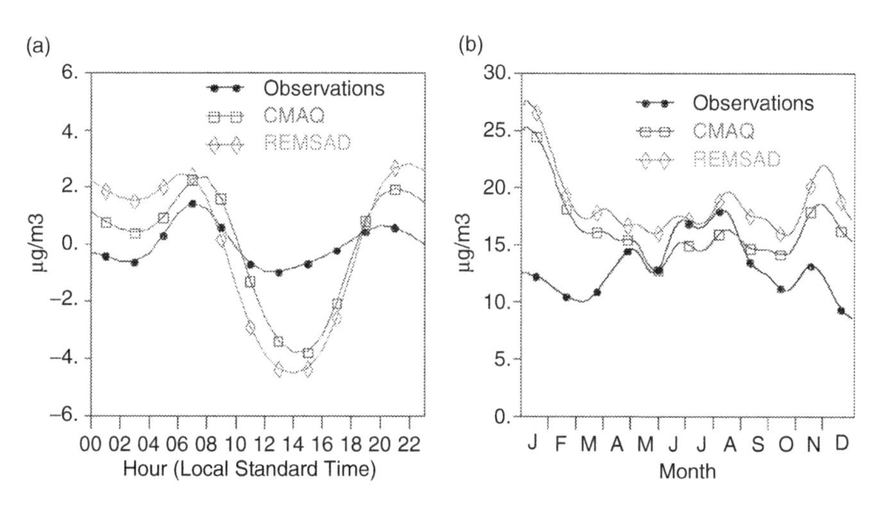

FIGURE 1. (a) Average observed and predicted diurnal cycles and (b) time series of the observed and predicted baseline components averaged over all TEOM monitors and corresponding model grid cells for $PM_{2.5}$.

overestimate observations during nighttime and severely underestimate observations during daytime hours. Plausible reasons for this discrepancy are misrepresentations of the strength of vertical mixing in the model or the magnitude of primary $PM_{2.5}$ emissions from area and mobile sources. Figure 1b illustrates that the low correlations on the baseline time scale stem from the higher $PM_{2.5}$ predictions by both models during wintertime while TEOM measurements show a decrease. Part of this decrease in TEOM measurements during wintertime is likely caused by the high operating temperatures of most of the currently-deployed TEOM instruments (30°C and 50°C). Volatilization losses can occur when the sample is heated from ambient temperature to the operating temperature, and such losses tend to be higher during colder ambient temperatures (Allen et al., 1997). On the other hand, CMAQ and REMSAD utilize MM5-simulated temperatures to calculate the partitioning between gas and particle phase. In other words, there is an inherent difference between measurement technique and modeling approach, and this difference exhibits seasonality, thereby affecting the baseline comparisons. Support for this explanation comes from the higher baseline correlations when CMAQ and REMSAD are compared against $PM_{2.5}$ filter observations from FRM monitors and the STN network. This highlights the importance of conducting $PM_{2.5}$ model evaluation on a network-by-network basis. In other words, data from different air monitoring networks should not be combined into a single dataset for the purpose of model evaluation.

Correlations between the synoptic and baseline components of sulfate measured by the IMPROVE, CASTNet, and STN networks, and predicted by CMAQ and REMSAD are consistently greater than 0.7, with baseline correlations exceeding 0.85 for CMAQ and 0.74 for REMSAD. It is noteworthy that there is relatively little difference in model performance across the different networks, a finding that is consistent with the regional-scale nature of sulfate concentrations in the eastern United States that has also been discussed by Gego et al. (2004). Furthermore, correlations for REMSAD are consistently lower than those for CMAQ for this pollutant across all networks. For nitrate, correlations on the baseline time scale are similar to those for sulfate, but correlations on the synoptic time scale are lower. For the baseline, CMAQ correlations are consistently higher than those for REMSAD.

In contrast to baseline correlations for sulfate and nitrate, correlations are relatively low for ammonium. A likely contributor to these lower correlations is the seasonal characterization of NH_3 emissions. The seasonality for NH_3 emissions is a well-known uncertainty that is currently being investigated from both bottom-up inventory development and from top-down estimation methods (Gilliland et al., 2003). An inverse modeling study is underway using this 2001 annual simulation to consider how the current seasonality estimates for NH_3 emissions should be modified to improve model predictions of ammonium aerosols and wet deposition.

Model predicted concentrations of elemental carbon (EC), organic carbon (OC), and crustal material are strongly influenced by emissions of primary PM since there is no secondary formation mechanism for EC and crustal material in

CMAQ and REMSAD. Consequently, the relatively weak correlations between the observed and predicted baseline components for these species point to potential problems in the temporal allocation of PM emissions during emission processing. To investigate this issue, we constructed the baseline component of EC observations, CMAQ and REMSAD predictions, and total $PM_{2.5}$ emissions at several STN monitoring locations. Examples of this analysis are shown in Figures 2a and b. The strong correlation between $PM_{2.5}$ emissions and model-predicted EC concentrations is clearly visible at the Decatur, GA monitor and, to a slightly lesser extent, at the Bronx, NY monitor. In both cases, the relatively poor correlation between observations and model predictions seems to be largely driven by temporal signature of the $PM_{2.5}$ emissions. Therefore, in order to improve model performance on longer time scales for primary species such as EC and crustal material, it is necessary to improve the temporal characterization of primary $PM_{2.5}$ emissions.

In summary, the results presented in Tables 1a-b illustrate that the models exhibit greatest skills at capturing longer-term (seasonal) fluctuations for temperature, wind speed, ozone, sulfate and nitrate. For total $PM_{2.5}$, ammonium, EC, OC and crustal $PM_{2.5}$, correlations are highest for the synoptic time scale, implying problems with factors other than meteorology in capturing the baseline fluctuations. For the variables for which hourly measurements were available, correlations were insignificant on the intraday time scale, suggesting that these models are not skillful in simulating the shorter-term variations in pollutant levels.

3.2. Analysis of the Weekday/Weekend Effect

Anthropogenic emissions of NO_x and VOC are reduced on weekends due to reduced traffic as well as industrial and commercial activities. The impact of these cyclical reductions of precursors on ozone has been the subject of numerous studies (e.g. Cleveland et al., 1974; Croes et al., 2003). Many studies found ozone

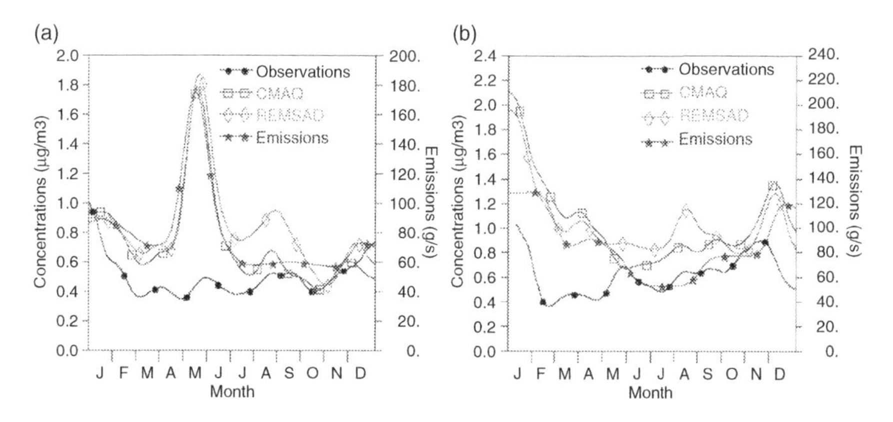

FIGURE 2. Baseline of EC observations, CMAQ and REMSAD predictions, and $PM_{2.5}$ emissions (a) Decatur, Georgia (b) Bronx, New York.

increases on weekends in urban core areas and laid out several hypotheses explaining this behavior, including reduced O_3 titration by NO and more efficient ozone production on weekends. One of the main motivations for such studies is to infer the likely response of ozone concentrations to emission control policies from these cyclical real-world emission reductions. While there is no direct way to evaluate photochemical model responses to hypothetical emission control scenarios, analysis of the weekday/weekend effect could provide a tool to evaluate the modeling system's ability to accurately reproduce the observed response to emission changes. While performing such analysis in sufficient detail is beyond the scope of this paper and will be presented in future work, we briefly outline some of the necessary steps in performing this analysis.

As a first step, it is necessary to establish that the weekday/weekend effect indeed exists during the period of analysis in both observations and model predictions. To this end, average weekly cycles need to be computed for each station, and the difference between average weekday (Monday-Friday) and weekend (Saturday-Sunday) concentrations needs to be determined. As an example, Figure 3 presents a scatter plot of the average CMAQ predicted difference between weekend daily maximum 1-hr ozone concentrations and weekday daily maximum 1-hr ozone concentrations versus the corresponding difference computed from observations at the same location. This figure illustrates that a weekday/weekend cycle of comparable magnitude is indeed present in observed and CMAQ-predicted ozone concentrations during the summer of 2001. Additional analyses are needed to determine the location of stations that show opposite magnitudes of the weekday/weekend differences, to perform analysis on early morning concentrations (when the difference in motor vehicle emissions between weekdays and weekends is most pronounced), to

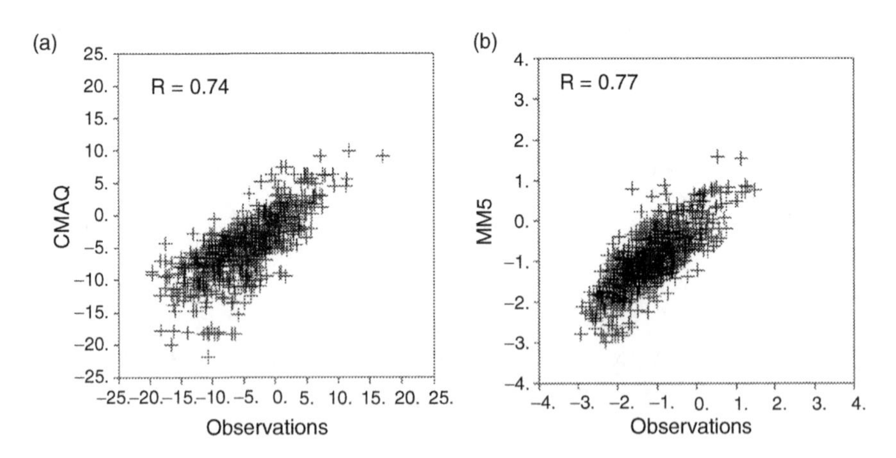

FIGURE 3. (a) Scatter plot of the average CMAQ predicted difference between weekend daily maximum 1-hr ozone concentrations and weekday daily maximum 1-hr ozone concentrations versus the corresponding difference computed from observations at the same location. (b) As in (a), but for daily maximum temperature.

restrict analysis to high ozone days only, and to include analysis of ozone precursors as well. Furthermore, it is important to ascertain whether the weekday versus weekend differences in ozone concentrations are caused by emission fluctuations or might be largely explained by meteorological effects when analysis is restricted to a single summer season only. Indeed, Figure 3b shows that for the summer of 2001 there was a distinct weekday/weekend fluctuation in temperature at most monitors in the eastern U.S. Most monitors show lower daily maximum temperatures on weekends than on weekdays, and MM5 captures this behavior. Because temperature both directly influences the rate of ozone formation and serves as a proxy for other meteorological parameters conducive to ozone formation, this figure illustrates that the existence of a weekday/weekend cycle in ozone concentrations for the summer of 2001 (Figure 3a) can not unequivocally be attributed to cyclical changes in precursor emissions. In the absence of longer time records of observed and predicted ozone concentrations, one might be able to apply statistical techniques to adjust observed and predicted ozone concentrations for meteorological variability to isolate the effect of emission reductions. Also, it is important to include an analysis of weekday/weekend differences in the processed emission files in the overall evaluation of the modeling system's ability to simulate this phenomenon.

Acknowledgments The Department of Commerce partially funded the research described here under contracts with Dr. E. Gego (EA133R-03-SE-0710), with the University of Idaho to Dr. P. S. Porter (EA133R-03-SE-0372), and with the State University of New York to Dr. C. Hogrefe (EA133R-03-SE-0650).

Disclaimer The research presented here was performed under the Memorandum of Understanding between the U.S. Environmental Protection Agency (EPA) and the U.S. Department of Commerce's National Oceanic and Atmospheric Administration (NOAA) and under agreement number DW13921548. Although it has been reviewed by EPA and NOAA and approved for publication, it does not necessarily reflect their policies or views.

References

Allen, G., C. Sioutas, P. Koutrakis, R. Reiss, F. W. Lurmann, and P. T. Roberts, 1997: "Evaluation of the TEOM method for measurement of ambient particulate mass in urban areas, *J. Air Waste Manage. Assoc.*, 47, 682-689.

Bey, I., D. J. Jacob, R. M. Yantosca, J. A. Logan, B. D. Field, A. M. Fiore, Q. Li, H. Y. Liu, L. J. Mickley, and M. G. Schultz, 2001: Global modeling of tropospheric chemistry with assimilated meteorology: Model description and evaluation, J. Geophys. Res., 106(D19), 23073-23096, 10.1029/2001JD000807.

Byun, D. W. and Ching, J. K. S. (eds.), 1999. Science algorithms of the EPA Models-3 Community Multiscale Air Quality Model (CMAQ) modeling system. *EPA/600/*

R-99/030, U. S. Environmental Protection Agency, Office of Research and Development, Washington, DC 20460.

Carolina Environmental Programs, 2003: Sparse Matrix Operator Kernel Emission (SMOKE) Modeling System, University of Carolina, Carolina Environmental Programs, Research Triangle Park, NC.

Cleveland W. S., T. E. Graedel, B. Kleiner, and J. L. Warner, 1974: Sunday and workday variations in photochemical air pollutants in New Jersey and New York. Science, 186, 1037-1038.

Croes, B. E., L. J. Dolislager, L. Larsen, and J. N. Pitts, 2003. Forum – the O3 "weekend effect" and NO_x control strategies – scientific and public health findings and their regulatory implications. Environ. Manag., July 2003, 27-35.

Gego, E., P. S. Porter, J. S. Irwin, C. Hogrefe, and S. T. Rao, 2004: Assessing the comparability of ammonium, nitrate and sulfate concentrations measured by three air quality monitoring networks, Pure Appl. Geoph., in press.

Gilliland A. B., R. L. Dennis, S. J. Roselle, and T. E. Pierce, 2003: Seasonal NH_3 emission estimates for the eastern United States based on ammonium wet concentrations and an inverse modeling method, J. Geophys. Res., 108 (D15), 4477, doi:10.1029/2002JD003063

Grell, G. A., J. Dudhia, and D. Stauffer, 1994: A description of the fifth-generation Penn State/NCAR Mesoscale Model (MM5). *NCAR Technical Note*, 138 pp., TN-398 + STR, National Center for Atmospheric Research, Boulder, CO.

Hogrefe, C., S. T. Rao, I. G. Zurbenko, and P. S. Porter, 2000: Interpreting the information in ozone observations and model predictions relevant to regulatory policies in the eastern United States; Bull. Amer. Meteor. Soc., 81, 2083-2106.

Hogrefe, C., S. T. Rao, P. Kasibhatla, W. Hao, G. Sistla, R. Mathur, and J. McHenry 2001: Evaluating the performance of regional-scale photochemical modeling systems: Part II – ozone predictions. Atmos. Environ, 35, 4175-4188.

ICF Consulting, 2002: User's Guide to the Regional Modeling System for Aerosols and Deposition (REMSAD) Version 7. Systems Applications International/ICF Consulting, San Rafael, CA 94903. 153 pages.

McNally, D., 2003: Annual application of MM5 for calendar year 2001, prepared for U.S. EPA by D. McNally, Alpine Geophysics, Arvada, CO, March 2003, 179 pp.

Pinder R. W., R. Strader, C. I. Davidson and P. J. Adams, 2004: A temporally and spatially resolved ammonia emission inventory for dairy cows in the United States, *Atmos. Env.*, 38, 3747-3756.

Rao, S. T., I. G. Zurbenko, R. Neagu, P. S. Porter, J.-Y. Ku, and R. F. Henry, 1997: Space and time scales in ambient ozone data. Bull. Amer. Meteor. Soc., 78, 2153-2166.

U.S. Environmental Protection Agency, 2003: User's guide to MOBILE6.1 and MOBILE6.2; EPA Office of Air and Radiation, EPA420-R-03-010, Assessment and Standards Division, Office of Transportation and Air Quality, U.S. Environmental Protection Agency. 262 pp.

61
Evaluation of CMAQ PM Results Using Size-resolved Field Measurement Data: The Particle Diameter Issue and Its Impact on Model Performance Assessment

Weimin Jiang, Éric Giroux, Helmut Roth, and Dazhong Yin†

1. Introduction

One uncertainty in evaluating particulate matter (PM) modelling results is caused by mismatches among regulated, measured, and modelled particle diameters. Current PM regulations are based on the aerodynamic diameter (EPA, 2004). PM measurements also target the aerodynamic diameter in order to address regulatory concerns. However, in air quality models, particles are often modelled on the basis of the Stokes diameter. To evaluate model performance using size–resolved measurement data, it seems logical that modelling results should be given in the aerodynamic diameter.

This paper focuses on the quantitative uncertainty associated with representing PM modelling results in these two different types of particle diameters. The objective is to understand differences between modelled PM concentrations expressed in the two types of diameters and the impact of the differences on model performance assessment.

For demonstration purposes, a modified version (Jiang and Roth, 2002) of the Community Multiscale Air Quality Model (CMAQ), which was originally developed by the Environmental Protection Agency of the United States (EPA) (Binkowski and Roselle, 2003), was used in this study. CMAQ simulations were conducted for an historical air pollution episode from July 31 to August 7, 1993 in the Lower Fraser Valley (LFV) domain, which covers southwest British Columbia, Canada, and northwest Washington State, United States, as shown in Fig. 1.

† Weimin Jiang, Éric Giroux, Helmut Roth, Dazhong Yin, National Research Council of Canada, 1200 Montreal Road, Ottawa, Ontario, Canada, K1A 0R6.

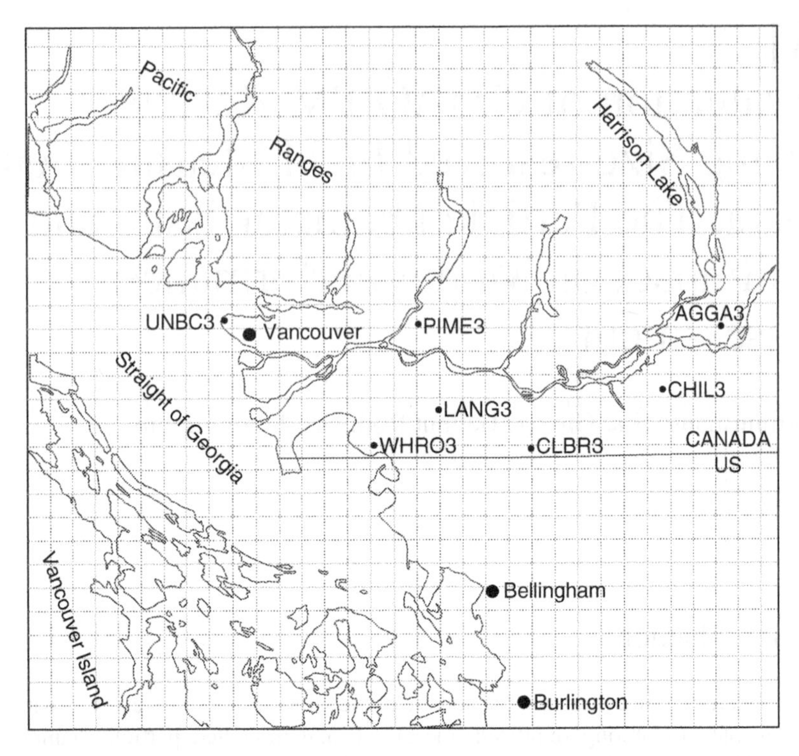

FIGURE 1. LFV modelling domain and 7 PM measurement sites. Horizontal grid resolution is 5 km × 5 km.

General model setup and preparation of emissions, meteorology, initial and boundary conditions for the model simulations are available in Jiang et al. (2004) and Yin et al. (2004).

2. Impact of the Diameter Type on PM_X Modelling Results

2.1. lnD_{st}– and D_{st}–based Particle Size Distribution Functions

CMAQ models particle number distributions using three superimposed log–normal distribution functions for nucleation (i), accumulation (j), and coarse (c) mode particles. Particles are treated as spheres with varying densities and physical diameters. The particle density in each mode at a particular time and location is determined by the chemical composition of the particles in the mode and the densities of the particle components (Jiang and Roth, 2003). For a spherical particle, the physical diameter is equivalent to the Stokes diameter, D_{st}, which is defined as the diameter of a sphere having the same terminal settling velocity and density as the particle (Seinfeld and Pandis, 1998).

For each mode, the $\ln D_{st}$–based particle number and mass distribution functions and the D_{st}–based particle mass distribution function are

$$n(\ln D_{st}) = \frac{dN}{d(\ln D_{st})} = \frac{N_0}{\sqrt{2\pi}\ln\sigma_g}\exp\left[-\frac{1}{2}\left(\frac{\ln D_{st} - \ln D_{g,st}}{\ln\sigma_g}\right)^2\right], \quad (1)$$

$$m(\ln D_{st}) = \frac{dMass}{d(\ln D_{st})} = \frac{\sqrt{\pi/2}\rho N_0}{6\ln\sigma_g} D_{st}^3 \exp\left[-\frac{1}{2}\left(\frac{\ln D_{st} - \ln D_{g,st}}{\ln\sigma_g}\right)^2\right], \quad (2)$$

$$m(D_{st}) = \frac{dMass}{d(D_{st})} = \frac{\sqrt{\pi/2}\rho N_0}{6\ln\sigma_g} D_{st}^2 \exp\left[-\frac{1}{2}\left(\frac{\ln D_{st} - \ln D_{g,st}}{\ln\sigma_g}\right)^2\right], \quad (3)$$

where N_0 is the total number concentration of all particles in the distribution, $D_{g,st}$ is the geometric mean diameter, σ_g is the geometric standard deviation, and ρ is the average density of particles in the mode.

CMAQ outputs mass concentrations of 21 particle components, number concentrations of the 3 particle modes, and surface area concentrations of 2 particle modes (Jiang and Roth, 2003). We calculate $D_{g,st}$ and σ_g from these quantities using mathematical formulas implemented in a postprocessor (Jiang and Yin, 2001; 2002).

As an example, Fig. 2 shows the modelled mass distribution functions based on $\ln D_{st}$ and D_{st} for the three individual particle modes and their superposition at 16:00, August 2, 1993, in the grid cell where the AGGA3 measurement site is located.

2.2. D_{st} as a function of D_{ca}

The aerodynamic diameter is defined as the diameter of a unit density (1 g·cm^{-3}) sphere that has the same settling velocity as the particle. It is called classical aerodynamic diameter, D_{ca}, in Seinfeld and Pandis (1998). D_{st} as a function of D_{ca},

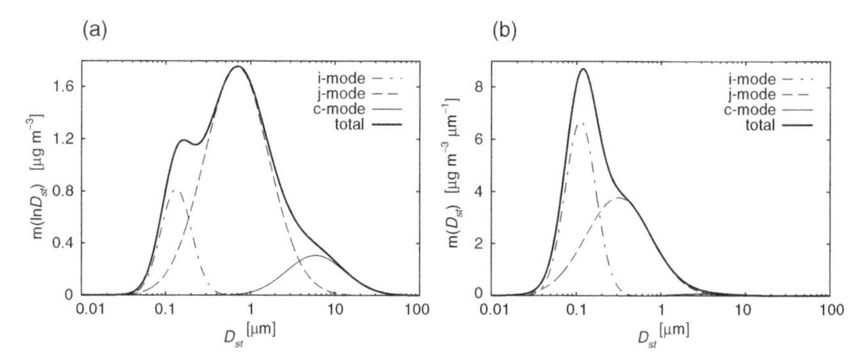

FIGURE 2. (a) $\ln D_{st}$– and (b) D_{st}–based mass distribution functions at AGGA3, 16:00, August 2, 1993, derived from CMAQ output by the PM_x post–processor of Jiang and Yin (2001).

$$D_{st} = \frac{1}{2}\left(\sqrt{B^2 + \frac{4\rho_0 D_{ca}(D_{ca}+B)}{\rho}} - B\right), \tag{4}$$

was obtained by combining Eq. (8.106) in Seinfeld and Pandis (1998) and an empirical relationship between the slip correction factor C_c and the particle diameter D_p

$$C_c = 1 + B/D_p, \tag{5}$$

where D_p is D_{st} or D_{ca} depending on the particle diameter type in question; ρ is particle density; ρ_0 is unit density (1 g·cm^{-3}); B is a function of the mean free path of air molecules. As an approximation, B is treated as a parameter and its best value of 0.21470 μm was obtained by fitting the C_c and D_p values in Table 8.3 of Seinfeld and Pandis (1998) with Eq. 5.

Fig. 3. shows graphically the relationship between D_{st} and D_{ca} for ρ = 1.5, 2.0, and 2.5 g·cm^{-3}. From the ratios of D_{st} and D_{ca} in the figure, it is clear that D_{st} is always smaller than D_{ca} for the same particle, since $\rho > 1$ g·cm^{-3} for all these cases.

2.3. Modelled PM_x Mass Concentrations Based on D_{st} and D_{ca}

For a mode k, the mass concentration of all particles smaller or equal to a D_{st}–based cut–off diameter x_{st} is (Jiang and Yin, 2001)

$$PM_{x,st,k} = \frac{\pi}{12}\rho_k N_{0,k} D^3_{g,st,k} \exp\left(\frac{9}{2}\ln^2\sigma_{g,k}\right) \left[1 + erf\left(\frac{\ln x_{st} - \ln D_{g,st,k}}{\sqrt{2}\ln\sigma_{g,k}} - \frac{3\sqrt{2}}{2}\ln\sigma_{g,k}\right)\right], \tag{6}$$

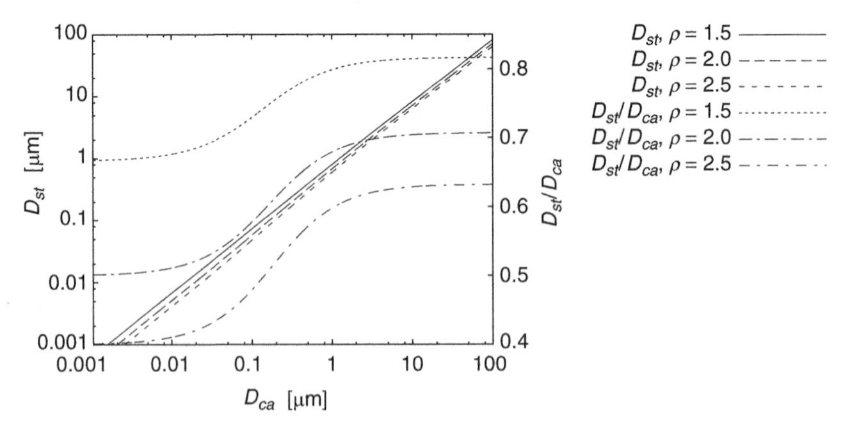

FIGURE 3. Relationship between D_{st} and D_{ca} for ρ = 1.5, 2.0, and 2.5 g·cm^{-3}.

where *erf* is the error function. Note that the subscript k indicates the mode.

To calculate the mass concentration, $PM_{x,ca,k}$, for all particles smaller or equal to a D_{ca}–based cut–off diameter x_{ca}, a value of x_{st} equivalent to x_{ca} is calculated using Eq. 4. Then $PM_{x,ca,k}$ is calculated using Eq. 6 and the equivalent x_{st} value.

Fig. 4(a) reveals graphically the differences between D_{st}– and D_{ca}–based mass concentrations under two example cut–off diameters x = 2.50 and 0.82 μm, respectively, for j–mode particles, at the same time and location as in Fig. 2. For x_{ca} = 2.50 and 0.82 μm, their equivalent x_{st} values are 1.87 and 0.60 μm, respectively, according to Eq. 4 (ρ = 1.746 g·cm^{-3} for j–mode particles at the time and location). Therefore, areas A and B in Fig. 4 represent the differences between D_{st}– and D_{ca}–based mass concentrations under the two example cut–off diameters, respectively, for the j–mode particles.

Applying the same calculations to all three particle modes, we obtain the differences in total mass concentrations caused by the diameter type change under the two cut–off diameters: $PM_{0.82,st}$ - $PM_{0.82,ca}$=0.57 μg·m^{-3}, and $PM_{2.5,st}$ -$PM_{2.5,ca}$ = 0.27 μg·m^{-3}. Note that the difference in $PM_{0.82}$ is substantially greater than the difference in $PM_{2.5}$. Relative differences between $PM_{x,st}$ and $PM_{x,ca}$, calculated as $(PM_{x,st}$-$PM_{x,ca})$ / $PM_{x,ca}$ × 100%, are 21.00% and 6.11% for x = 0.82 μm and 2.50 μm, respectively.

For the same example, Fig. 4(b) shows absolute and relative differences between $PM_{x,st}$ and $PM_{x,ca}$ as functions of the cut–off diameter x. The absolute difference curve has two peaks in the x range between 0.1 and 1 μm. The relative difference between $PM_{x,st}$ and $PM_{x,ca}$ increases when the cut–off diameter decreases. Note that the relative differences can reach hundreds and even thousands of percent at small particle size cut–off values.

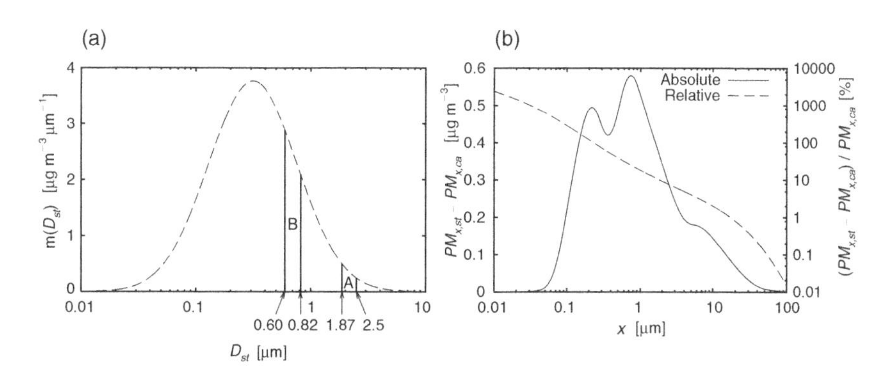

FIGURE 4. Differences between D_{ca}– and D_{st}–based particle mass concentrations at the same time and location as in Fig. 2. (a) conceptual difference at two cut–off diameters, x = 2.50 and 0.82 μm, for j–mode only; (b) absolute and relative differences between total $PM_{x,st}$ and $PM_{x,ca}$ for all three modes as functions of the cut–off diameter x.

3. Correlation and Differences Between $PM_{x\,st}$ and $PM_{x,ca}$ Throughout the Modelling Domain and Period

Fig. 5 presents linear correlations between $PM_{x,st}$ and $PM_{x,ca}$ at five cut–off diameters 0.01, 0.1, 1, 2.5, and 10 μm for all grid cells and modelling hours. In general, $PM_{x,st}$ and $PM_{x,ca}$ correlate well with very high R^2 values for larger x and more moderate R^2 for smaller x. Equations for the least–square fit lines indicate that $PM_{x,st}$ values are approximately 3 to 4 times the $PM_{x,ca}$ values for $x = 0.01$ and 0.10 μm. The ratios between $PM_{x,st}$ and $PM_{x,ca}$ approaches 1 when x becomes larger. This is consistent with the trend of relative differences between $PM_{x,st}$ and $PM_{x,ca}$ shown in the example of Fig. 4(b).

Differences between $PM_{x,st}$ and $PM_{x,ca}$ vary with time. Fig. 6 shows hourly values of relative differences between spatial averages of $PM_{x,st}$ and $PM_{x,ca}$ over all grid cells containing the measurement sites. The five curves correspond to $x = 0.01$, 0.1, 1, 2.5 and 10 μm. It appears that for smaller cut–off sizes, the relative differences generally increase along with the development of the pollution episode, which reached its peak on August 5. In contrast, the relative differences generally decrease over the same time period for larger cut–off sizes. In the last two days when the episode died down, the relative differences drop for the smaller and move higher for the larger cut–off sizes.

The variation of the relative differences between $PM_{x,st}$ and $PM_{x,ca}$ with location is relatively moderate. Fig. 7 shows the relative differences between time–averaged $PM_{x,st}$ and $PM_{x,ca}$ at each grid cell containing the measurement sites. Comparing with the sites close to the coast, the sites in and down the valley

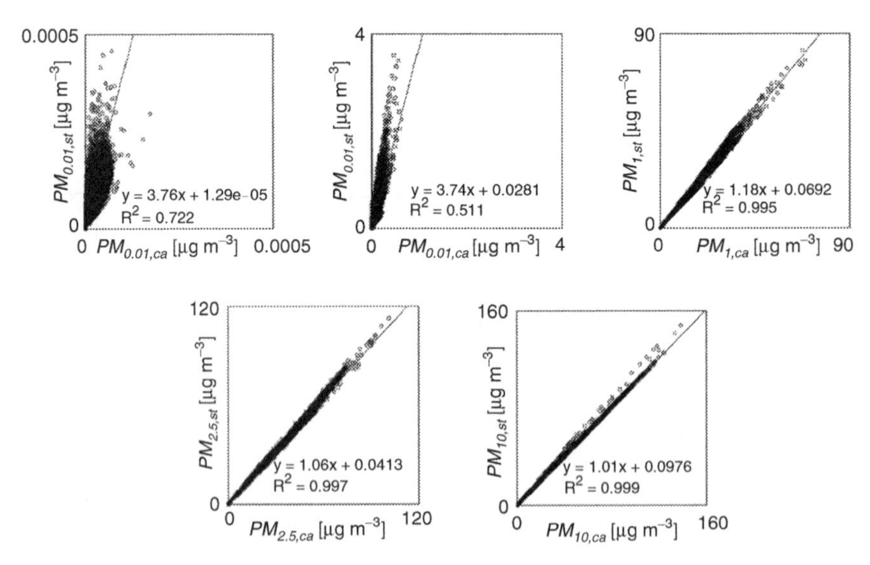

FIGURE 5. Linear correlation between $PM_{x,st}$ and $PM_{x,ca}$ at five cut–off diameters 0.01, 0.1, 1, 2.5, and 10 μm for all grid cells and modelling hours.

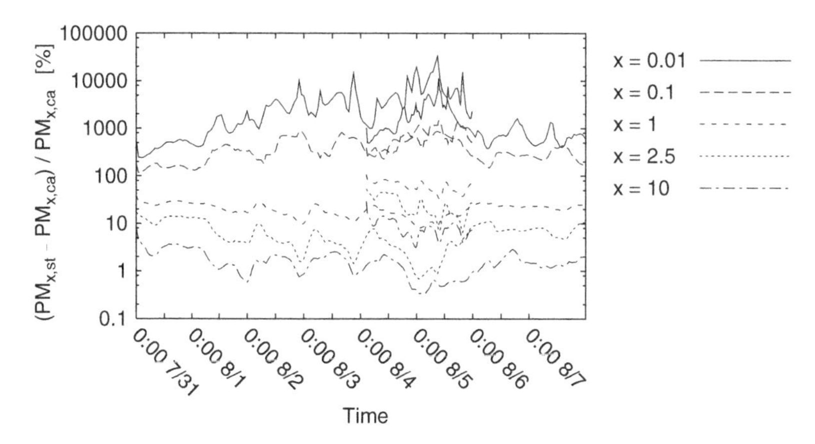

FIGURE 6. Hourly values of relative differences between spatial averages of $PM_{x,st}$ and $PM_{x,ca}$ over all grid cells containing the measurement sites.

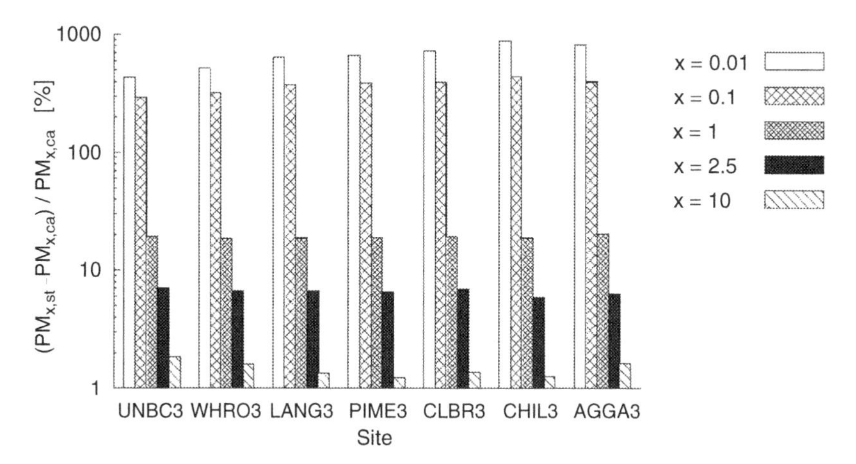

FIGURE 7. Relative differences between time–averaged $PM_{x,st}$ and $PM_{x,ca}$ at each grid cell containing the measurement sites.

appear to give higher relative differences for smaller cut–off sizes and lower relative differences for larger sizes.

4. Impact of the Diameter Type on Model Evaluation

Widespread differences between $PM_{x,st}$ and $PM_{x,ca}$ will directly affect conclusions on model performance. Extent of the impact will be dependent on particle sizes of interest, and time and location of modelling scenarios. The impact expressed in % differences will tend to be high when the particle cut–off sizes are small.

TABLE 1. $PM_{2.5}$ model performance statistics.

		$PM_{2.5,st}$	$PM_{2.5,ca}$	$PM_{2.5,st}-PM_{2.5,ca}$	PM_{I+J}
Bias (µg·m^{-3})	*Max*	21.08	18.60	2.48	30.78
	Min	−11.88	−12.08	0.20	−11.80
	Mean	−1.11	−1.78	0.67	0.53
Normalized Bias (%)	*Max*	177.74	156.83	20.91	259.57
	Min	−67.94	−69.08	1.14	−67.52
	Mean	−3.46	−9.67	6.21	13.63
Normalized Mean Bias (%)		−8.99	−14.42	5.43	4.29
Gross Error (µg·m^{-3})	*Max*	21.08	18.60	2.48	30.78
	Min	0.08	0.24	−0.16	0.11
	Mean	5.69	5.50	0.18	7.08
Normalized Gross Error (%)	*Max*	177.74	156.83	20.91	259.57
	Min	0.71	2.03	−1.32	1.28
	Mean	45.87	43.75	2.12	60.60
Normalized Mean Gross Error (%)		46.18	44.71	1.47	57.52

Complete quantitative analysis of the impact requires comprehensive size–resolved PM measurement data over wide size ranges, which are not available for the 1993 modelling scenario simulated in this study.

For the 1993 modelling scenario, PM measurements were mostly conducted for $PM_{2.5}$ on a 24-hour average basis. According to Figs. 4–7, the differences between $PM_{2.5,st}$ and $PM_{2.5,ca}$ are mostly less than 10%. Therefore, it is expected that using $PM_{2.5,st}$ or $PM_{2.5,ca}$ would cause moderate differences in $PM_{2.5}$ performance assessment. Table 1 lists some model performance statistics when modelled $PM_{2.5,st}$, $PM_{2.5,ca}$, and PM_{i+j} are compared with measured $PM_{2.5}$ mass concentrations at all measurement sites. PM_{i+j} is the total mass concentration of all fine mode particles, which has been used as an approximation for $PM_{2.5}$ by many other CMAQ applications (e.g., Mebust et al., 2003) and is listed here for reference. Note that although the average impact is moderate, using $PM_{2.5,ca}$ instead of $PM_{2.5,st}$ helps reduce maximum normalized bias and maximum normalized gross error by about 21%.

Using different particle diameter types affects model performance assessment not only for total particle concentrations, but also for individual PM species and particle compositions. Details on this subject are omitted here due to the length limitation of the paper.

5. Conclusions

For particles with greater than unit density, D_{st} is smaller than D_{ca}, while $PM_{x,st}$ is higher than $PM_{x,ca}$. $PM_{x,st}$ and $PM_{x,ca}$ generally correlate well. Quantitative differences between $PM_{x,st}$ and $PM_{x,ca}$ are functions of x, and are dependent on the shape of the particle distribution curves varying with time and location. Percentage differences between $PM_{x,st}$ and $PM_{x,ca}$ and their impact on model performance assessment tend to be high when the particle diameters of interest are small. At relatively high particle cut–off diameters, such as 2.5 µm, the average impact of varying modelled diameter type on PM performance assessment is generally moderate. However, it is still possible to obtain meaningful changes in maximum normalized biases and maximum normalized gross errors under such circumstances.

Acknowledgments The authors would like to thank the Pollution Data Branch, Air Quality Research Branch, and Pacific & Yukon Region of Environment Canada for the raw emissions and ambient measurement data that they provided.

References

Binkowski, F. S., and Roselle, S. J., 2003, Models–3 community Multiscale air quality (CMAQ) model aerosol component 1. Model description, *J. Geophys. Res.,* **108**(D6), 4183, doi:10.1029/2001JD001409.

EPA, 2004, Science Topics: Particulate Matter: Background (June 25, 2004); http://es.epa.gov/ncer/science/pm

Jiang, W., Giroux, É., Yin, D., and Roth, H., 2004, A modelling study on the impact of three sets of future vehicle emission standards on particulate matter loading in the Lower Fraser Valley: Interim report, Institute for Chemical Process and Environmental Technology, National Research Council of Canada, Ottawa, Ontario, Canada, 2004.

Jiang, W., and Roth, H., 2002, Development of a modularised aerosol module in CMAQ, 2002 Models–3 Workshop, Research Triangle Park, North Carolina, United States, October 21–23, 2002.

Jiang, W., and Roth, H., 2003, A detailed review and analysis of science, algorithms, and code in the aerosol components of Models–3/CMAQ I. Kinetic and thermodynamic processes in the AERO2 module, Report Number PET–1534–03S, Institute for Chemical Process and Environmental Technology, National Research Council of Canada, Ottawa, Ontario, Canada, March 2003.

Jiang, W., and Yin, D., 2001, Development and application of the PMx software package for converting CMAQ modal particulate matter results into size–resolved quantities, Report Number PET–1497–01S, Institute for Chemical Process and Environmental Technology, National Research Council of Canada, Ottawa, Ontario, Canada, June 2001.

Jiang W., and Yin, D., 2002, Mathematical formulation and consideration for converting CMAQ modal particulate results into size–resolved quantities, Paper #1.2, preprint volume, the Fourth Conference on Atmospheric Chemistry: Urban, Regional, and Global–Scale Impacts of Air Pollutants, Orlando, Florida, United States, January 13–17, 2002, American Meteorological Society.

Mebust, M. R., Eder, B. K., Binkowski, F. S., and Roselle, S. J., 2003, Models–3 community Multiscale air quality (CMAQ) model aerosol component 2. Model evaluation, *J. Geophys. Res.*, **108**(D6), 4184, doi:10.1029/2001JD0014010.

Seinfeld, J. H., and Pandis, S. N., 1998, *Atmospheric Chemistry and Physics : From Air Pollution to Climate Change*, John Wiley and Sons, New York.

Yin, D., Jiang, W., Roth, H., and Giroux, É., 2004, Improvement of biogenic emissions estimation in the Canadian Lower Fraser Valley and its impact on particulate matter modelling results, *Atmos. Environ.*, **38**, 507.

62
The U.K. Met Office's Next-Generation Atmospheric Dispersion Model, NAME III

Andrew Jones, David Thomson, Matthew Hort, and Ben Devenish*

1. Introduction

The impact of the Chernobyl power-plant accident in 1986 gave a major impetus to dispersion modelling activities around that time, especially in those countries directly impacted by the radioactive cloud. In the United Kingdom, the greatest contamination occurred in upland areas across the western half of the country where intense convective rainfall had intercepted the plume (most notably, in NW Wales, Cumbria and SW Scotland). Significant quantities of radionuclides were deposited locally in these upland grassland environments. The Met Office provided specialist forecasts during the incident (based essentially on trajectory techniques); however no operational long-range dispersion model was available for use at that time. Hence central government sanctioned us with developing an emergency-response modelling capability to provide detailed predictions of the transport and deposition of radioactive materials that might arise from any similar events in the future. The Met Office **N**uclear **A**ccident **M**od**E**l (abbreviated to NAME) was in use by 1988 with a major upgrade (NAME II) operational from 1994.

The Met Office's current operational dispersion model, NAME II, is a Lagrangian particle trajectory model (Ryall and Maryon, 1998). Emissions from pollutant sources are represented by particles released into a model atmosphere driven by the meteorological fields from our numerical weather prediction model, the Unified Model (Cullen, 1993). Each particle carries mass of one or more pollutant species and evolves by various physical and chemical processes during its lifespan. Although originally designed as an emergency-response nuclear accident model, subsequent development has greatly enhanced NAME's capabilities so that it is now used in a wide range of applications.

* Met Office, FitzRoy Road, Exeter, EX1 3PB, United Kingdom.

Today, the Met Office has international responsibilities as a Regional Specialist Meteorological Centre for environmental emergency-response modelling in the event of a serious atmospheric pollution incident in the European and African regions. Similarly, our role as a Volcanic Ash Advisory Centre to the aviation industry for the Icelandic/N.E. Atlantic region requires a volcanic ash cloud modelling capability. Real-time dispersion predictions are also needed in many other crisis situations (modelling of toxic releases, forest fires, dust plumes, etc.). For instance, the Met Office played an important role during the UK outbreak of foot-and-mouth disease in 2001, where our modelling work (Gloster et al., 2003) focused on investigating the likely mechanisms responsible for farm-to-farm transmission of the virus in the early stages of the crisis and any possible impacts of burning carcasses on large pyres.

The use of backwards-dispersion and attribution techniques allows source-receptor relationships to be studied. For instance, in the analysis of an individual pollution episode (U.K. Environment Agency, 2001) where the aim is to identify contributory sources responsible for the event, or in the estimation of source emission patterns based on observed monitoring data (Manning et al., 2003). The inclusion of a chemistry scheme in the dispersion modelling framework opens up an extended range of functionality, including air quality applications and source attribution of secondary pollutants.

Although supporting all of the applications discussed above, the current version of our NAME II model is intended primarily for predicting dispersion at medium-to-long ranges (that is, tens of kilometres up to global scales). Dispersion applications tend to use boundary layer averaging at longer ranges where plumes become well-mixed within the boundary layer. At short range, however, the smaller spatial and temporal scales involved are more computationally demanding. For instance, there can be significant vertical variation in a plume's structure that needs to be resolved, whereas shorter time-steps are also required to adequately represent the evolution. The increased cost will be more restrictive on the number of particles which might reasonably be followed in the short-range regime. Thus, a simple box-averaging scheme, which is used in NAME to derive concentrations from particle masses, is not especially well suited for obtaining detailed small-scale structure in the near-source region. Neither does the current NAME model explicitly resolve short-range aspects of the dispersion such as the influence of a building or small-scale terrain effects near to a source. Thus, in 2000, the Met Office initiated a project to develop a successor for NAME II providing improved modelling abilities and greater flexibility, which has resulted in our next-generation atmospheric dispersion model NAME III.

2. The Met Office's NAME III Dispersion Model

The **N**umerical **A**tmospheric-dispersion **M**odelling **E**nvironment NAME III, like its predecessor, is essentially a Lagrangian particle-trajectory model designed to predict the atmospheric dispersion and deposition of gases and particulates.

Unlike in NAME II however, particles can now acquire additional extent attributes that prescribe a spatial distribution around the particle's centre of mass and a temporal spread about its validity time. These particle-puff entities are used in NAME III to represent releases from pollution sources. One or more species are emitted on each individual particle-puff which can be subsequently tracked within a model atmosphere driven by appropriate meteorological data. Our particle-puff approach provides a unified framework for predicting atmospheric dispersion at both short ranges and long ranges, thus giving NAME III an all-range modelling capability (from a few metres to global scales).

A modular code design offers the user flexibility in configuring model runs and provides an extensible infrastructure onto which extra modules could be added. NAME III is capable of utilising meteorological data from a variety of sources: fields from a numerical weather prediction model, radar rainfall estimates, and single-site observations, with the available data used in a nested sense. Other effects, such as plume-rise (for buoyant or momentum-driven releases), radioactive decay of radionuclides, and in-situ chemical transformations, are also considered. At short ranges, NAME III functionality includes modelling of short-period concentration fluctuations and the effects of small-scale terrain or isolated buildings on dispersion.

2.1. A Particle-Puff Approach

The use of particles which have their mass concentrated at a single point is not especially well-suited to modelling dispersion at short ranges. The dimensions of a plume near to its source are usually small relative to the boundary layer depth and a fully three-dimensional representation of the plume has to be calculated. Furthermore, our output three-dimensional grid also needs to have a sufficiently fine resolution to adequately capture small-scale structure in the concentration field (since large spatial gradients would be smoothed out by averaging over too coarse of an output grid). However, this requires the release of a large number of particles (to avoid statistical noise), increasing the computational cost, so that such an approach could become prohibitively expensive. This issue can be resolved by defining a localised spatial distribution of the mass on each particle. That is, by spreading out the mass carried on a particle about its centroid (in our case using a Gaussian distribution), the particle can have a puff or cloud-like structure attached to it – and is consequently referred to here as a particle-puff.

The puff component of a particle-puff will grow, as time elapses, to represent the spread of material by turbulent diffusion. However such growth, if left unchecked, could result in separate regions of a puff entering significantly different flow regimes (in reality, this would create complex distortions of the puff shape which could not be modelled within a basic puff scheme). Our scheme will therefore allow a puff to grow to a certain size (the threshold value being flow-dependent) after which time the puff will split into multiple copies and subsequent growth is then represented by the random motion of these individual components (with further splitting occurring as is necessary). At longer ranges, pollutants usually become well-mixed within the boundary layer and the additional value afforded by the puff structure then diminishes; thus to improve computational performance, this information may be forgotten in later stages of the particle's evolution. Similarly, the puff scheme is not necessary when modelling long-range transport problems and so can be switched off for improved efficiency.

Other puff approaches have been adopted elsewhere. For instance, the puff-particle model of de Haan and Rotach (1998) in which Gaussian puffs are treated in a relative-diffusion framework while their centres of mass follow stochastic trajectories obtained from a Lagrangian particle-dispersion model. (NAME III is conceptually similar although our puffs are not evolved in relative-diffusion terms and the splitting of puffs is handled differently.) The turbulence-closure based SCIPUFF model of Sykes et al. (1993) represents a scalar concentration distribution as a collection of overlapping Gaussian puffs. Finally, a variety of kernel approaches have also been used in standard particle-dispersion models as a mechanism to locally smooth the mass associated with particles.

2.2. Nested Meteorological and Flow Information

NAME III incorporates a versatile flow processing system with the ability to use multiple sources of meteorological data and other flow information. By organising met and flow information in a modular way, a highly flexible and extensible environment is provided for managing such data. The dispersion calculation gets its flow information from one or more flow modules (for instance, there are separate flow modules delivering single-site met data or NWP met fields).

A flow module has a defined domain of applicability that prescribes the region in space and the interval in time where that flow module is valid. Multiple flow modules can be defined within a model run and their domains often overlap. If flow information is requested at a location where multiple flow modules are valid, a preferred priority ordering is used (typically retrieving data at the finest scale available). Similarly, flow modules can have a nested structure with information from the large-scale flow field feeding through as a forcing on the smaller scales. For example, our buildings module (which is designed to model principal features of the local flow around an isolated cuboid-shaped building) uses the ambient meteorology at the building location as its basic, unperturbed state. This basic flow could be provided as a single-site met observation (via a single-site flow module) or as input from an NWP model (via an NWP flow module).

Our principal source of NWP met data for NAME III is the Met Office's operational weather prediction model, the Unified Model (Cullen, 1993). Analysis fields and forecast data (for real-time predictions) are available with global coverage at ~60 km resolution and a limited-area mesoscale model covering the North Atlantic and European region at a resolution of ~12 km. In principle, gridded met data derived from other NWP modelling systems could be handled by a flow module in a similar manner. The effects of small-scale topography, such as isolated hills or ridges, on a local flow are modelled in NAME III by means of the LINCOM linear flow model (Astrup, 1996) courtesy of the Risø National Laboratory.

The LINCOM model is called from within NAME III, and access to the flow information is provided by an interface through the LINCOM flow module.

2.3. Chemistry Scheme

One of the principal differences between emergency-response roles and air quality applications, such as air quality forecasting and episode analysis, is the important role of chemical transformations in the latter. A comprehensive gas and aqueous phase chemistry scheme is included in the NAME III dispersion model. Our scheme, originally developed within NAME II (Redington and Derwent, 2002) and now imported into the NAME III environment, models both sulphate and nitrate chemistry. The set of reactions are derived from the Met Office global chemistry model STOCHEM (Collins et al., 1997) and account for 30 or so predominant atmospheric species. Concentrations of OH, HO_2, H_2O_2 and O_3 are initialised using monthly average 3-D background fields retrieved directly from STOCHEM. Primary pollutant species, such as sulphur dioxide, ammonia and nitric oxide, are emitted on particles from the various sources. Secondary species produced in the chemistry scheme are subsequently added to these particles.

Chemistry calculations are performed on a fixed three-dimensional chemistry grid. The initial species concentrations in a chemistry grid-box are obtained by summing the contributions from all particles occupying that box at the given time. Following completion of the chemistry calculations, the updated mass of each species in the chemistry box is reassigned back to these particles: primary pollutants being redistributed according to the relative proportion of the original contributions, and with secondary species being distributed among particles carrying the appropriate primary species in proportion with the original amount of primary pollutants. In a typical chemistry run, the scheme updates particle masses every 15 minutes (although an internal time-step of 100 seconds is generally used for most of the chemistry calculations).

2.4. Short-term Concentration Fluctuations

Short-duration fluctuations in the concentration of airborne substances can be significant in a wide variety of atmospheric dispersion problems, especially at short ranges downwind from a source. For instance, concentration fluctuations are important in assessing the ignition risk for flammable and explosive materials, in modelling the physiological response to some toxic compounds (Griffiths and Harper, 1985) and in influencing chemical reactions where reaction rates might be non-linear functions of the constituent concentrations. Fluctuations can also play a role in the design of source detection strategies, the perception of malodour and similar environmental-impact type problems, and the consideration of plume visibility for obscurant materials. Consequently in many dispersion scenarios it is insufficient to predict the mean concentration field alone – estimates of likely fluctuations about the mean value are also required.

A fluctuations scheme (Thomson, 2004) in NAME III, developed by combining turbulence theory with experimental data, can provide some guidance on the magnitude of fluctuations. The standard deviation of concentration is modelled by comparing parametrised statistics of particle pair separation (a measure of the instantaneous plume) with one-particle displacement statistics (a measure of the mean plume). In some cases, there is often a need to provide a fuller description of the probability distribution of the short-term concentration. Concentration fluctuations in dispersing plumes and puffs have been observed experimentally in field trials over many years, and such data have often been fitted to clipped-normal distributions (although other distributions have also been considered). Following this philosophy, the probability distribution of concentration fluctuations is modelled as a clipped-normal distribution with parameters calculated to be consistent with NAME III estimates of the mean concentration and the standard deviation of fluctuations. The user can request output as percentile concentrations (50th, 95th, etc.) or as exceedence probabilities of particular concentration thresholds which are deduced from the computed clipped-normal distribution. This approach to modelling fluctuation statistics is similar to that previously used in the ADMS 3 dispersion model.

3. Validation and Testing

Recent work has focused on the testing of NAME III, including comparisons against other dispersion models, and some initial validation studies.

3.1. Validating Against Data from the Kincaid Experiment

The Kincaid field experiment (Bowne and Londergan, 1983) was an extensive experimental campaign carried out during 1980 and 1981 around the Kincaid power plant in Illinois, USA. A buoyant plume containing the tracer SF_6 was released from a stack of height 187 m. Arc-wise maxima of the hourly-averaged ground level concentrations were measured at distances ranging from 500 m to 50 km from the stack. Meteorological variables were observed throughout the campaign on a 100 m tower at the power plant site and augmented by boundary layer soundings obtained from an on-site radiosonde ascent several times a day. The Kincaid experiment has been used extensively in the validation of dispersion models (e.g. ADMS, AERMOD, HPDM, and NAME II). In the validation study presented here, NAME III is compared against the Kincaid observations of quality 3 (as defined in the Model Validation Kit (Olesen, 1994)).

Results of this validation exercise for NAME III are presented in Table 1. The table also gives earlier validation statistics for our NAME II model. The NAME III predictions compare satisfactorily with observations and are a significant improvement on NAME II. The results show a small over-prediction of the mean concentration with NAME III (in contrast to the under-prediction experienced with NAME II). Spread in the predicted concentrations of NAME III is in good

TABLE 1. Validation statistics of NAME III and its predecessor NAME II for the Kincaid experiment: mean, bias, standard deviation (SD), normalised mean square error (NMSE), correlation (R), fractional bias (FB), fractional standard deviation (FS), and proportion of modelled values within a factor of 2 (FA2) of the arc-wise maximum concentrations.

	Mean $\mu g\ m^{-3}$	Bias $\mu g\ m^{-3}$	SD $\mu g\ m^{-3}$	NMSE	R	FB	FS	FA2
Observations	0.69	—	0.51	—	—	—	—	—
NAME III	0.74	−0.05	0.52	0.56	0.473	−0.072	−0.022	75.8%
NAME II	0.58	0.11	0.58	1.07	0.306	0.180	−0.113	66.7%

agreement with the observed spread. The normalised mean square error has been substantially reduced in the new model. Similarly, both correlation and the proportion of modelled concentrations within a factor of two of observed values have been improved in NAME III. The improvements seen in all of these statistical measures are partly a consequence of the new model formulation (i.e. using a puff scheme, etc.) but also partly due to inclusion of the meteorological pre-processor (which allows NAME III to determine its own values of boundary layer depth).

3.2. Comparison Runs with NAME II and HPAC

In preparation for its operational implementation, NAME III is being tested and compared against other dispersion models. One aspect of the testing involves running the NAME III model alongside its predecessor NAME II with the aim of identifying any potential deficiencies in the formulation or operation of our new model. These parallel runs are being performed using various test cases and also on a routine daily basis with real-time meteorological data. One such example is shown in Figure 1 which compares the dispersion predictions of NAME III and NAME II following a hypothetical eruption of Mt. Katla, Iceland on 2 July 2004. The charts depict a forecast at 12Z of mean volcanic ash concentrations within an altitude range 6,000–9,000 m following a six-hour (00Z-06Z) volcanic eruption throughout a column in the troposphere. The two models agree favourably on both the position and shape of the modelled ash cloud. This good agreement should be expected, to a certain extent, since each model uses the same source of meteorological data, although some differences are also evident owing to the different meteorological processing and dispersion calculations. NAME III is also currently being compared against the HPAC dispersion model (Defense Threat Reduction Agency, 2004).

4. Future Plans

Several themes have been identified for future development of our NAME III model, including the addition of a dust/sea-salt production model, support for ensemble forecasts of the meteorology, and a facility for statistical processing of multiple dispersion cases.

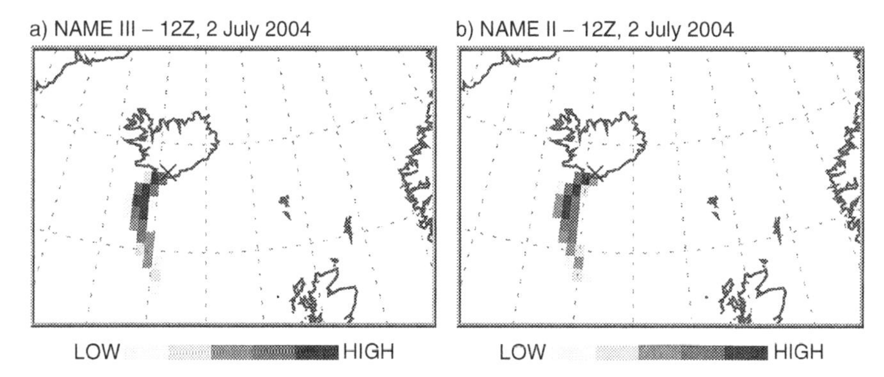

FIGURE 1. A comparison of NAME III and NAME II predictions of the Mt. Katla ash cloud at 12Z, 2 July 2004 following a hypothetical six-hour eruption commencing at 00Z on the same day. The plots show volcanic ash concentrations calculated as a six-hour mean over the slab at 6–9 km.

Progress has already been made in developing a scheme to represent the generation of dust by wind action over land surfaces and of sea-salt by ocean waves. Natural sources of airborne particulates, such as dust and sea-salt, can sometimes be responsible for a large fraction of measured atmospheric particulates (e.g. PM_{10}). Inclusion of these natural sources is therefore important from an air quality viewpoint (in applications such as air quality forecasting and the setting of regulatory objectives). Dust modelling also enables individual dust storm episodes to be studied. Such episodes, which typically see huge quantities of dust lifted into the atmosphere over vulnerable desert regions, are capable of transporting material over great distances. The common feature of wind-generated dust and sea-salt is their dependence on the meteorological conditions – that is, they can be regarded as met-dependent sources. Here the most significant meteorological parameter is wind speed (via the surface stress), although other variables can also play a role (e.g. soil type and vegetation cover are important in dust production).

There are typically many sources of uncertainty in dispersion modelling predictions arising from internal limitations of any particular dispersion model and from incomplete knowledge of the input data required to run that model. In regional-scale applications, such as long-range nuclear accident modelling, a major element of uncertainty is in forecasts of the synoptic-scale meteorology. Ensemble techniques are used to quantify uncertainty in a meteorological forecast and to provide a sample of forecast evolutions (such individual ensemble members can be obtained as multiple runs of a single "perturbed" NWP model, or as a collection of forecasts from separate NWP models in a so-called "poorman's ensemble"). The Met Office uses medium-range forecasts, out to ten days, from the European Centre for Medium-range Weather Forecasts (ECMWF) Ensemble Prediction System, and is also investigating the application of ensembles in short-range weather forecasts. In the area of dispersion, our aim is to configure NAME III to generate an ensemble of dispersion scenarios associated with multiple forecasts of the meteorology. NAME III also needs to perform statistical processing of this information to yield more relevant output quantities (ensemble-mean, "worst-case" scenario, probability forecasts, etc.). These developments will assist ensemble dispersion activities such as our involvement in the European Union's ENSEMBLES Project (Galmarini et al., 2004a, b).

Acknowledgments We are grateful to colleagues at the Risø National Laboratory, Roskilde, Denmark for their assistance in providing the LINCOM linear flow model. The NAME III development project has been supported by the NWP Research Programme and the Government Meteorological Research Programme of the Met Office.

References

Astrup, P., 1996, LINCOM user guide for version L1, Risø National Laboratory, DK-4000, Roskilde, Denmark, RODOS(WG2)-TN(96) 07:17.

Bowne, N. E., and Londergan, R. J., 1983, Overview, results and conclusions for the EPRI plume model validation and development project: plains site, EPRI report EA-3074.

Collins, W. J., Stevenson, D. S., Johnson, C. E., and Derwent, R. G., 1997, Tropospheric ozone in a global-scale three-dimensional Lagrangian model and its response to NOx emission controls, *J. Atmos. Chem.* **26**:223-274.

Cullen, M. J. P., 1993, The unified forecast/climate model, *Meteorol. Mag.* **1449**:81-94.

Defense Threat Reduction Agency, 2004, Hazard Prediction and Assessment Capability (HPAC), Consequences Assessment Branch, Defense Threat Reduction Agency, Fort Belvoir, Virginia, 22060-6201 (July 2004); http://www.dtra.mil/

de Haan, P., and Rotach, M. W., 1998, A novel approach to atmospheric dispersion modelling: the Puff-Particle Model, *Q. J. R. Meteorol. Soc.* **124**:2771-2792.

Galmarini, S., et al., 2004a, Ensemble dispersion forecasting, part I: concept, approach and indicators (submitted).

Galmarini, S., et al., 2004b, Ensemble dispersion forecasting, part II: application and evaluation (submitted).

Gloster, J., Champion, H. J., Sørensen, J. H., Mikkelsen, T., Ryall, D. B., Astrup, P., Alexandersen, S., and Donaldson, A. I., 2003, Airborne transmission of foot-and-mouth disease virus from Burnside Farm, Heddon-on-the-Wall, Northumberland during the 2001 epidemic in the United Kingdom, *Veterinary Record* **152**:525-533.

Griffiths, R. F., and Harper, A. S., 1985, A speculation on the importance of concentration fluctuations in the estimation of toxic response to irritant gases, *J. Haz. Mat.* **11**:369-372.

Manning, A. J., Ryall, D. B., Derwent, R. G., Simmonds, P. G., and O'Doherty, S., 2003, Estimating European emissions of ozone-depleting and greenhouse gases using observations and a modelling back-attribution technique, *J. Geophys. Res.* **108**:4405.

Olesen, H. R., 1994, Model Validation Kit for the workshop on 'Operational short-range atmospheric dispersion models for environmental impact assessments in Europe', NERI, DK-4000, Roskilde, Denmark.

Redington, A. L., and Derwent, R. G., 2002, Calculation of sulphate and nitrate aerosol concentrations over Europe using a Lagrangian dispersion model, *Atmospheric Environment* **36**:4425-4439.

Ryall, D. B., and Maryon, R. H., 1998, Validation of the UK Met. Office's NAME model against the ETEX dataset, *Atmospheric Environment* **32**:4265-4276.
Sykes, R. I., Parker, S. F., Henn, D. S., and Lewellen, W. S., 1993, Numerical simulation of ANATEX tracer data using a turbulence closure model for long-range dispersion, *J. Appl. Meteorol.* **32**:929-947.
Thomson, D. J., 2004, A practical model for predicting the fluctuation statistics of dispersing material (in preparation).
U.K. Environment Agency, 2001, Report into an air pollution episode: sulphur dioxide, September 2nd 1998, Midlands and South Yorkshire, United Kingdom Environment Agency report.

63
An Operational Evaluation of ETA-CMAQ Air Quality Forecast Model

Daiwen Kang, Brian K. Eder, Rohit Mathur, Shaocai Yu, and Kenneth L. Schere*

1. Introduction

The National Oceanic and Atmospheric Administration (NOAA), in cooperation with the Environmental Protection Agency, is developing an Air Quality Forecasting Program that will eventually result in an operational Nationwide Air Quality Forecasting System. The initial phase of this program, which couples NOAA's Eta meteorological model with EPA's Community Multiscale Air Quality (CMAQ) model, began operation since May of this year and has been providing forecasts of hourly ozone concentrations over the northeastern United States.

As part of this initial phase, an evaluation of the coupled modeling system has been performed in which both *discrete forecasts* (observed versus modeled concentrations) for hourly, maximum 1-hr, and maximum 8-hr O_3 concentrations and *categorical forecasts* (observed versus modeled exceedances / non-exceedances) for both the maximum 1-hr (125 ppb) and 8-hr (85 ppb) were evaluated. The evaluation encompasses one month (1 June – 30 June, 2004) and uses hourly O_3 concentration measurements from the EPA's AIRNOW network.

2. The Modeling Systems

The Eta-CMAQ air quality forecasting (AQF) system is based on the National Centers for Environmental Prediction's (NCEP's) Eta model (Black 1994; Rogers et al., 1996) and the U.S. EPA's CMAQ Modeling System (Byun and

* Daiwen Kang#, Brian K. Eder@, Rohit Mathur@, Shaocai Yu#, and Kenneth L. Schere@, NERL, U.S. Environmental Protection Agency, RTP NC, 27711, USA. # On Assignment from Science and Technology Corp., VA 23666, USA. @On assignment from Air Resources Laboratory, National Oceanic and Atmospheric Administration, RTP, NC 27711, USA.

Ching 1999). Otte et al. (2004) describe the linkage between the Eta and the CMAQ model. Here a brief summary relevant to this study is presented. The Eta model is used to prepare the meteorological fields for input to the CMAQ. The NCEP Product Generator software is used to perform bilinear interpolations and nearest-neighbor mappings of the Eta Post-processor output from Eta forecast domain to the CMAQ forecast domain. The processing of the emission data for various pollutant sources has been adapted from the Sparse Matrix Operator Kernel Emissions (SMOKE) modeling system (Houyoux et al., 2000) on the basis of the U.S. EPA national emission inventory. The Carbon Bond chemical mechanism (version 4.2) is used for representing the photochemical reactions.

The detailed information on transport and cloud processes in the CMAQ is given in Byun and Ching (1999). In this study for forecasting O_3 concentrations over the domain of the Northeast U.S., a 12-km horizontal grid spacing on a Lambert Conformal map projection is used. The vertical resolution is 22 layers, which are set on a sigma coordinate, from the surface to ~100 hPa. Vertically varying lateral boundary conditions for O_3 are derived from daily forecasts of the Global Forecast System (GFS). 3D chemical fields are initiated from the previous forecast cycle. The Eta 12 UTC cycles are used for the forecast cycle (Otte et al., 2004). The primary Eta-CMAQ model forecast for next-day surface-layer O_3 is based on the current day's 12 UTC Eta cycle, and products are issued daily no later than 1330 LDT. The target forecast period is local midnight through local midnight (04 UTC to 03 UTC for the Northeast U.S.). An additional 8-hr is required beyond midnight to calculate peak 8-h average O_3 concentrations. So a 48-hr Eta-CMAQ forecast is needed on the basis of the 12 UTC initialization to obtain the desired 24-hr forecast period. The model analysis period for this study is from June 1 to June 30, 2004. The model performances based on the 12 UTC run for the target forecast period are evaluated in this study.

3. O_3 Data

Hourly, near realtime, O_3 (ppb) data obtained from US EPA's AIRNOW program were used in the evaluation. Over 600 stations were available, mostly in urban areas, resulting in over 500,000 observations. In addition to the hourly data, both the maximum 1-hour and maximum 8-hour concentrations are calculated for each station and each day over the evaluation period. The calculation of 8-hour maximum is the same as model forecast using the forward calculation method and the calculation of the last 7 8-hour maximum concentrations including data from next day. The maximum 1-hr and 8-hr concentrations are considered missing if half of the hourly observation data is missing for the day (LST). If two or more monitoring stations are located within the same model grid cell, their average value is used as the representative measurement for that grid cell.

4. Results

Kang et al. (2003, 2004) present various statistical measures for model performance evaluation. The following equations summarize the discrete and categorical statistical measures used in this evaluation.

$$MB = \frac{1}{N}\sum_{1}^{N}(C_m - C_o) \tag{1}$$

$$NMB = \frac{\sum_{1}^{N}(C_m - C_o)}{\sum_{1}^{N} C_o} \cdot 100\% \tag{2}$$

$$RMSE = \sqrt{\frac{1}{N}\sum_{1}^{N}(C_m - C_o)^2} \tag{3}$$

$$NME = \frac{\sum_{1}^{N}|C_m - C_o|}{\sum_{1}^{N} C_o} \cdot 100\% \tag{4}$$

where C_m and C_o are modeled and observed concentrations, respectively.

$$A = \left(\frac{b+c}{a+b+c+d}\right) \cdot 100\% \tag{5}$$

$$CSI = \left(\frac{b}{a+b+d}\right) \cdot 100\% \tag{6}$$

$$POD = \left(\frac{b}{b+d}\right) \cdot 100\% \tag{7}$$

$$B = \left(\frac{a+b}{b+d}\right) \tag{8}$$

$$FAR = \left(\frac{a}{a+b}\right) \cdot 100\% \tag{9}$$

4.1. Discrete Evaluations

As seen in Table 1, discrete evaluations were performed for hourly, maximum 1-hr and maximum 8-hr O_3 forecast. To differentiate model performance at levels above typical background concentrations performance metrics were also computed using a 40 ppb observation threshold (Figure 1). The O_3 concentrations were low in June over the modeling domain with the mean observed O_3 concentration of

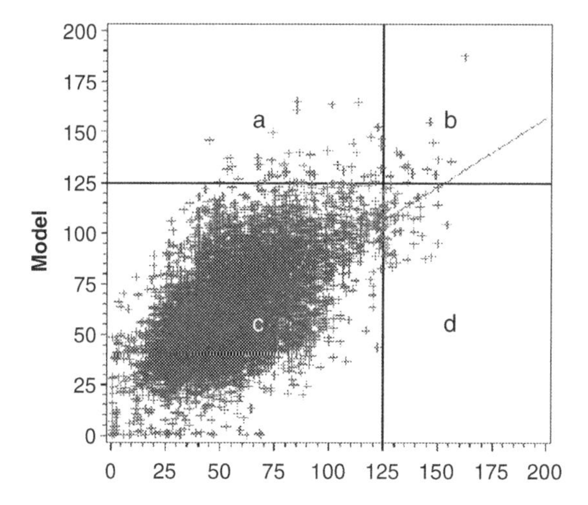

FIGURE 1. Example plot for categorical evaluation.

TABLE 1. Summary of discrete statistics for forecasts of hourly, maximum 1-hr, and 8-hr O_3 for all the concentration range and the range for all the observed concentrations greater than 40 ppb.

Metrics	Hourly		Max 1-hr		Max 8-hr	
	All	>40 ppb	All	>40 ppb	All	>40 ppb
N	536623	142913	18389	12218	18389	12218
Obs_mean	29.2	51.3	51.7	59.1	46.1	53.1
Mod_mean	41.6	52.3	57.5	60.3	53.9	56.5
MB (ppb)	12.4	1.0	5.7	1.2	7.8	3.4
NMB (%)	42.4	1.9	11.1	2.0	16.9	6.5
NME (%)	53.7	17.0	20.8	14.3	24.0	15.5
RMSE (ppb)	19.7	11.4	13.8	10.8	14.1	10.4
r	0.54	0.45	0.54	0.52	0.51	0.48

only 29.2 ppb. At this level of O_3 concentration, the model tends to over predict O_3 concentrations for all the categories in Table 1 with positive mean bias (MB) values. However, when only the points with greater than 40 ppb observed O_3 are considered, the over prediction becomes significantly smaller compared with the overall concentration range (with MB of 12.4 ppb versus 1.0 ppb for the hourly forecast, 5.7 ppb versus 1.2 ppb for maximum 1-hr forecast, and 7.8 ppb versus 3.4 ppb for the maximum 8-hr forecast, respectively). Other statistical metrics (NMB, NME, and RMSE) display the similar variation pattern and the RMSE values are all less than 20 ppb. The correlation coefficients range from 0.51 to 0.54 for overall forecast, and from 0.45 to 0.52 for the higher observed O_3 concentration range. **Because the air quality model (CMAQ) is tuned to forecast**

high O_3 episodes, the statistics suggest that during the relatively low O_3 period, the model performed fairly well.

Scatter plots of the model forecasts versus AQS observations (for both the maximum 1- and 8-hr ozone concentrations) are presented in Figure 2. In addition to illustrating the exceedance threshold areas (which were used in calculation of the categorical statistics), the plots also provide the 1:1.5 factor lines. As evident from the scatter plot, most of the over prediction occurs at the lower concentrations; when observed concentrations >40 ppb, majority of the model forecast are within a factor of 1.5.

Evaluation of model diurnal performance was also performed as seen in Figure 3, where boxplot (denoting 75th, 50th, 25th percentiles, max., min. and mean) of simple bias (Model-AIRNOW) over the entire analysis period of June is provided. As Figure 3 indicated, the model tends to over predict through all the time of day, however, the magnitude of over prediction during daytime is less than those during nighttime.

4.2. Time Series

Figure 4 displays the time series of correlation coefficient (a), normalized mean bias (b), and normalized mean error (c) for hourly, maximum 1-hr, and maximum 8-hr during June. As Figure 4 indicates, the correlation coefficients fluctuate from ~0.4 to ~0.8 except that on June 13th, slightly negative correlation coefficients for maximum 1-hr and maximum 8-hr were observed. However, the hourly correlation coefficient for that day is comparable with those on other days. The significant low correlation coefficients on June 13th were due to very low O_3 concentrations on that day across the domain, that is, no observable variation pattern exists. In fact, as Figure 4 (b) and (c) indicate, the normalized mean bias and

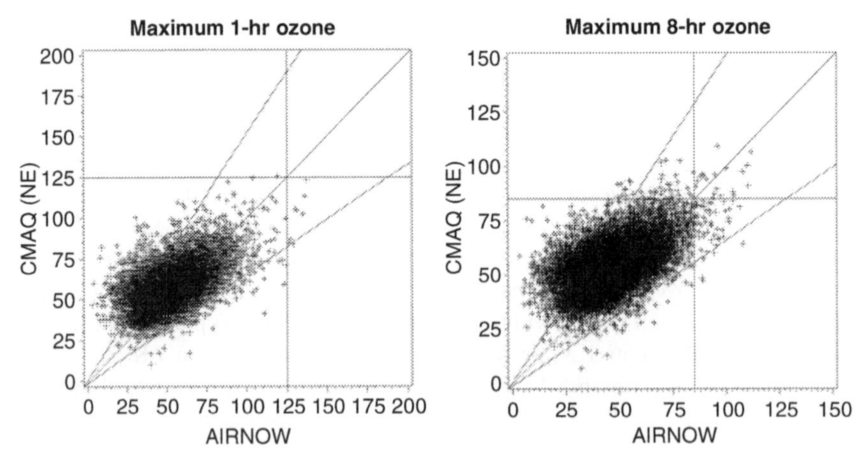

FIGURE 2. Scatter plots of the model versus AIRNOW for both 1- and 8-hour maximum ozone concentrations (ppb) with exceedance thresholds indicated.

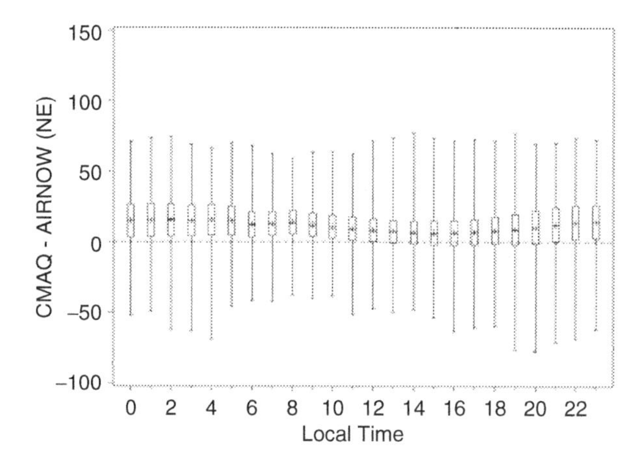

FIGURE 3. Boxplots of the diurnal variation (model – AIRNOW) for hourly ozone concentrations (ppb).

the normalized mean error on that day are relatively small. Both normalized mean bias and normalized mean error of hourly, maximum 1-hr and maximum 8-hr O_3 follow very similar day to day variations with maximum 8-hr values being the lowest and hourly values the highest. The day to day variations of biases are significant and are found to be associated with mesoscale meteorological conditions. Lower biases were always associated with high pressure system and clear skies such as those during June 2 to June 6, while higher biases were always associated with low pressure (front) system and cloudy skies (precipitation) such as on June 10th and June 25th.

4.3. *Categorical Evaluations*

As shown in Table 2, the **Accuracy** (A) for model prediction, which indicates the percent of forecasts that correctly predict an exceedance or non-exceedance, is close to 100%. However, care must be taken in interpretation of this metric, as it is greatly influenced by the overwhelming number of non-exceedances and for the period of this analysis, only a handful of exceedances were observed, especially for the maximum 1-hr forecast. To circumvent this inflation (which is common when evaluating the prediction of rare events like O_3 exceedances), the **Critical Success Index** (CSI) is often a better metric of model performance. The CSI provides a measure of how well the exceedances were predicted, without regard to the large occurrence of correctly predicted non-exceedances. For our evaluation, the CSI for the 8-hr exceedance is about 16%.

The **Probability Of Detection** (POD) metric is similar to the CSI, in that it measures the number of times a model predicted an exceedance when one actually occurred. In our evaluation, the POD for maximum 8-hr forecast is 25%.

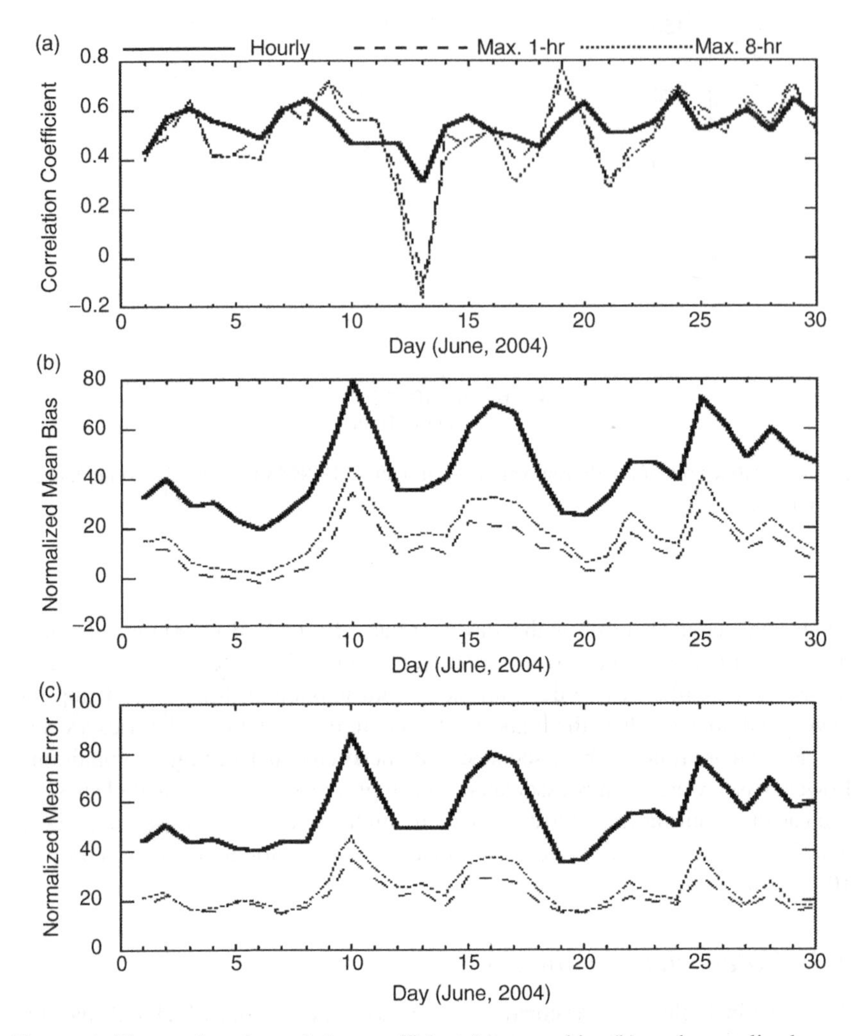

FIGURE 4. Time series of correlation coefficient (a), mean bias (b), and normalized mean bias (c) for hourly, maximum 1-hr, and maximum 8-hour ozone concentrations (ppb).

Measures of **Bias** (B), which for a categorical forecast indicates if forecast exceedances (1- and 8-hr) are under predicted ($B < 1$) or over predicted ($B > 1$), indicate that the model under predicts both maximum 1-hr and 8-hr exceedances.

The fifth categorical metric, the **False Alarm Rate** (FAR), indicates the number of times that the model predicted an exceedance that did not occur. During the analysis period, FAR for maximum 8-hr forecast is 67.8% and this is comparable with other similar regional scale forecast models (Kang et al., 2003).

TABLE 2. Summary of categorical statistics for forecasts of maximum 1-hr and 8-hr O_3.

	A (%)	CSI (%)	POD (%)	B	FAR (%)
Max 1-hr	99.95	N/A	N/A	0.13	100
Max 8-hr	99.48	16.25	25.33	0.79	67.8

5. Summary

The purpose of this research has been to evaluate the performance of the Eta-CMAQ air quality forecast system using O_3 observations obtained from EPA's AIRNOW program and a suite of statistical metrics evaluating both discrete and categorical forecasts developed from our previous research work.

Results from the discrete evaluation revealed that Eta-CMAQ tends to over predict O_3 concentrations (MB ranges from 1.0 ppb for hourly forecast with cut-off at observations greater than 40 ppb to 12.4 ppb hourly forecast with all forecast data) and that produces substantial errors (RMSEs range from 10.4 ppb max. 8-hr forecast with 40 ppb cutoff to 19.7 ppb for hourly forecast with all the forecast data). Only maximum 8-hr forecast generated sufficient exceedances for the categorical evaluation during June and the result revealed that the accuracy of the forecast is greater than 99%, Critical Success Index (CSI) is 16.3%, Probability of Detection (POD) is 25.3%, and False Alarm Rate (FAR) is 67.8%. The values of categorical measures for Eta-CMAQ are comparable with those of other models when they were operated during the high O_3 season in 2002 (Kang et al., 2003). The time series of correlation coefficients and biases (mean bias and normalized mean bias) displayed significant day to day variations in June. Higher correlation coefficients and low biases were generally associated with high pressure systems and low correlation coefficients and high biases generally occurred during days with low pressure systems. Even though the air quality forecast system is tuned to higher O_3 concentrations, both discrete and categorical statistics show that the mode still perform fairly well for the relatively low O_3 concentrations in June. With the progress of the summer O_3 season and normally high pressure systems dominate during summer in the modeling domain, the forecast system is expected to perform better than the results shown in this paper for the month of June.

Acknowledgements The authors would like to thank the model developers for providing results of their simulations.

Disclaimer This document has been reviewed and approved by the U.S. Environmental Protection Agency for publication. Mention of trade names or commercial products does not constitute endorsement or recommendation for use.

References

Black, T. The new NMC mesoscale ETA Model: description and forecast examples. Wea. Forecasting, 9, 265-278, 1994.

Byun, D. W. and J. K. S. Ching, Eds.,: Science algorithms of the EPA Models-3 Community Multi-scale Air Quality (CMAQ) modeling system, EPA/600/R-99/030, Office of Research and Development, U.S. Environmental Protection Agency, 1999.

Houyoux, M. R., J. M. Vukovich, C. J. Coats Jr., N. M. Wheeler, and P. S. Kasibhatla, Emission inventory development and processing for the seasonal model for regional air quality (SMRAQ) project. J. Geophys. Res. 105, 9079-9090, 2000.

Kang, D., B. K. Eder, and K. L. Schere, The evaluation of regional-scale air quality models as part of NOAA's air quality forecasting pilot program, Preprints, 26th NATO/CCMS International Technical Meeting on Air Pollution Modeling and its Application, 26-30 May 2003, Istanbul, Turkey, 404-411, 2003.

Kang, K., B. K. Eder, A. F. Stein, G. A. Grell, S. E. Peckham, and J. McHenry, The new England air quality forecasting pilot program: development of an evaluation protocol and performance benchmark. JAWMA, in press, 2004.

Otte, T. L., et al., Link the Eta model with the community multiscale air quality (CMAQ) modeling system to build a real-time national air quality forecasting system. Weahter and Forecasting, 2004 (in review).

Rogers, E., T. Black, D. Deaven, G. DiMego, Q. Zhao, M. Baldwin, N. Junker, and Y. Lin. Changes to the operational "early" Eta Analysis/Forecast System at the National Centers for Environmental Prediction. Wea. Forecasting, 11, 391-413, 1996.

64 AURAMS / Pacific2001 Measurement Intensive Comparison

P. A. Makar[1,*], V. S. Bouchet[2], W. Gong[1], M. D. Moran[1], S. Gong[1], A. P. Dastoor[2], K. Hayden[1], H. Boudries[3], J. Brook[1], K. Strawbridge[1], K. Anlauf[1], and S. M. Li[1]

1. Introduction

Research on the numerical prediction of chemically speciated gas and particle phase components of the atmosphere has been driven by public health studies linking both gases and particles to adverse health effects. Three dimensional air-quality models containing detailed chemical and physical processes for gas and particle formation (Meng et al., 1998; Dennis et al., 1996; Ackermann et al., 1998) provide a means of linking and describing the complex non-linear processes leading to these adverse air-quality health outcomes.

Measurement data that may be used for model comparison may be grouped into two broad classes, monitoring network and measurement intensive data. Data from monitoring networks usually is available from multiple locations across a wide spatial domain, and over long time periods, thus providing a means of evaluating model performance as an average over large spatial scales (Bouchet et al., 2004). The monitoring data at any given site are limited by factors relating to the cost of instrumentation and analysis, with the result being data that have limited chemical speciation, long time averages (typically a day or longer), and low measurement frequency (e.g. measurements may be made only one day in six). Recent network improvements (e.g. Airnow) aim to alleviate some of these concerns, but are still limited in terms of chemical speciation. The data from measurement intensives, in contrast, are highly speciated and time-resolved, but are usually made only at select sites for a limited total time, again due to expense concerns. While both forms of data are useful for model evaluation purposes, the level of detail in the measurement intensive data allows for a process-level evaluation of a model's results and thus provides an opportunity for model improvement.

[1] P. A. Makar, W. Gong, M. D. Moran, K. Hayden, J. Brook, K. Strawbridge, K. Anlauf, and S. M. Li, Meteorological Service of Canada, 4905 Dufferin Street, Downsview, Ontario, M3H 5T4, Canada.

[2] V. S. Bouchet and A. P. Dastoor, Meteorological Service of Canada, CMC, 2121 Route TransCanadienne, Dorval, Québec, H9P 1J3, Canada.

[3] H. Boudries, Aerodyne Research, Inc., 45 Manning Road, Billerica, MA 01821, USA.

This study describes and contrasts the comparison of the Meteorological Service of Canada's AURAMS model to (a) monitoring measurement data in a domain covering much of Western Canada and the North-Western United States of America, and (b) to intensive data collected in the Lower Fraser Valley of British Columbia, during the period August 25 – August 31, 2003. This time interval comprises the final week of a four-week intensive measurement campaign to study the processes leading to atmospheric particle formation that took place in the Greater Vancouver Region of Canada's west coast, "Pacific2001" (Li et al., 2002).

2. AURAMS: A Brief Description

A more detailed description of AURAMS is provided in the proceedings of the previous NATO conference (Makar et al., 2004, Bouchet et al., 2004, Gong et al., 2004) – only an brief overview will be provided here.

AURAMS (A Unified Regional Air-quality Modelling System) is comprised of five independent codes that are run in sequence in order to produce an air-quality simulation.

The first of these is the Canadian Emissions Processing System (Moran et al., 1997; Makar et al., 2003a), a set of codes that convert raw emissions data into temporarily varying, spatially gridded mass fluxes for each of the major emissions categories.

The second part of the system is one of two meteorological forecast codes (the Meteorological Service of Canada's MC2 (Benoit et al., 1997) or GEM (Côté et al., 1998) models), run in an "offline" mode to produce input meteorological input data for air-quality simulations. The simulations performed here make use of the GEM forecast model.

The third part of the system, the AURAMS Preprocessor, extracts and converts input data files of geophysical constants, land-use categories, meteorological information, and the emissions fields created by CEPS to unformatted binary files for the grid used in a given study.

The fourth component of the system is the Chemical Transport Model (CTM), the "core" of AURAMS that uses the above input data to predict the temporal and spatially varying concentrations of the gas and particle constituents of the atmosphere. The CTM of the AURAMS model is based on that of its predecessor, the CHRONOS model (Pudykiewicz et al., 1997), with extensive modifications for detailed particle processes. The configuration used in these runs included 42 gas-phase species, and 7 aerosol species (sulphate, nitrate, ammonium, sea-salt, black carbon, organic carbon, and crustal material) distributed over 12 sections ranging in size from 0.01 to 40.96 μm diameter.

The model makes use of a single forward operator splitting on its main processes (operator step of 900 seconds here). Descriptions of the operators and/or methodology may be found in Gong et al, 2003; Makar et al., 2003b; Zhang et al., 2001, 2002; Lurmann et al., 1986; Odum et al., 1996, and the proceedings papers noted above.

The fifth and final component of AURAMS is the post-processing code. The output of AURAMS is available on a highly time-resolved individual species and size-bin basis, but monitoring data are frequently available only on a coarser time and cut-size. In order to allow comparisons to monitoring data, the post-processor bins and time-averages data in the same manner as monitoring network observations to allow a direct comparison to those observations (Bouchet et al., 2004).

3. Model Domain, Network Data, Study Period Characteristics

Model simulations were performed on a 148 by 124 grid with a horizontal resolution of 21 km and 28 terrain-following layers in the vertical from the ground to 30 km. A comparison between AURAMS results for monitoring data from four networks (CAPMoN, NAPS, IMPROVE and AIRS) was provided in last year's proceedings, and will be repeated here with the most recent version of the model.

The model simulations encompassed the period August 25–August 31, 2001 (with a two day a priori spin-up period). This period was chosen due to observations made during the measurement intensive (Li et al., 2002) indicating rapid particle formation in the Greater Vancouver/Georgia Basin region of Canada's west coast, subsequent to the passage of a frontal system on August 23rd. The highest particle levels were recorded between the 27th -28th of August, with total PM levels reaching 20 $\mu g/m^3$.

4. Model Results and Analysis

The model domain, monitoring network stations employed for the comparisons, and the resulting model-measurement scatterplots, are shown in Figure 1.

The model's best fit to the measurements occurs for $PM_{2.5}$ SO_4, $PM_{2.5}$ NH_4, and PM_{total} NO_3, with slopes close to unity and correlation coefficients ranging from 0.595 to 0.404. Fine mode particle nitrate is being over-predicted (slope 2.84), while $PM_{2.5}$ OC and EC are being under-predicted. The latter are due at least in part to high observed values at five measurement stations – with the exception of a station at Portland, Oregon, these are all located in US National Parks (low anthropogenic emissions regions), and likely show the impact of forest fires on the measured OC and EC (forest fires having been observed in satellite observations during this time period). The correlation on fine-mode sea-salt is poor, while coarse mode sea-salt is being overpredicted. Crustal material is being overpredicted, with a low correlation, and likely results in an overestimate in the expected emissions of this portion of the model results. While further improvements are possible and likely, the model correlations and slopes are greatly improved over those presented at the previous NATO conference (Makar et al., 2004). These improvements are largely the result of the adoption of an improved mass conservation methodology

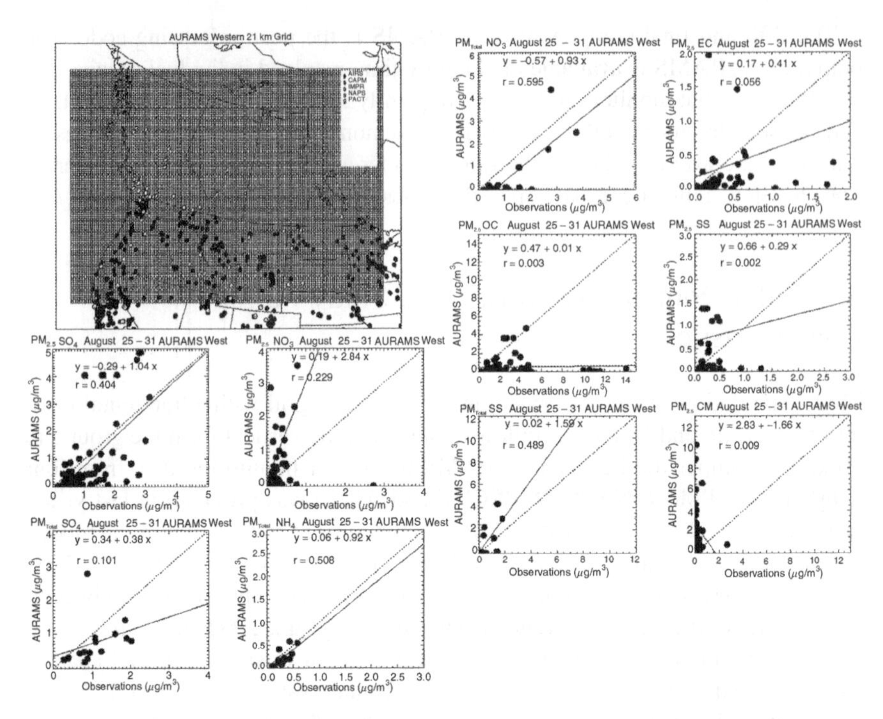

FIGURE 1. AURAMS Western Domain and monitoring stations, and Model – Measurement comparisons for $PM_{2.5}$ SO_4, $PM_{2.5}$ NH_4, $PM_{2.5}$ NO_3, PM_{total} NO_3, PM_{total} SO_4, $PM_{2.5}$ EC, $PM_{2.5}$ OC, $PM_{2.5}$ Sea-Salt, PM_{total} Sea-Salt, and $PM_{2.5}$ Crustal Material.

and the use of the Crank-Nicholson ($O(\Delta t^2)$) rather than Laasonen ($O(\Delta t)$) numerical method for vertical diffusion.

As noted above, the monitoring network data provide a generic estimate of model performance, but measurement intensive data are required to explain the scatter in the model results, and some of the model's biases. Here, measurements taken at two of the intensive sites (Langley and Slocan Park) will be used for model evaluation.

Figure 2 shows a comparison between modelled and measured (TEOM) total $PM_{2.5}$ at Langley. Similar features are shown between the two plots during the first few days of the comparison period, but on the last three days total particle concentrations are being overpredicted by the model, with model maxima on the order of 75 $\mu g/m^3$ and measured maxima on the order of 20μ g/m^3. A large portion of the model mass is particle nitrate: for example at 6:15 UT on August 30th, 52.8% of the total model-predicted $PM_{2.5}$ mass is nitrate.

Figure 3 shows a comparison between particle sulphate, nitrate and ammonium provided by the model and particle size and composition measurements taken using an Aerodyne Aerosol Mass Spectrometer. The effective size range for this instrument is up to 1.0 um, hence the smaller vertical scale for the measurement

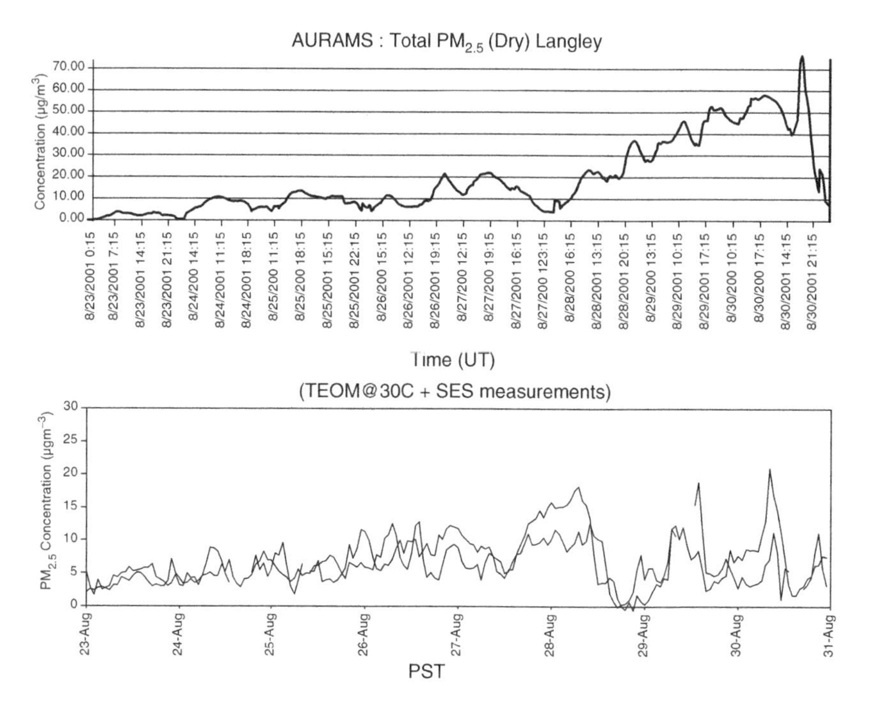

FIGURE 2. AURAMS Total $PM_{2.5}$ compared to TEOM measurements, Langley. Note difference in scales.

plots. The sulphate values (top) show similar magnitudes between model and measurements, with a difference in the timing of the peaks. The nitrate values show over-predictions, similar to those noted for Figure 1, and particle ammonium is both overpredicted and correlated with the nitrate.

The source of the NO_3 in particle nitrate is gas-phase nitric acid, which has in turn as its source NO_x. Nitric acid and NO_2 comparisons at this gridpoint are shown in Figure 4. The nitric acid over-prediction is greatest at the same time as the particle mass over-prediction. Large over-predictions in NO_2 are also apparent.

At this juncture in the analysis, it is worth pointing out that the Langley measurement site is downwind of the main emissions region in the area – urban Vancouver. The relative importance of the different operators on the solution (and more importantly, which one is responsible for the over-predictions) may be difficult to discern at a downwind site. For that reason, Figure 5 shows comparisons between model and measured NO_y at Slocan Park, a Pacific2001 intensive site located in a small municipal park in central Vancouver. At this site, the NO_y concentrations are likely dominated by fresh emissions of NOx, with little time for the reactive chemistry leading to HNO_3 and particle nitrate formation (compare NO_y and NO_x concentrations in model output, Figure 5). Once again, the model overpredicts NO_y. Of great importance to note, however, is that the model and

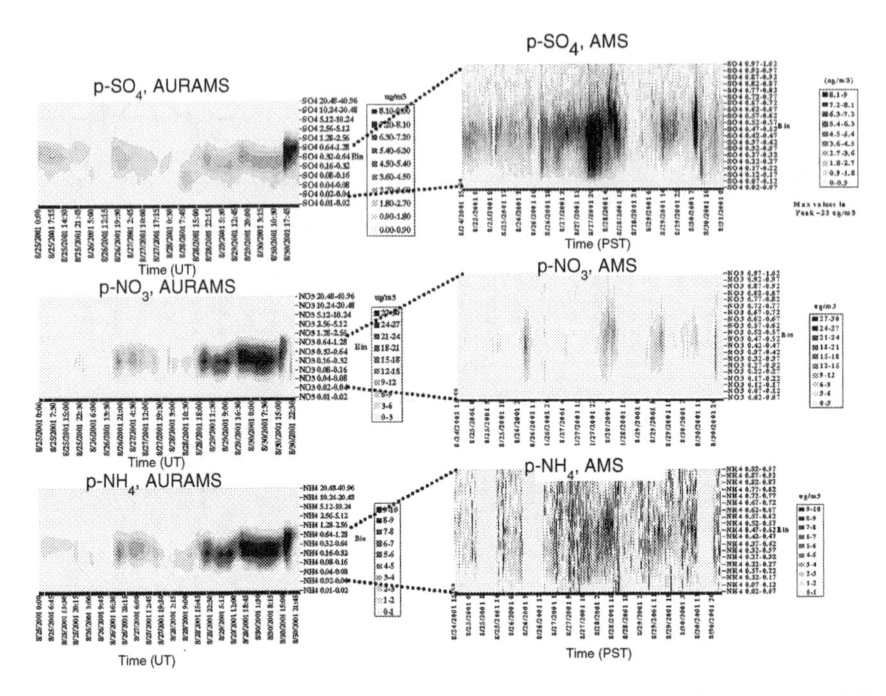

FIGURE 3. AURAMS p-SO_4, p-NO_3, p-NH_4 compared to AMS data. Note that the AMS size cutoff is 1 μm.

measured time series have the same behaviour with time – a double peak in concentrations, with the first peak near local midnight, and the second in the early morning. NO_y concentrations in both model and measurements maximize at night. A similar pattern is observed for measured and modelled VOCs (not shown); a double night-time peak in concentrations. The significance of Figure 5 is that it implies that the cause of the over-predictions is related to emissions or transport (chemistry not being a dominant factor so close to the source of NO_x emissions).

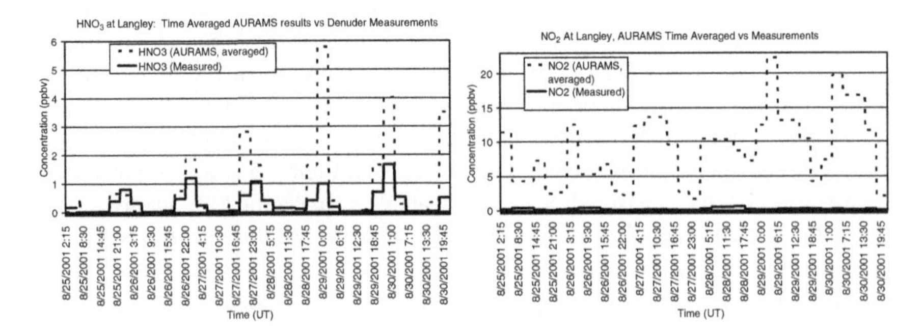

FIGURE 4. AURAMS HNO_3 and NO_2 at Langley, compared to observations.

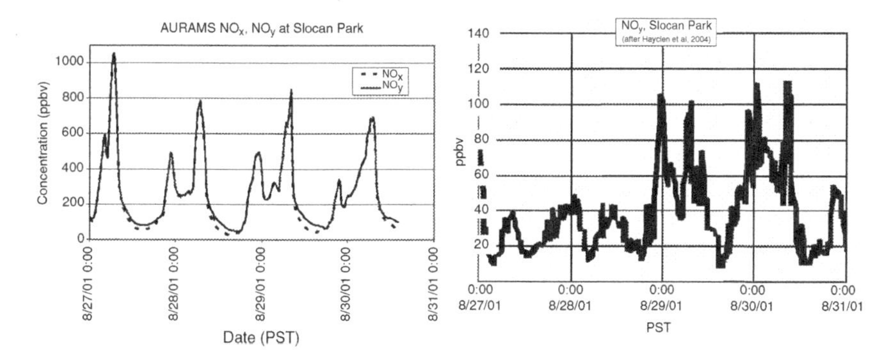

FIGURE 5. AURAMS NO_x at Slocan Park compared to observations.

In order to examine the transport effects, a comparison was made between (a) the winds observed at various stations in the LFV, and (b) the planetary boundary layer heights observed through the use of Lidar (Strawbridge et al., 2004). The former suggested that the horizontal winds were being accurately predicted by the GEM weather forecast model. The latter, however, showed that GEM had a systematic tendency to underpredict the boundary layer height at night near Langley, in turn indicating a tendency to overpredict the stability (underpredict the vertical diffusion transport) of the atmosphere under neutrally stable stratification. Figure 6 compares GEM-forecast and Lidar-measured PBL heights. The model minima in PBL height

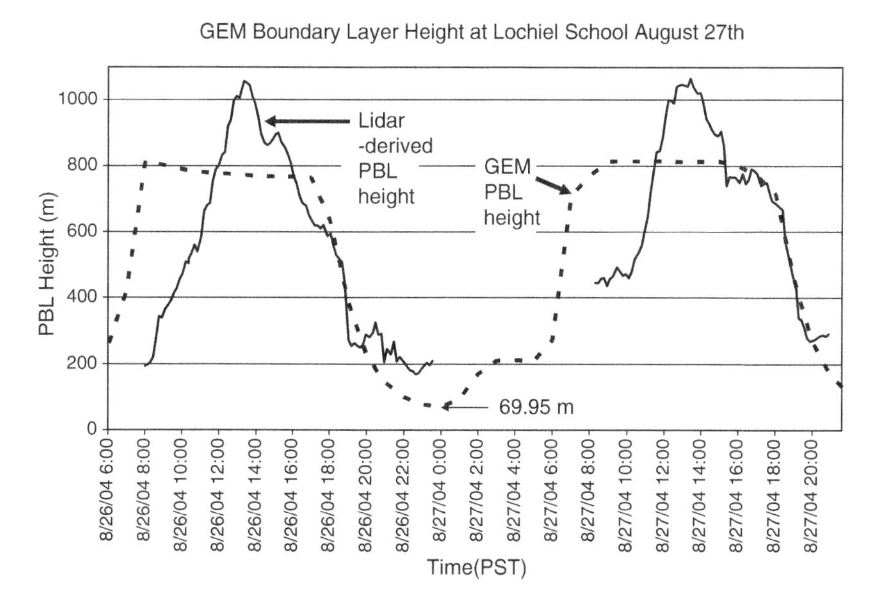

FIGURE 6. PBL Height at Langley: GEM vs Lidar-based observations.

is approximately 70 m, while the measured minima is approximately 200 m, a factor of 2.9 greater than the forecast. Figure 6 also suggests that the simulated morning increase in the PBL height occurs too early in the GEM forecast, and achieves a lower maximum value than the observations, under-predicting the maximum PBL height by approximately 250 m (30% under-prediction).

This in turn implies that the freshly emitted NO_x and VOCs in the AURAMS simulation will be mixed into a much smaller volume at night than occurred in the actual atmosphere. The mixing with the transition to instability in the morning takes place more slowly in the measurements than the model, but the daytime maxima imply that the model volume for effective vertical diffusion during the day will also be less than that in the observations.

5. Conclusions

The AURAMS model has been shown to have the most accurate predictions for $PM_{2.5}$ sulphate, and PM_{total} ammonium and nitrate, with OC and EC being biased low due to unresolved emissions, likely associated with forest fire smoke at certain locations in the model domain. The nitrate and likely the ammonium values, however, are overpredicted for $PM_{2.5}$. Comparisons between measurement intensive data and model results have suggested that the cause of these over-predictions is likely the result of night-time over-predictions of stability (under-predictions of the vertical diffusion constants) for the neutrally stable boundary layer. This in turn leads to an over-prediction in concentrations of NOx in source regions, with subsequent downwind over-predictions in HNO_3 concentrations. Current research work centers on an examination of the PBL parameterizations used in the GEM weather forecast model, for these conditions.

References

Ackermann, I. J., H. Hass, M. Memmesheimer, A. Ebel, F. S. Binkowski and U. Shankar, 1998: Modeal aerosol dynamics for Europe: Development and first applications. Atmos. Environ., 32, 2981-2999.

Benoit, R., M. Desgagné, P. Pellerin, S. Pellerin, Y. Chartier and S. Desjardins, 1997. The Canadian MC2: a semi-Lagrangian, semi-implicit wideband atmospheric model suited for finescale process studies and simulation. Mon. Wea. Rev., 125, 2382-2415.

Bouchet, V. S., M. D. Moran, L.-P. Crevier, A. P. Dastoor, S. Gong, W. Gong, P. A. Makar, S. Menard, B. Pabla, L. Zhang, 2004: Wintertime and summertime evaluation of the regional PM air quality model AURAMS, NATO-CCMS Conference, Istanbul Air Pollution Modeling and Its Application XVI, Plenum, New York, pp 97-104.

Côté, j., J.-G. Desmarais, S. Gravel, A. Méthot, A. Patoine, M. Roch and A. Staniforth, 1998. The operational CMC/MRB Global Environmental Multiscale (GEM) model. Part I: Design considerations and formulation. Mon. Wea. Rev., 126, 1373-1395.

Dennis, R. L., D. W. Byun, J. H. Novak, K. J. Galluppi, C. J. Coats, and M. A. Vouk, 1996: The next generation of integrated air quality modeling: EPA's Models-3. Atmos. Environ., 30, 1925-1938.

Gong, S. L., Barrie, L. A., Blanchet, J.-P., von Salzen, K., Lohmann, U., Lesins, G., Spacek, L., Zhang, L. M., Girard, E., Lin, H., Leaitch, R., Leighton, H., Chylek, P., and Huang, P., 2003: Canadian Aerosol Module: A size segregated simulation of atmospheric aerosol processes for climate and air quality models. 1. Module Development, *J. Geophys. Res.*, 108. D1, 4007.

Gong, W., P. A. Makar and M.D. Moran, 2004: Mass-conservation issues in modelling regional aerosol, NATO-CCMS Conference, Istanbul Air Pollution Modeling and Its Application XVI, Plenum, New York, pp 97-104.

Li, S.-M. and B. Thomson, 2002: PACIFIC2001 Air quality study – an overview. Proceedings, Fall AGU.

Makar, P. A., V. S. Bouchet, L.-P. Crevier, A. P. Dastoor, S. Gong, W. Gong, S. Menard, M. D. Moran, B. Pabla, S. Venkatesh, L. Zhang, 2004: AURAMS runs during the Pacific2001 time period – a model/measurement comparison, NATO-CCMS Conference, Istanbul Air Pollution Modeling and Its Application XVI, Plenum, New York, pp 97-104.

Makar, P. A., M. D. Moran, M. T. Scholtz, A. Taylor, 2003a: Speciation of volatile organic compound emissions for regional air quality modelling of particulate matter and ozone, *J. Geophys. Res.*, 108, D12: 10.1029/2001JD000797, ACH2-1, ACH2-51.

Makar, P. A., V. S. Bouchet, A. Nenes: 2003b. Inorganic Chemistry Calculations using HETV – A Vectorized Solver for the SO_4^{2-} -NO_3^- -NH_4^+ system based on the ISORROPIA Algorithms, Atmos. Environ., 37, 2279-2294.

Meng, Z., D. Dadub, and J. H. Seinfeld, 1998: Size-resolved and chemically resolved model of atmospheric aerosol dynamics, *J. Geophys. Res.*, 103, 3410-3435.

Moran, M. D., M. T. Scholtz, C. F. Slama, A. Dorkalam, A. Taylor, N. S. Ting, P. A. Makar, and S. Venkatesh, 1997. An overview of CEPS1.0: Version 1.0 of the Canadian Emissions Processing System for regional air-quality models, Proc., 7th AWMA Emission Inventory Symp., 28-30 Oct. Research Triangle Park, N.C., Air and Waste Manage. Assoc., Pittsburgh.

Odum, J. R., T. Hoffman, F. Bowman, D. Collins, R. C. Flagan, and J. H. Seinfeld, 1996. Gas/particle partitioning and secondary aerosol formation. Environ. Sci. Technol., 30, 2580-2585.

Pudykiewicz, J. A., A. Kallaur and P. K. Smolarkiewicz, 1997. Semi-Lagrangian modelling of tropospheric ozone, Tellus, 49B, 231-248.

Zhang, L., M. D. Moran, P. A. Makar, J. R. Brook, and S. Gong, 2001. Asize-segregated particle dry deposition scheme for an atmospheric aerosol module. Atmos. Environ., 35, 549-560.

Zhang, L., M. D. Moran, P. A. Makar, J. R. Brook, and S. Gong, 2002. Modelling gaseous dry deposition in AURAMS: a unified regional air-quality modelling system. Atmos. Environ., 36, 537-560.

65 Analyzing the Validity of Similarity Theories in Complex Topographies

Osvaldo L. L. Moraes[1], Otávio Acevedo[1], Cintya A. Martins[1], Vagner Anabor[1], Gervásio Degrazia[1], Rodrigo da Silva[1], and Domenico Anfossi[2]

1. Introduction

The dispersion of trace gases in the atmosphere depends on the state of the atmospheric boundary layer (ABL) and one of the most important parameters characterizing it is the intensity of turbulence within the ABL. Hence, the reliability of the atmospheric dispersion models depends on the way turbulent parameters are calculated and related to the structure of the ABL. Similarity theories are the usual tool used to study the ABL structure, and it is able to describe not only the distributions of turbulent statistical variables, but also profiles of the mean variables and spectra. Different kinds of similarity have different similarity scales, similarity relations and application ranges. There is a long history of field experiments contributing to the continuing development of similarity theories (Businger et al., 1971; Niewstadt, 1984; Sorbjan 1986; Mahrt et al. 1998) and the results of such studies lead to a well known description of the structure of turbulence over flat, homogeneous surfaces and various atmospheric conditions. However only in recent years attention has been devoted to the structure of the atmospheric turbulence over complex terrain. Kaimal and Finnigan (1994) addressed the problem of heterogeneity surface on several scales. On the smallest scale the effects are confined to the surface layer and it is related to local advection in micrometeorology, i.e. how far downwind of a change we must go to find a flow in equilibrium with the local surface. Over very complex terrain, however, this equilibrium may never be attained. Generally studies over non-homogeneous topography try to estimate surface fluxes from profile measurements once the one-dimensional conditions, upon which Monin-Obukhov similarity is predicted, are lost. On the other hand information on turbulence response to the surface changes is much less complete than that on the mean fields (Kaimal and Finnigan, 1994). Based on theory, Jackson and Hunt (1975) have suggested the existence of a two-layer to describe

[1] Laboratório de Física da Atmosfera, Departamento de Física, UFSM, Santa Maria, Brazil.
[2] Instituto di Scienze dell'Atmosfera e Del Clima, Torino, Italy.

the turbulence structure over hilltops, escarpments, etc. In the so-called inner layer (near the hill surface) of depth λ, turbulence is expected to be in equilibrium with the current boundary conditions, hence the turbulence structure may be predicted on the basis of surface layer laws. In the overlying outer layer, turbulence is expected to be modified by the rapid distortion effect (Townsend, 1972; Britter et al, 1981). Observations from different field studies (Mason and Sykes, 1979, Bradley, 1980; Taylor et al. 1983; Mason and King, 1985; Mickle et al, 1988) fail to provide a simple turbulence description for both the inner layer and to the outer layer. In this paper we analyse integral turbulence statistics in terms of usual Monin-Obukhov similarity parameters from data collect at two sites in the south of Brazil. Both sites are located in complex terrain, so that different wind directions represent different conditions regarding the equilibrium state.

2. Site and Instruments Details

We use in this study data collected at two observational campaigns conducted in southern Brazil. The effort of these observations has the main purpose of understanding the physical microclimatology of the regions, which will be partially flooded in the upcoming years, due to the construction of river dams. The first campaign was conducted at the Jacuí River valley, from August to September 2000, during Southern Hemisphere winter. The site (S 29° 28′, W 53° 17′ 9.2″) was located at the bottom of the valley where the ridges have an average elevation of 300 m above the valley. In this part the valley has a north-east to south-west orientation at a 50-km scale. However, on a smaller scale, the valley has a north-south orientation at the south location. At the floor the area is almost flat with weak slopes toward the ridges in the east-west direction. The second campaign (5 November 2001 to 15 December 2001) was conducted at the edge of a sharp cliff, at the bottom of which a river runs (Rio das Antas). The steep slope starts only a few meters from the tower, at its west side. Another sharp cliff exists at 1km, south of the tower. Figure 1 bellow display the topographic characteristics of these sites.

In both campaigns turbulence was measured by a three-dimensional Campbell sonic anemometer located at the 10-m level. Slow response sensors included Vaisala temperature and humidity probes, R. M. Young anemometers, Vaisala pressure and net radiometer sensors was also used. Further details on the site and measurements are described in Moraes et al. (2003) and Acevedo et al (2002).

Evolutions of the micrometeorological parameters were determined for every 3 minutes, using 30-minute data windows for the mean removal. Before the analyses the individual wind vectors were rotated into the mean wind direction in such a way that $\overline{v} = \overline{w} = 0$. In the second phase, the mean values, surface layer parameters ($u*$, $\overline{w'\theta'}$, L, $T*$, $q*$) and standard deviation of the turbulence variables (σ_i ($i = u$, v or w), σ_θ, σ_q) were calculated. A data quality analysis was applied to detect any spikes, kinks or missing portions in the data. In this paper we analyze only those sets for which $-2 \leq z/L \leq 2$. With these procedures the overall

FIGURE 1. Topographic view of the two sites. Left panel Jacuí River Valley. Right panel Antas River Valley. In both panels it is indicated the location of the micrometeorological towers.

number of thirty minutes test runs is 9802 (4940 in convective conditions, 4543 in stable conditions and 319 in neutral conditions) for the first campaign and 19200 (9115 in convective conditions, 8117 in stable conditions and 1968 in neutral conditions) for the second experiment. Each one of them was collected at a 10-Hz sample rate.

3. Results

As said in the introduction, the present paper intends to verify the applicability of the Similarity Theory, using tower-layer data in complex topography, to give the similarity relationships of the dimensionless velocity and temperature variances in terms of similarity scalings.

According to Monin-Obukhov (1954), the structure of surface layer turbulence is determined by a few key parameters. These are the height above ground (z), the buoyancy parameter (g / T_o, where g is the acceleration due to gravity and T_o is the mean surface temperature), the kinematic surface stress (τ / ρ, where ρ is the density of air), and the surface temperature flux $H_o / \rho c_p$ where H_o is the surface turbulent sensible heat flux and c_p is the specific heat at constant pressure). As proposed by Monin-Obukhov (1954) various atmospheric parameters and statistics, when normalized by appropriate powers of scaling velocity ($u*$) and scaling temperature ($T*$), become universal functions of the stability parameter (z / L as defined below).

For characterizing the surface layer structure, turbulent fluxes of heat and momentum were estimated using eddy correlation technique. The scaling of velocity has been evaluated from the expression

$$u_* = (\tau/\rho)^{1/2} = \left(\overline{w'u'}^2 + \overline{w'v'}^2\right)^{1/4}, \tag{1}$$

where $\overline{w'u'}$ and $\overline{w'v'}$ are the turbulent fluxes of momentum in the direction of u and v components of the wind vector velocity, respectively. Another important surface parameter, the Monin-Obukhov length scale, was obtained for each test runs using

$$L = -\frac{\rho c_p u_*^3}{\kappa \frac{g}{T_o} H_o}, \tag{2}$$

where κ is the von-Karman constant.

For each of the 30 min data series, mean wind speed, temperature, surface fluxes, stability parameters and velocity fluctuation standard deviations were determined. In the next step, all the results were block averaged for a range of z/L values. In order to analyze the effect of the ridges on the turbulence statistics, we separated the 30 min time series according to the wind direction. Data from the first campaign were separated in two groups: the first one has those time series for which the wind direction is $180° \pm 45°$ or $360° \pm 45°$. These time series are referenced as parallel. The second one has time series with other wind directions, and they are called transversal. Data from the second campaign were also separated in three groups. The first one has time series with wind direction is from the valley to the tower, i.e. $240° \pm 45°$. The second one has time series whose wind direction is from the tower to the valley, i.e. $60° \pm 45°$. The other group has time series with wind direction parallel to the valley.

Our analysis includes the standard deviation of wind velocity components normalized by the friction velocity and the standard deviation of temperature normalized by the scale of temperature.

Initially, we present in Figure 3 the scatter plots for the variation of σ_w/u^* in terms of z/L. Left panel presents this parameter for wind direction parallel to the valley axis while right panel display the results for wind direction transversal to the valley axis. In this situation and for the tower located at the edge cliff two situations are considered: flow from the valley and flow to the valley. In the figure each symbol represents these different wind directions. Figures 3 and 4 have the same analysis for the non-dimensional standard deviation of transversal and longitudinal turbulent velocity.

As we can see from the Figure 2 $\sigma_w/u*$, obtained at the bottom valley and at the border of the valley top are reasobly described by the stability parameter when the wind direction is parallel to the valley axis. In the first situation a good relationship can be obtained in the range $-2 \leq z/L \leq 1$ but at the border of the valley top this range is more limited ($-0.5 \leq z/L \leq 0.5$). When the wind direction is transversal to the valley axis (Figure 2 – right panel) we can observe the same tendency as observed in Figure 2 (left panel) but the scatter in the data does not permits to extrapolate any relationship. Panofsky and Dutton (1984)

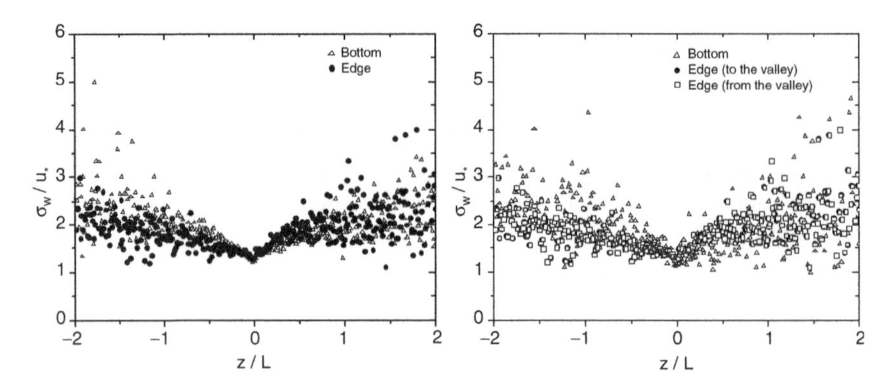

FIGURE 2. Standard deviation of vertical velocity component normalized by the friction velocity. Left panel display the results for wind direction parallel to the valley axis. Right panel is for wind direction transversal to the valley axis. For the tower located at the top of the valley two transversal directions are analyzed separately.

hypothesysed that the vertical fluctuations in the surface layer are produced by small eddies that rapidly adjust to terrain changes and they are never far from the equilibrium. However our results confirm this hypothesis only when the wind direction is parallel to the valley and only for a limited interval of z/L. In the neutral condition $\sigma_w/u* \approx 1.25$ for all situations in agreement with results obtained over homogeneous terrain (Dutton and Panofsky 1984).

Figures 3 and 4 show that a small scatter in the data points for $\sigma_u/u*$ and $\sigma_v/u*$ plots is observed only for the interval $-1 < z/L < 1$, i.e. for weakly stable and convective conditions and at the bottom of the valley and wind direction parallel to the valley axis. All other conditions can not be described by the stability parameter derived from the similarity theory.

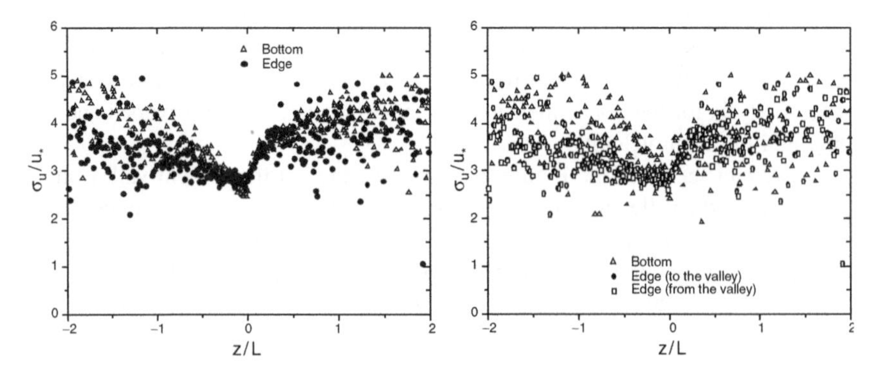

FIGURE 3. As Figure 2 but for the longitudinal velocity component.

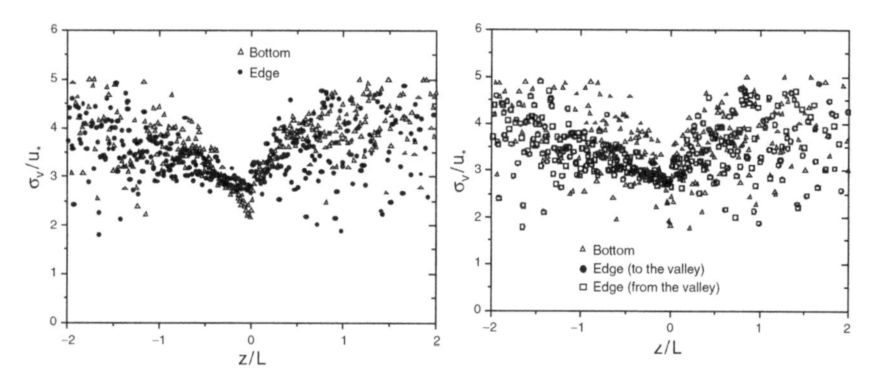

FIGURE 4. As Figure 2 but for the transversal velocity component.

4. Conclusions

In this paper observations of turbulence carried out in the bottom and in the top of a valley ridge were analyzed. The main purpose was to investigate the limit of the Monin-Obukhov Similarity Theory to describe the behavior of the non-dimensional standard deviation of the turbulence velocities components. The results indicate that this description is acceptable only when the wind direction is parallel to the valley axis and the observations were done at the bottom of the valley. For wind direction transversal to the valley, the parameters cannot be described in terms of the stability parameter. This is true for observations in the bottom as well as in the top of the valley. In this condition there is no local equilibrium and the turbulence flow does not adjust quickly to the distortion induced by topographic effects.

Acknowledgements This work was supported by the Brasilian funding agency CNPq and by two private companies: CERAN and DFESA.

References

Acevedo, O. C., Moraes, O. L. L., Silva, R., 2002, Turbulence Observations at the Edge of a Cliff, *15th Symposium on Boundary Layers and Turbulence,* Wageningen, The Netherlands, 592-595.

Bradley, E. F., 1980, 'An Experimental Study of the Profiles of the Wind Speed, Shearing Stress and Turbulence at the Crest of Large Hill', *Quart. J. Roy. Meteorol. Soc.*, **106,** 101-124.

Britter, R. E., Hunt, J. C. R., Richards, K. J., 1981, Airflow over two-dimensional hill: studies of velocity speedup, roughness effects and turbulence, *Quart. J. Roy. Meteorol. Soc.*, **107**, 91-110.

Businger, J. A., Wyngaard, J. C.; Izumi, Y., Bradley, F., 1971, Flux- Profile Relationships in the Atmospheric Surface Layer, *J. Atmos. Sci.*, **28**, 181-189.

Jackson, P. S., Hunt, J. C. R., 1975, Turbulent wind flow over a low hill, *Quart. J. Roy. Meteorol. Soc.*, **101**, 929-955.

Kaimal, J. C., Finnigan, J. J., 1994, *Atmospheric Boundary Layer Flows: Their Structure and Measurement*, Oxford University Press, New York, 289 pp.

Mahrt, L., Sun, J., Blumen, W., Delany, T., Oncley, S., 1998, 'Nocturnal Boundary- Layer Regimes', *Boundary-Layer Meteorology*, **88**, 255-278.

Mason, P. J., King, J. C., 1985, 'Measurements and Predictions of Flow and Turbulence over an Isolated Hill of Moderate Slope', *Quart. J. Roy. Meteorol. Soc.*, **111**, 617-640.

Mason, P. J., Sykes, R. I., 1979, 'Flow over an Isolated Hill of Moderate Slope', *Quart. J. Roy. Meteorol. Soc.*, **105**, 383-395.

Mickle, R. E., Cook, N. J., Hoff, A. M., Jensen, N. O., Salmon, J. R., Taylor, P. A., Tetzlaff, G., Teunissen, H. W., 1988, 'The Askervein Hill Project: Vertical Profiles of Wind and Turbulence', *Boundary-Layer Meteorology*, **43**, 143-169.

Monin, A. S., Obukhov, A. M., 1954, *Basic Laws of Turbulent Mixing in the Ground Layer of the Atmosphere.* Trans. Geophys. Inst. Akad. Nauk. USSR, **151,** 163-187.

Moraes, O. L. L., Acevedo, O. C., Silva, R., Magnago, R., Siqueira, A. C., 2003, Surface-Layer Characteristics over Complex Terrain. under Stable Conditions. *Boundary- Layer Meteorology*, **112**, 159-177.

Nieuwstadt, F. T. M., 1984, The Turbulence Structure of the Stable, Nocturnal Boundary Layer, *American Meteorological Society*, **41**, 2202-2216.

Sorbjan, Z., 1986, On similarity in the atmospheric boundary layer, *Boundary- Layer Meteorol.*, **34**, 377-397.

Taylor, P. A., Mickle, R. E., Salmon, J. R., Teunissen, H. W., 1983, 'The Kettles Hill Experiment – Site Description and Mean Flow Results', Internal Report Aorb-83-002-L, Atmos. Environm. Service, Downsview. Ont., Canada.

Townsend, A. A., 1972, Flow in a Deep Turbulent Boundary Layer over a Surface Distorted Water Waves, *J. Fluid Mech.*, **55**, 719-735.

66 Siting and Exposure of Meteorological Instruments at Urban Sites

Tim R. Oke[1]

1. Introduction

There is a growing need for meteorological data in urban areas in support of air pollution research and management, but measurement poses substantial challenges. Most densely-developed sites make it impossible to conform to the standard WMO Guidelines for site selection and instrument exposure (WMO, 1996) due to obstruction of airflow and radiation exchange by buildings and trees, unnatural surface cover and waste heat and water vapour from human activities. New guidelines (Oke, 2004) to assist in this task form the basis of the first part of this paper. Here emphasis is on those variables of greatest use in air pollution applications. Valid and repeatable results can be obtained despite the heterogeneity of cities, but it requires careful attention to principles and concepts specific to urban areas. Guidelines must be applied intelligently and flexibly, rigid 'rules' have little utility. It is necessary to consider exposures over non-standard surfaces at non-standard heights, splitting observations between more than one location, or being closer to buildings or anthropogenic heat and vapour sources than is normal WMO recommended practice.

1.1. Definitions and Concepts

1.1.1. Scales

The success of an urban station depends on an appreciation of the concept of scale. There are three scales of interest in urban areas (Oke, 1984):

(a) *Microscale* – typical scales of urban microclimates are set by the dimensions of individual elements: buildings, trees, roads, streets, courtyards, gardens, etc., extending from less than one to hundreds of metres. WMO guidelines for an open-country climate station are designed to avoid microclimate effects and to standardize, as far as is practical – a set height of measurement, single

[1] Department of Geography, University of British Columbia, Vancouver, BC Canada V6T 1Z2

surface cover, minimum distances to obstacles and little horizon obstruction. The aim is to achieve climate observations free of extraneous microclimate signals to characterize local climates. Avoiding anomalous microclimate influences is hard to achieve.

(b) *Local scale* – this scale includes climatic effects of landscape features, such as topography, but excludes microscale effects. In cities this means the climate of neighbourhoods with similar types of urban development (surface cover, size and spacing of buildings, activity). Typical scales are one to several kilometres.

(c) *Mesoscale* – a city influences weather and climate at the scale of the whole city, typically tens of kilometres in extent. A single station is not able to represent this scale.

An essential difference between the climate of urban areas and that of open-country sites is that the vertical exchanges of momentum, heat and moisture occur in a layer of significant thickness – called the urban canopy layer (UCL). The height of the UCL is approximately equivalent to the mean height of the main roughness elements (buildings and trees), z_H. Whilst the microclimatic effects of individual surfaces and obstacles persist for a short distance away from their source they blend, in the horizontal and vertical, by turbulence. The distance depends on the magnitude of the effect, the wind speed and the stability. Effects may persist up to a few hundred metres horizontally. In the vertical, individual element effects are discernable in the roughness sublayer (RSL) up to the blending height, z_r. Field measurements indicate z_r can be as low as $1.5z_H$ at densely built (closely spaced) sites, but greater than $4z_H$ in low density areas (Grimmond and Oke, 1999; Rotach, 1999; Christen, 2003). Instruments placed above z_r 'see' a blended, spatially-averaged signal representative of the local scale.

Each local scale surface type (e.g. distinct neighbourhood) generates an internal boundary layer that grows with fetch at a rate depending on the roughness and stability. In rural conditions height:fetch ratios vary from as small as 1:10 in unstable conditions to as large as 1:500 in stable cases (Garratt, 1992; Wieringa, 1993). Urban areas tend towards neutral stability, due to enhanced thermal and mechanical turbulence associated with the heat island and their large roughness, therefore, a ratio of about 1:100 is more typical. Given the nature of the UCL the height of the internal boundary layer is taken above the displacement height z_d (typically $z_d \sim 0.5 - 0.7z_H$; Grimmond and Oke 1999). So in a densely-built district ($z_H = 10$ m), it means $z_r \geq 15$ m and the fetch requirement over similar urban terrain is likely to be at least 0.8 km. This is a real site restriction because if the urban terrain is not similar out to at least this distance around the site, then observations will not be representative of the local surface type. At less densely developed sites, where heat island and roughness effects are less, the fetch requirements are likely to be greater. Above the blending height, but within the local internal boundary layer, measurements are within an inertial sublayer where boundary layer theory applies.

1.1.2. Source Areas ('Footprints')

A sensor placed above a surface 'sees' only a portion of its surroundings. This is its 'source area' which depends on the height and the process transporting the surface property to the sensor. For upwelling radiation and surface temperature viewed by an infrared thermometer the field-of-view of the instrument and the surface geometry set what is seen. By analogy sensors such as thermometers, hygrometers, gas analyzers, anemometers 'see' properties such as temperature, humidity, atmospheric gases, wind speed and direction that are carried from the surface to the sensor by turbulent transport.

The source area of a downfacing radiometer with its sensing element parallel to the ground is a circular patch with the instrument at its centre (e.g. Schmid *et al.* (1991). Depending on its field-of-view, a radiometer may see only a limited circle, or it may extend to the horizon. The instrument usually has a cosine response, so that towards the horizon it becomes increasingly difficult to define the actual source area seen, so the view factor is defined as the area contributing a set proportion of the instrument's signal (typically 50, 90, 95, 99, or 99.5%).

The source area of a sensor that derives its signal via turbulent transport is not symmetrically distributed about the sensor location. It is elliptical in shape and is aligned in the upwind direction from the tower. The influence of the ground area at the base of the mast is zero, because turbulence cannot transport the influence up to the sensor level. Further upwind the source starts to affect the sensor, this effect rises to a peak, thereafter decaying at greater distances (for the shape in both the x and y directions see Kljun *et al.*, 2004; Schmid, 2002). The position and shape of the ellipse source area ('footprint') vary considerably over time depending on the height of measurement (larger at greater heights), surface roughness, atmospheric stability (increasing from unstable to stable) and whether a turbulent flux or a meteorological concentration is being measured (larger for the concentration) (Kljun *et al.*, 2004).

Methods to calculate the dimensions of flux and concentration 'footprints' are available (Schmid, 2002; Kljun *et al.*, 2004) and best apply to instruments placed in the inertial sublayer, above the complications of the RSL and the complex geometry of the three-dimensional urban surface. Within the UCL the source areas of instruments cannot be evaluated reliably due to the obvious complications of the complex flow and radiation environments in the UCL. The immediate surroundings of the station will have the greatest effect and the extent of the turbulent source areas will grow with stability and the height of the sensor. The distance influencing screen-level (~1.5 m) sensors may be a few tens of metres in neutral conditions, less when it is unstable and perhaps more than a hundred metres when it is stable. At a height of three metres the equivalent distances probably extend up to about three hundred metres in the stable case. A rule-of-thumb is that the circle of influence on a screen-level temperature or humidity sensor has a radius of about 0.5 km, but this is likely to depend upon the building density.

1.1.3. Measurement Approaches

It follows from the preceding discussion that if the goal of an urban site is to monitor the local scale climate, there are two viable approaches:

(a) locate the site in the UCL at a location surrounded by average or 'typical' conditions for the urban terrain, and place the sensors at heights similar to those used at non-urban sites. This assumes the mixing induced by flow around obstacles is sufficient to blend properties to form a UCL average at the local scale; or
(b) mount the sensors on a tall tower above the RSL and measure blended values that can be extrapolated down into the UCL.

In general approach (a) works best for air temperature and humidity, and (b) for wind speed and direction and precipitation. For radiation the main requirement is an unobstructed horizon. Urban stations, therefore, often consist of instruments deployed both below and above roof-level and this requires that site assessment and description include the scales relevant to both contexts.

1.1.4. Urban Site Description

The dimensions of the morphometric features comprising the urban landscape confer the dimensions of urban climate scales. This emphasizes the need to adequately describe the properties of urban districts that affect the atmosphere. The most important features are the urban *structure* (dimensions of the buildings and the spaces between them, the street widths and street spacing), the urban *cover* (built-up, paved, vegetated, bare soil, water), the urban *fabric* (construction and natural materials) and the urban *metabolism* (heat, water and pollutants due to human activity). Proper description of a site should include measures of these descriptors. Then they can be used to select potential sites, and be incorporated in the site metadata to accurately describe the setting of the station.

These four features tend to cluster together to form characteristic urban classes. For example, in the central areas of cities many have tall buildings relative to street width that are densely packed (*structure*) so the ground is largely covered with buildings or paved surfaces made of durable materials such as stone, concrete, brick and asphalt (*cover*, *fabric*) where heat releases from furnaces, air conditioners, chimneys and vehicles are large (*metabolism*). Near the other end of the spectrum often are districts with low density housing of one- or two-storey buildings of light construction and considerable garden or vegetated areas with low heat releases, but perhaps large irrigation use.

There is no universally accepted scheme of urban classification for climatic purposes. The Urban Terrain Zone scheme of Ellefsen (1990/91) are a good start. They emphasize structure and indirectly reflect aspects of cover, fabric and metabolism because a given structure carries with it the type of cover, materials, and degree of human activity. Application of the scheme needs only aerial photography. A new simple scheme of Urban Climate Zones (UCZ) is forwarded (Figure 1). It incorporates groups of Ellefsen's zones, plus a measure of the

Urban Climate Zone, UCZ[1]	*Image*	*Roughness class*[2]	*Aspect ratio*[3]	*% Built (impermeable)*[4]
1. Intensely developed urban with detached close-set high-rise buildings with cladding, e.g. downtown towers		8	> 2	> 90
2. Intensely developed high density urban with 2 – 5 storey, attached or very close-set buildings often of brick or stone, e.g. old city core		7	1.2 – 2.5	> 85
3. Highly developed, medium density urban with row or detached but close-set houses, stores & apartments e.g. urban housing		7	0.5 – 1.5	70
4. Highly developed, low density urban with large low buildings & paved parking, e.g. shopping mall, warehouses		5	0.05 – 0.2	75 - 95
5. Medium development, low density suburban with 1 or 2 storey houses, e.g. suburban housing		6	0.2 – 0.5, up to > 1 With tall trees	35 - 65
6. Mixed use with large buildings in open landscape, e.g. institutions such as hospital, university, airport		5	0.1 – 0.5, depends on trees	< 40
7. Semi-rural development with scattered houses in natural or agri-cultural area, e.g. farms, estates		4	> 0.05, depends on trees	< 10

Key to image symbols: *buildings; vegetation; impervious ground; pervious ground*

[1] A simplified set of classes that includes aspects of the scheme of Ellefsen (1990/91) plus physical measures relating to wind, thermal and moisture controls (columns at right).

[2] Effective terrain roughness according to the Davenport classification (Davenport et al., 2000).

[3] Aspect ratio = z_H/W - related to flow regime types and thermal controls (solar shading and longwave screening). Tall trees increase this measure significantly.

[4] Av. fraction of ground covered by built features (buildings, roads, paved and other impervious areas) the rest of the area is occupied by pervious cover. Permeability affects the ability to store moisture and hence the moisture status of the ground.

FIGURE 1. Classification of distinct urban forms arranged in approximate decreasing order of their ability to impact local wind, temperature and humidity climate (Oke, 2004).

structure, z_H/W (W – element spacing or street width) known to be related to flow, solar access and the heat island, plus a measure of the surface cover (%Built) related to the degree of surface permeability. The importance of UCZ, is not their absolute accuracy to describe the site but their ability to classify areas of a settlement into districts with similar capacity to modify the local climate, and to identify potential transitions to different urban climate zones. Such classification is crucial when setting up an urban station to ensure that spatial homogeneity criteria are met for a station in the UCL or above the RSL. The number and description of classes may need adaptation to accommodate the special nature of some cities.

2. Choosing a Location and Site for an Urban Station

First, it is necessary to establish the purpose of the station. If there is to be only one station inside the urban area it must be decided if the aim is to monitor the greatest impact of the city, or of a more representative or typical district, or if it is to characterize a particular site. Areas having the highest probability of maximum effects can be judged from the ranked list of UCZ types in Figure 1. Similarly whether a station will be 'typical' can be assessed using the ideas behind Figure 1 to select extensive areas of similar urban development for closer investigation.

The search can be usefully refined in the case of air temperature and humidity by conducting spatial surveys, wherein the sensor is carried on foot, or mounted on a bicycle or a car and traversed through areas of interest to see if there are areas of thermal or moisture anomaly or interest. The best time to do this is a few hours after sunset or before sunrise on nights with relatively calm airflow and cloudless skies. This maximizes the potential for the differentiation of micro- and local climate differences.

If the station is to be part of a network to characterize spatial features of the urban climate then a broader view is needed informed by knowledge of the typical spatial form of urban climate distributions (e.g. isolines of urban heat and moisture 'islands'). It must be decided if the aim is to observe a representative sample of the UCZ diversity, or is it to faithfully reflect the spatial structure? The latter is usually too ambitious with a fixed-station network in the UCL because it requires many stations to depict the gradients near the periphery, the plateau region, and the nodes of weaker and stronger than average urban development. With sensors above the RSL, the blending action produces muted spatial patterns and fetch distance to the nearest UCZ transition, or the urban-rural fringe, is critical. In the UCL a distance to a change of UCZ of 0.5 to 1 km may be acceptable, but for a tower-mounted sensor the requirement is likely to be more like a few kilometres. Since the aim is to monitor local climate attributable to an urban area it is sensible to avoid locations extraneous microclimatic influences or non-urban local or mesoscale climatic phenomena that will complicate the urban record.

Once a UCZ type and its general location inside the urban area are chosen potential candidate sites are selected from map, imagery and photographic evidence and a foot survey. Areas of reasonably homogeneous urban development without large patches of anomalous structure, cover or materials, or a transition zone to a different UCZ are ideal. The precise definition of 'reasonably' however is not possible. For each candidate site the expected range of footprint areas should be estimated for radiation and turbulent properties. Key surface properties (e.g. mean height and density of obstacles, surface cover, materials) within the footprint areas should be documented. Their homogeneity should then be judged, 'by eye' or by statistical methods.

3. Exposure of Instruments

A curious legacy of open country standardization is that many urban stations are placed over short grass in open locations (parks, playing fields). As a result they monitor modified rural-type conditions, not representative urban ones. The guiding principle for the exposure of sensors in the UCL should be to locate them so they monitor conditions that are representative of the environment of the selected UCZ. The %Built category (Figure 1) is a crude guide to the recommended underlying surface. The most obvious requirement that cannot be met at many urban sites is the distance from obstacles. Instead it is recommended that the urban station be centred in an open space where the surrounding aspect ratio (z_H/W) is representative of the locality.

3.1. Temperature

Standard thermometry is appropriate for urban observations but radiation shielding and ventilation is even more necessary. In the UCL a sensor may be close to warm or highly reflective surfaces (sunlit wall, road, glass or hot vehicle). Hence shields must block radiation effectively. Similarly, the lower UCL may be so sheltered that forced ventilation of the sensor is essential.

In accord with the above the surface should be typical of the UCZ and the thermometer screen/shield centred in a space with approximately average z_H/W. In very densely built-up UCZ this might mean it is located only 5 to 10 m from buildings. If the site is a street canyon, z_H/W only applies to the cross-section normal to the axis of the street. The recommended open-country screen height of 1.25 to 2 m above ground level acceptable for urban sites but on occasion it may be better to relax this requirement to allow greater heights. Observations in canyons show slight air temperature gradients in the UCL (Nakamura and Oke, 1988), so as long as the sensor is >1 m from a wall error should be small, especially in densely built-up areas. Measurements at heights of 3 or 5 m are little different from those at the standard height. They even benefit by having larger source areas, the sensor is beyond the easy reach of vandals or the path of vehicles, and exhaust heat from vehicles is diluted.

Too often roofs are sites for meteorological observations. This may arise in the mistaken belief that at this elevation sensors are free from microclimates, such as those in the UCL. In fact roof tops have strongly anomalous microclimates. To be good insulators roofs are constructed of materials that are thermally extreme. In light winds and cloudless skies they become very hot by day, and cold by night, with sharp temperature gradients near the roof. Roofs design also ensures they are waterproof and shed water rapidly. This together with their openness to solar radiation and wind makes them anomalously dry. Roofs are also commonly affected by release of heat from roof exhaust vents.

Air temperatures above roof-level using towers, are influenced UCL and roof effects. Whilst there is little variation of temperature with height in the UCL, there is a discontinuity near roof-level both horizontally and vertically. Hence if meaningful spatial averages are sought sensors should be well above mean roof-level so that adequate blending is accomplished (>1.5 z_H if possible). Currently there are no methods to extrapolate air temperature data from above the RSL down into the UCL. Similarly, apart from statistical methods that require a large set of training data from a dense station network there is no scheme to extrapolate air temperatures horizontally inside the UCL.

3.2. Humidity

The guidelines for the siting and exposure of temperature sensors in the UCL, and above the RSL, apply equally to humidity sensors. Urban environments are notoriously dirty (dust, oils, pollutants) which means hygrometers are subject to degradation and require increased maintenance in urban environments. For example wet-bulb wicks become contaminated, hair strands disintegrate and the mirror of dew-point hygrometers and the windows of ultraviolet and infrared absorption hygrometers need to be cleaned frequently. Increased shelter in the UCL means forced ventilation is essential.

3.3. Wind Speed and Direction

The measurement of wind speed and direction is highly sensitive to distortion of the mean flow and turbulence by obstacles. Concerns arise at all scales, including the effects of local relief (hills, valleys, cliffs), sharp changes in roughness length (z_0) or the zero-plane displacement (z_d), clumps of trees and buildings, individual trees and buildings even the disturbance induced by the anemometer mast or mounting arm.

However, if a site is on reasonably level ground, has sufficient fetch downstream of major changes of roughness and is in a single UCZ without anomalously tall buildings nearby, then a mean log wind profile should exist in the inertial layer above the RSL. Within the RSL and UCL no one site can be expected to possess such a profile. Individual locations experience highly variable speed and direction shifts as the airstream interacts with individual buildings, streets, courtyards and trees. In street canyons the shape of the profile is different for along-canyon, versus across-canyon flow (Christen *et al.* 2002) and depends on position across and along the street (DePaul and Shieh, 1986). As an engineering approximation the profile in the UCL can be described by an exponential form (Britter and Hanna, 2003) merging with the log profile near roof-level.

The wind profile parameters z_0 and z_d can be measured using a vertical array of anemometers, or measurements of momentum flux or gustiness from fast-response anemometry in the inertial layer, but estimates vary with wind direction and are sensitive to errors (Wieringa, 1996). Methods to parameterize the wind profile parameters z_0 and z_d for urban terrain are also available (Grimmond and Oke, 1999; Davenport *et al.*, 2000). It is essential to incorporate z_d into urban wind profile assessments. Depending on the building and tree density this could set the base of the profile at a height between 0.5 and $0.8z_H$ (Grimmond and Oke, 1999).

The choice of height at which to measure wind in urban areas is a challenge, but if some basic principles are applied meaningful results can be attained. For rural observations the measurement height is set at 10 m above ground and the sensor should not be nearer to obstructions than ten obstacle heights. In most urban districts it is not possible to find such locations, e.g. in a UCZ with 10 m high buildings and trees a patch of at least 100 m radius is needed. If such a site exists it is unlikely to be representative of the zone. The RSL extends to a height of about $1.5z_H$ in a densely built-up area, and even higher in less densely developed sites. Hence in the example district the minimum acceptable anemometer height is at least 15 m, not the standard 10 m. When building heights are much taller, an anemometer at the standard 10 m height would be well down in the UCL, and given the heterogeneity of urban form and therefore of wind structure, there is little merit in placing a wind sensor beneath, or even at about, roof-level.

Laboratory and field observations show flow over a building creates strong perturbations in speed, direction and gustiness unlike the flow at an open site (Figure 2). These include modifications to the streamlines, recirculation zones on the roof and in the lee cavity behind the building, and wake effects that persist downstream for tens of building heights. Flat-topped buildings create flows on their roofs that are counter to the external flow and speeds vary from jetting to near calm. In general, roofs are very poor locations for climate observations unless the sensors are on tall masts.

There are many examples of poorly exposed anemometer-vane systems in cities. The data registered by such instruments are erroneous, misleading, potentially harmful if used to obtain wind input for wind load or dispersion applications, and wasteful of resources. The inappropriateness of placing anemometers and vanes on short masts on the top of buildings cannot be over-emphasized. Speeds and directions vary hugely in short distances, both horizontally and vertically. Results from instruments deployed in this manner bear little resemblance to the general flow and are entirely dependent on the specific features of the building itself, the mast location on the structure, and the angle-of-attack of the flow to the building. To get outside the perturbed zone wind instruments must be mounted at a considerable height, typically at a height greater than the maximum horizontal dimension of the major roof (Wieringa, 1996). This implies an expensive mast system for which it may be

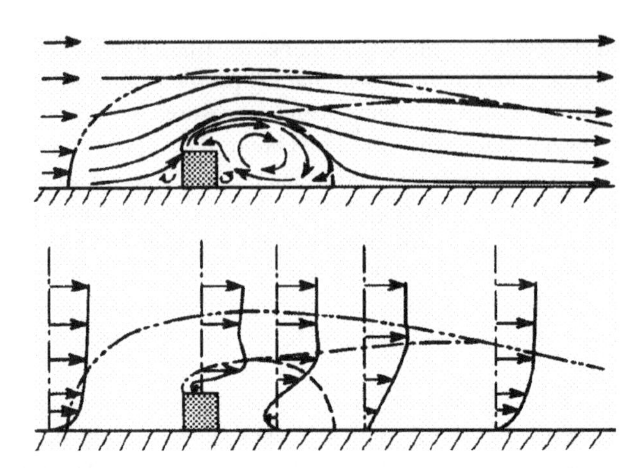

FIGURE 2. Flow (top) and the wind profile (bottom) around and over a sharp-edged building (Halitsky, 1963).

difficult to obtain permission. Nevertheless, this is the only acceptable approach if meaningful data are to be measured.

The following recommendations are made:

(a) in urban areas with low element height and density (UCZ 6 and 7) it may be possible to use the 'open country' exposure guidelines. To use the standard (10 m) height, obstacles should be < 6 m tall on average and > 10 times their height from the mast;
(b) in more densely built districts, with relatively uniform height and density of the elements (buildings and trees), the anemometer should be mounted on a mast of open construction at 10 m or 1.5 times the mean height of the elements, whichever is greater ;
(c) in urban districts with scattered tall buildings the recommendations are as in (b) but with special concern to avoid the wake zone of the tall structures; and
(d) it is not recommended to measure wind speed or direction in densely-built areas with multiple high-rise structures unless a very tall tower is used.
(e) instruments on open construction masts should be mounted on booms long enough to keep the sensors at least two, better three, tower diameters from the mast.
(f) sensors mounted on tall or isolated buildings must consider effects of the structure on the flow. This is likely to require analysis using wind tunnel, water flume or computational fluid dynamics models specifically tailored to the building in question, and including its surrounding terrain and structures.

The aim is to ensure all wind measurements are made at heights sufficient to ensure they are representative of the upstream surface roughness at the local scale and are as free as possible of confounding influences from micro- or local scale surface anomalies. The idea is to gain accurate measurements at whatever height is necessary to reduce error, rather than measuring at a standard height. This may mean the wind site is separate from the location of the other measurement systems. It may also result in wind observations at several different heights in the same settlement, necessitating extrapolation of the measured values to a common height using the log law. A suitable reference height may be 50 m above z_d.

Other exposure corrections for flow distortion, topography, and roughness effects may also have to be applied. If suitable wind observations cannot be made for a given urban site it is still possible to calculate the wind at the reference height using observations at another urban station or the airport using the 'logarithmic transformation' model of Wieringa (1986).

3.4. *Precipitation*

The measurement of precipitation is always susceptible to errors associated with the exposure of the gauge, especially due to the wind field in its vicinity. Given the highly variable wind field in the UCL and the RSL, there are concerns about: (a) the interception of precipitation during its trajectory to the ground by nearby collecting surfaces such as trees and buildings; (b) hard surfaces near the gauge causing splash-in to the gauge, and over-hanging objects dripping into the gauge; (c) the variable wind field around obstacles causing localized augmentation or the absence of rain- or snow-bearing airflow; and (d) the gustiness of the wind together with the turbulence around the gauge leading to under- or over-catch. The turbulent activity created by flow around sharp-edged buildings is more severe than that around natural obstacles and may last for greater distances in their wake. Again, the highly variable wind speeds and directions encountered on the roof of a building make it a site to be avoided.

It is recommended that precipitation gauges in urban areas are:

(a) located at open sites within the city where the standard exposure criteria can be met (e.g. playing fields, open parkland with a low density of trees, an urban airport); or
(b) located in conjunction with the wind instruments if a representative wind site is found. This may mean mounting the gauge above roof-level on a mast where it will be subject to greater than normal wind speed and hence the error of estimation will be greater than near the surface, and the gauge results must be corrected. It also means that automatic recording is favoured.;
(c) not located on the roofs of buildings unless exposed at sufficient height to avoid the wind envelope of the building.

Depth of snowfall should be made at an open site or, if made at developed sites, a large spatial sample must be obtained to account for the inevitable drifting around obstacles.

3.5. *Radiation*

Very few radiation flux measurements are conducted in urban areas. Most radiation sites are located in rural or remote locations specifically to avoid the aerosol and gaseous pollutants of cities that 'contaminate' their records. All short- and longwave fluxes are impacted by the properties of the atmosphere and surface of cities contributing to the net all-wave radiation balance that drives the urban energy balance (Oke, 1988). Incoming solar radiation is a fundamental forcing variable and its measurement should be given high priority when a station is established or upgraded. At automatic stations the addition of solar radiation measurement is simple and relatively inexpensive.

The exposure requirements for pyranometers and other incoming flux sensors are easily met in cities. What is required is that the sensor be level, free of vibration, free of any obstruction above the plane of the sensing element. So a high, stable and accessible platform like the roof of a tall building is often ideal. It is essential to clean the upper domes at regular intervals. In heavily polluted environments this may mean daily.

Outgoing fluxes of radiation (reflected solar, emitted and reflected longwave and the net short-, long- and all-wave radiant fluxes) are seldom monitored in cities. This means the albedo and the opportunity to invert the Stefan-Boltzmann relation and solve for the surface radiant temperature and the critical net radiation that supports warming/cooling of the fabric, and the exchanges of water and heat between the surface and the urban boundary layer are missing. The main difficulty is to ensure the field-of-view of a down-facing radiometer 'sees' a representative sample of the underlying urban surface including does it see both an adequate set of plan-view surface types, but also appropriate fractions of roof, wall, and ground surfaces, including the correct fractions of each that are in sun or shade? Soux *et al.*, 2004 developed a model to calculate these fractions for relatively simple urban-like geometric arrays. It is recommended that:

(a) down-facing radiometers be placed at a height at least as large as a turbulence sensor (i.e. a minimum of $2z_H$ is advisable) and preferably higher; and
(b) the radiative properties of the immediate surroundings of the radiation mast are representative of the urban district of interest.

3.6. Evaporation and other Turbulent Fluxes

Like radiation, evaporation observations in urban areas are almost non-existent at standard climate stations. The use of atmometers, evaporation pans or lysimeters to measure evaporation in the UCL is not recommended. Their evaporative surfaces are not representative of the surroundings and they are in receipt of micro-advection that is likely to force evaporation at unrealistically high rates. Micro-lysimeters can give the evaporation from individual surfaces, but are unsuitable for long-term observations.

Spatially-averaged evaporation and other turbulent fluxes (momentum, sensible heat, carbon dioxide) at the local scale can be observed using sensors above the RSL. Such fluxes are of practical interest in urban areas. The vertical flux of horizontal momentum, and integral wind statistics and spectra are central to questions of wind loading on structures and the dispersion of air pollutants. The turbulent sensible heat flux is required to calculate atmospheric stability (e.g. flux

Richardson Number or Obukhov length) and the depth of the urban mixing layer. Fast-response eddy covariance or standard deviation methods are recommended, rather than profile gradient methods. Appropriate instruments include sonic anemometers, infrared hygrometers and gas analyzers and scintillometers. Exposure should be as for wind sensors: above the RSL but below the internal boundary layer of the UCZ of interest. Accurate measurements rely on the flux 'footprint' being large enough to be representative of the local area of interest.

4. Metadata

The full and accurate documentation of station metadata is essential "to ensure the final data user has no doubt about the conditions in which data have been recorded, gathered and transmitted, in order to extract accurate conclusions from their analysis" (Aguilar *et al.*, 2003). It is even more critical for an urban station, because their sites possess both complexity and a greater propensity to change over time. Change means that site controls are dynamic so documentation must be updated frequently.

Urban stations may expose instruments both within and above the UCL, so site description must include both the micro- and local scales. Metadata should include: (a) a map at the local to mesoscale (~1:50,000) updated regularly to describe urban development changes and ideally an aerial photograph and a simple sketch map (at 1:500,000 or 1:1,000,000) to show the station relative to the rest of the urbanized region and any major geographic features. (b) a microscale sketch map (~1:5,000), according to metadata guidelines, updated each year (see Aguilar *et al.*, 2003). (c) horizon mapping using a clinometer and compass survey in a circle around the screen and a fisheye lens photograph looking at the zenith. (d) photographs in the cardinal directions taken from the instrument enclosure. (e) a microscale sketch of the instrument enclosure, updated when instruments are relocated or other significant changes occur. (f) repeat steps (b) to (d) above for any site where measurements of variables are made separate from the enclosure (on masts, roof-tops, more open locations).

5. Flux Measurements and Estimates Relevant to Dispersion

Until the 1980s there were very few measurements of local scale fluxes of heat, mass and momentum in urban areas. Those attempted were very short-term studies and the methods employed were experimental. By the 1990s questions regarding the relative merits of eddy covariance and gradient methods, the height of the RSL, the nature of turbulent and radiative source areas, and how to handle the unmeasured anthropogenic and storage heat fluxes were addressed. Today the emergence of robust, affordable, commercial flux instrumentation in combination with agreed field methods has meant that repeatable results, gathered over long periods, for several cities, have become available. This in turn has made it possible to conduct inter-site and inter-city comparisons of roughness, turbulence and both radiative and turbulent fluxes. The results have created databases suitable for the construction of parameterizations and testing the predictions of models (for reviews see Grimmond and Oke, 1999, 2002; Roth, 2000; Arnfield, 2003). Although further development of methods will occur, workers in urban meteorology now have a more substantial basis for comparison, testing and model development.

Nevertheless, most of this work on turbulence and flux determination remains in the research and experimental domain. To buy, install and maintain many of these sensors is a costly undertaking that is beyond the budget of most monitoring networks. The modest target of the WMO report (Oke, 2004) is progress toward better observation at ordinary climate stations, similar to those operated by national and other meteorological agencies, but located in urban environments. The most critical of those measures for air pollution applications (dispersion calculations or as model inputs) are wind speed, direction and gustiness. Here I stress that the exposure of wind instruments is sensitive to the effects of obstacles. It is hard to comprehend the number of studies in support of network monitoring, dispersion calculations or the 'validation' of flow or dispersion models that rely on poorly sited or incorrectly exposed wind sensors. This must arise either because of lack of understanding of scale concepts or because someone has deemed it to be too expensive or difficult to follow the protocols outlined here. However, it is no economy to expose good sensors on short masts near, or especially on, buildings where their readings are dominated by obstacle effects that are often totally at odds with the flow properties sought.

If sited and exposed correctly the relatively simple array of instruments at a standard station in a city are useful to estimate fluxes and more sophisticated turbulence variables of relevance to air pollution analysis. This is possible through use of a meteorological pre-processor scheme: a collection of algorithms to convert standard observations into the input variables required by models but that are not normally measured (e.g. atmospheric stability, fluxes of momentum, heat and water vapour, mixing height, dispersion coefficients, etc.). Examples include OLM (Berkowicz and Prahm, 1982; Olesen and Brown, 1992), HPDM (Hanna and Chang, 1992) or LUMPS (Grimmond and Oke, 2002). Such schemes typically require only spatially representative observations of incoming solar radiation, air temperature, humidity and wind speed, and estimates of average surface properties such as albedo, emissivity, roughness length and the fractions of the area vegetated or built-up or irrigated. Ideally the wind, air temperature and humidity measurements are taken above the RSL, but for temperature and humidity if only UCL values are available, they are usually acceptable because the schemes are not sensitive to these variables.

It is increasingly realistic to foresee the adoption of numerical models to generate fluxes and other meteorological properties which, in turn, will drive mesoscale flow, climate and air quality models of urban atmospheres. Probably the most practical of these, because of the relative simplicity of its input requirements and because it is the most widely adopted scheme so far, is the Town

Energy Balance (TEB) model of Masson (2000). TEB is also noteworthy because it has been tested against measured fluxes and climate variables for heavily developed sites in four cities (Masson *et al.*, 2002; Lemonsu *et al.*, 2004). Other models, usually more demanding in their input needs, are available or in development. What sets all these new models apart is their explicit recognition of the UCL, including its three-dimensionality. It is to be expected that in the future there will be an array of 'urban physics packages' that can be coupled to existing mesoscale models. They will vary in the scale they address and the demands they make with respect to input requirements. When validated against well-conducted urban observations such models will be able to complete the circle and provide valuable information to inform the optimal design of observation networks.

References

Aguilar, E., I. Auer, M. Brunet, T. C. Peterson and J. Wieringa, 2003, Guidance on metadata and homogenization, WMO-TD No. 1186, (WCDMP-No. 53), pp. 51.

Arnfield, A. J., 2003, Two decades of urban climate research: A review of turbulence, exchanges of energy and water and the urban heat island. *Int. J. Climatol.*, **23**, 1-26.

Berkowicz, R. and L. P. Prahm. 1982, Sensible heat flux estimated from routine meteorological data by the resistance method, *J. Appl. Meteorol.*, **21**: 1845-1864.

Christen, A., 2003, *pers. comm.*, Instit. Meteorol., Climatol. & Remote Sens., Univ. Basel.

Christen, A., R. Vogt, M. W. Rotach and E. Parlow, 2002, First results from BUBBLE: profiles of fluxes in the urban roughness sublayer, *Proc. 4th Symp. Urban Environ.*, Amer. Meteorol. Soc., Boston, pp. 105-106.

Davenport, A. G., C. S. B. Grimmond, T. R. Oke & J. Wieringa, 2000, Estimating the roughness of cities and sheltered country. *Proc. 12th Conf. Appl. Climat.*, Amer. Meteorol. Soc., Boston, pp. 96-99.

DePaul, F. T. and C. M. Shieh, 1986, Measurements of wind velocity in a street canyon, *Atmos. Environ.*, **20**: 455-459.

Ellefsen, R., 1990/91, Mapping and measuring buildings in the urban canopy boundary layer in ten US cities. *Energy and Build.*, **15-16**: 1025-1049.

Garratt, J. R., 1992, *The Atmospheric Boundary Layer*, Cambridge University Press, Cambridge, pp. 316.

Grimmond, C. S. B. and T. R. Oke, 1999, Aerodynamic properties of urban areas derived from analysis of urban form. *J. Appl. Meteorol.*, **38**: 1262-1292.

Grimmond, C. S. B. and T. R. Oke, 2002, Turbulent heat fluxes in urban areas: observations and a *L*ocal-scale *U*rban *M*eteorological *P*arameterization *S*cheme (*LUMPS*). *J. Appl. Meteorol.*, **41**: 792-810.

Halitsky, J., 1963, Gas diffusion near buildings. *Trans. Amer. Soc. Heat. Refrig. Air-condit. Engin.*, **69**, 464-485.

Hanna, Sr and J.C. Chang, 1992, Boundary layer parameterizations for applied dispersion modelling over urban areas. *Bound.-Layer Meteorol.*, **58**, 229-259.

Kljun, N., P. Calanca, M. W. Rotach and H. P. Schmid, 2004, A simple parameterization for flux footprint predictions, *Bound.-Layer Meteorol.*, **112**, 503-523.

Lemonsu, A., C. S. B. Grimmond, V. Masson, 2004, Modelling the surface energy balance of an old Mediterranean city core, *J. Appl. Meteorol.*, **43,** 312-327.

Masson, V., 2000, A physically-based scheme for the urban energy balance in atmospheric models. *Bound.-Layer Meteor.*, **94**, 357-397.

Masson, V., C. S. B. Grimmond and T. R. Oke, 2002, Evaluation of the *T*own *E*nergy *B*alance (*TEB*) scheme with direct measurements from dry districts in two cities", *J. Appl. Meteorol.,* **41**, 1011-1026.

Nakamura, Y. and T. R. Oke, 1988, Wind, temperature and stability conditions in an E-W oriented urban canyon, *Atmos. Environ.*, **22**: 2691-2700.

Oke, T. R., 1984, Methods in urban climatology. In *Applied Climatology, Zürcher Geograph. Schrift.*, **14,** 19-29.

Oke, T. R. (1988), The urban energy balance. *Prog. Phys. Geogr.*, **12**, 471-508.

Oke, T. R., 2004, *Urban Observations,* Instruments and Methods of Observation Programme, IOM Report, World Meteorol. Organiz., Geneva, in press.

Olesen, H. R. and N. Brown, 1992, The *OML* meteorological pre-processor: a software package for the preparation of meteorological data for dispersion models. MST LUFT-A, 122.

Rotach, M. W., 1999, On the influence of the urban roughness sublayer on turbulence and dispersion. *Atmos. Environ.*, **33**: 4001-4008.

Roth, M., 2000, Review of atmospheric turbulence over cities. *Quart. J. Royal Meteorol. Soc.*, **126**, 941-990.

Schmid, H. P., H. A. Cleugh, C. S. B. Grimmond and T. R. Oke, 1991, Spatial variability of energy fluxes in suburban terrain, *Bound.-Layer Meteorol.*, **54**: 249-276.

Schmid, H. P., 2002, Footprint modeling for vegetation atmosphere exchange studies: a review and perspective. *Agric. Forest Meteorol.*, **113**: 159-183.

Soux, A., J. A. Voogt and T. R. Oke, 2004: A model to calculate what a remote sensor 'sees' of an urban surface, *Bound.-Layer Meteorol.*, **111**: 109-132.

Wieringa, J., 1986, Roughness-dependent geographical interpolation of surface wind speed averages. *Quart. J. Royal Meteorol. Soc.*, **112**: 867-889.

Wieringa, J., 1993, Representative roughness parameters for homogeneous terrain, *Bound.-Layer Meteorol.*, **63**: 323-363.

Wieringa, J., 1996, Does representative wind information exist? *J. Wind Engin. Indus. Aerodynam.*, **65**:1-12.

World Meteorological Organization, 1996, Guide to Meteorological Instruments and Methods of Observation, 6th Edn., WMO-No. 8, World Meteorol. Organiz., Geneva.

Siting and Exposure of Meteorological Instruments at Urban Sites

Speaker: T. Oke

Questioner: R. Bornstein

Question: Is the RSL a blended a blending layer, i.e., is it a layer in which M-O theory can be used (i. e. blended) or is it still "blending" ?

Answer: It is a transition zone in which blending is taking place between the 'chaos' of the urban canopy layer below, in which the microclimatic effects of individual elements dominate, and the inertial layer above, in which blending permits M-O similarity to apply. Of course it is quite possible that horizontal inhomogeneity at the local scale does not permit an inertial layer to exist.

Name of Questioner: R. Yamartino

Question: In many cases, one only has airport wind data. Do the detailed urban tower data suggest that it is at all reasonable to profile the airport winds upward and then profile down over the city?

Answer: Yes they do. Naturally the usefulness also depends on whether the exposure at the airport site gives a well adjusted wind profile with which to work. Assuming that is the case, a technique such the 'logarithmic transformation' model of Wieringa (1986) [Quarterly Journal Royal Meteorological Society, 112, pp. 867-889] works well for practical applications.

67 The Effect of the Street Canyon Length on the Street Scale Flow Field and Air Quality: A Numerical Study

I. Ossanlis, P. Barmpas, and N. Moussiopoulos*

1. Introduction

The problems associated with urban atmospheric pollution have received increasing attention in the past decades. The residence time of a pollutant within a street canyon is a very important factor that affects both the study of hot spots and the chemical reactions that take place within the street canyon.

An example for such chemical reactions are the photocatalytic reactions that take place on the wall surfaces of a street canyon covered with TiO_2, like the ones studied in the PICADA project within EC's 5th Framework Programme. In the frame of this project, a field experiment took place in Guerville, France, which was designed in order to evaluate the performance of translucent TiO_2 coatings and also validate the microscale model MIMO which is suitable for street canyon applications. On this basis, a 3D study of the effect of the street canyon length on the flow field and the dispersion of pollutants within street canyons was undertaken. During this study, several numerical simulations for street canyons of the same height and width and therefore aspect ratio but different lengths have been performed, using MIMO as well as the commercial CFD code CFX TASCflow (URL 1). Additionally, a 2D simulation for a street canyon of the same aspect ratio was performed and the results obtained for the residence time and the flow field were compared.

2. Numerical Modelling

2.1. The Microscale Model MIMO

The increasing interest in the prediction of wind flow and pollutant dispersion over built-up areas and major advances in computer technologies have led to the development of efficient numerical models as an alternative or complementary approach to laboratory and field experiments.

* I. Ossanlis, Photios Barmpas and N. Moussiopoulos, P.O. Box 483, Aristotle University of Thessaloniki, Laboratory of Heat Transfer and Environmental Enginnering, GR 54124, Thessaloniki, Greece

The numerical model MIMO is a three-dimensional model for simulating microscale wind flow and dispersion of pollutants in built-up areas. It solves the Reynolds averaged conservation equations for mass, momentum and energy. Additional transport equations for humidity, liquid water content and passive pollutants can be solved. A staggered grid arrangement is used and coordinate transformation is applied to allow non-equidistant mesh size in all three dimensions in order to achieve a high resolution near the ground and near obstacles (Ehrhard et al., 2000).

2.2. Turbulence Modelling

Numerical modelling of turbulence plays a crucial role in providing accurate microscale wind fields, which are necessary to reliably predict transport and dispersion of pollutants in the vicinity of buildings. In MIMO the Reynolds stresses and turbulent fluxes of scalar quantities can be calculated by several linear and nonlinear turbulence models (Ehrhard, 1999).

In the frame of the present study, the standard k-e two-equation turbulence model by Jones and Launder was utilised because it is widely used and has been tested for various similar cases with very satisfactory results. Wall functions are also an important factor that may affect the results. Different laws of wall functions are implemented in MIMO. In the present study standard wall functions were used.

2.3. Solution Procedure

The governing equations are solved numerically on a staggered grid by using a finite volume discretisation procedure. For the numerical treatment of advective transport in the present study a 3D second-order total variation-diminishing (TVD) scheme is implemented which is based on the basic 1D scheme (Harten, 1986). Furthermore, a second-order flux-corrected transport (FCT) Adams Bashforth scheme can be applied (Wortmann et al., 1995). Both schemes are positive, transportive, conservative and they are characterized by a low level of numerical diffusion.

2.4. Calculation of the Residence Time

The residence time calculation has been the subject of several previous studies. As an example, heuristic calculations were performed with the aim to define and determine a typical penetration or dispersion time scale, which is strictly identical to the residence time in simple 2D street canyons (Sini et al., 1996).

In the present study, the residence time t of an inert pollutant is calculated using the formula:

$$t = \frac{m}{|\dot{m}|} \tag{1}$$

where $\dot{m}$ is the pollutant mass flux that escapes from the street canyon and m the steady-state mass of the pollutants in the canyon.

2.5. Code Validation

MIMO has been validated in previous studies for various test cases and it was applied successfully to air pollution problems. In particular, MIMO was tested against available experimental data in thermally neutral conditions for a) flow over a surface-mounted cube, b) flow in a single cavity and c) flow over a real site location in Hanover. These validation tests have been extensively presented and analysed by Sahm et al., 2002, Assimakopoulos, 2001, Ketzel et al., 2002 and Moussiopoulos, 2003. The results show that the standard k-ε model gives satisfactory predictions for all mean flow fields while providing a relatively good impression on the turbulent diffusion.

2.6. Geometric Specifications

A 3D numerical study was undertaken during which three cases were considered with street canyons of different length. In the first case the canyon length was assumed to be slightly above 6m (L = 6.096m; this is the length of one cargo container as those used in the Guerville experiment), while in the second and third cases the street canyon length was set equal to that of two and three cargo containers respectively (12.192m & 18.276m). The third configuration is the one that was chosen as the experimental street canyon configuration for the Guerville experiment (Figure 1). Additionally, a 2D simulation for a street canyon of the same aspect ratio was performed.

FIGURE 1. Configuration of the Guerville experiment.

2.7. Boundary Conditions

All numerical simulations were performed for identical boundary conditions. Due to lack of experimental measurements, all boundary conditions were imposed taking into account the results of previous street canyon studies.

An inlet logarithmic wind profile with a reference speed $U_\delta = 5$ m/s was assumed at the surface layer height $\delta = 35$m, with the wind direction perpendicular to the street canyon. In all cases the following assumptions were made: roughness length $z_o = 0.0125$m, inflow turbulence intensity 0.03, inlet temperature = 293K and massflow of passive pollutants at the mid-section of the street canyon $Q_s = 0.15$ mg/s.

3. Results and Discussion

For the 3D cases that were investigated with MIMO, results were extracted at horizontal planes at the heights of Y/H = 0.15, Y/H = 0.5 and Y/H = 1.0. As expected, the patterns are symmetrical with respect to the axis Z=0. Furthermore, graphs for the u and w velocity components for horizontal lines at the levels Y/H = 0.15 and 0.5 near the leeward and windward walls as well as at the centre of the street canyon are presented in Figures 5 and 6. For symmetry reasons, only half of the domain ($0 \leq Z/L \leq 1$) is displayed. In addition, calculations were made in order to estimate the residence time of the pollutants emitted from the source, which was placed along the entire length of the street canyon, at the street canyon ground level.

The calculated concentrations are non-dimensionalised via the following formula:

$$C^* = CU_\delta H / (Q_s / L) \tag{2}$$

where C^* is the non-dimensional concentration, C is the calculated pollutant concentration, U_δ is the reference wind velocity, H is the height of the street canyon, Q_s *is* the mass flow of the pollutant emission and L is the characteristic length of the source.

In the first case, at a horizontal level Y/H = 0.15, a system of two small counter-rotating vortices each located near the edges of the 2nd street canyon is predicted and as a result pollutants concentrate at an area near the edges of the windward wall. However, at the higher horizontal levels of Y/H = 0.5 and Y/H = 1.0, these vortices grow in size extending over the whole street canyon area at the rooftop level. As a consequence, pollutants are dispersed over areas extending over the largest part of the horizontal levels at these heights inside the street canyon (Figure 2).

In the second case, at all horizontal levels pollutants concentrate in an area extending around the vertical middle section of the street canyon. However, in this case this area reduces in size at heights of Y/H = 0.5 and Y/H = 1.0. Furthermore, at the rooftop level some pollutants are driven to the 1st street canyon, which lies upstream of the 2nd, probably due to the effect of a predicted

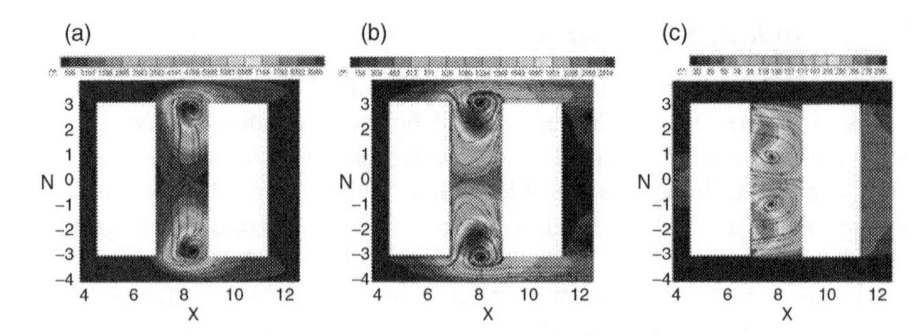

FIGURE 2. Concentration fields in and around the street canyon for the first case at (a) Y/H = 0.15 (b) Y/H = 0.5 and (c) Y/H = 1.0, as computed by MIMO.

recirculation zone which forms as the prevailing wind approaches the multi-street canyon configuration and extends over the roof top level of the 1st and 2nd street canyons (Figure 3).

In the third case pollutants also concentrate at an area extending around the vertical middle section of the street canyon. Similar to the previous configuration, this area reduces in size at heights of Y/H = 0.5 and Y/H = 1.0. Furthermore, at the rooftop level some pollutants are also driven to the 1st street canyon. Again this may be attributed to the above mentioned effect of the predicted recirculation zone above the roof top level (Figure 4).

From the u and w velocity components graphs it may be followed that for the first case MIMO predicts higher velocities at the centre of the canyon and at Y/H = 0.5. The same happens for the w-velocity (Figures 5 & 6). The negative sign means that the direction of u-velocity is from the windward side of wall to the leeward side of wall. The same sign means that the direction of the w-velocity is from the edges to the centre of the street canyon.

Figures 5 & 6 show also the u and w velocity components for the second and third cases at Y/H = 0.15 and Y/H = 0.5 respectively. Maximum values of w-velocity are predicted near the windward side of the wall while maximum

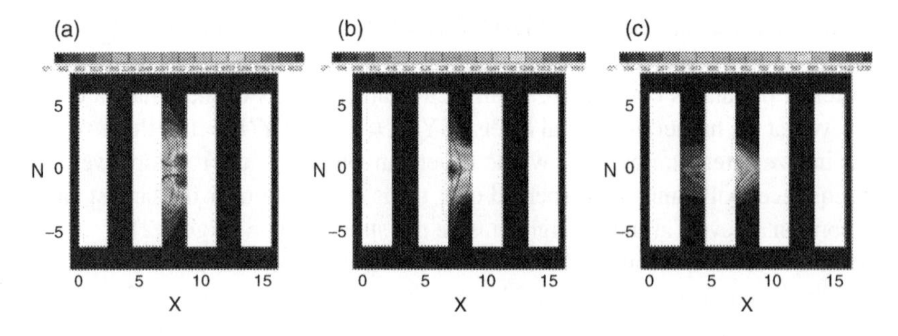

FIGURE 3. Concentration fields in and around the street canyon for the second case at (a) Y/H = 0.15 (b) Y/H = 0.5 and (c) Y/H = 1.0, as computed by MIMO.

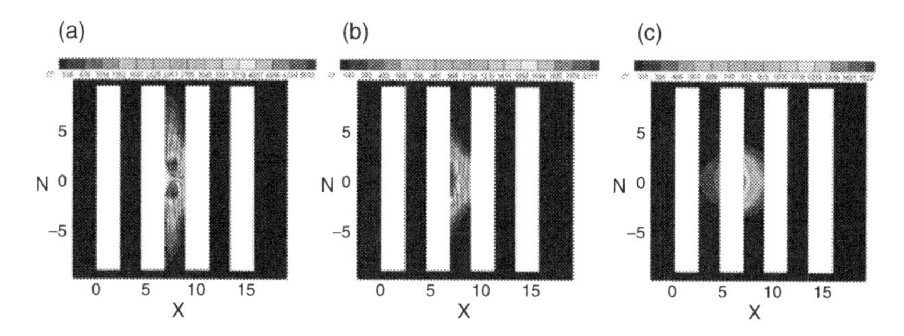

FIGURE 4. Concentration fields in and around the street canyon for the third case at (a) Y/H = 0.15 (b) Y/H = 0.5 and (c) Y/H = 1.0 as computed by MIMO.

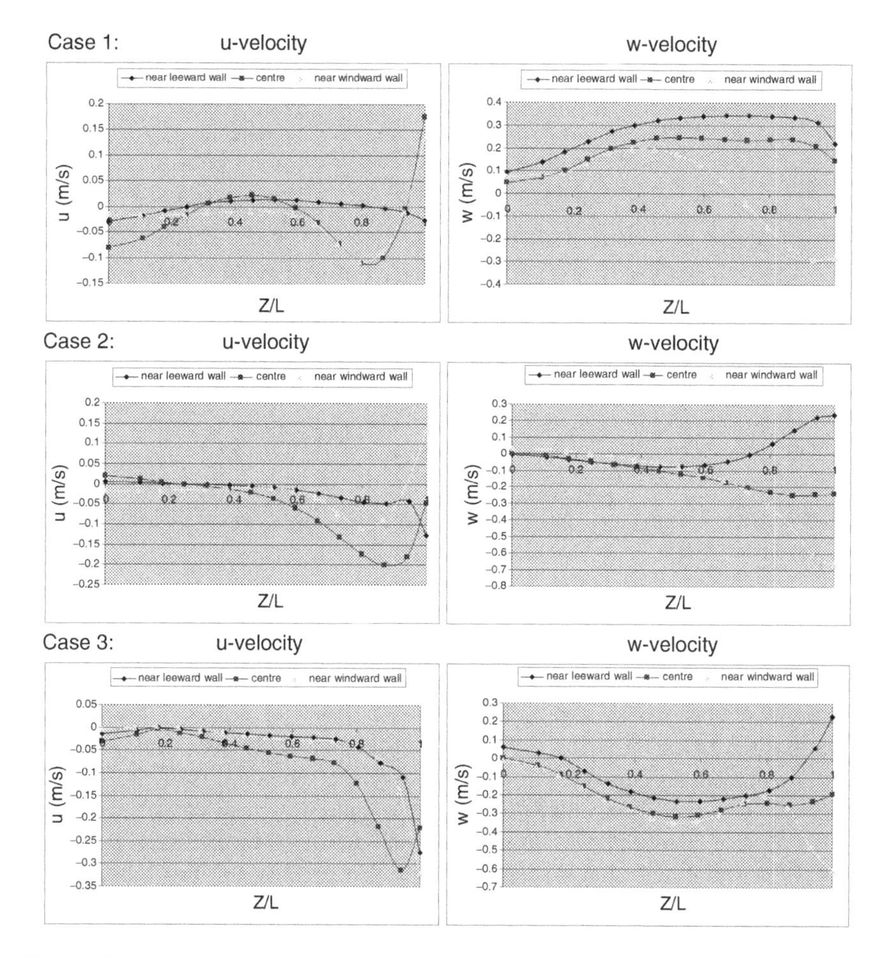

FIGURE 5. u and w velocity components along the street canyon for the three cases at the horizontal level Y/H = 0.15.

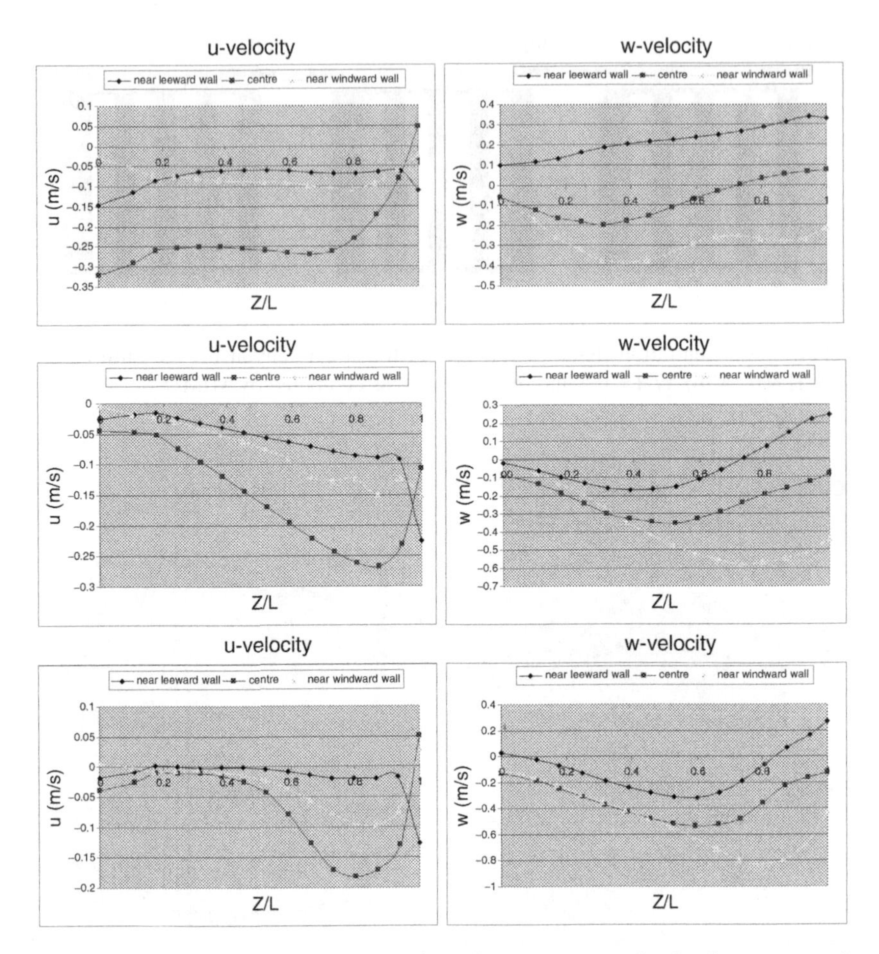

FIGURE 6. u and w velocity components along the street canyon for the three cases at the horizontal level Y/H = 0.5.

values for u-velocity are predicted near the centre line. The same conditions are also noticed for the third case.

In all three cases, the dispersion pattern of the pollutants is dominated by the coupling of vortices generated in the vertical and horizontal directions. Vortices generated in the horizontal direction act in the same way like the ones generated in the vertical direction that apparently dominate the flow fields in 2D street canyons in which only the u and w velocity components are important. In 3D cases, however, the component along the canyon axis is also important, as it dominates the inward – outward mass flux between the centre and the sides of the street canyon. As a result, in the 3D cases described in this section, in the first case the vortices at the two sides of the street canyon drive pollutants in a region

TABLE 1. Calculated pollutant residence time (s).

	MIMO	CFX-TASCflow
Case 1	2.79	2.52
Case 2	4.21	4.13
Case 3	4.86	4.48
2D case	6.12	5.99

near there, while the system of vortices that develop inside the street canyons in the other two cases results in an accumulation of the pollutants in a region near the vertical middle plane section of the street canyon.

The residence time calculated by both codes (MIMO and CFX-TASCflow) is presented in Table 1. Results for both codes are similar with very small deviations. Comparison between results from 2D and 3D simulations proves that the former lead to a noticeable overestimation of the residence time of pollutants in the street canyon. This happens because in 2D simulations there is only one pollutant exit while 3 exits exist for the 3D simulations. Consequently, a higher pollutant mass "escapes" in the 3D case and so the residence time is smaller.

The results presented in the above table indicate that the higher the length of a street canyon, the higher the residence time of a pollutant inside the canyon. This is the main reason that from the three cases investigated above, the third was chosen as the optimum experimental street canyon configuration for the experiment in Guerville: a longer pollutant residence time in the street canyon would lead to higher concentrations over a longer time period.

4. Conclusions

In this paper the microscale model MIMO was applied to simulate pollutant dispersion in street canyons with the same aspect ratio but different length. Both 2D and 3D numerical simulations were performed for assessing the importance of the three-dimensionality of the flow. Results for the residence time were compared with analogous results calculated with the commercial code CFX-TASCflow. The agreement between the two codes was in general fair.

The results of this study elucidate that the flow field developing inside a street canyon of a given aspect ratio and hence the dispersion of pollutants inside it, are strongly affected by the street canyon length. In particular, the residence time of a pollutant was found to increase drastically with street canyon length. Comparison between results from 2D and 3D simulations proves that the former lead to overestimation of the residence time of pollutants within a street canyon.

Aknowledgements The authors gratefully acknowledge funding from the PIC-ADA project (Contract N° GRD1-2001-40449) within GROWTH Measurements & Testing Infrastructures, EC's 5th Framework Programme.

References

URL1, http://www.software.aeat.com/cfx/products/cfx-tascflow

Ehrhard, J., 1999, Untersuchung linearer und nightlinearer Wirbelviskositatsmodelle zur Berechnung turbulenter Stromunger um Gebaude, Dissertation, University of Karsruhe, Department of Applied Thermodynamics.

Ehrhard, J., Khatib, I.A., Winkler, C., Kunz, R., Moussiopoulos, N. and Ernst, G., 2000, The microscale model MIMO: development and assessment, Journal of Wind Engineering and Industrial Aerodynamics, **85**, 163-176

Harten, A., On a large time-step high resolution scheme, Math. Comput. 46 (1986) 379-399.

Ketzel, M., Louka, P., Sahm, P., Sini, J.-F., Moussiopoulos, N. 2002, Intercomparison of Numerical Urban Dispersion Models -Part II: Street Canyon in Hannover, Germany. Water, Air and Soil Pollution: Focus, **2**, 603-613.

Moussiopoulos N. ed., 2003, Air Quality in Cities, SATURN/EUROTRAC-2, Subproject Final Report, Springer, Berlin, 298 pp.

Sahm, P., Louka, P., Ketzel, M., Guilloteau, E., Sini, J-F., 2002, Intercomparison of Numerical Urban Dispersion Models -Part I: Street Canyon and Single Building Configurations. Water, Air and Soil Pollution: Focus, 2, 587-601.

Sini, J.F., Anquentin, S. and Mestayer, P.G., 1996, Pollutant dispersion and thermal effects in urban street canyons. Atmospheric Environment, **30 (15)**, 2659-2677.

Wortmann-Vierthaler, M., Moussiopoulos, N., 1995, Numerical tests of a refined flux corrected transport advection scheme, Environ. Software 10, 157-176.

68 Limitations of Air Pollution Episodes Forecast due to Boundary-Layer Parameterisations Implemented in Mesoscale Meteorological Models

Leiv H. Slørdal[1], Sandro Finardi[2], Ekaterina Batchvarova[3,4], Ranjeet S. Sokhi[3], Evangelia Fragkou[3], and Alessio D'Allura[2]

1. Introduction

Dispersion models require information on the turbulence characteristics in the planetary boundary layer (PBL). This information is most often extracted from either meteorological measurements or from numerical (prognostic or diagnostic) models, and the requested turbulence parameters are then estimated using a PBL pre-processor.

Traditionally, Monin-Obukhov (M-O) similarity theory is applied when estimating the surface turbulent fluxes and the various vertical profiles of averaged quantities in the surface layer of the PBL (Beljaars and Holtslag 1991; Hanna and Chang 1992; Zilitinkevich et al. 2002b). In this similarity approach several simplifying assumptions are made, among which the requirement of quasi-stationary and horizontally homogeneous flow, and constant (independent of height) turbulent fluxes are the most crucial (Arya 1988). In urban areas and in complex terrain these assumptions are obviously not fulfilled. The theory is particularly questionable in very stable conditions, i.e. under conditions typically prevailing during pollution episodes in winter. In very stable conditions turbulence tends to be sporadic, and wave-turbulence interaction becomes increasingly important as well as drainage effects due to even small terrain slopes (Högström 1996). Moreover, observational data suggest that developed turbulence can exist in the stably stratified surface layer at much larger Richardson numbers than the

[1] Norwegian Institute for Air Research (NILU), Kjeller, Norway.
[2] ARIANET, Milan, Italy.
[3] University of Hertfordshire (UH), Hatfield, UK.
[4] National Institute of Meteorology and Hydrology (NIMH), Sofia, Bulgaria.

classical M-O theory predicts (Zilitinkevich, 2002). The M-O theory is nevertheless applied in most models even in these cases, mostly due to lack of other practical formulations (Mahrt 1999).

In the following study the output from a M-O based meteorological pre-processor, METPRO (Slørdal et al., 2003; van Ulden and Holtslag 1985; Bøhler 1996), is analysed for a particularly stable period in January 2003 in the city of Oslo, Norway. During this period extraordinarily high pollution levels were measured at several sites within the city. A similar pre-processor SURFPRO (Silibello C., 2002; Finardi et al., 1997) is applied to analyse a winter air pollution episode, characterised by strong nighttime stability, in Milan. Since no turbulence measurements exist for these episodes, the sensitivity of the pre-procrssors output on some of the model assumptions are investigated in this study.

2. The Meteorological Pre-Processors METPRO and SURFPRO at Very Stable Conditions

The M-O turbulence parameters estimated by the preprocessors are based on either measured meteorological input data or on model output from meteorological circulation models. Typical pre-processor outputs are the friction velocity (u_*), the temperature scale (θ_*) and the Monin-Obukhov length (L):

$$u_* \equiv \left(\overline{u'w'}\right)_0^{1/2} = (\tau_0/\rho)^{1/2}, \quad \theta_* \equiv -\frac{(\overline{\theta'w'})_0}{u_*} - \frac{H_0}{\rho C_p u_*}, \quad L \equiv -\frac{(\overline{u'w'})_0^{3/2}}{\kappa \frac{g}{\theta}(\overline{\theta'w'})_0} \equiv \frac{u_*^2}{\kappa \frac{g}{\theta}\theta_*} \tag{1}$$

Here τ_0 is the surface momentum flux, H_0 is the surface sensible heat flux, ρ is the air density and C_p is the specific heat at constant pressure. Based on an estimate of the surface roughness, z_0, measurements of the wind at one height and the temperature difference between two heights, all made within the surface layer (inertial sub layer), u_*, θ_* and L are computed by an iterative solution of the profile equations for wind speed and temperature. In these expressions κ is the von Karman constant, with a prescribed value of 0.41. The Monin-Obukhov length, L, is a measure of the buoyant stability of the air. Small positive and negative values of L indicate stable and unstable conditions, respectively. The neutral regime is found for large positive or negative values.

In very stable situations, $z/L > 1$, convergence of the above mentioned iterative procedure is not ensured. In this regime, Holtslag (1984) proposed an analytical solution in which, an upper bound is imposed on the surface sensible heat flux through specifying a constant positive limit value for the temperature scale, i.e. $\theta_* = 0.08$ K. The analytical solution gives real (physical) values for L as long as its value is greater than a minimum value, L_0, given by $L > L_0 \equiv 5z/\ln(z/z_0)$.

Thus, if the wind speed is measured at $z = 25$ m and $z_0 = 0.5$ m, this minimum value of L will be about 32 m. According to Holtslag (1984) lower values of L correspond to very stable conditions, in which there is little or no turbulence. Holtslag (1984) also presents a "practical" solution for $L < L_0$, which is given as

$$L=\left(L_0\frac{L_n}{2}\right)^{1/2}, \textit{ where } L_n=\kappa U(z)^2 T/2g\theta_* \left\{ln\left(\frac{z}{z_0}\right)\right\}^2 \tag{2}$$

Equation (2) gives values of L that are continuous at $L = L_0$ and which results in $L = 0$ for $U(z) = 0$. Equation (2) is just an interpolation formula. The M-O similarity theory predicts that the turbulence has ceased when $L < L_0$.

In many urban applications a minimum allowed positive value of the Monin-Obukhov length is employed directly. This is a simple way of accounting for the effects of the urban heat island and the increased mechanical turbulence induced by the urban canopy. The minimum values predicted by $L > L_0 = 5z/\ln(z/z_0)$ coincide rather well with tabulated values of minimum L in urban areas of Hanna and Chang (1992).

3. Application of METPRO During a Peak Pollution Episode in the City of Oslo

In order to investigate the pre-processors performance under stable conditions, METPRO has been applied, with input data from an urban meteorological measurement station in Oslo (Hovin), to one of the most recent (and most severe) pollution episodes in Oslo. This episode occurred in the second week of January 2003.

The city of Oslo is located at the northern end of the Oslo fjord, surrounded by several hills up to 600 m height and with three main valleys emanating from the city basin, the largest to the northeast, one to the north and one to the northwest. During low wind conditions with strong ground based or slightly elevated inversions the pot-formed topographical features of the area contributes to worsen the dispersion conditions, thereby capturing pollutants emitted within the urban airshed. These effects are of particular importance during the wintertime season.

The meteorological station at Hovin measures the wind at a height of a 25 m mast. The station is located in a park area surrounded by what must be characterised as an urban environment. The surface roughness for momentum for this area has been estimated to be about 0.5 m. On the 7th of January hourly NO_2 values up to about 600 $\mu g/m^3$ were observed at monitoring station Alna, the highest NO_2 values ever observed in Oslo. High $PM_{2.5}$ values were observed as well (hourly value of 152 $\mu g/m^3$ at Kirkeveien). However, relatively large variability in the concentrations at the different measurement sites within the central city area indicate that sporadic periods of enhanced turbulent mixing occurred at various areas throughout the 3 day pollution episode from the 7th to the 9th of January.

Prior to the episode the measured ground temperatures were low, about –20°C. An area of high pressure prevailed over the north Atlantic on the 3rd of January, and it was transported to the East, arriving at the sea area between UK and Southern Norway on the 7th. On the 6th, relatively warmer air masses were transported at higher altitudes from the northwest; this led to the formation of a strong inversion that lasted from the 7th to the 10th of January. The ground surface was covered with snow or ice during the course of the episode. **The ambient (2m) temperature varied from – 18 to + 1°C at the** station of Hovin.

The radiosonde vertical temperature profiles for the period 3-10 January measured at 00 UTC at the meteorological station **Blindern were analyzed**. The main reason for the formation of the long lasting ground based inversion was found to be the advection of warm air above the lowest atmospheric layers. In Figure 1a and 1b time series of the wind speed (measured at 25 m) and the ΔT-value (measured between 25 m and 8 m) at Hovin are shown. In the same plots (see right axis) the measured NO_2 values at the Løren station are presented as well. This is the AQ-station closest to the Hovin meteorological station, located less than a km north of Hovin.

As seen from Figures 1a and 1b, the highest NO_2 concentrations at Løren starting on the afternoon of the 6th lasting throughout the 9th are coinciding with relatively low wind speeds and a build up of a strong surface inversion that lasts for the whole period. The hourly NO_2 concentrations at the Løren station reach slightly above 400 μg/m³ and as much as 13 hourly values above 200 μg/m³ are measured during the 7th and 8th. It can be noted from Figure 1b that $\Delta\theta$ maintains its positive value throughout the daytime hours on the 7th, 8th and 9th of January. The term "long lived stable boundary layer" (Zilitinkevich, 2002) therefore seems to describe these conditions better than the traditional term of the "nocturnal boundary layer".

By applying the observed wind speed and ΔT values at Hovin as inputs, i.e., the data presented in Figures 1a and 1b, the M-O based meteorological pre-processor

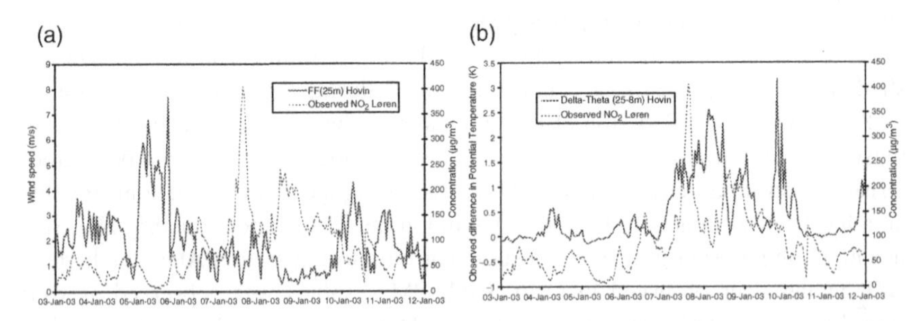

FIGURE 1. (a) Time series of hourly values of the observed wind speed (m/s) at Hovin and observed hourly NO_2 concentrations (μg/m³) at Lørenduring the January 2003 pollution episode. (b) Time series of hourly values of the observed ΔT (converted to $\Delta\theta$; K) at Hovin and observed hourly NO_2 concentrations (μg/m³) at Løren during the January 2003 pollution episode.

METPRO can be applied to estimate various dispersion parameters for this pollution episode. Furthermore, different methods of parameterisation can be compared. In the following the measurement height of the wind speed ($z = 25$ m) is applied as reference height when estimating the dimensionless value of the stability parameter z/L.

In Figure 2a three estimates of the inverse Monin-Obukhov length are presented. The lower bound on L, applied in METPRO for stable conditions in urban areas, is clearly revealed by the constant maximum value on the $1/L$-curve during the entire three day pollution period from the afternoon of the 6th to the end of the 9th, see curve a in Figure 2a. The corresponding minimum value of L is 32 m in this case. When allowing for lower values of L by applying the interpolation formula, Eq.(2), curve b in Figure 2a is computed. A continuous variation of the $1/L$-curve is now found with maximum values reaching about 0.40 m^{-1}, implying minimum values of L of about 2.5 m. The sensitivity of applying $\theta_* = 0.08$ K in the very stable regime ($z/L > 1$) was investigated by increasing this value to $\theta_* = 0.15$ K, and then recalculating $1/L$ without applying a lower limit on L. The resulting 1/L-curve is shown as curve c in Figure 2a. This rather large increase in θ_* leads to somewhat larger peak values of 1/L as compared to the curve b results in Figure 2a. 1/L now reach values of 0.54 m^{-1}, giving minimum values of L as low as 1.9 m. L is thus marginally influenced by the constant chosen for θ_* in the very stable case. Moreover, the turbulent diffusivities estimated from the above M-O parameters remain extremely low, i.e. in the range 0.01–0.25 m^2/s, throughout the three-day episode. This is the case even when the lower limit is applied for L.

The corresponding effect on the estimated PBL heights of allowing for lower values of L in the very stable regime is shown by curve a–c in Figure 2b. The PBL heights are estimated by use of Zilitinkevich (1972) stable layer parameterisation, i.e. $h = 0.4\sqrt{u_* L/f}$. Very low PBL heights are estimated throughout the three-day pollution episode and therefore only heights below 200 m are shown.

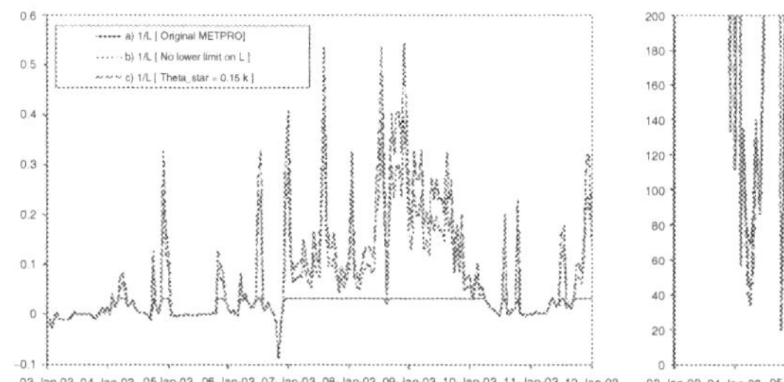

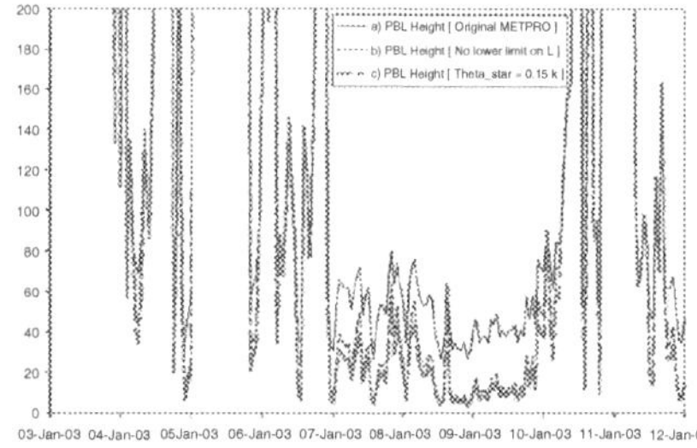

FIGURE 2. (a) Values of the inverse M-O length (1/L): a) L estimated from original METPRO, b) with no lower limit on L, c) with no limit on L, but with $\theta* = 0.15$ K instead of $\theta* = 0.08$ K when $z/L > 1$. (b) Estimated values of the PBL height (h): a) L estimated from original METPRO, b) with no lower limit on L, c) with no limit on L, but with $\theta* = 0.15$ K instead of $\theta* = 0.08$ K when $z/L > 1$.

Without limitations on L, the PBL height reaches unrealistic values as low as 4 to 5 m, i.e. values much less than the average height of the roughness elements in urban areas. However, when the lower bound on L is applied, PBL heights down to about 25 m (i.e. slightly higher than the average building height in Oslo) are found, indicating that a lower limit on L is indeed needed for urban applications. Furthermore, the results are not very much influenced by the choice of the constant for the temperature scale, θ_*. The period with very low estimated PBL heights coincides rather well with three radio-soundings profiles (7th, 8th and 9th), which exhibit the deep surface based inversion.

4. Application of SURFPRO During a Winter Pollution Episode in the City of Milan

SURFPRO has been applied to a severe pollution episode that happened in Milan during December 1998, and has been analysed in Kukkonen et al (2004) in comparison with episodes recorded in other European cities.

Milan city is located in the central part of the Po river basin, in Northern Italy, in a flat area. The atmospheric circulation of the Po valley is characterised by the strong modification of synoptic flow due to the high mountains that surround the valley on three sides. Calm conditions and weak winds occur frequently. The most severe winter episodes are commonly associated with high pressure, weak winds and elevated temperature inversions.

In this work we used meteorological observations from the radiosoundings of the suburban airport of Linate and from the urban surface station of Juvara, where the air quality station is located at the road pavement level, while the corresponding meteorological station is located on a building roof at a height of approximately 30 m.

A period of elevated PM_{10} and NO_2 concentrations is clearly distinguishable from 14 to 19 December 1998 (Figure 3). Exceedancees of UE limits for both

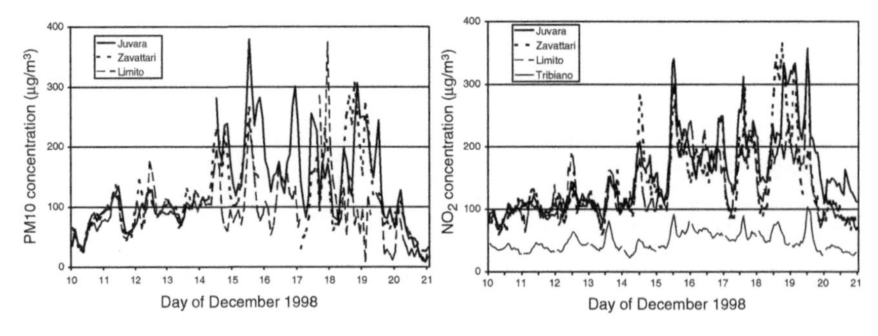

FIGURE 3. PM10 (left) and NO_2 (right) concentrations recorded by some of the stations located in Milan Province during December 1998 episode.

pollutants were recorded by all the stations of Milan city air quality network during the episode.

The synoptic weather conditions were characterised by a high-pressure ridge arriving at the Mediterranean basin on December 13th and remaining quasi-stationary, centred over the western Mediterranean for the whole of the episode duration. The local surface wind speed was low or it was calm. The temperature of the surface layer was rather low for Milan climatology and did not change substantially during the course of the episode (Figure 4). An intense, slightly elevated temperature inversion was formed on 13 December that reached its maximum depth and magnitude (with a temperature growth of about 15°C in the first 1500 metres height) on 15 December and prevailed until 19 December according to radiosonde data at Linate airport. The observed inversions were caused by the advection of warm air aloft carried by the incoming high-pressure ridge.

The M-O based meteorological pre-processor SURFPRO has been applied to estimate turbulence scaling parameters, mixing height and Eulerian dispersion parameters (K_Z) for this pollution episode. A roughness length of 1 metre has been considered to describe urban conditions. In Figure 5 two estimates of the inverse Monin-Obukhov length are presented, with and without the limiting value on L. If applied, this limiting value affects all the periods during which stable conditions were recorded.

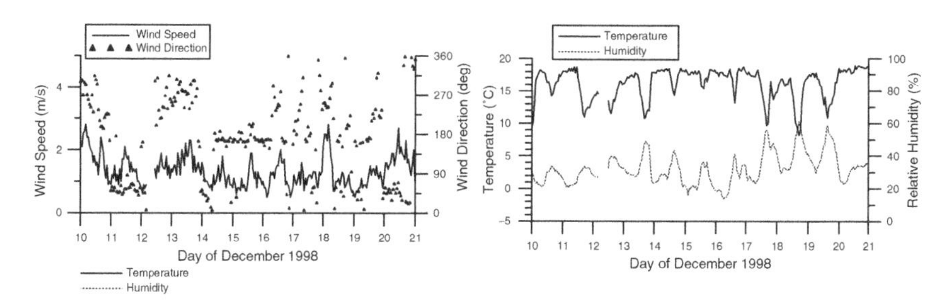

FIGURE 4. Wind speed and direction (left), temperature and relative humidity (right) observed at the urban station of Juvara during the December 1998 episode.

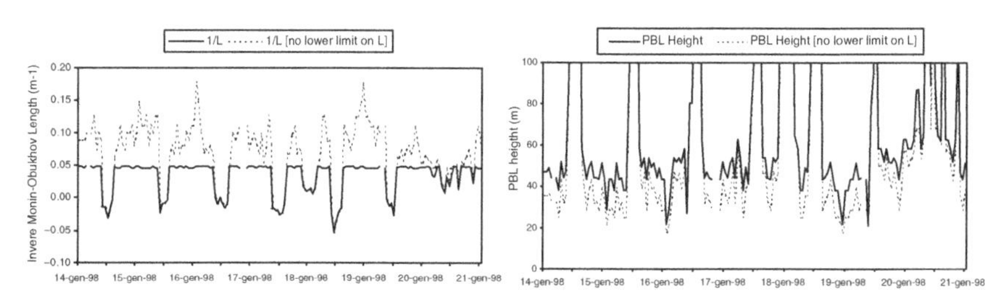

FIGURE 5. (a) Estimated values of the inverse M-O length (1/L). (b) Estimated values of the PBL height (h) with and without considering a lower limit to MO length (L) – zoom on the first 100 metres.

For the considered application this limit turned on when the wind speed was lower than about 1.8 m/s, a limit that was nearly never exceeded in stable night time conditions. The corresponding effect on the estimated PBL heights is shown in Figure 5. Very low PBL heights are estimated during the whole episode. Daytime values (estimated using a simple encroachment method basen on observed temperature profiles and estimated sensible heat flux) rarely reach 200 metres, while during stable conditions the estimated boundary layer depth is always lower than 50 metres. Values in the range 20-30 metres are obtained without lower limit to the MO lenght.

5. General Discussion and Conclusions

Two meteorological pre-processors (METPRO and SURFPRO), applying traditional Monin-Obukhov similarity theory, have been applied in the parameterisation of dispersion parameters for winter pollution episodes in Oslo and Milan. The boundary layer depths diagnosed for both cities were extremely low during stable conditions. These values roughly varied in the range 25-50 meters. Very low values, rarely > 0.1 m^2s^{-1}, were obtained for the vertical eddy diffusivity.

The episode in Oslo was characterized by very stable atmospheric conditions lasting for several days. Extremely low PBL heights and almost zero turbulent dispersion was predicted during a three-day pollution episode.

By applying a lower (urban) limit to the estimated Monin-Obukhov length, estimated PBL heights lower than the average roughness elements were avoided. Nevertheless, this simple way of urbanizing the scheme did not have a profound impact on the estimated vertical turbulent diffusivities. In reality the turbulent exchange may have been stronger than predicted by the applied M-O theory, at least sporadically in time and space, because of wave – turbulence interactions and drainage effects (Högström 1996). This is also supported by the available air quality measurements that show relatively large variability in concentration levels throughout the 3 day pollution episode.

Instead of using pre-processors of the above type, output from numerical weather prediction models or mesoscale circulation models could be applied. However, lots (if not most) of these models also apply M-O similarity theory in their PBL schemes, and direct extraction of surface fluxes from these models will produce similar results as shown above.

Even though the considered pollution episode in Oslo was quite severe, the meteorological conditions are rather characteristic for pollution episodes in Norwegian, and probably other Nordic, cities. Under such conditions the PBL is termed “long lived stable boundary layers” by Zilitinkevich (2002) as opposed to the traditional “nocturnal boundary layer”. Zilitinkevich et al., (2002) propose a practical method for including the effects of internal gravity waves in the description of such long-lived SBLs, thereby enhancing surface layer turbulence. However, in order to apply this method, information on the Brunt-Väisälä frequency in the adjacent layers of the free atmosphere is needed. Observations of the Brunt-Väisälä frequency is normally not available, but estimates of this quantity may be extracted from NWP or mesoscale circulation models. This type of PBL parameterisations together with related expressions on the PBL height (Zilitinkevich et al., 2002a) should be tested further on long-lived episodes, as in Sodeman and Foken (2004).

Acknowledgements The study has been supported by the European Commission research project FUMAPEX, and cluster CLEAR. At the NIMH the work was related to NATO CLG (979863) and the BULAIR Project EVK2-CT-2002-80024.

References

Arya, S. P. S., 1988, Introduction to Micrometeorology. Academic Press, Inc., San Diego, Caliofornia 92101. pp. 307.

Beljaars, A. C. M., and Holtslag, A. A. M., 1991, Flux parameterisation over land surfaces for atmospheric models. *J. Appl. Meteorol.*, **30,** 327-341.

Bøhler, T., 1996, MEPDIM. The NILU Meteorological Processor for Dispersion Modelling. Version 1.0. Model description. Kjeller (NILU TR 7/96).

Finardi S., Morselli M. G., Brusasca G. and Tinarelli G., 1997, A 2D Meteorological Pre-processor for Real-time 3D ATD Models", International Journal of Environmental Pollution, Vol. 8, Nos. 3-6, 478-488.

Hanna, S. R. and Chang J. C., (1992, Boundary-layer parameterizations for applied dispersion modeling over urban areas. *Boundary-Layer Meteorol.*, **58**, 229-259.

Högström, U., 1996, Review of some basic characteristics of the atmospheric surface layer. *Boundary-Layer Meteorol.*, **78**, 215-246.

Kukkonen, J. M., Pohjola, Sokhi, R. S., Luhana L., Kitwiroon, N., Rantamäki, M., Berge, E., Odegaard, V., Slørdal, L. H., Denby B., and Finardi, S., 2004, Analysis and evaluation of local-scale PM10 air pollution episodes in four European cities: Oslo, Helsinki, London and Milan. Atmospheric Environment, in press.

Mahrt, L., 1999, Stratified atmospheric boundary layers. *Boundary-Layer Meteorol.*, **90**, 375-396.

Silibello C., 2002, SURFPRO (SURrface-atmosphere interFace PROcessor) User's guide. Report Arianet 2002.

Slørdal, L. H., Solberg, S., and Walker, S. E., 2003, The Urban Air Dispersion Model EPISODE applied in AirQUIS$_{2003}$. Technical description. Norwegian Institute for Air Research, Kjeller (NILU TR 12/03).

Sodeman, H. and Foken, T., 2004, Empirical evaluation of an extended similarity theory for the stably stratified atmospheric surface layer, *Q. J. R. Meteorol. Soc.* (2004), **130**, pp. 2665-2671.

van Ulden, A. P. and Holtslag, A. A. M., 1985, Estimation of Atmospheric Boundary Layer parameters for Diffusion Application. *J. Appl. Meteorol., 24,* 1196-1207.

Zilitinkevich, S. S., 1972, On the determination of the height of the Ekman boundary layer. *Boundary-Layer Meteor.* **3**, 141-145.

Zilitinkevich, S. S., 2002, Third-order transport due to internal waves and non-local turbulence in the stably stratified surface layer. *Q. J. R. Meteorol. Soc.*, **128**, 913-925.

Zilitinkevich, S. S. and Calanca, P., 2000, An extended similarity theory for the stably stratified atmospheric surface layer. *Q. J. R. Meteorol. Soc.*, **126,** 1913-1923.

Zilitinkevich, S. S, Baklanov, A., Rost, J., Smedman, A.-S., Lyksov, V. and Calanca, P., 2002a, Diagnostic and prognostic equations for the depth of the stably stratified Ekman boundary layer. *Q. J. R. Meteorol. Soc.*, **128**, 25-46.

Zilitinkevich, S. S., Perov, V. L. and King, J. C., 2002b, Near-surface turbulent fluxes in stable stratification: Calculation techniques for use in general-circulation models. *Q. J. R. Meteorol. Soc.*, **128**, 1571-1587.

Posters
Role of Atmospheric Models in Air Pollution Policy and Abatement Strategies

69
Use of Lagrangian Particle Model Instead of Gaussian Model for Radioactive Risk Assessment in Complex Terrain

Marija Zlata Božnar and Primož Mlakar*

1. Introduction

In this work an experience of using Lagrangian Particle model Spray (in AriaIndustry package from Arianet, Italy) for risk assessment at Krško Nuclear Power Plant (NPP) in Slovenia is described. Krško NPP is placed in a semi-open basin sourrounded with very complex terrain. The area is characterised by very low wind speeds and thermal inversions.

2. Description of the System

The model calculates dilution coefficients in automatic mode (every half hour) without operator assistance. It is using meteorological data from several automatic stations covering local complex orography. Model results[1] are used by NPP experts for dose projection algorithms in case of an emergency situation. The application described is an example of good practice of automatic use of highly efficient numerical model for routine risk assessment calculations in complex terrain.

3. Modelling Limitations

There were several practical programming problems and limitations that have to be solved when planing to use the Lagrangian particle model and 3D mass consistent wind field model for automatic batch runs without operator assistance.

* AMES d.o.o., Na Lazih 30, SI-1351 Brezovica pri Ljubljani, Slovenia

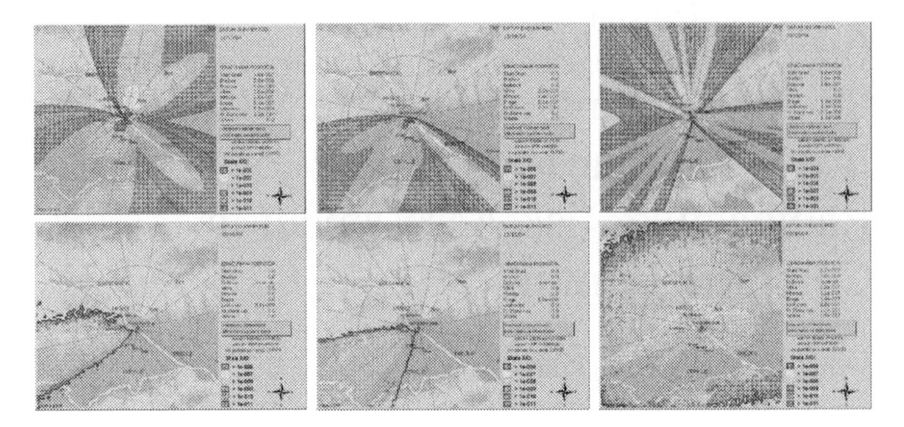

FIGURE 1. Comparison of daily dilution coefficients (averages made of half hour modelling results) for Gaussian model (upper three figures) and corresponding ones for Lagrangian particle model.

In the second part of the presentation air pollution modelling experiences from one year of operation are given. Previously a Gaussian model (according to U.S. Regulatory Guide 1.145) was used. A detailed analysis of differences and agreements for the results of both models[2] will be given to show the extents of complex terrain influence.

4. Results

Figure 1 shows examples of results of Gaussian and Lagrangian models for three days in autumn 2004. The basic difference in the meteorological data used is the wind. Due to the existence of nearby buildings, the Gaussian model according to U.S. R.G. 1.145 should therefore use the ground level wind. The Lagrangian model uses the wind measured at the height of probable exhaust. The ground level wind direction is fluctuating due to very low wind speeds. In the upper layers wind direction is fluctuating less. This situation is only one example of meteorological situations in complex terrain that the Gaussian model can not cope with properly.

5. Conclusions

The analysis prove the need of using advance numerical models in highly complex terrain applications like the one at Krško.

References

1. Breznik, B., Božnar, M., Mlakar, P., Tinarelli, G. Dose projection using dispersion models. V: Batchvarova, E. (ed.), Syrakov, D. (ed.). Eighth International Conference on Harmonisation Within Atmospheric Dispersion Modelling for Regulatory Purposes, Sofia, Bulgaria, 14-17 October 2002. Proceedings. [Sofia]: National Institute of Meteorogy and Hydrology: Bulgarian Academy of Sciences, 2002, pp. 409-413. [COBISS.SI-ID 17175079]
2. Božnar, M., Mlakar, P., Breznik, B. Advanced modelling of potential air pollution dispersion around Krško NPP using 3D wind field reconstruction and Lagrangian particle model. V: Ravnik, M. (ed.), Žagar, T. (ed.). International Conference Nuclear Energy for New Europe 2003, September 8-11, 2003, Portorož, Slovenija. Proceedings. Ljubljana: Nuclear Society of Slovenia, 2003, pp. 602-1-602.8. [COBISS.SI-ID 18041127]

70
Study of Air Pollutant Transport in Northern and Western Turkey

Tayfun Kindap, Alper Unal, Shu-Hua Chen,Yongtao Hu,
Talat Odman, and Mehmet Karaca

1. Introduction

While most regulation and research of air pollution episodes are focused on Europe, transport of air pollutants from Europe to Northern and Western Turkey hasn't been studied sufficiently, although unusually high air pollution episodes occur in these regions. As an example during 05-12 of January, 2002, it has been identified from observational data that particulate matter levels were almost 4-5 times (364 $\mu g/m^3$) higher than the World Health organization standards for Europe (70 $\mu g/m^3$). For the same period, other pollutants were significantly higher than the standards as well. During this period, analysis of the meteorological data indicate that there were significant westerly-southwesterly winds. These findings indicate that in addition to local emissions, long-range transport might be contributing to air pollution episodes in Turkey. In this study we aim to examine the potential sources of pollution in western Turkey and its surrounding regions. We developed a framework to model air quality in Europe using MM5 (i.e., for meteorological modeling) and CMAQ (i.e., for air quality modeling). It should be noted that we developed our own emissions modeling to process gridded EMEP emissions. In the next sections we give more information on these methods.

2. Methodology

2.1. Meteorology

The MM5 horizontal grid resolution is 50 km with 137 cells along the east-west direction and 116 cells in the north-south direction covering all Europe, and it consists of 37 vertical levels. Model physics options include the RRTM (Rapid Radiative Transfer Model) radiation scheme, Kain-Fritsch scheme, MRF boundary layer scheme, and Simple Ice microphysics scheme. Although the model didn't use a high resolution due to the resolution of the emission inventory. The meteorological model produced reasonable results as indicated from the comparison between the predictions and observational data.

2.2. *Tracer and Backward Trajectory*

In order to identify movement in the weather pattern we conducted tracer and backward trajectory analysis. The results of these analyses are given in Figure1.a and b. at the beginning of the tracer simulation, air was emitted from different cities of Europe. Figures 1a and 1b show propagating air parcels after 12 hours and 72 hours respectively. Although the western European region (e.g., London, Paris or Berlin) is far from Turkey, these tracers can propagate to Turkey and nearby regions after a 60-h integration. Meanwhile, tracers from nearby regions such as Warsaw, Kiev, and Bucharest reach Turkey in less than 12 hours. The tracer study indicates that aerosols can be transported from western and northwestern Europe to eastern and southeastern Europe during the simulation period. Besides the tracer approach, trajectories might help to see any source areas for interested places in term of transport (Figure1.c) A study of the trajectories indicates that some places from western and northwestern as well as inside Turkey were affected by air masses traveling from northwestern Europe during the simulation.

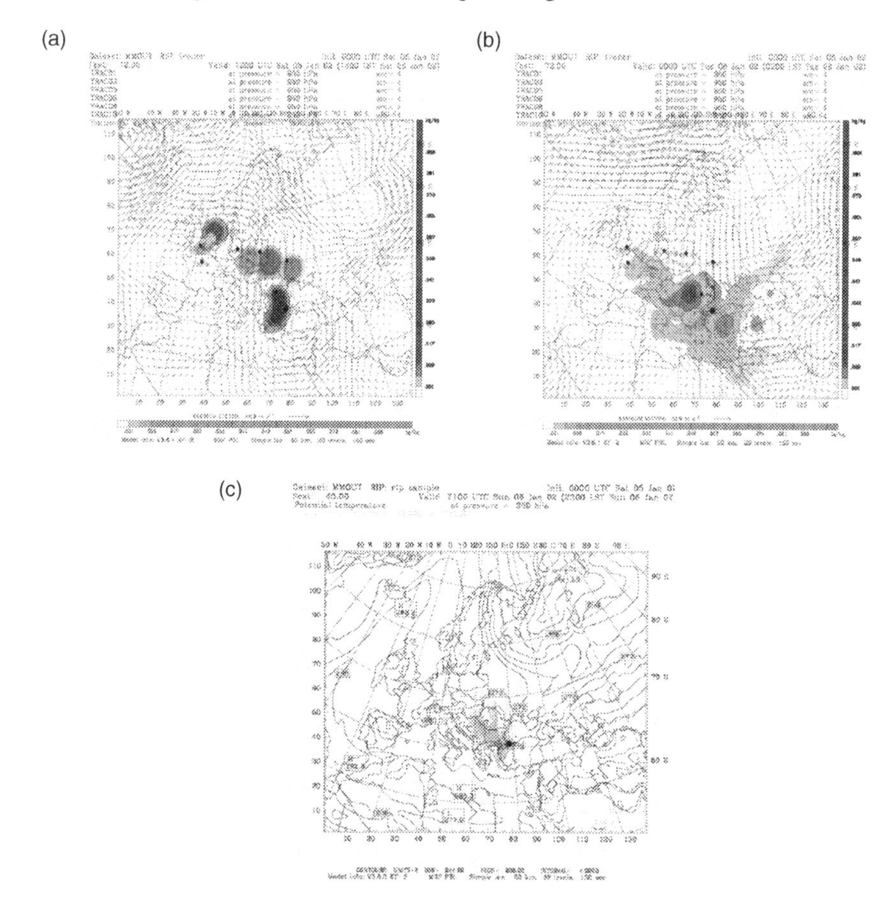

FIGURE 1. Backtrajectory was run from 42 to 72 hrs.

3. Emission Modeling

An algorithm was developed to generate an emissions database to be used in the air quality study. A grid annual emission data was obtained from EMEP (The Cooperative Programme for Monitoring and Evaluation of the Long-range Transmission of Air Pollutants in Europe) data base (EMEP, 2004) for anthropogenic sources (50 km resolution). Biogenic emissions were not available in EMEP grid emissions database. For this reason, methodology developed by Guenther (1996) was utilized for biogenic sources. For each data sources, speciation, temporal allocation, and vertical distribution were performed. For emissions data Quality Assurance/Quality Check (QA/QC) methods were also conducted.

4. Air Quality Modeling

After preparing the necessary inputs, including meteorological and emissions data, we ran the CMAQ model for our domain for January 5-12 2002. The CMAQ horizontal grid size was set to 50 km with 132 cells along the east-west direction and 111 cells in the north-south direction covering all of the Europe. There are 20 layers in the vertical direction. The CB4 chemistry mechanism was selected in this study.

4.1. Brute-Force Sensitivity

To demonstrate long-range aerosol contributions to Turkey, as indicated by tracer and trajectory analyses, we have conducted brute-force sensitivities. For this purpose we increased emissions of individual countries by 50 percent, which is the reported uncertainty associated with emissions (EMEP, 2004). This approach enables us to determine the contribution from individual countries to Turkey. As a result of the brute-force runs, it has been found that contributions of Eastern European countries to air pollution in Northern/Western Turkey range from 4 to 10 percent for the selected episode.

References

EMEP/CORINAIR, 2001, Atmospheric Emission Inventory Guidebook Third Edition (2001). UNECE-Convention on long-range trans-boundary air pollution & European Environmental Agency.

Guenther, A., 1997, Seasonal and spatial variations in the natural volatile organic compound emissions, *Ecol. Appl.*, **7(1)**,34-45.

WHO guidelines for Europe; http://www.worldbank.org/html/fpd/em/power/standards/airqstd.stm#who

71
Source Term Assessment from Off-Site Gamma Radiation Measurements

Bent Lauritzen and Martin Drews*

During a nuclear accident in which radionuclides are released to the atmosphere automatic off-site gamma radiation measurements may be used to assess the source strength and the dispersion of radionuclides. In the following, a newly developed sequential data assimilation method for such on-line source term estimation is presented and results are shown using both real and simulated radiation data.[1]

The method is based on a discrete-time, stochastic state space model describing the dynamics of the significant plume parameters (the state variables) and their coupling to measurements through a non-linear atmospheric dispersion model. A general system equation describes the time evolution of the state,

$$X_t = f(X_{t-1}, u_t) + w_t, \tag{1}$$

while the physical observables are linked to the state variables X_t through a static measurement equation,

$$Y_t = h(X_t) + v_t. \tag{2}$$

Here w_t and v_t are taken as uncorrelated white noise Gaussian processes with covariance matrices Q_t and R_t, respectively; f describes the dynamics of the state variables, and u_t is an optional forcing term. The set of observables, Y_t, may comprise both radiation and meteorological measurements taken during the accident.

In the state space model (1-2), the expectations of the state variables X_t conditioned on the measurement data Y_t are obtained via an extended Kalman filter. The embedded parameters of the model, *i.e.* the covariance matrices Q and R, adjustable parameters of the atmospheric dispersion model $h(\cdot)$, as well as the initial state, $X_{t=0}$, may be determined through maximum likelihood estimation, that is by repeated applications of the Kalman filter, thereby making the method essentially free of external parameters.

---

* Risø National Laboratory, DK-4000 Roskilde, Denmark

The method has been tested against simulated dose rate data from a fabricated release of a radionuclide and against measurement data obtained during an ^{41}Ar atmospheric dispersion experiment. In both cases, the atmospheric dispersion and radiation field is calculated within the Gaussian plume model, applicable up to a few kilometers from the release point. To keep the calculations computationally simple, the dynamics of the plume parameters are modeled as a random walk process,

$$X_t = X_{t-1} + w_t, \tag{3}$$

and the covariance matrices Q and R assumed to be constant and diagonal.

For the simulation, the fictitious release and dispersion of a radionuclide is modeled as a Gaussian plume with the time evolution of the plume parameters given by Eq. (3). The state variables are provided by the radionuclide release rate divided by the wind speed (q/u), the plume height (h), and the main advection direction (θ), while the Gaussian plume dispersion parameters are prescribed externally. The "measurement" vector Y_t consisting of a set of four simultaneous dose rate measurements and one wind direction measurement was calculated within the Gaussian plume model, with Gaussian random errors added according to Eq. (2).

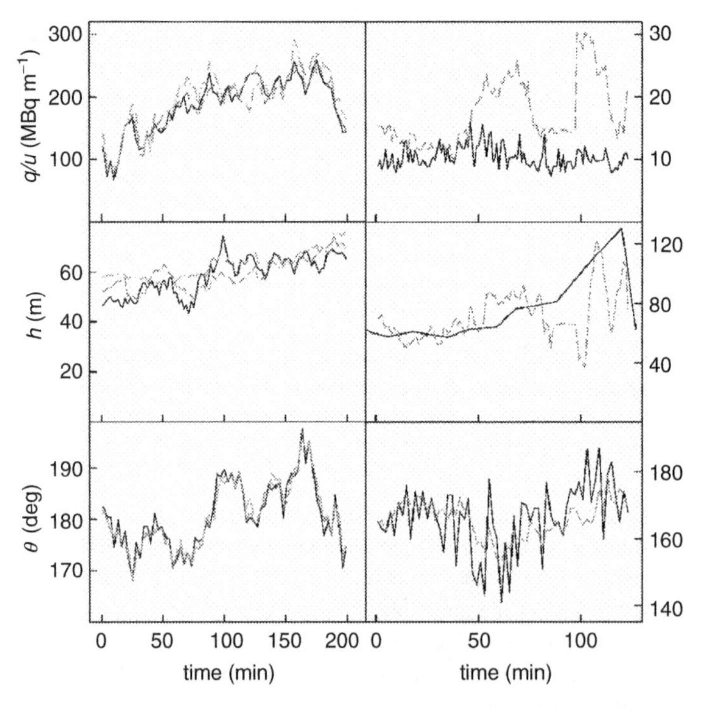

FIGURE 1. Left column: Kalman filtered state variables for Setup 1 (dashes) and Setup 2 (dot-dash) compared to the simulated release and dispersion data (solid lines). Right column: Kalman filtered state variables (dashed curves) compared to measured ^{41}Ar release and dispersion data (solid curves).

In Figure 1 (left column), the Kalman filtered state variables are shown for two different setups of gamma radiation detectors: Setup 1 having the four detectors spanning the plume perpendicular to the main advection direction, and Setup 2, having the detectors covering only roughly half of the plume cross section. In both cases, the Kalman filter recovered both the hypothetical dispersion parameters (q/u, h, θ) and the embedded state space parameters (Q, R). Best agreement with the simulated data is found for Setup 1, where the detectors span the plume cross-section.

In the ^{41}Ar atmospheric dispersion experiment, gamma-spectra from the decay of ^{41}Ar released from a nuclear research reactor were recorded by an array of four NaI(Tl) scintillation detectors. A visible aerosol tracer was added to the plume, allowing for direct measurement of the plume height by a Lidar scanning technique. In Figure 1 (right column), the Kalman filter results are compared to the measured values. The method is seen rather accurately to reproduce the measured source term, plume height and main advection direction. The estimated ^{41}Ar source term is found in most cases to differ from the measured values by less than a factor of two.

Using data from automatic gamma monitoring networks located in the vicinity of many nuclear installations, the proposed method may be used for real-time estimation of the source term in case of an accidental release of radionuclides to the atmosphere. The source term estimation will directly improve assessment of the radiological consequences of an accident in the neighborhood of the nuclear installation and may also in turn be used as input for both long- and short-range atmospheric dispersion model calculations. The proposed method proves computationally efficient and therefore has the potential of becoming an efficient operational assimilation tool for nuclear emergency management.

References

1. Drews, M., Lauritzen, B., Madsen, H., Smith, J.Q. 2004. Kalman filtration of radiation monitoring data from atmospheric dispersion of radioactive materials, Rad. Prot. Dos., *to be published.*

72 Determination of the Impact of Different Emission Sources in the Air Quality Concentrations: The Teap Tool

R. San José, J. L. Pérez*, and R. M. González**

1. Introduction

The advances examined by the Air Qulity Modelling Systems have been quite important in the last five years. The tirad generation of Air Quality Modelling Systems such as MM5 (PSU/NCAR) and CMAQ (EPA, USA) is a reliable software tool to simulate the air quality concentrations. Advances in computer capabilities in recent years have been also considerable and the capability to use a set of PC's (clusters) to run complex models such as MM5 and CMAQ is nowadays a real issue. The Eulerian non-hydrostatic mesoscale meteorological models and dispersión models including chemistry with clouds and aerosols are today reliable software tools which can be used in real-time and forecasting mode. The TEAP (A tool to evaluate the impact of indutrial emissions) Eureka project has been running in the last 2 years and the results and experience is shown in this contribution. Moreover, we show an extended version of this tool which is used for having complete air quality information in real-time and forecasting mode of the impact of different (5) industrial sources. The results show the importance of the impact on air concentrations of the different industrial emission plants used in the simulation in a mesoscale domain centered in an area located in the south of Madrid (Spain). The results show that the system can be used to determine the importance of 5 different industrial plants (simulated for this research experiment) on the emissions in the air concentrations in real-time and forecasting mode.

* Environmental Software and modelling Group, Computer Science School, Technical University of Madrid (UPM) (Spain), Campus de Montegancedo, Boadilla del Monte 28660 Madrid (Spain) http://artico.lma.fi.upm.es.

** Department of Geophysics and Meteorology, Faculty of Physics, Universidad Complutense de Madrid (Spain).

2. Results

The system took 140 CPU hours in a Pentium IV, 3,06 Ghz, 1 Gbyte RAM. The 12 CMAQ simulations – two emission reduction scenarios (50 and 100%) for each virtual industrial plant and one for ON mode (all virtual industrial plants emitting) and CAL mode (the five virtual industrial plants are not emitting at all) for 9 km and 3 km spatial resolutions took 70% of the total CPU time. Post-processing analysis took 25% of the total CPU time. The system could be mounted in a PC cluster with 20 PCs to obtain about 10 times speed-up in the CPU time which will make the system "operational" under daily basis operations since the total CPU time will be about 14 hours.

The system generated a total of 134139 files which mostly are hourly images of the different combinations between OFF, 50% OFF and ON (excluding switching off more than one virtual industrial plant). A total of 2,91 Gbytes were generated.

We have built a tool which implements an adaptation of the MM5-CMAQ modelling system named OPANA V3. The system shows an extraordinary capability to simulate all changes produced when five different virtual industrial plants are implemented in the 3 km resolution and 9 km resolution model domains. We have analysed more 11 scenarios corresponding to 0% and 50% emission reduction for each virtual industrial plant and CAL scenario (all virtual industrial plants are switched off). The results show that the system can be used as a real-time and forecasting air quality management system for industrial plants. The system requires it to be mounted in a cluster platform in order to handle the CPU times required for such daily simulations.

Acknowledgements We would like to thank Professor Dr. Daewon Byun formerly at Atmospheric Modeling Division, National Exposure Research Laboratory, U.S. E.P.A., Research Triangle Park, NC 27711 and currently Professor at the University of Houston, Geoscience Department for providing full documentation of CMAQ and help. We also would like to thank to U.S.E.P.A. for the CMAQ code and PSU/NCAR for MM5 V3.0 code.

73 Advanced Atmospheric Dispersion Modelling and Probabilistic Consequence Analysis for Radiation Protection Purposes in Germany

Harald Thielen, Wenzel Brücher, Reinhard Martens, and Martin Sogalla*

1. Introduction

Currently within the licensing procedure of nuclear facilities in Germany, atmospheric dispersion calculations are still predominantly based on the simple Gaussian plume approach. Advanced model systems consisting of a diagnostic flow model together with a Lagrangian particle simulation model, as e.g. described in a guideline of the German Association of Engineers (VDI 3945 part 3)[1], have less restrictions with respect to terrain effects, source configuration and non-stationary conditions.

An upgraded standard model system based on the Lagrangian approach was recently adopted for the German licensing procedure of non-radioactive airborne pollutant emissions (Technical Instruction on Air Quality Control, TA-Luft, AUSTAL2000)[2]. This advanced mesoscale model chain consists of a particle simulation model linked to a diagnostic flow model. In order to reach the same level of accuracy and flexibility for the standard model for nuclear facilities, it is intended to adapt AUSTAL2000 to the additional requirements for radiation protection purposes.

2. Planned Model Sytem

The AUSTAL2000 model can take into account horizontally and vertically variable meteorological fields of wind and turbulence, arbitrarily shaped source geometry, gravitational settling of aerosol particles and dry deposition. To provide consistent realistic atmospheric dispersion modelling of both radioactive and non-radioactive airborne pollutants, this model chain will be extended for nuclear

* H. Thielen, W. Brücher, R. Martens, and M. Sogalla, Gesellschaft für Anlagen-und Reaktorsicherheit (GRS) mbH, Cologne, Germany D-50667

regulatory purposes by including γ-cloudshine, radioactive decay, wet deposition and an appropriate database of radioactive substance properties.

The theoretical basis for the extension with a γ-cloudshine module is established in VDI 3945, part 3[1]. A corresponding module for a particle model was developed and successfully verified with respect to γ-cloudshine calculations currently applied in Germany.

In combination with a probabilistic analysis which takes into account the variability of the weather situations, a refined and realistic assessment of the radiological consequences of either incidental or accidental releases can be performed. By means of cumulative complementary frequency distributions (CCFD) the probability of reaching or exceeding a specified dose value at any point of interest can be analysed. Figure 1 shows a typical result of a long term dispersion calculation including a probabilistic analysis at four points of interest.

The extension of the AUSTAL2000 model for radiation protection purposes will harmonise the standard methods for atmospheric dispersion of radioactive and non-radioactive on a modern and sophisticated level. The resulting model chain will offer deterministic as well as probabilistic analyses. In combination with the explicit consideration of terrain or building effects based on pre-calculated flow and turbulence fields an in-depth analysis of radiological consequences in the vicinity of a release will be enabled.

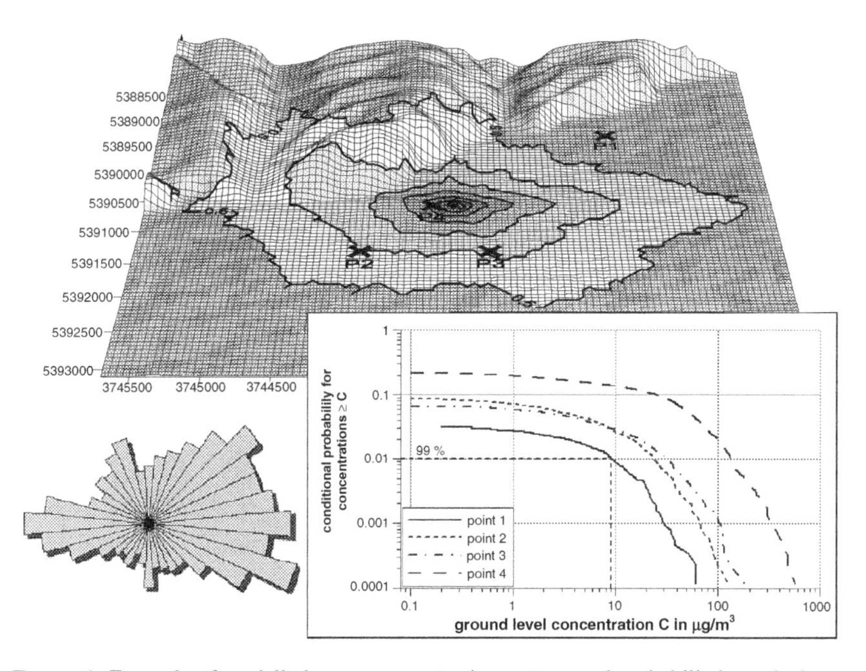

FIGURE 1. Example of modelled mean concentration pattern and probabilistic analysis.

References

1. VDI, Verein Deutscher Ingenieure: Environmental meteorology, Particle model. Kommission Reinhaltung der Luft im VDI und DIN, VDI 3945, Blatt 3, Beuth Verlag (Berlin, September 2000).
2. Erste Allgemeine Verwaltungsvorschrift zum Bundes-Immissionsschutzgesetz (Technische Anleitung zur Reinhaltung der Luft – TA Luft) vom 30. Juli 2002, GMBl. 25-29, S. 511/605. Corresponding computer program is available via: http://www.austal2000.de

Integrated Regional Modeling

74 Comparison of Different Turbulence Models Applied to Modelling of Airflow in Urban Street Canyon and Comparison with Measurements

Miroslav Jicha and Jiri Pospisil*

1. Introduction

Residential buildings in cities are ventilated using different systems for ventilation. Except for systems that consist in opening windows, there are more sophisticated systems that are controlled by different manners. Among them, hybrid systems are recently gaining more attention as they promise to be energy effective with a good control of indoor air quality. Operation of such ventilation systems can be significantly influenced by airflow in street canyons. Thus, correct prediction of airflow is a prerequisite for efficient functioning of such systems. Authors of this paper choose Agiou Fanouriou street canyon in Athens as a test canyon for comparison of modeling and field measurements.

2. Mathematical Model of the Studied Street Canyon

The Agiou Fanouriou Street canyon is located in residential part of city of Athens. A regular net of perpendicularly intersecting street canyons forms this urban area. Information about the geometry of the urban canyon was obtained from Niachou, (2002). The studied canyon has a SSE orientation and its main axis is 150 degrees from the North direction. The canyon's length is 76 m, width is 8 m and the average height of buildings is 21 m. Balconies and shades in the form of awnings disturb facades thus forming the street canyon. Buildings have two floors above the ground with plane facades. Terraces form another three floors of the building. The roofs of the buildings are flat. Buildings of the same geometry form the studied street canyon. The canyon is modelled without trees and without standing cars. A numerical model of a calculated area involves not only the studied street canyon but also surrounding buildings.

* Miroslav Jicha and Jiri Pospisil, Brno University of Technology, Faculty of Mechanical Engineering, Technicka 2, 616 69 Brno, Czech Republic.

A series of calculations were done for three different wind directions, namely longitudinal, perpendicular and oblique (45°) wind direction. The wind direction orientation is related to the longitudinal axis of the studied street canyon. Velocity fields were solved for wind velocities 3 m/s and 6 m/s as obtained from the measurements at a height of 25 m above ground level.

In total, three different turbulence models were tested. As a first one, K-ε Hi-Re model was used as the most common model used in CFD. As a second, K-ε RNG model of turbulence was taken into account. From different studies it follows that this model provides good results of air flow in urban areas. Third, the K-ε Lo-Re model of turbulence (Lien, F.S., Chen, W.L., and Leschziner, M.A., 1996) was used. Three different wind directions and two different wind velocities were tested with all models of turbulence.

3. Conclusions

The most intensive movement of air appeared with oblique wind direction to the canyon axis. This wind direction forms a more complex air velocity field showing two vortexes with opposite rotation in the cross section. Velocity fields obtained for longitudinal wind direction shows lower air velocity movement in the studied canyon compared with the oblique wind direction. This results from buildings configuration in front of the studied canyon in the direction of a longitudinal axis of the canyon. The front side (entrance) of the canyon forms a border of a perpendicularly oriented highway. A continuous block of buildings was modeled on the opposite side of the highway. This block of buildings lifts the air above the studied canyon in situations of parallel wind direction. Perpendicular wind direction creates one vortex in the upper part of the canyon. This wind direction causes very different velocity fields around the ends of the canyon compared with central part of the canyon.

Comparison of predicted and measured wind velocity values shows the best agreement for K-ε RNG model of turbulence.

Acknowledgment This work is a part of COST Action 715 and Eurotrac-2 subproject SATURN and was financially supported by the Czech Ministry of Education under the grants OC715.80 and OE32/EU1489, respectively and the Brno University of Technology Research Plan No. MSM 262100001.

References

1. A. Niachou, K. Papakonstantinou, M. Santamouris, 2002: RESHYVENT ClusterProject - WP10-Urban Impact in the EU, *Report No. RESHYVENT-WP10-TR-2-June* 2002.
2. A. Niachou, M. Santamouris, 2003, RESHYVENT ClusterProject -WP10 -Urban Impact in the EU, *Report No. RESHYVENT-WP10-TR-10-March* 2003.
3. Lien, F. S., Chen, W. L., and Leschziner, M. A. 1996, 'Low-Reynolds-Number Eddy-Viscosity Modelling Based on Non-Linear Stress-Strain/Vorticity Relations', *Proc. 3rd Symp. on engineering Turbulence Modelling and Measurements,* Crete, Greece.

75 Pollutant Dispersion in a Heavily Industrialized Region: Comparison of Different Models

Maria R. Soler, Sara Ortega, Cecilia Soriano, David Pino, and Marta Alarcón

1. Introduction

The objective of this study is to analyse the meteorological fields and the dispersion patterns of different pollutants released by several types of industries on a heavily industrialized region located in the NE part of Spain. Petrochemical industries and several chemical plants are concentrated in an industrial park placed near the shoreline of the Mediterranean Sea and close to a relatively high mountain range, facts which increase the complexity of the wind fields in the region and therefore of the resultant pollutant dispersion patterns. Numerical modelling has been carried out with two different models: the 3-D Urban Airshed Model with variable grid (UAM-V) (Biswas et al., 2001), implemented to MM5 meteorological model (Grell et al. 1994); and an Australian model TAPM (Luhar and Hurley, 2003), which has its own photochemical module. The models have been run from the 7th to 9th of August 2003, during a summer ozone episode mainly characterized by a synoptic situation of high pressures which favour the development of mesoscale circulations forced by the topography. First results show how meteorological fields are certainly a critical component of the dispersion modeling systems because in this area local wind circulations are the main cause of the plume dispersion. In consequence, its evaluation is considered as a preliminary and an important point of this study. Afterwards a comparison between the dispersion patterns given by the two different photochemical is carried out and agreements and differences are analyzed.

Maria R. Soler and Sara Ortega, Department of Astronomy and Meteorology, University of Barcelona, Barcelona, SPAIN.
Cecilia Soriano, Department of Applied Mathematics I, Universitat Politècnica de Catalunya, Barcelona, SPAIN.
David Pino, Department of Applied Physics, Universitat Politècnica de Catalunya, Barcelona, SPAIN.
Marta Alarcón, Department of Nuclear Physics and Engineering, Universitat Politècnica de Catalunya, Barcelona, SPAIN.

2. Meteorological Model Evaluation

The meteorological simulations were performed using the nesting capabilities of the two models. Four nested domains were used at horizontal resolutions of 27, 9, 3 and 1 km. The dimensions of each domain are 31×31 for the two outer domains, and 46×40, 67×40 grid cells for the two inner domains, respectively. In the case of TAPM, all domains had the same number of cells (67×40). Table I shows the statistics of the comparison between measurement and the output of the two meteorological model's for different surface stations.

The results above show that both models give good agreement with observations, especially for wind velocity. TAPM model tends to overestimate the wind velocity, while the Bias value for MM5 model depends of the station. For wind direction both models simulate correctly the sea breeze entrance, while during night time the wind velocity is low or calm and the models behaviour is worse.

3. Dispersion Simulation

Since the intention of this study is to evaluate the contribution of the plume emitted by the main industries in the region, and to study its dispersion patterns, neither initial nor boundary conditions have been considered in the simulations.

Previous to the photochemical simulation, the models have been executed in a non-reactive mode using CO as a tracer in order to analyze the plume dispersion pattern. Results indicate that during daytime the plume is dispersed inland to the mountain range due to the sea breeze flow, while during nighttime, the plume is dispersed towards the sea due to the prevailing drainage winds.

The photochemical simulation shows a similar behaviour for ozone and CO although the previous day accumulated precursors cause some differences in the dispersion. As a consequence, ozone is generated in wider regions which are sometimes disconnected from the direct impact area of the plume.

TABLE 1. Statistics of the comparison between measured and simulated winds.

Stations	Bias (ms^{-1})		Accuracy (ms^{-1})		RMSE (ms^{-1})		Direction Accuracy (°)	
	MM5	TAPM	MM5	TAPM	MM5	TAPM	MM5	TAPM
Alcover	0.8	1.7	1.1	1.7	1.8	1.0	52	29
Botarell	–0.3	1.0	0.9	1.1	1.1	0.9	37	32
Constantí	1.1	1.9	1.3	1.9	1.3	1.0	51	35
Espluga de F.	–0.03	0.7	1.6	1.3	1.9	1.4	42	69
Reus	1.0	1.8	1.1	1.8	1.1	0.7	38	32
Torredembarra	0.02	0.1	0.9	0.8	1.2	1.0	43	37
Vila-Seca	0.5	1.0	1.5	1.0	0.8	0.6	50	43

References

Biswas, J., Hogrefe, C., Rao, S. T., Hao, W., Sistla, G., 2001: Evaluating the performance of regional-scale photochemical modeling systems,. Part III-Precursor predictions. *Atmospheric Environment*, 35, 6129-6149.

Grell, G. A., 1993: Prognostic Evaluation of Assumptions used by Cumulus Parameterization. *Monthly Weather Review*, 121, 764-787.

Luhar A., Hurley P., 2003: Evaluation of TAPM, a prognostic meteorological and air pollution model, using urban and rural point-source data', *Atmospheric Environment.*, 37, 2795-2810.

76 Study of Odor Episodes Using Analytical and Modeling Approaches

Cecilia Soriano, F. Xavier Roca, and Marta Alarcón

1. Introduction

Odor episodes are of particular concern in Catalonia, a region in NE Spain with an important industrial sector whose facilities, on many occasions, are located very close to population areas. In this contribution we show a methodology that has been applied to investigate the origin of bad odors episodes. The procedure combines, on one hand, an analytical approach, based on the acquisition of samples, and which requires the participation of the affected population; and on the other hand, a modeling approach, based on the use of mesoscale meteorological models to track back in time the origin of the air mass responsible of the discomfort. The system has been applied to the investigation of odor episodes suspected to be caused by a landfill facility.

2. Analytical Approach

The Laboratory of the Environment Center (LCMA) has been working for several years on the development of odor maps for several regions. The group has developed an analytical system which detects the VOCs responsible for bad odors. Since odor nuisance is a result of several factors (both physiological and psychological) that determine the behavioral response of the individual, the system relies on the collaboration of members of the affected community, which turn on the sampler device whenever they experience the smell problem.

Measurements are acquired with a VOC sampler developed at LCMA (and whose main characteristics are: remote activation, operating range 40-200 ml/min,

---

Cecilia Soriano, Department of Applied Mathematics I, Universitat Politècnica de Catalunya (UPC). Av. Diagonal 647, 08028 Barcelona, SPAIN.
F. Xavier Roca, Laboratory of the Environment Center (LCMA), Universitat Politècnica de Catalunya (UPC), Barcelona, SPAIN.
Marta Alarcòn, Department of Nuclear Physics and Engineering, Universitat Politècnica de Catalunya (UPC), Barcelona, SPAIN.

constant flow model, solid adsorbent multi-bed tubes Carbotrap + Carbopack X+ Carboxen 569). Samples are taken for short-term air sampling (2-6 hour) during odor episodes by affected residents. Representative VOC standards (C_6-C_{10} aliphatic hydrocarbons, C_6-C_8 aromatic hydrocarbons, C_5 esters, C_2 aliphatic halocarbons, C_6 aromatic halocarbons and C_{10} monoterpenes) are used for calibration and validation purposes. An optimized thermal desorption-gas chromatography-mass spectrometry (TD-GC-MS) method is used to quantify these analyses down to 0.01–100 mg/Nm^3 in outdoor air samples.

The analytical system is an effective tool for the identification of sources responsible for the odor events. Identification of the VOC component, correlated with the meteorological records prevailing during the hours of the episodes (wind direction), can lead to the identification of the wind sectors where the odor originated.

3. Mesoscale Modeling Approach

To further confirm the causes of the episodes, a modeling approach has been added to the methodology explained above, by using a 3-D mesoscale meteorological model at high resolution. The ideal situation for this kind of study would

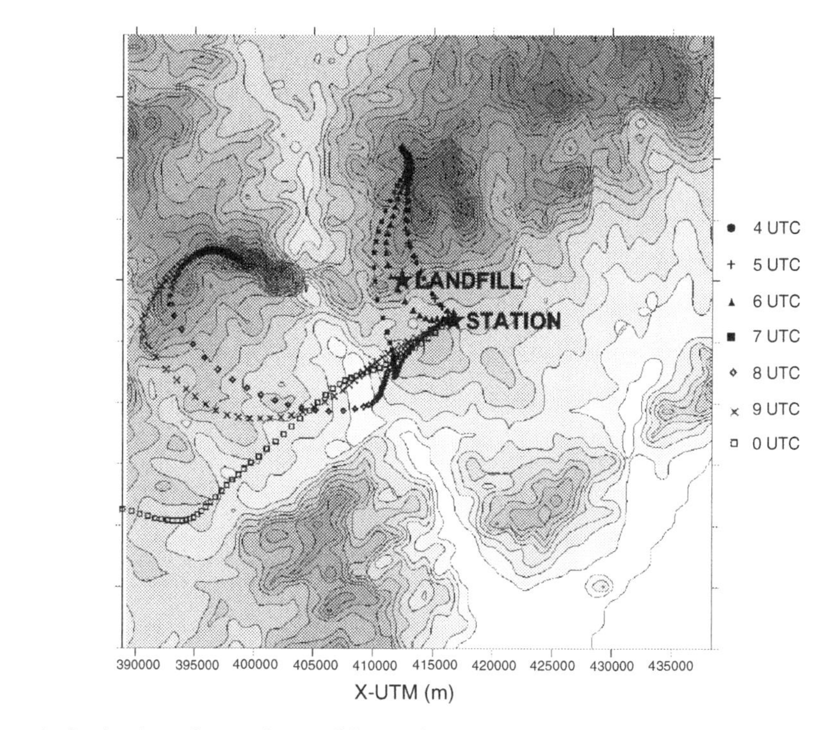

FIGURE 1. Backtrajectories starting at different times on June 22, 2003, showing that the air mass reaching the station that detected typical VOCs of landfill activity came from the landfill area at the times of the episode.

be to perform a full dispersion simulation of the suspected source. However, this is not always possible since most of the time the industrial activity under investigation is not willing to provide information about its emissions. For this reason the developed methodology makes use only of the meteorological module of the mesoscale model. The high-resolution wind fields provided by the model are used to track back in time the air-mass that has caused the bad odor episode at a given moment in a certain location.

The model used for the simulation was the Australian model TAPM, and four nested domains were defined, consisting of 50×50 cells of 12, 6, 3, 1.5 and 1-km horizontal resolution and 30 vertical σ-layers, with the lowest one situated at 10-m AGL. Figure 1 shows kinematic backtrajectories starting at different times and calculated using the winds in the first σ-level predicted by the mesoscale model over the inner domain. Backtrajectory calculations have proven to be useful to confirm the origin of the episode, as the ones reaching the monitoring station during the times of the episode (reported from 5:30 UTC to 7:30 UTC) passed above the questioned source a few hours before (in this case, the landfill facility).

77
Application of Back-Trajectory Techniques to the Characterization of the Regional Transport of Pollutants to Buenos Aires, Argentina

Ana Graciela Ulke*

Backward trajectories from Buenos Aires, Argentina, were used in an exploratory study to characterize the airflow and the relationship with synoptic patterns and to contribute to the identification of air pollution source regions. Buenos Aires is the most populated city of the country. Air pollution climatology studies have shown that there is an important frequency of reduced ventilation conditions, especially during the cold season (Ulke, 2000). Case studies have demonstrated the transport of biomass burning products from Brazil during the tropical dry season toward the region (Longo et al, 1999).

Three-dimensional trajectories were calculated using the Hybrid Single-Particle Lagrangian Integrated Trajectories Model (HYSPLIT) developed at NOAA's (National Oceanic and Atmospheric Administration) Air Resources Laboratory (ARL) (Draxler and Hess, 1997, Draxler and Rolph, 2004). The final run 6 hourly meteorological data from NCEP Global Data Assimilation System for South-America (FNL-SH) were used (http://www.arl.noaa.gov). The trajectories were computed at five altitudes, twice a day, for a one-year period. A convergence criterion was applied to assess the representativeness of individual trajectories. A cluster analysis was performed to help characterize transport patterns in association with synoptic conditions on a seasonal and annual basis (Stohl, 1998). The year 2003 was selected due to the availability of some aerosol measurements at ground-based sites in Buenos Aires (34° 34'S, 58° 31'W).

At lower altitudes, during winter, in general, cluster mean back-trajectories reached Buenos Aires from the W-SW sector, associated with the path of cold fronts. The most representative clusters differ only in the travel distance, being more frequently related to the relatively slow-moving cold fronts. Another cluster, although coming from the SW, is associated with a counter-clockwise motion due to the anticyclone behind the cold front. During summer, more clusters were

* Department of Atmospheric Sciences, Faculty of Sciences, University of Buenos Aires, Pabellón II, Piso II, 1428 Ciudad Universitaria, Buenos Aires, Argentina. Email: ulke@at.fcen.uba.ar.

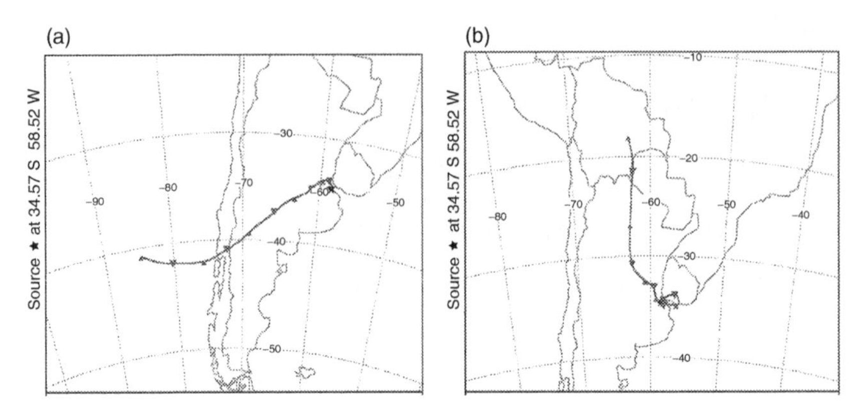

FIGURE 1. Back-trajectories ending at 12 UTC on 7 July near the surface (a); 8 February at higher altitude (b).

identified, because of the greater variability of the weather patterns. The most representative cluster is related to slow-moving air parcels and anticyclonic circulation very near the trajectory endpoint. The following one is associated with the S-SW sector and greater travel distances. The transient seasons exhibited an intermediate behavior. Among the trajectories arriving in Buenos Aires at higher levels, the most important component is that associated with the South American low-level jet east of the Andes. This continental to regional scale pattern plays an important role in moisture and energy exchange between the tropics and extratropics. It can also transport biomass-burning products from Amazonia to southeastern South America. Situations of interest were selected and analyzed. Figure 1 shows the related trajectories reaching Buenos Aires during winter (a) and summer (b).

The results show a satisfactory behavior in the characterization of the contribution from different source regions in association with synoptic conditions and seasonal circulation in South America. Some source regions were in good agreement with aerosol measurements. Further studies using at least a 5-year database will be carried out in order to obtain a climatology of the transport patterns of air parcels arriving in Buenos Aires.

Acknowledgments NOAA Air Resources Laboratory is acknowledged for the provision of the FNL-SH data and the HYSPLIT model. This work was supported by University of Buenos Aires Grant X170.

References

Draxler, R. R. and Hess, G. D, 1997, Description of the HYSPLIT_4 modeling system. NOAA Technical Memorandum ERL ARL-224, December, 24 pp.

Draxler, R. R. and Rolph, G. D., cited 2004, NOAA ARL HYSPLIT Model. NOAA Air Resources Laboratory (http://www.arl.noaa.gov/ready/hysplit4.html).

Longo, K. M., Thompson, A. M., Kirchhoff, V. W. J. H., Remer, L. R., Freitas, S. R., Silva Dias, M. A. F., Artaxo, P., Hart, W., Spinhirne, J. D., Yamasoe, M. A., 1999, Correlation between smoke and tropospheric ozone concentration in Cuiaba during Smoke, Clouds and Radiation-Brazil (SCAR-B), Journal of Geophysical Research, 104, D10, 12113-12129.

Stohl, A., 1998, Computation, accuracy and applications of trajectories. A review and bibliography. Atmospheric Environment, 32, 947-966.

Ulke, A. G., 2000, Air pollution potential in Buenos Aires city, Argentina. International Journal of Environment and Pollution 14, 1-6: 400-408.

Effects of Climate Change on Air Quality

78
Application of Source-Receptor Techniques to the Assessment of Potential Source Areas in Western Mediterranean

Marta Alarcón, Anna Avila, Xavier Querol, and M. Rosa Soler*

1. Introduction

Receptor-oriented methodologies have been recently used to provide maps of potential sources for particular target species. Here, a receptor-oriented methodology has been applied to the chemistry of precipitation data collected at a background site in NE Spain (La Castanya rural station, LC) in order to identify the likely sources of anthropogenic pollutants (SO_4^{2-}, NO_3^-, NH_4^+) and alkaline species to this site. The concentration fields were obtained on the basis of the residence time of the air mass trajectories in each source region.

2. Methodology

The methodology consists in a statistical approach that combines concentration data for selected species in rain at a regional background site with backward trajectories ending at this site.

The sampling station is at 700 m asl in a clearing of the holm oak forest at La Castanya (LC) in the Montseny Mountains (41°46′N, 2°1°′E) located 40 km to the NNE of Barcelona, and 25 km west from the Mediterranean Sea (Figure 1). A total of 157 weekly precipitation samples were collected from July 1995 to September 2000, and analyzed for the major ions. For all the samples, the 12 UTC 96-hour isentropic back-trajectory at 1500 m was computed for each rainy day.

* Marta Alarcón, Department of Physics and Nuclear Engineering, Universitat Politècnica de Catalunta (UPC). Avda. Víctor Balaguer s/n, 08800 Vilanova i la Geltrú, SPAIN.
Anna Avila, Center for Ecological Research and Forestry Applications (CREAF), Universitat Autònoma de Barcelona, Bellaterra, SPAIN.
Xavier Querol, Institut Jaume Almera (IJA), Consejo Superior de Investigaciones Científicas, Barcelona, SPAIN.
M. Rosa Soler, Department of Astronomy and Meteorology, Universitat de Barcelona (UB), SPAIN.

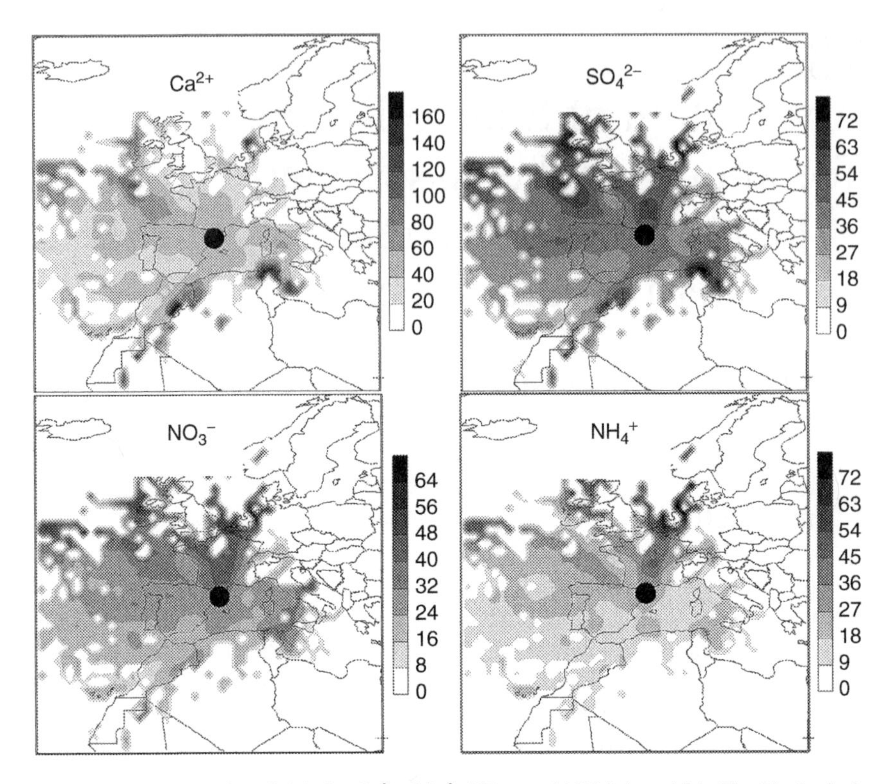

FIGURE 1. Concentration fields for Ca^{2+}, SO_4^{2-} NO_3^- and NH_4^+ (μmol/L). The black circle shows the receptor site LC.

Trajectories were computed from the meteorological data of the U.S. National Climate Data Center (1997-2003) and the NCEP/NCAR Reanalysis (1983-1996), using the HYSPLIT (Hybrid Single-Particle Lagrangian Integrated Trajectories) model (Draxler and Hess 1997). Each trajectory was associated to the corresponding sample concentration. Superimposing a 1° × 1° grid to the trajectory domain, a logarithmic mean concentration for each grid cell was calculated following the Seibert et al. (1994) methodology.

3. Results

Figure 1 shows the result of the model application for Ca^{2+}, SO_4^{2-}, NO_3^- and NH_4^+. The cells with less than 5 trajectory steps were not taken into account.

The highest concentrations of anthropogenic pollutants (SO_4^{2-}, NO_3^-, NH_4^+) originated in North Europe (France, Belgium, The Netherlands and western Germany), a widely reported result in the literature. Sulfate, besides this, shows also predominant contributing sources in the Western Sahara, South Morocco and

Tunisia, in coincidence with source areas for Ca^{2+}, and Na^+ and Cl^- (not shown). Two hypothesis may account for this high sulfate and chloride contents: a) interactions of regionally emitted SO_2 and calcium carbonate from dust (to form calcium sulfate), together with the simultaneous transport of marine aerosol; and b) contribution of dust and salt from salty soils and dry lakes (shebkas) in these regions of North Africa.

References

Draxler R. R. and Hess G. D. 1997 Description of the HYSPLIT modeling system, NOAA Tech. Memo., ERL, ARL-224, 24 pp., Natl. Oceanic and Atmos. Admin., Boulder, Colorado.

Seibert P., Kromp-Kolb H., Balterpensger U., Jost D. T., Schwikowski M., Kasper A. And Puxbaum H. 1994 Trajectory analysis of aerosol measurements at high alpine sites. In Transport and Transformation of Pollutants in the troposphere. Ed. Borrel, P.M., Borrell P., Cvitas T. and Seiler W., 689-693. Academic Publishing, Den Haag.

New Developments

79
Influence of the Autocorrelation Function in the Derivation of Fundamental Relationship $\varepsilon \propto \sigma_v^2/C_0 T_{Lv}$

G. A. Degrazia, O. C. Acevedo, J. C. Carvalho, A. G. Goulart,
O. L. L. Moraes, H. F. Campos Velho, and D. M. Moreira*

1. Introduction

The lateral dispersion parameter σ_y is a statistical quantity of great interest both for dispersion modelling and for the derivation of fundamental relations associated to the turbulence study. The purpose of the present investigation is to derive a new expression for the turbulence dissipation rate. The development consists on a binomial expansion of an algebraic relation for the lateral dispersion parameter originated from the fitting of esperimental data.

2. Derivation of the Turbulence Dissipation Rate

Tennekes (1982) used an equation for the lateral dispersion parameter based on exponential autocorrelation function and inertial subrange similarity arguments (Hinze, 1975) to derive the following fundamental expression for the turbulence dissipation rate ε:

$$\varepsilon = \frac{2}{C_0} \frac{\sigma_v^2}{T_{Lv}} \tag{1}$$

where C_0 is the Kolmogorov constant.

A deep review of the specific literature from the past 20 years, shows that the following simple algebraic relation has been exhaustively used to fit observed dispersion parameters (σ_y, σ_z) in the planetary boundary layer under different stability conditions:

* G. A. Degrazia, O. C. Acevedo, A. G. Goulart, O. L. L. Moraes, Departamento de Física, Universidade Federal de Santa Maria, Santa Maria, RS Brazil.
J. C. Carvalho, D. M. Moreira, Engenharia Ambiental, PPGEAM, Universidade Luterana do Brasil, Canoas, R. S, Brazil.
H. F. Campos Velho LAC, Instituto Nacional de Pesquisas Espaciais, São José dos Campos, SP, Brazil.

$$\sigma_y = \frac{\sigma_v t}{\left(1 + \frac{1}{2}\frac{t}{T_{Lv}}\right)^{1/2}} \tag{2}$$

At this point, we perform a binomial series expansion of the empirical expression (2). In this case, for $t < T_{Lv}$:

$$\sigma_y^2 = \sigma_y^2 t^2 \left(1 - \frac{t}{2T_{Lv}} + \ldots\right) = \sigma_y^2 t^2 - \sigma_y^2 \frac{t^3}{2T_{Lv}} + \ldots \tag{3}$$

It is plausible to associate the term $\sigma_v^2 t^3/2T_{Lv}$ to the inertial subrange high-frequency eddies (this negative term in the right hand side of (3) shows that the small eddies of the inertial subrange are removed as time proceeds). This relationship can be established through the use of the Lagrangian structure function D_{Lv} (the ensemble average of the square of the change in Lagrangian velocity), the Lagrangian autocorrelation function ρ_{Lv} and the inertial subrange Lagrangian turbulent spectrum (Monin and Yaglom, 1975):

$$D_{Lv}(\tau) = 2\sigma_v^2\left[1 - \rho_{Lv}(\tau)\right] = 2\int_0^\infty \left[1 - \cos\left(2\pi n\tau\right) S_{Lv}(n)\, dn\right] \tag{4}$$

where $S_{Lv}(n)$ is the inertial subrange Lagrangian turbulent spectrum, given by (Hanna, 1981):

$$S_{Lv}(n) = \frac{B_0}{2\pi}\varepsilon n^{-2} \tag{5}$$

where B_0 is a constant.

Substitution of (5) in (4) leads to:

$$D_{Lv}(\tau) = 2\sigma_v^2\left[1 - \rho_{Lv}(\tau)\right] = C_o \varepsilon\tau \tag{6}$$

where $C_0 \equiv B_0\, \pi$.

Equation (6) establishes the relationship for the Lagrangian autocorrelation function $\rho_{Lv}(\tau)$ in terms of inertial subrange parameters:

$$\rho_{Lv}(\tau) = 1 - \frac{C_o \varepsilon\tau}{2\sigma_v^2} \tag{7}$$

Substitution of (7) in the classical Taylor statistical diffusion theory results $\sigma_y^2 = \sigma_v^2\, t^2 - C_0\, \varepsilon t^3/6$, which after comparison to the empirically originated expression (3), leads to:

$$\varepsilon = \frac{3}{C_0}\frac{\sigma_v^2}{T_{Lv}} \tag{8}$$

Therefore, this result suggests that the widely used coefficient 2 in equation (1) is not an universal one. In fact, it does not even agree with the also widely used expression (2), which leads to a coefficient 3.

3. Conclusions

The main result of the present study is equation (8), which represents an alternative expression for the dissipation rate, derived from a functional form obtained from experimental σ_y data rather than from an exponential form of the autocorrelation function. The difference from this result to the classical, widely accepted expression for the dissipation rate (equation 1) is the numerical coefficient: 3 for the new function, 2 for the classical one. As a consequence, the analysis developed here showed that the dissipation rate functional form that is consistent with the observationally derived equation (2) for σ_y has the higher coefficient.

References

Hanna, S. R., 1981. Lagrangian and Eulerian time-scale in the daytime boundary layer. Journal of Applied Meteorology 20, 242-249.

Hinze, J. O., 1975. Turbulence. Mc Graw Hill, 790 pp.

Monin, A. S., Yaglom, A. M., 1975. Statistical fluid mechanics: Mechanics of turbulence, 2, MIT press, 874 pp.

Tennekes, H., 1982. Similarity relation, scaling laws and spectral dynamics. In: Nieuwstadt F. T. M. and Van Dop H. eds.. Atmospheric Turbulence and Air Pollution Modeling. Reidel, Dordrecht, 37-68.

80

A Model for Describing the Evolution of the Energy Density Spectrum in the Convective Boundary Layer Growth

Antonio Goulart*, Haroldo F. C. Velho, Gervásio Degrazia, Domenico Anfossi, Otávio Acevedo, Osvaldo L. L. Moraes, Davidson Moreira, and Jonas Carvalho

1. Introduction

The literature regarding the Convective Boundary Layer (CBL) at early morning transition is scarce. Most papers are focused on the growth after convection is fully established. Tennekes (1973) proposes a conceptual model for the time required to erode the nocturnal inversion, while Kaimal et al. (1976) show a 'typical' diurnal cycle of surface heat flux and Convective Boundary Layer depth, but this begins after the surface heat flux is already positive. Wyngaard (1983) discusses the morning ('neutral-convective') transition briefly, and makes the point that only a relatively small buoyancy flux is needed to cause the transition. In this work, starting from the spectral form of the turbulent energy equation, a model for the CBL growing is described.

2. Budget for the Spectral Form of the Turbulent Energy Equation

It is possible to derive an equation for the energy spectrum function in a turbulent flow from the momentum conservation law, expressed by Navier-Stokes equations. In the case of a homogeneous turbulent flow, where the buoyancy and

* Antonio Goulart, URI – Dep. de Ciências Exatas, Santo Ângelo, Brazil; UFSM – Dep. de Física, Santa Maria, Brazil.
Haroldo F. C. Velho, INPE – LAC, São José dos Campos, Brazil.
Gervásio Degrazia, UFSM – Dep. de Física, Santa Maria, Brazil.
Domenico Anfossi, CNR – ICG, Turin, Italy. Otávio Acevedo, UFSM – Dep. de Física, Santa Maria, Brazil.
Osvaldo Moraes, UFSM – Dep. de Física, Santa Maria, Brazil.
Davidson Moreira, ULBRA – PPGEEAM, Canoas, Brazil.
Jonas Carvalho, ULBRA – PPGEEAM, Canoas, Brazil.

inertial energy transfer effects are important, the TKE Fourier Transform of the dynamic equation for the energy density spectrum reads (Hinze, 1975):

$$\partial/\partial t \; E(k,t) = W(k,z,t) + (g/T_0)H(k,t) - 2\nu k^2 E(k,t) \tag{1}$$

where: $(g/T_0)\,H(k,t)$ is the buoyancy term and $W(k,t)$ is the energy-transfer-spectrum function that represents the contribution due to the inertial transfer of energy among different wave-numbers. Is assumed that $H(k,t)$ is,

$$H(k,t) = c_1 \gamma_c \epsilon_0^{-2/3} k^{-2/3} E_0(k) \sin(\pi t/2\tau_f) \tag{2}$$

where c_1 is a constant to be determined from experiments or model simulations, and τ_f is the time at which the height of CBL becomes constant.

Pao (1965) parameterized the term $W(k,t)$ on the basis of dimensional analysis, as follows:

$$W(k,t;z) = -\frac{\partial}{\partial k}\left(\alpha^{-1} \varepsilon^{1/3} k^{5/3} E(k,t)\right) \tag{3}$$

where α is the Kolmogorov constant and ε is the rate of molecular dissipation of kinetic energy.

Substituting equations (3) and (2) in equation (1) yields an expression for the energy spectrum function $E(k,t)$. During the growth of the CBL, the length (z_i) and velocity scales (w_*) vary from zero up to the maximum value that occurs when the CBL reaches the steady-state behavior. We suggest the following expressions for the length and velocity scale,

$$Z_i(t) = 2\sqrt{\frac{\tau_f \overline{(w\theta)}_{0i}}{\gamma_0 \pi}\left(1 - \cos\left(\frac{\pi}{2}\frac{t}{\tau_f}\right)\right)}, \qquad w_*(t) = \left[\frac{g}{T_0} Z_i(t)\overline{(w\theta)}_{0i} \sin\left(\frac{\pi}{2}\frac{t}{\tau_f}\right)\right]^{1/3} \tag{4}$$

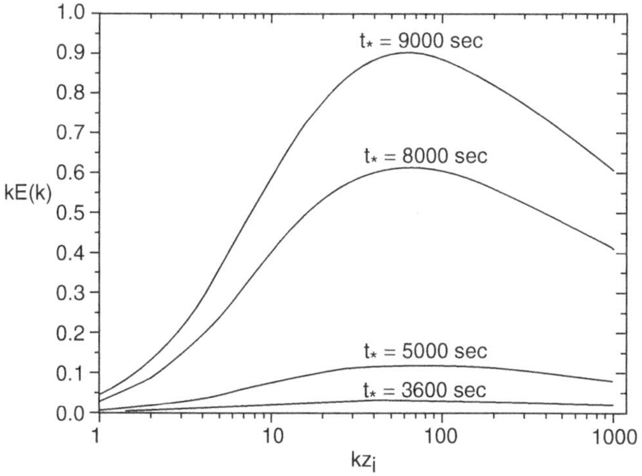

FIGURE 1. The evolution of the 3-D energy density spectrum in the morning.

Figure 1 displays the time evolution of turbulent kinetic energy. We promptly observe a great variability of this energy at early morning. To this point, it is important to mention that, again to our knowledge, there are no experimental works to quantify the turbulent kinetic energy variation at early morning when the CBL begins to form.

References

Hinze, J. O., 1975, Turbulence, Ed. Mc. Graw Hill. 790p.

Kaimal, J. C., Wyngaard, J. C., Haugen, D. A., et al., 1976, Turbulence Structure in the Convective Boundary Layer, *Journal Atmospheric Science*, **33**: 2169.

Pao, Y. H., 1965, Structure of Turbulent Velocity and Scalar Fields at Large Wavenumbers, *The Physics of Fluids,* **8**:1063.

Tennekes, H., 1973, A Model for the Dynamics of the Inversion above a Convective Boundary Layer, *Journal Atmospheric Science*, **30**: 567.

Wingaard, J. C., 1983, Lectures on the Planetary Boundary Layer, in D. K. Lilly and T. Gal-Chen (eds.), Mesoscale Meteorology – Theories, Observations, and Models, D. Reidel, Dordrecht, 650.

81
Simulation of the Dispersion of Pollutants Considering Nonlocal Effects in the Solution of the Advection-Diffusion Equation

D. M. Moreira, C. Costa, M. T. Vilhena, J. C. Carvalho,
G. A. Degrazia, and A. Goulart*

1. Introduction

A way to solve the problem of closing the advection-diffusion equation is based on the transport hypothesis by gradient that, in analogy with molecular diffusion, assumes that the turbulent flux of concentration is proportional to the magnitude of the gradient of medium concentration. This work differs from the traditional method; a generic equation is used for the turbulent diffusion being considered that the flux more yours derived is proportional to the medium gradient. This way, an equation is obtained that takes into account the asymmetry in the process of dispersion of atmospheric pollutants. Therefore, the proposal of this work is to obtain an analytic solution of this new equation using the Laplace Transform technique, considering the Convective Boundary Layer (CBL) as a multilayer system. To investigate the influence of the nonlocal effects in the turbulent dispersion process, the model is evaluated against atmospheric dispersion experiments that were carried out in Copenhagen under unstable conditions.

2. Model

Using the generic equation for turbulent diffusion sugered by van Dop and Verver (2001), where the vertical turbulent contaminant flux can be written as (time independent):

* Davidson Martins Moreira, Jonas da Costa Carvalho, ULBRA – PPGEEAM, Canoas, Brazil.
Camila Costa, Marco Túllio Vilhena, Universidade Federal do Rio Grande do Sul, Instituto de Matemática, Porto Alegre, Brazil.
Antonio Goulart, Gervásio Annes Degrazia, UFSM – Departamento de Física, Santa Maria, Brazil.

$$\overline{w'c'} = -K_z \frac{\partial \bar{c}}{\partial z} + \left(\frac{S_k \sigma_w T_l u}{2} \right) \frac{\partial \bar{c}}{\partial x} \tag{1}$$

where S_k is the skewness, σ_w is the vertical turbulent velocity and T_l is the Lagrangian time scale.

In this work, the Eq. (1) is introduced in the advection-diffusion equation and the equation that arise for to be solved is:

$$u \frac{\partial \bar{c}}{\partial x} = \frac{\partial}{\partial z}\left(K_z \frac{\partial \bar{c}}{\partial z} \right) - \frac{\partial}{\partial z}\left(\beta \frac{\partial \bar{c}}{\partial x} \right) \tag{2}$$

where $\beta = 0.5S_k\, \sigma_w T_l u$. Applying the Laplace transform in equation (2) in variable x, results the well-known solution of Vilhena et al. (1998):

$$\bar{c}_n(x,z) = \sum_{j=1}^{8} \frac{p_j}{x} w_j \left[A_n e^{(F_n^* - R_n^*) z} + B_n e^{(F_n^* + R_n^*) z} + \frac{Q}{2R_n^* K_n} \left(e^{(F_n^* - R_n^*)(z - H_s)} - e^{(F_n^* + R_n^*)(z - H_s)} \right) H(z - H_s) \right] \tag{3}$$

where $H(z - H_s)$ is the Heaviside function and

$$R_n^* = \pm \frac{1}{2} \sqrt{\left(\frac{\beta_n}{K_n} \frac{p_j}{x} \right)^2 + \frac{4u_n}{K_n} \frac{p_j}{x}} \quad ; \quad F_n^* = \frac{\beta_n}{2K_n} \frac{p_j}{x}$$

The classical statistical diffusion theory, the observed spectral properties and observed characteristics of energy-containing eddies are used to estimate the turbulent parameters (Degrazia et al., 1997, 2001).

3. Results

Figure 1 shows the effect of nonlocal transport for different skewness in the ground-level concentration. It is verified quickly that alters the peak of the maximum concentration quantitatively (in spite of not altering the position of the maximum). The countergradient term in the turbulence closure made an additional term to appear in the advection-diffusion equation. This term takes get information on the asymmetrical transport in the CBL. The character nonlocal in previous works was introduced in the eddy diffusivity, but now as a new term of the differential equation. We verified that the skewness was more effective in the distance of the peak concentration, altering the value of the maximum. This result is very expressive because the determination of the maximum in the ground-level concentration is one of the most important aspects to be considered in the control of the quality of the air. Additionally, the incorporation of the countergradient term didn't generate larger effort computational in relation to the original problem of Vilhena et al. (1998).

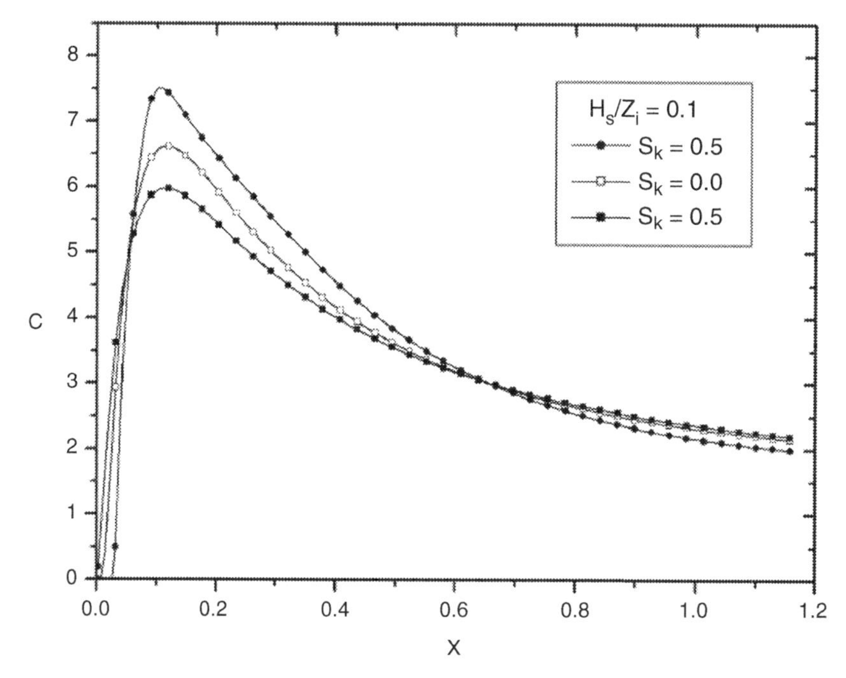

FIGURE 1. Plot of nondimensional ground-level concentration in function of the nondimensional distance for the experiment 8 of Copenhagen. $(\overline{C^y} = \bar{c}uz_i/Q)$; $(X = xw_*/uz_i)$.

References

Degrazia, G. A., Campos Velho, H. F., Carvalho, J. C., 1997. Nonlocal exchange coefficients for the convective boundary layer derived from spectral properties. Contr. Atmos. Phys., 57-64.

Degrazia, G. A., Moreira, D. M. and Vilhena, M. T., 2001. Derivation of an eddy diffusivity depending on source distance for vertically inhomogeneous turbulence in a convective boundary layer. J. Appl. Meteor. 40, 1233-1240.

van Dop, H. and Verver, G., 2001. Countergradient transport revisited. J. Atmos. Sci. 58, 2240-2247.

Vilhena, M. T., Rizza, U., Degrazia, G. A., Mangia, C., Moreira, D. M. and Tirabassi, T., 1998. An analytical air pollution model: development and evalution. Contr. Atmos. Phys. 71, 315-320.

82 Concentration Fluctuations in Turbulent Flow

L. Mortarini[1] and E. Ferrero[2]

1. Introduction

The aim of this work is the formulation of a mathematical model for reacting turbulent flows and its validation through comparison with other models and experimental data. The model is based on the Lagrangian statistical description of the turbulent diffusion, developed firstly for the turbulent dispersion of one particle in homogeneous flows and later gradually extended to non homogeneous flows (Van Dop, 1985) and two-particle dispersion (Durbin, 1980, Thomson, 1990, Borgas and Sawford, 1994).

The Lagrangian stochastic models are based on the Langevin equation for the turbulent velocities. In order to specify the different term of the equation two coefficients need to be determined which should reflect the properties of the turbulent flow. The first one is obtained according to the Kolmogorov's theory of locally isotropic turbulence in the inertial sub-range and the second is a solution of the Fokker-Planck equation with a specified probability density function of the process, or in other words the Eulerian probability density function of the turbulent velocities.

2. The Model

The Lagrangian Stochastic model is used to calculate the mean concentration and its fluctuations in isotropic homogeneous turbulence. The model is based on the Thomson (1990) theory and the three-dimensional solution is considered. The concentration fluctuations are computed through a numerical integration based on the probability density function of the particle's separation. The results are compared firstly with the theoretical behaviour prescribed by the classical Taylor dispersion theory and then experimental data are considered.

[1] Università di Torino, Torino, Italy.
[2] Università del Piemonte Orientale, Alessandria, Italy.

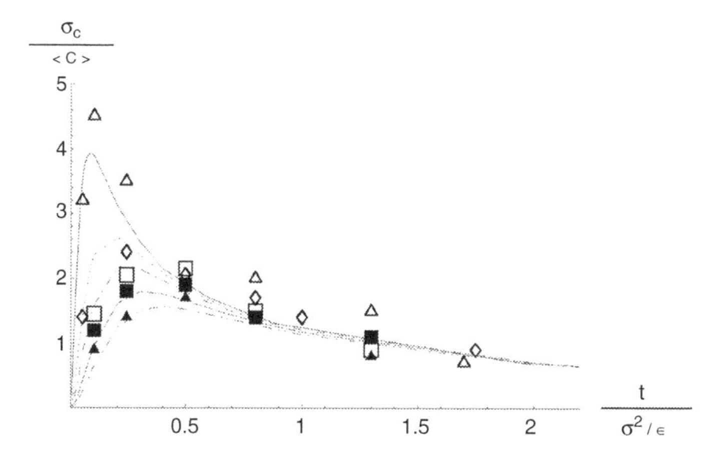

FIGURE 1. Comparison between model simulations (lines) and experimental data (symbols) for different source sizes.

3. Results

In order to test the model we compared the RMS of the particle separation with the Richardson t^3-law first. We found three different ranges inside the inertial range. As a matter of fact, an intermediate range exists:

$$\begin{cases} \langle \Delta^2 \rangle = \frac{1}{3} C_0 \varepsilon t^3 & t_\eta \ll t \ll t_0 \\ \langle \Sigma^2 \rangle = \frac{1}{3} (C_0 - \gamma) \varepsilon t^3 & t_0 \ll t \ll T_L \end{cases}$$

where t_η is the Kolmogorov length scale and t_0 depends on the initial separation $t_0 = (\Delta_0^2 / \varepsilon)^{1/3}$.

Then we compared the models results against the wind tunnel experiment carried out by Fackrell and Robins (1982). The results depend on the Kolmogorov constant, which is known assuming sligtly different values in the literature. The best results were found with $C_0 = 2$ (Figure 1).

References

Borgas M. S. and Sawford B. L., 1994, A Family of Stochastic Models For Particle Dispersion in Isotropic Homogeneous Stationary Turbulence. *J. Fluid Mech.* **279**: 69-99

Durbin P.A., 1980, A stochastic model of two-particle dispersion and concentration fluctuation in homogeneous turbulence, *J. Fluid Mech.*, **100**, 279-302

Fackrell J. E. and Robins A. G., 1982, Concentration fluctuations and fluxes in plumes from point sources in a turbulent boundary, *J. Fluid Mech.*, **117**:1-26

Thomson D. J., 1990, A stochastic model for the motion of particle pairs in isotropic high Reynolds number, and its application to the problem of concentration variance. *J. Fluid Mech.* **210**: 113-153, 1990

Van Dop H., Nieuwstadt F.T.M. and. Hunt J.C.R, 1985, Random walk models for particle displacements in in-homogeneous unsteady flow, *Phys. Fluids*, **28**, 1639-1653

Model Assessment and Verification

83
Skill's Comparison of Three Canadian Regional Air Quality Models Over Eastern North America for the Summer 2003

David Dégardin[1], Véronique Bouchet[2], and Lori Neary[3]

1. Introduction

Three Canadian numerical air quality models: CHRONOS (**C**anadian **H**emispheric and **R**egional **O**zone and **NO**x **S**ystem), AURAMS (**A** **U**nified **R**egional **A**ir quality **M**odelling **S**ystem) and GEM-AQ (**G**lobal **E**nvironmental **M**ultiscale – **A**ir **Q**uality model) present different modelling approaches as well as different degrees of complexity in the way they represent the physicochemical interactions of the atmosphere. Large differences also exist in the way these three models are used or the kind of evaluations they have been subjected to. In order to establish a benchmark comparison, the three models will be evaluated over a one month period, starting in August, 2003. Ozone concentrations measured during the same period by air quality monitoring networks will constitute the evaluation database for this work. The three models will be compared in their native mode.

2. Models Description

Both CHRONOS and AURAMS are off-line, three-dimensional semi-Lagrangian chemical tracer models based on Pudykiewicz's formulation (1997). They involve a non-oscillatory transport algorithm and a comprehensive oxidant chemistry module modified to allow secondary organic aerosol formation and heterogeneous

[1] David Dégardin, UQAM, Dépt. des sciences de la Terre et de l'Atmosphère, Case Postale 8888, Succ. Centre Ville, Montréal, Québec, Canada, H3C 3P8.
[2] Véronique Bouchet, Meteorological Service of Canada, CMC, 2121 Route Transcanadienne, Dorval, Québec, Montréal, H9P 1J3.
[3] Lori Neary, York University, Dept. of Earth and Atmospheric Science, 134 Petrie Bldg., 4700 Keele Street, Toronto, Ontario, Canada, M3J 1P3.

chemistry and deposition modules. As detailed in Bouchet et al. (2003), Makar et al. (2003) and Gong et al. (2003), AURAMS makes use of a detailed sectional particulate matter (PM) representation and aqueous processes. On the other hand, GEM-AQ is an on-line model based on the Canadian meteorological model GEM (Côté et al., 1998a, b) which simulates the transport of the chemical tracers using the same advection scheme as the meteorological fields. The chemical components are the same as the one used in AURAMS, although there are fewer PM components. For this study, CHRONOS is run with a horizontal resolution of 21 km and 25 vertical levels describing the tracer paths in the first 4 kilometres of the atmosphere. AURAMS and GEM-AQ are run on a horizontal grid with a resolution of 42 km and 28 levels from the ground to 30 kilometres.

3. Preliminary Results

Non-validated real-time observations, provided by Canadian and American networks, are taken as reference data. The selected observation sites respect the three following criteria: located in the simulation domain; qualified either as a rural, suburban or urban air quality station; and present at least 75% of available data for the simulation period and after a first degree of validation step. More than 700 qualified sites as shown on Figure 1 met these conditions. The hourly mixing ratios measured at each site are compared with the one derived at the nearest grid point by the models.

The comparisons presented here, are based on hourly averages from August 10th, 2003 to September 10th, 2003 over the whole simulation domain. Two techniques are used here. The first based on Taylor (2001) evaluates the degree of correspondence between two fields by representing, with a single point, the RMS difference, the correlation coefficient between two fields and their standard deviations. The second technique presented by Schmidt et al. (2001) estimates the quality of the models to reproduce the right mixing ratio as a function of the simulated pollution level.

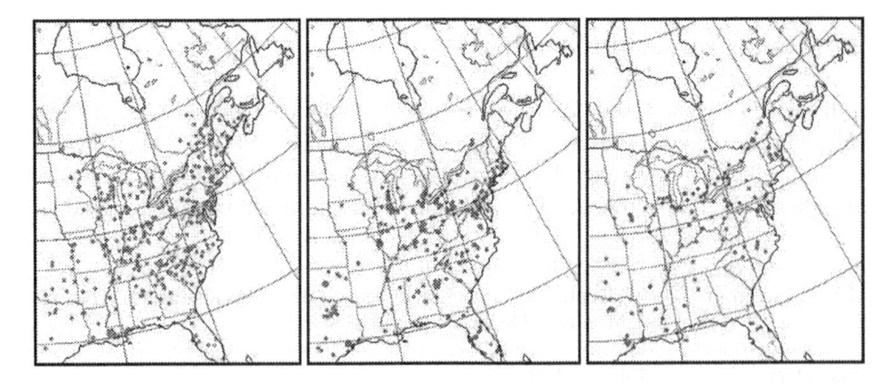

FIGURE 1. Distribution of the rural (a), suburban (b) and urban (c) sites which are available over the simulation domain.

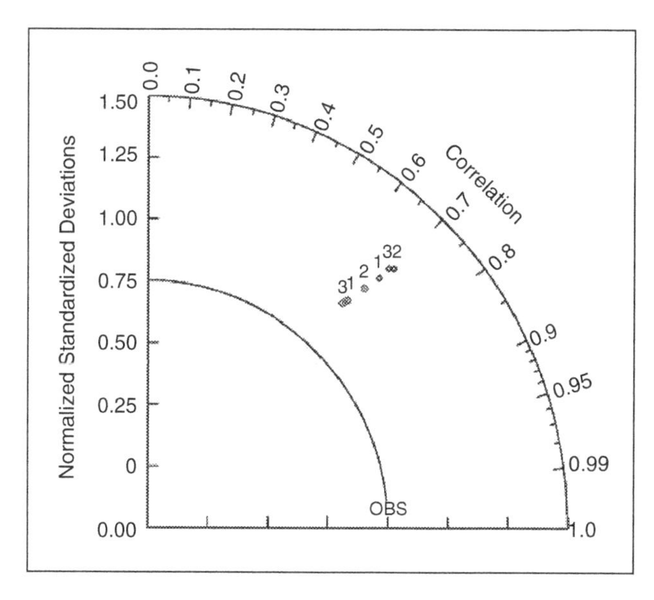

FIGURE 2. Taylor diagram according to the urban (1), suburban (2) and rural (3) typology sites, averaged over the whole simulation domain. Statistics have been derived from the hourly data base.

The Taylor diagram, presented in Figure 2, shows that, on average and for all typology, both CHRONOS (diamonds) and AURAMS (centered dots) have correlation coefficients close to 0.69. Their RMS errors are spread around 1.00. The similarity in performance of both models is understandable as they share the same atmospheric chemistry module (ADOM IIb). On the other hand, for all typology, the standard deviations derived from AURAMS, which are around 1.27, are systematically smaller than those derived from CHRONOS, which averaged at 1.45. The standard deviations computed from both models indicate that they tend to slightly overestimate the amplitude of the time variation of the mixing ratio. Finally, the spatial averages of the correlation coefficients and the RMS differences are not clearly influenced by the typology of the reference sites.

Figures 3 and 4 present the distribution of the difference between simulated and observed mixing ratios as a function of three pollution classes. Limits of the pollution classes are based on the Canadian Standards for ozone. For each pollution class, five bins are defined in order to categorize the difference amplitude. From these graphics, we should note that the first class represents about 80% of the simulation period and the two others classes share the remaining 20%. The pollution class distributions show that, on average over the whole domain, both CHRONOS (Figure 3) and AURAMS (Figure 4) clearly overestimate, by more than 15 ppb, mixing ratios larger than 65 ppb. 70 to 90% of the modeled data fall in that category. On the other hand, for the class where mixing ratios are less

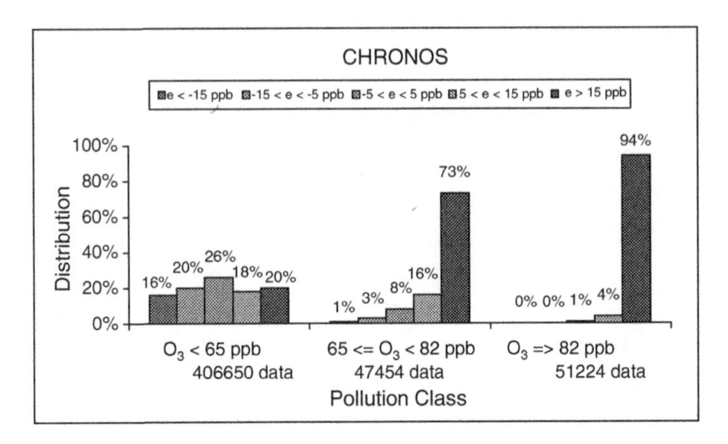

FIGURE 3. CHRONOS distribution of the difference $e = c_{sim} - c_{obs}$ as a function of the pollution classes.

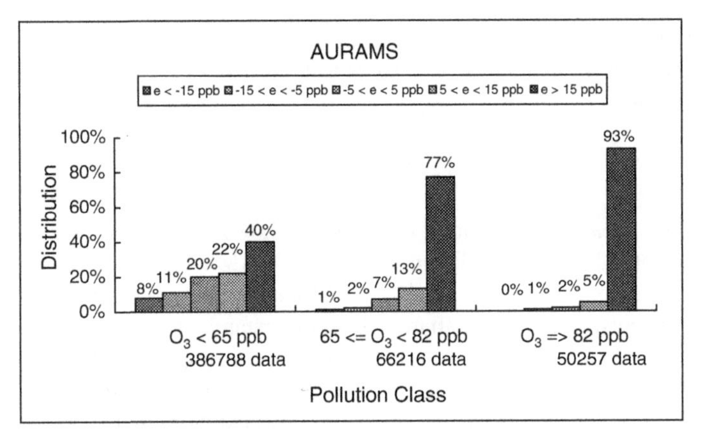

FIGURE 4. AURAMS distribution of the difference $e = c_{sim} - c_{obs}$ according to pollution classes.

than 65 ppb, 50% of the observed mixing ratios are relatively well simulated by the two models. AURAMS clearly overestimates below 65 ppb. This overestimation is twice as frequent for CHRONOS.

4. Conclusion

Due to time constraints, results from GEM-AQ are not presented in this proceeding. The next step of the survey consists of finalising comparisons with observation including the performances of GEM-AQ. Preliminary results illustrates that both models overestimate the hourly ozone mixing ratios as well as the amplitude of their variations. Besides, these results do not specifically depend on the typology of the reference sites. Further investigations will be undertaken in order to identify the potential sources of the difference between simulations and observations.

References

Bouchet, V. S., M. D. Moran, L.-P. Crevier, A. P. Dastoor, S.-L. Gong, W. Gong, P. A. Makar, S. Menard, B. Pabla and L. Zhang, 2004, Wintertime and summertime evaluation of the regional PM air quality model AURAMS, in: *Air pollution modeling and its application XVI,* C. Borrego and S. Incecik, Kluwer Academic / Plenum Publishers, New York, pp. 97-104.

Côté, J., J.-G. Desmarais, S. Gravel, A. Méthot, A. Patoine, M. Roch and A. Staniforth: 1998a. The operational CMC-MRB Global Environmental Multiscale (GEM) model; Part I-Design considerations and formulation. *Mon. Wea. Rev.*, **126**, 1373-1395.

Côté, J., J.-G. Desmarais, S. Gravel, A. Méthot, A. Patoine, M. Roch and A. Staniforth: 1998b. The operational CMC-MRB Global Environmental Multiscale (GEM) model; Part II-Results. *Mon. Wea. Rev.*, **126**, 1397-11418.

Gong, W., P. A. Makar, and M. D. Moran: 2003. Mass-conservation issues in modelling regional aerosol. in: *Air pollution modeling and its application XVI,* C. Borrego and S. Incecik, Kluwer Academic / Plenum Publishers, New York.

Makar, P. A., V. S. Bouchet, L.-P. Crevier, S.-L. Gong, W. Gong, S. Menard, M. Moran, B. Pabla and S. Venkatesh: 2003. AURAMS runs during the Pacific 2001 time period – a Model/Measurement Comparison. in: *Air pollution modeling and its application XVI,* C. Borrego and S. Incecik, Kluwer Academic / Plenum Publishers, New York, pp. 153-160.

Pudykiewicz, J. A., A. Kallaur and P.K. Smolarkiewicz: 1997. Semi-laghrangian modelling of tropospheric ozone. *Tellus*, **49B**, 231-248.

Schmidt, H., C. Derognat, R. Vautard, M. Beekmann: 2001. A comparison of simulated and observed ozone mixing ratios for the summer of 1998 in Western Europe. *Atm. Env.* **35**, 6277-6297.

Taylor, K. E., 2001, Summarizing multiple aspects of model performance in a single diagram, *JGR*. **106**:D7, 7183-7192.

84
Region-Based Method for the Verification of Air Quality Forecasts

Stéphane Gaudreault, Louis-Philippe Crevier, and Michel Jean*

1. Introduction

A fundamental problem in the development of an air quality forecast system is the implementation of an evaluation protocol. Traditionally, statistics are computed to compare the model output to the observations. These methods are limited in that they are generally insensitive to location and timing error. In this paper, we describe a framework to address these limitations. This framework adopts a region-based approach and encompasses both formalism and a software tool that is under active development. More specifically, the framework permits the specification and manipulation of invariants associated with topological elements of an air quality forecast.

2. Algorithm Description

Recently, a region (or object-oriented) based approach to numerical weather prediction model verification was presented by Bullock et al. (2004; Chapman et al., 2004). They provided an advanced diagnostic method for the evaluation of quantitative precipitation forecasts. Forecast and observed precipitations patterns are compared using geometric invariants, rather than on a grid point by grid point basis. This paper modifies and extends this approach for n-dimensional fields in general and air quality forecasts in particular.

Region-based methods attempt to identify significant structures and patterns within fields. The algorithm will mark grid cell belonging to specific regions from data fields and calculates a set of invariants (characteristics) for each region. Current calculated characteristics include size, axis orientation, (major and minor) axis length, area, centroid location (both geometric and value-weighted)

* Meteorological Service of Canada, Canadian Meteorological Centre, 2121 Route Trans-Canadienne, Dorval, Québec, Canada, H9P 1J3.

and compactness. The algorithm can be applied to both the forecast field and analysis valid at the same time. Comparison between the invariants from both fields will allow matching regions from the model output to regions from the analysis.

The algorithm can be generalized to work with n-dimensional fields. Let $G \subset \Re^n$, an n-dimensional grid. We define a scalar k-field on G as an injective function $f:G^k \rightarrow \Re$ $(k \leq n)$. In theory, invariants in both time and space could therefore be calculated for 4-dimensional fields. Two- and three-dimensional fields have been tested up to now. These fields can come from model output, analysis or observations. However, it is important to note that the same grid must be used to compare different fields.

The algorithm consists in two broad steps: region extraction and region matching. To extract the regions from the desired field, one first needs to choose a criterion (e.g. grid cell values greater than a specific threshold) for the field. This criterion is then applied to the field to extract individual grid cells matching it. The resulting fields may consist of large well-organized regions, of a large number of small regions or a mix of the two. A filter can be applied to noisier fields to smooth out the regions. A binary mask is computed from the filtered field. This mask is then applied to the original field and makes it possible for a region to consist in a grouping of smaller regions. From then on, such a group is analyzed as a single region. All invariants are calculated to characterize the regions.

Once invariants have been calculated in the two fields to compare, regions must be matched based on combinations of characteristics. Two regions with the same area and orientation but with different locations may be matched. In initial tests, this step is performed by the user to obtain better knowledge of the relative weight to give to each characteristic in the matching process. Artificial intelligence algorithms are being investigated to perform this procedure automatically and effortlessly in the future.

3. Perspectives and Future Work

A new tool for region-based analysis of air quality model fields is being developed at the Canadian Meteorological Centre. This tool will extract significant features or regions from model fields and calculate a set of invariants (characteristics) for each region and make it practical to compare, in the near future, air quality model output to analysis or other models on a daily basis. Qualitative insight on the performance and value of region-based methods can then be gained. As knowledge and experience of these methods accumulates, it is our hope that more quantitative metrics may be developed from this approach.

References

Bullock, R., B. G. Brown, C. A. Davis, M. Chapman, K. W. Manning and R. Morss, An Object-Oriented Approach to the Verification of Quantitative Precipitation Forecasts:

Part I – Methodology, 20th Conference on Weather and Forecasting, Seattle, Washington (2004); http://ams.confex.com/ams/pdfpapers/71819.pdf

Chapman, M., R. Bullock, B. G. Brown, C. A. Davis, K. W. Manning, R. Morss and A. Takacs, An Object-Oriented Approach to the Verification of Quantitative Precipitation Forecasts: Part II – Examples, 20th Conference on Weather and Forecasting, Seattle, Washington (2004); http://ams.confex.com/ams/pdfpapers/70881.pdf

85 On the Comparison of Nesting of Lagrangian Air-Pollution Model Smog to Numerical Weather Prediction Model ETA and Eulerian CTM CAMX to NWP Model MM5: Ozone Episode Simulation

Tomas Halenka[1], Krystof Eben, Josef Brechler, Jan Bednar, Pavel Jurus, Michal Belda, and Emil Pelikan*

The spatial distribution of air pollution on the local scale of parts of the territory in Czech Republic is simulated by means of Charles University Lagrangian puff model SMOG nested in NWP model ETA. The results are used for the assessment of the concentration fields of ozone, nitrogen oxides and other ozone precursors. A current improved version of the model based on Bednar et al. (2001) covers up to 18 groups of basic compounds and it is based on trajectory computation and puff interaction both by means of Gaussian diffusion, mixing and chemical reactions of basic species. Results of summer photochemical smog episode simulations are compared to results obtained by another couple adopted in the framework of the national project as a basis for further development of data assimilation techniques, Eulerian CTM CAMx nested in NWP model MM5. There are measured data from field campaigns for some episodes as well as air-quality monitoring station data available for comparison of model results with reality.

Usually, there is a problem with emission data for the simulations and definitely they are far from actual instantaneous data. Both the couples have rather older databases of emissions available with many uncertainities, for

[1] Regular associate of the Abdus Salam ICTP, Trieste, Italy.
* Tomas Halenka, Josef Brechler, Jan Bednar, Michal Belda, Charles University in Prague, Faculty of Math. and Physics, Prague, Czech Republic.
Krystof Eben, Pavel Jurus Emil Pelikan, Institute of Computer Science, Academy of Science, Czech Republic.

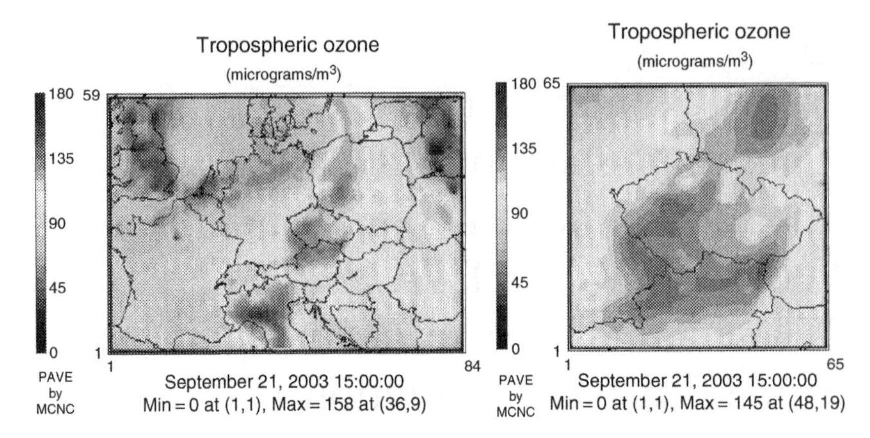

FIGURE 1. Surface concentration of O_3 ($\mu g.m^{-3}$) simulated with CAMx-MM5 for 21 September 2003, Central Europe (left panel), Czech Republic (right panel).

CAMx-MM5 with some problems in resolution and proper description of significant point sources, in case of SMOG-ETA there is a problem with huge database of emission sources taken into account individually, both point and area sources, transportation being taken in terms of area sources. The latest development involves the incorporation of biogenic emissions in these studies. To avoid the problem with emission uncertainities, assimilation techniques are adopted in CAMx-MM5 nesting.

The model simulation of a photochemical smog episode in September 2003 using MM5-CAMx couple on regional and local scale with a sample of the results is presented in Figure 1. Assimilation techniques are tested now to improve the results. ETA-SMOG couple was run for the remote location of region of Hruby Jesenik in framework of a local project for June 2000 in simplified manner (see Halenka et al., 2004), in Figure 2 the results of full day by day run are presented in terms of day-time averages. The agreement between simulation and measurement is pretty good, total shift probably due to absence of Polish sources and possible biogenic emissions. New attempts are carried out to include biogenic emissions into the model chemistry.

Acknowledgment This work has been performed in the framework of the project VaV/740/2/01 funded by the Ministry of Environment of the Czech Republic, the project 205/02/1488 funded by the Grant Agency of the Czech Republic and the project 1ET400300414 funded by Academy of Science, Czech Republic.

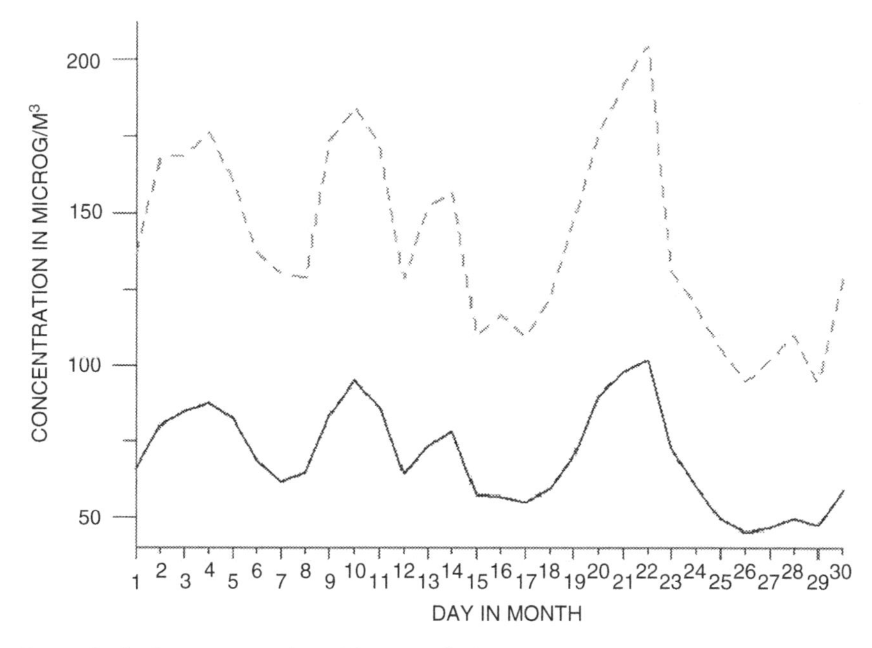

FIGURE 2. Surface concentration of O_3 ($\mu g.m^{-3}$) simulated with SMOG-ETA for June 2000 (solid bottom curve) and measured data from Cervenohorske sedlo site (dashed curve above).

References

Bednář, J., Brechler, J., Halenka, T. and Kopářek, J., 2001: Modelling of Summer Photochemical Smog in the Prague Region, *Phys. Chem. Earth* (*B*), **6**, 129–136.

Halenka, T., Brechler, J., Bednar, J., 2004: Modelling activity in the framework of the national project "Transformation of air-pollution, modelling its transport and dispersion". In: *Air Pollution Modeling and its Application,* C. Borrego, S. Incecik (eds.), Kluwer Academic/Plenum Publishers, New York, pp. 629–631.

86 High Resolution Air Quality Simulations with MC2-AQ and GEM-AQ

Jacek W. Kamiński, Lori Neary, Alexandru Lupu, John C. McConnell*,
Joanna Strużewska, Małgorzata Zdunek, and Lech Łobocki**

1. Introduction

MC2-AQ is a multiscale meteorological model combined with an atmospheric air quality module. It is capable of simulating meteorological phenomena and atmospheric chemistry in a wide range of scales, from the regional to the agglomeration scale. High-resolution models, containing comprehensive descriptions of atmospheric chemistry are nowadays applied at spatial resolutions of one kilometre or even higher, despite numerous deficiencies in the available emissions inventories. Hence, a highly important question – quite apart from the general performance assessment of a given model – regarding the results' reliability limitations due to the quality of the emission database.

The ESCOMPTE field measurement campaign (Cros *et al.*, 2004) provided a valuable database for evaluation of air quality models. The observations gathered during the second intensive observation period (IOP2) have been used here to test the MC2-AQ model, and through simulations accomplished with various spatial resolutions, to address the issues of the high-resolution emission database.

2. Modelling Framework

A three-dimensional air quality model (MC2-AQ) with on-line oxidant chemistry has been used to study the formation of ground-level ozone and other atmospheric pollutants on regional and urban scales. Processes which have been traditionally

* Jacek W. Kamiński, Lori Neary, Alexandru Lupu and John C. McConnell: Department of Earth and Space Science and Engineering, York University, Toronto, Ontario, Canada, M3J 1P3.
** Joanna Strużewska, Małgorzata Zdunek and Lech Łobocki: Institute of Environmental Engineering Systems, Warsaw University of Technology, Warsaw, Poland, 00-653.

treated by a separate chemical transport model have been included within an existing non-hydrostatic mesoscale model, the Mesoscale Compressible Community Model (MC2). The chemical tracers are calculated on the same grid and use the same advection algorithm as is used by the meteorological model, and the physical parameterizations affecting the chemical tracers have, to the greatest extent possible, been harmonized with those used in the meteorological calculations. A full description of the meteorological model MC2 is given in, for example, Tanguay *et al.* (1990), Benoit *et al.* (1997), and Mailhot *et al.* (1998); the description of the air quality module may be found in Plummer *et al.* (2001) and Kaminski *at al.* (2002). The specific MC2-AQ components, horizontal domains and emission fields which were used for the purpose of this study are summarized in the next two subsections.

2.1. Model Domains

MC2-AQ model (dynamics version 4.9.7 with physics library version 4.0+), with chemistry, has been run at a horizontal resolution of approximately 100 km. Initial and boundary conditions for the meteorological variables were taken from the CMC Global Objective Analysis. Initial and boundary conditions for chemical tracers were taken from a global CTM model available at York University (Pathak, 2002).

In order to control the growth of errors in the model representation of synoptic-scale features, a simplified four-dimensional data assimilation (FDDA) scheme was applied. The method of Newtonian relaxation or 'nudging' (Stauffer and Seaman, 1990) has been applied to the horizontal wind components and temperature. The model fields have been nudged towards the gridded objective analyses. Linear interpolation was used to generate analyzed fields for nudging at intermediate times. No additional observational information was introduced to the nudging scheme and the value of the nudging coefficient was set to 2×10^{-4} s^{-1} for the horizontal wind, and 1×10^{-4} s^{-1} for temperature.

Output from the MC2-AQ simulation on the 100 km grid was used to provide meteorological and chemical boundary conditions for a subsequent run of the MC2-AQ model at 10 km resolution. The simulation on the nested 10 km grid was run without interruption for the study period, and nudging of model variables was not applied. Given the smaller geographic extent of the domain, the influence of the boundary conditions has been relied upon to control the growth of errors in the model meteorology (Anthes et al., 1989). Further domain nesting at horizontal resolutions of 3 and 1 km have been done in a similar way (Figure 1). Domain information is summarized in Table 1.

TABLE 1. Model domains.

Horizontal resolution	Time step	Meteorological boundary conditions	Chemical boundary conditions	Grid dimensions
100 km	900 s	OA-3D-Var	CTM	56×56
10 km	200 s	100 km run	100 km run	207×207
3 km	60 s	10 km run	10 km run	151×151
1 km	20 s	10 km run	10 km run	227×207

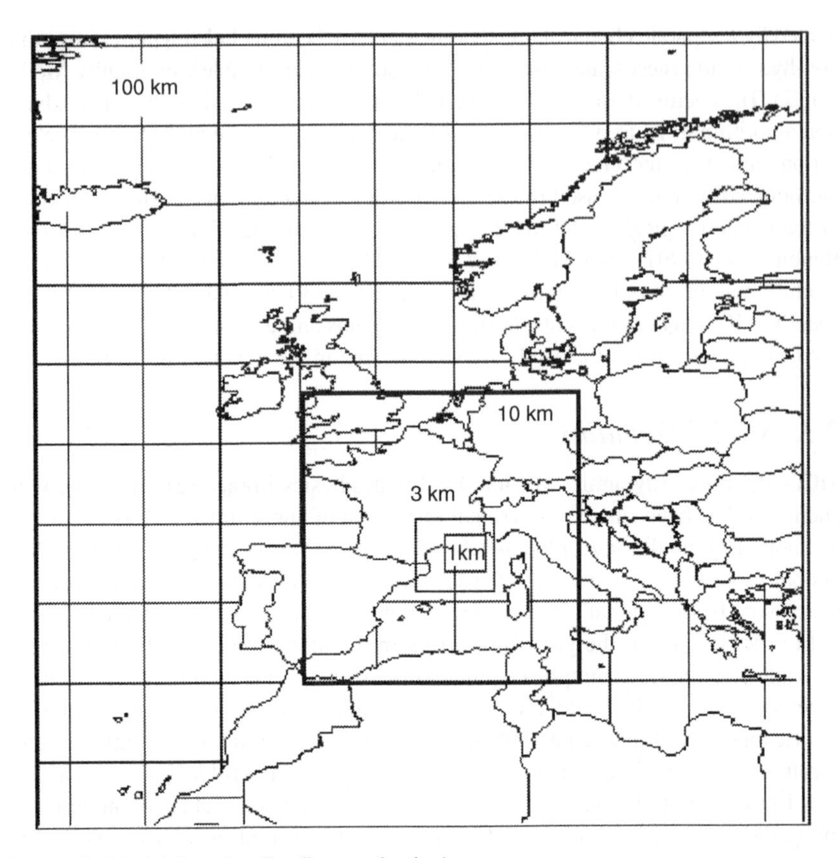

FIGURE 1. Model domains (Lat/Lon projection).

A constant set of vertical levels has been used for model runs on all domains. The model top was set at 20 km with a total of 44 levels. There were 17 model levels specified in the lowest 1.5 km of the atmosphere, allowing for good resolution of the planetary boundary layer.

2.2. *Emission Data*

The emission database consisted of the EMEP inventory blended with the detailed ESCOMPTE data. The EMEP inventory includes estimates for SO_2, CO, NO_x and total non-methane VOC (total and for individual SNAP sectors) given in tonnes of pollutant per year per EMEP grid square. Emission processing for use in MC2-AQ included spatial interpolation to the MC2-AQ grid and VOC speciation according to the MC2-AQ chemical module. For VOC splitting the methodology based on the UK VOC speciation has been applied (Passant, 2002). The total VOC profile for the UK includes 227 species, given for 9 SNAP sectors.

The hydrocarbons included in the chemical mechanism were lumped according to the method used in the MC2-AQ model. For the 100 km run only EMEP area emissions were used. It was assumed that emission fluxes were distributed within the lowest model layer.

The model domain for the 10km resolution run includes the western part of the Mediterranean Sea basin. Since the detailed inventory available for ESCOMPTE covers only a small part of the modelled area, the EMEP emissions (the same as for the 100 km resolution run) were used. Within the ESCOMPTE area, EMEP data was replaced with detailed inventory. It includes the hourly set of data, on a 1 km^2 resolution grid, for NO_x, CO, SO_2 and speciated VOC. A total of 21 inorganic species and 126 hydrocarbons information is available, and is categorized according to the SNAP 97 classification system.

3. Model Simulations and Results

The MC2-AQ model was run for IOP 2A (20 to 24 June, 2001, 00 UTC) in a cascade mode with the final resolution of 1 km. The 10 km run was delayed by 12 hours with respect to the 100 km run, and the 1 km / 3 km runs were delayed by 6 hours with respect to the 10 km run. While the general synoptic features were correctly modelled by the MC2, there was a systematic underestimation of the amplitude of the surface air temperature variation during the diurnal cycle. While some of these errors are due to the sea / land mask inaccuracy at the seashore, the others are apparently due to inadequate soil moisture / soil temperature fields, adopted from the CMC climatological data. Further work will include efforts to address these issues.

In general, ozone values produced by MC2-AQ were underestimated, although at some stations the ozone maxima are reasonably well reproduced. It is also apparent that there is too much NO_x in the system which can be directly attributed to the over-estimation of NO_x emissions. The overestimation of the NO_x concentration is particularly noticeable in the lowest model layer.

Time series analysis at several measurement points (Rognac-les-Brets data shown here as an example) reveals an interesting feature: the underestimation of the ozone concentrations is larger for the 1 km resolution run than for the 3-km one (Figure 2). This behaviour may be attributed to the lack of information on the release height for the area sources in the emission database, which may produce bigger errors at a higher spatial model resolution. The overestimated NO_x mixing ratios in the lowest model layer indicate that the area emission fluxes should be distributed over several model layers. Due to very high model resolution in the boundary layer, the thickness of the lowest layer is ~15 m. Area emissions data include some sectors such as individual heating, ship emissions and emissions from airports, for which injection height should be taken into consideration but it is not provided in the inventory. At the same time, comparisons at other points (Marignane Ville shown here) display better agreement and lack of such systematic difference in the 3 km and 1 km results comparison (Figure 3).

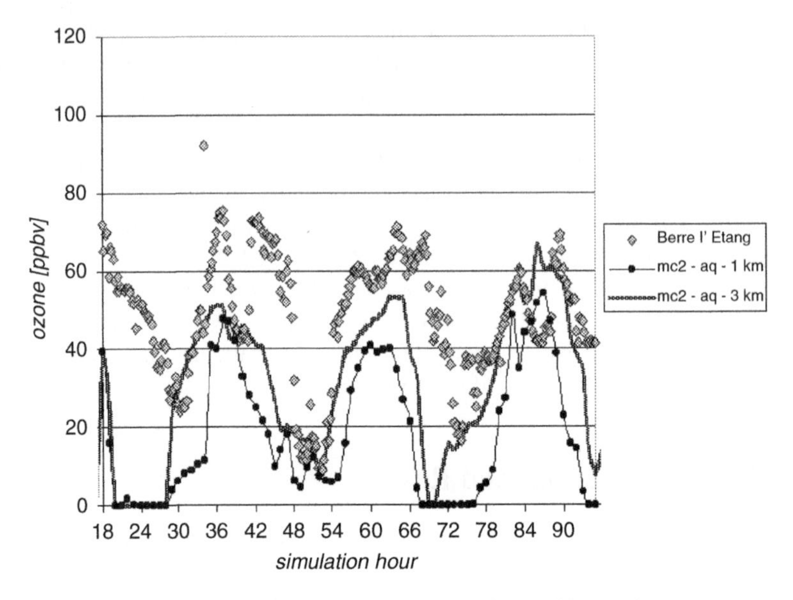

FIGURE 2. Ozone mixing ratio (at 5 m) time series for Berre l'Etang site.

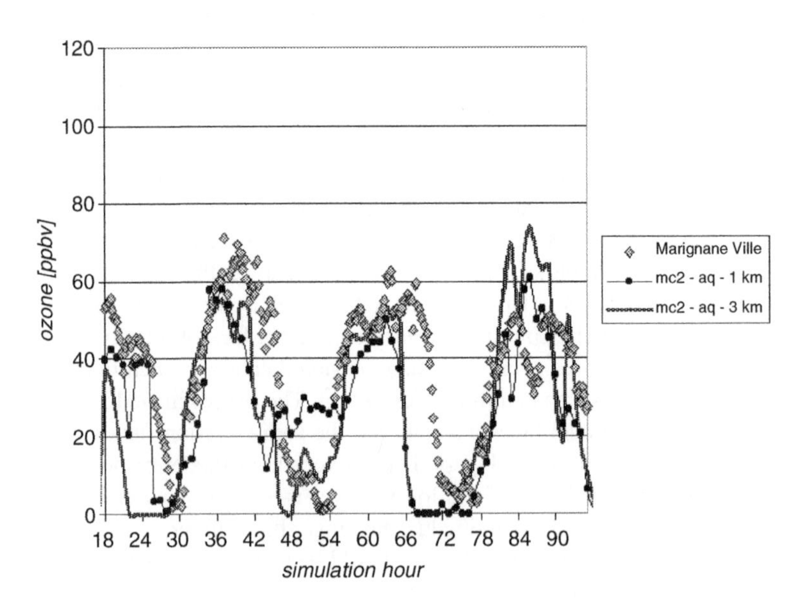

FIGURE 3. Ozone mixing ratio (at 5 m) time series for Marignane Ville site.

An interesting outcome of the present analysis relates to the estimation of ship emissions. It was found that during the sea breeze events, the ozone produced over the sea was not able to reach the shore. While moving across the ship corridor, it was almost completely titrated by NO_x, which concentrations were apparently overestimated. It is likely that this was due to the overrating of ship emissions. This hypothesis lead to a review of the ship emissions by the ESCOMPTE team.

4. Conclusions and Further Work

The overall objectives of this work were to determine the ability of the MC2-AQ modelling system to reproduce the measurements and to investigate the sensitivity of the modelling system to improved resolution of emission fields. To accomplish the latter, MC2-AQ was run at 1 and 3 km horizontal resolutions for IOP-2A using original ESCOMPTE emission fields and 3×3 grid averaged fields respectively. It is apparent that while 1 km runs provide better representation of the fine scale structures directly related to the topography and the emission fields, the evaluation of time series at least at some measurement points yields dubious results. It seems that the deterioration of the forecast quality at a higher resolution may be due to the emission data quality, in particular to the lack of information of the release height structure of certain groups of data classified as 'area' sources.

Evaluation of the meteorological fields produced by MC2 for the IOP-2A scenario shows an under-estimation of surface temperature. This is most likely due to a lack of high resolution geophysical and climatological fields provided to MC2. Therefore, it is difficult to assess the impact of high resolution emissions on the system. In addition to the problems with meteorological fields, the analysis of MC2-AQ model results suggest that the emissions provided in the ESCOMPTE inventories are overestimated for NOx. Furthermore, inventory information classified as 'area source' emissions contains data from 'elevated sources', i.e. aircraft, stacks, ships. Since MC2-AQ was run at relatively high vertical resolution, the injection height of 'surface' sources will play an important role in ozone generation. It appears that only mobile and biogenic emissions should be released to the lowest model layer. However, height and sector information cannot be recovered from the present inventories.

In view of the mentioned problems, the presented results should be regarded as preliminary. Further simulation experiments are needed to confirm hypotheses and to deliver corrected results. The temperature amplitude problem and the apparent NO_x emission overestimation must be resolved. Further analysis will involve an odd oxygen (NO_2+O_3) balance to examine the correlation with observed ozone. It is also anticipated that improving geophysical and land use fields could improve the dry deposition and, consequently, the predictive value of the model.

Acknowledgement The authors acknowledge support from the Canadian Foundation for Climate and Atmospheric Sciences and from the Meteorological Service of Canada, Environment Canada.

References

Anthes, R. A., Kuo, Y.-H., Hsie, E.-Y., Low-Nam, S. and Bettge, T. W., 1989: Estimation of skill and uncertainty in regional numerical models. *Quart. J. Roy. Meteorol. Soc.*, **115**, 763-806.

Benoit, R., Desgagne, M., Pellerin, P., Pellerin, S., Chartier, Y., 1997, The Canadian MC2: a semi-Lagrangian, semi-implicit wideband atmospheric model suited for fine scale process studies and simulation. *Mon. Wea Rev.* **125**: 2382-2414.

Cros, B., Durand, P., Cachier, H., Drobinski, Ph., Frejafon, E., Kottmeier, C., Perros, P. E., Peuch, V.-H., Ponche, J.-L., Robin, D., Saýd, F., Toupance, G. and Wortham, H., 2004, The ESCOMPTE program: an overview. *Atmos. Res.* **69**: 241-279.

Kaminski, J. W., Plummer, D. A., Neary, L., McConnell, J. C., Struzewska J. and Lobocki, L., 2002, First Application of MC2-AQ to Multiscale Air Quality Modelling over Europe, *Phys. Chem. Earth*, **27**: 1517-1524.

Mailhot, J., Belair, S., Benoit, R., Bilodeau, B., Delage, Y., Fillion, L., Garand, L., Girard C. and Tremblay, A. 1998, *Scientific description of the RPN physics library – Version 3.6.* Recherché en Prevision Numerique, Atmospheric Environment Service, Dorval, Quebec, 188 pp.

Pathak, J., 2002: *3D global modelling of tropospheric oxidants*, PhD thesis, York University.

Passant, N.R., 2002, *Speciation of UK emissions of NMVOC*, AEAT/ENV/R/0545 report, (February 2002).

Plummer, D. A., McConnell, J. C., Neary, L., Kaminski, J., Benoit, R., Drummond, J., Narayan, J., Young, V., and Hastie, D. R., 2001: Assessment of emissions data for the Toronto region using aircraft-based measurements and the air quality model. *Atmos. Env.* **35**: 6453-6463.

Stauffer D. R., Seaman, N. L. Use of four-dimensional data assimilation in a limited-area mesoscale model. Part I: experiment with synoptic scale data. *Mon. Wea. Rev.*, **118**, 1250 -1277, 1990.

Tanguay, M., Robert, A. and Laprise, R., 1990, A semi-implicit semi-Lagrangian fully compressible regional forecast model, *Mon. Wea. Rev.*, **118**, 1970-1980.

87 Nonlinear Models to Forecast Ozone Peaks

Carlo Novara, Marialuisa Volta, and Giovanna Finzi*

1. Introduction

In this paper, the problem of forecasting tropospheric ozone concentration is approached by means of nonlinear black-box modeling techniques[1]. In particular Neuro-Fuzzy, Auto Regressive Ciclostationary and Nonlinear Set Membership models are identified and tested in the one day ahead forecast of daily maximum ozone concentration in Brescia, a highly populated and industrialized area in the Po Valley (Northern Italy). The model performances are assessed by means of indexes and statistical indicators suggested by the European Environment Agency.

2. Models and Methods

Four model classes have been implemented and identified: **neuro-fuzzy systems** (NF)[2,3], **Nonlinear Set Membership** (NSM) method[4], **Auto Regressive Ciclostationary with eXogenous inputs models** (ARCX)[5] and the **persistent model** (PM), used as a lower bound. In order to test the capabilities of the predictors to foresee if the O_3 concentration will overcome an assigned threshold, European Environment Agency[6] has defined the following skill parameters: **SP** is the *fraction of correct forecast smog events* (range from 0 to 100 with a best value of 100); **SR** is the *fraction of realized forecast smog events* (range from 0 to 100 with a best value of 100); assuming an equal weight to the correct forecasting of smog events and of a non-smog event, the scoring parameters SP and SR can be combined to the *success index* **SI,** ranging from −100 to 100 with a best value of 100; finally **S** represents the so-called *skill score*, evaluating how much a given prediction model is globally superior to the *persistent model.*

* Carlo Novara, D.AU.IN., Politecnico di Torino, Italy.
Marialusia Volta and Giovana Finzi D.E.A. Università di Brescia, Italy.

The city of Brescia is located in the Po Valley in Northern Italy and is characterized by high industrial, urban and traffic emissions and continental climate. The examined data records consist of O_3, CO, NO and NO_2 hourly concentrations measured by the urban air quality monitoring station. Local temperature monitored and forecasted data are available from the meteorological office. The models have been identified on 1995-1998 (identification set); the best performing models have been selected evaluating the 1999 (first validation set) indexes and validated on 2000-2001 summer season data (second validation set). The forecast models process chemical and meteorological measured data until 8 p.m. and forecast the maximum expected hourly concentration value expected during the day after. The models process the maximum ozone concentration until 8 pm, the mean nitrogen dioxide concentration from 4 pm to 8 pm, the mean ozone concentration from 4 pm to 8 pm, the maximum temperature until 8 pm, the estimate of maximum temperature for the following day. The model inputs and the lag values have been chosen and handled by means of statistical techniques[7].

3. Prediction Results and Conclusions

The prediction performances of the identified models have been evaluated on the base of the performance indices for 130 $\mu g/m^3$ threshold. In Table 1 the performance indexes obtained on first and second validation set are shown.

Neuro-Fuzzy and Nonlinear Set Membership models exhibit alike performances for both validation sets; they correctly predict 6-7 exceeding concentrations out of 10, providing 3-5 false alarms out of 10. ARCX model shows a lower capability to forecast ozone peaks, mainly during the second validation set, stressing the less robustness of this technique to different conditions. The NF and NSM models process the same input data sets and show similar performances. This fact underlines that the information content of the input data can't totally explain the complex accumulation tropospheric ozone processes. In fact the lack of synoptic meteorological data, driving the persistence of the accumulation processes, strongly restricts the skill of the models.

TABLE 1. Prediction performances of identified models.

	1999				2000-01			
Model	SP [%]	SR [%]	SI [%]	S [%]	SP [%]	SR [%]	SI [%]	S [%]
NF	63.5	74.1	51.8	27.8	66.7	55.3	60.2	22.1
NSM	72.4	73.1	52.8	28	71.8	56	63.5	5.5
ARCX	61.9	75	51.1	23.7	35.9	48.3	31.3	22.7
PER	65.1	66.1	47.6	0	41.5	42.5	34.4	0

References

1. Schlink U., Dorling S., Pelikan E., Nunnari G., Cawley G., Junninen H., Greig A., Foxall R., Eben K., Chatterton T., Vondracek J., Richter M., Dostal M., Bertucco L., Kolehmainen M., and Doyle M. 2003. A rigorous inter-comparison of ground-level ozone predictions Atmospheric Environment, 37, 3237-3253.
2. Shing J. and R. Jang, 1993. Anfis: Adaptive-network-based fuzzy inference system, *IEEE trans. on System, Man and Cybernetics*, 23, 665-683.
3. Babuvska R. and H.Verbruggen, 2003. Neuro-fuzzy methods for nonlinear system identification, *Annual Reviews in Control*, 27, 73-85.
4. Milanese M. and C. Novara, 2004. Set membership identification of nonlinear systems, *Automatica, vol.40.*
5. Schlink U., M. Volta, 2000. Grey Box and Component Models to Forecast Ozone Episodes: A Comparison Study, *Environmental Monitoring and Assessment*, 65 (1/2), 313-321.
6. Van Aalst R. and F. de Leeuw, 1998. National ozone forecasting system and international data exchange in northwest europe, Tech. Rep.9, *European Environmental Agency*.
7. Granger C. 1980. Testing for causality, a particular viewpoint, *Journal of Economic Dynamic Control*, 2, 329-352.

88
Evaluation of MC2 Profile Data During the Pacific2001 Field Study

Bradley J. Snyder and Xin Qiu*

1. Introduction

The University of British Columbia (UBC) has been using a variety of mesoscale models as part of their research endeavours over the past decade. One such model is the Mesoscale Compressible Community Model (MC2) developed by Reserche Prevision Numerique (RPN) (Benoit et al 1997) which UBC has run at horizontal resolutions as fine as 2km. This model was used operationally as forecast guidance and decision support during the Pacific 2001 Field Study. One of the goals of this field study was to use the rich dataset to develop and evaluate air quality models over the Lower Fraser Valley (LFV) of BC. Post-field study, a concerted effort has been made to use MC2 as the driving model for daily CMAQ runs at UBC.

One way to improve air quality modelling is to improve the driving meteorological model. It is believed that further improvements in MC2 model physics will lead to higher quality simulations in the boundary layer, thereby improving air quality forecast capability (Qiu, 2002). The goal of the present study is to report on comparisons made of various MC2 model configurations with observations from Pacific 2001.

2. Data and Methodology

Observational data consisted of radiosonde measurements at Langley and Chilliwack (situated approximately 50km apart) in the LFV from 13-31 August 2001.

Model data was comprised of 13 sensitivity tests using a recent version (v4.9.7) of the MC2 model at horizontal resolutions of 4km and 2km. This version of MC2, run at RWDI, incorporated 24km Global Environmental Multiscale (GEM) model initial conditions. Original MC2 output from UBC

* Bradley J. Snyder, Environment Canada. Brad.snyder@ec.gc.ca. Xin Qiu, RWDI West Inc, xq@rwdi.com

was also used to compare with the 3 scenarios. Preliminary results consisted of three scenarios. Model evaluation was based on two parameters which have important impacts on air quality forecasting: the afternoon mixing height (MH) and the morning inversion strength. Mixing heights were calculated based on two techniques (Holzworth, 1967 and Heffter, 1980). Inversion strength was calculated based on the temperature gradient between the surface and 200m.

Three sensitivity tests have been performed to date. These include the following:

- Control Run. Using 20 km MC2, nesting run at 4 km using Force-Restore scheme + default model settings**
- Change Geophysical Fields. Change all geophysics for 4 km run based on control (4km ISBA run).
- Change cloud physics. From 20 km Kain Fritch, run 4 km MC2 using Kong Yau scheme based on Change-all geophysics at 4 km

3. Results

Standard performance statistics performed on the MH data produced differing results for the two sites. Generally, values of Root Mean Squared Error (RMSE) were in the range of 300-400m at Langley and 400-600m at Chilliwack.

In terms of RMSE, the scenario which showed the worst performance was that using the Kong Yau scheme (KY); the scenario which showed the best performance was the control run. It is believed that, especially during wet days, the KY scheme may overestimate moisture field and generate weaker convective conditions without a well-defined capping inversion. When so-called wet days were removed from the analysis, the results changed. The control run performed least well. Meanwhile, performance from the other scenarios improved.

The three scenarios examined outperformed the original MC2 runs in most cases, but only marginally. This may be related to the objective methods of MH calculation; a subjective assessment of temperature profile differences (not presented here) indicated the appearance of some finer details in PBL with these scenarios.

With respect to inversion strength, model performance results for the two sites differed. There was no clear best configuration, although the results RMSE from Chilliwack did suggest that the Force Restore configuration generated the least error. In terms of bias, this setup also was the only one which showed a tendency to over-forecast the inversion strength at both sites whereas the rest had a negative bias.

**Default model settings are based on coarse resolution geophysical and climatological data as MC2 input. Details can be found from http://weather.eos.ubc.ca/wxfcst/STATUS/ABOUT/MC2_details.html

4. Future Work

Three scenarios are not enough to draw definitive conclusions on the best meteorological set up. A complete suite of 13 sensitivity tests shall be completed during the summer of 2004, providing a more complete set of data in which to evaluate changes to model physics and vertical resolution. The expectation is that a 'best' scenario will be found and that this can be implemented in order to improve CMAQ output.

References

Benoit, R., M. Desgagne, P. Pellerin, S. Pellerin, Y. Chartier and S. Desjardins, 1997: The Canadian MC2: A semi-lagrangian, semi-implicit wide-band atmospheric model suited for fine-scale process studies and simulation. Mon. Wea. Rev., Vol 125.

Heffter, J. L., 1980. Air Resources Laboratories Atmospheric Transport and Dispersion Model (ARL-ATAD), Tech. Memo. ERL ARL-81, Natl. Oceanic and Atmos. Admin., Rockville, MD.

Holzworth, G. C., 1967: Mixing depths, wind speeds and air pollution potential for selected locations in the United States. J. Appl. Meteor., 6, 1039-1044.

Qiu, X., 2002: High-resolution Mesoscale meteorological modelling in Southern Ontario. Ph.D. Thesis, Faculty of Earth and Space Science, York University.

27th International Technical Meeting on Air Pollution Modelling and Its Application

October 24–29, 2004

Bruce Ainslie	University of British Columbia Department of Earth & Ocean Sciences 6339 Stores Road Vancouver, BC V6T 1Z4 Canada Email: bainslie@eos.ubc.ca
Tara Allan	University of Calgary 708 Harris Place, NW Calgary, AB T3B 2V4 Canada Email: tamoran@ucalgary.ca
Sebnem Andreani Aksoyoglu	Paul Scherrer Institute Ofla/012 Villigen Psi, CH-5232 Switzerland Email: sebnem.andreani@psi.ch
Domenico Anfossi	CNR-ISAC Corso Fiume 4 Torino, 10133 Italy Email: anfossi@to.infn.it
Jeff Arnold	Atmospheric Science Modeling 1200 6th Avenue 9th Floor, OEA-095 Seattle, WA 98101 USA Email: jra@hpcc.epa.gov

Jose M. Baldasano
Universidad Politecnica de Cataluna
Etseib-upc, Dept Proyectors
Ingenieria, Avda Diagonal 647 Office 1023
Barcelona, 08028
Spain
Email: jose.baldasano@upc.es

Photios Barmpas
Aristotle University Thessaloniki
University Campus
Thessaloniki, 54124
Greece
Email: fotishb@aix.meng.auth.gr

Ekaterina Batchvarova
National Institute of Meteorology & Hydrology
66, Blvd Tzarigredsko Chaussee
Sofia, 1784
Bulgaria
Email: ekaterina.batchvarova@meteo.bg

Nadjet Bekhtaoui
Ecole Point du Jour
129/3 Cite 500 Log, Seddikia
Oran, 31025
Algeria

Prakash Bhave
NOAA - Atmospheric Sc. Modeling Div
U.S. Epa, Office of Research
& Development Mail Drop E243-03
Research Triangle Park, NC 27711
USA
Email: bhave.prakash@epa.gov

Robert Bornstein
SJSU
Department of Meteorology
1 Washington Square
San Jose, CA 95192
USA
Email: pblmodel@hotmail.com

Carlos Borrego	Universidade de Aveiro Dept De Ambiente E Ordenamento Aveiro, 3810-193 Portugal Email: borrego@ua.pt
Veronique Bouchet	Environment Canada 2121 Trans Canada Highway Dorval, QC H9P 1J3 Canada Email: veronique.bouchet@ec.gc.ca
Marija Zlata Boznar	AMES D.O.O. Na Lazih 30 Brezovica Pri Ljubijani, 1351 Slovenia Email: marija.boznar@ames.si
Josef Brechler	Charles University Department of Meteorology & Environment Protection V Holesovickach 2 Prague 8, 180 00 Czech Republic Email: josef.brechler@mff.cuni.cz
Peter Builtjes	TNO-MEP PO Box 342 Apeldoorn, 7300 Netherlands Email: p.j.h.builtjes@mep.tno.nl
Anabela Carvalho	Universidade de Aveiro Dept De Ambiente E Ordenamento Campus Univ De Santiago Aveiro, 3810-193 Portugal Email: avc@dao.ua.pt

Eric Chaxel
Laboratoiare des Ecoulements
Geophysiques et Industriels
BP43
Grenoble, 38041
France
Email: eric.chaxel@hmg.inpg.fr

Lawrence Cheng
Alberta Environment
4th Floor, Oxbridge Place
9820-106 Street
Edmonton, AB T5K 2J6
Canada
Email: lawrence.cheng@gov.ab.ca

Jesper H. Christensen
National Environment Research Institute
PO Box 358
Frederiksborgvej 399
Roskilde, 4000
Denmark
Email: jc@dmu.dk

Andrei Chtcherbakov
Ontario Ministry of the Environment
125 Resources Road
Etobicoke, ON M9P 3V6
Canada
Email: andrei.chtcherbakov@ene.gov.on.ca

Daniel Cohan
Georgia Tech
School of Earth & Atmospheric Sciences
311 Ferst Drive
Atlanta, GA 30332
USA
Email: dcohan@eas.gatech.edu

Ana Margarida Costa
Universidade de Aveiro
Dept De Ambiente E Ordenamento
Campus Univ De Santiago
Aveiro, 3810-193
Portugal
Email: anamarg@dao.ua.pt

Sreerama Daggupaty
Meteorogical Service of Canada
Environment Canada, Arqi
4905 Dufferin Street
North York, ON M3H 5T4
Canada
Email: sam.daggupaty@ec.gc.ca

Richard Daye
EPA VII
21607 NE 178th Street
Holt, MO 64048
USA
Email: daye.richard@epa.gov

David Degardin
Universite du Quebec a Montreal
Dept Des Sciences De La Terre
Et De l'atmosphere UQAM
Cp 888 Succ Centre Ville
Montreal, QC H1M 1R7
Canada
Email: degardin@sca.uqam.ca

Luca Delle Monache
University of British Columbia
6339 Stores Road
Vancouver, BC V6T 1Z4
Canada
Email: lmonache@eos.ubc.ca

Robin Dennis
U.S. EPA/NOAA
Mail Drop E243-01
Atmospheric Modeling Division
Research Triangle Park, NC 27711
USA
Email: dennis.robin@epa.gov

Stephen Dorling
University of East Anglia
School of Environmental Sciences
Earlham Road
Norwich, NR47TJ
United Kingdom
Email: s.dorling@uea.ac.uk

Adolf Ebel
University of Cologne, RIU
Aachener Str. 201-209
Cologne, DE-50931
Germany
Email: eb@eurad.uni-koeln.de

Brian Eder
NOAA & EPA
Mail Drop 243-01
Research Triangle Park, NC 27709
USA
Email: eder@hpcc.epa.gov

Shannon Fargey
University of Calgary
119 Tuscany Ravine Road NW
Calgary, AB T2L 3L3
Canada
Email: sefargey@ucalgary.ca

Giovanna Finzi
Universita di Brescia
Dept Di Elettronica Per L'automazione
Via Branze 38
Brescia, 25123
Italy
Email: finzi@ing.unibs.it

Bernard Fisher
Environment Agency
Kings Meadow House
Kings Meadow Road
Reading, RG1 8DQ
United Kingdom
Email: bernard.fisher@environment-agency.gov.uk

Stefano Galmarini
European Commission
Joint Research Center
TP 441
Ispra, 21020
Italy
Email: stefanco.galmarini@jrc.it

Edith Gego
308 Evergreen Drive
Idaho Falls, ID 83401
USA
Email: e.gego@onewest.net

Wanmin Gong
Meteorological Service of Canada
4905 Dufferin Street
Downsview
North York, ON M3H 5T4
Canada
Email: wanmin.gong@ec.gc.ca

Arno Graff
Umweltbundesamt
Postfach 330022
Bismarckplatz 1
Berlin, 14193
Germany
Email: arno.graff@uba.de

Allan Gross
Danish Meteorological Institute
Lyngbyvej 100
Copenhagen, DK-2100
Denmark
Email: agr@dmi.dk

Tomas Halenka
Charles University
Department of Meteorology & Environment
Protection, Faculty of Math & Physics
V Holesovickach 2
Prague 8, 180 00
Czech Republic
Email: tomas.halenka@mff.cuni.cz

Menouar Hanafi
The University of Sciences and Technology
129/3 Cite 500 Log, Seddikia
Oran, 31025
Algeria

Christian Hogrefe	Atmospheric Science Research Center NYSCDEC Centre 625 Broadway Albany, NY 12233-3259 USA Email: chogrefe@dec.state.ny.us
Selahattin Incecik	Istambul Technical University Department of Meteorology Maslak Istambul, 34469 Turkey Email: incecik@itu.edu.tr
Trond Iversen	University of Oslo PO Box 1022, Blindern Moltke Moes V. 35, 11th Floor Olso, N-0315 Norway Email: trond.iversen@geo.uio.no
Michel Jean	Canadian Meteorological Centre 2121 Trans-canada Highway Dorval, QC H9P 1J3 Canada Email: michel.jean@ec.gc.ca
Weimin Jiang	National Research Council of Canada Room 233, M2 1200 Montreal Road Ottawa, ON K1A 0R6 Canada Email: weimin.jiang@nrc-cnrc.gc.ca
Miroslav Jicha	Brno University of Technology Faculty of Mechanical Engineering Technicka 2 Brno, CZ-61669 Czech Republic Email: jicha@eu.fme.vutbr.cz

Andrew Jones

Met Office
Exeter
Fitzroy Road
Exeter, EX1 3PB
United Kingdom
Email: andrew.jones@metoffice.com

Ieesuck Jung

Texas A&M University-Kingsville
MSC 213
917 Ave B
Kingsville, TX 78363
USA
Email: kaij000@tamuk.edu

George Kallos

University of Athens, School of Physics
University Campus
Bldg Phys-5
Athens, 15784
Greece
Email: kallos@mg.uoa.gr

Daiwen Kang

Science and Technology Corporation
US EPA, NERL/AMD
Mail Drop E243-01
Research Triangle Park, NC 27711
USA
Email: kang.daiwen@epa.gov

Andreas Kerschbaumer

Freie Universitaet Berlin
Institut Fuer Meteorologie
Carl-Heinrich-Becker-Weg 6-10
Berlin, DE-12165
Germany
Email: kerschba@zedat.fu-berlin.de

Oswald Knoth

Institute for Tropospheric Research
Permoserstrasse 15
Leipzig, 04318
Germany
Email: knoth@tropos.de

Sonya Lam
Alberta Environment
111 Twin Atria Bld.
4999-98 Avenue
Edmonton, AB T6B 2X3
Canada
Email: sonya.lam@gov.ab.ca

Bent Lauritzen
Riso National Laboratory
PP Box 49
Roskilde, 4000
Denmark
Email: bent.lauritzen@risoe.dk

Naishi Lin
University of Calgary
305, 703-14th Avenue SW
Calgary, AB T2R 0N2
Canada
Email: phillip988@hotmail.com

Deborah Luecken
U.S. EPA, MD E243-03
86 T.W. Alexander Drive
Research Triangle Park, NC 27711
USA
Email: luecken.deborah@epa.gov

Jeff Lundgren
RWDI West Inc.
222-1628 West First Avenue
Vancouver, BC V6J 1G1
Canada
Email: david.gregory@rwdiwest.com

Larry Mahrt
Oregon State University
Coas
Corvallis, OR 97331
USA
Email: mahrt@coas.oregonstate.edu

Alberto Martilli
University of British Columbia
6339 Stores Road
Vancouver, BC V6T 1Z4
Canada
Email: amartilli@eos.ubc.ca

Clemens Mensink
VITO
VITO-IMS
Boeretang 200
Mol, B-2400
Belgium
Email: clemens.mensink@vito.be

Primoz Mlakar
AMES D.O.O.
Na Lazih 30
Brezovica Pri Ljubljani, 1351
Slovenia
Email: primoz.mlakar@ames.si

Osvaldo Moraes
Federal University of Santa Maria
UFSM - Departmento De Fisica
Campus Universitario
Santa Maria, 97119.900
Brazil
Email: moraes@mail1.ufsm.br

Michael Moran
Meteorlogical Service of Canada
4905 Dufferin Street
North York, ON M3H 5T4
Canada
Email: mike.moran@ec.gc.ca

Luca Mortarini
University of Turin, ISAC-CNR
Corso Fiume 4
Torino, 10100
Italy
Email: luca.mortarini@fastwebnet.it

Lori Neary
York University
4700 Keele Street
Department of Earth & Space Science & Engineering
North York, ON M3J 1P3
Canada
Email: lori@yorku.ca

Trevor Newton
Levelton Consultants
#151 808-4th Avenue SW
Calgary, AB T2P 3E8
Canada
Email: tnewton@levelton.com

Ann-lise Norman
The University of Calgary
2500 University Drive NW
Calgary, AB T2N 1N4
Canada
Email: annlisen@phas.ucalgary.ca

Talat Odman
Georgia Institute of Technology
Environmental Engineering Department
311 Ferst Drive
Atlanta, GA 30332
USA
Email: talat.odman@ce.gatech.edu

Tim Oke
University of British Columbia
Deptartment of Geography
1984 West Mall
Vancouver, BC V6T 1Z2
Canada
Email: toke@geog.ubc.ca

Koray Onder
Golder Associates Ltd.
1000-940 6th Avenue SW
Calgary, AB T2P 3T1
Canada
Email: koray_onder@golder.com

Soon-ung Park	Seoul National University San 56-1 Shinlim-dong Kwanak-ku Seoul, 151-742 South Korea Email: supark@snu.ac.kr
Steven Porter	University of Idaho 308 Evergreen Drive Idaho Falls, ID 83401 USA Email: psp@srv.net
Jiri Pospisil	Brno University of Technology Faculty of Mechanical Engineering Technicka 2896/2 Brno, 61669 Czech Republic Email: pospisil@eu.fme.vutbr.cz
S.T. Rao	US Environmental Protection Agency 109 T.W. Alaxander Drive (MD-E-243-02) Research Triangle Park, NC 27711 USA Email: rao.st@epa.gov
Eberhard Renner	Institute for Troposheric Research Permoserstrasse 15 Leipzig, D-04318 Germany Email: renner@tropos.de
Pietro Salizzoni	Politecnico di Torino-ECL Lyon Via Delle Orfane 30 Torino, 10121 Italy Email: pietro.salizzoni@infinito.it

Roberto San Jose
Technical University of Madrid
Campus De Montegancedo
Boadilla Del Monte, 28660
Spain
Email: roberto@fi.upm.es

Astrid Sanusi
University of Calgary
Department of Physics & Astronomy
2500 University Drive NW
Calgary, AB T2N 1N4
Canada
Email: sanusi@phas.ucalgary.ca

Guy Schayes
University of Louvain
Inst D'astronomie de Geophysique
Chemin Du Cyclotron 2
Louvain-la-neuve, B-1348
Belgium
Email: schayes@astr.ucl.ac.be

Francis Schiermeier
U.S. NOAA
303 Glasgow Road
Cary, NC 27511
USA
Email: schiermeier@msn.com

Howard Schmidt
Lockheed Martin/ REAC
2890 Woodridge Avenue
Bldg 209
Edison, NJ 08837
USA
Email: howard.d.schmidt@lmco.com

Pilvi Siljamo
Finnish Meteorological Institute
PO Box 503
Vourikatu 24
Helsinki, 00101
Finland
Email: pilvi.siljamo@fmi.fi

Bradley Snyder
Environment Canada
201-401 Burrard Street
Vancouver, BC V6C 3S5
Canada
Email: brad.snyder@ec.gc.ca

Mikhail Sofiev
Finnish Mateorological Institute
Sahaajankatu 20 E
Helsinki, 00880
Finland
Email: mikhail.sofiev@fmi.fi

Maria Rosa Soler
University of Barcelona
Department of Astonomy & Meteorology
Avinguda Diagonal 647, 7a Plan
Barcelona, 08028
Spain
Email: rosa@am.ub.es

Cecilia Soriano
Universitat Politecnica de Catalunya
UPC-ETSEIB Dept Math Aplicada1
Diagonal 647
Barcelona, 08028
Spain
Email: cecilia.soriano@upc.es

Wayne Speller
Golder Associates Ltd.
1000, 940-6th Avenue SW
Calgary, AB T2P 3T1
Canada
Email: wspeller@golder.com

Douw Steyn
The University of British Columbia
Department of Earth & Ocean Sciences
Vancouver, BC V6T 1Z2
Canada
Email: douw.steyn@ubc.ca

Peter Suppan

IMK-IFU
Institute for Meteorology & Climate
Research, Kreuzeckbahnstr 19
Garmisch-parternkirc, 82467
Germany
Email: peter.suppan@imk.fzk.de

Benjamin Terliuc

NRC-Negev
PO Box 9001
Beer Sheva, 84190
Israel
Email: terluic@zahav.net.il

Harald Thielen

GRS
Gesellschaft Fur Anlagen Und
Reaktorsicherheit Schwertnergasse 1
Cologne, 50667
Germany
Email: thi@grs.de

Gianni Tinarelli

Arianet S.R.L.
Via Gilino 9
Milan, 20128
Italy
Email: g.tinarelli@aria-net.it

Daniel Q. Tong

Princeton University
Science Technology & Environment
Policy Program
Woodrow Wilson School
Princeton, NJ 08544
USA
Email: quansong@princeton.edu

Sema Topcu

Istambul Technical University
Department of Meteorology
Maslak
Istambul, 34469
Turkey
Email: stopcu@itu.edu.tr

Silvia Trini Castelli
ISAC-C.N.R.
Corso Fiume 4
Torino, 10122
Italy
Email: trini@to.infn.it

Dean Vickers
Oregon State University
College of Oceanic & Atmosph Sciences
Ocean Admin Bldg 104
Corvallis, OR 97331
USA
Email: vickers@coas.oregonstate.edu

Vivian Wasiuta
University of Calgary
#104, 4500, 39th Street NW
Calgary, AB T3A 0M5
Canada
Email: vlwasiut@ucalgary.ca

Brian Wiens
Environment Canada
Room 200, 4999-98 Avenue
Edmonton, AB T6B 2X3
Canada
Email: brian.wiens@ec.gc.ca

Ralf Wolke
Institute for Tropospheric Research
Permoserstrasse 15
Leipzig, Saxony 04318
Germany
Email: wolke@tropos.de

Robert Yamartino
Integrals Unlimited
509 Chandler's Wharf
Portland, ME 04101
USA
Email: rjy@maine.rr.com

Gregory Yarwood

ENVIRON Internation
101 Rowland Way, Suite 220
Novato, CA 94945-5010
USA
Email: gyarwood@environcorp.com

Shaocai Yu

U.S. EPA
Atmospheric Modeling Division
E243-01 NERL US EPA
Research Triangle Park, NC 27711
USA
Email: yu.shaocai@epa.gov

CPSIA information can be obtained at www.ICGtesting.com
Printed in the USA
LVOW01*2045271013

358777LV00005B/371/A